Important Tables

Chemistry

A World of Choices

Chemistry
A World of Choices

Second Edition

Paul B. Kelter
University of North Carolina at Greensboro

James D. Carr
University of Nebraska

Andrew Scott

Boston   Burr Ridge, IL   Dubuque, IA   Madison, WI   New York   San Francisco   St. Louis
Bangkok   Bogotá   Caracas   Kuala Lumpur   Lisbon   London   Madrid   Mexico City
Milan   Montreal   New Delhi   Santiago   Seoul   Singapore   Sydney   Taipei   Toronto

McGraw-Hill Higher Education

*A Division of The **McGraw-Hill** Companies*

CHEMISTRY: A WORLD OF CHOICES
SECOND EDITION

Published by McGraw-Hill, a business unit of The McGraw-Hill Companies, Inc., 1221
Avenue of the Americas, New York, NY 10020. Copyright © 2003, 1999 by The McGraw-Hill
Companies, Inc. All rights reserved. No part of this publication may be reproduced or distributed
in any form or by any means, or stored in a database or retrieval system, without the prior written
consent of The McGraw-Hill Companies, Inc., including, but not limited to, in any network or other
electronic storage or transmission, or broadcast for distance learning.

Some ancillaries, including electronic and print components, may not be available to customers
outside the United States.

 This book is printed on recycled, acid-free paper containing 10% postconsumer waste.

International   1 2 3 4 5 6 7 8 9 0 KGP/KGP 0 9 8 7 6 5 4 3 2
Domestic   1 2 3 4 5 6 7 8 9 0 KGP/KGP 0 9 8 7 6 5 4 3 2

ISBN 0–07–231590–3
ISBN 0–07–119937–3 (ISE)

Publisher: *Kent A. Peterson*
Developmental editor: *Shirley R. Oberbroeckling*
Marketing manager: *Thomas D. Timp*
Project manager: *Sheila M. Frank*
Senior production supervisor: *Laura Fuller*
Coordinator of freelance design: *David W. Hash*
Cover/interior designer: *Rokusek Design*
Cover image: © *Corbis Stock Market/George B. Diebold*
Senior photo research coordinator: *John C. Leland*
Photo research: *Toni Michaels/PhotoFind LLC*
Supplement producer: *Brenda A. Ernzen*
Senior media project manager: *Tammy Juran*
Media technology lead producer: *Steve Metz*
Compositor: *GAC—Indianapolis*
Typeface: *10/12 Times Roman*
Printer: *Quebecor World Kingsport, Inc.*

The credits section for this book begins on page 627 and is considered an extension of the copyright page.

Library of Congress Cataloging-in-Publication Data

Kelter, Paul B.
      Chemistry : a world of choices / Paul B. Kelter, James D. Carr, Andrew Scott. — 2nd ed.
          p.   cm.
      Includes bibliographical references and index.
      ISBN 0–07–231590–3 (acid-free paper) — ISBN 0–07–119937–3 (acid-free paper)
      1. Chemistry. I. Carr, James D. II. Scott, Andrew, 1995–. III. Title.

QD33.2 .K45   2003
540—dc21   2002019615
                                                                                                         CIP

INTERNATIONAL EDITION ISBN 0–07–119937–3
Copyright © 2003. Exclusive rights by The McGraw-Hill Companies, Inc., for manufacture and export. This book
cannot be re-exported from the country to which it is sold by McGraw-Hill. The International Edition is not
available in North America.

www.mhhe.com

To Norma, and Phil and Faye, for all the love that they give.
Paul B. Kelter

To Rosalind, who put up with my spending so much time on the writing of this book; to Rebecca and Louise, the next generation.
James D. Carr

To my family, and the friends on both sides of the Atlantic, who have given so much help and encouragement.
Andrew Scott

About the Authors

Paul Kelter is M. F. Rourk Professor of Chemistry and Biochemistry at the University of North Carolina at Greensboro. Professor Kelter received his B.S. in chemistry from the City College of the City University of New York in 1976, and his Ph.D. in analytical chemistry (1980) from the University of Nebraska–Lincoln (UNL), where his research mentor was Jim Carr, the second author of this book! He did postdoctoral research at the University of Georgia (1980–81). Dr. Kelter has been honored with many teaching awards, including the 1990 University of Wisconsin–Oshkosh Distinguished Teaching Award, the University of Nebraska Outstanding Teaching and Instructional Creativity Award (1999), and his personal favorites, the inaugural *and* second University of Nebraska Student Body Outstanding Teacher of the Year Awards (1996 and 1997). Dr. Kelter is a lifetime member of the University of Nebraska Academy of Distinguished Teachers.

James D. Carr is Professor of Chemistry and Coordinator of the general chemistry program at the University of Nebraska–Lincoln. He received his B.S. degree in chemistry at Iowa State University in 1960 and his Ph.D. in analytical chemistry at Purdue University in 1965. He did postdoctoral research at the University of North Carolina in 1965–66, after which he joined the faculty of the Department of Chemistry at the University of Nebraska. He was awarded a University Distinguished Teaching Award in 1981, the University of Nebraska Outstanding Teaching and Instructional Creativity Award in 1996, and the Distinguished Teaching Award sponsored by the Nebraska Teaching Improvement Council in 2001. Additionally, he has been recognized five times by the University of Nebraska Parents Association for Services to Students. During 1999, while in York, England, he participated in curricular revisions on preuniversity chemistry in the United Kingdom.

Andrew Scott is a writer and lecturer. He received a B.Sc. in biochemistry from Edinburgh University, Scotland, and a Ph.D. in chemistry from Cambridge University, England. His books have been translated into many languages and he has written for a wide variety of newspapers and magazines. He has taught chemistry and general science for several universities and colleges.

Brief Contents

Contents

Chapter 5

The Role of Energy in Chemical Reactions   177

Chapter 6

Creating with Carbon— The Importance of Molecular Structure   205

Chapter 7

Properties of Water   243

Chapter 8

Acids and Bases   275

Chapter 9

Acid Rain   305

Applications

Preface

We live in a world of choices in which chemistry plays a central role. Chemistry is a critical part of our society. Modern medicines are a result of chemistry. So are computer chips. Refining metals for use in so many products is a result of chemistry. We make plastics via chemical processes. We increase yields of all kinds of foods via chemistry. The chemistry that is so beneficial puts stresses on the environment, and we must understand and address these concerns. We live in a world of risks and benefits—*A World of Choices*. And chemistry can be used to help us understand and deal with these choices.

This textbook is written for students who like to think about ideas and have the hunger to understand how the ideas of chemistry relate to our individual and social day-to-day decisions. While chemistry is a challenging subject, it is also understandable and vital.

Chemistry: A World of Choices focuses the writing and content on interesting topical applications that are the central context from which the chemistry is understood. It also stresses the importance of asking the right questions to yield answers that help you to make meaningful choices.

New and Improved for the Second Edition

Organization

- Chapter 18—*Chemistry at Home* is a **NEW** chapter discussing topics that include the chemistry needed in building a home and keeping the house clean, chemicals in the kitchen, personal care chemistry, and pharmaceuticals.
- Chapter 16—*The Chemistry of Life* contains **NEW** topics on carbohydrates, lipids, chemicals in the body, and genetic engineering.
- Chapter 17—*The Chemistry of Food* now handles the concepts of vitamins and minerals in separate headings, with **NEW** topics that include food additives, choosing a balanced diet, energy needs, genetically modified food, and the critical choices about what we eat.
- Chapter 4 has a **NEW** section (4.6) on Green Chemistry.
- Chapter 6 has been rewritten to include an entirely **NEW** section (6.5) on functional groups. Chapter 6 also includes **NEW** content on stereoisomers, optical isomers, and geometric isomers.
- Chapter 8 has a **NEW** section (8.2) on molarity and (8.4) on oxides as acids and bases.
- Chapter 10 has a **NEW** focus on water quality.
- Chapter 11 has a **NEW** section (11.3) on Dalton's Law of Partial Pressures.
- Chapter 12 has a **NEW** focus on air quality, and contains **NEW** content on ozone destruction and bromine.
- Chapter 13 has a **NEW** section (13.6) on corrosion.
- Chapter 15 has a **NEW** section (15.3) on spectroscopy.

Application-Rich

- *Green Chemistry* is discussed in Chapter 4 and integrated throughout the text. An icon in the margin highlights areas where Green Chemistry is particularly applicable.
- *Case in Point* boxes discuss recent scientific developments, current news items, and hot topics that have been explained within the context of the chapter content.
- *Consider This* poses questions that require the student to use critical thinking skills on a variety of topics.
- *Pro/Con Discussions* provide two sides to a social issue regarding uses of chemistry or chemical technology. There are gray areas in many public policy decisions and those that involve chemistry are no exception.

- *Hands-On* boxes present fun with household chemistry. These are simple experiments that can be done at a desk or any sink and allow you to have hands-on experiences.

Problems

- Many "worked examples" have been revised to more clearly show the problem-solving process.
- There are many **NEW** end-of-chapter problems for *every* chapter.
- *Food for Thought* questions at the end of every chapter are created to provoke thoughtful discussion about topical issues related to chemistry.

Pedagogy

- To provide better accessibility for the student, we have included more titled headings illustrating the nature of the content presented.
- Excellent and accessible student readings are given at the end of every chapter.

Media

- **NEW** Online Learning Center with information, instructor's tools, and study tools for students. The students have access to material presented in the in the Websites section at the end of every chapter, expanding their knowledge of the material.

Instructor Resources

Online Learning Center—is a comprehensive, book-specific website (http://www.mhhe.com/physsci/chemistry/kelter/) offering excellent tools for both the instructor and the student. Instructors can create an interactive course with the integration of this site, and a secured Instructor Center stores your essential course materials to save you prep time before class. This center offers PowerPoint lecture outlines, PowerPoint images, a table library, and more. The Online Learning Center content has been created for use in PageOut, WebCT, or Blackboard course management systems.

Test Bank—Edited by Conrad Bergo of East Stroudsburg University, the test bank contains over 1500 multiple choices, true–false, and matching questions. The questions are comparable to the problems in the text.

Computerized Test Bank—This test bank contains all of the questions in the Test Bank, while allowing instructors to edit or create their own test templates. The Test Bank is formatted for easy integration into the course management systems PageOut, WebCT, and Blackboard.

Digital Content Manager—is a multimedia collection of visual resources allowing instructors to utilize artwork from the text in multiple formats to create customized classroom presentations, visually-based tests and quizzes, dynamic course website content, or attractive printed support materials. The Digital Content Manager is a cross-platform CD containing an image library, photo library, and a table library.

Transparency Set—Over 100 four-color illustrations from the text are reproduced on acetate for overhead projection.

Instructors' Solutions Manual—contains the solutions to all end-of-chapter questions in the text. Answers to all the questions in the lab exercises as well as tips on setting up the labs and getting the best results for the students are also included.

ChemSkill Foundations—is developed by James D. Spain and Harold J. Peters. Chem-Skill Builder software challenges your students' knowledge of introductory chemistry with an array of individualized problems. Organized to accompany any introductory chemistry text, this student-oriented software generates questions for students in a randomized fashion with a constant mix of variables. No two students will receive the same electronic homework problems—ensuring an accurate test of the students' knowledge. This unique software program records grades on the quizzes. These grades can easily be transferred to an instructor's record-keeping file.

Student Resources

Online Learning Center—is a comprehensive, exclusive website that provides the student access to the web-related activities in the end-of-chapter websites sections. The website also includes quizzing and other study tools for the students. Log on at http://www.mhhe.com/physsci/chemistry/kelter/.

Student Study Guide/Selected Solutions—by Ken Hughes, Professor of Chemistry (Emeritus), University of Wisconsin–Oshkosh, reviews essential points from the text and presents problems and hints to assist the student in solving them. Solutions to selected problems in the end-of-chapter problem sets are included.

Laboratory Manual—20 lab experiments geared to complement the content in the text, and designed with the nonscience major in mind. The manual includes sections on lab safety. The manual is edited by Jerry Walsh and Dennis Burnes at the University of North Carolina at Greensboro.

Walkthrough

This section highlights pedagogy and the electronic media that support *Chemistry: A World of Choices,* Second Edition, and shows how the entire package fits together. Use the materials that work best for you, or use them all!

Worked exercises in every chapter walk you through the problem solving process. Many of the problems have been revised to more clearly show the problem solving process. The problem is stated, a solution is worked through and a check is presented to ensure the solution makes chemical sense.

Green Chemistry

Green Chemistry is discussed in Chapter 4 and integrated throughout the text. An icon in the margin highlights areas where here chemistry is particularly applicable

exercise 3.3

Lewis Structures—Ions and Polyatomic Ions

Problem

Draw the Lewis structure for these ions and polyatomic ions:

(a)	Cs^+	**(b)**	Cl^-	**(c)**	CN^-
(d)	NH_4^+	**(e)**	S^{2-}	**(f)**	NO_3^-

Solution

(a) Cesium (Cs) is a metal from Group 1A. It has one valence electron in its outer shell, which it loses when the ion is formed:

$$Cs\cdot \longrightarrow Cs^+ + e^-$$

(b) Chlorine (Cl) is a nonmetal from Group 7A. It has seven valence electrons in its outer shell, and so it will gain one electron to form the chloride ion:

$$e^- + :\ddot{\underset{..}{C}l}\cdot \longrightarrow :\ddot{\underset{..}{C}l}:^-$$

(c) Carbon (C) is from Group 4A, with four valence electrons; nitrogen (N) is from Group 5A, with five valence electrons. The bond that forms between them is covalent, a triple bond. For nitrogen, this creates a stable octet. Carbon, however is still shy one electron and so must gain an electron to form an octet, creating a polyatomic ion with a charge of -1:

$$e^- + \cdot\dot{C}\cdot + \cdot\dot{N}: \longrightarrow :C\!\equiv\!N:^-$$

(d) The single valence electrons of the four hydrogen atoms form single bonds with nitrogen. Nitrogen gives up an electron in the process, creating a polyatomic ion with a charge of $+1$:

$$4H\cdot + \cdot\dot{N}: \longrightarrow \left[\begin{array}{c} H \\ | \\ H\!-\!N\!-\!H \\ | \\ H \end{array}\right]^+ + e^-$$

(e) Sulfur (S) is a nonmetal from Group 6A. It has 6 electrons in its outer shell, and will gain 2 electrons to form the sulfide ion:

$$2e^- + :\ddot{S} \longrightarrow :\ddot{\underset{..}{S}}:^{2-}$$

(f) Nitrogen is from Group 5A, with 5 valence electrons. Oxygen is from Group 6A, and has 6 electrons in its outer shell. The double and single bonds that form between the nitrogen and oxygen atoms create stable octets around all atoms except one oxygen, in which an extra electron is needed.

$$e^- + \cdot\dot{N}: + 3\,\ddot{\underset{..}{O}}: \longrightarrow \left[\begin{array}{c} :\!\ddot{O}\!: \\ | \\ :\!\ddot{\underset{..}{O}}\!-\!N\!=\!\ddot{\underset{..}{O}} \end{array}\right]^-$$

Check

In the cases shown here the atoms attain a stable configuration. For cesium and chlorine, the two resulting ions can react to form the ionic compound cesium chloride (CsCl).

Hands-On

The *Hands-On* boxes present household chemistry. These are simple experiments that can be done at a desk or any sink and allow you to have hands-on experiences.

Case in Point

Case in Point boxes discuss recent scientific developments, current news items, and hot topics that have been explained within the context of the chapter content.

Consider This

The boxes entitled *Consider This* pose questions that require the student to use critical thinking skills on a variety of topics.

Pro/Con Discussion

Two sides to a social issue regarding uses of chemistry or chemical technology are presented by two of the authors. There are gray areas in many public policy decisions and those that involve chemistry are no exception.

hands-on

Like Oil and Water

When two people just cannot "get along" we sometimes say that they are "like oil and water," meaning they just will not mix effectively.

Try this simple activity to explore the inability of oil and water to mix.

We use mineral oil, which contains hydrocarbons such as $C_{16}H_{34}$.

Collect:

- a small, colorless soda bottle (glass or plastic) with the cap
- enough mineral or baby oil to fill 1/2 of the bottle
- enough water to fill the other 1/2 of the bottle
- some food coloring
- a funnel (optional)

Put a couple of drops of food coloring into the water and swirl to mix. Add the water to the bottle, using the funnel to avoid making a mess. Add the oil to the bottle (again, using the funnel) and fill to the top. Seal the bottle tightly.

A photograph of what you should see is shown in Figure 6.17. Why are there two separate layers? The answer, and its implications for much more global issues, relates to our theme of *structure dictating function*. Before we explore the answer, however, give your bottle a very vigorous shake for about 30 seconds in an attempt to force the oil and water to mix (but do make sure the cap is on tightly!). Look at the bottle immediately after shaking. What happens? For a second or so you may think you have succeeded; but very soon the two layers separate once again. They just will not stay together.

Food for thought: Try the activity again, adding a squirt of liquid dishwashing detergent. What happens? Why? What does the observed behavior say about the structure of food coloring? We discuss the behavior of detergents in Section 7.3.

Figure 6.17 Why do the oil and water layers separate in the wave bottle?

Use of Oxygenates in Gasoline

Green Chemistry

Most compounds in gasoline are hydrocarbons, that is, they consist only of hydrogen and carbon. **Oxygenates** are compounds which contain oxygen and are added to gasoline in large percentage (5–20%) in order to lower the amounts of pollutants given off by the automobile. The two classes of compounds usually considered are the alcohols, methyl alcohol (methanol)

Problems

Each chapter has a variety of problems and tools for the student.

The text offers a wide range of end-of-chapter problems to solve based upon the knowledge gained in the content and worked examples of the chapter. Some are quite simple while others require a high level of critical thinking skills. **Food for Thought** problems provide thoughtful discussions about topical issues related to chemistry.

Readings are sources for further information on the chapter topics and also for research. **Websites** provide the Internet address and a quick review of information the student will see when going to the site. The website material is written for this text by Paul Kelter.

52. Are the components of petroleum used only for fuels, solvents, and lubricants? Explain.
53. What are the differences between sand, clay, silt and humus?
54. A 40.0-g soil sample is acidified and the carbon dioxide that was liberated was recovered and found to weigh 0.73 g. What percent of this soil was limestone?
55. A second 40.0-g sample of the soil mentioned in Exercise 54 was heated to several hundred degrees in the presence of an excess of oxygen. The carbon dioxide that was liberated was trapped and found to weigh 3.55 g. What percent of the soil was humus (about 50% carbon)?

food for thought

56. When the Statue of Liberty was refurbished, there was discussion that she should be polished to look like shiny copper. It was decided not to do this. What do you think?
57. In the text, we listed the primary way of obtaining pure gold from rock, that of combining it with sodium cyanide. Look up some other ways that gold is isolated. What are the economic, social, and health advantages and disadvantages of each?

readings

1. Matthews, Samuel W. Nevada's mountain of invisible gold. *National Geographic*, May 1968, p. 668.
2. Davidson, Keay, and Williams, A. R. Under our skin: Hot theories on the center of the Earth. *National Geographic*, January 1996, p. 100.
3. Jarnoff, Leon. Iceman. *Time*, October 26, 1992, pp. 62-66.
4. Hall, Alice J. Liberty lifts her lamp once more. *National Geographic*, July 1986, p. 2.
5. *Minerals Yearbook*, U.S. Department of Interior, Bureau of Mines, Washington, D.C., published annually since 1932.
6. Alternative Energy Institute, Inc. January 1999 Coal Fact Sheet http://www.altenergy.org/2/nonrenewables/fossil_fuel/facts/coal.html.

websites

www.mhhe.com/Kelter   The "World of Choices" website contains activities and exercises including links to websites for: The American Petroleum Institute; a discussion of corrosion from "The Corrosion Doctors"; the LaMotte Corporation, which makes soil test kits; and much more!

Media Walkthrough

INSTRUCTOR MEDIA

Online Learning Center (OLC) is a secure, book-specific website. The OLC is the doorway to a library of resources for instructors.

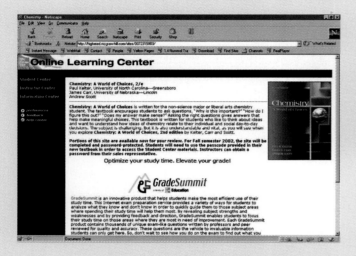

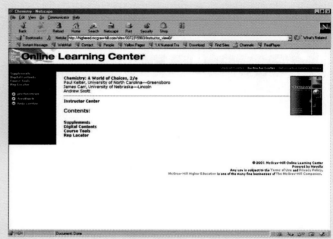

- The Instructor Center will host the Test Bank, the transparency list, as well as other tools for instructor use.
- PowerPoint Presentation—is organized by chapter and ready for the classroom, or the instructor can customize the lecture to reflect teaching style.
- Computerized Test Bank—contains questions with algorithms and over 200 algorithms-based questions that the instructors can edit to create his or her own test templates.
- Course Management Systems—PageOut, WebCT, and Blackboard. All of the following tools are available on the Online Learning Center or in a cartridge for your course delivery system:
  1. Computerized Test Bank
  2. PowerPoint Lecture Presentation
  3. Images from the text
  4. Tables from the text
- Digital Content Manager—CD-ROM set includes electronic files of all full-color images in the text. Import the images into your own presentation, or use the PowerPoint presentation written by the lead author, Paul Kelter, and provided for each chapter.

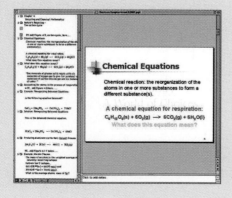

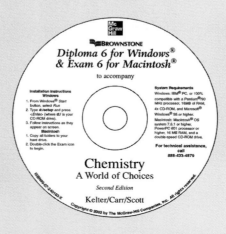

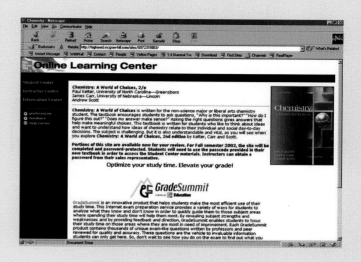

**Create a custom course website with PageOut,
free to instructors using a McGraw-Hill textbook.**

To learn more, contact your McGraw-Hill publisher's
representative or visit www.mhhe.com/solutions.

STUDENT MEDIA

- Online Learning Center is the doorway to access most of the media for *Chemistry: A World of Choices,* Second Edition.
- The OLC includes self-assessment quizzes (**Test Yourself**) and
- Online Web Studies (**Study Tools**) for each section in the text.

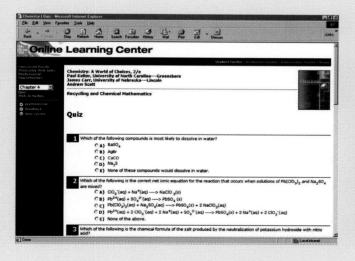

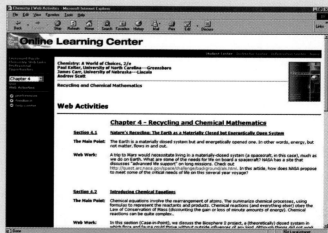

Acknowledgments

As we continue our voyage in this second edition, and boldly go where no book has gone before, we thank the Captain and First Officer who guide our ship so very smoothly—Kent Peterson and Shirley Oberbroeckling, thank you for all that you do. We are grateful to the entire crew, notably Sheila Frank and Toni Michaels, for making our ideas look so beautiful in print. These are trying times. Traveling and learning together, may our future journeys in our "World of Choices" bring freedom and peace.

Besides the publishing team, many dedicated professionals within the chemistry community helped us refine and direct the progress of this book.

In March of 2000, a select group of individuals came together at the first Liberal Arts Chemistry Summit to share ideas about liberal arts chemistry, the professor needs and the student needs. Their insight has been very valuable in understanding this discipline area.

Mary Evenson, *St. Cloud University*

Sue Godbey, *Eastern Kentucky University*

Ralph Jacobson, *California State Poly–San Luis Obispo*

Brian Laird, *University of Kansas*

Jason Ribblett, *Ball State University*

Wayne Riley, *Texas Tech University*

Zhihuz Shen, *Utah State University*

We would like to thank the many reviewers who contributed to the accuracy and presentation of information. The reviewers have given comments based on the first edition and the various drafts of the second edition.

Susmita Acharya, *Cardinal Stritch University*

Bob Allen, *Arkansas Technological University*

Karen Anderson, *Madison Area Technical College*

Wanda D. Bailey, *John Jay College of Criminal Justice*

S. Basu-Dutt, *State University of West Georgia*

Conrad H. Bergo, *East Stroudsburg University*

William F. Berkowitz, *Queens College*

Ildy Boer, *County College of Morris*

Simon Bott, *University of Houston–Houston*

Brenda Braaten, *Framingham State College*

William Burns, *Arkansas State University*

Vincenzo Cammarata, *Auburn University*

Jeff Charonnat, *California State University–Northridge*

Ron Choppi, *Chaffey College*

Kent Clinger, *Lipscomb University*

Wheeler Conover, *Southeast Community College*

Sharon Cruse, *University of Louisiana–Monroe*

Thomas D. Crute, *Augusta State University*

Kim Davis, *Mississippi Delta Community College*

Kay Davis, *Garden City Community College*

Charles Dickson, *Catawba Valley Community College*

Rodney Alvin Dixon, *Towson University*

Jerry Easdon, *College of the Ozarks*

R. Max Ferguson, *Eastern Connecticut State University*

Ted Fickel, *University of Judaism*

Kate Flickinger, *Maui Community College*

Peter D. Frade, *Wayne State University*

Lawrence Fuller, *SUNY–Oswego*

Brian Glaser, *Black Hawk College*

Terry Gleason, *Saddleback College*

Susan E. Godbey, *Eastern Kentucky University*

Garth Grantier, *Alfred State College*

Stan Grenda, *University of Nevada–Las Vegas*

Ryan Harden, *Central Lakes College*

Vahak Harutunian, *Moorpark College*

Charlene Haskin, *McMurry University*

Alton Hassell, *Baylor University*

Noralynn Hassold, *Kettering University*

Sally R. Harms, *Wayne State College*

Barbara Hillery, *SUNY–Old Westbury*

Rita Hoots, *Yuba College–Woodland*

Paul Horton, *Indian River Community College*

Kathleen House, *Illinois Wesleyan University*

Xiche Hu, *University of Toledo*

Lidija Kampa, *Kean University*

Paul Karr, *Wayne State College*

Sidney A. Katz, *Western Texas College*

Cindy L. Kepler, *Bloomsburg University*

Jaeju Ko, *Indiana University of Pennsylvania*

Roger Kugel, *St. Mary's University*

Patricia L. Lang, *Ball State University*

Richard H. Langley, *Stephen F. Austin State University*

Mihaela Leonida, *Fairleigh Dickinson University–Teaneck*

Jackie C. Lindbeck, *Florida Hospital College of Health Sciences*

Dahong Lu, *Fitchburg State College*

Ralph Martinez, *Humboldt State University*

Ruhlullah Massoudi, *South Carolina State University*

Mark B. Masthay, *Murray State University*

Dan McNally, *Bryant College*

William P. Metzar, *Broome Community College*

Carl E. Minnier, *CCBC–Essex*

Michael D. Mosher, *University of Nebraska–Kearney*

Huggins Z. Msimanga, *Kennesaw State University*

Ruth Ann Murphy, *University of Mary Hardin Baylor*

William Nelson, *Mineral Area College*

Lynda P. Nelson, *Westark College*

Larry Neubauer, *University of Nevada–Las Vegas*

Mohammad A. Omary, *Colby College*

Mark Ott, *Western Wyoming Community College*

Gerri Ottewill, *Portsmouth University*

Richard A. Paselk, *Humboldt State University*

David Peitz, *Wayne State College*

Laura E. Pence, *University of Hartford*

Richard Pendarvis, *Central Florida Community College*

H. Mark Perks, *University of Maryland–Baltimore*

Joanna Petridou, *Spokane Falls Community College*

Gary O. Pierson, *Central State University*

Robert D. Pike, *College of William & Mary*

Wendy Pogozelski, *SUNY–Geneseo*

Bruce R. Prall, *Marian College of Fond Du Lac*

Laura Pytlewski, *Moraine Valley Community College*

Jeffrey A. Rahn, *Eastern Washington University*

Rajendra P. Raval, *Virginia Union University*

Michael Rodgers, *Southeast Missouri State University*

Ruth Russo, *Whitman College*

Kathryn M. Rust, *Tennessee Technological University*

Mahin Sadrai, *St. Cloud State University*

Elsa C. Santos, *Colorado State University*

Traci Santos, *Stonehill College*

Elaine L. Schalck, *Alvernia College*

Kerri Scott, *University of Mississippi*

Shirish Shah, *College of Notre Dame of Maryland*

Brenda R. Shaw, *University of Connecticut*

Ralph Shaw, *Southeastern Louisiana University*

Steven M. Socol, *McHenry County College*

Cynthia Somers, *Red Rocks Community College*

Gordon Sproul, *University of South Carolina–Beaufort*

David A. Stanislawski, *Chattanooga St. Tech Community College*

Douglas Strout, *Alabama State University*

James Swartz, *Thomas More College*

Erach R. Talaty, *Wichita State University*

Joseph C. Tausta, *SUNY–Oneonta*

Larry A. Teter, *Three Rivers Community College*

Robert L. Thomas, *Angelo State University*

John S. Thompson, *Texas A&M University–Kingsville*

Robert C. Troy, *Central Connecticut State University*

Eric Trump, *Emporia State University*

Priscilla J. Tuttle, *Suffold Community College–Eastern*

John B. Vincent, *The University of Alabama–Tuscaloosa*

Christy L. Vogel, *Cabrillo College*

Yan Waguespack, *University of Maryland–Eastern Shore*

Robert W. Wallace, *Bentley College*

Willis Weigand, *Pennsylvania State University–Altoona*

Michael J. Welsh, *Columbia College–Chicago*

James E. Whinnery, *West Texas A&M University*

Yuefeng Xie, *Pennsylvania State University–Harrrisburg*

Martin G. Zysmilich, *George Washington University*

On The Need to Know

Sister Sara sings to Sky Masterson:
"I'll know when my love comes along . . . "
Sky croons in response:
"Mine will come as a surprise to me . . .
Mine, I'll leave to chance . . . and chemistry . . . "
Sister Sara (rather sharply):
"Chemistry?"
Sky (with great confidence):
"Yeah, Chemistry!"

From the 1950 Broadway musical, *Guys and Dolls,*
based on the book by Damon Runyon.

As you may have guessed, Sky Masterson and Sister Sara fall in love, sing the show's theme song down Broadway, and live happily ever after as a result of the "chemistry" between them. After all, the show is a love story. Real life is not a Broadway show. It is actually far more interesting, and it depends on chemistry to make everything happen, not just happy endings.

What makes us fall in love, in real life? What attracts you to someone else? Why do we so often attribute our social attractions and antagonisms to "chemistry"? From a physical standpoint, the feelings other people arouse in us, and the feelings we arouse in them, truly are the result of chemistry.

Interactions between humans depend on the chemistry within our bodies. Chemical reactions in our eyes and nerves allow us to see other people and feel their touch. Chemical reactions in our brains allow us to decide what to say to people; and the chemistry within nerves and muscles allows us to say it.

Chemists and biochemists already understand much of the basic chemistry involved in letting us see, talk to, and touch other people, but they are a long way from understanding the special "chemistry" that makes some relationships grow and others wither. Yet it is fun to try to learn the answers to these more subtle mysteries of chemistry, because it is so human to wonder, to find out about our world. In short, it is human to do the thing we call "science."

Studying "love" and attractions can be a very scientific process. A psychologist might study what we think

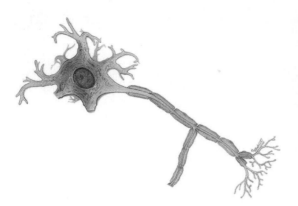 

The human nervous system consists of a vast array of neurons (nerve cells) like the ones shown here. The neurons communicate with our "central processing unit," the brain. Every thought, every sensation, every feeling we have travels along these neural pathways. The signals that neurons send are chemical. Would you like to touch someone's hand? Waves of sodium and potassium ions trade places across cell membranes generating an action potential of 100 millivolts in a matter of a few milliseconds. It doesn't sound very romantic, but it works. The neurochemical processes involved in social interactions are far more complex and, as a result, not so readily understood, but this we know: chemistry is central to all biology.

about when we meet that special someone. An anthropologist might study ethnic or racial patterns in people's choices of partners. A chemist, however, would be interested in the chemical changes in your body when you see someone you love or to whom you are attracted.

For many years, chemists have been looking at the chemical signals that initiate and control the sexual interactions of certain insects. These chemicals, known as **pheromones,** are produced by one member of a species to affect the behavior of another. For example, just a single molecule of the pheromone bombykol emitted by a female silkworm moth can, under certain conditions, elicit a response from the male silkworm moth. That's powerful chemistry. Pheromones are known to play a part in the behavior of some mammals as well.

Some research suggests that humans may also produce and respond to pheromone-like signals, which may influence our social interactions. This, however, is a controversial and much debated possibility. In September 2000 the first gene possibly involved in human pheromone activity was discovered. We are just in the very early stages of discovering the chemical foundations of human feelings and behavior. Such debate and controversy is part of the excitement of chemistry. Chemists are still learning about the world, still feeling their way, constantly discovering more about the influence of chemicals on all aspects of our lives.

P.1 | Why Do We Need to Understand Chemistry?

Where might the chemical study of love and attraction lead us? We will certainly learn more about how we and the rest of the world work, and in some scientific studies, that is sufficient reward. Often, however, new chemical understanding leads to an important practical application for the chemistry. In the case of insect pheromones, it is now possible to isolate the pheromones, manufacture them, and use them in chemical traps in the battle to control insect populations (Figure P.1). Any such application of scientific knowledge to serve some practical purpose is called **technology.**

Every day we benefit from countless scientific and technological advances that have greatly improved the quality of our lives (see Case in Point: Life without Chemical Technology). Do you need to understand chemistry to enjoy a glass of homogenized vitamin-fortified milk? Not really. Would you perform even better if you knew how the synthetic microfibers in your running suits were designed to keep moisture away from the skin? Perhaps not. Yet there will be times when knowledge of chemistry will serve you well.

If the study of the chemistry of human attractions generates potential new technologies, the possibilities will be controversial and stimulate fierce debate. Such debates have already raged about the contraceptive chemicals used to control human reproduction, about the use of the chemical techniques involved in genetic engineering, and about the use of chemical technology to alter our food, to give just three prominent examples. To be informed about such debates, to participate in them, and to be ready to make educated decisions on such issues for yourself, you will need to know some chemistry. *The intimate connection between chemistry, technology, and life is the essential reason for our need to understand chemistry.* The connection between chemistry and everyday life is a major theme of this book.

Figure P.1   This trap makes use of insect pheromones to attract and capture insects: in this instance the Japanese Beetle. The trap targets the beetle without endangering the surrounding plants, soil, or other living organisms.

What Is Chemistry All About?

Before we proceed any further, we should briefly consider what we mean by "chemistry" and even the term "science" itself. **Chemistry** is the systematic study of all substances. This includes their composition, the properties they exhibit, and the changes they undergo when they react with other substances. The change resulting from the interaction of one substance with another is called a **chemical reaction,** and the substances themselves are called *chemicals*. Chemistry is sometimes formally defined as "the systematic study of matter," which means that it is the study of everything that occupies space and possesses mass (and consequently has some weight). In other words, it is the systematic study of *virtually everything*, because almost

 ## case in point

Life without Chemical Technology

It is seven o'clock and your roommate wakes you up. Time to crawl out of bed and face the day: a sociology lecture, an accounting test, and three hours of chemistry lab. . . . Things could be worse. Imagine what our relatively pampered lifestyles would be like if we could not take advantage of chemical technology.

You want to have a cup of "decaf?" Sorry, the decaffeinating process involves interactions between methylene chloride or carbon dioxide and the coffee beans. You go to turn on the lights. Sorry! Incandescent light bulbs involve heating a tungsten filament in an evacuated glass bulb. Making the glass and extracting the tungsten from its ore both depend on chemical processes.

Time to take a shower and get ready. You will have to get clean without modern soap, shampoo, or toothpaste because all are products of industrial chemistry. You want to put on your pantyhose or cotton/polyester socks? You can't do that either because in a world without chemical technology, polymers such as nylon and polyester do not exist. Neither do the synthetic dyes that color your clothing.

Time to go to your first class. Hop in your car—wait! There is no car! No modern plastics, metals, or glass are available without chemical production. And gasoline is one product of the chemical processing of crude oil.

You run to class and arrive almost on time. The teacher is teaching, but you can't take notes because the ink for your pen is a combination of synthetic chemical dyes, which in a world lacking chemical technology would not exist. Nor can you have books, which depend on chemically treated paper, ink, and glue.

This whole experience is giving you a headache. Time for an aspirin. But . . . you guessed it, aspirin is a chemically manufactured product. Well, if you can't have anything else, how about a drink of water? Well, yes, water is available, but almost all water is now made safe to drink by treatment with chemical disinfectants.

Perhaps the only bright spot in a rapidly deteriorating day is that you can take a relaxing walk around the neighborhood when you get home. Then again, concrete and asphalt are just two more products of the technological application of chemistry—to say nothing of your sneakers.

everything that we can see and touch and feel is composed of matter. This description also incorporates the definition of science: a systematic study—a study of the physical world and how it works.

The whole Earth is made of chemicals, as are all the other planets and the stars (Figure P.2). All the living things on Earth are made of chemicals, and chemical reactions sustain every single thing you do, every thought that you think, every opinion you form. All of the materials and machines we construct are made out of the chemicals available on Earth, often transformed by chemical reactions into forms that meet our specific needs.

So the study of chemistry is the study of the Sun and stars, the Earth, the sea, all life, and all of our materials and machines. To fully understand these things, you need to understand some chemistry.

Making Choices: How Human Activity Affects the Chemistry of Our World

Although our world is a world of chemicals that sustain us and are exploited by us, it is also a world of choices. Throughout your life you will make choices that will essentially be decisions that affect how you experience and manipulate the chemical world. When you decide what to eat, you are choosing chemicals that will become a part of your body and will release the energy to power your activities. When you decide what to wear, you are choosing chemical materials that will give you the combination of comfort, warmth, protection, and appearance that you desire. When you decide on a method of transport, you are choosing chemicals gathered from the environment that will power your journey, chemicals that will be released to the environment as a result of it, and chemicals that may well affect the chemistry of your body. A ride by bicycle, for example, will have a very different effect on the chemistry of both you and your environment than a ride by car. Even when you decide where to live, you are choosing

Figure P.2 Chemistry is nothing less than the study of the universe. The Earth, the other planets and their satellites, the Sun and all other stars are composed of chemicals. Here we see the gaseous raw material from which stars are born. The Earth and its moon were formed from such materials over 4 billion years ago. Why they are so different from one another is a matter of physics and chemistry.

whether to be surrounded by the chemistry of plants or the chemistry of brick and concrete; whether to breathe in the chemical scents of the city or the chemical scents of the suburbs or country (Figure P.3). Life is mainly a series of chemical choices!

The Impact of Chemistry

Here's a challenge: Name five things (processes, products, etc.) that do not, in some fashion, involve chemistry. Are you having a hard time coming up with *any* good answers? Chemistry has an impact on us from the time we wake up until the time we go to sleep. In fact, chemical processes occur in our bodies when we sleep!

Large organizations such as industrial corporations and governments are also fully involved in the business of making chemical choices. They choose which chemical raw materials should be taken from the environment and in which ways these chemicals should be used. They choose which chemicals can be released as "wastes" to the environment, and which chemicals should be reused rather than disposed of as waste.

The practice of medicine is largely a matter of altering the chemistry of the body to change the chemistry of sickness into the chemistry of good health. Regardless of whether synthetic drugs or natural remedies are chosen, the purpose of all medicines is to manipulate the chemistry of the body. The need to choose and use medicines to help you when you are ill is strongly influenced by your own choices about which foods you eat, how much exercise you do, and whether or not you smoke cigarettes. So chemical choices can keep you well or make you ill in addition to being used to help you get better when you are ill.

Many people make all of their chemical choices without much knowledge of chemistry, and without even realizing that their choices have anything to do with chemistry. Some knowledge of chemistry, however, can be a great help in making appropriate choices; whereas a lack of chemical knowledge can increase the risks of making inappropriate and perhaps even dangerous choices. Chemical knowledge makes you aware of the benefits of some chemicals, the

A.

B.

Figure P.3   Although you may not be fully aware of the fact, the decisions you make concerning what to eat, how to travel, and where to live affect the chemistry in your life. **A.** Life in the big city may expose your body to high levels of air pollution. When you drive your car in the city, you add to that pollution. Riding your bike may give you greater mobility and be more environmentally friendly, yet you may expose your lungs to even greater levels of air pollutants. **B.** Life in the country may offer fewer big-city conveniences, but the air you breathe may well be cleaner. The pollution from automobile exhaust will dissipate quickly. There may be other hazards however. What about the water you drink? Are there any safeguards to protect you from agricultural or industrial run-off in the groundwater?

drawbacks and dangers of others, and of more complex situations in which the balance between risks and benefits depends on particular circumstances.

Chemical knowledge enables you to realize that a chemical that is beneficial in small amounts may be harmful, or even lethal, in large amounts. Common table salt, for example, is essential for life, but can increase the risk of high blood pressure and heart disease if consumed to excess over long periods of time. In very large doses, table salt could even kill you within minutes. Chemical knowledge enables you to understand how cholesterol can be an essential part of the body, but can also be regarded as a health hazard when consumed in excessive amounts. Chemical knowledge enables you to understand why ozone is a "good thing" high up in the atmosphere (where it screens us from potentially damaging radiation from the Sun), but a "bad thing" at street level (where it is an irritating and hazardous pollutant). Throughout this course, you will repeatedly be introduced to situations in which chemical knowledge enables you to understand the choices you make throughout your life and, we hope, help you to make more appropriate choices. You will also discover another equally fulfilling reason for learning about chemistry, namely that it is fascinating and great fun to discover the web of chemical events that sustains everything within and around you.

P.2 | How We Find Out about Chemistry

Chemistry in Practice

Chemists, like all scientists, find out about the world by making observations and measurements. Sometimes those observations are of contrived situations, called **experiments,** and sometimes they are observations of nature at work. By and large, chemists do most of their work in laboratories rather than in nature. Nature is so complex that we are not usually able to figure out what is going on by simply observing what happens. Instead we do experiments that bring substances together under *controlled* conditions to observe and measure what happens. Typically we try to identify what new substances have been formed during a reaction, how fast this has happened, whether any minor products form in addition to the principal product, and how much energy change has occurred during the reaction. This often requires the use of complicated equipment to try to understand what is going on at the atomic and molecular level. A fundamental principle of chemical understanding is that we can explain many readily observable features of our world by figuring out what is happening on an atom-by-atom basis.

There is no rigid formula that all chemists (and other scientists) follow to find out more about the world, but the general pattern involves the following steps:

1. **State the Problem:** Be clear about what you want to know (what *questions* you want answered) or what you want to do.
2. **Design Experiments:** Set up equipment and substances in such a way that clear measurements and observations can result.
3. **Gather Data:** This is the process of making observations and measurements and writing them down. An alert scientist will make a note of unexpected results, as well as results that were expected. The unexpected results can often lead to the most significant new discoveries. There is a certain satisfaction if an experiment confirms what you had expected at the beginning but considerable excitement when the experiment shows that something else is happening instead.
4. **Interpret the Data:** Figure out what the experiment reveals to you about nature.
5. **Plan Future Work:** This frequently means just going back to step 1 again, either to investigate a new problem or simply to repeat your work. Repetition is very important: *to be convincing, the results of scientific experiments must be repeatable.*
6. **Publish Your Results:** Scientists must always, eventually, set out their results in a very clear form for other scientists to read, comment on, and try to confirm. The history of science is full of examples of published results which later had to be retracted when other scientists could not confirm the results and went on to identify errors in the original work that led to the wrong conclusions being drawn. So the publication of results is important, not

only as a means of spreading information, but also as part of the process of checking that results are repeatable.

The goal of the experiments is often to test the viability of a particular idea, a **hypothesis.** A hypothesis is a statement of what we think will happen during an experiment, so it embodies some *prediction* and possibly a tentative explanation of why it happens as it does. A hypothesis is not a guess, but rather it results from our best understanding of the materials involved. It is frequently said that we can never prove that a hypothesis is true; the best we can do is prove that it is false. Therefore the best experiments are designed to do just that, in other words, to show that what the hypothesis predicts will happen does not in fact occur.

A hypothesis that stands up to such testing gains in credibility and may eventually be spoken of as law, an idea or explanation that carries the weight of a body of supporting physical evidence. Scientists also seek to understand the fundamental behavior giving rise to such laws. These explanations are called **theories.** Ultimately, the goal of the scientific community is to discover the underlying **scientific laws** that describe fully how matter in the universe behaves.

 ## hands-on

Here is an experiment for you to try. You will need vinegar, water, baking soda, baking powder, household ammonia, dish-washing liquid, as well as a narrow-necked bottle such as a soda bottle, and a candle. Some of the activities will be messy, so do them in a place where the mess can be easily cleaned up.

- Put a tablespoon of baking powder into the bottle and add about 4 ounces of water. Observe what happens. Try to keep the contents inside the bottle by pressing your hand over the top of the bottle. Are you able to do so? Allow the escaping gas to flow over a burning candle flame. What happens?
- Repeat the previous experiment using baking soda instead of baking powder.
- Repeat the experiment with baking soda, but use vinegar instead of water.
- Repeat both the baking powder and baking soda experiments using household ammonia instead of water.
- Try adding a generous squirt of dish-washing liquid to the water first, before adding the baking powder. Predict what you think will happen before you do this experiment.
- Design your own experiment using these substances and predict what you expect to happen. Were you correct?
- Write down the chemical names of the ingredients in the baking powder, baking soda, vinegar, and ammonia. As the semester goes along, find out what chemical processes happen when these substances are brought together.

A lot of household products, especially cleaners, are powerful chemical agents. Some can become dangerous when mixed together. For instance, when chlorine bleach (Clorox®, for example) is added to ammonia, the mixture is potentially lethal. Your experiments should be done in a controlled environment, and only if you have an idea of what the hazards are.

The **scientific method** is what we have just been discussing. It essentially involves looking at nature, asking questions, and learning from the answers we obtain. The scientist's job is to think of appropriate questions and devise ways to test them. (Case in Point: The Pheromone and the Spawning Goldfish details one such endeavor.) The most essential principle of the scientific method is to be accepting of physical evidence, even when that evidence is not as expected. A willingness to rethink an idea in light of new evidence lies behind all advances in scientific knowledge.

Learning from nature is not always easy. It often involves following and then eventually abandoning false leads, and constantly modifying our view of things in the light of what

■ case in point

The Pheromone and the Spawning Goldfish: An Application of the Scientific Method

Like many species of fish that live in temperate climates, individual female goldfish spawn (release their eggs) only once or twice a year. Spawning typically takes place in the low light of early morning in turbid ponds and lasts only a few hours. Spawning females are generally greatly outnumbered by males, whose season for spawning (release of sperm) is much longer than that of the females. How does an individual male goldfish "know" when a female is ready to spawn? And once he does know, how does he maximize his chances of spawning successfully in this highly competitive environment? There has been much research into these questions over the years and Michelle DeFraipoint and Peter Sorensen suggest that pheromones (chemical signals) play an important role. Their 1993 paper* explores the impact of a particular pheromone, 17α,20β-dihydroxy-4-pregnen-3-one (or 17,20βP for short), on the reproductive behavior and physiology of male goldfish. Their work is an excellent illustration of how the scientific method is used to answer questions about how the world works.

When a female fish enters a reproductive cycle, her sex hormone levels rise to stimulate maturation of her eggs. DeFraipoint and Sorensen have discovered she also releases several of these hormones to the water where they function as pheromones. One of these pheromones is 17,20βP. The concentration of this pheromone is low, equivalent to about 1 molecule of 17,20βP for every 10 *billion* molecules of water in a 20-gallon tank. The male detects this (using his sense of smell) and responds with a dramatic increase in his own hormones. This much was known prior to conducting this study. What was not known, however, was *how* 17,20βP works to spur the male's reproductive success. This was the focus of the study.

The two scientists first posed three specific questions that they wanted to answer:

1. Does exposure to 17,20βP actually change the behavior of male goldfish?
2. Does overnight exposure to the pheromone enhance the success of fertilization?
3. Does exposure to 17,20βP enhance the amount and quality of the sperm released by the male goldfish?

The next step was to design experiments to address the questions. The experiments were designed to collect

The female goldfish emits a pheromone that stimulates the male goldfish into mating.

meaningful data (information) that could be used to test a variety of hypotheses.

Proper experimental design for this study involved having what is known as a *control* population of fish. The control population consisted of goldfish that were not exposed to the pheromone but were otherwise treated the same as the goldfish exposed to the pheromone. Controls are used in all sorts of experiments to set up two situations, differing in only one respect (such as presence of the pheromone). This allows any differences between what happens in the two situations to be attributed to the single original difference between them.

Interpretation of the data is where the authors gave meaning to the data. In this and subsequent studies, the data support the observations that males who were exposed to 17,20βP "nudged, chased and spawned with females two to five times more frequently" than the control males (question 1). Further, those exposed to 17,20βP overnight had better spawning success than the control group (question 2). Their sperm were more active and thus more fertile (question 3).

Yet the authors also raise some questions for future study. The actual physiological mechanisms by which 17,20βP enhances spawning and related behavior are not yet known. The interactions of 17,20βP and other pheromones are not yet understood. The effect of being in a natural waterway, rather than a laboratory aquarium under very tightly controlled conditions, is also food for thought for the future.

Recently, two advances have been made to the understanding of how 17,20βP functions as a pheromone in goldfish. First, DNA fingerprinting (see Chapter 16) was used to follow the reproductive success of male goldfish. As predicted, exposure to the pheromone greatly

*DeFraipoint, Michelle, and Sorensen, Peter W. Exposure to the Pheromone 17α,20β-dihydroxy-4-pregnen-3-one Enhances the Behavioural Spawning Success, Sperm Production and Sperm Motility of Male Goldfish. *Animal Behaviour*, 46 (1993): 245–246.

enhanced the ability of males to gain access to females and fertilize their eggs. Second, Sorensen and his group discovered that the actions of 17,20βP are enhanced by a second, related chemical that is released by ovulatory females. This suggests that the pheromone may be a mixture of two substances.

New work by other scientists has shown that plants emit a set of compounds when they are infested with caterpillars. These compounds send a message to moths to stop laying eggs on the plant. They also signal wasps that there is a feast waiting!

The specific processes involved in asking and answering questions may differ from scientist to scientist and from study to study. Yet the general approach of asking questions about nature and seeking out answers in a rational, systematic, observable, repeatable fashion is what makes the scientific method such a powerful tool to learn about our world.

Zheng, W., Strobeck, C. & Stacey, N. E. 1997. The steroid pheromone 17a,20b-dihydroxy-4-pregnen-3-one increases fertility and paternity in goldfish. *J. Exp. Biol.* 200 (1997): 2833–2840.

Sorensen, P. W., Scott, A. P., Stacey, N. E. & Bowdin. 1995a. Sulfated 17,20b-dihydroxy-4-pregnen-3-one functions as a potent and specific olfactory stimulant with pheromonal actions in the goldfish. *Gen. Comp. Endocrinol.* 100 (1995a): 128–142.

nature reveals to us (Figure P.4). So the scientific method is not a linear process that always takes us in a straight line from a question about nature to the discovery of the answer. We must go back and forth between the key aspects of the scientific method. We must ask some questions, make observations, test hypotheses, ask new questions, devise new hypotheses, make more observations—and so on. Being human, we also make mistakes, but the scientific method corrects our mistakes when we discover that our supposed results cannot be repeated (or, more embarrassingly, someone else discovers they cannot be repeated).

As we apply the scientific method, our understanding of the world becomes ever more thorough because we are always answering questions. Yet the need to know more will always exist, because new questions are always being raised, often more quickly than the old questions are answered. Science is a fascinating, never-ending cycle of inquiry and discovery, inquiry and discovery. . . . The more we learn about nature through this process, the more we realize how much remains unknown. By the end of the 19th century, many scientists believed humanity was near to a complete understanding of nature, with almost nothing of importance left to be discovered. Nowadays we realize just how false such an attitude was and is. It has been said that one of the most important scientific discoveries of the 20th century is the realization of how little we really know. Modern science is full of mysteries, ranging from some of the most fundamental properties of matter up to the nature of consciousness and thought.

Scientists Are People

Scientists are subject to the same personal and social pressures as anyone else. For most scientists, their interest in science itself is of primary importance. Yet they all want to be well-regarded by colleagues. They want to get invited to prestigious conferences. They want to do well financially. The overwhelming majority of scientists do their science rationally, but the desire to get ahead is something that everyone, no matter what they do for a living, has to grapple with. So, like any other profession, science contains its share of frauds and scandals, muddles, jealousies, stupidities, and simple honest mistakes.

Scientists are not perfect. However, science, when properly conducted, is above all else a rational process. Science in its purest form is also a completely open process. Through the publication of results and open discussion of their meaning, everyone should ideally have access to what scientists are claiming to have found out about nature, and how they are going to apply their newly acquired knowledge. Everyone should be free to challenge the conclusions of scientists or to argue that their plans are misguided. In these disagreements, the ultimate authority is a carefully conducted experiment. In this free and open pure form, science is perhaps the most glorious expression of the importance of the freedom of thought and speech.

In the long run science is an amazingly efficient self-correcting activity. In other words, mistakes or deceptions can never survive unchallenged forever (and usually do not survive for

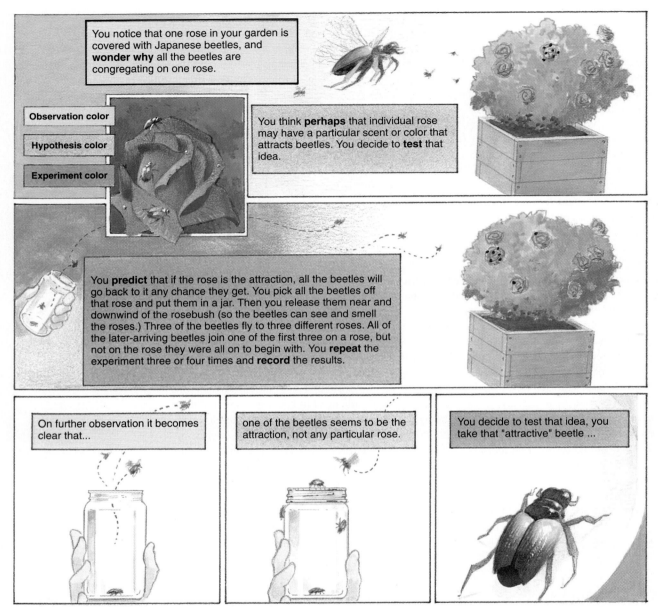

Figure P.4   Science is not an activity that must be confined to professional scientists in laboratories. It is a very human endeavor, accessible to any curious person.

long) because one principle of the scientific method is that results must be seen to be consistently repeatable by many different people before they can be fully accepted. To gain respect and belief in science, you must, eventually, allow your results to face the scrutiny of many other scientists, who are often quite keen to discover that you are wrong. This essential feature of the scientific method can be summarized in very simple everyday terms: the mistakes get fixed, eventually.

You now know that the scientific method is not some deep, dark secret that separates scientists from the rest of society. On the contrary, it is our goal that you should understand and adopt the scientific method in your day-to-day thinking. Perhaps the most interesting part of the scientific method is that you can and should raise your own questions about results that scientists present to you. As soon as you learn to apply the scientific method, it becomes possible for you to be one of the people who fixes past mistakes or prevents mistakes in the future.

P.3 | Chemical Technology: Risks and Benefits

In Section P.2 we focused our attention on how scientists learn things about nature. One reason for learning things about nature is simple curiosity. The other major reason is to be able to use what we learn to *modify* what nature has supplied in ways intended to improve our lives, often through technology (Figure P.5A,B,C,D). Science and technology frequently advance together and sometimes are done by the same individuals, but they are separate enterprises with different goals. The goal of science is to learn; the goal of technology is to make use of what we learn. Frequently in this book we will show how scientific knowledge has directly affected technology and our everyday lives.

The modern application of chemical technology is big business. In 1999, academic institutions spent $915 million on chemical research and development; chemical firms spent almost $5 billion. The fact that so much money is spent on chemistry means that this chemistry is doing an awful lot of good for many people. The advantages—the benefits—that the industrial use of chemistry brings us include new medicines; effective pesticides and fertilizers; synthetic materials and fabrics; new types of industrial materials, paints, cosmetics, and artificial sweeteners; and much more. See, for instance, all that technology can glean from an ear of corn in the Case in Point: In Your Ear. The fruits of this labor are all around us, and make

A. Smelting.

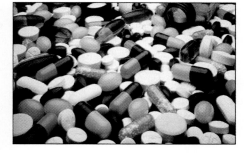

B. Pills.

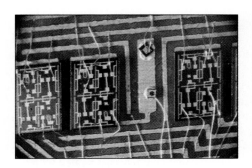

C. Computer chip.

D. Astronaut suit.

Figure P.5
Science and technology provided these advances to our society. But some of these benefits bring risks as well.

case in point

In Your Ear: The Chemistry and Technology of Corn

What do sugar, ethanol, and golf tees have in common? The connection is corn. This nutritionally and economically important crop can be processed into a startling array of products. What is the fate of an ear of corn? As shown in the figure, most of it is used to feed people and animals. Yet with our new chemical understanding of the substances that make up corn, technology is opening up new horizons.

Modifying food to meet (or even create!) consumer demand is not at all new. Aztec, Zuni, and other cultures began growing corn domestically about 4700 years ago. Since the start of farming in the New World, over 300 varieties of corn have been developed. In the new millennium, however, our understanding of how things work chemically has advanced to the point where we can take an ear of corn, analyze its structure for specific pest or disease resistance, and create new pest-resistant corn varieties. We can process that corn into packaging that will chemically fall apart much faster than styrofoam "peanuts," and therefore be less environmentally harmful. Corn can even be processed to make golf tees.

Why is this important to know? The beauty and richness of chemistry is that we are now at a point where we can *chemically modify* what nature has supplied. The changes are sometimes good (disease control), sometimes bad (air pollution), and occasionally fairly meaningless (flexible polymers that kids can squeeze out of a toy monster). Yet the *overall* influence of chemistry in our lives is profound. Your interaction with chemistry will not end with this course because, as a chemistry-wise Pogo surely would have said, "We have met the chemistry, and it is us!" We want you to enjoy the subject for a lifetime. After all, it is what you are!

Rhoades, Robert E. The Golden Grain, Corn. *National Geographic* June 1993: 92–117.

Wilford, John Noble. Corn in the New World: A Relative Latecomer. *New York Times* 7 March, 1995.

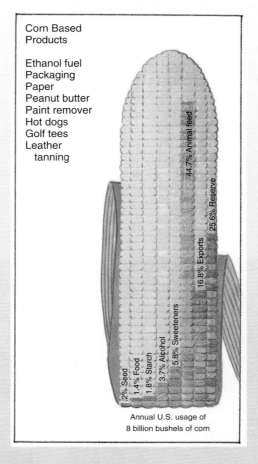

Corn Based Products

Ethanol fuel
Packaging
Paper
Peanut butter
Paint remover
Hot dogs
Golf tees
Leather
   tanning

2% Seed
1.4% Food
1.8% Starch
3.7% Alcohol
5.8% Sweeteners
16.8% Exports
44.7% Animal feed
25.6% Reserve

Annual U.S. usage of
8 billion bushels of corn

Corn represents the ultimate confluence of chemistry, technology, and history. It has been used for thousands of years as a food source. It can now be processed into many different products, ranging from packaging to golf tees.

possible many of the things we associate with a high standard of living. This sounds good! Yet, life is never simple. The modern use of chemistry is like a double-edged sword: it cuts both ways.

The benefits that result from our exploitation of chemistry are accompanied by drawbacks and dangers—the risks. The chemical processes used by the chemical industry generate a wide variety of hazardous wastes. These can pollute the soil, water, and air. Some can combine with other chemicals in the atmosphere to fall back to Earth as acid rain. The pesticides and fertilizers used by farmers can make their way into groundwater, creating the potential for contamination of water supplies. These few examples illustrate that there are risks associated with our use of chemical technology. These risks must always be weighed against the benefits. The field of study that weighs the pros and cons of a process is called **risk-benefit analysis.** Many of the choices that face us in our interaction with the world of chemistry involve deciding when the benefits outweigh the risks and when the risks outweigh the benefits.

Another form of risk analysis takes into account the cost of avoiding risks and addresses the practical side of the financial resources available. This is known as **cost-benefit analysis.** It is cost-benefit analysis that comes into play when a community or government agency must decide its degree of tolerance for certain levels of toxins in the environment. If a community has scarce monetary resources, it may decide that it needs a new water treatment plant more than the removal of well-covered asbestos in the city hall.

Assessing Risks, Assessing Benefits

Any choice we make usually has its consequences. To make appropriate choices we must consider what there is to be gained from a particular choice (the benefits) and what there is to be lost (the risks). We make decisions every day based on our own personal risk-benefit analyses. Do you eat that extra piece of cake after dinner? What are the risks? Roughly 400 extra calories might be a problem. Perhaps the sugar will make you sleepy or possibly hyperactive. The fat may increase your long-term risk of heart disease. What about the benefits? The taste and texture of the cake will give you pleasure, and if the added calories force you to exercise to assuage your guilt, then they could count as a benefit rather than a risk.

consider this:

Hot Button Issues

Although the scientific laws that govern the behavior of chemicals are constant, the way in which we use chemistry is always changing, as we learn more about the science. There are a number of controversial (so called "hot-button") issues relating to chemistry in our new millennium. One example is the use of the pesticide atrazine in farm fields. Another is the genetic modification of crops. Can you name five other current "hot-button" issues?

Many of the risk-benefit analyses involved in making choices in chemistry tend to be much more complex than the simple "yes or no" evaluation of whether to eat a piece of cake, but the underlying logic is the same. As we examine chemistry and society throughout this book, we will look at many issues in terms of risk assessment, using both risk-benefit and cost-benefit considerations. A few things to keep in mind during your own evaluation of the issues are:

- There is virtually never an important issue that involves only science. Social concerns and politics are almost always also involved. So risk assessment that involves scientific matters must also consider the interactions between science and society.
- Chemical processes range from almost entirely beneficial and harmless (the synthesis of safe vaccines, for example) to the obviously repugnant (such as the manufacture of chemical weapons). Most, however, lie somewhere in between. The ones that lie between the extremes are the most challenging to assess.
- Something that can be viewed as a benefit to one group of people may pose a risk to another group. The workers in a factory may view the chemical processes that give them employment as benefits, whereas the people whose land and water are polluted by the factory waste may view the same chemistry as posing only risks.
- When considering risks and benefits directly, we should try to compare only those things that are directly comparable and can be stated in the same terms. For example, lives saved versus lives lost is a valid comparison. Time saved by using a new robot-based manufacturing system versus time lost while installing the system is also a valid comparison. However, comparing the additional expense of an emission-control device to the tons of pollutants released in automobile exhausts is very difficult, because money and weight, like apples and socks or money and lives, are not the same. We are often forced to attempt such comparisons, however, when making the difficult choices of real life.

Each chemical process that we perform and then use, must be evaluated on the basis of its risks and its benefits. This is a judgment that you will be asked to contemplate many times in this course and beyond.

P.4 | Chemical Information: Where Can You Get It and How Accurate Will It Be?

A vital part of the scientific method is the publication of results, to enable other scientists to become aware of them, to comment on them, to learn from them, and to check their accuracy. So a vital part of all scientists' work is the spreading of new information, usually in the form of research papers published in specialized journals.

We truly live in the information age. Figure P.6A and B shows that between 1907 and 2000, the number of chemistry-related articles published per year grew at an exponential rate. Although the growth has lately become less dramatic, there is far too much information available for even the most diligent of readers.

Chemists read only a small fraction of the available science literature, so how can nonscientists keep abreast of important developments while remaining confident that what they read will be reasonably accurate? There is a fairly reliable rule of thumb about science information: the closer the source of information is to the researcher, the more detailed and

Figure P.6 The number of chemistry articles published per year has grown from less than 8000 (1907) to over 573,000 (2000). **A.** One of the authors (JC) is shown here alongside all of the abstracts of chemistry-related articles published in 1910, 1920, and so forth. **B.** The increase in the number of scientific articles published in the 20th century is astounding. Still, the relationship between science and society is ever-present. Note the drop in number of chemistry-related articles during the first and second world wars!

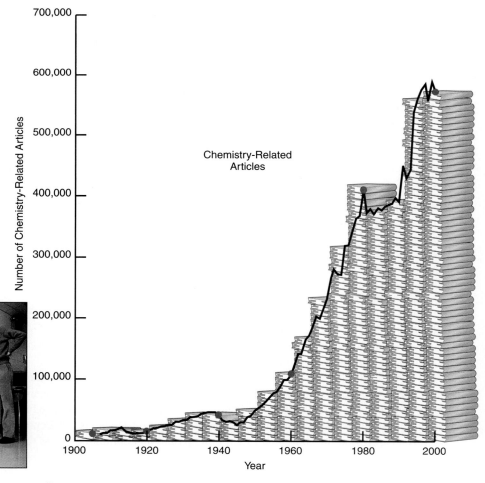

A.

B.

accurate that information is likely to be. A look at various reports about a skin-cream product will help make the point.

In 1988, Retin-A was touted on the evening news and in newspapers as an anti-aging skin cream. The popular press said that Retin-A could remove wrinkles and huge consumer demand for the product ensued. The study (the *primary* reference) that served as the basis for the unbridled optimism of the news reports was published by Dr. Jonathan Weiss in the *Journal of the American Medical Association* (*JAMA*).

The goal of Weiss's research was to study the effect of Retin-A on sun-damaged skin. Of the 40 people tested, 37 developed dermatitis to some degree as a result of the application of Retin-A. Patients also reported burning, tingling, and severe itching. Such symptoms, which lasted from 2 weeks to 3 months after the initial treatment, decreased when Retin-A was used less often. After the initial dermatitis (an inflammation of the skin), many patients did show improvement in skin quality, such as fewer wrinkles. These conclusions, drawn from Weiss's own study, are far less uniformly favorable than the news media reports of his study.

Weiss and his co-authors made it clear in their *JAMA* report that the number of participants in the study was very small and that more studies were necessary. But this caution and the negative aspects of the treatment itself were not reported in the mass media. As we go from the primary reference to more commonly available popular resources, something often gets lost in the translation. It is a bit like the children's game "operator" in which a message is verbally passed from one child to the next. The message gets more and more garbled the farther along it goes.

There are four levels of available information about science (Figure P.7):

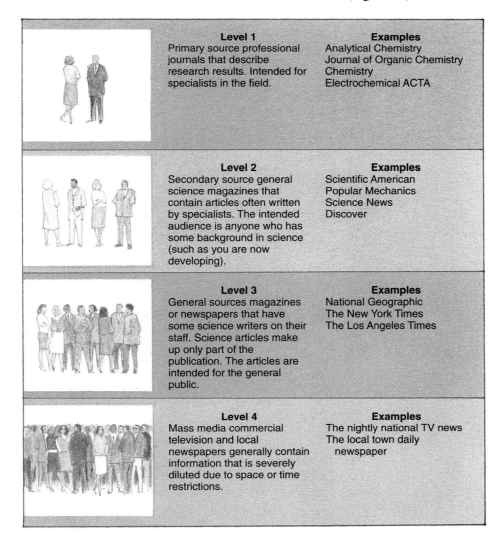

Level 1
Primary source professional journals that describe research results. Intended for specialists in the field.

Examples
Analytical Chemistry
Journal of Organic Chemistry
Chemistry
Electrochemical ACTA

Level 2
Secondary source general science magazines that contain articles often written by specialists. The intended audience is anyone who has some background in science (such as you are now developing).

Examples
Scientific American
Popular Mechanics
Science News
Discover

Level 3
General sources magazines or newspapers that have some science writers on their staff. Science articles make up only part of the publication. The articles are intended for the general public.

Examples
National Geographic
The New York Times
The Los Angeles Times

Level 4
Mass media commercial television and local newspapers generally contain information that is severely diluted due to space or time restrictions.

Examples
The nightly national TV news
The local town daily newspaper

Figure P.7 The more information is condensed and simplified, the more people it can reach and influence.

pro con discussions

Is Chemical Technology a Good Thing?

pro

Without modern chemical technology we would be living like people did back in the Middle Ages, and it would *not* be idyllic.

Diseases that we scarcely bother to think about in the modern world would be killing millions due to the lack of modern vaccines and drugs. Our food supply would be much less varied and less secure, because modern pesticides and herbicides would not be there to help us. In fact, without fertilizers the land might not be able to support all the people it does now, so our civilization might collapse amid food riots, local land wars, and trade wars.

Even if we could still get along with one another reasonably well, most of the things we do for entertainment would be denied to us. Forget television, forget cinema, forget glossy colorful magazines, forget computers, forget cars, forget airplanes—forget just about everything that makes modern life interesting!

We would have to build from wood and stone, wear uncomfortable clothes that would be dull to look at and very difficult to clean. There would be lots for us to do because just staying alive would take so much more effort, but there would not be much that a modern person would consider worth living for.

Sure, there are the problems caused by chemical technology, such as industrial pollution, the use of limited supplies of raw materials, and the fact that the chemicals we make can cause new illnesses and not just cure old ones. But chemical technology also provides the answers to the problems it creates. It will give us the means to recycle materials more efficiently and give us ways to clean up the messes we have made. We may even learn to trap the Sun's energy in an efficient and dependable way, allowing us to forget about oil and coal as sources of fuel. What the future needs is more chemical technology, not less!

con

They call us "greens," "ecology-nuts," and a hundred other derogatory names, but it all boils down to one bottom line—they think that those of us who are truly concerned about the negative impact of chemical technology are crackpots who worry too much. They think we don't appreciate the economic benefits of chemistry-based manufacturing. We *do* appreciate them. We see the jobs, the money, the products—we see them all. But it is sometimes hard to see them *clearly* through the haze of pollution that is a result of our industry-based lifestyle.

It is only within the last 20 years that Lake Erie can once again support edible fish—and then only after a 50-year battle to stem industrial dumping into the lake as if it were a huge toilet bowl. There are cities such as Los Angeles that until recently exceeded safe levels of ozone in the air over 100 days each year. Industrially produced herbicides and insecticides flow from farmers' fields into municipal water systems. Oil spills occur somewhere in the world's oceans more than once every day!

Who supplies the oil and other raw materials to feed our voracious appetites for manufactured material goods? Mother Earth is the source, and her supply is not endless. Some observers estimate that we have as little as a 25-year supply of crude oil remaining, though others put the figure closer to 200 years. Whichever is correct, we will eventually run out, and those things that come from the oil—the energy, the plastics, the medicines—will all be gone. The minerals that supply metals do not exist in endless supply. Even the trees that supply cellulose for paper are, in many parts of the world, being ravaged by acid rain and other forms of human-based pollution.

We manufacture because we consume. We pollute because we manufacture. And we recycle and look for energy alternatives only because we "greens" of the world see that there is only so much pressure you can put on nature before she pushes back—hard.

further discussion

Try to develop your own arguments to support first the PRO position and then the CON position. Then write down a few thoughts about what you think is a sensible and balanced view about the benefits and drawbacks of modern chemical technology.

- Level 1: primary source—written accounts intended for professionals.
- Level 2: secondary source—written accounts intended for the science-literate public.
- Level 3: general source—written accounts on a variety of subjects intended for any audience.
- Level 4: mass media—briefly written audio or audiovisual reports intended for any audience.

At each level the information gets less detailed and less technical than at the previous level. That is not necessarily bad, because it is important to express results in ways that everyone can understand. Yet, as the Retin-A example demonstrates, we do have to remember that attempts at simplification and selective quotation can easily amount to *mis*information. If you are really interested in any media report about science, follow up your interest by consulting other sources that are closer to the original source of the information (or tackle the original source itself).

Where you get your information depends on how much time you have and how complete you want the information to be. Very few of us have the time, or the need, to go to level 1 (primary) sources. Yet relying strictly on the mass media causes us to sacrifice detail and quality. As a compromise, you may opt for secondary and general sources. You will find references to many primary, secondary, and general sources throughout this book.

Our introduction to the importance of chemistry—its relevance in everyday life, and the methods used to find out about chemistry—is complete. We can now begin to look in more detail at what chemicals are and how they influence our lives. In Chapter 1 we will begin by considering what we know (or suspect) about the origins of the chemicals we now find around us and within us.

main points

- Everything tangible on Earth, as well as everything in the universe, is made of chemicals.

- Chemical reactions in our bodies keep us alive and allow us to move, see, hear, talk, and think.

- Even if we don't realize it, we make chemical choices every day of our lives.

- Chemists and other scientists find out about nature in a systematic way, by asking questions, designing and performing experiments, gathering data, interpreting the data, and checking that their results can be repeated.

- The application of scientific knowledge to solve problems and improve our standard of living is known as technology.

- Modern civilization relies very heavily on chemical technology.

- The application of chemical knowledge brings both risks and benefits, which can be assessed in a procedure known as risk-benefit analysis.

- It's important for you to keep informed about the latest advances in science and technology. However, be careful about accounts of scientific studies reported in the mass media. Sometimes the reporters oversimplify the details and draw unwarranted conclusions.

important terms

Chemical reactions are the changes that occur when different chemicals interact. (p. 3)

Chemistry is the systematic study of all matter: its composition and properties, and the way all matter interacts. (p. 3)

Cost-benefit analysis is a form of risk-benefit analysis that occurs when we are weighing the benefits of something against its financial costs. (p. 13)

Experiments are actions in which substances are brought together under controlled conditions in order to observe and measure what happens. (p. 6)

A **hypothesis** is a statement of what we think will happen based on past observations, therefore it embodies some prediction about what will happen. A hypothesis is not a guess; rather, it results from our best understanding of the materials we are working with. (p. 7)

Pheromones are chemical signals that initiate and control interactions between animals of the same species. (p. 2)

Risk-benefit analysis is the field of study that weighs the pros and cons of a process. (p. 12)

Scientific laws describe the behavior of material in the physical world, often in terms of mathematical relationships. (p. 7)

Scientific Method: The rational approach to learn about nature. (p. 7)

Technology is the application of scientific knowledge to serve some practical purpose. (p. 2)

A **theory** is an explanation of why things happen as they do, based on an accumulation of data. (p. 7)

exercises

1. Use your own words to define each of these terms:
   a. science
   b. matter
   c. chemistry
   d. risk-benefit analysis
   e. primary source of information
   f. secondary source of information
2. Reread the Case in Point: Life Without Chemical Technology (Section P.1). There are many activities that you do during your day that we did not mention that are dependent on chemical technology. List five of these activities or processes.
3. State whether each of these is a valid risk-benefit comparison and justify your answer:
   a. Money saved versus money spent.
   b. Acidity of a river versus number of fish killed.
   c. Concentration of sulfur dioxide in the air from dirty coal versus cost to burn clean coal.
   d. Cost for industrial pollution controls versus cost of a government fine for having no such controls.
4. The latter part of May brings with it the end of the school year, thoughts of summer (work or play?), and, for some fair-skinned students, the desire to get a golden tan. Some sit in the sun. Others go to tanning salons. List some risks and benefits to consider when judging whether a tan is worth having.
5. Many science classes discuss the scientific method as a series of specific steps to be followed. List the steps in the scientific method. After doing so, close the book, take a few moments to have a soda, come back, and, without looking at the book, make another list of the steps. Do the lists differ? If so, what can you conclude about the idea of the scientific method as a series of predetermined, sequential steps?
6. The information given in the box to the right is similar to what you might find on a box of cold cereal.

a. What information is fact?
b. What does the cereal manufacturer want you to conclude about this product?
c. Are the comparisons fair? Why or why not?
7. A recent study examined the effect of the long-term absence of gravity on the bone density of space shuttle astronauts. After finding some calcium loss for one male astronaut after 7 days in Earth orbit, the study concluded, "long-term living in zero-gravity will cause substantial loss of bone density in space travelers." Was this conclusion valid? What could have been done to make the

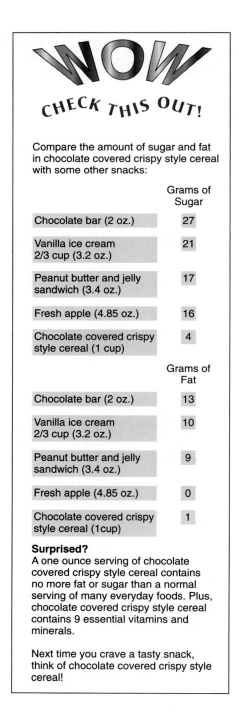

WOW
CHECK THIS OUT!

Compare the amount of sugar and fat in chocolate covered crispy style cereal with some other snacks:

	Grams of Sugar
Chocolate bar (2 oz.)	27
Vanilla ice cream 2/3 cup (3.2 oz.)	21
Peanut butter and jelly sandwich (3.4 oz.)	17
Fresh apple (4.85 oz.)	16
Chocolate covered crispy style cereal (1 cup)	4

	Grams of Fat
Chocolate bar (2 oz.)	13
Vanilla ice cream 2/3 cup (3.2 oz.)	10
Peanut butter and jelly sandwich (3.4 oz.)	9
Fresh apple (4.85 oz.)	0
Chocolate covered crispy style cereal (1cup)	1

Surprised?
A one ounce serving of chocolate covered crispy style cereal contains no more fat or sugar than a normal serving of many everyday foods. Plus, chocolate covered crispy style cereal contains 9 essential vitamins and minerals.

Next time you crave a tasty snack, think of chocolate covered crispy style cereal!

experiment more meaningful? Why might such a study not be done?

8. Characterize each of these as a type of information source (primary, secondary, etc.). If you are not sure, look up an issue at the campus library.
   a. *The Washington Post*
   b. *The Winneconne News* (or any small-town newspaper)
   c. *The Journal of the American Chemical Society*
   d. *Good Housekeeping*
   e. *Popular Science*

9. Try to list several materials or processes related to chemistry which you have been in contact with today.

10. Continue to develop the series of observations and experiments shown in Figure P.4. What important step or steps of the scientific process is/are not included in this garden experiment?

11. What variables would you have to control to set up an experiment to determine the effect of a new fertilizer on the growth of beans? Describe what experiments you would recommend the fertilizer manufacturer do to test the worth of such a new product. Describe what experiments you would recommend a strict environmentalist do to test the worth of such a new product.

12. Based on your results of the series of experiments with baking soda, baking powder, and so on, what similarities and differences are between baking soda and baking powder? What seems to be necessary to get gas evolution from these substances?

13. Diabetes is a serious genetic disease characterized by the inability to regulate the amount of glucose (sugar) in one's blood. Suppose that a new medicine is proposed to treat diabetes. The medicine is shown by experiments to have these features. The new medicine causes the glucose level in the blood of most persons to return to normal values within an hour after eating and to relieve most of the symptoms associated with diabetes. Suppose though that a small percentage of diabetics react occasionally by having their blood sugar level drop far below normal so that severe symptoms or even death can occur. How large a fraction of the population must respond favorably before you would recommend that the drug be put on the market?

food for thought

14. An important theme in this text is that science is absolutely intertwined with all aspects of our lives. It can get particularly sticky when one of the aspects is politics.

    The various parts of the Job Creation and Wage Enhancement Act of 1995 were passed by Congress and enacted into law between 1995 and 1997. This act was aimed at curbing excessive government regulation of business and offered risk-benefit analysis and peer review as one way to do this. The federal government has been in the business of consumer protection since 1906, with the creation of the Pure Food and Drug Act, which regulates the shipment of adulterated food in interstate commerce. Since then, organizations such as the Food and Drug Administration, the Department of Agriculture, and the Environmental Protection Agency have looked after the general welfare of the public.

    One part of the Job Creation and Wage Enhancement Act of 1995 deals with risk assessment. Let's say that an agency would like to regulate the use of a solvent that is known to be lethal with continual exposure. The act requires an estimate of "at least three other risks with which the public is familiar." This might equate the risk of the solvent with smoking a certain number of cigarettes per day, the likelihood of being killed in a highway crash, and the likelihood of being killed in a workplace accident: that is, measure risk against risk.

    Another part of the legislation requires that, "each agency (will) submit its scientific and economic findings to outside panels for peer review" and that these panels "shall not exclude peer reviewers merely because they represent entities that may have a potential interest in the outcome."

    Let us assume that you are a member of Congress. Do you support or oppose such an act? Why? Whom does the act protect? What might be the long-term effect of the act? Look up some current consumer protection decisions. Would this act strengthen or denude these decisions? You make the call!

15. Determining risk is often no simple matter. For instance, if a public-health risk is considered to be statistically insignificant, that is, it is very unlikely to happen, public perception of that risk may still drive government action or reaction. Nothing better illustrates this than the story that took the international spotlight in March of 1996: British Beef Banned Because of Mad Cow Disease.

    The actual facts of the story are that the British Government announced that a possible "link" existed between ten reported cases of a fatal human brain disease known as Creutzfeldt-Jakob Disease and a similar disease in cattle known as bovine spongiform encephalopathy (B.S.E.)—the now infamous Mad Cow Disease. B.S.E. was known to have affected some 11 million cattle in Britain in the late 1980s, but measures had been taken and the epidemic was largely under control. During all those years, British health officials reassured the beef-eating public that there was no danger of transmission of B.S.E. to humans. Then came 10 cases of Creutzfeldt-Jakob, typically a disease of the elderly, that had affected younger people and showed some unusual characteristics. The expert medical committee studying the issue stated that in the "absence of a credible alternative" the most likely explanation was exposure to B.S.E.-infected beef in the years before 1989. This they said "is a cause for great concern."

The British beef industry, worth some $6.5 billion a year to the British economy, was brought to its knees. Britons boycotted beef at home and the beef was banned in Europe. To regain public confidence, the British government assumed the responsibility of overseeing the killing of millions of cattle. The costs were in the billions of dollars. Yet, as of this writing, absolutely no scientific evidence exists to show a definite link, across species, between the two brain diseases.

Consider the issue from the point of view of the British Government, the farmers, and then the general public. The government cannot force a wary public to buy a food product that they have lost confidence in. Who should pay for the consequences of the action taken? To what extent do you entrust government agencies to protect the public health? Is any degree of risk unacceptable?

Do you remember this incident? From what sources did you get your information (primary, secondary, general)? Were you well enough informed that you could make a personal decision as to what to do if you were faced with the same situation?

16. A fly called a mediterranean fly (medfly) causes severe damage to fruit in California. In recent years there have been extensive programs to kill these insects by widespread spraying of insecticides. Suppose that a pheromone were discovered that would bring medflies in to a central point from 500 meters away and that 90% of the flies are killed by a small amount of insecticide when they reach the central location. Consider and compare the costs and benefits of the two treatments. Would you recommend that the state require the pheromone treatment? Why or why not?

17. Go to the chemistry section of your college or university library. Pick up a chemistry journal and glance through it. Read the abstract and the conclusions section of an article whose title catches your eye. Can you make any sense of the writing? What makes it difficult to read?

18. Look in a weekly news magazine and find an article dealing with chemistry. Ask a chemistry instructor to help you interpret what it means. Decide whether you think proper public decisions have been made regarding this issue. Make an outline of the article and your reactions to it.

readings

1. Sexual Chemistry. *U.S. News and World Report* (19 July 1993): 57–64.
2. Selinger, Ben. *Chemistry in the Marketplace,* 4th ed. Orlando, Florida: Harcourt Brace Jovanovich, 1988: 458.
3. Weiss, Jonathan A., et al. A Topical Treatment for Sun-Damaged Skin. *JAMA* (29 January 1988): 527–532.
4. Baum, Stuart J. *An Introduction to Organic and Biological Chemistry,* 4th ed. New York: MacMillan, 1987: 424–425.
5. Hileman, B., Long, J. R., and Worthy, W. Funding Continues to Grow at a Slower Pace. *Chemical & Engineering News* (19 August 1991): 32–70.
6. Brennan, M. B. and Long, J. R. Facts and Figures for Chemical R&D. *Chemical & Engineering News* (17 August 1992): 38–61.
7. Rutowski, Ronald L. Mating Strategies in Butterflies. *Scientific American* (July 1998): 64–69.
8. Rodriguez, Ivan, Greer, Charles A., Mok, Mai Y., and Mombaerts, Peter. A Putative Pheromone Receptor Gene Expressed in Human Olfactory Mucosa. *Nature Genetics* 26, No. 1 (2000): 18–19.

resources

Bronowski, Jacob. *A Sense of the Future.* Cambridge, Mass.: MIT Press, 1977.

Brunk, Conrad F.; Hayworth, Lawrence; and Lee, Brenda. *Value Assumptions in Risk Management: A Case Study of the Alachlor Controversy.* Waterloo, Ontario: Wilford Laurier University Press, 1991.

Goodstein, David. *Scientific Elites and Scientific Illiterates, Ethics, Values and the Promise of Science.* Forum Proceedings, Sigma Xi, The Science Research Society (25–26 February 1993): 61.

websites

www.mhhe.com/kelter   The "World of Choices" website contains activities and exercises including links to websites for: National Science Foundation's Office of Legislative and Public Affairs; Oak Ridge National Laboratories; Cable News Network, CNN; and much more!

Origins

1

Every tradition grows ever more venerable—the more remote is its origin, the more confused that origin is. The reverence due to it increases from generation to generation. The tradition finally becomes holy and inspires awe.

—Friedrich W. Nietzsche *Human, All Too Human*

The Earth is our home, and it is a beautiful place. As we live our lives on the surface of the Earth, we can appreciate the beauty of the landscape, the sea, and the sky. From an aircraft, we can appreciate a wider view, enjoying the curves of coastlines, the mixed patterns of farmland, the majesty of great mountains, and the grandeur of cities. We have also now become familiar with a more distant view of our home, as a gorgeous jewel-like sphere set in the blackness of space. All of the beauty, in all of the views, is due to the chemicals of the Earth and the interactions among them.

As we emphasized in the Prelude, everything that we can see, touch, or feel is either composed of matter or produced as a result of the activities of matter. Matter is the stuff of chemistry, biology, physics, and geology: the touchable stuff of the universe. In this chapter we look more closely at the form in which matter occurs and consider where it came from.

Every culture has its own tradition of understanding origins—a wonderful and ancient story about the creation of the universe and all its matter. In this chapter, we describe the scenario that is based on scientific observation. It is not a complete story, or a perfect one, but it is one that brings order to our understanding of the remote origins of the universe. Many doubts and debates remain, which is not surprising because we are describing events which nobody witnessed. It is, however, the most accurate account available, based on the interpretation of all the existing scientific evidence.

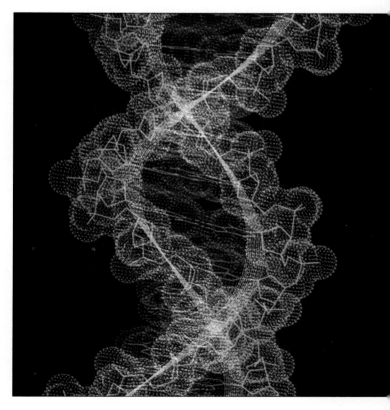

1.1 | The Origin of Matter

The universe is unimaginably large. More remarkable is the fact that it is still expanding. This means that in the past it must have been much smaller. The best evidence available leads most scientific observers to agree that the origin of our expanding universe occurred about 12–15 billion years ago, with a cataclysmic event known as the **Big Bang.** This idea is derived by considering what we know about the universe today and using our knowledge of physics to work backward in time. It is

When we admire the beauty of the Earth from any viewpoint, we are admiring the beauty of chemistry.

not just some speculative flight of the imagination. It is the product of a rigorous application of mathematical physics. It suggests that at the time of the Big Bang, space, energy, and matter underwent an explosive expansion, growing at almost the speed of light, cooling and changing as it went (Figure 1.1).

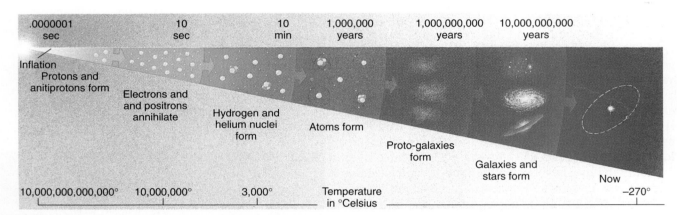

Figure 1.1   Key events in the history of the universe as we currently understand it, from the Big Bang to the present.

The ability of scientists to make credible statements about nature depends on their ability to measure, to quantify, and, by doing so, to discover and define underlying relationships. Nowhere is this more evident than in the study of the origin of the universe. Scientists must answer questions such as how fast or slow, how big or small, how hot or cold, how weak or strong. Then to share this information, certain standard units of measure have to be agreed on and used by the international scientific community so that everyone is speaking in the same

consider this:

Energy in Our World

Your hot water heater burns natural gas to heat water. This releases energy which actually does the heating. The natural gas came from the Earth as a result of the decay of plant material many, many years ago. The plants grew in response to sunlight. The sunlight is a result of nuclear reactions on the Sun. Therefore, nuclear energy has been used to heat your hot water!

Please think of some other everyday processes that require or release energy. Ultimately, where does this energy come from?

Table 1.1

Units of Measure

Physical Quantity	Unit	Abbreviation
Base units		
Mass	kilogram	kg
Length	meter	m
Time	second	s
Related units		
Volume	cubic meters	m^3
Density	kilograms per cubic meter	kg/m^3
Speed	meters per second	m/s
Force	newton ($kg \cdot m/s^2$)	N
Energy	joule ($N \cdot m$)	J

language. In this way hypotheses can be tested by many scientists and any necessary corrections can be made. As we go through this chapter, we will introduce you to some of these standard units. There are, in fact, just seven base units of measure from which all others can be derived. They are referred to as SI units, from the International System of Units. Table 1.1 lists three of these units; the related units are derived from the base units.

Matter and Energy

Immediately after the Big Bang, the universe was tiny and incredibly hot. There would have been no real distinction between what we think of as matter and energy for the two were continually being *interconverted*.

   Matter is the physical material of the universe: it occupies some space and can be said to have **mass**—a measure of its quantity. **Energy** is a more subtle phenomenon that can be defined as "the capacity to do work" (this will be explained in more detail in Chapter 5). Energy is essentially a property of physical systems, like light from the Sun or heat from the flame of a candle. It is energy that can make change happen.

   The intimate relationship between matter and energy is embodied in Albert Einstein's famous equation $E = mc^2$. The E stands for energy (measured in units called joules), the m stands for mass (measured in kilograms), and the c stands for the speed of light (300,000,000 meters per second). This equation indicates the amount of mass that corresponds to a given amount of energy and vice versa.

   You can see from the equation that the value of c is squared, meaning it must be multiplied by itself. The result is a number consisting of 17 digits, 16 of which are zeros. Here you see another difficulty facing scientists in taking the measure of the universe: the scale of the quantities involved, both large and small. They use a form of mathematical shorthand known as *scientific notation*. In scientific notation

$$300,000,000 \text{ becomes } 3 \times 10^8$$

by moving the decimal point 8 places to the left ($\times 10^8$) and using only those digits needed to give an accurate sense of the number (3). The value c^2 is written 9×10^{16}. (For more about the numbers used in scientific notation, see the appendix at the end of this chapter: Significant Figures.) Figure 1.2 shows how scientific notation can extend the usefulness of any unit of measure, from the very small to the very large.

   Given that matter possesses mass, any piece of matter must embody an amount of energy as summarized by Einstein's equation. The amount of energy held within matter is very large as we can demonstrate with a simple example. Using the equation $E = mc^2$, we can calculate that the conversion of just 1 gram of matter (the mass of about four frosted flakes) would release 9×10^{13} joules of energy. If this energy were used to heat an ice cube the size of a typical city block, the ice would immediately turn into boiling water

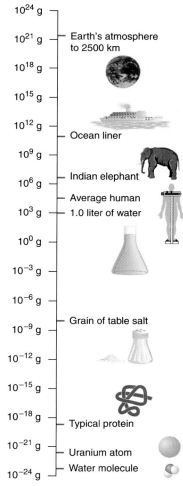

Figure 1.2   Quantities of mass, measured in grams, using scientific notation. A gram is 1/1000 of a kilogram.

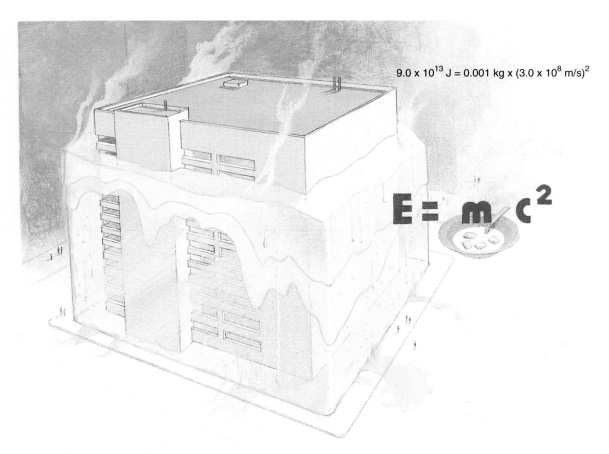

$$9.0 \times 10^{13} \text{ J} = 0.001 \text{ kg} \times (3.0 \times 10^8 \text{ m/s})^2$$

$$E = mc^2$$

Figure 1.3   The conversion of matter with a mass of only 1 gram to energy would release enough heat to turn a cube of ice the size of a city block into boiling water. This example is meant only to give you some sense of the energy content of matter. Such a conversion is a practical impossibility.

(Figure 1.3). A small fraction of the energy embodied within matter actually can be released in nuclear bombs and (less violently and more usefully) in nuclear power plants, but these are highly specialized circumstances.

Unique circumstances existed one millionth of a second after the Big Bang. The temperature of the universe was probably around 10 trillion °C (degrees Celsius). At this temperature, and under the conditions of the Big Bang, matter and energy were readily interconvertible. To begin with, radiant energy, such as light, predominated over matter. As the universe expanded and cooled, however, increasing amounts of matter began to emerge out of the original energy. To examine what form the matter took and what happened next, we need to consider the form in which matter occurs today.

"But How Do You Know?"

When we discuss the origin of the universe in our classes, students invariably ask, "How do you know what happened, you weren't there!" The first and most honest answer to such a question should be "We don't have absolute proof, but our rational view of the evidence leads to some likely scenarios." Remember that a scientific theory not only incorporates the data and observations of the past and present; it also makes **definite predictions** about the results of observations to come. Every new observation that supports the theory indicates that the basic concept is correct—and it is in this way that we "know." And so science progresses.

The issue of how we know about events we cannot ourselves observe (via **direct evidence**) is an important one. The key to understanding such events is **indirect evidence.** This means looking at clues that, when taken together, provide substantial evidence that *an event has occurred without our having seen it.* That evidence is interpreted in light of the way we know matter to behave when we have been able to observe or measure it directly (Figure 1.4).

A.

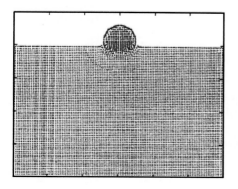

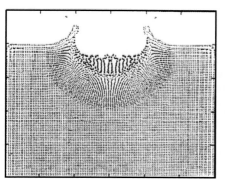

B.

C.

Figure 1.4   The important thing about indirect evidence, **A.** is that it is based on our direct experience of the physical world, and the way we know matter and energy to behave. If direct experience of a similar event is not possible, then we can still apply our knowledge of the physical laws of matter and energy to create a model that validates our hypothesis. **B.** This computer model simulates the creation of a crater formed as the result of a meteor-like impact. **C.** A photo of the Arizona meteor crater. Some 50,000 years ago a meteorite approximately 50 meters in diameter is thought to have hit the Earth just 40 miles east of what is now Flagstaff, Arizona. It wasn't Flagstaff back then.

We infer information from indirect evidence all the time. For example, you've had a raw egg sitting in boiling water for 20 minutes. Is the egg still raw or is it hard-boiled? *How do you know?* Have you actually seen the inside of the egg? Of course not. You do, however, *infer from the accumulation of evidence* that something you cannot observe directly has, in fact, occurred. Simply crack open the shell and you will have all the direct evidence you need. For scientists studying the origin of the universe, the evidence is there, and accumulating, although the payoff will never be so direct as a simple peek inside. The best scientific models explaining our origins rely heavily on many different pieces of indirect evidence, which you can explore further by reading some of the references at the end of this chapter. The Case in Point: Star Light, Star Bright gives you one such example.

Subatomic Particles

There are just three principal building blocks in all matter familiar to each of us: **protons, neutrons,** and **electrons.** They are the *subatomic particles* (the word particle really just means "little bit"). For our purposes, the most important characteristics of these particles are mass, size, and electrical charge.

The **mass** of a particle is a measure of how much matter it contains. Mass is different from weight, although the two terms can often be used

exercise 1.1

Mass-Energy Conversions

Problem

In a nuclear power station, fuel such as uranium is converted into other elements and particles of smaller mass. Calculate the energy released when 1.000 kg of uranium is converted into 0.999 kg of other elements.

Solution

The mass (m) converted into energy is equal to $1.000 - 0.999 = 0.001$ kg; in scientific notation that's 1×10^{-3} kg. The decimal point is moved three places to the *right* so the exponent is negative. We use Einstein's equation (Equation 1), substitute numbers (Equation 2), square c (Equation 3), then group similar components (Equation 4), and solve:

$$E = mc^2 \qquad \textbf{(Equation 1)}$$
$$E = (1 \times 10^{-3}\,\text{kg}) \times (3 \times 10^8\,\text{m/s})^2 \qquad \textbf{(Equation 2)}$$
$$E = (1 \times 10^{-3}\,\text{kg}) \times (9 \times 10^{16}\,\text{m}^2/\text{s}^2) \qquad \textbf{(Equation 3)}$$
$$E = (1 \times 9) \times (10^{-3} \times 10^{16}) \times (\text{kg} \cdot \text{m}^2/\text{s}^2) \qquad \textbf{(Equation 4)}$$
$$E = 9 \times 10^{13}\,\text{J}$$

Check

We have here the same mass as our example of four frosted flakes: 1 gram is equivalent to 0.001 kg. (The special characteristics of uranium, which we study in more depth in Chapter 14, make such a mass-energy conversion practical.) When multiplying, the exponents of 10 are added together to determine the final value: $10^{(-3 + 16)} = 10^{13}$. Table 1.1 lists joules (J) as N · m. Substituting for N gives (kg · m/s²) · m, and ultimately the kg · m²/s² we see in Equation 4; the multiplication sign ($\times$) has been replaced by a multiplication dot ($\cdot$). You will learn more about *unit* conversions in Chapter 4.

interchangeably without confusion. An object's **weight** depends on how strong a force of gravity the object is exposed to. An astronaut who walked on the Moon, for example, weighed less than when on Earth, because the Moon's gravity is less than the Earth's. The astronaut's mass remained unchanged, however, because the amount of matter present in his body does not depend on where he was. Despite this distinction, the unit of mass used in the everyday world, the gram, is also used as a unit of weight. This causes no confusion provided we do all our measuring on Earth, because the weight of an object on Earth (the force with which it is attracted to Earth) is directly proportional to its mass.

Because subatomic particles are so much smaller than everyday objects, we do not usually measure them in everyday units like centimeters and grams (or inches and pounds), just as we would not measure the width of a hair in miles. The units that are used, however, can be converted into more familiar units if we wish. The masses of protons, neutrons, and electrons, and ultimately the atoms that are made from them, are measured in **atomic mass units (amu).** There are 6.02×10^{23} atomic mass units in 1 gram, so the atomic mass unit is a very small unit of mass indeed.

Protons and neutrons both have masses that are very close to 1 amu, so close that we can treat the mass of both particles as being the same—equal to 1 amu. Electrons contain much less matter than protons or neutrons, and the mass of an electron is approximately 0.00055 amu (Figure 1.5).

Distance on the atomic scale is often measured in units of picometers (pm). There are 1 trillion (1×10^{12}) pm in 1 meter. Protons and neutrons have similar sizes; each has a radius

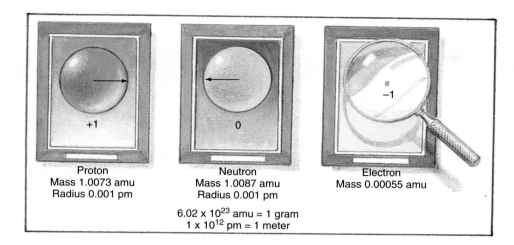

Proton
Mass 1.0073 amu
Radius 0.001 pm

Neutron
Mass 1.0087 amu
Radius 0.001 pm

Electron
Mass 0.00055 amu

6.02×10^{23} amu = 1 gram
1×10^{12} pm = 1 meter

Figure 1.5   A portrait of the subatomic particles protons, neutrons, and electrons, and their key characteristics. Electrons are so incredibly tiny that it is barely meaningful to assign any radius to them. The electrons in atoms, however, move around within regions of space (the electron "clouds") that are much larger in size than protons or neutrons. In the hydrogen atom, the electron cloud is a sphere 30 pm in radius. (Rendering not to scale.)

case in point

Star Light, Star Bright: The Beauty of Indirect Evidence

We have never touched a star or another planet. A trip to our nearest planetary neighbor, Mars, would take at least 1.5 years, assuming it were possible to attend to the human needs of long-term space travel. Yet we say with confidence that our largest planet, Jupiter, some 400 million miles away, contains the element hydrogen, and that this is also the main component of the stars, including our Sun. How do we learn about the planets and stars without actually visiting them to take samples (as we have done with our Moon)?

The power of indirect evidence shines through here. While we have never visited other celestial bodies, we receive energy from them, in the form of *electromagnetic radiation*. This includes visible light, as well as ultraviolet radiation, microwaves, radio waves, and so on. Visible light can be broken down into its constituent wavelengths to reveal a distinct pattern of various colors. The spectrum of a star, when adjusted for its movement away from the Earth, contains a pattern matching that from hydrogen that we can sample directly here on Earth. We can therefore conclude with certainty that stars contain hydrogen, even though we cannot actually visit the stars.

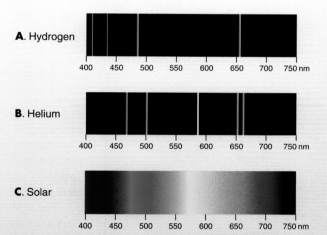

A. Hydrogen

B. Helium

C. Solar

The Sun is 71% hydrogen **(A)**, 27% helium **(B)**, with 2% heavier elements, such as carbon and iron. The planet Jupiter has a similar composition. Such information is derived in part by finding within the solar spectrum **(C)** the individual spectra of the different elements.

of approximately 0.001 pm. Electrons are much smaller, so small that it is barely meaningful to quote a value for their sizes. As we shall soon see, however, the influence of electrons is felt throughout large volumes of space.

Electric Charge and the Electric Force

The third vital characteristic of the subatomic particles (in addition to mass and size) is the type of electric charge they carry. You may already know that there are two kinds of electric charge—positive and negative. Protons carry a positive electric charge (which has arbitrarily been assigned a value of +1), whereas electrons carry an equivalent amount of negative

exercise 1.2

Mass versus Weight

Problem

The gravitational force on the planet Jupiter is 2.54 times that of Earth's. Given that fact, how much would you expect a 280-pound linebacker to weigh on Jupiter if his team played all its home games there? How would his mass change from Earth to Jupiter?

Solution

Recall in our discussion that the weight of an object is proportional to the gravitational pull on that object. This means that the weight of the football player on Jupiter will be 2.54 times that on Earth, or 280 × 2.54 = 711 pounds. The mass, however, is constant, because mass is a measure of the amount of substance, and there is as much of the player on Jupiter as there is on Earth.

Check

The calculated weight is greater on Jupiter than on Earth. This makes sense because Jupiter is more massive than Earth. Therefore, it's gravitational pull is greater, resulting in a greater weight for the football player.

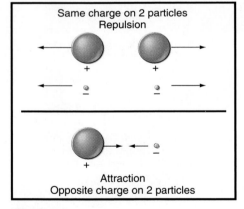

Figure 1.6
A. Objects with the same type of electric charge are repelled from each other by the electric force. **B.** Objects with opposite electrical charges are attracted to each other by the electric force.

charge (assigned a value of −1). The positive and negative charges of protons and electrons are the source of all the much larger, cumulative positive and negative charges found in everyday life. The "negative" end of a battery, for example, is negative because it acts as a source of electrons. When a flashlight is turned on, electrons flow from the negative end of the battery through the filament in the bulb to the positive end of the battery.

The neutron is an electrically neutral particle (hence its name)—it does not carry a net electrical charge. This makes the neutron a minor player in the drama of chemistry because, as we shall see, chemistry is all about the interactions of protons and electrons that result from their electrical charges.

Electric charge is a major influence on what happens during chemical reactions because the motion of an electrically charged particle is influenced by the presence of other charged particles. This influence is what we call a force, in this case an *electric force*. Particles with the same type of charge *repel* one another. Particles with opposite types of charge *attract* one another. These attractions and repulsions do all the "pushing" and "pulling" that makes chemical reactions happen (Figure 1.6).

It takes energy to move anything in the direction opposite to which a force would cause it to move. This idea is central to understanding the involvement of energy in chemical reactions. Moving a negatively charged electron away from a positively charged proton requires an input of energy. Similarly, pushing two electrons together requires energy, as does pushing two protons together. These ideas are explored more fully in Chapter 5.

Fundamental Forces

The electric force is one of four **fundamental forces** that play a part in the workings of the universe. It is more appropriately known as the **electromagnetic force,** because it is responsible for magnetism as well as electric repulsions and attractions.

case in point

The Ups and Downs of Sub-Subatomic Particles Known as Quarks

For many years protons and neutrons were thought to be indivisible particles, with no "smaller bits" within them. In recent years, however, physicists have widely accepted the view that protons and neutrons are themselves composed of even more fundamental particles called *quarks.* There seem to be quarks with electric charges of $+\frac{2}{3}$ and $-\frac{1}{3}$. A proton is composed of two of the $+\frac{2}{3}$ quarks (termed "up" quarks) and one of the $-\frac{1}{3}$ quarks (termed "down" quarks). Neutrons are composed of only one of the "up" quarks and two of the "down" quarks.

The physicist Murray Gell-Mann appropriated the name quark from a line in James Joyce's *Finnegans Wake:* "Three quarks for Muster Mark," which Gell-Mann theorizes is actually a call to the bar: "Three quarts for Mister Mark." Since the recipe for making a neutron or proton is essentially "take three quarks" as Gell-Mann puts it, he felt the name was justified: "the number three fitted perfectly the way quarks occur in nature." Quarks also have "flavors" and "colors." Who says physicists don't have a sense of humor?

Quarks have never been isolated or "split" from the larger particles they form, but their presence within protons and neutrons has been detected indirectly. Quarks are thought to have been formed in the first thousandth of a second after the Big Bang. The existence of the last predicted quark, the so-called "top quark," was confirmed in March 1995.

Gell-Mann, Murray. *The Quark and the Jaguar: Adventures in the Simple and the Complex.* New York: W.H. Freeman and Co., 1994: 180–181.

We've also mentioned one other, more familiar, fundamental force: that of **gravity.** Gravity causes all objects with mass to be attracted toward one another, so gravity is not associated with repulsion, only attraction.

The two other fundamental forces are the **strong nuclear force** which binds protons and neutrons together in the nuclei of atoms, and the **weak nuclear force** responsible for subtle transformations of the *subatomic* particles within some atoms.

Together, these four fundamental forces do all the pushing and pulling and changing that makes things happen.

1.2 | The Origin of Atoms

The **atom** is the basic particle of matter in which an element's characteristic properties are expressed. There are 92 naturally occurring **elements** in the Earth, the basic material from which all substances are derived. It is the number of *protons* in an atom of any one element that defines its uniqueness (Figure 1.7). Some of the names of the elements will be familiar to you, such as iron (composed of iron atoms) and copper (composed of copper atoms). Others will probably be completely unknown to you, such as lutetium (composed of lutetium atoms) or terbium (composed of terbium atoms).

The essential features of atomic architecture are outlined in Figure 1.7, using three different atoms as examples. The protons and neutrons are clustered together within a central **nucleus,** while the electrons move around within defined regions of space surrounding the nucleus. This movement of electrons allows them to occupy a significant space in much the same way as a tiny fly can buzz around a large room and annoy all of its occupants. This is why, when we depict an atom, we show the **electron cloud** around the nucleus, not the electrons themselves.

Although the number of protons within an atom of an element does not vary, the number of neutrons can. For example, all atoms containing only one proton are known as hydrogen atoms; but different forms of hydrogen exist. The most

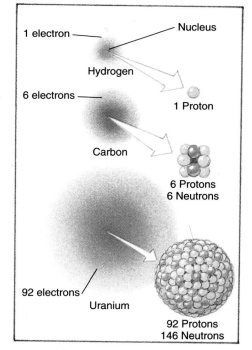

Figure 1.7   The structure of three atoms (hydrogen, carbon, uranium), illustrating the basic principles of atomic structure. An impression of true relative size can be gained by appreciating that the nucleus of a hydrogen atom located in the center of the electron cloud, is so small that on the scale of this sketch it would not be visible. If a hydrogen atom were a sphere the size of the New Orleans Superdome, the nucleus would be the size of a small pea, with the much smaller electron moving around within all the rest of the space.

common hydrogen atom has no neutron in its nucleus. Less common are the "heavy" atoms of hydrogen, atoms with one neutron (deuterium) and two neutrons (tritium). Atoms of the same element that have a different number of neutrons are known as **isotopes.** We will learn more about the isotopes of hydrogen later.

Where did all these elements and, therefore, all these kinds of atoms come from? The source initially was the Big Bang, and later from less massive explosions called supernovae (plural of supernova). First we need to consider how the protons, neutrons, and electrons that formed very shortly after the Big Bang combined to form the atoms of the modern universe. The essential principle behind the origin of the atoms is straightforward: the smallest, simplest atoms formed first, then the bigger atoms were built up by combinations of the nuclei of simpler atoms.

■ case in point

Isotopes: The Power of the Atom Unleashed and Harnessed

One way to generate atomic energy, either for producing electricity or for use in weapons, is *nuclear fission*. The nucleus of an atom is split apart, releasing a large amount of energy in the process. A neutron strikes the nucleus of a heavy element causing it to break (fission) into two smaller nuclei and a few extra neutrons, and release a lot of energy. The key to producing such power is to separate the rare fissionable isotope $^{235}_{92}U$ from the dominant isotope $^{238}_{92}U$, which does not undergo fission. This is the so-called "weapons-grade" uranium that you hear of in the news media. This was first done in 1943 at Oak Ridge, Tennessee, where the smaller mass of the 235-isotope was separated from the slightly heavier 238-isotope. $^{235}_{92}U$ proved to be fissionable and allowed the construction of the first atomic bomb which many believe led to the end of the Second World War. Just a few kilograms of $^{235}_{92}U$ powered the explosion "heard around the world" on August 6, 1945.

Chemically, both isotopes behave the same, but the nuclear properties differ significantly. When a neutron strikes a $^{235}_{92}U$ nucleus, it breaks into pieces and releases enormous amounts of energy. When a neutron strikes a $^{238}_{92}U$ nucleus, the neutron is captured by the nucleus to form $^{239}_{92}U$; then a transformation occurs of a neutron into a proton, and the resultant atom becomes neptunium, $^{239}_{93}Np$. This process represents the first earthly synthesis of an element with an atomic number greater than uranium.

Isotopes have their humanitarian uses as well, in particular in the field of medical diagnosis. For example, subatomic particles splitting off from radioisotopes can act as tracers. Their presence can be detected and captured on photographic film, creating an image of organs in the body.

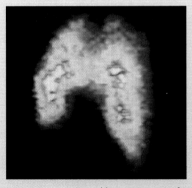

The enormous energy possessed by matter was forcefully demonstrated by the atom bomb (top). Isotopes are used every day in the medical world to aid in diagnosis. The radioisotope iodine, for example, is used to assess the condition of the thyroid gland (bottom).

Origin of the Light Elements

As the universe cooled after the Big Bang, protons, neutrons, and electrons emerged from the chaotic blend of primordial energy and matter. As we just mentioned, a single proton forms the nucleus of an hydrogen atom. A little more than 2 minutes after the Big Bang, approximately one quarter of the hydrogen nuclei in the universe combined to form helium nuclei, which contain two protons and two neutrons. This process is an example of **nuclear fusion** (the merging of atomic nuclei). The neutrons of the helium nuclei are actually formed from the hydrogen nuclei—matter can be interconverted among different types of particles (in this case, protons to neutrons) during nuclear fusion. Reactions that involve changes to the nucleus of an atom, and thus the makeup of the atom itself, are a special category of processes known as **nuclear reactions.** We will be looking at nuclear chemistry in much greater depth in Chapter 14.

When the nuclei of smaller atoms fuse to make larger atoms, vast amounts of energy can be released. This is the same process that drives the explosive power of the hydrogen bomb. A small part of the matter of the hydrogen nuclei is actually converted into energy and so the total mass of the resulting nuclei and particles is a little bit less than the total mass of the nuclei that underwent fusion. The fusion process generates two neutrons from preexisting protons and also involves the release of particles called positrons. **Positrons** are essentially "electrons" that carry a positive charge rather than a negative charge. These exotic "creatures" have no role to play in present-day chemistry, although they and the isotopic elements that release them do play a major role in medical diagnosis (see the Case in Point: Isotopes).

Nearly all the different naturally occurring atoms in the universe today have been formed by nuclear fusion reactions, in which the atoms of "lighter" elements combined to form new elements with "heavier" atoms. Some isotopes of some atoms are formed by various processes by which the nucleus decays. Nuclear fusion only occurs in situations of very high energy, such as within the first few minutes after the Big Bang or within stars (as we shall see).

For at least the first 1 billion years of the universe's existence, the only types of atomic nuclei that existed in abundance were hydrogen and helium nuclei. For the first 500,000 years or so of that time, the nuclei existed alone, without any associated electrons. Electrons eventually became associated with the nuclei, to form true hydrogen and helium atoms, about 500,000 years after the Big Bang, when the temperature had fallen to 10,000°C. Before that time, the electrons and nuclei would have possessed too much energy of motion to become bound together to form atoms. Energy, remember, allows electrons and nuclei to resist the force of attraction between them.

Atomic Number and Mass Number

As you will learn throughout your study of chemistry, the identity as well as the behavior of different atoms is directly related to the makeup of the atomic nucleus. Chemists take this information and use it to identify specific atoms in a special kind of chemical notation, as shown in Figure 1.8. Each atom is represented by a symbol of one or two letters, representing the element concerned (H for hydrogen, He for helium, etc.). These letters are sometimes preceded by two numbers:

The lower number (or *subscript*) is called the **atomic number** of the element, which is the number of protons in the nucleus. The top number (or *superscript*) is the **mass number,** which is the total of the number of protons added to the number of neutrons. (Refer to the Periodic Table of Elements printed

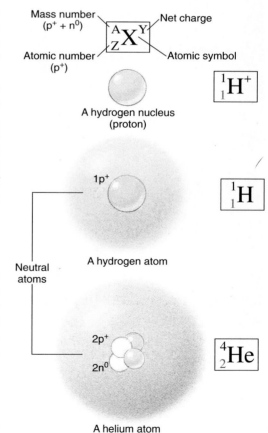

Figure 1.8   Atomic nuclei are represented here using the special notation, a kind of "chemist's shorthand."

Figure 1.9 Clumps of gaseous matter gather and heat up as stars are born in this photo taken by the Hubble Space Telescope. Courtesy NASA.

on the inside back cover of this book to find the atomic number of each element; use the table of elements to find a listing of element names and symbols).

We know that every atom (or nucleus) of a specific element must have the same atomic number. In other words, the number of protons in an atom *defines* the type of atom. All hydrogen atoms, for example, have one proton in the nucleus and an atomic number of 1; all carbon atoms have six protons in the nucleus and an atomic number of 6; and so on. We know too that atoms of any one element can have different numbers of neutrons and so can have different mass numbers. These are the different isotopes of the element. Some elements exist in nature as a single isotope but others have many, up to five or ten. Deuterium, an isotope of hydrogen with one proton and one neutron in its nucleus, has a mass number of 2 and is represented in "chemist's shorthand" by $_1^2$ H. The isotope of carbon having eight neutrons is called carbon-14 and is represented as $_6^{14}$ C. The subscript "6" is the atomic number (which always equals the number of protons) and the superscript "14" is the mass number (which is the total number of protons plus neutrons).

Stars and Galaxies

Everything became relatively quiet in the universe about 500,000 years after the Big Bang when electrons became attached to hydrogen and helium nuclei, to form hydrogen and helium atoms. The next major change did not occur until about 1 billion years (10^9 years) after the Big Bang. At this time, the effects of the force of gravity—the mutual force of attraction between particles of matter—began to pull the existing matter into distinct and growing clumps. As these first accumulations of matter grew larger, their ever-increasing net gravitational attraction caused yet more material to be drawn into them.

As the clumps formed, they began to warm up, due to the increasingly frequent collisions between the particles of matter within them. Eventually, the temperature within some regions of the clumps became high enough to drive the electrons off the atoms again and initiate nuclear fusion reactions between the nuclei present (almost entirely nuclei of hydrogen and helium atoms). These nuclear fusion reactions released large amounts of energy, causing the materials participating in them to form stars (Figure 1.9). Stars are essentially massive nuclear fusion reactors, in which the fusion of atomic nuclei generates all of the heat and light released by a star.

Figure 1.10 M31 is the nearest large galaxy to our Milky Way.

Several billion years after the Big Bang, countless new stars were igniting throughout the universe, and these stars became clustered together in the vast collections of stars that we call galaxies (Figure 1.10). At this point, the universe would have first begun to resemble the place we live in today: a seemingly limitless region of dark space speckled with the bright lights of countless galaxies and individual stars.

Our star, the Sun, is now about 5 billion years old and is still composed of about 71% hydrogen and 27% helium, with only small amounts of all the other elements. This is despite the fact that the main process that has kept the Sun "burning" for 5 billion years has been the conversion of hydrogen into helium by nuclear fusion reactions. So the Sun's supply of hydrogen "fuel" will therefore last for a long time yet (Figure 1.11).

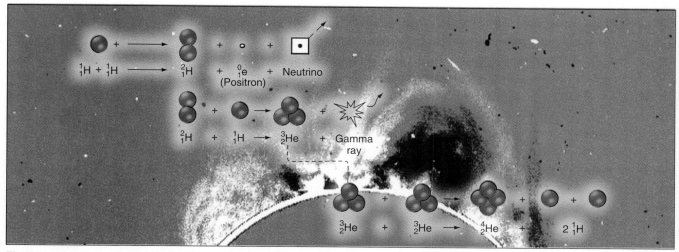

Figure 1.11 Our Sun has been shining for billions of years and is expected to continue for several billion more. For the whole history of the Earth the Sun has been an ever-present source of energy, released by the nuclear fusion reactions within it.

Origin of the Heavy Elements

So far we have described the origin of only the first two elements: hydrogen and helium. The larger atoms of other elements are also formed in stars, or during the explosive death throes of some stars, by different nuclear fusion reactions. After a long period during which hydrogen is steadily converted into helium, stars undergo changes in size and structure which make other fusion reactions possible. Helium nuclei can fuse to form the nuclei of lithium or beryllium atoms. Fusion between beryllium and another helium nucleus can generate the nucleus of a carbon atom, containing six protons and six neutrons. Fusion of a further hydrogen nucleus to a carbon nucleus can produce a nitrogen nucleus, with seven protons and six neutrons. In this way a wide variety of different nuclei can be formed, each with the release of nuclear energy, during the later stages of a star's long existence.

The fusion of nuclei heavier than iron (atomic number, 26) to make still heavier nuclei does not release energy but, instead, takes in energy. Nevertheless, the fusion of helium and other light nuclei releases sufficient energy to allow an aging star to continue to release energy overall and keep shining although its temperature and its size change dramatically.

Eventually, however, a star's "fuel supply" of light nuclei runs out. At this point it stops shining and can face a range of fates, depending on its precise size and composition. Some stars simply "fizzle out," slowly cooling and dimming. In others however, something much more dramatic can happen. The force of gravity, no longer being adequately resisted by the outward pressure caused by the release of energy from fusion reactions, pulls all the materials violently inward, causing the entire star to collapse in a dramatic *implosion*. The initial implosion is then followed by a devastating rebound explosion, as the star is destroyed, and much of its materials are flung outward. This is known as a supernova event (Figure 1.12).

A supernova explosion scatters a star's harvest of atomic nuclei (plus electrons and other subatomic particles) out into space. The very high energies associated with a supernova event cause additional nuclear fusion reactions to add new elements to the harvest. Once out in the low-energy domain of outer space, however, electrons are able to become associated with the nuclei to form neutral atoms. Some of the nuclei that are the first products of nuclear fusion reactions are unstable, meaning they are liable to split up (decay) into other nuclei. These unstable nuclei are called radioactive isotopes. The nuclei of radioactive isotopes decay to form nuclei of other atoms, which may themselves be stable or unstable. Eventually, through a series of decays, stable, long-lived atoms remain.

The atoms scattered from supernovae may eventually fall into other stars or may be mixed with other atoms in interstellar space until they gather together within the clumps that will form new stars. Thus, a second generation of stars can be born, followed by a third

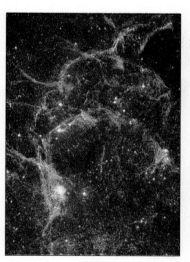

Figure 1.12 A supernova event scatters a star's harvest of atoms into space.

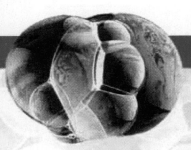

exercise 1.3

Protons, Neutrons, Electrons, and Symbols

Problem

Use the tables on the inside back cover to help you fill in the missing information:

	Symbol	Element	Protons	Neutrons
(a)	$^{15}_{7}\text{N}$	_____	_____	_____
(b)	$^{239}_{94}$ ____	_____	_____	_____
(c)	$^{}_{}$ ____	W	_____	110
(d)	$^{55}_{25}$ ____	_____	_____	_____

Solution

Remember that the upper number (always the larger of the two, except in hydrogen) is the mass number, equal to the sum of the protons and neutrons in the nucleus. The atomic number (always the lesser of the two) is the lower number and is equal to the number of protons in the atom.

(a) $^{15}_{7}\text{N}$ *Element:*   The element is given: **N,** which is **nitrogen.**

     Protons: The atomic number equals the number of protons: **7.**

     Neutrons: The mass number (15) minus the atomic number (7) equals the number of neutrons: $15 - 7 =$ **8.**

(b) $^{239}_{94}\text{Pu}$ *Element:*   The atomic number of this element is 94, which corresponds to **plutonium (Pu).**

     Protons: The atomic number equals the number of protons: **94.**

     Neutrons: The mass number (239) minus the atomic number (94) equals the number of neutrons: $239 - 94 =$ **145.**

(c) $^{184}_{74}\text{W}$ *Element:*   The element is given: **W,** which is **tungsten.**

     Protons: The periodic table shows that the atomic number of tungsten is **74,** which equals the number of protons.

     Neutrons: The number of neutrons is given. You get the mass number by adding the given number of neutrons (110) to the atomic number (74): $110 + 74 =$ **184.**

(d) $^{55}_{25}\text{Mn}$ *Element:*   The atomic number of this element is 25, which corresponds to **manganese (Mn).**

     Protons: The atomic number equals the number of protons: **25.**

     Neutrons: The mass number (55) minus the atomic number (25) equals the number of neutrons: $55 - 25 =$ **30.**

Check

If you've done the work correctly, the mass number (the superscript number to the left of the element symbol) will be equal to the number of protons and neutrons combined. Because different isotopes of each element exist, in each of these exercises the number of neutrons had to be identified for you. The atomic mass given in the periodic table, appearing below the chemical symbol, is actually a weighted average of the different atomic masses of the isotopes of that element.

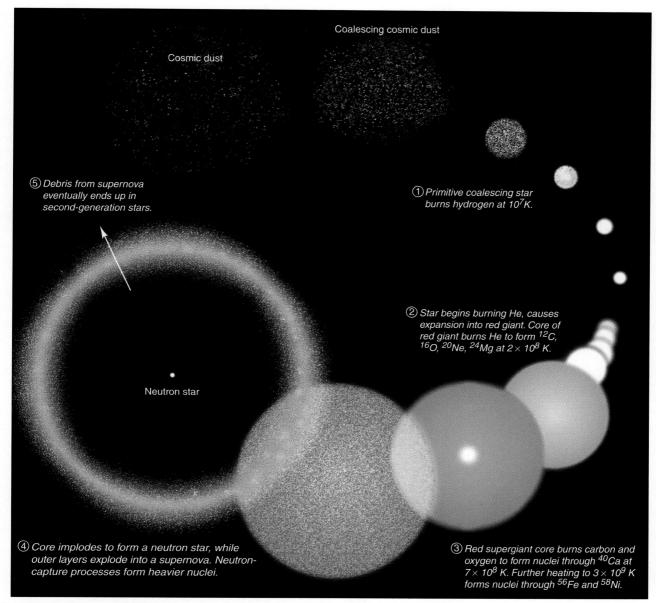

Figure 1.13   The life cycle of a star brings new elements into existence.

generation of stars, and so on (see Figure 1.13). In each generation, new fusion reactions can occur, generating atoms of other, larger elements. Our Sun is probably a second- or third-generation star.

So, all of the naturally occurring atoms of the universe, perhaps as many as 92 different kinds of atoms found on Earth, have been formed either in the immediate aftermath of the Big Bang, within the nuclear "furnaces" of stars, during the violent supernova events that accompany the death of some stars, or by nuclear decay of other, larger nuclei (see Chapter 14). This is the history of every atom on Earth and every atom in your body. We truly are children of the stars, made, in a very real sense, from star dust.

Despite the fact that new atoms have been forming for at least 9 billion years, there are still plenty of atomic raw materials available, namely hydrogen and helium, that will be here for a long time to come. Even today, about 92.7% of the atoms of the universe (and more than 70% of the atoms in the galaxy) are hydrogen atoms and 7.2% of the atoms in the universe are helium atoms, so the atoms of the other elements we know of account for only about

0.1% of all the atoms in existence. "The Universe" appears to be a show that will run for a long time yet.

1.3 | The Origin of the Earth and Solar System

We have so far discussed the origin of atoms in the universe but have not considered their physical form. There are three **states of matter** (Figure 1.14). A **solid** (*s*) is matter that has a definite shape and volume: a block of wood, a piece of stone. A **liquid** (*l*), water for example, is matter that has a definite volume but no definite shape. It takes the shape of whatever it fills, whether it's a glass or a lake bottom. A **gas** (*g*), just as our atmosphere, has no definite shape or volume. It will occupy the full volume of the container that holds it. Helium in a tank is more greatly compressed than the helium that fills a balloon for instance. Stars, like our Sun, are essentially balls of hot gas. The intense heat within provides the energy needed to keep the atoms moving and well apart in space. Now as we consider the formation of the Earth, and by extension the other planets of the solar system, we need to consider what happens to gaseous matter as it cools and condenses.

Astronomers hypothesize that the Sun and planets formed from the collapse of a huge, slowly spinning cloud of gas and dust. Most of the matter was drawn inward by the force of gravity and became the Sun. The little remaining dust and gas clumped together to form planets; a large body of evidence suggests that this happened 4.6 billion years ago.

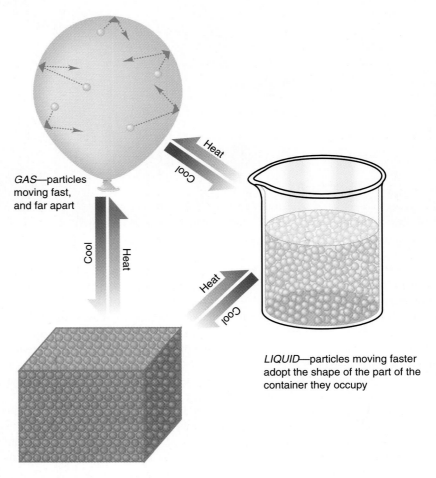

GAS—particles moving fast, and far apart

LIQUID—particles moving faster adopt the shape of the part of the container they occupy

SOLID—jostling particles close together within a fixed shape

Figure 1.14 The three states of matter—solid, liquid and gas

As a gas cools, it loses the energy that keeps the atoms apart, and so atoms can adhere to one another to form liquids and then solids. Much of the material of the "newborn" Earth, still very hot, took liquid form. The denser elements, mostly molten iron and nickel, began to sink toward the center of the Earth, while the less dense materials, consisting mostly of silicon and oxygen, floated toward the surface. **Density** is a property of matter that relates its mass to a given volume (kg/m^3 or g/cm^3). The denser a material is, the more matter it packs into a cubic meter or centimeter (try the Hands On on the following page). Table 1.2 shows the densities of some substances. The differing densities of certain materials caused the Earth to form into distinct layers as it cooled (Figure 1.15). The *average* density of the Earth is 5.52 g/cm^3.

The molten material at the Earth's surface eventually cooled to form the solid rock of the Earth's crust. The oldest rocks found on the Earth today are over 4 billion years old. Even with all the time that has passed, the interior of the Earth is still very hot and much of it is still liquid. At the very center, there is a solid core composed of iron and nickel. The inner core is solid because the enormous pressure there locks the atoms together; they no longer have the mobility of the atoms in the outer liquid core, despite the high temperature. The pressure affects the state of matter of the iron and nickel in the inner and outer core (solid versus liquid) but not their densities. Recent evidence shows that the solid iron core of the Earth rotates slightly faster than does the mantle. The liquid outer core acts like a lubricant between the two solid layers.

There is a lot of debate about the likely composition of the atmosphere of the early Earth. To begin with, there may have been very little atmosphere, with hydrogen and helium gases having been lost to space rather than held by the rather weak gravitational attraction between these gases and the Earth. Volcanic activity may then have spewed out vast quantities of various gases to form the primeval atmosphere. Another theory suggests that frequent collisions between the Earth and comets brought in large quantities of the water and gases that formed both the oceans and atmosphere of the young Earth. There would have been no ocean or other surface water to begin with, because the surface was so hot that any water on it would have immediately evaporated. Eventually, however, the cooling of the Earth must have initiated a phase of great rains, which formed the oceans and rivers and streams.

Table 1.2

Densities

Material	Density (g/cm^3)
Hydrogen	0.000089
Oxygen	0.0014
Cork	0.24
Wood, pine	0.35–0.50
Diethyl ether	0.71
Water	1.00
Carbon tetrachloride	1.6
Table salt	2.16
Silicon dioxide	2.6
Aluminum	2.70
Iron	7.8
Copper	8.9
Nickel	8.9
Silver	10.5
Lead	11.3
Mercury	13.5
Gold	19.3
Osmium	22.61

consider this:

Explaining Atmospheric Concentrations

Gravitational attraction is proportional to the mass of a body. Based on this principle, can you explain why Jupiter and Saturn have large amounts of hydrogen in their atmospheres but Earth, Mercury, and Mars do not?

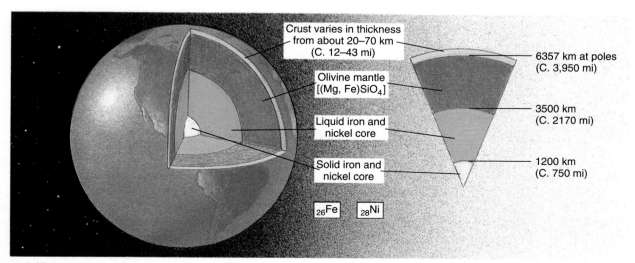

Figure 1.15 The internal structure and composition of the Earth.

■ hands-on

You've heard that oil and water don't mix. This fact does aid those who must clean up an oil spill at sea, in that the oil will tend to pool together. Oil also has a lesser density than water, which means it will float on top of the water.

Do you ever give much thought to the subject of density? You do although you may not know it. Let's suppose you want to pack a box of breakables for shipping. What material do you use to pad your delicate items: styrofoam "peanuts" or sand? You'll be paying by the pound, so let's consider the weight of each. A Styrofoam peanut weighs around 50 milligrams, roughly 0.0001 pound. A grain of sand weighs much less. So why do you pack your collection of Steuben glass with Styrofoam peanuts? Why do they work so well as a packing material? The answer has to do with density.

Density is not one of the fundamental SI units. Rather, it is derived from the base units of mass (grams) and length (volume, which is length cubed, centimeters × centimeters × centimeters): g/cm³. To get a good sense of what density means, let's try the following activity. You will need about 1 pound of sand, 1 pound of Styrofoam peanuts or a similar packing material, a large garbage bag, a plastic shopping bag, and a kitchen scale.

• Use the kitchen scale to weigh out 1 pound of sand, using a plastic bag to contain it. Put this aside.
• Without using the scale, fill the garbage bag with what you guess to be 1 pound of Styrofoam peanuts. Weigh these on the scale. Add more peanuts as necessary. Now consider these questions:

1. Which weighed more, the bag of sand or the bag of peanuts?
2. Which was less dense? Why?
3. Based on your observations, can you estimate the density of sand and that of the peanuts? (Use pounds per cubic foot.)
4. Based on this activity, can you describe why styrofoam peanuts work better than sand as a packing material? Include in your discussion the importance of the gas in the peanut.

This activity is courtesy of a wonderful educator, Linda Woodward.

Figure 1.16 We have evidence of early chemical activity in these ancient beds of banded iron as they appear today in Michigan; the formations are known as red beds. The iron formed a compound with the oxygen in the atmosphere upon reaching the Earth's surface, and, in essence, rusted.

1.4 | The Formation of Molecules and Ions

Since the time of its formation, conditions have changed dramatically on the surface of the Earth, where life is found. The atoms of perhaps 92 elements have reacted with each other to form a staggering variety of **compounds**—chemical substances composed of two or more elements chemically bonded together. Most of the chemicals on Earth are compounds, containing specific combinations of elements (Figure 1.16).

ELEMENTS—contain only one kind of atom
COMPOUNDS—contain different kinds of atoms chemically bonded together
MIXTURES—a blending of substances not chemically bound to each other
(see Figure 1.17)

Now we have reached the real core of chemistry. **Chemistry** is the study of the ways in which atoms react with each other to make new combinations. These reactions are very different from the fusion of atoms—the *nuclear reactions*—described before, in which new elements are created. In ordinary chemical reactions, the nuclei of atoms remain intact, but the electrons which surround the nuclei become rearranged. The interaction of these electrons is what forms the **chemical bonds** between atoms, the grouping of atoms due to several kinds of attractive forces. Chemistry is all about the rearrangement of electrons within and between the particles we know as atoms, and the *molecules* and *ions* that derive from atoms.

Figure 1.17 Elements,
Compounds and Mixtures

Elements
contain only
one kind of atom,
copper atoms (Cu)
in this case.

Compounds
contain different kinds
of atoms chemically
bonded together,
water molecules (H_2O)
in this case.

Mixtures have different
components combined
together, such as this
mixture of copper sulfide
and water. The suspended
solid can be separated from
the liquid by filtering

Solutions are
homogeneous mixtures
such as this solution of
copper sulfate pentahydrate
in water. The components in
the solution cannot be
separated by filtering

Figure 1.18   The great recycling of matter on Earth—it keeps going, and going. . . .

It is because the atomic nuclei retain their identity throughout chemical changes that the atoms available for chemical reactions on Earth today are the same ones that have been here since the origin of the Earth. Atoms of carbon that once were part of Hannibal's elephants are now in pine trees in Colorado; atoms of sulfur that once rained down from the eruption of Mt. Vesuvius are in the hairs on your head; and an atom of argon in the next breath you take could once have been inhaled by Ghengis Khan or Alexander the Great (Figure 1.18). The Big Bang, the stars, and supernova explosions made the atoms of Earth. By and large, all of the subsequent chemistry involving these atoms has involved the rearrangement of their electrons, rather than the creation or destruction of new atomic nuclei.

Molecules

A **molecule** is a distinct particle composed of two or more atoms linked together by the sharing of electrons between the nuclei concerned. The electrons no longer belong to only one atom but swarm around the nuclei of the various atoms within the molecule, forming wonderful patterns. The nuclei within a molecule are chemically *bonded* together by their mutual attraction to the shared electrons (Figure 1.19). The chemical bonds formed by the shared electrons are known as **covalent bonds.**

The covalent bonds holding molecules together are very strong, compared to the much weaker forces of attraction that exist among different molecules. This allows a molecule to behave as a distinct particle, moving about and bouncing off other particles. When we say a covalent bond is "strong," we mean that it requires a lot of energy to disrupt the forces of attraction between the nuclei and the electrons that are shared between the nuclei. These shared negatively charged electrons can be thought of as a kind of "electric glue," holding together all the positively charged nuclei that are attracted to them.

The simplest molecule is the hydrogen molecule, composed of two hydrogen atoms (the simplest type of atom) bonded together by electron sharing (Figure 1.20). We can represent a hydrogen molecule by the formula H_2. This simply indicates the type and number of atoms present using the standard symbols for the atoms rather than their names. When two individual hydrogen atoms come together under appropriate conditions of temperature and pressure, they can undergo a chemical reaction which leaves their electrons shared between the two nuclei, forming a covalent bond. The bond, and the molecule as a whole, forms a vibrant, dynamic partnership, rather than a rigid structure. The nuclei oscillate in and out, averaging an optimum distance known as the *bond length* of the H—H bond. When the nuclei move closer to each other than the optimal bond length, electric repulsion builds up between them, pushing them apart. If the nuclei move farther apart than the optimal distance, the attraction of the nuclei for the shared electrons pulls them together again. The continual inward and outward

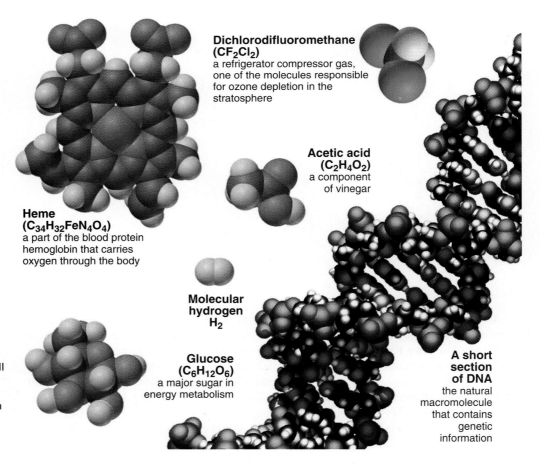

Dichlorodifluoromethane (CF_2Cl_2)
a refrigerator compressor gas, one of the molecules responsible for ozone depletion in the stratosphere

Acetic acid ($C_2H_4O_2$)
a component of vinegar

Heme ($C_{34}H_{32}FeN_4O_4$)
a part of the blood protein hemoglobin that carries oxygen through the body

Molecular hydrogen H_2

Glucose ($C_6H_{12}O_6$)
a major sugar in energy metabolism

A short section of DNA
the natural macromolecule that contains genetic information

Figure 1.19 A selection of common molecules, ranging from the simplest (the hydrogen molecule) to some of the most complex (a molecule of DNA, only a small part of which is shown here). The common feature of all molecules is that they contain two or more atoms bonded together by electron-sharing within covalent bonds.

movement of the nuclei involved in covalent bonds means that the bonds can best be visualized as tiny vibrating "springs" rather than static rigid rods.

When chemists carry out chemical reactions, they often heat the chemicals which they wish to react together. When anything is heated, its constituent particles move about more quickly and therefore collide more violently with one another, causing their bonds to oscillate more violently. If the collisions are sufficiently violent (or technically speaking, occur with sufficient energy), they can cause existing chemical bonds to break and new chemical bonds to form. That is one reason why chemistry laboratories have Bunsen burners—to break chemical bonds! Chemical reactions are processes in which preexisting chemical bonds are broken, and new chemical bonds are formed, with electron rearrangement being the fundamental process involved in breaking and making bonds.

Notice that the hydrogen molecule (H_2) is a molecule of a single element—only one type of atom is involved (Figure 1.21A). Most molecules formed, however, are molecules of compounds containing atoms of different elements bonded together, a simple one being carbon monoxide (CO, Figure 1.21B). The different neighboring atoms within the more complex molecules of compounds (like the molecules shown in Figure 1.19) are linked by bonds of differing lengths and strengths. The precise lengths and strengths of the bonds depend on the elements involved and the chemical surroundings provided by the rest of the molecules concerned.

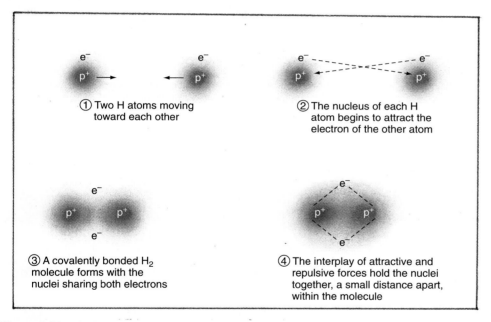

① Two H atoms moving toward each other

② The nucleus of each H atom begins to attract the electron of the other atom

③ A covalently bonded H_2 molecule forms with the nuclei sharing both electrons

④ The interplay of attractive and repulsive forces hold the nuclei together, a small distance apart, within the molecule

Figure 1.20   The formation of a hydrogen molecule (the simplest possible molecule) in which two hydrogen atoms become bonded together by the mutual sharing of their electrons. Chemical bonds that hold nuclei together due to the *sharing* of electrons are known as covalent bonds.

Ions

There is another way in which electron rearrangement can lead to the formation of compounds. When the atoms of elements come into contact, one or more electrons can be transferred from one atom to the other. This happens when the nucleus of one of the atoms has a substantially stronger attraction for electrons than the nucleus of the other atom. The resulting particles, known as **ions,** have a charge imbalance. The ion is *positive* if it has *given up an electron* (lost a negative charge, hence it has a positive charge). The ion is *negative* if it has *taken on an electron* (has an extra negative charge). This imbalance creates an attraction between the resulting ions. The attraction is called an **ionic bond.** Ionic bonds are in the same range of strengths as covalent bonds.

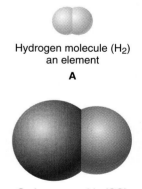

Hydrogen molecule (H_2) an element

A

Carbon monoxide (CO) a compound

B

Figure 1.21   **A.** The hydrogen molecule is made up of just a single element. **B.** Carbon monoxide is made up of two elements (carbon and oxygen). Carbon monoxide is considered a compound because of its different constituents, the hydrogen molecule is not.

Ionic bonding occurs when the positive charge on one atom's nucleus is strong enough to attract the electrons away from the other atom. The elements sodium and chlorine, for example, can react to form the compound sodium chloride (NaCl, common table salt), as shown in Figure 1.22. During this reaction, one electron is pulled off each sodium atom and added to a chlorine atom. This converts the sodium atoms into sodium ions, each with a charge of $+1$, meaning that it has one more proton than electron. Each chlorine atom becomes an ion of chlorine, called chlor*ide* ion, with a charge of -1, meaning it has one more electron than protons. The chemical symbol for the sodium ion is Na^+ and that for chloride ion is Cl^-.

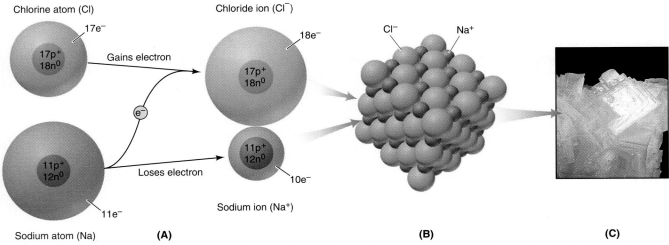

Figure 1.22 Sodium chloride **(C)**, table salt, is formed when sodium atoms and chlorine atoms become ions by the loss and gain of electrons, respectively **(A)**. The oppositely charged ions are attracted to one another, to become tightly held together **(B)** by what is known as an ionic bond. (Nuclei are not drawn to scale.)

exercise 1.4

Subatomic Makeup of Sodium and Chloride Ions

Problem

How many protons, neutrons, and electrons are in the sodium ion ($^{23}Na^+$) and chloride ion ($^{35}Cl^-$)?

Solution

The $+1$ charge on the sodium ion means that there is one more proton than electrons—the atom has given up an electron. Check the periodic table at the back of the book to find the atomic number of sodium. An element is defined by its proton count (atomic number), so sodium will *always* have **11 protons.** It must therefore have **10 electrons** if the charge is $+1$. The number of neutrons is the mass number (23) minus the number of protons (11) = **12 neutrons.**

Similar thinking goes into determining the particle count for the chloride ion, which has **17 protons, 18 neutrons,** and **18 electrons.**

Check

For a positively charged ion such as Na^+, there will be more protons than electrons (11 protons versus 10 electrons). For a negatively charged ion such as Cl^-, there will be fewer protons than electrons (17 protons versus 18 electrons).

A single positive or negative sign indicates a charge of $+1$ or -1; larger charges are indicated by a number followed by a sign, such as Ca^{2+} or P^{3-}.

The oppositely charged ions of sodium and chlorine are attracted to one another, allowing them, under suitable conditions, to settle into a regular crystalline structure shown in Figure 1.22. The ionic bonds that form the solid ionic compound NaCl are strong; however, they can be broken quite readily when the salt is dissolved in water. The ions separate and drift around in the water. The positive and negative ions embark on a virtually independent existence, moving around as free ions, although surrounded by a "cage" of water molecules. The sea contains substantial quantities of dissolved sodium chloride (in addition to other dissolved ionic compounds), which accounts for the characteristic salty taste of seawater. Many rocks also contain ionic compounds, with the ions locked within solid crystalline structures.

Exercises 1.4 and 1.5 will reinforce your understanding of proton, neutron, and electron counts in ions. The electron count is of particular importance, as we will begin to see in Chapter 2, because through this we can begin to *predict the chemical behavior* of atoms, ions, and molecules.

From the earliest days of its existence, the Earth has contained a great variety of ions and molecules derived from the basic set of naturally occurring elements. Perhaps the single-most astonishing result of all this chemical complexity is the fact of life itself.

1.5 | Origin of Life on Earth

We have examined the origin of atoms, the origin of the Earth composed of these atoms, and the way in which atoms can combine to form compounds composed of molecules or ions. Atoms, molecules, and ions are the simple raw materials of natural chemistry. Living creatures such as humans are built out of these raw materials, but how did the necessary building process occur? How did the simple chemicals of the early Earth give rise to the gloriously intricate and adaptable chemistry of life?

The origin of life is a subject that generates tremendous interest and controversy. Nevertheless, scientists raise questions, examine the evidence, carry out experiments, construct whatever theories seem to best fit the evidence, and continually modify and refine these theories as new evidence becomes available.

It is generally assumed that the life currently found on Earth is derived from earlier forms of life that actually originated on the Earth. There is no direct evidence to support this idea, however. Some scientists argue instead that the first life forms on Earth were "colonists" in the form of single-celled organisms or at least the raw materials from which life evolved, brought to Earth on the surface of meteorites or comets (Figure 1.23). Somewhere, somehow, life first began, presumably due to the interactions of mere chemicals. Explaining how it could have begun is one of the most fascinating challenges facing chemists.

What Is Life?

In discussing the origin of life, it helps to be clear about what we mean by "life." Unfortunately life is not an easy concept to define. Few scientists would dispute, however, that the living organisms found on Earth are *complex chemical assemblies that are capable of giving rise to other chemical assemblies similar to themselves.*

All of the living organisms we know of fit within that chemical definition, and all are also based on a branch of chemistry known as **organic chemistry**—the chemistry of carbon. By "the chemistry of carbon," we mean the chemistry of compounds which contain carbon atoms as key parts of their structures. We also know that all life on Earth is sustained by the same basic chemical processes involving genes made of the chemical called DNA (deoxyribonucleic acid). Genes control the production of complex molecules called proteins, which in turn oversee all the chemical reactions of life. These chemical foundations of life are examined in Chapter 16. The fact that the same central chemical processes lie at the heart of all life on Earth—from

Figure 1.23 Scientists have speculated that single-celled "colonists" from within the solar system may have landed on Earth via meteors.

exercise 1.5

Protons, Neutrons, and Electrons in Ions

Problem

Fill in the missing information in this table:

	Symbol	Protons	Neutrons	Electrons	Charge
(a)	$^{14}_{7}\text{N}^{3-}$	_____	_____	_____	_____
(b)	$^{=}_{=}\underline{\ \ }^{-}$	34	45	36	_____
(c)	$^{=}\text{F}^{-}$	_____	10	_____	-1
(d)	$^{52}_{24}\underline{\ \ }^{3+}$	_____	_____	_____	_____

Solution

(a) $^{14}_{7}\text{N}^{3-}$ This is an ion of nitrogen, referred to as *nitride*.

Protons: The atomic number equals the number of protons: **7.**

Neutrons: The mass number (14) minus the atomic number (7) equals the number of neutrons: $14 - 7 = \textbf{7.}$

Electrons: The charge is -3. This means that there are 3 more electrons than protons, so there must be

$$7 + 3 = \textbf{10 electrons}$$

Charge: The charge is given: **−3.**

(b) $^{79}_{34}\text{Se}^{2-}$ This is an ion of selenium.

Protons: The number of protons is given. The atomic number is the subscript to the left of the symbol: **34.** Using the periodic table and the list of elements at the back of the book, you can identify the element as selenium (**Se**).

Neutrons: The number of neutrons is given. The mass number can be calculated by adding the number of protons (34) and neutrons (45): $34 + 45 = \textbf{79.}$ The mass number is the superscript to the left of the symbol.

Electrons: The number of electrons is given: 36.

Charge: There are 34 protons and 36 electrons. This means that there are two more electrons than protons: $36 - 34 = 2$. Because there are more electrons, the charge is negative: **−2.**

(c) $^{19}_{9}\text{F}^{-}$ This is an ion of fluorine, referred to as fluoride.

Protons: The periodic table shows that fluorine's atomic number is 9, therefore it has **9** protons. The atomic number is the subscript to the left of the symbol.

Neutrons: The number of neutrons is given. The mass number is calculated by adding the number of protons (9) and neutrons (10): $9 + 10 = \textbf{19.}$ This is the superscript to the left of the symbol.

Electrons: The charge is -1. This means that there is one more electron than protons: $9 + 1 = 10$. There are **10 electrons.**

Charge: The charge is given: **−1.**

exercise 1.5 *(continued)*

(d) $^{52}_{24}Cr^{3+}$ This is an ion of chromium.

Protons: The atomic number (subscript to left of symbol) equals the number of protons: **24.** Looking at the periodic table we see that this corresponds to chromium (**Cr**).

Neutrons: The mass number (52) minus the atomic number (24) equals the number of neutrons: $52 - 24 = $ **28.**

Electrons: The charge is $+3$. This means that there are three more protons than electrons, so the number of protons would equal the number of electrons if you added three more electrons. Mathematically this can be represented as:

$$24 \text{ (\# of protons)} = \text{\# of electrons} + 3$$

Subtracting 3 from both sides gives us:

$$24 - 3 = \text{\# of electrons} + 3 - 3$$
$$\mathbf{21 = \text{\# of electrons}}$$

Charge: The charge is given: $+3$.

bacteria to trees to humans—is a strong piece of evidence suggesting that all life on Earth shares a common origin. All life on Earth, in other words, is derived from the same ancient early life forms, which, when they appeared perhaps 4 billion years ago, must have been much simpler than anything we would today recognize as life.

Recreating the Steps Toward Life

A classic series of experiments to determine whether the most basic chemicals of life could have formed from nonliving materials were carried out in Chicago in 1951 by Harold C. Urey and his student Stanley Miller. They mixed water (H_2O), representing primordial oceans or pools, with the gases methane (CH_4), ammonia (NH_3), and hydrogen (H_2)—gases believed to have predominated in the early atmosphere. This simple chemical mixture was then heated within a closed system of flasks and tubes (Figure 1.24) causing water vapor to circulate throughout the apparatus. The gaseous mixture was subjected to electrical discharges intended to simulate lightning in the atmosphere, while the heating and a separate cooling phase caused a constant cycling of the water vapor between the simulated "ocean" and "atmosphere." After about a week, the originally colorless water had turned a deep reddish brown. When the apparatus was eventually opened up for analysis, a large number of very complex molecules were found dissolved in the water and mixed into the vapor. Many of these compounds were common constituents of modern life forms, including several amino acids, the building blocks of proteins (discussed in Chapter 16).

The Miller-Urey experiment dramatically demonstrated that some of the most fundamental chemical compounds of life could readily be formed in conditions believed to simulate the conditions on the early Earth. This was taken to be firm support for the idea that life could have originated on Earth as a result of spontaneous chemical reactions. Many subsequent, similar simulations (often using different approximations of the composition of the early atmosphere) have generated the same or similar mixtures of the key chemicals of life.

It is also very significant, however, that many of the most essential chemicals of life (including many generated in the Miller-Urey experiment) have now been identified in space. The Murchison meteorite was found to contain a mixture of organic chemicals very similar to those produced in the Miller-Urey apparatus, including important amino acids. Such discoveries rekindle interest in the possibility that life might have originated somewhere other than on Earth. Combined with the results of the simulation experiments, it certainly indicates

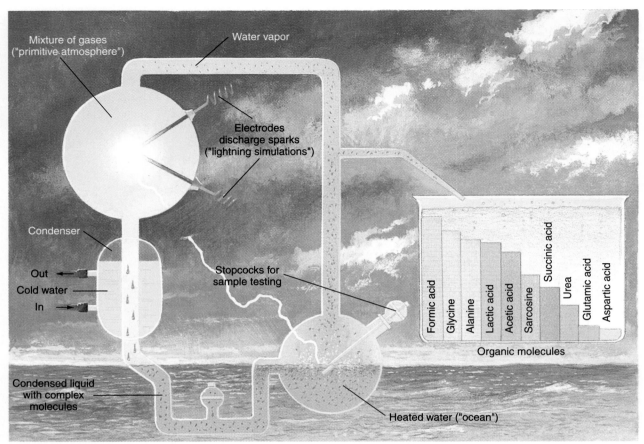

Figure 1.24 An illustration of the apparatus used by Stanley Miller and Harold Urey to mimic the basic chemistry of the waters and atmosphere of the primordial Earth. The heat caused the water to evaporate and cycle through the apparatus. The electrical discharges simulated the effects of lightning on the mixture of water, methane, ammonia, and hydrogen in the "atmosphere." The condenser cooled the gas mixture to cause droplets of "rain" to return to the "ocean" (water phase) to be heated and recycled. After several days, the simple chemical starting materials yielded a variety of the most important chemicals found in living things.

that some of the basic chemical raw materials of life can form readily throughout the universe and were probably abundant on the early Earth.

Of course, generating the raw materials of life is very different from generating life itself. We must clearly state that nobody has succeeded in recreating the origin of even the simplest form of life under conditions resembling those of the primordial Earth. This is perhaps hardly surprising, since nobody knows the precise combination of chemicals and conditions that presumably led to the origin of life. Slow but steady progress is being made, however, in recreating a series of plausible steps through which simple chemical systems capable of reproducing themselves could have emerged.

The most excitement at present is centered on a chemical called RNA (ribonucleic acid) which (as discussed in Chapter 16) is a key chemical in all modern life forms. Chemists are very close to demonstrating how molecules of RNA might have emerged spontaneously and might then have been able to participate in the formation of copies of themselves. Such primitive chemical reproduction might then, through a series of seemingly plausible steps, have led to the origin of simple living organisms based on the same "chemical logic" as modern life.

Research into the origins of life, however, remains a very controversial field, with heated debates proceeding among the various scientists involved. At present nobody knows exactly how life began, and this represents an important field of continued scientific study.

1.6 | Origins of Chemistry as a Science

According to the scientific account of our origins:
- The physical processes operating in the universe produced atoms.
- The atoms reacted with one another to make molecules and ions.
- Atoms, molecules, and ions reacted to create the first simple life forms.
- The chemical processes of life allowed evolution eventually to generate us—thinking beings able to contemplate and manipulate the chemistry that created us and sustains us.

Viewed from this perspective, the origins of chemistry as a science also depended on chemistry, and are a part of chemistry. They are part of the chemistry that sustains the human brain, makes it capable of puzzlement and experimentation, and makes it capable of creating organized societies in which different humans can share their accumulated knowledge.

Long before anyone knew about atoms, there were people who were "doing chemistry." They lit fires, cooked food, used heat to release metals from ores, and mixed molten metals to generate new kinds of metal alloys. They were curious about the chemical properties of the plants and rocks around them. They developed ideas, or theories, to try to make some sense of the chemistry of their world, such as the early notion of the four "elements," namely earth, air, water, and fire. The people who devised such ideas were moving along the right path, although we now recognize 92 naturally occurring fundamental substances called elements, rather than just four.

As human history moved along through the Stone Age, Bronze Age, Iron Age, and into modern times, increasing numbers of curious people investigated an increasing variety of interesting properties of the substances they encountered. When they found practical uses for what they had learned, new technologies resulted. The purposeful starting of fires was one of the first steps along the path of chemical technology that would eventually lead to many of the innovations that sustain our modern lifestyles. We have even reached the stage where we can manufacture new elements. We can mimic the alchemy of the stars to realize, at last, the ancient alchemist's dream of changing one element into another, even if it is the element plutonium (atomic number = 94) that is produced from uranium (atomic number = 92) rather than gold from lead.

Chemistry is not an activity reserved for just a small community of scientists. We are *all* chemists, in one way or another. We always have been and always will be because our world is built of chemicals. From chemicals we originated, and the continuing interactions between chemicals are inextricably intertwined with our future.

main points

- The universe is probably 12–15 billion years old, and was probably formed by an event known as the Big Bang. At the time of the Big Bang, the distinction between matter and energy was very blurred. Even today, matter and energy can be interconverted.

- All chemicals are composed of particles of matter, which can be atoms, molecules, or ions. Molecules and ions are derived from atoms.

- Nuclear fusion was (and is) responsible for the origin of elements. The process of nuclear fusion also releases incredible amounts of energy, such as the energy released by stars, including our Sun.

- Atoms are composed of subatomic particles called protons, neutrons, and electrons. Protons and neutrons themselves are made up of quarks. Protons have a positive electric charge. Electrons have a negative electric charge. Neutrons are electrically neutral.

- The electric force causes the attraction between protons and electrons (and between any objects with opposite signs of charge). The electric force also causes repulsion between objects with the same sign of charge (such as two electrons or two protons). Elements are the simplest type of substance, each one characterized by a unique number of protons. Compounds are substances made up of two or more elements chemically bonded together.

- Molecules are made of two or more atoms that are held together by the sharing of electrons; each molecule behaves as an independent unit.

- Ions are electrically charged particles, derived from atoms or molecules when the atoms or molecules either lose or gain one or more electrons. Life was probably formed on Earth as a result of natural chemical processes.

- Our understanding of chemistry is helping to shed light on the mysteries of the origin of the universe and of life.

important terms

amu is the abbreviation for **a**tomic **m**ass **u**nit, also called "dalton." (p. 26)

Atom is the smallest part of an element to retain the properties of the element. (p. 29)

Atomic mass unit (amu) is approximately the mass of one hydrogen atom; exactly 1/12 the mass of one atom of carbon having mass number 12. (p. 26)

Atomic number is an integer equal to the number of protons in an atom's nucleus. (p. 31)

Big Bang is the event, occurring 12–15 billion years ago, when the universe was created. (p. 22)

Chemical bond involves electrons holding two or more atoms together in a molecule or crystal. (p. 38)

Chemistry, at the atomic level, is study of the interaction of atoms as they react to form different combinations and make new substances. (p. 38)

Compound is a substance made of specific ratios of different chemical elements in a particular arrangement. (p. 38)

Covalent bond is a chemical bond formed by the sharing of electrons between two atoms. (p. 40)

Density is the ratio of the mass of a substance to its volume. (p. 37)

Direct evidence is obtained from actual observation. (p. 24)

Electromagnetic force is the force associated with light, electricity, and magnets. (p. 28)

Electron is the subatomic negatively charged particle surrounding the atomic nucleus. (p. 25)

Electron cloud is the region of space surrounding a nucleus where the electrons are located. (p. 29)

Element is one of over 92 naturally occurring substances of which the matter of the solar system is made. (p. 29)

Energy is the capacity for doing work. (p. 23)

Fundamental forces are the four forces responsible for all interactions between matter in the universe: strong and weak nuclear forces, electromagnetism, and gravity. (p. 28)

Gas (*g*) is one of three states of matter; it fills the volume of the container that holds it regardless of the shape and size of the container. (p. 36)

Gravity is the force of gravitational attraction between two bodies. (p. 29)

Indirect evidence is inferred from observation of related phenomena. (p. 24)

Ion is an atom or molecule possessing an electric charge. (p. 41)

Ionic bond is a bond formed by attraction of positively and negatively charged ions. (p. 41)

Isotopes are atoms with the same number of protons but a different number of neutrons. (p. 30)

Liquid (*l*) is one of three states of matter; it fills the bottom of the container that holds it to the extent of its own volume. (p. 36)

Mass is a measure of an amount of matter. (pp. 23, 25)

Mass number is the integer which is the sum of the number of protons and neutrons in a nucleus. (p. 31)

Matter is anything that occupies space and has mass. (p. 23)

Molecule is a group of atoms bonded together by shared electrons in a specific arrangement, which behaves as an independent unit. (p. 40)

Neutron is a subatomic nuclear particle that has no electrical charge and so contributes only to the mass of the nucleus. (p. 25)

Nuclear fusion is the formation of a larger nucleus by the merging of two smaller nuclei. (p. 31)

Nuclear reactions are those reactions in which the makeup of an atomic nucleus changes and new elements or isotopes are produced. *Chemical* reactions, in comparison, are changes in which one set of substances is converted into another set but the nuclei of atoms involved remain unchanged. (p. 31)

Nucleus (plural nuclei) is the positively charged center of an atom. (p. 29)

Organic chemistry is the chemistry of compounds of carbon. (p. 43)

Positron is a short-lived particle with the mass of an electron but a positive charge. (p. 31)

Proton is a positively charged subatomic particle in the nucleus of atoms; its number determines atomic number. (p. 25)

Solid (*s*) is one of three states of matter; it has a fixed shape. (p. 36)

States of matter are the three forms in which matter appears: gas, liquid, and solid. (p. 36)

Strong nuclear force is the force that holds protons and neutrons together in an atomic nucleus. (p. 29)

Weak nuclear force is responsible for the transformations of sub-subatomic particles in the nucleus of some atoms. (p. 29)

Weight is a measure of force equal to the product of an object's mass times the local gravitational acceleration. (p. 26).

exercises

1. When a woman is pregnant, we cannot actually see the fetus growing inside of her. Give three examples of indirect evidence that she is indeed pregnant.
2. What are the four fundamental forces? How do you imagine each of these forces acting on a raindrop?
3. A quantity of hydrogen atoms having a mass of 4.032 grams undergoes fusion to form 4.004 g of helium. The lost mass has become energy. How much energy is released in this fusion process?
4. The kinetic energy of an object equals one-half that object's mass multiplied by the square of its velocity ($KE = \frac{1}{2} mv^2$). Calculate the kinetic energy of a 0.142-kg baseball moving at a speed of 40.0 meters per second.
5. Calculate the energy released if a baseball, with a mass of 0.142 kg, were transformed into energy.
6. If a hydrogen atom were the size of a baseball, how big would the nucleus be if it were in the same proportion as is true for real hydrogen atoms?
7. How many hydrogen atoms would need to lie side by side to make a line of hydrogen atoms 1 cm long?
8. If an electron had a mass of 1 gram, how heavy would a hydrogen atom be if it were in the same proportion as is true for real hydrogen atoms?

9. Read an article on the origin of the universe which was published after this textbook. Compare what it says to what we say in this book.
10. Sketch an atom of oxygen, showing the nucleus, protons, neutrons, and electrons.
11. Sketch an atom of nitrogen (N), showing the protons, neutrons, and electrons.
12. C-12 and C-13 are the two stable isotopes of carbon. What are the similarities and differences between these isotopes in terms of protons, neutrons, and electrons?
13. A hydrogen bomb produces an explosion resulting from the nuclear fusion of deuterium and tritium, (two isotopes of hydrogen). What causes an atomic bomb to explode?
14. Beginning with the Big Bang, list the significant chemical events in the formation of the universe.
15. Give two examples of the use of isotopes in medical diagnosis.
16. Describe the formation of a star. Why do stars release light and heat?
17. Why don't we have to worry about the sun burning out?
18. Suggest a theoretical fusion reaction to produce a bromine-81 atom.
19. Suggest a theoretical fusion reaction to produce an atom of element 117 having an atomic weight of 291.
20. Explain why we are really "made from stardust."
21. An iron atom has an atomic number of 26 and a mass number of 56. How many neutrons must the atom contain?
22. Suppose a plutonium atom (atomic number 94) splits apart (nuclear fission) and one of the products is krypton (atomic number 36). What must be the remaining element?
23. How many helium atoms are needed to fuse into a silicon atom having 14 neutrons? Would this transformation give off energy or require energy?
24. List several nouns that describe things not made of matter.
25. Boron (atomic number 5) consists of two different isotopes and has an average atomic mass of about 10.8. What are reasonable guesses for the mass number of the two isotopes? Which isotope is present in greater amount?
26. Chlorine has an atomic number of 17. The two common isotopes of chlorine have mass numbers 35 and 37. How many neutrons are there in one atom of each of these two isotopes?
27. Identify each of these elements:
    a. $^{91}_{40}X$ b. $^{186}_{75}X$ c. $^{33}_{16}X$
    d. $^{14}_{6}X$ e. $^{51}_{23}X$ f. $^{131}_{54}X$
28. How many protons and neutrons are in each of these elements?
    a. ^{91}Zr b. ^{73}Ge c. $^{24}Mg^{2+}$
    d. ^{209}Bi e. $^{35}Cl^{7+}$ f. $^{65}Zn^{2+}$

29. How many protons and neutrons are in each of these elements?
    a. ^{227}Ac   b. ^{70}Ga   c. ^{40}K
    d. ^{259}No   e. ^{239}Np   f. ^{56}Fe

30. How many protons, neutrons, and electrons are there in one atom of each of the following elements?
    a. ^{13}C   b. ^{35}Cl   c. ^{60}Co   d. ^{235}U

31. The three states of matter, solids, liquids, and gases, together form our surroundings. Name the solids, liquids, and gases found in a lake.

32. A very muscular person and a person with 35% body fat and very little muscle are attempting to float in a pool. Explain why the person with more body fat will have an easier time floating.

33. Suppose that you did this experiment to identify the metal used in a coin found in an Egyptian pyramid. You determined the mass of the coin to be 17.85 grams. When you put the coin in a graduated cylinder (used for measuring volumes) filled with water, the volume went up from 19.8 mL to 21.7 mL. What is the density of the coin?

34. Which more closely resembles a covalent bond: (a) two people standing side by side or (b) two people holding hands. Why?

35. In potassium bromide, individual potassium ions have 18 electrons and bromine ions have 36 electrons. Which atom is positive and which is negative?

36. What is the difference between a covalent and an ionic bond?

37. Look up (in Table 1.2) the densities of diethyl ether, water, mercury, and carbon tetrachloride. If these were placed into a drinking glass, what sequence would there be from top to bottom?

38. Look up (in Table 1.2) the densities of wood, lead, and osmium. Would these float or sink in each of the liquids mentioned in question 37?

39. How many protons, neutrons, and electrons are in each of the following ions?
    a. ^{23}Na$^+$ b. ^{81}Br$^-$ c. ^{48}Ti^{2+} d. ^{90}Sr^{2+}

40. How many protons, neutrons, and electrons are in each of the following ions?
    a. ^{56}Fe^{6+} b. ^{40}K$^+$ c. ^{19}F$^-$
    d. ^{14}N^{3-} e. ^{127}I$^-$ f. ^{127}I^{7+}

41. How many protons, neutrons, and electrons are in each of the following?
    a. ^{195}Pt$^+$ b. ^{93}Nb   c. ^{37}Cl$^-$
    d. ^{35}Cl^{7+} e. ^{122}Sb^{2+} f. ^{56}Fe^{2+}
    g. ^{182}W   h. ^{85}Rb$^+$ i. ^{29}Si^{4-}

42. Fill in the missing information in this table:

	Symbol	Protons	Neutrons	Electrons	Charge
a.	$^{126}_{53}$I$^-$				
b.	—$^{5+}$	35	55		+5
c.	^{137}— $^-$	56		54	
d.	$^{62}_{29}$Cu^{2+}				
e.	$^{51}_{23}$—$^{5+}$				
f.	—Ni$^-$		31		+2

43. Fill in the missing information in this table:

	Symbol	Protons	Neutrons	Electrons	Charge
a.	^{27}—$^-$	13		10	
b.	$^{88}_{38}$—$^-$				+2
c.	—$^+$	30	35		
d.	—$^-$	19	18	19	
e.	—Sb^{3+}		76		
f.	$^{133}_{55}$Cs$^-$				+1

For the following questions, refer to this attached table of atomic composition of typical atoms. Not all isotopes are mentioned here, only the most abundant one for the first 20 elements.

Element	Proton	Neutron	Electron
H	1	0	1
He	2	2	2
Li	3	4	3
Be	4	5	4
B	5	6	5
C	6	6	6
N	7	7	7
O	8	8	8
F	9	10	9
Ne	10	10	10
Na	11	12	11
Mg	12	12	12
Al	13	14	13
Si	14	14	14
P	15	16	15
S	16	16	16
Cl	17	18	17
Ar	18	22	18
K	19	20	19
Ca	20	20	20

44. How many protons, neutrons, and electrons are there in one molecule of carbon dioxide, CO_2?

45. How many protons, neutrons, and electrons are there in one molecule of hydrogen fluoride, HF?

46. How many protons, neutrons, and electrons are there in one molecule of phosphorus pentafluoride, PF_5?

47. Organic chemistry is the basis for many types of compounds: pharmaceuticals, agricultural products, food preservatives, etc. These products are constantly researched and developed to improve our lives. Using a web search, find a recent news event involving organic chemistry.

48. Why do we refer to DNA as a "building block of life?"

49. Name one way you used chemistry today.

50. What would a job description be for a chemist at a company that makes running shoes?

food for thought

51. Ice has a density less than that of water and steel has a density greater than that of water. How is it that both the ocean liner Titanic (made mostly of steel) and the iceberg which it struck floated on water? What caused that to change after the impact?

52. The atomic bomb releases energy by fission (splitting of atoms) of uranium or plutonium and the hydrogen bomb releases energy by fusion of hydrogen nuclei. Why do both of these processes release energy?

53. In the years immediately after the discovery of atomic energy, it was widely believed that atomic power generation would be so cheap that electricity would be nearly free. What problems are associated with atomic power plants? How many of these problems are technical or chemical? How many of these problems are attitudinal or psychological?

54. The Miller-Urey experiment showed that amino acids found in living organisms could be synthesized from simple compounds, given "primordial conditions." You are a reviewer on a scientific panel that wants to judge the validity of the Miller-Urey experiment. What additional evidence would you like to see before you are convinced that life actually *was* derived from simple molecules?

55. It has been said that detection of oxygen gas in the atmosphere would be a good indicator of life on that planet. Why is this a reasonable statement?

56. Theories that describe the origin of atoms in the universe are undergoing constant scrutiny and refinement. There is ongoing debate about whether and how long ago the Big Bang actually occurred, although most experts support the idea. Is it possible to determine exactly how the universe began? Is it appropriate for textbooks such as this one to even raise the issue if the experts themselves are not sure?

57. The final Main Point of this chapter says that "Our understanding of chemistry is helping to shed light on the mysteries of the origin of the universe and of life." Is this really useful for us to know? Why should we care—or should we?

58. Draw a graph with proton number on the x-axis and neutron number on the y-axis. Locate on this graph 15 stable nuclei. Refer to the *CRC Handbook of Chemistry and Physics* "Table of the Isotopes" for your information. (Check the reference section of the library.) Save this graph so you can extend it after you have studied Chapter 14.

59. Suppose you are asked to reconcile the description of the beginning of the universe found in this book with the Genesis description. What would you say?

60. Suppose you are asked to show that the description of the beginning of the universe is *irreconcilable* with the Genesis description. What would you say?

readings

1. Peebles, P., James E., Schramm, David N., Turner, Edwin L., and Kron, Richard G. The Evolution of the Universe. *Scientific American,* 271 (October 1994): 53.

2. Hawking, Stephen. *A Brief History of Time: From the Big Bang to Black Holes.* New York: Bantam Books, 1988: 117.

3. Kirshner, Robert P. The Earth's Elements. *Scientific American,* 271 (October 1994): 59.

4. Kirshner, Robert P. Supernova: Death of a Star. *National Geographic,* (May 1988): 618.

5. Broad, William J. Earth's Inner Core Found to Be Spinning on Its Own. *New York Times* (18 July 1996): A1.

6. Wilford, John Noble. Signs of Primitive Life on Mars Are Found in Ancient Meteorite. *New York Times* (7 August 1996): A1.

7. Wilford, John Noble. Replying to Skeptics, NASA Defends Claims About Mars. *New York Times* (8 August 1996): A1.

8. Browne, Malcolm W. Planetary Experts Say Mars Life Is Still Speculative. *New York Times* (8 August 1996): D20.

9. Orgel, Leslie E. The Origin of Life on Earth. *Scientific American,* 271 (October 1994): 77.

10. Gore, Rick. How Did Life Begin? *Time* (October 11, 1993): 68–74.

11. Mojzsis, S. J., Arrhenius, G., McKeegan K. D., Harrison, T. M., Nutman A. P., and Friend R. L. Evidence of Life on Earth Before 3,800 Million Years Ago. *Nature* (7 November 1996): 55–59.

12. Chaboyer, Brian C. Rip Van Twinkle—The Oldest Stars Have Been Growing Younger. *Scientific American* (May 2001): 44–53.

resources

Lemonick, Michael D. and Nash, J. Madeleine, Unraveling Universe. *Time* (6 March 1995): 77–84.

The Once and Future Universe. *National Geographic,* 163 (June 1983): 704.

websites

www.mhhe.com/kelter   The "World of Choices" website contains activities and exercises including links to websites for: The National Aeronautics and Space Administration (NASA) site called "Origins"; Fermilab National Accelerator Laboratory Site; The journal *Scientific American,* and much more!

Appendix to Chapter 1

Significant Figures

Maybe the most significant figure to you is the actual amount of money you get to take home from your part-time job after the federal and state governments have taken their share. That figure, whether it's $50 or $75.47, represents a *counted* number, an actual number of dollars and cents. No matter how paltry it might seem, that's exactly how much, or how little, you have to spend. To scientists, the idea of significant figures has special meaning, having to do not with something that is counted, but is *measured.* If you think about it, you'll realize any measurement must have some degree of uncertainty about it, for the devices used to measure are made by humans and read by humans. Most measurements can be inferred to have an uncertainty of 1 in the last significant place. A measure of 5.25 inches has an *uncertainty* of 0.01 inches (or 5.25 ± 0.01); the true value of the length is between 5.24 and 5.26 inches. The question for the scientist is, in any given measurement, which digits are important (or *significant*) and which can be ignored?

Let's consider an example that may well relate to the source of your paycheck: the fast-food chains in town. Most major fast-food corporations have "Nutritional and Ingredient Information" sheets available. The following data are from one popular chain that sells Mexican food:

Soft taco: serving size = 78 grams; calories = 180
Bean burrito: serving size = 198 grams; calories = 390

A leading hamburger chain gives the following data for its products:

Small fries: serving size = 68 grams; calories = 220; protein = 3.13 grams
Large hamburger with secret sauce: serving size = 215 grams; calories = 560; protein = 25.2 grams

What can you infer about the consistency of the serving size based on these data? That is, *based on the information that you are given,* should you assume that one taco is nearly identical to any other? Can you assume that every serving of small fries has precisely 3.13 grams of protein?

If your own experience is anything like that of your textbook authors, then you know that sometimes the burger comes with secret sauce dripping from all sides of the sandwich. Other times, the location of the secret sauce is truly a secret. On occasion, perhaps the mass of beans in a burrito is about the same as that of a feather. Another burrito seemingly contains enough beans to mortar all the bricks in the student union building.

When the nutritional information sheet says that a serving of fries has 3.13 grams, is that number "right on"? Is it better to say "about 3 grams," to indicate that there is a lot of variability from sample to sample? Does the ".13" in the 3.13 really have meaning, or is it *insignificant?*

If data are to have meaning in science (or to the general public, for that matter), the data must be as close to "the truth" as possible. They must mean what the scientists intend for them to mean. One way to interpret what data mean is to look at the number of **significant figures** (or "sig. figs.") in given values. We saw with the food data that just because data are reported to a certain number of figures, it does not mean that the figures are all significant. And as with the fast-food examples, we see that unless the scientists understand how to use significant figures properly, silly conclusions can result. Similarly, you need to know how to interpret and to report the proper number of significant figures, so your results have your intended meaning when they are transmitted to others.

Here are some simple rules to follow:

1. **Nonzero digits are significant.** For example, 3.25 newtons (a unit of force, N) has three significant figures; 12.657 newtons has five significant figures.
2. **Zeros that are sandwiched in between nonzero digits are significant.** There are four significant figures in 250.4 milliliters (a measure of volume, mL) as well as in 10.01 milliliters.
3. **Zeroes to the right of the decimal point are significant.** When we say that 2.0 inches has two significant figures, we are implying that the uncertainty is the same as 2.4 inches. Thus, both 5.00 inches and 5.25 inches have three sig. figs.
4. **Zeroes to the left of the decimal point (whether shown or not) are ambiguous as to their significance.** The zeroes in the value 500 pounds may be there simply as place-holders or they may represent actual measured numbers. A zero is just as likely to show up in a measurement as any of the other nine digits. From the number 500 pounds, we can't tell whether the writer meant that the number has an uncertainty of 100 pounds, 10 pounds, or just 1 pound. This is a very important justification for using scientific (or exponential) notation. Writing 500 as:

5×10^2 indicates there is 1 sig. fig.
5.0×10^2 indicates there are 2 sig. figs.
5.00×10^2 indicates there are 3 sig. figs.

This makes it unambiguous as to how many significant figures there are.

5. **Zeros that precede the first nonzero digit are *not significant*.** A measurement of mass recorded as 0.0160 kilograms (kg) has only three significant figures. The zero to the left of the decimal point is there simply to emphasize the position of the decimal point. The zero between the decimal point and the number 1 serves as a place holder. In scientific notation, this would be written as 1.60×10^{-2} kg. Now, it is clear that the number has three sig. figs.

When working with values in mathematical computations, we must decide if the answer makes sense not only with regard to the magnitude of the number, but also the number of significant figures. The most important overall rule to follow when doing mathematical computations is to *complete all calculations before dropping any insignificant digits*. Again, we will look at a few simple rules and examples to clarify the choices.

Addition and Subtraction

When adding and subtracting, proceed as normal until all calculations are carried out. The final answer is determined to the correct number of significant figures by counting the number of digits after the decimal point in each of the measurements being added or subtracted. The sum or the difference should contain *the same number of digits to the right of the decimal point as the measurement with the least number of decimal digits* (that is, the crudest measurement).

Multiplication and Division

When multiplying or dividing numbers, the answer should contain no more significant figures than the fewest number of significant figures in any of the numbers being multiplied or divided in the calculation.

Connections

By taking these two things (the gold elixir and the gold) man can refine his body so that he never grows old and he never dies. Seeking for these external substances to fortify and strengthen oneself is like the feeding of the flame with oil so that it does not die out.

—Ko Hung (Chinese Physician and Scholar), 320 A.D.

Gold is perfection: create it, and you have created perfection; ingest it, and you will live forever—so many of our ancestors thought. The desire to change ("transmute") base metals into gold was the driving force behind the study of chemistry for over 2000 years. From about 600 B.C. to 1600 A.D., the world of chemistry was vibrant in Greece, the Middle East, and China. The desire to produce gold was the common thread that linked the chemical efforts of these societies.

Why did this interest occur at these places during those times? One of the key points in the Prelude was that there is an intimate relationship between science and society. To understand the origins and development of chemical thought, we must examine aspects of the origins and development of society itself.

At the end of the last Ice Age, about 8000 B.C., the climate of many regions of the Earth became more temperate. A critical development in the history of humankind occurred at this time: the first known strains of cultivable wheat and barley appeared in the Middle East, in an area aptly described as the Fertile Crescent. The cultural consequences of this development have been overwhelmingly important.

Until the end of the last Ice Age, much of the world's population had been nomadic, constantly traveling to find food. Imagine yourself as a hunter-gatherer, having to pack up all of your belongings every few weeks and move on. You would need approximately 250 square miles of land just to keep you and a group of your 25 closest friends fed for the year. Now imagine that you could cultivate a crop of wheat and barley on an acre of land, crops that would yield enough grain to keep your family fed for a year. You could stay put, build a house. If you had a steady and reliable source of water, you might even produce more grain than your family needed. You could use that surplus to *barter*. So it happened in

the Middle East. With the ability to sow and harvest grain came a truly significant change in lifestyle: a change from the nomadic life to one centered on organized agricultural villages. These permanent settlements eventually permitted the development of many chemical technologies related to farming. Among them were an understanding of how to work with natural resources such as metals; the development of potteries, glazes, and pigments; the use of fertilizer; and even the fermentation of grain into alcohol.

As resources were developed, trade routes for exchanging resources with other settlements became important, thus, much of the world's early economic activity centered on Greece, especially Ionia and, later, Athens. This region was in the center of a system of trade routes that connected Asia, Egypt, and Europe (see map). Relative stability and prosperity brought with it the opportunity for reflection. Philosophers were given to speculate about the nature of life and the world. Influential thinkers such as Socrates, his student Plato, and then Plato's student, Aristotle, all came from Greece. If we continue the line of students one step further, we come to Aristotle's student Alexander the Great who, although he only lived 32 years, conquered many countries, allowing for the interplay of scientific ideas among civilizations including Greek, Egyptian, Mesopotamian, Persian, and Indian. The link with India served as a conduit for interaction with China, a country with already well-developed ideas about the world. In this way, much of the world began to share ideas about the goals and processes associated with the earliest practice of chemistry. Interestingly, many of these ideas connect to centuries of attempts to make gold, to create "perfection" in metallic form; they connect to a philosophical ideal that could be attained through working with matter, a practice called *alchemy.*

As early as 8000 B.C. the area around the Fertile Crescent was home to some of the world's first farmers. The ancient city of Jericho was particularly blessed with fertile land, temperate climate, and a plentiful source of water. Fields could yield two or three crops a year of wheat and barley. This supported the farm families as well as traders who might barter salt or semiprecious gems for foodstuffs.

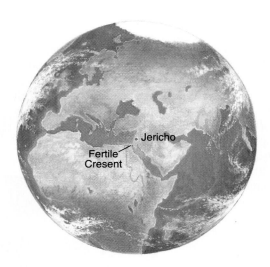

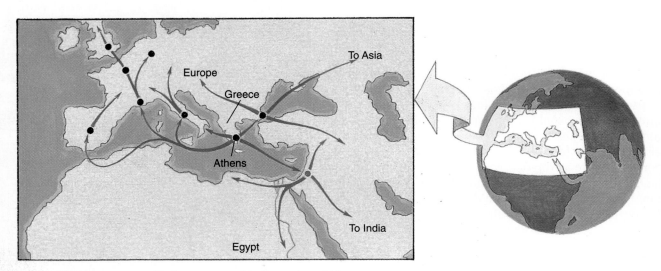

Trade routes 2000 years ago enabled commerce of both goods and ideas.

2.1 | Alchemy and Its Connection to Modern Chemistry

Was It Silly to Try to Change Other Substances into Gold?

Chemists have known for over 200 years that chemical reactions cannot be used to change atoms of one element into atoms of another element. Two thousand years ago, however, attempts to produce gold from other substances seemed perfectly reasonable. Ko Hung, the fourth-century scholar quoted on the opening page of this chapter, was the foremost Chinese authority on the subject of alchemy. He justified attempts to change "baser" substances into gold, a process called **transmutation,** by noting that many examples of transmutation are found in the natural world: "caterpillars [turn] into moths, snakes turn into dragons, oysters into frogs, rotting grass into fireflies, and old bears become foxes." This may sound odd to us, but there is an important realization here: that matter, living and nonliving, does change over time. Just as humans grow old and infirm, most materials in some way deteriorate. What Ko Hung knew to be true was that humans could act as agents of change: copper was combined with zinc to make brass, iron was made into steel, and silver was extracted from ores. So why shouldn't it be possible to make gold?

Alchemy was in essence a philosophy—the idea that humans could strive and attain an ideal: life without pain, suffering, or death. It was also a belief that through chemistry this ideal could be achieved, since chemistry was recognized as a tool that could be used to change matter from one form into another. Alchemists were not, by and large, simply greedy crackpots trying to make a fortune from gold. Rather, they were trying to reach this philosophical goal of perfection using their best available understanding. This was a philosophy common to many cultures for hundreds of years (Figure 2.1). The work of the alchemists created a foundation upon which we built our modern understanding of chemical transformations.

The attempts to make gold were confused by misleading appearances: if something looked like gold, felt like gold, weighed as much as gold, and conducted heat as gold does, it was easy to believe it was gold (Figure 2.2). Also working against the alchemists was their

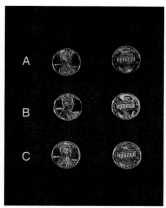

Figure 2.2 Take a good look at the three pennies shown here. What do they look like? Most students tell us they look like **A.** a copper penny, **B.** a silver penny, and **C.** a gold penny. Actually, the first is a copper penny, the second is a penny chemically coated with a mixture of zinc and copper known as *white brass,* and the third is a penny after heating to form the more common yellow brass. Could you tell the difference between the golden penny and real gold?

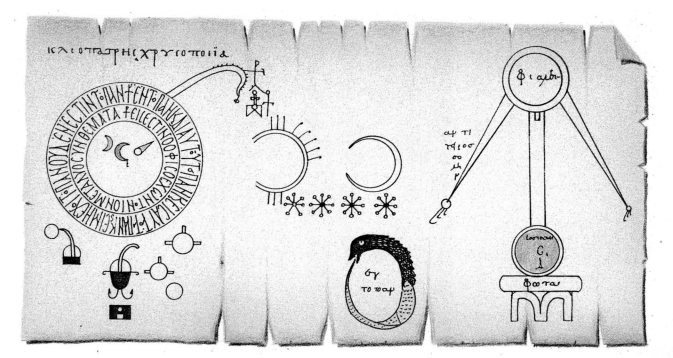

Figure 2.1 A re-creation of a page from a fourth-century (A.D.) alchemist's book on gold making. This is the formula and equipment used by an alchemist known as Cleopatra.

imperfect understanding of the nature of matter itself. Bernard Trevisan was a 15th-century alchemist who had spent his life and a sizable fortune in search of the secret of turning base matter into gold. The secret he realized only as he lay dying: "To make gold, one must start with gold."

The Diverse Origins of the Word "Chemistry"

Many recipes for transmutation were based on reasonably careful observations of nature, which led to some less reasonable assumptions. One such assumption held that when a living thing dies, the body leaves a seed that, impelled by the spirit, undergoes many changes to become, ultimately, perfect. The metal equivalent of perfection is gold. Therefore, it was first necessary to "kill" the starting metal, so that it would lose its metallic properties. The most visible metallic property is color, so killing the metal meant making it black (the absence of all color). It could then be made into gold. The Egyptian word for black is *khem.* This is one possible origin of the word *chemistry.* However, the Greek word for casting or pouring metals is *cheo,* another possibility. The Chinese too might have coined the word, from *chin-I* meaning "gold-making juice."

Whatever the origin of the word that came to name the science of chemistry, it is known that much knowledge of Chinese philosophy and transmutation was brought to the West via Middle Eastern countries such as Persia (the ancient name for the country that now includes Iran). The Arabic definite article "al" was added to the Chinese name "chin-I," giving rise to the original term: *alchemy.*

Although the alchemists never actually managed to make gold from other substances, their experimentation led to a better understanding of many chemicals. The English philosopher Francis Bacon (1561–1626) compared alchemy "to the man who told his sons he had left them gold buried somewhere in his vineyard; where they by digging found no gold, but by turning up the mould about the roots of the vines, procured a plentiful vintage." For chemists the harvest included a practical knowledge of processes such as distillation, fermentation, putrefaction, and discovery of the elements antimony, bismuth, zinc, arsenic, cobalt, and phosphorus (Figure 2.3).

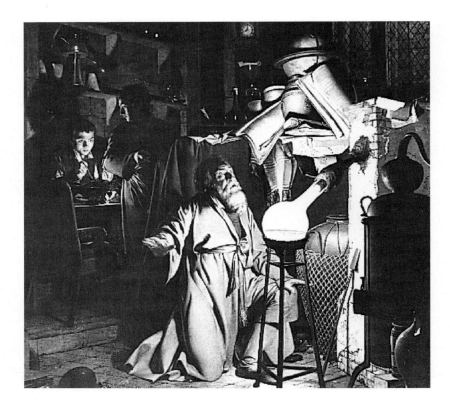

Figure 2.3 A portrait of "The Alchymist." This is a portion of a painting by the Englishman Joseph Wright. Some suggest the subject of the painting is the German alchemist Hennig Brand, shown in his laboratory lit by the glow of phosphorus, which he discovered in 1669. The apparatus shown here is used for distillation, a process used today to separate substances.

2.2 | The Advent of Modern Chemistry

Rethinking the Elements

Modern scientific chemistry began to gain momentum in the late 18th century. The ideal sought now was a better understanding of matter itself; philosophy would be left to the philosophers. But one important aspect of philosophy still shaped the debate, and that was the idea that all matter is made up of just a few essential "elements," each with its own distinctive property. Any material represents some combination of these elements. The philosophers of Ancient Greece thought in terms of earth, air, fire, and water. The alchemists, whose time was spent in laboratories, had discovered that there was more, not less, to matter than first meets the eye. The focus of those who now studied the science of chemistry was not gold, but air.

Curiosity, careful observation, experimentation, and publication (or the exchange of letters) led to both a complete description of what air is made of and to the development of an accepted scientific process and community. The discovery process was initiated by the Scotsman Joseph Black who in 1756 discovered that air consists of more than one type of gas. This was no easy task. Until about 1750, air had been thought to be composed of a single substance. Joseph Black determined that air contained a distinct component he called "fixed air," a gas that turned samples of lime-water cloudy. This gas was carbon dioxide. Henry Cavendish discovered "inflammable air" (hydrogen) in 1766. Then in 1772, one of Black's students, Daniel Rutherford, reported on another component of air, "noxious air," which could not support life nor combustion. This was later determined to be nitrogen gas, another major component of air.

Perhaps the most interesting component of air is oxygen. Its discovery, and the revelation of its importance, is usually connected with three names. These are Karl Scheele (Sweden), Joseph Priestley (England), and Antoine Lavoisier (France). It was Scheele who first prepared and described oxygen in 1771, calling it "fire air" because of the ease with which things burned in it. He sent the manuscript describing this work to a publisher who delayed publication for years. By the time the work was published, others had done the same experiments, published their results, and gained the principal credit for the discovery.

Joseph Priestley is more popularly credited with discovering oxygen. Using sunlight and a large magnifying glass, he heated a red-orange substance we now call mercury(II) oxide in a closed container and observed that droplets of pure mercury formed and a gas was given off. Others had decomposed mercury(II) oxide to give pure mercury, but no one had noticed the gas before (Figure 2.4). Priestley collected the gas and began a series of careful observations of its behavior. He quickly discovered that smoldering wood splints and candles burst into brilliant flame in this new gas. Mice placed in the gas became very active. When Priestley himself breathed the gas, he claimed it made him feel very "light and easy." He reported his work in 1774.

Among all the distinguished scientists who were working during the late 18th century, it is Antoine Lavoisier, the French chemist, who is often said to be the "Father of Modern Chemistry." Not only was he meticulous in his observations and systematic in his approach, but he made accurate measurements of the quantities of material being observed. He recognized that combustion is the reaction of a fuel with oxygen in the air.

By carefully weighing everything in his experiments, Lavoisier was able to demonstrate that the increase in weight of a metal when it oxidizes to form an oxide is just the same as the decrease in weight of the air in which it reacts (Figure 2.5). When a metal oxide is heated with charcoal (carbon), the decrease in weight is caused by the removal of oxygen from the compound to form carbon dioxide, whose weight in turn is the weight of the charcoal plus that of the oxygen. For example, 1.00 gram of tin will react with 0.27 gram of oxygen to form 1.27 grams of tin oxide. The 1.00 gram of tin will be recovered if the tin oxide is reacted with carbon.

As sometimes happens in matters that are familiar, even obvious, to us, we might overlook how crucial Lavoisier's statement of the **Law of Conservation of Mass** is to the practice of chemistry. Although matter might change its form during a chemical reaction, it cannot be created or destroyed—mass and therefore matter is *conserved*. The accepted way of doing

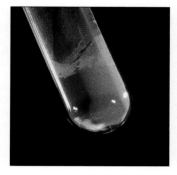

A. Before

B. Heating

C. After

Figure 2.4 A. The mineral cinnabar, a rough red earthy substance now known to have the formula HgO and named mercury(II) oxide. **B.** By heating the mineral, Priestly showed that it decomposed into **C.** elemental mercury and gaseous oxygen. He also showed that mercury would react with oxygen to reform mercury(II) oxide.

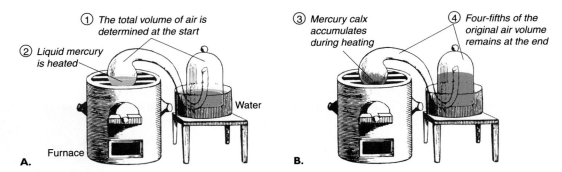

① The total volume of air is determined at the start

② Liquid mercury is heated

③ Mercury calx accumulates during heating

④ Four-fifths of the original air volume remains at the end

Water

Furnace

A.

B.

Figure 2.5 Lavoisier's experimental setup. **A.** Heating mercury in air creates a compound of mercury and oxygen (mercuric oxide or "calx"). **B.** The volume of oxygen taken up can be measured by the rise of water in the water-enclosed glass bell. Lavoisier then demonstrated (not shown here) that when the process was reversed and the mercuric oxide heated, the volume of oxygen released (the water level in the bell goes down) is equal to that originally taken up by the heating of mercury.

chemistry changed abruptly with Lavoisier's emphasis on measuring the amounts of materials entering and leaving chemical systems.

The next critical step forward came in 1799 with Joseph Louis Proust, a Bohemian teaching chemistry in Spain. He published his finding that stated that when substances come together to form compounds, they always do so in fixed proportions, by mass. The **Law of Definite Proportions** underscored the fact that not only did matter react in a consistent and predictable way, but also with mathematical precision. Thus, from our earlier examples: 1.00 gram of tin will *always* react with just 0.27 gram of oxygen to form tin oxide; 2.30 grams of magnesium will always react with just 1.51 grams of oxygen to form magnesium oxide.

consider this:

When Is a Law Really a Law?

No matter how carefully you test the Law of Conservation of Mass in a chemistry laboratory, you will never find it broken. It is an ancient and powerful principle of chemical understanding, reminding us that all the atoms present at the start of a chemical reaction must also be present at the end of the reaction.

For many years, however, we have known from the work of Albert Einstein that "energy has mass," and so if energy is released during a chemical reaction, that energy carries away some mass with it. Alternatively, if energy is absorbed by chemicals during a reaction, the energy brings in some mass with it. So it is not strictly true that the chemicals produced by a chemical reaction have the same mass as those at the start of the reaction. The amount of mass associated with the energy changes during chemical reactions can be calculated using Einstein's famous equation: $E = mc^2$. The "c^2" part of that equation is the speed of light squared, a value that is so large, 9×10^{16} m^2 s^{-2}, it means that tiny amounts of mass correspond to large amounts of energy. For this reason, no chemistry laboratory balance will ever come close to detecting the vanishingly small mass changes due to the energy lost or gained during chemical reactions. So the Law of Conservation of Mass remains a true and very important observation for practical chemistry.

This subtlety demonstrates two important points about the scientific method. First, everything we learn by applying the *scientific method* is always subject to challenge and revision by later discoveries, even the statements we are so sure of we call them *laws*. Second, laws and theories that we know are not precisely true, can nevertheless prove very useful to us because they remain true in many practical circumstances.

Atomic Theory

By the end of the 18th century, a vast amount of data had been collected on a wide variety of substances and their properties. The challenge facing chemists of the day was to make some sense out of all this. It seemed to be increasingly clear that a certain underlying order existed in the material world. The Law of Conservation of Mass supported this idea as did the Law of Definite Proportions.

exercise 2.1

The Law of Conservation of Mass

Problem

Magnesium is a light metal that when reacted with oxygen forms an extremely adherent high-melting layer of magnesium oxide. Magnesium is used in metal alloys that need to be structurally strong but lightweight—like those in aircraft bodies, car engines, cameras, and luggage. A piece of magnesium weighing 2.30 grams reacts with oxygen to form 3.81 grams of magnesium oxide. How much oxygen reacted with the magnesium?

Solution

Because mass is neither created nor destroyed in a chemical reaction, the increase in weight is the weight of oxygen which reacted.

$$\text{3.81 grams} - \text{2.30 grams} = \textbf{1.51 grams of oxygen reacted}$$

Check

The weight of the magnesium (2.30 grams) and that of the oxygen (1.51 grams) totals the weight of the magnesium oxide (3.81 grams).

$$2.30 + 1.51 = 3.81$$

The profile of John Dalton (1766–1844) is not one that suggests a person of revolutionary ideas. He was a self-taught mathematician, scientist, and philosopher who spent much of his life in Manchester, England. Dalton's foremost contribution to chemistry was a kind of *general* scientific discovery. He knew that compounds contain definite proportions, by mass, of their constituent elements. No matter how a specific compound is prepared, the percentage composition of the masses involved is always the same. Water is always 11% hydrogen and 89% oxygen, by mass. Other people had noticed the same thing, but Dalton went further in his thinking. He realized that the same pair of elements could combine in different proportions and thus form different compounds having very different properties—the **Law of Multiple Proportions.** For instance, carbon and oxygen can form two different compounds: one which consists of 42.9% carbon and 57.1% oxygen, and another very different compound having 27.3% carbon and 72.7% oxygen. The first of these two compounds (carbon monoxide) is highly toxic to animal life whereas the other (carbon dioxide) is a harmless gas present in every breath we take. The names of these compounds make their difference clear: carbon **mono**xide (CO), is composed of molecules containing one carbon atom and **one** oxygen atom (mono = 1); carbon **di**oxide (CO_2) is composed of molecules containing one carbon atom and **two** oxygen atoms (di = 2) (see Figure 2.6).

What was Dalton's revolutionary idea? Actually it was an old idea first associated with the philosophers of ancient Greece—that there existed in nature tiny indivisible building blocks of matter called *atoms.* What was new was Dalton's realization that these atoms existed

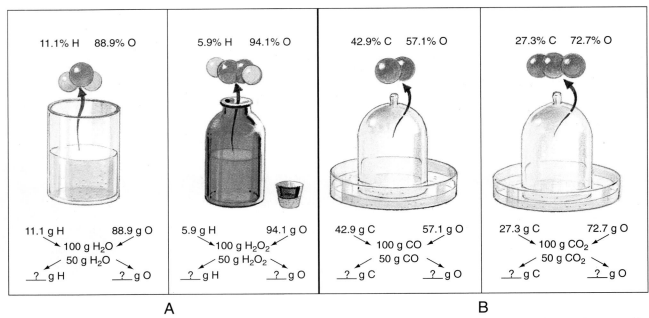

Figure 2.6 An underlying order exists in the way certain elements combine to form compounds. Quantitative measurements show that the elements combine in definite proportions, in a quantitatively consistent ratio of one element to another. Dalton realized that two elements can form more than one compound: **A.** hydrogen and oxygen and **B.** carbon and oxygen can each combine to form two different compounds, each in distinct proportions.

exercise 2.2

The Law of Definite Proportions

Problem

A sample of methane, sometimes referred to as swamp gas, is found to contain 12.0 grams of carbon and 4.03 grams of hydrogen. If a second sample of methane is found to contain 30.0 grams of carbon, how many grams of hydrogen does it contain?

Solution

The ratio of the masses of elements is constant. Therefore, since the amount of carbon is increased by a factor of 2.5 (30.0 grams ÷ 12.0 grams = 2.50), then the same must hold true for hydrogen.

$$\text{Grams of hydrogen} = 4.03 \times 2.5 = \textbf{10.1 grams}$$

Check

Use the data you have to show that the ratio of carbon to hydrogen is essentially the same for both samples:

$$\frac{12.0}{4.03} = \textbf{2.98} \text{ and } \frac{30.0}{10.1} = \textbf{2.97}$$

not as some philosophical concept but as a physical reality, particles that had a unique identifiable mass and specific physical characteristics. What Dalton proposed was a "modern" theory of the atom that could account for the law of conservation of mass, the laws of definite

■ case in point

Science and Politics: A Sometimes Volatile Mixture

You may know the old expression, "nothing in this world is certain except death and taxes." The quote needs to be changed for Antoine Lavoisier. Taxes caused his death. To understand the tragedy, you need to know that up until the French Revolution, France was ruled by a succession of kings, the last one being Louis XVI. The public desire for change spawned the French Revolution of 1789, which resulted in the creation of the First Republic followed by the Reign of Terror during which the King and Queen, Lavoisier, and hundreds of others were executed. The First Republic lasted until 1804, when, after a coup d'etat, Napoleon was crowned Emperor.

Before the revolution, Lavoisier participated in many aspects of French government. He was the Director of Gunpowder, improving the method of its manufacture and increasing France's supply. He worked on the commission that devised the metric system. To aid French farmers, he established a model farm and devised improved methods of soil cultivation. It was his involvement in an altogether different kind of "farm" that ultimately got him into trouble.

Tax collection in France was done through a private company called the Ferme Gènèrale, or Tax Farm. Lavoisier invested much of his family fortune in the Tax Farm and was therefore a tax collector in a very corrupt process that exempted the wealthy and, to use current vernacular, "soaked the poor". Lavoisier most likely did this strictly as a business venture, however, it cost him his life.

French revolutionaries viewed the Tax Farmers as agents for the repressive prerevolution conditions. The Farmers were arrested in 1793 and many were put to death. Lavoisier was beheaded by the guillotine on May 8, 1794, after the court ruled "The Republic has no need of scholars." He was 51 years old.

Lavoisier's scientific career began with working out an improved method for manufacturing potassium nitrate, KNO_3 (an important component of gunpowder) and devising ways of bringing fresh water and lighted streets to the people of Paris. By age 25 he was admitted to the Royal Academy of Sciences. His work on street lighting led him to consider the whole problem of combustion—what happens when something burns?

Laviosier's tremendous impact on chemistry continued up until his death. What additional discoveries might have been possible if Lavoisier had lived for another 20, perhaps 30 years? What might the effect of these discoveries have been on the social, economic, and scientific history of the world? Of course, we cannot know. The famous mathematician Joseph Louis Lagrange, a contemporary of Lavoisier, said this of his fellow scientist: "It took them only an instant to cut off that head, but France may not produce another like it for a century."

Gould, Stephen J. The Passion of Antoine Lavoisier. *Natural History,* June 1989.

Jaffe, Bernard. Lavoisier. *Crucibles: The Story of Chemistry,* 4th ed. New York: Dover Publications, 1976: 69–83.

and multiple proportions, and the difference between elements and compounds. Dalton's **atomic theory** states:

- *All matter consists of atoms.* Atoms are tiny indivisible particles of an element that cannot be created or destroyed. (In fact, atoms can be further broken down into subatomic particles, but it is the atom itself that makes an element unique.)
- *All the atoms of one element are identical in mass and other properties and are different from atoms of any other element.* (Isotopes of a given element are now known to vary in mass because of differing numbers of neutrons.)
- *Atoms of one element cannot be converted into atoms of another element.* (This does happen in nuclear reactions but never in a chemical reaction.)
- *Compounds are the chemical combination of two or more elements in a specific ratio.*

Chemical Symbols

Dalton devised a shorthand language for describing atoms and their compounds that consisted of circular symbols (Figure 2.7). This was swiftly replaced by a simpler one; the present system of naming and assigning symbols to elements and compounds was devised by Jöns Jakob Berzelius, of Sweden, in the 1820s. He assigned elements symbols of one or two letters, usually the first letter of the Latin name of the element and a characteristic second letter. The first letter is always capitalized and the second is lower case. Latin was used because it was then a common language among chemists of the four major scientific countries of the middle 19th

Figure 2.7 John Dalton proposed a system of circular symbols for the elements, which could then be combined to show compounds (molecules). The symbols rapidly proved to be terribly difficult to use in any context other than as a code. Chemical compounds and reactions were very ponderous to write out. (Berzelius spoke very forcefully in favor of his alternative system of letter symbols, which any printer could publish easily.) Interestingly, Dalton's symbols look surprisingly similar to our modern space-filling models of molecules. Note that some of the weights proposed by Dalton are quite close to modern values but others are off a lot, often about half or double the modern values.

ELEMENTS

	wt			wt
HYDROGEN	1	STRONTIUM	46
NITROGEN	5	BARIUM	68
CARBON	5,4	IRON	50
OXYGEN	7	ZINC	56
PHOSPHOR	9	COPPER	56
SULFUR	13	LEAD	90
MAGNESIUM	20	SILVER	190
CALCIUM	24	GOLD	190
SODIUM	28	PLATINUM	190
POTASSIUM	42	MERCURY	167

MOLECULES

WATER
AMMONIA
METHANE
CARBON MONOXIDE
CARBON DIOXIDE
SULFUR TRIOXIDE

century: Germany, France, Britain, and Sweden. Fortunately for us, many elements have the same name in English as in Latin, so their symbols are easy for us to learn. For example, the symbol for hydrogen is H, chlorine is Cl, and calcium is Ca. Some names in English do not match the Latin names, so their symbols are more of a problem for us. The symbol for sodium, for example, is Na (natrium), iron is Fe (ferrum), and potassium is K (kalium). The back endsheet of this book lists the modern symbols and full names of all the atoms.

Chemical Formulas

Chemists summarize the elements present in chemical compounds and the proportions in which they are present, using **chemical formulas.** In writing the formula of a compound, the symbols of all its elements are written together. For example, a combination of nitrogen and oxygen is NO. Hydrogen and chlorine is HCl. If the ratio in which the atoms are present is not 1:1, subscript numbers are used to indicate the ratios in which they are present. A compound containing one atom of carbon for every two atoms of oxygen is thus written as CO_2 (which is carbon dioxide); a compound containing one iron atom for every three chlorine atoms is $FeCl_3$; and so on.

Ions, derived from atoms or groups of atoms which have lost or gained electrons, are written with the charge as a superscript after the relevant symbol or symbols. For example, Na^+ is a sodium ion, Ca^{2+} is a calcium ion, and Cl^- is a chloride ion. Examples of **polyatomic ions** (containing several atoms covalently bonded together) are sulfate (SO_4^{2-}), hydroxide (OH^-), and ammonium (NH_4^+) ions (Table 2.1). These ions are also included in formulas of compounds such as calcium sulfate, $CaSO_4$; ammonium chloride, NH_4Cl; and calcium hydroxide $Ca(OH)_2$. Note the use of parentheses when a subscript number must refer to a *group* of atoms before it, rather than just a single atom before it. We consider polyatomic ions to be a group because they often stay together when a compound is added to water. When sodium hydroxide, NaOH, is added to water, two ions form. One is a positively charged ion, Na^+. The other is the polyatomic ion, OH^-. In the same way, when copper(II) nitrate, $Cu(NO_3)_2$, is added to water, the compound falls apart to form Cu^{2+}, the copper ion, and 2 nitrate ions, NO_3^-. The nitrogen and oxygen remain as a group. The best way to know the polyatomic ions in Table 2.1 is to memorize them. We will be using them enough that they may well become quite familiar to you.

Table 2.1

Polyatomic Ions

Formula	Name
NH_4^+	Ammonium
OH^-	Hydroxide
CN^-	Cyanide
NO_3^-	Nitrate
NO_2^-	Nitrite
SO_4^{2-}	Sulfate
SO_3^{2-}	Sulfite
CO_3^{2-}	Carbonate
HCO_3^-	Hydrogen carbonate or bicarbonate
ClO^-	Hypochlorite
ClO_2^-	Chlorite
ClO_3^-	Chlorate
ClO_4^-	Perchlorate
PO_4^{3-}	Phosphate
HPO_4^{2-}	Hydrogen phosphate
$H_2PO_4^-$	Dihydrogen phosphate

exercise 2.3

Writing a Compound's Formula

Problem

Write the formula for the compound which comes from the combination of:

    (a)   1 atom of carbon and 4 atoms of hydrogen
    (b)   1 atom of carbon, 2 atoms of fluorine, 2 atoms of chlorine
    (c)   6 atoms of carbon, 12 of hydrogen, and 6 of oxygen
    (d)   1 calcium ion and 2 chloride ions
    (e)   2 copper ions, 2 hydroxide ions, and 1 carbonate ion
    (f)   1 sodium ion and 1 bicarbonate ion

Solution

The number of atoms are written as subscripts following any element having more than one atom per molecule. Numbers always follow the symbol they modify. The number 1 is assumed and not written, so numbering begins with 2. The table on the back endsheet will provide the element's symbol if you do not know it.

(a) CH_4 This is the formula for methane (natural gas or "swamp gas").

(b) CF_2Cl_2 This is Freon 12, once commonly used as a refrigerant (often referred to as a CFC, chlorofluorocarbon, in the news media).

(c) $C_6H_{12}O_6$ There are several compounds having this formula. Three important ones are glucose, fructose, and galactose, sugar compounds that are important in our diets.

(d) $CaCl_2$ This compound is calcium chloride. Among other uses, it is used to spread on dusty roads to cause water to be absorbed out of the air to settle the dust.

(e) $Cu_2(OH)_2CO_3$ or $Cu(OH)_2 \cdot CuCO_3$ This is the greenish-black coating that covers copper or brass objects that have been out in the open air for a long time.

(f) $NaHCO_3$ Sodium bicarbonate or sodium hydrogen carbonate is sold as baking soda.

Check

There are very specific, and not always obvious rules regarding which atom goes first in a written formula. Methane is never written "H_4C." Experience, along with some key rules, will be your guide as we work our way through the text. Note that "CFC" is not itself a formula but just an abbreviation, which is definitely not the same thing.

Looking again at $Cu(NO_3)_2$, how many oxygen, nitrogen and copper atoms does it contain? The **formula unit** of the substance contains the group of atoms of each kind indicated by the formula. There are 2 nitrate ions, each with 3 oxygen atoms. The compound therefore contains a total of 6 oxygen atoms. Each nitrate ion has 1 nitrogen atom, so there are 2 nitrogen atoms in the entire compound, to go along with 1 copper atom in the form of an ion.

exercise 2.4

Atoms in the Compound

Problem

Ammonium phosphate, $(NH_4)_3PO_4$, is an important crop fertilizer. How many of each kind of atom are in the formula unit?

Solution

Remember that the parentheses mean that everything within gets multiplied by the sub-script immediately following. This means that the number of each kind of atom gets multiplied by 3.

(a)   *Hydrogen atoms:* Each of the 3 ammonium ions in the formula unit contains 4 hydrogen atoms. There are $3 \times 4 = 12$ hydrogen atoms in the compound.

(b)   *Nitrogen atoms:* Each ammonium ion in the formula unit contains 1 N, so there are 3 nitrogen atoms in the compound.

(c)   *Phosphorus atoms:* There is only 1 phosphorus atom in the formula unit.

(d)   *Oxygen atoms:* There are no parentheses around the 4 oxygen atoms, so the total is 4.

Table 2.1 will be a nice resource as you learn to name compounds late in Chapter 3.

exercise 2.5

Polyatomic Ions in a Compound

Problem

Write the formula of the compound containing 2 iron ions and 3 sulfate ions (see Table 2.1).

Solution

Two parts of iron makes that portion of the compound "Fe_2." Three sulfate groups are within parentheses, "$(SO_4)_3$." The entire compound has the formula $Fe_2(SO_4)_3$. In Chapter 3, we will learn enough about the placement of elements in the periodic table to see why the combination of iron and sulfate ions gives this particular formula.

Atomic Masses

Dalton and others tried to work out how much atoms weigh, at least relative to one another. We would now say that they tried to measure the atomic masses of each of the elements. They did this by carefully analyzing the composition of a dazzling variety of compounds. The biggest stumbling blocks encountered were removed much later when it was realized that many elements such as hydrogen, oxygen, nitrogen, and chlorine exist as molecules containing two atoms of the element. Also, water has two atoms of hydrogen for every atom of

exercise 2.6

Ions or Uncharged Substances?

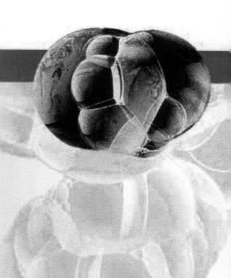

Problem

Ions are especially important in the human body, where they play a central role in maintaining the balance of electric charge in living cells. You may remember from the Prelude that a nerve impulse in part consists of an exchange of ions across cell walls. Which of the following are ions?

a. K^+ **b.** SO_3^{2-} **c.** SO_3 **d.**   Sb

Solution

(a) K^+ and **(b)** SO_3^{2-} are ions because they carry an overall electrical charge. The charge is written as a superscript following the final element shown. If the charge is 1, then a single + or − sign is shown; if the charge is greater than 1, the value is shown before the sign of the charge (for example 2+).

(c) SO_3 shows no charge sign and so is a neutral compound, in this case a gas that can react with water to form acid rain.

(d)   Sb is the symbol for the element antimony. The number of capital letters in a chemical formula signal the number of elements present. Thus, the beautiful black crystal *stibnite*, SbS_3 (which is used as an industrial lubricant), is a compound consisting of 1 atom of antimony (Sb) and 3 atoms of sulfur (S).

oxygen, instead of just one as had been presumed for a long time. Once a table of reasonably accurate atomic masses had been compiled, done first by Berzelius in 1828, a better understanding of how atoms combine became possible.

The modern values of atomic mass use the most common isotope of carbon (containing six protons and six neutrons) as the reference standard. The mass of this "carbon-12" isotope is assigned a value of exactly 12 atomic mass units (that is, 12.000 . . .). This effectively *defines* the atomic mass unit (amu) as corresponding to one twelfth of the mass of an atom of carbon-12.

For some elements, later intensive investigations led to major changes in the accepted values of atomic mass. For instance, in 1828 Berzelius listed the atomic mass of sodium (Na) as 46.62 units and potassium (K) as 78.51 units, instead of the modern values of 22.99 amu and 39.10 amu. This large discrepancy arose because Berzelius based his values on analysis of the compounds formed between these elements and oxygen, which he believed to have the formulas NaO and KO. We now know that the true formulas are Na_2O and K_2O. With our present knowledge of the correct formulas, we would divide each of Berzelius's values by 2 to give 23.31 and 39.26, which are both close to modern values. Misunderstandings of this kind bedeviled a generation of chemists in the 19th century but were finally all sorted out by about 1850. The modern values of the atomic masses are listed in Table 2.2. We will use these values at many points throughout this text.

The determination of atomic masses was, as we shall see, one factor that allowed the elements to be organized into what is now known as the Periodic Table of the Elements (look ahead to Figure 2.11 or see the inside back cover). This structured list of the elements is a treasure trove of chemical and mathematical information. However, the treasure can only yield its gold (and hydrogen, carbon, etc.) if we understand its periodic structure.

Table 2.2

The Elements

Element	Symbol	Atomic Number	Atomic Mass†	Element	Symbol	Atomic Number	Atomic Mass†
Actinium	Ac	89	(227)	Europium	Eu	63	152.0
Aluminum	Al	13	26.98	Fermium	Fm	100	(257)
Americium	Am	95	(243)	Fluorine	F	9	19.00
Antimony	Sb	51	121.8	Francium	Fr	87	(223)
Argon	Ar	18	39.95	Gadolinium	Gd	64	157.3
Arsenic	As	33	74.92	Gallium	Ga	31	69.72
Astatine	At	85	(210)	Germanium	Ge	32	72.61
Barium	Ba	56	137.3	Gold	Au	79	197.0
Berkelium	Bk	97	(247)	Hafnium	Hf	72	178.5
Beryllium	Be	4	9.012	Hassium	Hs	108	(269)
Bismuth	Bi	83	209.0	Helium	He	2	4.003
Bohrium	Bh	107	(267)	Holmium	Ho	67	164.9
Boron	B	5	10.81	Hydrogen	H	1	1.008
Bromine	Br	35	79.90	Indium	In	49	114.8
Cadmium	Cd	48	112.4	Iodine	I	53	126.9
Calcium	Ca	20	40.08	Iridium	Ir	77	192.2
Californium	Cf	98	(251)	Iron	Fe	26	55.85
Carbon	C	6	12.01	Krypton	Kr	36	83.80
Cerium	Ce	58	140.1	Lanthanum	La	57	138.9
Cesium	Cs	55	132.9	Lawrencium	Lr	103	(262)
Chlorine	Cl	17	35.45	Lead	Pb	82	207.2
Chromium	Cr	24	52.00	Lithium	Li	3	6.941
Cobalt	Co	27	58.93	Lutetium	Lu	71	175.0
Copper	Cu	29	63.55	Magnesium	Mg	12	24.31
Curium	Cm	96	(247)	Manganese	Mn	25	54.94
Dubnium	Db	105	(262)	Meitnerium	Mt	109	(268)
Dysprosium	Dy	66	162.5	Mendelevium	Md	101	(258)
Einsteinium	Es	99	(252)	Mercury	Hg	80	200.6
Erbium	Er	68	167.3	Molybdenum	Mo	42	95.94

†All atomic masses are shown to four significant figures. Values in parentheses represent the mass number of the most stable isotope of radioactive elements.

2.3 | The Meaning of Periodic: Putting the Elements on the Table

When we say that something behaves *periodically,* we mean that some aspect or property repeats itself in a recognizable pattern. The seasons can be said to have *periodicity* because they come and go in a regular way each year. A swing, or any other kind of pendulum, behaves periodically as it moves back and forth in a continuous and regular manner. Even the elements can be grouped together to reveal a pattern of periodic behavior. This is a direct result of the connection between an atom's structure and its behavior.

Looking for Patterns

As increasing numbers of elements were discovered and as reliable values were determined for their atomic masses, people began to notice periodic relationships among the properties of elements. Just as we can organize jumbled playing cards by suit (Figure 2.8), we can also

Table 2.2

The Elements—continued

Element	Symbol	Atomic Number	Atomic Mass†	Element	Symbol	Atomic Number	Atomic Mass†
Neodymium	Nd	60	144.2	Seaborgium	Sg	106	(263)
Neon	Ne	10	20.18	Selenium	Se	34	78.96
Neptunium	Np	93	(237)	Silicon	Si	14	28.09
Nickel	Ni	28	58.69	Silver	Ag	47	107.9
Niobium	Nb	41	92.91	Sodium	Na	11	22.99
Nitrogen	N	7	14.01	Strontium	Sr	38	87.62
Nobelium	No	102	(259)	Sulfur	S	16	32.07
Osmium	Os	76	190.2	Tantalum	Ta	73	180.9
Oxygen	O	8	16.00	Technetium	Tc	43	(98)
Palladium	Pd	46	106.4	Tellurium	Te	52	127.6
Phosphorus	P	15	30.97	Terbium	Tb	65	158.9
Platinum	Pt	78	195.1	Thallium	Tl	81	204.4
Plutonium	Pu	94	(242)	Thorium	Th	90	232.0
Polonium	Po	84	(209)	Thulium	Tm	69	168.9
Potassium	K	19	39.10	Tin	Sn	50	118.7
Praseodymium	Pr	59	140.9	Titanium	Ti	22	47.88
Promethium	Pm	61	(145)	Tungsten	W	74	183.9
Protactinium	Pa	91	(231)	Uranium	U	92	238.0
Radium	Ra	88	(226)	Vanadium	V	23	50.94
Radon	Rn	86	(222)	Xenon	Xe	54	131.3
Rhenium	Re	75	186.2	Ytterbium	Yb	70	173.0
Rhodium	Rh	45	102.9	Yttrium	Y	39	88.91
Rubidium	Rb	37	85.47	Zinc	Zn	30	65.39
Ruthenium	Ru	44	101.1	Zirconium	Zr	40	91.22
Rutherfordium	Rf	104	(261)			110	(269)
Samarium	Sm	62	150.4			111	(272)
Scandium	Sc	21	44.96			112	(285)
						114	(289)

Figure 2.8   When we first get our hand of cards from the dealer, we establish *periodic relationships* as we classify them by suit and number.

group together elements with similar or systematically changing properties. For instance it was clear that lithium, sodium, potassium, rubidium, and cesium

[Li, Na, K, Rb, Cs]

were very similar in their physical and chemical properties. All five are soft, metallic, highly reactive elements. Each reacts vigorously with water. One atom of each of them combines with one half a molecule of chlorine gas (Cl_2), that is, with one chlorine atom, to make a white, water-soluble salt. Sodium chloride ($NaCl$), what we know as table salt, is an example. But it was also clear that the five of them were distinct from one another; for one thing their atomic masses varied significantly.

Similar if less dramatic trends were seen for other sets of elements, such as:

[Mg, Ca, Sr, Ba]   [O, S, Se, Te]   and   [F, Cl, Br, I]

In 1829 Wolfgang Döbereiner noted that for certain groups of three elements (or **triads**) the atomic mass of the middle element is just about equal to the average value of the mass of the light and the heavy elements of the group. Take, for example, atomic masses of the triad of sulfur (S), selenium (Se), and tellurium (Te):

S, 32.07 amu   Se, **78.96 amu**   Te, 127.60 amu

The mass of selenium is equal to just about half of the masses of sulfur and tellurium combined:

$$\frac{32.07 + 127.60}{2} = \frac{159.67}{2} = \textbf{79.84 amu}$$

Several of these triads were noted, but there were so many exceptions that not much came of this relationship. For instance, the group O (oxygen), S, Se doesn't work, but in a roundabout way even that was a clue to the form of the underlying periodicity.

In 1864, John Newlands found that when he arranged elements in increasing order of their atomic masses, similarities in chemical properties occurred every eight elements. Unfortunately, he called this the "Law of Octaves." He was widely ridiculed for such a proposal, which sounded like a relationship to musical scales.

Many chemists were on the trail, but the German Julius Lothar Meyer (1830–1895) and the Russian Dimitri Mendeleev (1834–1907) played central roles in solving the mystery of chemical periodicity. Meyer concentrated on the *physical* properties of the elements, especially properties related to density. He identified repeating properties of the elements when they were arranged according to atomic mass and made some proposals to explain the patterns he observed. Mendeleev concentrated more on the *chemical* properties, especially the way the elements combined with oxygen, hydrogen, and chlorine. Again, he was able to find repeating chemical properties of the elements when they were ordered by atomic mass.

Predictions Based on Patterns

Meyer (Figure 2.9A) was a leading chemist in Germany, one of the preeminent countries in chemical research, so his work on chemical periodicity initially received much more attention than Mendeleev's. The Russian Mendeleev (Figure 2.9B) was from a country where little or no recognized chemical research was being done. It is Mendeleev, though, to whom history has given the greater honor.

We now think of Mendeleev as the principal figure in the discovery of chemical periodicity, because he took a giant step beyond Meyer. Not only did he use his understanding of chemical periodicity to correct what were then accepted but inaccurate measures of atomic mass for some elements, but he actually *predicted* the existence and properties of the yet undiscovered elements germanium, gallium, technetium, and scandium. This audacious step seemed foolhardy. It was difficult enough to measure the properties of an element once it was discovered, but to predict chemical reactivities and physical properties before even the existence of an element had been demonstrated was a leap of faith that seemed to border on madness. Yet, within 4 years, the first of Mendeleev's predicted elements was discovered, and the agreement between the observed properties and those predicted by Mendeleev was nearly perfect. Mendeleev's "madness" was now recognized as "genius," and all chemists quickly came to accept his system of periodicity. Table 2.3 summarizes Mendeleev's predictions for the properties of the element that he called "ekasilicon" (meaning "one more than" or one after silicon), what we now call germanium.

A.

B.

Figure 2.9   Portraits of **A.** Meyer and **B.** Mendeleev. Julius Lothar Meyer and Dimitri Mendeleev played central roles in understanding the periodic nature of the elements.

Table 2.3

Comparison of Mendeleev's Ekasilicon vs. Germanium

Ekasilicon (Ek)	Germanium (Ge)
Predicted in 1871	Discovered in 1886
Atomic mass = 72	Atomic mass = 72.6
Density = 5.5 g/cm³	Density = 5.47 g/cm³
Color: dirty gray	Color: grayish white
Reacts on heating in air	Reacts on heating in air
to give white solid EkO_2	to give white solid GeO_2
Density of EkO_2 = 4.7 g/cm³	Density of GeO_2 = 4.703 g/cm³
Oxide does not change upon high heating	Oxide does not change upon high heating
Decomposes steam with difficulty	Does not decompose steam
$EkCl_4$ will be a liquid	$GeCl_4$ is a liquid
Density of $EkCl_4$ will be 1.9 g/cm³	Density of $GeCl_4$ is 1.887 g/cm³
Boiling point of $EkCl_4$ will be less than 100°C	Boiling point of $GeCl_4$ = 86°C

To appreciate Mendeleev's insight, you must realize that at the time there existed published data on 63 known elements. Atomic mass was one distinguishing characteristic and it made sense to order the elements numerically, using their atomic masses. Each element also had associated with it certain physical properties and characteristic reactivities with other elements. Of particular interest was that among the elements, certain groups of elements shared similar characteristics. To continue with our example of a deck of cards, Hearts would form one such group, Clubs the next, and so on. The cards could be arranged numerically, with numbers repeating (periodically) from suit to suit. This was essentially the approach employed by Mendeleev (Figure 2.10). When he came to a spot in his table which cried out for an element of a certain mass and with certain characteristics, even if no such element was known to exist, he left a "blank."

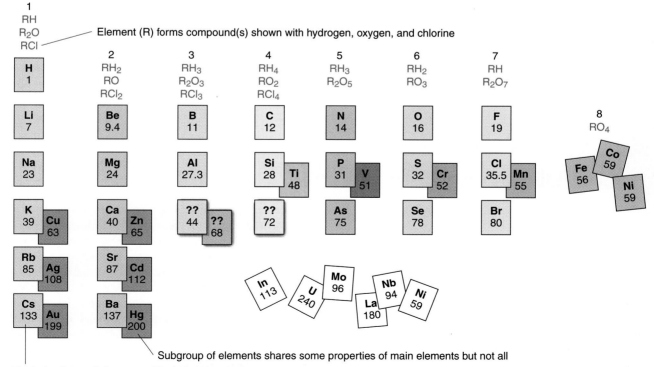

Figure 2.10 A portion of the periodic table published by Mendeleev in 1872. His original version had groups running horizontally, not vertically as shown here. The format here anticipates the orientation used in the modern periodic table. The numbers are the atomic masses Mendeleev was working with at the time.

Mendeleev, like Newlands before him, recognized that chemical properties seemed to repeat with every eighth element. (At this time none of the unreactive gaseous elements—He, Ne, Ar, Kr, Xe, and Rn—that belong to what we now call Group 8 was known.) The similarities were very strong for the first 18 elements but then odd things happened. Mendeleev was forced to group copper (Cu), silver (Ag), and gold (Au) with the first group that included lithium and sodium. These elements fit in the sense that just as sodium forms Na_2O and $NaCl$, two atoms of copper, silver, or gold combine with one atom of oxygen to form Cu_2O, Ag_2O, and Au_2O, and one atom of copper, silver, or gold combines with one atom of chlorine to form $CuCl$, $AgCl$, and $AuCl$. Copper, silver, and gold, however, are much less **reactive** (that is, much less likely to undergo chemical reactions) than sodium or other elements in the sodium group. The sodium group elements all react explosively with water but copper, silver, and gold are completely unreactive with water. These differences would ultimately be taken account of by separating the periodic elements into A and B subgroups.

Mendeleev had some wild cards in his deck. For a time, considerable confusion was caused by the discovery of several elements that should fit in with Group 3B; all were functionally so similar to lanthanum that it was hard to distinguish among them. Fortunately, very few of these elements had been discovered in 1872 when Mendeleev published his work. Two decades of very careful work revealed 13 elements where Mendeleev had expected only one! A 14th element of this set was later found in 1937. In the modern periodic table they are now listed separately as the *lanthanide series,* along with a second series known as the *actinide series.*

Another surprise came when, from 1894 to 1898, the elements argon, krypton, neon, and xenon were discovered as completely unreactive components of air. None of these elements could be made to react with any other element or compound. The atomic masses of these elements suggested they belonged between the Group 7A and Group 1A elements. A whole new family of elements had been discovered. These gases were collectively known as the inert or "noble" gases, due to their inability to react with any other elements (unless subjected to extreme "encouragement" by modern chemical techniques). Mendeleev could not have predicted these elements, but he lived to see this entire new family of elements entered into his table and was apparently delighted. Figure 2.11 shows the periodic table of the known elements, with the atomic number and atomic mass of each element shown.

The Final Hurdle: Introducing the Atomic Number

As the periodic table grew, revelations of patterns among the elements accumulated. The essential structure of the periodic table was taking shape, but one vital discovery lay ahead.

■ case in point

Double Discoveries

Meyer and Mendeleev's near simultaneous discovery of chemical periodicity is but one example of a "double discovery." This often occurs in scientific research when sufficient facts have become known to make a certain discovery nearly inevitable.

Scheele's and Priestley's near simultaneous discovery of oxygen is another good example of scientists working in separate countries doing very similar work. In 1886, Charles Hall (United States) and Paul Héroult (France) both invented a very similar method for commercial production of aluminum, which serves as the focus of our discussion on chemical mathematics in Chapter 4 and Earth as a natural resource in Chapter 13. Double discoveries can

happen when two independent scientists have the same scientific interests, information, and available technology.

A modern example of a double discovery is that of the virus that causes acquired immunodeficiency syndrome (AIDS), which was discovered almost simultaneously in France and America in 1983. The leading scientists in each group, Dr. Luc Montagnier at the Pasteur Institute in France and Dr. Robert Gallo at the National Institutes of Health, both claimed credit and the glory that so often accompanies being first. The debate about who really discovered the AIDS virus continued for more than a decade. History will credit Dr. Montagnier's group as the first to make the discovery.

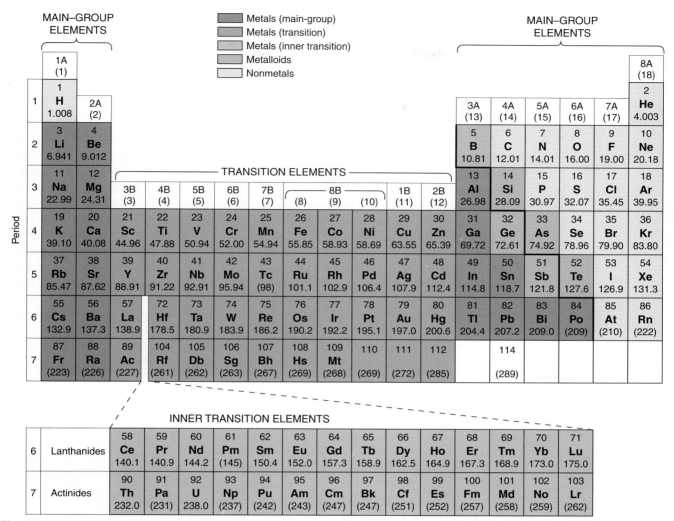

Figure 2.11   The periodic table with A and B subgroups shown in their proper position. The chemically inert "noble" gases are positioned to the far right as Group 8A. The lanthanide and actinide series are placed at the base of the table. This table shows both the naturally occurring elements found on Earth plus those that have been produced only in the laboratory.

Mendeleev and Meyer relied on the atomic masses of the elements to determine proper sequencing of elements within the periodic table. The difficulty was that a few elements had to be placed out of sequence (in terms of atomic mass) in order to be placed in the "correct" position dictated by chemical properties.

All the pieces of the puzzle came together in 1914 with Henry Gwyn Jeffreys Moseley working in Manchester, England (the city in which Dalton had worked 100 years before). He knew that elements emitted X-rays when they were bombarded by high-energy electrons; he believed that the X-rays for each element would be distinct, a sort of X-ray "fingerprint." Upon examining the X-ray profiles of different elements he discovered a new pattern. The energies of the X-rays were not only distinct but they were mathematically related to atomic mass: they increased in proportion to what he called the *atomic number* of an element (Figure 2.12). What he used as the atomic number was the position of an element in the table. Thus, hydrogen was 1, helium 2, and so on. This in fact is the number of protons in the nucleus of each atom of the element.

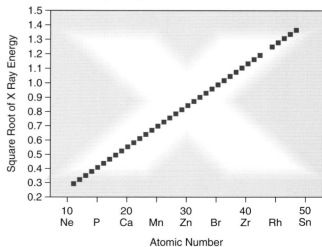

Figure 2.12   Moseley's results plotting X-ray emission energies vs. atomic number, showing a straight-line relationship. The missing point at atomic number 43 is because this element was unknown in Moseley's time.

When the elements are put in order of their atomic number, rather than atomic mass, the systematic properties that Mendeleev observed can shine through. For example, the elements cobalt and nickel, which Mendeleev originally thought to have the same mass, each find their proper place. Using atomic numbers Moseley was able to create order among the elements of the lanthanide series. The solution was not only useful, it was elegant.

The result of the careful work of many individuals appears as the modern form of the **periodic table,** a systematic arrangement of the elements, shown in Figure 2.13. What we now call individual **groups** are aligned vertically, into *columns;* elements within a group share similar chemical properties and reactivities. The horizontal *rows* are called **periods;** a particular set of chemical reactivities is represented by each position within a row. The table is subdivided into Groups A and B. Groups that have an "A" after their number are known as the **main groups,** and the elements they contain as **main-group elements.** Groups with "B" after their name, in the middle of the periodic table, are known as the **transition elements.** If we "read" through the periodic table like the page of a book, moving along the top period (or row) from left to right, then along the second period from left to right, and so on, we encounter all the elements from atomic number 1 through 57. To find element 58 we must jump down to the **lanthanide series** (elements 58 through 71, *inner* transition elements). We then return to where we left the main table, to find elements 72 through 89, then jump to the **actinide series** for elements 90 through 103, before going back to where we left the main table to find elements 104 to 114 (the largest of the human-made elements produced so far). Some science

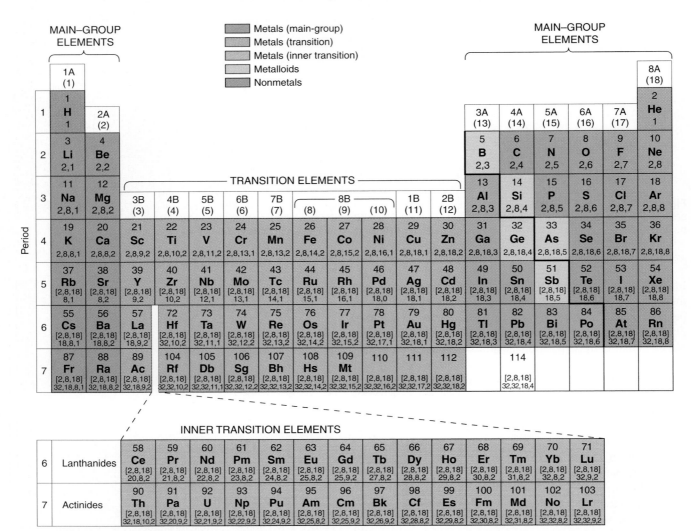

Figure 2.13   The periodic table of the elements. This table includes the electron configuration of each atom, which as we will see is the key factor in determining the periodic properties of the elements. To find the mass of each atom, see Table 2.2 or Figure 2.11.

organizations use a "1 to 18" designation for the A and B groups, as shown in Figures 2.11 and 2.13. We will use the A and B designations throughout this text.

2.4 | The Electronic Logic of the Periodic Table

The discussion so far reveals only some of the organizational logic behind the structure of the periodic table. We know that elements in the same group share broadly similar chemical reactivities—they tend to participate in the same general kinds of reactions. Other aspects of the periodic table's structure may seem rather illogical to you, however. When we follow the sequence of atomic numbers from 1 to 114, why are there only two elements in the first period, eight in the second and third periods, eighteen in the fourth and fifth periods; and why do we have to make the "jumps" to the lanthanide and actinide series?

The reasons for these seemingly illogical peculiarities are largely due to the *arrangement of the electrons* within atoms. You may have noticed that the periodic table of Figure 2.13 includes a series of numbers identified as the electron *configuration*. To find the underlying order of periodic behavior among elements, scientists had to look deeper—at the connection between the inner structure of the atom and its outward behavior.

Modeling the Atom

While working on a problem, scientists try to visualize what's happening. They develop a working **model,** a simplified representation of reality that allows them to test their ideas. Dalton's model of the atom was that of an indestructible particle with a unique mass. The model was useful in the sense that it focused the energies of several generations of chemists on the search for all the elements, identifying the chemical and physical properties of each. However, the model did nothing to explain why the elements behaved as they did. By the end of the 19th century it was becoming increasingly clear that the atom was not the final word in matter. There were certain subatomic particles common to all the elements. The first to be discovered was the negatively charged electron, next came the positively charged proton, and finally the neutron. An atom full of oppositely charged particles required a new model.

In 1911, physicist Ernest Rutherford (Figure 2.14) offered the model of an atom with all of its positive charges concentrated in a tiny but massive central nucleus, with electrons occupying the relatively large outer reaches. The Danish physicist Niels Bohr would refine this model even further. He envisioned the electrons moving around the nucleus like tiny satellites orbiting the Earth. It is the Bohr model that you often see in logos and even in popular literature, the symbol of the atomic age. This model is in fact an oversimplification. A more realistic picture arises through the deep and dark waters of theoretical physics. We will need to "dip our toes" into these waters to understand more but will not wade in too deep.

Figure 2.14 Ernest Rutherford proposed the model of an atom with all of its positive charges concentrated in a tiny but massive central nucleus.

Energy Levels

What is so striking about Bohr's model of the atom is his realization that electrons surrounding the nucleus are restricted to certain **energy levels.** Electrons in low energy levels are those closest to the nucleus. They possess, by definition, less energy than those in higher energy levels. The electrons in lower energy levels therefore occupy a smaller volume of space as they travel around the nucleus. Farther out, the higher energy electrons occupy a greater volume. The energy levels occupied by electrons are also known as **electron shells.** (To learn more about the work of Niels Bohr and his model of the atom, see the Case in Point: Quantum Leap.)

Only a certain number of electrons can occupy each energy level (each shell). If we number each shell, starting with "1" as the innermost level, the maximum number of electrons in

case in point

Quantum Leap: Niels Bohr and the Bohr Atom

In the early years of the 20th century, our ideas about the nature of the atom were changing dramatically. The first hint that the atom was actually composed of smaller, more fundamental particles, came with the discovery of the electron in 1897. Discovery followed discovery: radioactivity (alpha rays, beta rays, gamma rays), X-rays, protons. Yet with all this activity involving some of the most famous names in chemistry and physics, the result was more confusion, not less. Evidence suggested an atom with a positively charged nucleus surrounded by negatively charged electrons. No one could figure out how the negatively charged electrons managed to stay away from the positively charged protons in the nucleus. It was equally mysterious why all the positively charged protons stuck together so closely in the nucleus when "everyone knew" that like charges repel each other and opposite charges attract. Now enter Niels Bohr (1885–1962), a young man from Denmark with a gift for theoretical physics.

Bohr concentrated his energies on the simplest atom, hydrogen. Consider the visible light spectrum produced when hydrogen gas is subjected to an electric charge. If the electrons in a sample of hydrogen gas are moving to and fro all through the space surrounding the nuclei, then the hydrogen spectrum would depict gradual shifts in energy that would show as a continuous spectrum of changing color. In fact the visible light spectrum of hydrogen consists of a pattern of discrete colored lines (see the figure). Others had already worked out numerical formulas that predicted the position of the spectral lines and related their specific wavelengths to a set of integers (1, 2, 3 . . .), but no one knew what this relationship meant.

Bohr incorporated into his thinking the ideas of physicist Max Planck, who argued that energy is emitted not in one continuous stream but in tiny finite bundles (*quanta*). What if electrons exist only at certain energies in relationship to the nucleus, and not at intervening energies. Might this account in some way for the separate lines in the spectrum of hydrogen? In this sense Bohr took the "quantum leap"—such thinking ran completely counter to accepted ideas about electric charges. Thus, was born "quantum" mechanics.

When we say our thinking represents a quantum leap, we mean that we have jumped over intervening possibilities to settle on a different level of understanding. When we say that something is *quantized,* we mean that it can exist in certain stages, but not in intervening stages. Let's try some analogies:

- A ladder is quantized but an inclined plane is not. One must stand on one step or another of the ladder, not in between, but one can stand at any elevation on the ramp.

- Among musical instruments, the notes of trumpets and flutes are quantized but those of trombones and violins are not. Piano keys are quantized: one can play any black or white key but not the note "in the crack" in between.

A. White light consists of a wide range of different wavelengths and frequencies—what you see as the colors of the rainbow when you view white light through a prism. **B.** A gaseous sample of hydrogen is first excited by an electric discharge. The resulting light, when passed first through a slit and then a prism, shows a distinctive pattern of light of definite wavelengths (measured in nanometers, nm). **C.** Niels Bohr identified the energy levels that electrons occupy as "quantum numbers," symbolized by the letter "*n.*" The release of energy when an electron "falls" into a lower energy level results in the spectral lines of hydrogen. **D.** The relationship between energy changes and wavelength in the hydrogen atom.

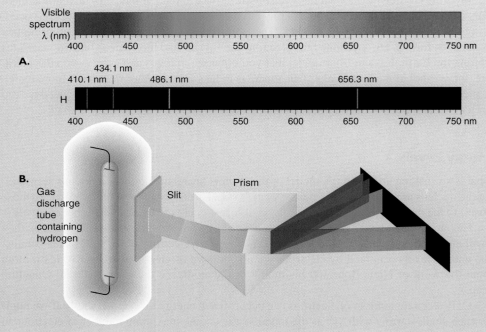

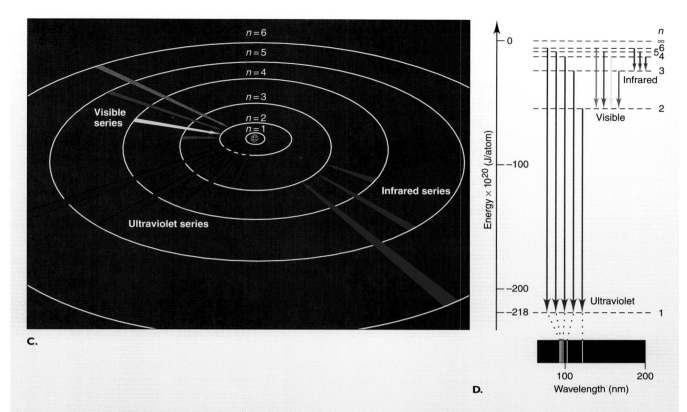

C.

D.

In an unexcited atom of hydrogen, the electron occupies a certain quantized energy level that is close to the nucleus, but if excited that electron can "jump" rapidly to higher energy levels. Bohr identified these energy levels by what he called *quantum numbers* using the symbol *n*. The energy level closest to the nucleus is numbered 1, and the levels increase in *n* value at greater distance from the nucleus. Energy is required for an electron to jump away from the nucleus (from a low *n* to a higher *n*) but is given off when an electron "falls" from a distant energy level to one closer to the nucleus (*n* going from high to low). It is the release of this energy that appears in the spectral lines of hydrogen, each line indicative of a change between energy levels. Even Einstein praised Bohr's idea and wished that he had thought of it himself. This is illustrated in Figure C.

The quantized energy levels of the hydrogen atom are arranged like a strange ladder. The higher up the ladder the electron climbs, the closer together the rungs become. The bottom rung of the ladder, *n* = 1, is at ground level and the atom is said to be in its *ground state*. In the ground state, the electron has the lowest energy it can attain. Most free hydrogen atoms exist in the ground state when at moderate temperatures. At higher temperatures, the atom can take in energy and the electron can rise to higher energy levels. The atom is in an *excited state*. After about a nanosecond, the electron falls back to a lower level, giving off the energy as light or some other form of electromagnetic radiation. This is shown in Figure D.

Bohr was asked why the electron didn't fall into the nucleus. He was forced to say that he didn't know why but that electrons in atoms simply do behave in the way he described. From 1913 to 1935 Bohr and others refined these ideas. These were years that turned physics on its head and led to a much richer understanding of atoms and thus to a more fundamental view of chemical behavior.

each shell can be calculated by multiplying the shell number (call it "*n*") by itself ($n \times n = n^2$) and then by 2.

$$\textbf{Maximum number of electrons per level} = n \times n \times 2 = \textbf{2}\textbf{\textit{n}}^{\textbf{2}}$$

This means that the maximum number of electrons in each energy level is

First energy level (*n* = 1): $1 \times 1 \times 2 = \textbf{2}$
Second energy level (*n* = 2): $2 \times 2 \times 2 = \textbf{8}$

The element that has both the first and second energy levels filled is neon with a total of 10 electrons.

We now know the maximum number of electrons possible in *any* given energy level. The periodicity of the elements, as recognized by Meyer and Mendeleev, is intimately connected to the ways electrons fill energy levels. Notice that the first shell is filled by only two electrons. Thus, hydrogen (one electron) and helium (two electrons) fall within that level and make up the entire first period of the periodic table (Figure 2.13). Then the next shell (with a capacity of 8 electrons) starts to fill, and when full, makes up the second period. Here are the electron arrangements of the eight elements in the second period:

Li (2,1)   Be (2,2)   B (2,3)   C (2,4)   N (2,5)   O (2,6)   F (2,7)   Ne (2,8)

The **electron configuration** of an element is simply a listing of how many electrons are in each energy level of the atoms of a particular element. With the third period the configurations get more complicated, but there is still a logic to the order, as we shall see.

One very common way to visualize electron configuration is to use diagrams in which circles of different diameters represent the electron shells, with dots or crosses representing electrons in these energy levels (Figure 2.15). These diagrams are very useful, but you should be careful not to take them too literally. They do not give an accurate picture of reality. They are simply a useful way of keeping track of the electron arrangements within atoms.

Now that you know about energy levels and electron arrangements, you are better able to appreciate the logic of the periodic table. Choose any of the first seven main groups of the periodic table (Groups 1A through 7A) and examine the electron configuration of each element in the group. You should notice a striking consistency, namely that all elements in any of these main groups have the same number of electrons in their highest (outer) energy level. In fact, the number of these outer electrons is the same as the main group number. So elements in Group 1A have one electron in their outer energy level, elements in Group 2A have two electrons in their outer energy level, and so on through to the elements in Group 7A, which all have seven electrons in their outer shell. The outer energy level is also called the *valence shell,* which relates to the atom's reactivity. The electrons in this outer level, often referred to as **valence electrons,** are the ones that will enter into bonding arrangements and are directly related to an element's reactivity.

Now look at the one remaining main group, Group 8A. The rule that the group number equals the number of valence-level electrons also applies to elements in this group, apart from helium. The rule linking main-group number with the number of outer electrons largely explains why elements in any one main group share very similar chemical reactivities. We have

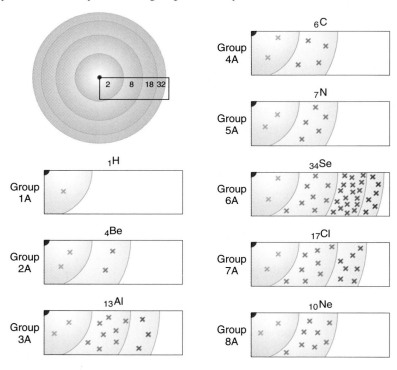

Figure 2.15 To help you make an accounting of all the electrons in an atom, we will artificially restrict all electrons to a cutaway section. This "model" does *not* give an accurate representation of the position of electrons in actual atoms, but is rather a form of electron bookkeeping. For the main-group elements note that as you move through a period, from Group 1A through Group 8A, the number of electrons in the outer level increases by one, always matching the group number.

exercise 2.7

Energy Level Calculations

Problem

Among the actinide series is the element uranium which has all of its first four electron shells full. Calculate the number of electrons in its third and fourth shells.

Solution

Because these shells are full, calculate the maximum number of electrons possible for each. Use the formula $2n^2$, with n being the number of the energy level:

Third energy level ($n = 3$): $2 \times 3^2 = 2 \times 9 = \mathbf{18}$
Fourth energy level ($n = 4$): $2 \times 4^2 = 2 \times 16 = \mathbf{32}$

For about 150 years, uranium was the element with the greatest known atomic mass (238). Uranium was the first element to demonstrate radioactivity, and study of its behavior was intimately tied up with the development of atomic energy.

Check

Remember that the energy levels are associated with a three-dimensional volume of space, so we can expect to see number of electrons increase significantly the farther from the nucleus we go.

already emphasized that chemical reactions are essentially processes involving electron rearrangements. The *outer* electrons of atoms are the most crucial electrons involved in these rearrangements because they are the ones most directly exposed to other atoms, molecules, and ions, so atoms with the same number of outer electrons tend to participate in the same kind of chemical reactions.

The first connection we pointed out between the structure of the periodic table and the structure of atoms is that the elements are arranged in order of their atomic numbers. This second connection between the groups formed, and the electron configuration of the elements is actually more fundamental, and we have not yet finished exploring it.

Energy Levels and Period Numbers

To discover another crucial connection between the periodic table and electron configuration, look across a few of the periods (rows) of the table, in search of some unifying link among all the elements in any one period. Elements in the first period (hydrogen and helium) have electrons in only one energy level; elements in the second period (lithium, beryllium, boron, etc.) have electrons in two energy levels; elements in the third period (sodium, magnesium, aluminum, etc.) have electrons in three energy levels; and so on. So the rule is that the period number of an element is the same as the number of electron-occupied shells in the atoms of that element. The only apparent exceptions to this rule are the lanthanides and actinides, but (for reasons of electronic structure that we need not explain here) the lanthanides should really be regarded as an insertion in the middle of Period 6; whereas the actinides should really be regarded as an extension of Period 7 (this is indicated by the arrows in Figure 2.13). The periodic table in Figure 2.16 was prepared with the lanthanides and actinides inserted into their "proper" places. This form is too large and inconvenient for general use.

There is a different way of expressing the link between period number and the number of occupied energy levels. Work through the periodic table in order of increasing atomic number. You will find that each time you jump to a new period, you jump to an atom containing one

1A

| 1 | 1 H | **2A** | | | | | | | | | | | | | | | | | | | **8A** |

Periodic table layout:

- Period 1: 1 H; 2 He
- Period 2: 3 Li, 4 Be; 5 B, 6 C, 7 N, 8 O, 9 F, 10 Ne
- Period 3: 11 Na, 12 Mg, **3B**; 13 Al, 14 Si, 15 P, 16 S, 17 Cl, 18 Ar
- Lanthanides / Actinides
- Groups (transition): 4B, 5B, 6B, 7B, ⌐8B⌐, 1B, 2B
- 3A, 4A, 5A, 6A, 7A
- Period 4: 19 K, 20 Ca, 21 Sc, 22 Ti, 23 V, 24 Cr, 25 Mn, 26 Fe, 27 Co, 28 Ni, 29 Cu, 30 Zn, 31 Ga, 32 Ge, 33 As, 34 Se, 35 Br, 36 Kr
- Period 5: 37 Rb, 38 Sr, 39 Y, 40 Zr, 41 Nb, 42 Mo, 43 Tc, 44 Ru, 45 Rh, 46 Pd, 47 Ag, 48 Cd, 49 In, 50 Sn, 51 Sb, 52 Te, 53 I, 54 Xe
- Period 6: 55 Cs, 56 Ba, 57 La, 58 Ce, 59 Pr, 60 Nd, 61 Pm, 62 Sm, 63 Eu, 64 Gd, 65 Gd, 66 Dy, 67 Ho, 68 Er, 69 Tm, 70 Yb, 71 Lu, 72 Hf, 73 Ta, 74 W, 75 Re, 76 Os, 77 Ir, 78 Pt, 79 Au, 80 Hg, 81 Tl, 82 Pb, 83 Bi, 84 Po, 85 At, 86 Rn
- Period 7: 87 Fr, 88 Ra, 89 Ac, 90 Th, 91 Pa, 92 U, 93 Np, 94 Pu, 95 Am, 96 Cm, 97 Bk, 98 Cf, 99 Es, 100 Fm, 101 Md, 102 No, 103 Lr, 104 Rf, 105 Db, 106 Sg, 107 Bh, 108 Hs, 109 Mt, 110, 111, 112, 114

Figure 2.16 Periodic table with the lanthanides and actinides inserted into their proper position, as dictated by the electron configuration.

more occupied energy level than the previous atom. So moving from one period to another corresponds to moving to atoms whose outer electrons are in the next higher energy shell.

The organization of the periodic table does successfully predict what types of chemical reactions the different elements will participate in and the degree of reactivity they exhibit in doing so. In discovering the laws of chemical periodicity Mendeleev, Meyer, and the others working in the late 19th century, years before the discovery of the electron, unknowingly laid the groundwork for the discovery of this central chemical truth: electron configuration of an atom determines the chemical reactivity of an element.

exercise 2.8

Electron Patterns

Problem

Look at the elements of Group 3A in Figure 2.13. Boron has 3 electrons in its outer shell ($n = 2$). Aluminum, gallium, indium, and thallium each have 3 electrons in the outer shell, but that outer shell has a larger value of n as we go down through the group, 3, 4, 5, and 6 respectively. Identify the elements that have 5 electrons in their outer shells and their corresponding values of n.

Solution

The main group number is the key to the number of electrons in the outer shell, so we are looking for the elements that make up **Group 5A.** You can determine the number of electron shells by counting the number of values given in the electron configuration. Check Table 2.2 for the full name of each element:

nitrogen ($n = 2$)	phosphorus ($n = 3$)	arsenic ($n = 4$)
antimony ($n = 5$)	bismuth ($n = 6$)	

Check

Because the $n = 1$ level has a maximum of 2 electrons, the elements must have at least 2 energy levels. A quick look down the column of Group 5A shows 5 as the value in the valence level throughout. The electron configuration shows the number of levels (n). For example, bismuth (Bi) has the configuration (2, 8, 18, 32, 18, 5) and so $n = 6$.

As we will see later, this group epitomizes the wide range of different appearances possible within the same group. Nitrogen is a gas (N_2 makes up 78% of air), whereas the other elements are all solids. Bismuth is in fact a very dense metal commonly used in remedies for upset stomachs, like Pepto *Bis*mol.

exercise 2.9

Finding Elements, Finding Groups

Problem

Use the electron configuration given here to identify the element and its group:

(a)   (2,8,5)   **(b)**   (2,8,18,8,2)   **(c)**   (2,8,6)   **(d)**   (2,7)

Solution

(a)   The element has a total of 15 electrons (2 + 8 + 5 = 15) and can be identified as the element **phosphorus (P)** with atomic number 15. It has 5 electrons in its outer shell (2,8,**5**) and is therefore in **Group 5A.**

(b)   There are 38 electrons in this element, which is **strontium (Sr),** atomic number 38. Strontium is a **Group 2A** element with 2 electrons in its outer shell (2,8,18,8,**2**).

(c)   **Sulfur (S),** atomic number 16, is a yellow solid. It belongs to **Group 6A.**

(d)   **Fluorine (F),** atomic number 9, is a highly reactive pale yellow gas. It belongs to **Group 7A.**

Orbitals

We have said that diagrams of electrons in energy levels such as those in Figure 2.15 should not be taken too literally. The main reason for this is that electrons do not circle the nucleus of an atom in regular "orbits," like satellites orbiting the Earth. These figures are meant only to give you a clear idea of the different and discrete *energies* possessed by an atom's electrons; they are not meant to represent an actual picture of electrons in motion, in space. So how do electrons really move and fill those relatively wide open spaces surrounding the nucleus? To be honest, nobody really knows for sure; the evidence says that electrons move as waves within specific *regions* of space.

The regions of space within which electrons move are known as electron **orbitals.** Unlike the orbits of satellites, electron orbitals are three-dimensional volumes of space which can be spherical (Figure 2.17A), dumbbell shaped (Figure 2.17B), or have even more exotic shapes (Figure 2.17C). There are four general types of orbitals: *s, p, d,* and *f* **orbitals.**

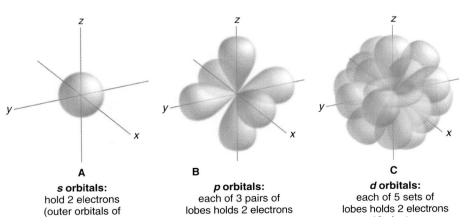

A

***s* orbitals:**
hold 2 electrons
(outer orbitals of
Groups 1A and 2A)

B

***p* orbitals:**
each of 3 pairs of
lobes holds 2 electrons
= 6 electrons
(outer orbitals of
Groups 3A to 8A)

C

***d* orbitals:**
each of 5 sets of
lobes holds 2 electrons
= 10 electrons
(found in elements
with atomic no. of 21
and higher)

Figure 2.17   The general shapes of *s, p,* and *d* orbitals, the regions of space around the nuclei of atoms in which electrons are likely to be found. (The *f* orbital is too difficult to depict.)

exercise 2.10

Predictive Value of Electron Patterns

Problem

What might you surmise about the ability of Group 1A elements to give up their outer electrons to form corresponding positive ions as we travel *down* the group? What might be the impact of this trend on chemical behavior?

Solution

Let's start with lithium [Li (2,**1**)], in Period 2. The outer level (valence) electron, in energy level 2, is very close to its nucleus. Attraction of the positively charged nucleus for its negatively charged electrons is fairly strong, so considerable energy is required to form the lithium cation (Li$^+$). The valence electron for sodium [Na (2,8,**1**)] is already farther away from its nucleus, with 10 electrons in between. So even though the sodium nucleus has a greater positive charge (11 protons), the electron is lost more easily from sodium than from lithium.

This trend continues down the group, so that the cesium valence electron [Cs (2,8,18,18,8,**1**)] is very easily lost. This capability is made use of in devices called phototubes, which detect light intensity. Cesium is often used in combination with other metals because of the ease with which cesium releases electrons. Light shining on the detector surface causes the cesium atoms to be ionized and their valence electrons sent into an electric circuit, where the electron flow is read as light intensity.

Check

The tendency of Group 1A elements to react *increases* as we move down through the group. This corresponds to the increasing distance of the valence electron from the nucleus.

Each electron orbital can contain a maximum of two electrons, so each shell must have enough orbitals to accommodate the maximum number of electrons. The shape and orientation of each orbital type determines how many will fit into a given energy level. A distinct pattern emerges (the numbers in red are the maximum number of electrons possible):

2
First shell: 2 electrons, **1 orbital** (*s*)

2 + 6
Second shell: 8 electrons, **4 orbitals** (*s* + 3-*p*)

consider this:

Paradigms

Thomas Kuhn, a philosopher of science, suggests that because any scientist is a product of society, he or she can be held back by commonly accepted ideas or *paradigms* shared by that society. Dalton's model of the indestructible atom was one such paradigm. When evidence mounts against a particular concept and new ideas take its place, that is a *paradigm shift*. It frees a scientist to go forward, to explore other options. Given the level of scientific understanding at the time, Mendeleev could not have predicted the reason for chemical periodicity, even though his work was crucial in directing other scientists toward that goal. Can you think of other paradigm shifts, past or present, that represent a significant change in the way the natural world is or was viewed?

Third shell: 18 electrons, **9 orbitals** (*s* + **3**-*p* + **5**-*d*)
2 + 6 + 10

Fourth shell: 32 electrons, **16 orbitals** (*s* + **3**-*p* + **5**-*d* + **7**-*f*)
2 + 6 + 10 + 14

The different-level orbitals of any one type, such as all the *s* orbitals, all have the same general shape (as shown in Figure 2.17A), but their volume increases as the energy level increases (Figure 2.18). Every energy level possesses one *s* orbital. Shells from level 2 upward each have three *p* orbitals (in addition to their single *s* orbital). Shells from level 3 upward each have five *d* orbitals (in addition to one *s* and three *p* orbitals) and from level 4 upward each have seven *f* orbitals (in addition to one *s* orbital, three *p* orbitals, and five *d* orbitals). The different orbitals are numbered according to which shell they are in: 1*s*; 2*s*; 2*p*; 3*s*, 3*p*, 3*d*; etc.

The question still remains—*what are orbitals?* It is important to realize that the orbitals that can be occupied by electrons are not physical "things." They are not empty "boxes" or "balloons" sitting around atoms ready to be filled by electrons. The orbitals have no physical reality because they are really a mathematical description of the *regions of space* where electrons are *likely* to be found. Only the actual electrons have any physical reality. The same thing applies to the electron shells; the wording might imply some physical structure, but there is no such structure. The only things actually present are the electrons themselves, which are said to "occupy" the shells.

At first this can be a difficult idea to grasp, so it helps to consider an everyday example. Air-traffic controllers and pilots often talk about "flight paths" taken by aircraft. You may have flown along such flight paths yourself, but were any paths visible in the sky when you looked out of the window? No. "Flight path" is a name given to a set of coordinates which guide pilots, and the aircraft they fly, to a destination. It represents the airspace in which the aircraft is most likely to be found at any given moment. In the same way, electron orbitals are volumes of space in which electrons are likely to be found. At any one time, there is only a 90% probability that an electron will be found within the boundary of the orbital that it is currently occupying.

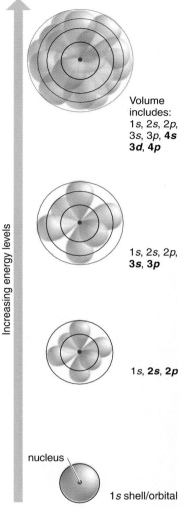

Increasing energy levels

Volume includes:
1*s*, 2*s*, 2*p*, 3*s*, 3*p*, **4s 3d, 4p**

1*s*, 2*s*, 2*p*, **3s, 3p**

1*s*, **2s, 2p**

nucleus

1*s* shell/orbital

Figure 2.18 The orbital with the lowest energy is the one with the smallest diameter: the *s* orbital shown at the bottom of the figure. As electrons are added to higher energy orbitals, the *s* shell and its *p* orbitals increase in volume. Here, from the bottom to the top of the figure, you see the gradually increasing volume of ever larger atoms as electrons are added to orbitals increasingly far from the nucleus.

2.5 | A Tour of the Periodic Table

Now that all the connections have been made (atomic mass → periodicity → atomic number → electron configuration → electron orbitals), let's go on a quick tour of the periodic table, concentrating on the key characteristics and reactivities of the elements it contains. We will begin by taking an aerial look at the chart, as we "fly in" for our more detailed tour. From high above we can clearly see the groups (columns) and periods (rows) we have discussed earlier, but we also notice that the table is subdivided by color into three distinct sections (Figure 2.19).

- Elements in the purple section such as iron, sodium, and mercury, are the elements we call **metals.** These elements share a range of metallic characteristics, including being shiny, malleable (that is, able to be hammered into shape), and able to conduct electricity. The metals tend to form positive ions.
- The pale pink section, including oxygen, hydrogen, and nitrogen, are the **nonmetals.** These elements tend to form negative ions and so will form ionic bonds with metals; they form covalent bonds with other nonmetals. Nonmetals don't conduct electricity.
- The thin pale green band running diagonally down the chart contains elements, including silicon and germanium, that are difficult to classify as either metal or nonmetal. They have some properties associated with the metals and other properties associated with the nonmetals. They are called the **semimetals** or **metalloids.**

Figure 2.19 The metals (purple) are found on the left side of the periodic table and the nonmetals (pale pink) are in the upper right-hand corner. The metalloids (semimetals in pale green) lie diagonally between the metals and the nonmetals.

1 H																	2 He
3 Li	4 Be											5 B	6 C	7 N	8 O	9 F	10 Ne
11 Na	12 Mg											13 Al	14 Si	15 P	16 S	17 Cl	18 Ar
19 K	20 Ca	21 Sc	22 Ti	23 V	24 Cr	25 Mn	26 Fe	27 Co	28 Ni	29 Cu	30 Zn	31 Ga	32 Ge	33 As	34 Se	35 Br	36 Kr
37 Rb	38 Sr	39 Y	40 Zr	41 Nb	42 Mo	43 Tc	44 Ru	45 Rh	46 Pd	47 Ag	48 Cd	49 In	50 Sn	51 Sb	52 Te	53 I	54 Xe
55 Cs	56 Ba	57 La	72 Hf	73 Ta	74 W	75 Re	76 Os	77 Ir	78 Pt	79 Au	80 Hg	81 Tl	82 Pb	83 Bi	84 Po	85 At	86 Rn
87 Fr	88 Ra	89 Ac	104 Rf	105 Db	106 Sg	107 Bh	108 Hs	109 Mt	110	111	112		114				

Metals
Non-metals
Metalloids

58 Ce	59 Pr	60 Nd	61 Pm	62 Sm	63 Eu	64 Gd	65 Tb	66 Dy	67 Ho	68 Er	69 Tm	70 Yb	71 Lu
90 Th	91 Pa	92 U	93 Np	94 Pu	95 Am	96 Cm	97 Bk	98 Cf	99 Es	100 Fm	101 Md	102 No	103 Lr

Figure 2.20 The reaction of Group 1A metals with water can be quite vigorous!

Group 1A

Let's now go to Group 1A, and head directly "south" down the group, looking at the elements it contains. Based simply on Mendeleev's work, we know that these elements will have properties in common. True enough. Lithium (Li), the top metal of Group 1A, reacts with water. Remember that by "reacts" with water, we mean chemically interacts with water, leading to the formation of new substances. Sodium (Na) also reacts with water, and the reaction is more violent. Potassium (K) appears to burst into flames when added to water, although strictly speaking it is hydrogen gas given off by the reaction that is burning, rather than the potassium itself (Figure 2.20). The members of Group 1A are all soft metals, with relatively low melting points. Cesium (Cs), for example, would melt like butter in the heat of a hot summer day. The elements of Group 1A are often called alkali metals because when added to water they react to form an *alkaline* (basic) solution (characteristically bitter, slippery, caustic, and reactive with acids).

As you can see from Figure 2.21, the common electronic structural feature of the Group 1A elements is that the atoms of each element have just one electron in their outermost (highest) energy level—the valence electron. As we have already discussed, an atom's outer, or valence electrons are the most significant ones because they interact with other valence electrons on other atoms during chemical reactions. The inner ("core") electrons are much less directly involved in the rearrangements of electrons that accompany chemical reactions. Group 1A elements tend to lose the outer electron, forming ions with a +1 charge. This represents a more stable electron configuration.

Group 8A

Let's now jump all the way across the table to Group 8A. The members of this group are the inert (or noble) gases: helium, neon, argon, krypton, xenon, and radon. They do not normally combine with any other atoms and so they all exist in the form of individual atoms and are gases at everyday temperatures and pressures. We have emphasized that chemical reactivity is determined by electronic arrangements, so what is so special about the electron configuration of the inert gases? We find that all have eight valence electrons, apart from helium, which has two valence electrons (Figure 2.22). It is very significant that the two valence electrons of a helium atom occupy a *completely full* 1s orbital, amounting to a completely full first energy level. So, having *either* eight valence electrons or enough valence electrons to fill the first energy level must reflect *stable* electron arrangements. By stable we mean very resistant to change and therefore unlikely to be altered by chemical reactions. When atoms have eight electrons in their valence shell, we say that they have an **octet.** This is an especially stable arrangement and one that is frequently the result of chemical reactions.

Even if we could not explain *why* the valence electron arrangements of the noble gases are very stable, the discovery that they are would be very important. In fact, chemists can

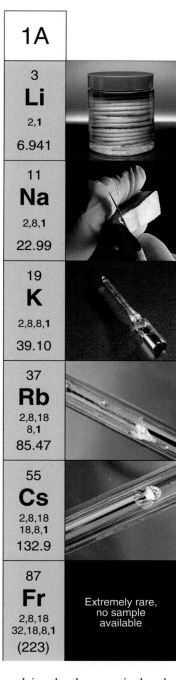

Figure 2.21
Group 1A.
(The energy-level
configuration
is shown.)

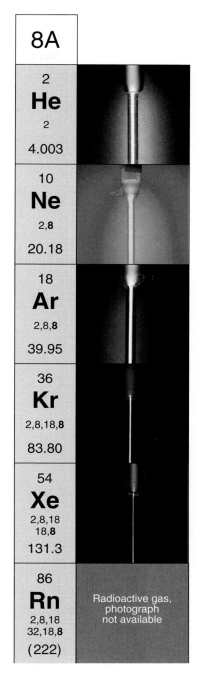

Figure 2.22
Group 8A. This
shows the light
emitted by each
element after
being energized
with a large
electric field.
The gases
themselves
are invisible.
The energy-level
configuration
is shown.

explain why these particular electron arrangements *are* stable. An octet is stable in the sense that a lot of energy is required either to add an electron to or remove an electron from an atom having such an octet.

For instance, sodium has one electron in its $n = 3$ level but eight electrons in $n = 2$. Removing the $n = 3$ electron requires energy of only 490 kilojoules per mole. On the other hand, removing one of the $n = 2$ (inner) electrons requires 4560 kJ/mol. Removing one of neon's eight $n = 2$ electrons requires 2080 kJ/mol. Adding an electron to fluorine to form an octet actually *gives off* 322 kJ/mol.

You will be able to appreciate the significance of stable energy configurations more fully by the time you reach the end of this book. In the meantime, it is important for you to simply recognize that the outer electron arrangements of the inert gases represent very stable configurations; that is why these gases are chemically inert. This experimentally discovered stability of the inert gases will be very important when we consider, later in this chapter, *why* elements other than the noble gases participate in chemical reactions.

Group 7A

As the tour continues, we head "west" from Group 8A to find Group 7A, elements, whose atoms contain 7 valence electrons (Figure 2.23). These elements tend to form ions by adding a single electron to their outer shell, so the ions formed from Group 7A elements typically have a -1 charge. The most important elements in this group are fluorine, chlorine, bromine, and iodine. At everyday temperatures and pressures these elements exist in the form of *diatomic* molecules (molecules made up from two atoms of the same element): F_2, Cl_2, Br_2, and I_2. Fluorine and chlorine are gases at room temperature and pressure, bromine is a liquid and iodine is a solid. These elements are highly reactive nonmetals. Fluorine is the most reactive member of the group and is actually the most chemically reactive of all the non-metals.

Here is something to think about: any member of the Group 7A elements will react vigorously with any member of the Group 1A elements to form compounds that have a 1:1 ratio of Group 7A to Group 1A atom. The mathematics of the valence electrons ($7 + 1 = 8$) ensures an octet—the number of valence electrons found in inert elements. (Remember this, it is intended to plant a seed in your mind that we will nourish later!) For example, chlorine (Group 7A) can react with sodium (Group 1A) to form sodium chloride, NaCl (table salt). Group 7A elements are often found in nature as salts, formed when they react with Group 1A elements. The Greek word for salt is "hals," so the Group 7A elements are known collectively as **halogens** (sodium chloride is often referred to by geologists as halite). Compounds composed of only two elements, such as sodium chloride, are called **binary compounds.**

exercise 2.11

Group 1A

Problem

How many protons, neutrons, and electrons are there in one sodium atom (Na in Figure 2.21)? How about in one sodium ion?

Solution

Sodium (Na) has an atomic number of 11, the number of protons or positive charges. Because an atom is electrically neutral, a sodium atom must also have 11 electrons. The atomic mass of 23 (rounded up from the atomic mass of 22.99) is the sum of protons and neutrons: $23 - 11 = 12$ neutrons. With only 1 valence electron, sodium will give up this electron to form an ion with 10 electrons (Na^+). This creates a configuration of two full electron shells. The nucleus is not affected so the 11 protons and 12 neutrons remain unchanged.

Check

You can check the electron configuration in Figure 2.13 to see how many valence electrons an element has. The lower the number, the more likely that the element will give up electrons rather than take them in.

The fact that an atom and ion are distinct is clear from a comparison of Na and Na^+. Sodium (a solid metal) is dangerously reactive. Sodium ion, a constituent of table salt, is important to our bodies functioning correctly. A high concentration of Na^+ is present in blood serum.

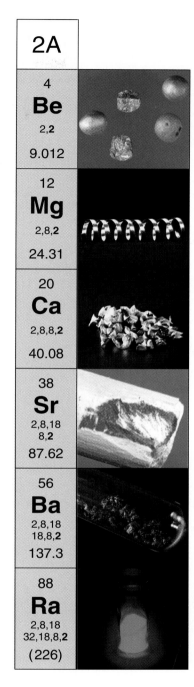

Figure 2.23
Group 7A.
(The energy-level
configuration
is shown.)

7A	
9 **F** 2,**7** 19.00	Photograph not available
17 **Cl** 2,8,**7** 35.45	
35 **Br** 2,8,18,**7** 79.90	
53 **I** 2,8,18 18,**7** 126.9	
85 **At** 2,8,18 32,18,**7** (210)	Extremely rare, no sample available

Figure 2.24
Group 2A.
(The energy-level
configuration
is shown.)

2A	
4 **Be** 2,**2** 9.012	
12 **Mg** 2,8,**2** 24.31	
20 **Ca** 2,8,8,**2** 40.08	
38 **Sr** 2,8,18 8,**2** 87.62	
56 **Ba** 2,8,18 18,8,**2** 137.3	
88 **Ra** 2,8,18 32,18,8,**2** (226)	

Group 2A

At opposite sides of the periodic table, in Groups 1A and 7A, we have found elements whose atoms react vigorously with one another, in a 1:1 ratio. Let's take one step over from Group 1A, to Group 2A, to see what we find.

What we find are elements that have only two electrons in the outer shell (Figure 2.24); their tendency will be to share or give up the two valence electrons. Group 2A elements will react with Group 7A elements to form compounds that have a 1:2 ratio ($2 + 7 + 7 = 16$, two octets). For example, calcium will react with chlorine to form calcium chloride, $CaCl_2$, a chemical used to melt snow and ice on roads and sidewalks.

Remember that the Group 1A elements were reactive metals. The elements in Group 2A are also reactive metals, because they can react with oxygen and water, although they react less vigorously than the adjacent Group 1A metals. For instance, whereas potassium produces a flame when added to water, calcium reacts more gently, forming bubbles of hydrogen gas.

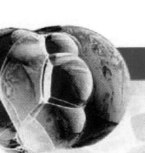

exercise 2.12

Interpreting Energy Changes in Atoms

Problem

Which would require more energy:

(a)   Removal of an electron from a potassium atom (K) or a potassium ion (K$^+$)?
(b)   Removal of an electron from a sodium atom (Na) or a neon atom (Ne)?
(c)   Removal of an electron from a sodium atom or chlorine atom (Cl)?

Solution

(a)   Looking back at Figure 2.13, you can discern that the K$^+$ ion forms from the K atom by the loss of the valence electron in its $n = 4$ level. This creates an ion with an electron configuration of 2,8,8 having an octet in the $n = 3$ level. Any further loss of an electron from the remaining octet would require a great deal more energy.

(b)   As in (a), the key issue is whether an octet is formed or destroyed. The process that results in an octet will require much less energy than that in which an octet is destroyed. This means that removal of an electron from the sodium atom (2,8,1) requires much less energy than the removal of an electron from neon (2,8).

(c)   The removal of an electron from a sodium (2,8,1) forms an octet, but removal of an electron from chlorine (2,8,7) does not, in fact this leaves us further removed from an octet. Therefore, more energy is required to remove an electron from chlorine than from sodium.

     Notice, then, that the reaction of sodium with chlorine results in sodium giving up an electron to chlorine, so that each ends up with an octet. The resulting ionic compound is sodium chloride, more commonly known as table salt.

exercise 2.13

Group 8A

Problem

How can it be that helium, which is the second most abundant element in the universe, after hydrogen, is very rare on Earth? Try to explain why.

Solution

Helium is an inert gas and as such does not enter into reactions; therefore, it will not be bound to rocks or anything else in the solid part of the Earth (unlike hydrogen which is readily found in water, H$_2$O). The helium that existed in the early days of the Earth was in the atmosphere. Its low atomic mass means that it has a very low density and so will rise to the top of the atmosphere. This is why helium-filled balloons fly. Much of the helium in the early atmosphere was lost to outer space. It essentially evaporated away from the Earth. The helium we have with us today has been formed deep within the Earth by the radio-active decay of uranium and similar elements.

The compounds of these metals with oxygen (called *oxides*) form alkaline solutions when added to water and they are therefore called *alkaline earth* metals. The term "earth" was used by the alchemists to refer to all metal oxides.

Group 3A

A brief stopover at Group 3A will serve to introduce us to aluminum, Al, the most important member of this group, owing to its great abundance in nature as the aluminum-containing ore known as bauxite, from which aluminum is obtained. Aluminum is used to make cans, foil, etc. (Figure 2.25). We will introduce aluminum again when we discuss recycling and chemical mathematics in Chapter 4 and later in Chapter 13.

Group 4A

It's just a short stroll over to Group 4A (Figure 2.26), which contains what is perhaps the most significant element of all: carbon. Why do we think that carbon is so special? In a word, life.

Carbon is the central essential element in all living things. Without carbon there would be no bones, blood, muscles, skin, brains; no trees or grasses, no tigers, no butterflies—no life as

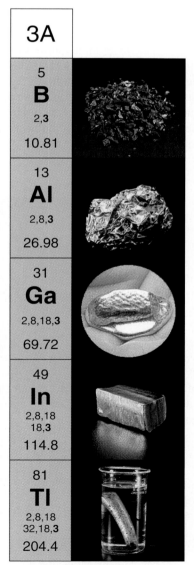

Figure 2.25   Group 3A. (The energy-level configuration is shown.)

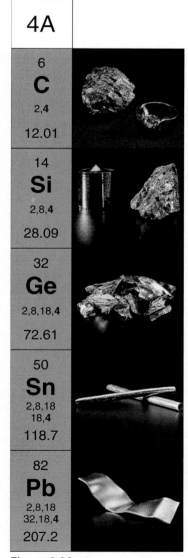

Figure 2.26   Group 4A. (The energy-level configuration is shown.)

we know it. Carbon is so central to life because its atoms can link up into long chains and small rings, with other atoms attached to the chains or sometimes incorporated within them. The chemistry of carbon is so important and diverse that a specialized branch of chemistry, known as *organic chemistry,* is devoted to its study. We will explore organic chemistry and the unique attributes of carbon compounds more in Chapter 6.

When living organisms die, they chemically decay. Given suitable circumstances and millions of years, such material becomes coal, petroleum, and natural gas. So carbon is also the foundation of the petroleum industry, including products such as oils, plastics, pharmaceuticals, and fuels.

In plants, carbon dioxide gas (CO_2) is chemically combined with water (H_2O) to produce carbon-containing chemicals called sugars and release oxygen gas (O_2) into the atmosphere. This reaction is powered by the energy of the Sun, and is known as *photosynthesis.* It is the mechanism by which carbon gets into plants, and so, indirectly, gets into animals, including humans. The food which supplies our carbon is either plant material (a direct source) or animal material (an indirect source that has at some point along the food chain derived carbon from plant matter). Animals (including humans) are able to release energy from food by performing what is essentially the reverse process to photosynthesis. This energy-releasing process is know as *respiration,* in which oxygen is combined (indirectly) with molecules of food to generate carbon dioxide (which we exhale), water, and energy (which our bodies use to power the energy-requiring processes of life).

Silicon is another Group 4A element of great importance. It is the second most abundant element in the Earth's crust (oxygen being the most abundant). Over three fourths of the crust is composed of silica (SiO_2) and compounds containing the silicate ion (SiO_4^{4-}). Silica is the basic constituent of sand and is the main component of glass products. Silicates are found in rocks and clays. Silicon is also the basis of the semiconductor business, as we shall see in Chapter 3. The elements in Group 4A (like those in other groups) become more metallic in character as we move down the group. Below silicon in Group 4A, we find the semimetal germanium (Ge), used in semiconductors, and the metals tin (Sn) and lead (Pb).

exercise 2.14

Group 4A

Problem

Silicon carbide (carborundum) is a compound composed only of an equal number of silicon and carbon atoms. It is nearly as hard as diamond and so is used to sharpen and grind steel objects. What fraction (and percentage) of the mass of carborundum is silicon?

Solution

Carbon and silicon atoms have atomic masses of 12.01 and 28.09, respectively. When determining a ratio, the mass unit involved is not important, it just has to be the same unit for both. The combined mass, one atom each, is $12.01 + 28.09 = 40.10$. To determine the fraction of silicon present, you simply compare the part to the whole (divide):

$$\frac{28.09}{40.10} = 0.7005 \text{ or } 70.05\%$$

Even though the number of the two kinds of atoms is equal, the mass of silicon is more than twice that of carbon.

Group 5A

Stepping over to Group 5A, we find nitrogen, another critical element. Molecular nitrogen, N_2, is the major component of air, accounting for 78% of the volume. It is an extremely unreactive molecule but can be made to combine with molecular hydrogen (H_2) to form ammonia,

exercise 2.15

Group 5A

Problem

Write the formulas for some very different compounds of nitrogen and oxygen, using the ratios shown:

	Nitrogen (N)	Oxygen (O)
(a)	1	1
(b)	2	1
(c)	1	2
(d)	2	4

Solution

(a) **NO.** Nitric oxide is formed during thunderstorms and carried down into the soil, acting as a natural fertilizer. In the formula, nitrogen is written first because it is closer to the metal side of the periodic table. When only one atom of an element is in the formula, the "1" is not written.

(b) **N_2O.** The compound nitrous oxide is the dental anesthetic commonly referred to as "laughing gas." It is also used as the propellant in canned whipped cream.

(c) **NO_2.** The compound nitrogen dioxide (NO_2) is definitely not a laughing matter. This noxious brown-yellow gas, a constituent of automobile exhaust, contributes to air pollution. Examples (b) and (c) show how crucial is the placement of subscripts: the subscript 2 must be placed *following* the element it modifies.

(d) **N_2O_4.** The compound is known as dinitrogen tetroxide. Despite the same ratio of nitrogen to oxygen (1:2), N_2O_4 is *definitely not* the same as NO_2. With each element having more than one atom, subscripts must be shown for both.

Check

Two molecules of NO_2 can react to make a single molecule of N_2O_4:

$$2NO_2 \rightarrow N_2O_4$$

Both substances react with water to make acidic solutions.

NH_3. This is accomplished industrially, using the Haber-Bosch process, or naturally, with the aid of nitrogen-fixing bacteria. Ammonia is used in many industrial processes, including the manufacture of fertilizers. Nitrogen and its fellow group-member phosphorus are (along with carbon, hydrogen, and oxygen) essential components of deoxyribonucleic acid, or DNA, which forms the genetic material of almost all forms of life (Figure 2.27). Nitrogen is also an essential part of all proteins, chemicals which direct the function of most of the chemistry of life (Chapter 16).

Like nitrogen, phosphorus is used in fertilizers because soils are often deficient in this vital nutrient, which enters plants through their roots. Phosphorus is also an essential component in some laundry detergents, such as $Na_5P_3O_{10}$ (sodium tripolyphosphate), partly because phosphate ions combine with calcium in "hard water." In this way the calcium ions are removed, preventing the formation of the calcium-based deposits, which build up in pipes and on utensils through the use of untreated hard water. The use of phosphorus-based detergents is decreasing because of the understanding that overfertilization of lakes and rivers results when such detergents enter the natural environment from the washing machine. A group "portrait" of Group 5A is shown in Figure 2.28.

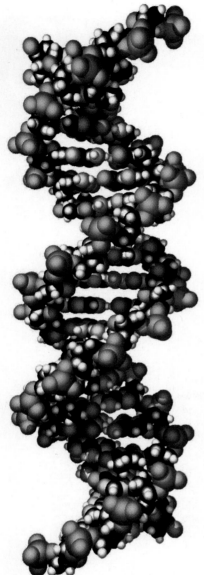

Figure 2.27 The double-helix of DNA (above), deoxyribo-nucleic acid—the "stuff of our genes." DNA molecules control most of the chemistry of life, yet they are composed of only five different atoms: carbon (black), hydrogen (pale blue), nitrogen (dark blue), phosphorus (purple), and oxygen (red). It is DNA that passes down inherited characteristics from one generation to the next.

Group 6A

Our final close-up look on our tour is at Group 6A (Figure 2.29). By far the most important element in the group is oxygen. Molecular oxygen (O_2) makes up about 21% of the volume of air. Scientists think that O_2 in a planet's atmosphere is so chemically reactive that it can exist only where there is life to constantly regenerate it, as plants do on Earth. Astronomers searching for extraterrestrial life would regard finding molecular oxygen as evidence (although not proof) of the existence of present or past life on a distant planet. Oxygen is also found as *ozone* (O_3) in the upper atmosphere and, far too often, in the air we breathe. Ozone, which is very chemically reactive, is an important component of the upper atmosphere because it filters out potentially harmful ultraviolet radiation (we shall have more to say about this in Chapter 12). It is, however, a dangerous contributor to ground-level pollution because it irritates the lungs, eyes, nasal passages, and throat. Summer brings with it "ozone alerts" in big cities. When the ozone level is too high, people with lung problems such as asthma are warned to stay indoors and not to exert themselves, in order to minimize their exposure to ozone.

Figure 2.28 Group 5A. (The energy-level configuration is shown.)

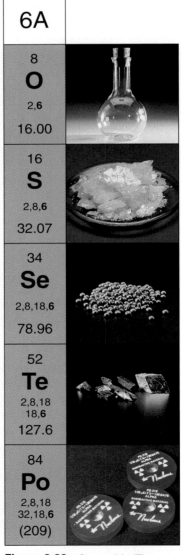

Figure 2.29 Group 6A. (The energy-level configuration is shown.)

exercise 2.16

Group 6A

Problem

Sulfur reacts with oxygen to make a compound in which there are about equal masses of sulfur and oxygen.

(a)   What must be the formula of this compound?
(b)   This compound then reacts with still more oxygen to form a second compound with one more oxygen atom. What is its formula?

Solution

(a)   Each sulfur atom has a mass of about 32 amu and each oxygen atom has a mass of about 16 amu. For the masses to be equal, there must be two oxygens for each sulfur ($2 \times 16 = 32$), so the compound must be SO_2 (sulfur dioxide).
(b)   One more oxygen atom gives sulfur trioxide, SO_3. We will learn more in Chapter 9 about how sulfur and oxygen compounds from smokestack emissions react with water to contribute to a problem known as acid rain.

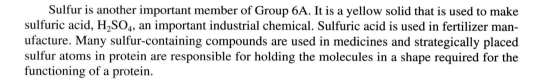

   Sulfur is another important member of Group 6A. It is a yellow solid that is used to make sulfuric acid, H_2SO_4, an important industrial chemical. Sulfuric acid is used in fertilizer manufacture. Many sulfur-containing compounds are used in medicines and strategically placed sulfur atoms in protein are responsible for holding the molecules in a shape required for the functioning of a protein.

The Transition Elements

Up to this point, we have toured the main groups of the periodic table. These groups are the ones that have the letter A as part of their designation (1A, 5A, etc.). The transition elements are metals situated between Groups 2A and 3A on the periodic table. The transition metals have group numbers ending in B, ranging from 1B to 8B (in a strange order, for reasons we need not consider). These elements have the typically metallic properties of high melting point (mercury is the exception being a liquid at room temperature), high density, hardness, and electrical conductivity. The best-known transition metals—iron, copper, nickel, and zinc—are used in commercial products from buildings to coins. Other, less well-known transition metals are also important. Palladium (Pd), for example, is used in automobile catalytic converters to help convert noxious carbon- and hydrogen-based exhaust fumes to more benign carbon dioxide and water.

   At the bottom of the periodic table lie the inner transition elements (the lanthanides and actinides). Most of these are extremely rare. The best known of them are uranium and plutonium, which are used as nuclear fuels in power plants, spacecraft, some military vehicles, and as explosives in nuclear warheads. Many of these elements are not found in nature, but created synthetically, mostly for the purposes of research. All of the elements with atomic numbers larger than 92 are made by bombarding existing atoms with neutrons, or with other atoms. Naming these newer heavy elements has become something of a problem.

   As we prepare to leave our brief tour of the periodic table, let's remember what we have seen. The table is divided into groups, based both on electronic structure and chemical and physical properties. Elements become more metallic as we head down a group. We have been introduced to all of the groups of elements. Some are highly reactive metals (Groups 1A and 2A); some are highly reactive nonmetals such as oxygen in Group 6A and the halogens (Group 7A). Some are chemically inert (Group 8A), and some form the basis of life (nitrogen, carbon, hydrogen, oxygen, phosphorus, and sulfur). We reviewed some very special elements along the way, and some others that play a smaller part in science, technology, and life.

We have come a long way, roughly 2000 years, in one chapter! We have covered the origins of modern chemical thought, the discovery of the gases, atomic theory, and the periodic arrangement of the elements. We called this chapter "Connections" because modern chemistry *is connected* to the knowledge of the ancients by the work of thousands of people gaining insights about their world through experimentation, communication, and correction of misunderstandings. It is about all the connections they discovered between the outer world we experience every day and the inner world of the atom. Francis P. Venable, the long-time Chair of the Chemistry Department at the University of North Carolina, in the early years of this century described the process of scientific discovery in this way:

> "There is first the groping after causes, and then the struggle to frame laws. There are intellectual revolutions, bitter controversial conflicts, and the crash and wreck of fallen philosophies."

consider this:

Transition Metals

The relative abundances of the transition metals in the Earth's crust and the year in which they were discovered are listed below. There is an enormous disparity in the abundances of these elements in the Earth's crust. The two transition elements known to the ancients, iron and copper, are very different in abundance. These two elements are also well known to us today because they are both extremely important in commerce and technology and will be studied in Chapter 13.

Element	Abundance (ppm)	Discovered
Scandium (Sc)	5	1879
Titanium (Ti)	4,400	1791
Vanadium (V)	150	1830
Chromium (Cr)	200	1797
Manganese (Mn)	1,000	1774
Iron (Fe)	50,000	~3000 B.C.
Cobalt (Co)	23	1735
Nickel (Ni)	80	1751
Copper (Cu)	70	~3000 B.C.
Zinc (Zn)	132	1746

(The term ppm stands for parts per million. One part per million is one gram in a million grams. A million grams is fairly close to a ton.)

With which of these metals are you most familiar? Least familiar? Rank the metals in decreasing order of your familiarity with them.

main points

- The development of chemistry was first stimulated by the establishment of organized and stable villages.

- Ancient alchemy, focused on the desire to produce gold, was a reasonable endeavor, given the understanding of the times. It led to a greatly improved knowledge of the behavior of matter.

- Tremendous advances in chemical understanding took place from the late 18th century onward, beginning with the identification of the gases in air.

- Dalton's Atomic Theory made significant strides toward an appreciation of the nature of matter. It focused attention on the key fact that atoms combine in specific fixed proportions to form compounds.

- Chemical formulas list the elements present in a compound and the proportions in which they are present.

- The periodic table lists the elements in order of atomic number. Its structure is based on periodic relationships between the chemical properties of different elements, which are due to their electron configuration.

- The atoms of the noble gases (Group 8A) have very stable electron arrangements. Apart from helium, all of the noble gases have a stable octet of outer electrons. Helium has a completely full first (and only) electron shell.

- When atoms of elements react together to form compounds, the products of those reactions tend to have electron configurations similar to those of the inert gases.

important terms

Actinide series is a series of 14 inner-transition elements (atomic numbers 90–103) that are all similar in properties and result from filling of the $5f$ orbitals. (p. 74)

Alchemy was the study of chemical and physical properties of substances with goals that included conversion of base metals into gold. (p. 57)

Atomic theory stresses that matter is composed of atoms that combine in simple proportions. (p. 63)

Binary compound is a compound of only two elements. (p. 86)

A **chemical formula** is a shorthand method of describing the composition of pure substances, giving the identity and number or ratio of each element in the substance. (p. 64)

A **d orbital** is one of a set of five orbitals (four are shaped like three-dimensional four-leaf clovers, one like an hourglass surrounded by a doughnut); it appears in level $n = 3$ and higher. (p. 81)

Electron configuration is the electron arrangement and pattern of orbitals in an atom. (p. 78)

Electron shells are sets of energy levels corresponding to a single value of n. (p. 75)

Energy levels are the discrete energies of electrons in an atom. (p. 75)

An **f orbital** is one of seven orbitals of complex shape; it appears in level $n = 4$ and higher. (p. 81)

Formula unit is the number of atoms of each kind in a molecule or the simplest formula of an ionic compound. (p. 65)

Group, as used in the periodic table, refers to the chemical elements found in a vertical column; the elements all share the same outer-shell electron configuration. (p. 74)

Halogen is any one of the elements in Group 7A (F, Cl, Br, I, At). (p. 86)

Lanthanide series is a series of 14 inner-transition elements (atomic numbers 58–71) that are similar in properties and result from filling of the $4f$ orbitals. (p. 74)

Law of Conservation of Mass states that matter is neither created nor destroyed during chemical processes. (p. 59)

Law of Definite Proportions recognizes that a pure compound always has the same fractions (proportions), by mass, of its constituent elements. (p. 60)

Law of Multiple Proportions recognizes that the same elements can combine to form different compounds when the constituents have different ratios of mass. (p. 61)

Main group, as used in the periodic table, consists of groups of elements whose column is headed by the letter A. (p. 74)

Main-group elements are those that belong to the "A" groups in the periodic table. (p. 74)

Metalloids are semimetals, elements that have properties intermediate between metals and nonmetals. (p. 83)

Metals are elements that tend to form positive ions, conduct electricity, be malleable, and form ionic compounds with nonmetals. (p. 83)

Model is a representation of reality (an image) used to relate unfamiliar phenomena to something familiar for ease of understanding. (p. 75)

Nonmetals are elements that tend to form negative ions, don't conduct electricity, and form covalent bonds with other nonmetals and ionic bonds with metals. (p. 83)

An **octet** consists of eight electrons having the same maximum n value in an atom. (p. 84)

Orbitals are regions of space in which electrons of certain energy levels are usually found. (p. 81)

A **p orbital** is one of three orbitals shaped like an hourglass; it appears in level $n = 2$ and higher. (p. 81)

Period, as used in the periodic table, refers to the elements arranged in horizontal rows. (p. 74)

Periodic table is a systematic arrangement of elements based on electron configuration and periodic behavior. (p. 74)

Polyatomic ions consist of ions composed of more than one atom. (p. 64)

Reactive describes elements or compounds that readily undergo chemical reactions. (p. 72)

An *s* **orbital** is a spherical orbital centered on the nucleus; it appears in all levels, starting with $n = 1$. (p. 81)

Semimetals are metalloids, elements that have properties intermediate between metals and nonmetals. (p. 83)

Transition elements are elements that belong to the "B" groups in the periodic table. (p. 74)

Transmutation is the conversion of one element into another. (p. 57)

A **triad** is a set of three elements having similar properties. (p. 70)

Valence electrons are the electrons in the outermost energy level of the atom (highest n value), those most likely to be involved in bonding. (p. 78)

exercises

1. The molecular formula of glucose is $C_6H_{12}O_6$. Glucose is oxidized in the body to provide energy. How many atoms of carbon, hydrogen, and oxygen will be present in the products of the oxidation of one molecule of glucose, according to this equation?

$$C_6H_{12}O_6 + 6\,O_2 \rightarrow 6\,CO_2 + 6\,H_2O$$

2. A 2-pound bag of marbles contains 40 red, 60 black, and 10 white marbles. Following the Law of Definite Proportions, how many of each color marble will a 6-pound bag contain?

3. Do you think the chemical properties of the compounds NO and NO_2 are similar or different? Why?

4. How many atoms of each element are present in these compounds?
   a. $CaCl_2$
   b. H_3PO_4
   c. $Na_2Cr_2O_7$
   d. H_2O

5. What makes a carbon atom different from a nitrogen atom?

6. Chemists have discovered many types of chemical reactions. Is it possible, using a specific chemical reaction, to convert carbon to nitrogen? Why or why not?

7. Give the chemical symbols of these elements:
   a. carbon
   b. bromine
   c. tin
   d. potassium
   e. hydrogen

8. Name the elements represented by these chemical symbols:
   a. Ag
   b. Fe
   c. O
   d. Na
   e. He

9. Explain the Law of Conservation of Mass, the Law of Definite Proportions, and the Law of Multiple Proportions in terms of the Atomic Theory.

10. What is the atomic mass of each of these elements?
    a. oxygen
    b. cesium
    c. nitrogen
    d. chromium
    e. argon

11. Chloroform is one of millions of *organic* compounds. Organic compounds are essentially carbon-containing compounds derived from living organisms or manufactured by humans. Organic compounds include plastics; oils; the proteins, fats, and carbohydrates in foods; the cellulose in trees (and paper); and most of the key compounds in that great organic machine: the human body.

    A sample of chloroform was found to contain 12.0 grams of carbon, 106.4 grams of chlorine, and 1.01 grams of hydrogen. If a second chloroform sample contains 30.0 grams of carbon, how many grams of chlorine and how many grams of hydrogen does it contain?

12. Two binary compounds of phosphorus and chlorine are known. One contains 3.43 grams of chlorine per gram of phosphorus and the other contains 5.72 grams of chlorine per gram of phosphorus. What must be the formulas for these two compounds?

13. Figure 2.6A shows two different binary compounds of hydrogen and oxygen. How many grams of hydrogen and oxygen would there be in a 50.0-gram sample of each? Figure 2.6B shows two different binary compounds of carbon and oxygen. How many grams of carbon and oxygen would there be in a 50.0-gram sample of each?

14. Sand is SiO_2 and a typical clay is $Al_2Si_2O_7$. These two classes of substances account for most of the silicon in the Earth's crust. Calculate the percent of silicon in each of these substances.

15. Write the formulas of compounds having these sets of atoms:
    a. 2 atoms of carbon and 4 atoms of hydrogen
    b. 1 atom of phosphorus and 5 atoms of fluorine
    c. 2 atoms of hydrogen, 1 atom of sulfur, and 4 atoms of oxygen

d. 2 atoms of iron and 3 atoms of oxygen

16. Write the formulas of compounds having these sets of atoms:
    a. 2 atoms of sodium, 1 atom of carbon, and 3 atoms of oxygen
    b. 6 atoms of carbon, 12 atoms of hydrogen, and 6 atoms of oxygen
    c. 1 atom of hydrogen, 1 atom of nitrogen, and 3 atoms of oxygen
    d. 7 atoms of carbon, 6 atoms of hydrogen, and 2 atoms of oxygen

17. Write the formulas of compounds having these sets of atoms:
    a. 5 atoms of carbon, 10 atoms of hydrogen, and 1 atom of oxygen
    b. 2 atoms of boron and 6 atoms of hydrogen
    c. 3 atoms of silicon, 2 atoms of oxygen, and 8 atoms of hydrogen

18. Write the formulas of compounds having these sets of atoms:
    a. twice as many bromine atoms as sodium atoms
    b. two thirds as many aluminum atoms as oxygen atoms
    c. four times as many chlorine atoms as lead atoms
    d. equal numbers of potassium and manganese atoms but four times as many oxygen atoms
    e. three times as many ammonium ions as phosphate ions
    f. twice as many potassium ions as sulfate ions

19. Nitrogen and oxygen form several compounds having formulas N_2O, NO, NO_2, N_2O_3, N_2O_4, and N_2O_5. Which two of these compounds have identical percent compositions?

20. Carbon and hydrogen form thousands of compounds having no other elements. A few of the formulas of these compounds are C_4H_{10}, C_4H_8, C_6H_6, C_6H_{12}, and C_6H_8. Which two of these compounds have identical percent compositions?

21. Explain the relationship between a deck of cards and the periodic table.

22. Why is the table of elements named the "periodic table"?

23. Why are elements in the periodic table listed in order of atomic number rather than atomic mass?

24. Which structural component of the atom ultimately accounts for the periodic behavior of the elements of the periodic table?

25. Logically arrange these foods into four groups. Add foods to each group so that each has the same number of items. Give an example of additional information you would need to further classify the members of each group.

    carrot, chicken, banana, bread, bagel, fish, cucumber, orange, pasta, pork chop, apple, rice, potato

26. What are the similarities and differences between a monthly calendar and the periodic table?

27. Which of these are *not* noble gases?

a. chlorine
b. argon
c. helium
d. hydrogen

28. Answer each of these questions by supplying the correct word(s), number, symbol, or response.
    a. How many protons are in a nucleus of a sodium atom?
    b. How many neutrons are there in a ^{19}F atom?
    c. How many electrons are there in a potassium ion?
    d. List two elements that tend to form $+2$ ions.

29. Answer each of these questions by supplying the correct word(s), number, symbol, or response.
    a. How many protons are in a nucleus of a potassium atom?
    b. How many neutrons are there in a ^{31}P atom?
    c. How many electrons are there in a chloride ion?
    d. List two elements that tend to form -1 ions.
    e. Nuclei having the same number of protons but different number of neutrons are _____.

30. Answer each of these questions by supplying the correct word(s), number, symbol, or response.
    a. Atoms with the same number of protons but different number of electrons are different _____.
    b. Most of the mass of an atom is comprised of its _____.
    c. The only common ion with zero electrons is _____.
    d. The atomic mass of antimony is _____.
    e. The atomic number of sulfur is _____.

31. Answer each of these questions by supplying the correct word(s), number, symbol, or response.
    a. An ion has the same number of _____ as the original atom but a different number of _____.
    b. Isotopes of an element have a different number of _____ but the same number of _____.
    c. The atomic number of oxygen is _____.
    d. The only element whose atom has zero neutrons is _____.

32. Answer each of these questions by supplying the correct word(s), number, symbol, or response.
    a. What is the name of the element whose symbol is F?
    b. What is the name of the element that has an atomic number of 14?
    c. What is the mass of an atom of the element having 33 protons in its nucleus?
    d. What is the expected charge of an ion of selenium (Se)?
    e. What is the symbol of the element potassium?
    f. How many neutrons are in the nucleus of a typical phosphorus atom?

33. Answer each of these questions by supplying the correct word(s), number, symbol, or response.
    a. What is the name of the element whose symbol is B?

b. What is the name of the element that has an atomic number of 8?

c. What is the mass of an atom of the element having 30 protons in its nucleus?

d. What is the expected charge of an ion of barium (Ba)?

e. What is the symbol of the element sodium?

f. What process is responsible for the release of energy on the Sun?

g. How many neutrons are in the nucleus of a typical sodium atom?

34. Döbereiner knew of the elements silicon and tin but not germanium. What atomic mass would he have predicted for germanium?

35. Assume that element 113 will soon be discovered or synthesized. Under what element will it be placed in future periodic tables?

36. Newlands, Meyer, and Mendeleev all thought of every eighth element being similar. Sodium is the ninth element from lithium and is similar. Why the difference?

37. Use Figure 2.12 to identify the lighter-weight elements not known at the time of Moseley.

38. The new owners of an old house find some hand-painted crockery in the attic, 17 pieces in all that almost make up four complete table settings:

a. If you were the new owners, how might you arrange the crockery to determine what was missing, that is, make up a "table of tableware" based on the "properties" of the pieces found?

b. A local antique dealer mentions that she has such a piece of crockery (pictured here) in her shop. Is it a missing piece? Where does it fit in your table? Is anything else missing?

39. What letter of the alphabet is the initial letter of the greatest number of chemical elements?

40. What letters of the alphabet are used as single letter elemental symbols?

41. Write a sentence using only chemical symbols. You may ignore distinctions of capital and lowercase letters.

42. What chemical symbols are English words in themselves?

43. Fill in the crossword puzzle outline with one letter in each box so that each pair of boxes: 12, 13, 14, 34, 24, and 32 are all chemical symbols.

1	2
3	4

44. Fill in the crossword puzzle outline with one letter in each box so that each of the letters are single letter elemental symbols and each pair of boxes: 12, 13, 14, 34, 24, and 32 are all chemical symbols.

1	2
3	4

45. Find sets of elemental symbols that are valid written both forward and backward.

46. How many electrons are in each shell for each of these elements:
   a. Li   b. N   c. Si   d. Ca

47. How many electrons are in each shell for each of these elements?
   a. K   b. Ar   c. As   d. Mn

48. Using just the electron configuration of these main-group elements, identify the column of each in the periodic table:
   a. Be (2,2)   b. Al (2,8,3)
   c. Cl (2,8,7)   d. Kr (2,8,18,8)

49. Using just the electron configuration of these transition elements, identify the column of each in the periodic table:
   a. Sc (2,8,9,2)   b. Fe (2,8,14,2)
   c. Mo (2,8,18,12,2)

50. Name all the elements that have all their electron shells totally filled.

51. Name all the metals whose names do not end with the letters "-ium." Do you see a pattern to these names?

52. The $n = 2$ energy level (shell) has eight electrons when filled. How many orbitals must make up this shell?

53. How many electrons are there in the p orbitals of each of these elements?
   a. Be   b. C   c. Ne   d. Al

54. How many electrons are there in the s orbitals of each of these elements?
   a. Li   b. F   c. K   d. Fe

55. What formula of a chemical compound formed from Group 1A and 7A elements is a palindrome?

56. Use the electron configurations to identify the element and the group to which it belongs:
   a. 2,3   b. 2,8,2
   c. 2,8,8,2   d. 2,8,18,4

57. In this cartoon drawing of the atom arsenic, draw in the appropriate number of electrons for each energy level.

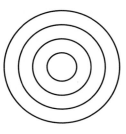

58. Give the number of valence electrons for each of these elements. Do you notice a pattern?

a. boron
b. carbon
c. nitrogen
d. oxygen
e. fluorine
f. neon

59. Which element would you expect to be more reactive, chlorine or argon? Why?

60. Predict the number of shells occupied by electrons for these elements. Do you notice a pattern?
    a. hydrogen
    b. lithium
    c. sodium
    d. potassium
    e. rubidium

61. Describe the element sodium in terms of atomic number, atomic mass, protons, neutrons, electrons, valence electrons, number of occupied energy levels, and electron configuration. What are the similarities and differences between sodium and magnesium? Sodium and potassium?

62. Fill in the appropriate number of electrons for each subshell:
    a. $1s$
    b. $2s2p$
    c. $3s3p3d$
    d. $4s4p4d4f$

63. Use the expanded electron configuration to identify each of these elements and the group to which it belongs:
    a. $1s^2 2s^2 2p^6 3s^2 3p^6 4s^2 3d^8$
    b. $1s^2 2s^2 2p^6 3s^2 3p^6 4s^2 3d^{10} 4p^4$

64. Write out the electron configuration of these:
    a. sodium atom   b. phosphorus atom
    c. titanium atom

65. What noble gas would have the same electron configuration as a:
    a. sodium ion   b. sulfide ion
    c. lithium ion

66. Match these following descriptions with the correct group of the periodic table.
    a. alkali metals
    b. contains the second most abundant element in the Earth's crust
    c. halogens
    d. stable/inert
    e. alkaline earth metals
    f. contains the major component of air
    g. bauxite
    h. most important element makes up about 21% of air

67. Name the family to which each of these elements belongs:
    a. Co   b. F   c. Xe   d. Ca
    e. K   f. Pr

68. Are these elements metals or nonmetals?
    a. Mg   b. Si   c. Ge
    d. Cl   e. O   f. Hg

g. Co   h. Fe   i. Rn

69. Name the family to which each of these elements belongs:
    a. Fm   b. I   c. Pt   d. Lu
    e. Kr   f. Cs   g. Sr

70. What atomic number would fall under polonium in the periodic table?

71. What appearance would you predict for astatine? Why do you suppose that no photo is available for use in Figure 2.23?

72. What groups of the periodic table will react with each other in a 1:1 ratio to form a complete octet?

73. Give some properties and uses of the transition metals.

74. What fraction of the Earth's crust is made up of transition metals?

75. Would you expect a sodium atom or sodium ion to be larger? Explain your answer.

76. Would you expect a chlorine atom or chloride ion to be larger? Explain your answer.

77. List ions having $+3, +2, +1, -1, -2,$ and -3 charges which have the same electron configuration as argon.

78. How many electrons must be lost from a calcium atom to leave an octet of valence electrons? What is the value of "n" of this octet of electrons? What must be the formula of the compound of oxygen with calcium?

food for thought

79. Mendeleev correctly predicted the existence and properties of element 43 even though it would not be discovered during his lifetime. He used the traits of the elements Mn, Mo, and Ru that were known in his time. Following in his footsteps, look up the

    • atomic mass
    • density
    • appearance
    • formulas of oxides and chlorides
    • melting point

    for each of these compounds and use the information to predict the corresponding properties of element 43 (technetium). After making your predictions, look up these properties in the *CRC Handbook of Chemistry and Physics* and see how well you did.

80. Research the elements that had been discovered by 1840 and highlight their symbols on a periodic table. Do you think Mendeleev could have figured out the periodic law based on that information? The *CRC Handbook of Chemistry and Physics* is a good source for this information.

81. Much later in life, Mendeleev also predicted element 84 but was much less sure of its properties. Why do you suppose that he was so unclear about this element when he had been so successful in his earlier predictions?

82. For hundreds of years during the alchemical period it was assumed that one metal could be transmuted into another. For about 100 years during the 19th century it was assumed that no such thing was possible. In the 20th century it became clear that some elements spontaneously change into other elements. Describe how the change in worldviews (*paradigm shifts*) can cause these changes in attitude. What specific people were responsible for these changes in outlook? What convincing evidence caused others to accept these radically new outlooks?

83. Group 7A elements will react with each other to make compounds such as ICl, ICl_3, ICl_5, and ICl_7. Why do you suppose that there are only even numbers of atoms in these molecules. In other words, why do ICl_2 and ICl_4 not exist?

readings

1. Bronowski, Jacob. *Ascent of Man.* Boston: Little, Brown and Co., 1973:68.
2. *The Human Dawn.* TimeFrame Series. Alexandria, Va.: Time-Life Books, 1990:97.
3. Needham, Joseph. *Science and Civilization in China.* vol. 5, pts. 2 and 3. London: Cambridge University Press, 1980.
4. Jaffe, Bernard. *Crucibles: The Story of Chemistry.* 4th ed. New York: Dover Publications, 1976:12.
5. Leicester, Henry M. *The Historical Background of Chemistry.* New York: Dover Publications, 1971.
6. Ihde, Aaron J. *The Development of Modern Chemistry.* New York: Dover Publications, 1984.
7. Service, Robert F. Element 107 Leaves the Table Unturned. *Science* (25 August 2000): 1270.
8. Armbruster, Paul, and Hessberger, Fritz Peter. Elementary Matters: Making New Elements. *Scientific American* (September 1998): 72–77.
9. Scerri, Eric R. The Evolution of the Periodic System. *Scientific American* (September 1998): 78–83.

websites

www.mhhe.com/kelter   The "World of Choices" website contains activities and exercises including links to websites for: "Webelements," with information on the elements, Los Alamos National Laboratory, The Lawrence Berkeley National Laboratory that discusses the formation and detection of "transfermium" elements, and NASA, with a focus on the solar spectrum.

Bonding

In our happiness, we should form many sweet bonds.

—Marcel Proust, The Past Recaptured

College life is filled with many pressures—papers due, tests to take, chemistry textbooks to read. Time seems to fly by as the days melt into each other. The strains are often so great that we sometimes forget that the world is changing at a frenetic pace. There is no better reflection of this change than the intertwining of computers with the fabric of our daily lives.

Computers have changed the nature of our world. Computer-generated weather forecasts, digital pictures from satellites, computer labs in high schools, and computerized data acquisition in chem labs are now routine. Automatic teller machines (ATMs), scanners at the grocery store, and bank-card approvals via huge computer networks are integral parts of our culture.

We recognize the extraordinary detail that goes into miniaturizing billions of electric circuits so that they fit on the tiniest silicon-based computer chip. The electrons travel through the gold paths in predictable ways. They do not stray onto the chip itself or flow through the plastic coating on the motherboard. The papers that you write for your English class are stored on a floppy disk that contains iron(III) oxide, Fe_2O_3, which can store the information.

Why is it that electrons can travel through metal but not most types of plastic? Why can Fe_2O_3 store data? How is it that billions of circuits can be etched into one computer chip the size of a fingertip? It all has to do with the way in which atoms combine to form compounds. This interaction of atoms with each other to form discrete groups, whether molecules or ions, is called **bonding.** Understanding the basic principles of bonding can help us explain the uses of the essential materials in a computer. It can also help us understand many other things, such as the special properties of water (Chapter 7), the behavior of organic substances (Chapter 6), and how soaps and detergents clean pots and pans (Chapter 7). Let's talk first about the basics.

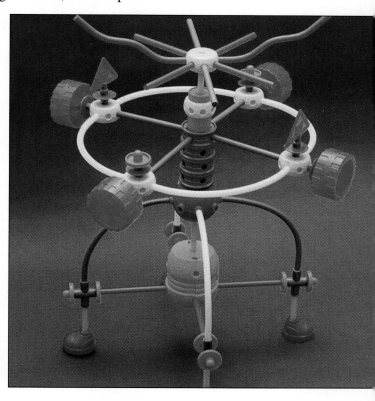

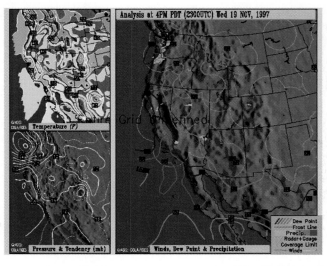

Computer-controlled networks speed up our access to goods and services on a daily basis.

3.1 | The Principles of Bonding

Why do atoms become bonded to one another (Figure 3.1)? The arrangement of the electrons that surround the nuclei of atoms, especially the outer-shell electrons, is the key to chemical bonding. In considering electron arrangements and rearrangements, we must always remember these essential facts:

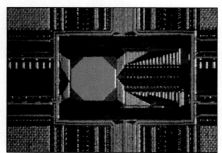

This section of a computer chip has been magnified 300 times to enable us to see the intricate detail in its circuits.

- There is a *force of attraction* between negatively charged electrons and positively charged nuclei.
- There is a *force of repulsion* between electrons.
- There is a *force of repulsion* between nuclei.

All chemical bonding is a reflection in one way or another of these interactions between electrons and nuclei. You can remember it this way—

opposites attract $(-/+)$ and like repels like $[(-/-), (+/+)]$. The Hands On activity at the bottom of this page provides a simple demonstration of how electrical charges affect the way matter interacts, even on the tiny scale of atoms.

The Trap of Stability

Strange as it may seem, the most significant clue to the reasons for chemical bonding lies with those atoms that *do not* ordinarily form chemical bonds, the noble gases. Why don't the atoms of the noble gases (in Group 8A) participate in chemical reactions? The short answer is that they already have very stable, balanced, electron arrangements, so there is no tendency for their electrons to become rearranged further.

When atoms collide with one another, as they do constantly, the turmoil induced by the collisions allows the electrons to "explore" new arrangements. If this allows the electrons to shift into new arrangements which are more stable (that is, less likely to change) than the original arrangements, then the electrons can become *trapped* in these more stable arrangements.

The electrons of atoms of the noble gases are *already in very stable arrangements*—so they do not participate in any further rearrangement. During chemical reactions, the electrons of *other* types of atoms tend to "fall into" and become trapped in new arrangements much closer to the stable electron arrangements of the noble gases. This is one of the central principles of chemistry and bonding: *chemical reactions are very often guided by the tendency of atoms to acquire an electron arrangement similar to the noble gas that is closest to them in the periodic table.*

Elementary chemistry textbooks often summarize this principle by saying that atoms react "in order to" attain the electron arrangement of the nearest noble gas (usually a stable octet of electrons). This is misleading. It implies that atoms have the ability to try to do things.

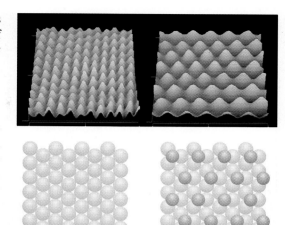

Figure 3.1   This micrograph of gold was made with one of the most recent methods—scanning-tunneling microscopy. We can see the surface of the gold before (left) and after (right) a single layer of copper was overlaid. Note that atoms align in a staggered pattern.

■ hands-on

Try this quick activity: Place two 15-centimeter (6-inch) strips of cellophane tape on a table top, leaving about 2 centimeters of the tape hanging over for you to grab. Grab one strip of tape between each thumb and fore-finger, then lift the two pieces of tape off the table top quickly and simultaneously.

*   Slowly bring the sticky sides close together. What happens? Why?
*   Slide your thumb gently along *one* of the strips and bring both the strips close together again. What happens? Why?

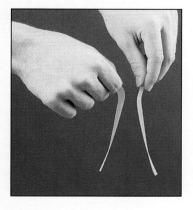

Positive charge causes tape strips to repel one another.

When you ripped the tape strips off of the table, their surfaces became positively charged because some surface electrons were left on the table! When you brought the strips together, the like positive charges repelled one another, so the strips were pushed apart.

When you slid your thumb along one piece of tape, you reversed the charge on that strip, because some electrons were transferred from your thumb to the tape. When you brought the pieces together again, their opposite charges attracted one another.

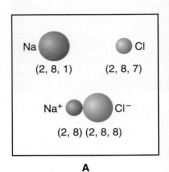

A

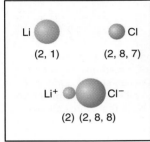

B

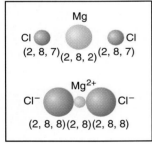

C

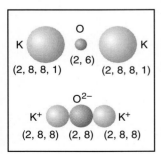

D

Figure 3.3 Ionic bonding in **A.** sodium chloride, **B.** lithium chloride, **C.** magnesium chloride, and **D.** potassium oxide. Notice that nonmetals get larger when they gain an electron and metals get smaller when they lose an electron.

Figure 3.2 The patterns formed by fallen leaves reflect the effect of the environment on their path. They collect in corners and snag on outcroppings.

They do not. Atoms are unthinking particles of matter. As they bump about in their random dance their electrons get rearranged and often become trapped in more stable arrangements.

For an everyday comparison, consider the tendency of fallen leaves to accumulate in corners, and hollows, sheltered from the wind (Figure 3.2). Do the leaves move about "in order to find" hollows and corners? Of course not, but they automatically and inevitably do accumulate in these places because they get trapped there, when blown into them by the wind.

The electron arrangement of the noble gases must be very stable. By stable we mean very resistant to change and therefore unlikely to be altered by chemical reactions. Atoms other than those of Group 8A can achieve a noble gas electron arrangement by gaining, losing, or sharing electrons.

3.2 | Classifications of Bonding

Ionic Bonding—Electron Transfer

The simplest way to explore the tendency of atoms to acquire electron arrangements similar to noble gases is to look again at ionic bonding. In Chapter 1, we examined the ionic bonds formed between sodium and chloride ions, derived from the transfer of one electron from a sodium atom to a chlorine atom (Figure 3.3A). You can now appreciate why this happens. When the sodium atom loses an electron it becomes a sodium ion with the stable electron arrangement of the chemically inert gas neon (2,8). When the electron lost by the sodium atom is gained by the chlorine atom, a chloride ion is generated with the stable electron arrangement of argon (2,8,8). In both cases the ions have acquired a stable **octet** of outer electrons.

The formation of lithium chloride (Figure 3.3B) emphasizes that gaining the stable electron arrangement of a noble gas does not always involve gaining a stable octet of outer electrons. The lithium atom donates its outer electron to a chlorine atom to become a lithium ion with the stable electron arrangement of helium (2).

Figure 3.3C introduces further subtleties, by showing you what happens when magnesium atoms react with chlorine. Each magnesium atom must lose two electrons before attaining the stable electron arrangement of neon (2,8). This means that each magnesium atom will form a "double-charged" Mg^{2+} ion, and will react with *two* chlorine atoms rather than just one. So the formula for magnesium chloride will be $MgCl_2$, and a crystal of magnesium chloride will contain two chloride ions for every magnesium ion.

How about potassium oxide (Figure 3.3D)? Here, each potassium atom loses one electron to form a K^+ ion (electron arrangement 2,8,8); while each oxygen atom must gain two electrons (from two separate potassium atoms) to form an oxide ion O^{2-} (electron arrangement 2,8). So the formula for potassium oxide is K_2O.

Triple-charged ions are also possible. An aluminum atom, for example, would form an Al^{3+} ion when reacting to form an ionic compound by losing three electrons and so attaining the stable electron arrangement of neon (2,8).

case in point

Storing the Data: The Case of a Lone Electron

The most important ionic substance in computers is iron(III) oxide, Fe_2O_3. This material is known in nature as the mineral hematite and is an important ore for production of iron and steel. In computers, a thin coating of iron oxide is used on floppy disks to store information. The same material is used on audio and video tapes to store sound and images.

There are three common oxides of iron: FeO, Fe_2O_3, and Fe_3O_4. Notice that the three compounds have different *oxidation numbers* for iron, which is a different way of saying that the three compounds have different numbers of oxygen per iron atom. There is 1.000 oxygen per iron in FeO, 1.500 oxygens per iron in Fe_2O_3, and 1.333 oxygens per iron in Fe_3O_4. The iron(III) oxide used for information storage is the so-called gamma-Fe_2O_3 instead of the more common, naturally occurring mineral hematite, which is alpha-Fe_2O_3. This distinction of alpha versus gamma means that although the chemical makeup of the two substances is the same, the arrangement of the atoms in the crystal is different. These two forms of iron oxide are said to be *polymorphs*.

The structure of gamma-Fe_2O_3 is a cube with oxide ions at the corners and on the center of each face of the cube. Some of the Fe^{3+} ions are in the interior of the cube surrounded by four of the oxide ions, including one corner oxide and the three oxides on faces adjacent to this corner. Other Fe^{3+} ions are at the center of the edges of the top and bottom of the cube.

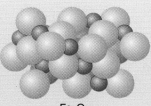

Fe2O3

Some of the electrons in iron(III) oxide are unpaired. This means that some of the orbitals in iron(III) oxide have only one electron instead of two. Electrons behave as though they spin about their own axis. The result of this is that they act like tiny magnets. When an orbital has two electrons, they spin in opposite directions and the magnetic effects of the two electrons cancel out. Atoms or molecules with unpaired electrons have unpaired spins and hence have an overall magnetic effect. An external magnet can align the magnetic direction of the electrons in

a piece of Fe_2O_3 and this direction is retained for a long time. Recording tapes and computer disks are made by coating a suitable plastic material with a thin layer of very tiny granules of Fe_2O_3. Each granule can be oriented either north pole pointed up or down.

The magnetic direction of these tiny regions of iron oxide makes possible a form of binary coding, and thus the use of binary arithmetic. Binary arithmetic uses only zeroes and ones to represent numbers (a base 2 system) instead of the base 10 system we commonly use. Computers do their arithmetic and logic using binary arithmetic and they discern 1 from 0 by whether the magnetic direction of a small region of the iron oxide is aligned north pole up or down. (Actually, "up or down" means pointed toward the center of the disk or the edge of the disk, but up and down are the words that are usually used.) For purposes of word processing, letters and symbols are given their own binary code (known as ASCII).

The orientation of the magnetic pole of this iron oxide is determined by the previous application of a magnetic field to a tiny region of the disk. The region influenced is very tiny (about 2 square micrometers) but *much* larger than the iron oxide granules themselves, which are slender and needlelike with a length of about 0.5 micrometers and a width of about 0.1 micrometer. About 1000 of these particles with their magnetic poles aligned together make up one "bit" of information. Present technology allows about 30 million bits per square centimeter but industry is now closing in on 6 billion bits per square centimeter.

Gamma iron(III) oxide has been used in commercial recording-tape manufacture since 1937 and is still used extensively today. New materials are continually being tried in attempts to make tapes and disks which will store more information with greater reliability. Some of these newer tapes use cobalt oxide and chromium oxide, metallic iron particles, and barium and titanium compounds with iron oxide.

We see then, that the basis of data storage—and, in fact all of the computer logic that begins with "yes" or "no" as 1 or 0, and ends with desktop publishing and "surfing the web"—really hinges on electron spin orientation in iron(III) oxide!

You should now begin to appreciate *why* chemical compounds contain their constituent elements in the proportions that they do. Another principle of bonding is this: *the ratios in which atoms combine to form compounds are determined by the electron arrangements of the atoms concerned.* The differing alterations in electron arrangement allow the atoms to attain the electron arrangement of the nearest inert gas.

exercise 3.1

Forming Ionic Compounds

Problem

Based on your understanding of the charges that atoms will likely form when they ionize, predict the formulas of the compound formed when each of these sets of atoms combines:

    (a)   Ba and F   **(b)**   Ca and O   **(c)**   K and P

Solution

The problem-solving strategy here requires a good understanding of how to interpret the periodic table. The ion that each atom is likely to form is dictated by the group to which it belongs.

(a)   **BaF_2**   Barium belongs to Group 2A, so it will lose two electrons, forming Ba^{2+}. Fluorine is in Group 7A, so it will gain one electron to complete its valence octet, becoming F^-. In order to preserve electronic neutrality, two fluorines will combine with each barium, giving **BaF_2**: $(+2) + 2(-1) = 0$.

(b)   **CaO**   Calcium is in Group 2A and will form a Ca^{2+} ion. Oxygen is in Group 6A, and will form O^{2-}. Electrical neutrality requires there to be one oxide ion for each calcium ion, giving **CaO**, $(+2) + (-2) = 0$.

(c)   **K_3P**   This is a particularly interesting case because although potassium will *always* form K^+, phosphorus can gain differing numbers of electrons (the reasons relate to electron configuration that will not be dealt with here). The P^{3-} is a common ion, leading to the electrically-neutral compound **K_3P**, $3(+1) + 1(-3) = 0$.

Check

The elements forming positive ions are the metals Ba, Ca and K. If you discount the transition elements in the middle, then these three elements are found to the left of the main-group elements shown in the periodic table at the back of your book. The elements forming negative ions are the nonmetals: F, O, and P.

In the case of ionic compounds, the ions always combine in ratios that will result in an electrically-neutral compound overall. So single-charged positive and negative ions will combine in a 1:1 ratio, the most common example being NaCl, table salt (Figure 3.3A). A double-charged positive ion and a single-charged negative ion will form in a 1:2 ratio, so that Mg^{2+} can combine with Cl^- to form magnesium chloride, $MgCl_2$ (Figure 3.3C), and so on. As we shall see, covalently-bonded compounds are also electrically-neutral. So it is also a general rule of chemistry that *chemical compounds are electrically neutral overall.*

Ionic bonding is found in many compounds that form when a metal reacts with a non-metal. In ionic compounds it is the metals that form the positive ions and the nonmetals that form the negative ions. In general, the atoms concerned tend to come from opposite sides of the periodic table. The ions formed from elements on the left of the periodic table tend to be positively charged ions because these atoms need to *lose* electrons to reach a noble gas electron configuration. The ions formed from elements on the right of the periodic table tend to be negatively charged ions because the atoms need to *gain* electrons to reach a noble gas electron configuration.

The most fundamental principle of chemical reactivity is that electrons become re-arranged to adopt stable configurations. The tendency of atoms to react in ways that leave them with the electron arrangement of the nearest noble gas (usually a stable octet) is very important and very common, *but it is not universally true.* Later in this book you will encounter examples of atoms reacting in ways that *do not* leave them with an electron arrangement of a noble gas. Instead, the atoms adopt other stable arrangements.

Covalent Bonding—Sharing of Electrons

Another way for atoms to bond is by covalent bonding, in which electrons become shared between neighboring atoms. This type of bonding was also introduced in Chapter 1, and in many cases it too can be explained in terms of electrons rearranging toward the stable arrangements found in the noble gases.

Consider the simplest molecule of all, the hydrogen molecule. Figure 3.4A illustrates how each hydrogen atom in a hydrogen molecule has a share of two electrons, rather than just one, giving each atom an electron arrangement much more like the stable arrangement of helium.

In water molecules (Figure 3.4B), each of the hydrogen atoms again attains a share of two electrons (approaching the stable arrangement of helium), while the oxygen atom gains a share of eight outer electrons, moving its outer electron arrangement much closer to that of neon (the nearest noble gas to oxygen in the periodic table).

Remember that the electrons in atoms will either be arranged as individual electrons occupying an orbital by themselves, or arranged into pairs of electrons occupying an orbital. Each orbital can accommodate a maximum of two electrons, so the electrons must occupy the orbitals either alone or as pairs, but never as triplets, or anything larger.

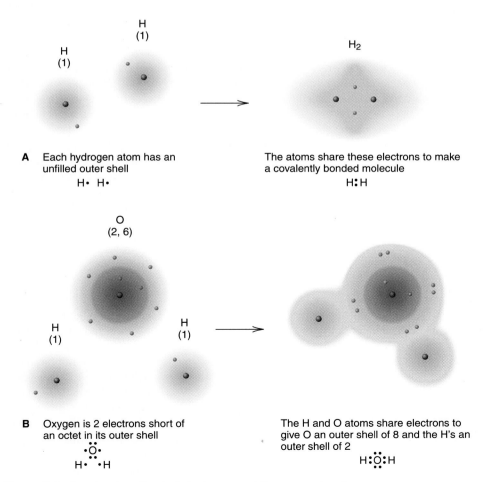

A   Each hydrogen atom has an
    unfilled outer shell
        H• H•

    The atoms share these electrons to make
    a covalently bonded molecule
            H**:**H

B   Oxygen is 2 electrons short of
    an octet in its outer shell
        •Ö•
        H• •H

    The H and O atoms share electrons to
    give O an outer shell of 8 and the H's an
    outer shell of 2
            H**:**Ö**:**H

Figure 3.4   Covalent bonding in **A.** hydrogen and **B.** water molecules.

Covalent bonds are formed when nonmetals react with one another. Recall that the non-metals are in the top right corner of the periodic table.

3.3 | Representations of Chemical Compounds

The fact that electrons move around singly or in pairs lies at the heart of a very simple way of visualizing the electron arrangements within compounds. The technique, which was developed by Gilbert N. Lewis in 1916, uses dots to represent the valence electrons of atoms, with the relevant element symbol being used to represent the inner electrons and the nucleus. The dots are drawn singly or in pairs, depending on how the valence electrons are arranged in each atom concerned. Figure 3.5A shows the **Lewis structures** of the elements from hydrogen through neon.

Lewis Structures

Lewis structures offer an easy way to visualize the way in which the electrons within compounds are often arranged to give each atom the number of valence electrons (usually eight) carried by the nearest noble gas. The Lewis structures of molecules of hydrogen, oxygen, nitrogen, water, ammonia, formaldehyde, and ethanoic (acetic) acid shown in Figure 3.5B should help to make this clear.

Each pair of shared electrons is regarded as a single covalent bond. Thus, atoms held together by four shared electrons (such as the carbon and oxygen atoms of the CH_2O molecule) are held by a **double bond.** Atoms held by six shared electrons (such as the nitrogen atoms of the N_2 molecule) are held by a **triple bond.**

Rather than drawing all the dots, we often use a line to indicate each pair of shared electrons. This alternative way of drawing the structures of molecules is also illustrated in Figure 3.5B. Lewis structures are often drawn using a mixture of the line and dot notation, in whatever way is appropriate to the molecule in question and the aspects of its structure that we are interested in. In the Lewis structure, any pair of electrons that is *not* a part of a bond is referred to as a **lone pair.** You can see a lone pair in the structure of ammonia, for example.

Real molecules have specific *three-dimensional shapes.* Most are not flat. Figure 3.6 gives you some idea of the real three-dimensional shapes of some simple molecules. The shapes of molecules play a very important part in determining how they interact with and react with other chemicals. We will see several examples of this later in the book. Included in Figure 3.6, for comparison, are simple structural formulas of the compounds. A **structural formula** is a simplified diagram that shows just the arrangement of atoms and bonds within a molecule.

The formation of ions can also be represented by Lewis structures. Here the Lewis structures depict a loss or gain of valence electrons rather than a sharing.

Valence

Due to their electron arrangements, the atoms of each element usually (although not always) participate in a fixed number of covalent or ionic bonds. Atoms of carbon and silicon (Group 4A elements), for example, usually form four covalent bonds. This is because each of the four electrons in the outer shell can combine with an electron from another atom to form the shared electron-pair that corresponds to a bond. So, forming four bonds will bring in four new electrons, giving the carbon atom a stable octet. For similar reasons, oxygen will usually form two covalent bonds, and hydrogen and fluorine will usually form one covalent bond. Nitrogen will usually form three bonds, sulfur will often form two, and the other halogens (Group 7A) will usually form just one bond. In some cases, however, the "bonds" we are referring to are part of multiple bonds (such as double or triple bonds) rather than always being in the form of single bonds. In the ionic compounds shown in Figure 3.3 we see that the number of ionic bonds formed corresponds to the charge of the ions.

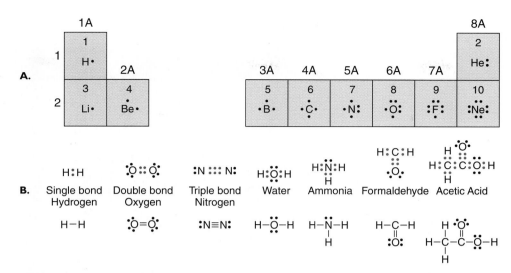

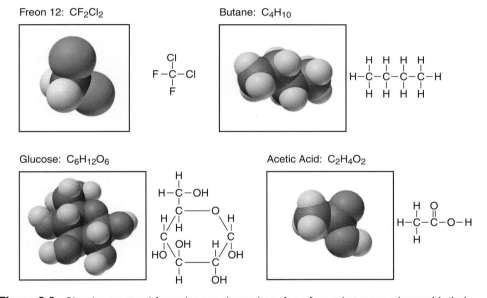

Figure 3.5 **A.** Lewis structures of the atoms of the elements from hydrogen through neon. **B.** Lewis structures of selected molecules. Covalent bonds are shown with both dot and line symbols. Note that each atom in every molecule has attained a stable configuration of electrons similar to a noble gas.

Figure 3.6 Simple structural formulas are shown here for a few substances, along with their corresponding space-filling models. The models give you a sense of the shapes of the molecules. Each bond line represents a pair of shared electrons. Unlike the Lewis structures in Figure 3.5B, however, those electrons not involved in bonding are omitted. This is a common practice in writing structural formulas. Just because our illustration omits them don't forget that they are really there.

Silicon dioxide is a good example of a covalent compound which has trillions of atoms of silicon and exactly twice that number of oxygen atoms in a huge lattice (a tightly packed cubic array). The lattice is not ionic; rather, each silicon-to-oxygen bond is a pair of shared electrons and SiO_2 is termed a **network covalent** compound. Such compounds, mostly of mineral origin, are extremely hard and have very high melting and boiling points. Contrast network covalent silicon dioxide (melting point of 2230°C) with molecular covalent carbon dioxide (melting point of −56.6°C).

The number of bonds that an atom usually forms is known as the atom's **valence** (or "combining power"). The valence number equals the number of electrons an atom would need to gain or lose (whichever is the smaller) in order to attain the stable outer electron arrangement

exercise 3.2

Lewis Structures—Covalent Compounds

Problem

Draw the Lewis structure and structural formula of these molecules:

(a)   HF   Hydrogen fluoride is used to etch glass and in the manufacture of silicon integrated circuit chips.

(b) CH_4 Methane, as we've mentioned before is a form of natural gas, often referred to as swamp gas.

(c) CO_2 Carbon dioxide is an important constituent of our atmosphere; it is the source of carbon for plant growth.

Solution

(a)   As you can see from Figure 3.5, hydrogen belongs to Group 1A and has a single (valence) electron. Fluorine belongs to Group 7A, with seven valence electrons. A single bond forms between the unpaired atoms in hydrogen and fluorine, leaving three lone pairs. The reaction can be represented with Lewis structures as:

$$ H\cdot \ + \ \cdot\ddot{\underset{\cdot\cdot}{F}}: \ \longrightarrow \ H-\ddot{\underset{\cdot\cdot}{F}}: $$

The structural formula of hydrogen fluoride is depicted simply as **H—F.**

(b)   The single valence electrons of the four hydrogen atoms bond with the four unpaired valence electrons of the carbon atom (Group 4A) creating four single bonds. No lone pairs remain, so the Lewis structure and structural formula are identical:

$$ 4H\cdot \ + \ \cdot\dot{C}\cdot \ \longrightarrow \ H-\overset{\displaystyle H}{\underset{\displaystyle H}{C}}-H $$

(c)   Oxygen is a Group 6A element with six valence electrons. Its two unpaired electrons will pair up with two of the unpaired electrons of carbon, each oxygen atom forming a double bond with carbon:

$$ 2\cdot\ddot{O}\cdot \ + \ \cdot\dot{C}\cdot \ \longrightarrow \ \ddot{O}{=}C{=}\ddot{O} $$

The structural formula of carbon dioxide is depicted simply as **O=C=O.**

Check

The question to ask is has the rearrangement of electrons in the compounds formed resulted in a more stable configuration. A stable arrangement for hydrogen is similar to the inert gas helium, having two electrons in the outer shell. That configuration is attained in the case of both **(a)** hydrogen fluoride and **(b)** methane. A more stable arrangement of electrons for the larger atoms would be that of an octet. The fluorine in hydrogen fluoride **(a),** the carbon in methane **(b),** and both the oxygens and the carbon in carbon dioxide **(c)** all have the configuration of a stable octet, in this case that of neon. Notice that the carbon in carbon dioxide must form two double bonds in order to achieve an octet.

exercise 3.3

Lewis Structures—Ions and Polyatomic Ions

Problem

Draw the Lewis structure for these ions and polyatomic ions:

(a)	Cs^+	**(b)**	Cl^-	**(c)**	CN^-
(d)	NH_4^+	**(e)**	S^{2-}	**(f)**	NO_3^-

Solution

(a)   Cesium (Cs) is a metal from Group 1A. It has one valence electron in its outer shell, which it loses when the ion is formed:

$$Cs \cdot \longrightarrow Cs^+ + e^-$$

(b)   Chlorine (Cl) is a nonmetal from Group 7A. It has seven valence electrons in its outer shell, and so it will gain one electron to form the chloride ion:

$$e^- + \ddot{\underset{..}{Cl}} \cdot \longrightarrow \ddot{\underset{..}{Cl}} {:}^-$$

(c)   Carbon (C) is from Group 4A, with four valence electrons; nitrogen (N) is from Group 5A, with five valence electrons. The bond that forms between them is covalent, a triple bond. For nitrogen, this creates a stable octet. Carbon, however is still shy one electron and so must gain an electron to form an octet, creating a polyatomic ion with a charge of -1:

$$e^- + \cdot \overset{..}{C} \cdot + \cdot \ddot{N} {:} \longrightarrow {:}C \equiv N {:}^-$$

(d)   The single valence electrons of the four hydrogen atoms form single bonds with nitrogen. Nitrogen gives up an electron in the process, creating a polyatomic ion with a charge of $+1$:

$$4H \cdot + \cdot \ddot{N} {:} \longrightarrow \left[\begin{array}{c} H \\ | \\ H-N-H \\ | \\ H \end{array} \right]^+ + e^-$$

(e)   Sulfur (S) is a nonmetal from Group 6A. It has 6 electrons in its outer shell, and will gain 2 electrons to form the sulfide ion:

$$2e^- + {:}\ddot{S} \cdot \longrightarrow {:}\ddot{\underset{..}{S}} {:}^{2-}$$

(f)   Nitrogen is from Group 5A, with 5 valence electrons. Oxygen is from Group 6A, and has 6 electrons in its outer shell. The double and single bonds that form between the nitrogen and oxygen atoms create stable octets around all atoms except one oxygen, in which an extra electron is needed.

$$e^- + \cdot \ddot{N} {:} + 3 \cdot \ddot{\underset{..}{O}} {:} \longrightarrow \left[\begin{array}{c} {:}\ddot{O} {:} \\ | \\ {:}\ddot{\underset{..}{O}} - N = \ddot{O} \end{array} \right]^-$$

Check

In the cases shown here the atoms attain a stable configuration. For cesium and chlorine, the two resulting ions can react to form the ionic compound cesium chloride (CsCl).

of the nearest noble gas. Here is the last principle of chemical bonding that we will mention: *the formulas of chemical compounds can often be worked out very easily by considering the valences of the atoms present.*

There are four basic principles to consider when considering how different atoms will interact:

1. Atoms involved in chemical reactions tend to acquire a more stable electron arrangement.
2. The proportions in which atoms form compounds is often determined by the outer-shell arrangement of the atoms involved.
3. Compounds formed during chemical reactions tend to be electrically neutral overall.
4. The number of bonds that atoms form, and hence the chemical formula of the resulting compound, correlates to the valences of the atoms present.

3.4 | The Forces of Bonding

In Section 3.3 we discussed ionic and covalent bonding, the two major classifications of bonding in the formation of chemical compounds. An often asked question is "How do the atoms know whether they should undergo ionic or covalent bonding?" As we stated earlier, the atoms don't "know" anything, but they do interact so as to achieve the lowest energy state available.

It usually works out that when a metal reacts with a nonmetal, the lowest energy situation involves a transfer of electrons so that an ionic compound is formed. Conversely, when nonmetals react with other nonmetals, the lowest energy situation involves the sharing of electrons so that covalent compounds are formed. This distinction is important because the properties of the compounds formed depend on the bonding that holds them together. Covalent bonding results in the formation of molecules. Molecular compounds are composed of individual molecules, which can contain anything from two up to several thousand atoms. In some compounds, however, including ionic solids, billions of ions bond together in a three-dimensional lattice.

The most obvious differences between molecular compounds and lattice compounds are in their physical properties. Molecular substances tend to be soft and to melt or even boil at relatively low temperatures. Compounds with crystal lattice structures tend to be hard and to melt only at very high temperatures and to boil at even higher temperatures. A compound's **melting point** is the temperature at which the solid form of the compound becomes liquid. Likewise, a compound's **boiling point** is the temperature at which the liquid boils and transforms into a gas. Table 3.1 lists melting and boiling points for the chlorine compounds of elements across two rows of the periodic table. Notice that the compounds of chlorine (a nonmetal) with metals have very high melting and boiling points. But the compounds of chlorine and other nonmetals have much lower melting and boiling points. The chlorine compound of aluminum is an interesting middle ground. It behaves in many ways like an ionic compound but in other ways it behaves like a covalent compound. Nature is more variable than our simple classifications. There is really a continuum of bonding types but it is very useful even to experienced chemists to think of ionic and covalent classifications of chemical bonds. Now lets look further at chemical bonding in order to explain chemical and physical properties of chemical substances.

Electronegativity

The ability of an atom that is forming a bond to attract electrons from its bonding partner can be indicated by a number known as the atom's **electronegativity.** The higher the electronegativity value, the stronger is the atom's "electron-attracting" ability. You will notice from Figures 3.7 and 3.8 that the nonmetals have much higher electronegativities than the metals, and that the elements at the top of the columns in the periodic table are more electronegative than those below. (Electronegativities are not well defined for the noble gases.) Based on these patterns, fluorine is the most electronegative element. This means that it has the greatest tendency to attract electrons when forming a compound.

Ionic compounds have very high melting points and boiling points because the temperature must be very great to break up the crystals. Covalent compounds, on the

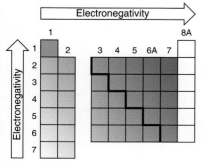

Figure 3.7 A portion of the periodic table showing the electronegativity tendencies of the main-group elements.

Table 3.1

Comparison of Ionic and Covalent Bonding

Group*	1A	2A	3A	4A	5A	6A	7A

Chlorine compounds formed with Period 2 elements

	LiCl	BeCl$_2$	BCl$_3$	CCl$_4$	NCl$_3$	OCl$_2$	ClF
Melting point (°C)	614	405	−107.3	−23	<−40	−20	−154
Boiling point (°C)	1325	520	12.5	76.8	<71 (explodes)	3.8 (explodes)	−100.8

Chlorine compounds formed with Period 3 elements

	NaCl	MgCl$_2$	AlCl$_3$	SiCl$_4$	PCl$_3$	SCl$_2$	Cl$_2$
Melting point (°C)	801	708	192‡	−70	−112	−78	−101
Boiling point (°C)	1413	1412	183 (sublimes @ 180)	57.6	75.5	59 (decomposes)	−34.6

*Elements shown in **boldface** belong to the groups listed at the top of the table. Chlorine is itself a Group 7A element.

‡When AlCl$_3$ melts, it changes from an ionic solid to a covalent liquid, Al$_2$Cl$_6$. The stark change in the nature of the bonding leads to the unusual melting and boiling point relationships seen here.

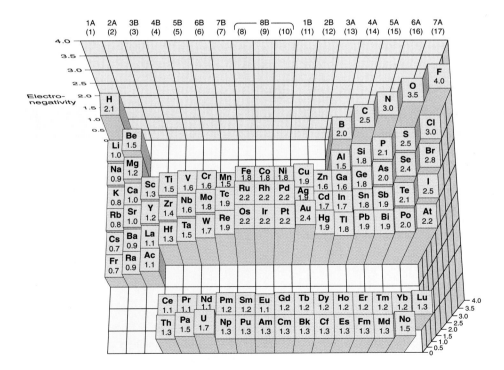

Figure 3.8 Elements with the highest electronegativity values have the strongest attraction for the electrons within a chemical bond. Note that electronegativity of atoms in a group decreases moving down the periodic table, and electronegativity of atoms in a period increases from left to right.

other hand, have much lower melting and boiling points because relatively little energy is needed to break the intermolecular attractions between molecules. For example, sodium chloride (NaCl) and acetic acid (CH$_3$COOH) have nearly the same *formula weight* but very different properties because NaCl is an ionic compound and CH$_3$COOH is a covalent compound.

NaCl: m.p. 601°C   b.p. 1413°C
CH$_3$COOH: m.p. −95.35°C   b.p. 56.2°C

Table 3.1 compares the melting and boiling points of the chlorides of the elements in the first two full periods of the periodic table. The metals from the left side of the periodic table form ionic bonds; the nonmetals from the right side form covalent bonds. Note the change in values as you go from one side of Table 3.1 to the other. Boron chloride (BCl_3) and aluminum chloride ($AlCl_3$) are interesting cases, and these elements are often listed as semimetals.

A covalent bond consists of a pair of electrons. Usually, each atom involved in the bond donates one electron to create the pair. This however, is not a hard and fast rule. Consider what happens when a molecule of NH_3, with an unshared pair of electrons, encounters a molecule of BF_3, which has only six electrons around the central boron atom.

$$
\begin{array}{ccc}
\text{F} & \text{H} & \text{F} \quad \text{H} \\
| & | & | \quad | \\
\text{F—B} \quad + \quad \text{:N—H} \quad \longrightarrow \quad & & \text{F—B—N—H} \\
| & | & | \quad | \\
\text{F} & \text{H} & \text{F} \quad \text{H}
\end{array}
$$

The boron-to-nitrogen bond in the product is an ordinary covalent bond but both of the electrons in this bond originated in the nitrogen atom. Such a bond, formed when both electrons originated in the same atom is called a *coordinate covalent bond.*

Polar Covalent Bonding—Unequal Sharing of Electrons

When atoms become covalently bonded, the shared electrons involved in the bonds will only be shared *equally* if the atoms are identical. This means that equal sharing of electrons only occurs in covalent bonds within molecules of like elements: O_2, O_3, H_2, Cl_2. In the much greater number of covalent bonds within compounds, the electrons are shared *unequally* between the different atoms. The reason for this lies in the differing structures of the individual atoms: their differing number of protons and electrons. These differences mean that some atoms attract the shared electrons more strongly than other atoms.

exercise 3.4

Electronegativity and Bonding, Part I

Problem

If you wanted to form a binary compound (a compound of two different atoms) with the greatest electronegativity difference, which two atoms would you choose? What would be the formula of this compound? *What is that nature of the bond formed?* What force might be holding the atoms of this compound together?

Solution

The greatest electronegativity is shown by fluorine, 4.0 (Figure 3.8); the smallest electronegativity is shown by cesium or francium, both 0.7. Francium does not exist in appreciable amounts, so **cesium fluoride (CsF)** is usually said to be the compound with the greatest electronegativity difference in its atoms. This is an example of ionic bonding. The force that holds the ions together is called the **electrostatic force.**

Check

Cesium, in Group 1A, forms a $+1$ ion, and fluorine, in Group 7A, forms the -1 ion. The combination of Cs^+ and F^- gives CsF. The positive charge of the cesium ion attracts the negative charge of the fluorine ion.

exercise 3.5

Electronegativity and Bonding, Part II

Problem

Carbon, which has an intermediate electronegativity of 2.5, is a constituent of millions of different compounds, *organic* compounds. Think of a covalent carbon compound in which carbon is bonded to atoms **(a)** more electronegative and **(b)** less electronegative than itself.

Solution

(a) CCl_4

The compound, carbon tetrachloride, has been used as a dry cleaning agent and fire extinguisher.

(b) CH_4

This is methane, "swamp gas."

Check

The electronegativity of C (2.5) is less than Cl (3.0) but greater than H (2.1).

When atoms with substantially different electronegativities are involved in covalent bonds, the bonds become *polarized* due to the resulting unequal distribution of electrons. By polarized we mean that the bonds have a slightly negative "pole" around the atom that attracts the shared electrons more strongly and a slightly positive "pole" around the atom that attracts the electrons less strongly. The bonds associated with this type of unequal sharing of electrons are called **polar covalent bonds.**

The partial electrical charges at the poles of polar covalent bonds can be indicated by using the Greek letter "delta" (δ) followed by a plus or minus sign, to denote the δ^+ and δ^- ends of the bonds. Figure 3.9 illustrates this with some common polar covalent bonds. Hydrogen chloride is a good example of polar covalent bonding. The δ^+ and δ^- regions within molecules held together by polar covalent bonds have a great influence on the chemical reactions these molecules can participate in.

Figure 3.9 Some molecules containing polar covalent bonds.

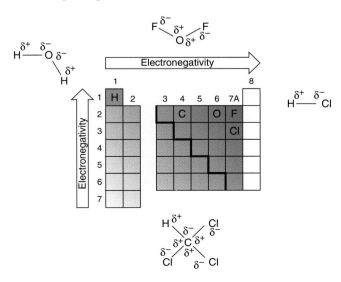

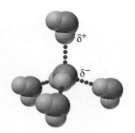

Figure 3.10 Water has a relatively high boiling point because of the attractive forces among molecules of water.

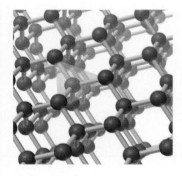

Figure 3.11 Carbon atoms in diamond bond to each other such that the four neighbors are at the corners of a tetrahedron surrounding each atom. All the atoms in a single diamond are covalently bonded in a huge network.

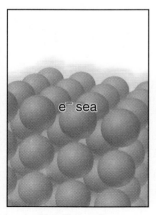

Metallic bonding

Figure 3.12 Metallic bonding in sodium. Many atoms of sodium come together. The nuclei and inner electrons of sodium metal are packed closely together, surrounded by a "sea" of mobile shared outer electrons (e⁻).

Polar and Nonpolar Molecules

The polar covalent bonds within a molecule are often arranged in a way that gives the molecule, overall, a δ^+ end and a δ^- end. Any such molecule is said to be a **polar molecule.** Polar molecules behave noticeably differently from nonpolar molecules. Water (H_2O) is one of the most polar of all molecules, as we will discuss in Chapter 7, and its unique properties have shaped the way life developed on Earth. One dramatic example of the significance of water being a polar molecule is its **boiling point** (the temperature at which it boils) of 100°C. When water boils, all of the forces of attraction between the individual water molecules must be broken by supplying heat energy. The partial charges on water molecules cause forces of attraction to exist not just between oxygen and hydrogen atoms within the molecule but between the molecules themselves (Figure 3.10). Nonpolar molecules have much weaker molecule-to-molecule attraction. If water molecules were not polar, and therefore not attracted to each other, the energy needed to separate them would be much less and so the boiling point of water would be much lower. With a lower boiling point, all of the water on Earth would have evaporated into the form of gas (water vapor). The Earth has oceans, rivers, lakes, and probably life because water molecules are polar!

The covalent substances we have considered have all been composed of fairly small molecules. As we will see in Chapter 4, compounds such as polyethylene (the plastic used in milk bottles), and proteins and starches are covalent molecules with thousands of atoms. Larger still are the covalently-bonded substances such as diamond, quartz, carborundum, and elemental silicon. These do not really form molecules. All the carbon atoms in a single diamond are covalently bonded in a huge network (Figure 3.11). There are not little diamond molecules packed together. These "network covalent solids" have in common the properties of high melting point and chemical stability. In the case of the silicon network solid, this makes it ideal for use in computers, as discussed in the Case in Point, Silicon: From Little Chips Comes Big Money.

Metallic Bonding—A "Sea" of Electrons

The bonding of metal atoms within a metal is somewhat distinct from the types of bonding we've discussed thus far. In **metallic bonding,** the atoms share their outer electrons *throughout the structure of the metal.* Thus, the outer electrons are not shared with specific partners as in covalent bonding, or transferred to a specific atom as in ionic bonding. Instead, they are free to move from atom to atom. The electrons "belong" equally to each of millions of atoms.

A good way to visualize the structure of a metal is to imagine the atoms' "cores" (their nuclei and inner electrons) packed together as closely as possible within a "sea" of mobile outer electrons (Figure 3.12). This sea of mobile electrons within metals explains why metallic elements are good conductors of electricity. An electrical current is simply a *flow of electrons* from one place to another. If one end of a piece of metal (such as a copper wire) is made negatively-charged and the other end is made positively-charged, electrons will flow through the metal (from negative end to positive end) in an electrical current. Nearly all electrical wires are copper wires, but the electrical contacts that attach the components of a computer are gold-coated to give greater corrosion resistance.

We have now explored the three types of chemical bonding: ionic, covalent, and metallic. We can find all of these inside just about every living and nonliving thing! Even though we classify bonding types in these three seemingly distinct ways, they are actually closely related in that they all are the result of electrons leaving their home atom and entering into combinations with other atoms.

We now have a reasonable start in our understanding of chemical bonding. Bonding is the combining of atoms into larger groups called compounds. Just as we name all the elements, so we also name the millions of compounds. Ionic compounds are named by a slightly different system than are covalent compounds, and organic compounds are named by a different system still. In order that we understand each other for the rest of the book, we now need to spend some time talking about how we name the substances that we will discuss. The set of rules for naming compounds is called *chemical nomenclature.*

case in point

Silicon: From Little Chips Comes Big Money

Silicon Valley is the home of California's computer industry, where computers and related products are designed and manufactured. At the heart of the industry is a tiny chip of silicon whose manufacture is big business, an economic juggernaut with year 2000 worldwide sales of just over $200 billion, most of which was generated from the United States and Japan. Why is silicon so important to the computer industry that the whole area is nicknamed after this Group 4A element?

Pure elemental silicon is a very hard, solid material. It is isolated from sand, which is mostly silicon dioxide, SiO_2. Silicon is a semimetal or *metalloid*. Metals, such as copper, conduct electricity extremely well and nonmetals, like sulfur, essentially don't conduct at all. As a semimetal, silicon is a rather poor conductor of electricity, but not so poor as true insulators; silicon is therefore termed a *semiconductor*.

For a substance to conduct electricity well, it must have electrons which are not tied down to a fixed position. In pure silicon, each atom is bonded to four other silicon atoms; all the available electrons are used up in the bonding so are unable to move around. However, if the pure silicon is *doped* with small amounts of an element in Group 5A (P, As, or Sb), the atom fits into the crystal structure of the silicon to make a solid solution but brings along one more electron than would a silicon atom. This extra electron is not fixed to one place and so can move around to conduct electricity. Specifically, this free electron can move toward electrically positive places. *Doping* is the process in which small amounts of the doping element are added into previously very pure silicon.

Silicon can alternately be doped with a Group 3A element such as B, Al, or Ga. Such an atom also fits into the crystal structure of the silicon but is short one electron to bond properly. The place where this electron would go is called a *positive hole*. This "hole" can also move toward negative charge and conduct electricity. Silicon with positive holes is termed a "*p*-type" semiconductor; silicon with mobile, negatively charged electrons is termed "*n*-type" silicon.

Silicon that has been reacted with oxygen to make SiO_2 (as in sand) is not an electrical conductor at all—rather it is an electrical insulator. Manipulation of silicon to make

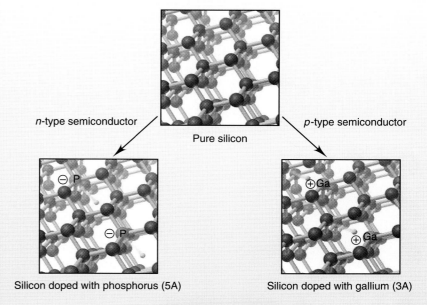

n-type semiconductor   Pure silicon   *p*-type semiconductor

Silicon doped with phosphorus (5A)   Silicon doped with gallium (3A)

SiO_2

Doping of silicon increases its conductivity. *n*-Type (or negative-type) doping of silicon with phosphorus (purple) adds additional valence electrons to increase conductivity. With *p*-type (or positive-type) doping of silicon with gallium (orange) nearby silicon electrons can enter the unoccupied orbitals of gallium atoms ("holes") and thereby increase conductivity.

n-type and *p*-type silicon and silicon dioxide is crucial to the semiconductor industry.

Silicon was first used in the electronics industry to make transistors. Transistors are small pieces of *n*-type silicon in contact with *p*-type silicon. These very small devices did the same job as much larger, more power-demanding vacuum tubes.

After a few years it became clear that entire electrical circuits could be laid out on a single piece of silicon by judicious doping in specific places on the piece of silicon. Now the entire operating chip of a computer with millions of individual electrical units is made on a single silicon wafer.

Computer chip.

hands-on

For this activity you will need a generous supply of spherical candies or marbles, all the same size, and a shallow container with a flat bottom, such as a box.

- Arrange several of these spheres or balls on the flat surface. If necessary, put pieces of cardboard inside the container to wedge the balls together into as small an area as possible.

Notice that the closest packing occurs when the balls are arranged with six neighbors touching a central ball.

- Lay a second layer of balls onto the first layer.

Notice that you can only put a ball into half of the triangular-shaped low spots in the bottom layer. When this is done, another layer of balls that looks like the bottom layer has been formed, except there are gaps left in the second layer. A third layer will fill into these lower open spots or on top of the second layer of balls. This arrangement is similar to the ways in which metal atoms pack together in forming a solid.

Imagine bending a wire of metal whose atoms are arranged in this way and whose valence electrons are swarming around not attached to any specific atom or pair of atoms. The atoms at the bend simply take up new positions with new neighbors and the electrons rush around to new positions. Metals can be deformed or bent by physical force (they are *malleable*) because their atoms are not tied to specific neighbors, as is the case with covalent and ionic substances.

3.5 | What's in a Name?
Chemical Nomenclature

What's in a name? In Korea, the family name comes first. In the United States, the surname comes first and the family name last. In Mexico, both families are represented fully in a child's name.

Throughout the world, different cultures have all kinds of ways of maintaining connections with family histories. We know something about families by the names that their members have. In the same way, we know about chemical substances by the names that they have. One key difference between the names of people and those of chemicals is that we really cannot know much about how people behave based solely on their name (for example, someone named "Rocky" is not necessarily a fighter—she may be a school teacher). Knowing the name of a chemical, however, can tell us an extraordinary amount about that substance's composition. Our understanding of composition and bonding can then tell us about its behavior. The names of some chemicals are very long and complex. To begin with, however, we will stick to the basics with substances that have a first and a last name. If these compounds contain two elements they are called **binary compounds.**

3.5   What's In a Name? Chemical Nomenclature

119

Binary Salts

Binary salts are examples of binary compounds. They are ionic compounds composed of positively charged ions and negatively charged ions. The positively charged ions are known as **cations,** while the negatively charged ions are known as **anions.**

Sodium chloride, NaCl, is the best-known binary salt because it is the "table salt" we use to season food. Its cations are sodium ions (Na^+) and its anions are chloride ions (Cl^-). Calcium fluoride (CaF_2) is another binary salt. Its cations are calcium ions (Ca^{2+}), while its anions are fluoride ions (F^-).

The name of the cation always comes first in the name of a binary salt and is simply the name of the element concerned (sodium or calcium in the examples just given). The name of the anion always comes last in the compound's name. It is formed by adding "ide" to the first part of the element concerned (hence chloride and fluoride, from chlorine and fluorine). So what would we call the binary salt formed by the combination of lithium and oxygen? You are able, we hope, to suggest "lithium oxide."

Sodium always forms ions with a +1 charge, called the +1 **oxidation state** (Figure 3.13). **Oxidation** is a term used to describe a process involving the loss of one or more electrons, so sodium atoms are *oxidized* when they become ions. Calcium always forms ions with a +2 charge (the +2 oxidation state).

Many of the transition elements (Groups 1B–8B) can form ions in different oxidation states, depending on the circumstances. Iron atoms, for example, can form Fe^{2+} and Fe^{3+} ions (with oxidation states of +2 and +3, respectively). When we name compounds containing iron ions, we need a way to indicate which oxidation state the iron ions are in. We do this using Roman numerals, as follows:

$$Fe^{2+} \text{ is iron(II) and } Fe^{3+} \text{ is iron(III)}$$

So the compound called iron(III) chloride must contain Fe^{3+} ions and have the formula $FeCl_3$. Could you have predicted that formula? You would be able to because chloride ions are always Cl^-, so we need three chloride ions to counterbalance the +3 charge on each iron(III) ion, to create an electrically neutral compound overall.

Copper(II) oxide will contain Cu^{2+} ions, and since oxide ions are always O^{2-} its formula will be CuO.

Salts with Polyatomic Ions

Polyatomic ions were discussed in Chapter 2 (see Table 2.1). When naming a compound with a polyatomic anion or cation, use the entire name of the polyatomic ion. For example:

$$Ca(OH)_2 \text{ is calcium hydroxide and } NH_4Cl \text{ is ammonium chloride}$$

Binary Covalent Compounds

Binary covalent compounds are *covalently-bonded* compounds that contain two different types of atoms. These compounds usually contain two different nonmetal elements. We usually begin the names of such compounds with the name of the element with the lower electronegativity (carbon would precede sulfur in CS_2, carbon disulfide). This is usually, *though not always,* the element in the lower numbered group (the one closest to the left of the periodic table, as with carbon and sulfur). We then add the first part of the name of the other element, followed by "ide" as before. We also often use the prefixes shown in Table 3.2 to indicate the ratios in which the atoms are present:

We usually use "mono" for "one" only when it refers to the second element in a compound, otherwise we use the element's name alone if only one atom is present. Also, the final "a" of the prefixes is dropped if it comes before a vowel (for example, "pentoxide" rather than "pentaoxide"). "Organic" compounds, composed largely of carbon and hydrogen atoms, are named using a special set of rules, described in Chapter 6.

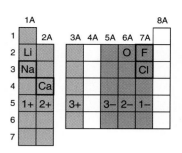

Cations (+)	Anions (−)
+1 Lithium	Ox<u>ide</u> −2
+1 Sodium	Chlor<u>ide</u> −1
+2 Calcium	Fluor<u>ide</u> −1

Figure 3.13 Electronegativity table with cations and anions.

Table 3.2

Prefixes Commonly Used in Chemical Nomenclature

Prefix	Value	Example	Name of Compound
mono	1	CO	carbon monoxide
di	2	SO_2	sulfur dioxide
tri	3	NI_3	nitrogen triiodide
tetra	4	CCl_4	carbon tetrachloride
penta	5	PCl_5	phosphorus pentachloride
hexa	6	Si_2Cl_6	disilicon hexachloride
hepta	7	IF_7	iodine heptafluoride
octa	8	Si_3H_8	trisilicon octahydride
nona	9	B_5H_9	pentaboron nonahydride
deca	10	P_4O_{10}	tetraphosphorus decoxide

Many compounds are not named in this systematic way; for instance, B_5H_9, listed in Table 3.2, is actually called pentaborane, and Si_3H_8 is technically named trisilane. One final example of this is the compound that might be known as dihydrogen monoxide but is always called water.

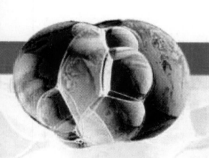

exercise 3.6

Naming Binary Salts

Problem

Name these binary salts. Note that titanium and iron are transition elements which can react to form ions in various oxidation states. You will need to work out the relevant state and use Roman numerals to identify it. (*Hint:* Work out the charge on the anions, and remember that compounds are electrically neutral overall.)

   (a)   KCl   **(b)** TiO_2 **(c)** Fe_2O_3

Solution

(a)   Potassium, a Group 1A element, is the metal in this salt and is named first. Chlorine, the anion, is named second and is given the ending "ide" in place of the element's ending "ine." The name of the binary salt is therefore **potassium chloride.** It is potassium chloride that is used in low-sodium table salt.

(b)   Titanium is a Group 4B metal, which is more abundant in the Earth's crust than nickel and copper! Titanium can exist in a number of oxidation states, even when combining with oxygen. For example, it can exist at Ti^{2+} in TiO, Ti^{3+} in Ti_2O_3, and Ti^{4+} in TiO_2. Because several oxidation states are possible, we must specify which oxidation state we are dealing with. Therefore, we name TiO_2, which is used as a white pigment in paints and paper, **titanium(IV) oxide.**

(c)   Iron is a Group 8B transition metal. Most transition metals have more than one nonzero oxidation state. Because of iron's multiple oxidation states, we must be careful to use Roman numerals in our naming. This compound is **iron(III) oxide.** Fe_2O_3 is the best iron ore, as well as the rust that forms when we leave iron tools exposed to moisture.

exercise 3.7

Compounds With Polyatomic Ions

Problem

Name these compounds:

(a) $BaSO_4$ (b) $Fe(OH)_3$ (c) $(NH_4)_2Cr_2O_7$

Solution

As in our naming rules for binary salts, the cation part of the name comes before the anion part. The keys here are to know (memorize, if necessary) the charges that the polyatomic ions have. These will dictate the charges of the remaining atoms.

(a)   The sulfate ion has a -2 charge. Barium, a Group 2A metal, will always have a $+2$ charge when it is part of a compound. The combination gives the name **barium sulfate.** Medical patients who are to have X-rays made of the stomach or intestines drink barium sulfate, which will trace out the internal structure of these soft organs.

(b)   Here we have a -1 charge on the hydroxide anion. Because there are three such anions, the total negative charge to be balanced is $-1 \times 3 = -3$. The iron must therefore have a $+3$ charge. This is perfectly reasonable for this transition metal. The compound is **iron(III) hydroxide,** used for purifying water and as a red-orange pigment.

(c)   In this case, we have two polyatomic ions. Even though the formula is awkward in that it seems long, the name of the compound is not as awkward, because each ion has only one possible charge: $+1$ for the ammonium, and -2 for the dichromate anion. The name of the compound is **ammonium dichromate.** It is used in photoengraving and pyrotechnics.

Getting Formulas from Names

A good strategy for getting formulas from the names of compounds is based on recognizing that chemical names contain the names of the elements that make up the compounds. The prefixes or numerals represent the number of each kind of atom. You can use this information to find the formula.

For example, let's determine the formula of magnesium nitride. Magnesium is in Group 2A and so will form the Mg^{2+} ion (by losing two electrons). Nitrogen is in Group 5A and so would form an N^{3-} ion (by gaining three electrons). The simplest way for Mg^{2+} to combine with N^{3-} to form a neutral compound is as Mg_3N_2 (in which a total charge of $+6$ would be contributed by the magnesium ions, counterbalancing a total charge of -6 from the two nitride ions). You could actually have arrived at that formula by a neat trick called the **cross-over method:**

1.   Calculate the values of the charges on the ions.
2.   "Cross them over" by swapping them between the elements.
3.   Use these values as the subscript numbers in the formula.

exercise 3.8

Naming Binary Covalent Compounds

Problem

Name these compounds:

       (a) CO_2 **(b)** P_2O_5 **(c)** BrF_3

Solution

Your knowledge of electronegativity is critical here to understand why the elements are listed in the order they are given:

(a)   Carbon is less electronegative than oxygen, so it is listed first. The oxygen gets the prefix "di" for 2, giving the name **carbon dioxide.** CO_2 is a constituent of air and necessary for plant growth.

(b)   Oxygen is more electronegative than phosphorus, and there are five oxygen atoms combined with two oxygen atoms, giving the name **diphosphorus pentoxide.** Actually, the formula is P_4O_{10}, but the simplest ratio of P_2O_5 is frequently cited. This substance is formed when any phosphorus-containing substance is burned.

(c)   Fluorine is the most electronegative element, so it is placed after the bromine, even though they are both in Group 7A. The name of this compound, which contains three fluorine atoms and one bromine atom, is **bromine trifluoride.** The substance is used as a solvent for other fluoride compounds.

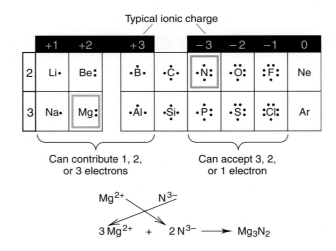

What you are really doing here is calculating the *valences* of the atoms involved (then crossing them over). That is why it works for covalently-bonded compounds too, because, even though they do not form ions, they follow the same rules of valence as ionic compounds. The valence of a polyatomic ion just equals the value of the charge on the ion, regardless of whether it is positive or negative.

A *word of warning,* however: the cross-over method does not always predict the correct formulas of compounds, but it will work in most cases.

The best way to develop naming skills is to try to name the compounds you see in the text and write formulas as often as possible. Knowing substances is like knowing people. You

exercise 3.9

Determining Formulas from Names

Problem

Work out the formula for each of these compounds:

(a)   sodium chloride
(b)   calcium fluoride
(c)   iron(III) nitrate

Solution

The first word in the formula represents the cation, often a metal. The second word in the formula represents the anion part, often a nonmetal or polyatomic anion. Your job is to determine the oxidation states of the anion and cation part. If the cation or anion can exist in one of several oxidation states, you need to determine which one is correct in the formula.

(a)   Sodium, a Group 1A metal, can only exist in compounds as Na^+. Therefore, the chlorine has an oxidation state of -1, CL^-. The formula of the neutral compound is **NaCl,** common table salt.
(b)   In calcium fluoride, both elements only have one possible oxidation state when in compounds Ca^{2+} and F^-. The resulting compound is **CaF_2,** used in huge amounts in the production of steel.
(c)   This compound contains iron(III), or Fe^{3+}. The nitrate anion exists with a -1 oxidation state, meaning that there must be three nitrate ions to form an electrically neutral compound. The formula is therefore **$Fe(NO_3)_3$.** The compound is used to aid certain dyes to bind to fabric, especially silk.

find that when you see them often, they become familiar. You may even start to like being with them!

Summing Up

The structure and function of the computer utilizes ionic, covalent, and metallic substances. The chemical bonding is directly responsible for the ability to store data (Fe_2O_3), carry electric current (copper and gold), and perform the mathematical operations that are the basis of computer logic (silicon-based semiconductors.) We have examined the nature of chemical bonds and the basic rules for naming chemical compounds. You have now learned enough of the vocabulary and principles of chemistry to begin your exploration of chemistry of the modern world. Our first exploratory journey will examine the issue of recycling and reveal some other important chemical principles and methods as we go along.

main points

- There are three types of chemical bonding: ionic, covalent, and metallic.

- Each type of bonding results in substances that have different physical properties.

- The chemical behavior of compounds can be explained in terms of their bonding.

- Lewis structures are the chemist's shorthand for representing atoms with their valence electrons.

- Chemical nomenclature deals with the rules for naming and assigning formulas to compounds and ions.

important terms

An **anion** is an atom or group of atoms having a negative charge. (p. 119)

Binary compounds are compounds composed of only two elements. (p. 118)

Binary salts are compounds containing one metal and one nonmetal. (p. 119)

Boiling point is the temperature at which a substance boils; above this temperature the substance is a gas. (pp. 112, 116)

Bonding is a collective term used to describe ways in which atoms are held together in compounds or crystals. (p. 101)

A **cation** is an atom or group of atoms having a positive charge. (p. 119)

Cross-over method is a strategy for writing the correct formula for an ionic compound. (p. 121)

Double bond is a covalent bond formed by two pairs of shared electrons. (p. 108)

Electronegativity is a measure of the ability of an atom to attract available electrons to itself. (p. 112)

Electrostatic force is the attraction of opposite electric charges and mutual repulsion of like charges. (p. 114)

A **lattice** is a tightly packed regular array of particles. (p. 112)

Lewis structure is a diagram showing atoms and valence electrons in an ion or molecule. (p. 108)

Lone pair describes a pair of valence electrons which are not shared but rather retained by a single atom. (p. 108)

Melting point is the temperature at which a substance melts; above this temperature the substance is a liquid. (p. 112)

Metallic bonding is chemical bonding in metals, neither ionic nor covalent, in which atoms share their valence electrons communally. (p. 116)

network covalent compound is a three-dimensional lattice of covalently-bonded atoms. (p. 109)

Octet is a set of eight valence electrons associated with an atom. (p. 104)

Oxidation is the loss of electrons by a substance. (p. 119)

Oxidation state is the actual charge of an ion or the apparent charge on a covalently bonded atom. (p. 119)

Polar covalent bond is a bond in which the electrons are shared unevenly between two atoms. (p. 115)

Polar molecule is a molecule in which the center of negative charge is not at the same place as the center of positive charge. (p. 116)

Structural formula is a molecular formula that shows how atoms are bonded together. (p. 108)

Triple bond is a covalent bond involving the sharing of three pairs of electrons. (p. 108)

Valence refers to the number of bonds which an atom can form. (p. 109)

exercises

1. How many electrons must these atoms gain or lose to obtain a noble gas configuration? Li, P, I, O, Mg, B, H
2. Describe the differences between the bonding of ionic compounds and of covalent compounds. Include typical examples and describe classes of compounds likely to utilize both types of bonding.
3. Make a suggestion of why six chloride ions surround each sodium ion in NaCl, but eight chloride ions surround each cesium ion in CsCl.
4. Which of these elements would be expected to form positive ions and which negative ions in ionic compounds?
   a. Na   b. Cl   c. P   d. S
   e. Mg   f. Fe   g. Cu   h. Sr
   i. La   j. F
5. Which combinations of atoms are expected to form covalent bonds?
6. Which combinations of atoms are expected to form ionic bonds?
7. Identify the principal kind of bonding of each of these compounds:
   a. KCl   b. SO_2 c. HNO_3 d. MgO
8. Identify the principal kind(s) of bonds in each of these substances:
   a. Cu   b. $CaCO_3$ c. RbBr
9. Which element loses an electron most easily?
10. Why do the Group 1A metals lose an electron so easily?
11. Describe the relative size of ions with respect to the atoms from which they are made.
12. Rank these species in order of increasing size:
    a. Ca   Sr
    b. Na   Al
    c. Na   K Na^+
    d. P^{3-} Cl^- Ca^{2+}
13. Draw Lewis dot structures for each of these ions:
    a. Na^+ b. Br^- c. Cl^-
    d. OH^- e. CN^-

14. Draw Lewis dot structures for each of these molecules:
    a. H_2O b. NH_3 c. HCl   d. CH_2O

15. Draw Lewis dot structures for each of these poly-atomic ions.
    a. CO_3^{2-} b. NO_3^- c. NH_4^+ d. SO_4^{2-}

16. Draw as many Lewis dot arrangements as possible of a molecule having a formula:
    a. $C_3H_6Cl_2$ b. C_4H_{10} c. C_3H_6 d. $C_2H_6O_2$

17. Draw as many Lewis structures as possible of a molecule having a formula:
    a. C_2H_6O b. C_3H_7F
    c. $C_2H_4Cl_2$ d. C_3H_4

18. Draw Lewis dot structures for each of these atoms:
    a. Li   b. C   c. F
    d. N   e. O

19. The elements P, S, and Cl and the nonmetal elements below them frequently form bonds to five or six other atoms. Draw a Lewis dot structure of each of these compounds:
    a. PCl_5 b. SF_6 c. ClF_5

20. Structural chemists were surprised when the correct structure of B_2H_6 was shown to have two of the hydrogen atoms shared between the two boron atoms. Draw an electron dot structure of this substance.

21. Of this group of atoms, create three stable molecules by forming covalent bonds between the correct atoms. Use all of the atoms, and each only once. H, H, Cl, Cl, Cl, Cl, N, Br, O, C, I, F

22. Draw the Lewis dot structures of the ions in these compounds: NaCl, LiCl, $MgCl_2$, K_2O, $AlCl_3$.

23. Based on electron arrangements, how many of each atom are found in these molecules (that is, fill in "x" for each molecule)? CCl_x, $CHCl_x$, CH_xCl_x

24. Which of these following compounds are covalently bonded? Which have ionic bonds? HCl, NaCl, CCl_4

25. How many electrons are found in a single covalent bond?

26. How many covalent bonds are formed by each of these atoms? H, B, C, N, S, I, He

27. Is the compound CCl_5 possible? Why or why not?

28. Of the atoms H, Na, C, B, Cl, Ar, Mg, N, and Br, which will generally *not* form covalent bonds?

29. Draw Lewis structures of these molecules: BF_3, H_2O_2, KBr, H_2S.

30. Draw the neutral Lewis structures for calcium and chlorine. By moving electrons, draw the Lewis structure of calcium chloride with correct charges (you may need more than one of each atom). Highlight the electrons you moved.

31. Which would you expect to have a higher melting point, NaCl or glucose ($C_6H_{12}O_6$). Why?

32. Predict the formula of the compound of calcium and fluorine. What type of bonding is involved in this compound?

33. Predict the formula of the compound of sodium and oxygen. What type of bonding is involved in this compound?

34. Predict the formula of the compound of chlorine with one carbon. What type of bonding is involved in this compound?

35. Draw an electron dot structure of a molecule of hydrogen cyanide (HCN).

36. Draw an electron dot structure of chloroform ($CHCl_3$).

37. What weight of arsenic must be doped into 100 grams of pure silicon if one atom out of 1000 is to be arsenic?

38. Of these pairs, which atom is more electronegative?
    a. Na or C
    b. P or O
    c. H or He
    d. Cl or F

39. In the molecule HCl, would you expect the electrons of the covalent bond to be found closer to H or Cl? Why?

40. Identify two pairs of atoms that form polar, diatomic molecules. Do the same for nonpolar diatomic molecules.

41. Draw a molecule containing δ^+ and δ^- charges. Draw a molecule that has no partial charges.

42. Explain why water has a higher boiling point than butane, C_4H_{10}.

43. You need to create an electrical conductor. What element of the periodic table might you choose to work with? Why?

44. Write the formulas for the ionic compounds containing these set of ions:
    a. K^+ and SO_4^{2-} b. Mg^{2+} and NO_3^-
    c. Fe^{3+} and O^{2-} d. Li^+ and PO_4^{3-}

45. Write the formulas for the ionic compounds containing these sets of ions:
    a. Ca^{2+} and PO_4^{3-} b. K^+, Al^{3+}, and SO_4^{2-}
    c. K^+ and $Cr_2O_7^{2-}$

46. Name these compounds:
    a. Cl_2O b. SO_3 c. N_2O_3

47. Work out the formula for each of these compounds:
    a. copper(I) chloride   b. tin(IV) oxide
    c. dinitrogen tetroxide   d. ammonium sulfate

48. Supply either the correct name or formula.
    a. BF_3
    b. $CaCO_3$
    c. sodium nitrite
    d. cobalt(III) chloride
    e. carbon disulfide

49. Supply the correct formula.
    a. potassium carbonate
    b. iron(III) bromide
    c. sodium nitrate
    d. cobalt(III) oxide
    e. carbon dioxide

50. Supply the correct formula.
    a. calcium oxide
    b. iron(III) sulfate
    c. dinitrogen monoxide
    d. calcium carbonate
    e. iron(II) chloride

food for thought

51. A tiny cube of gamma-iron oxide is described in the Case-in-Point Storing the Data as being a cube with oxide ions at the corners and the centers of each face. Draw such a cube or make one out of gumdrops and toothpicks. Use red gumdrops to represent oxide ions in a pattern with red gumdrops at the corners of a cube and also at the centers of each face of the cube. This cube represents a *unit cell of cubic closest packed* ions. Put a white gumdrop on the end of another toothpick and maneuver it around inside the cube until you find a position in which the white gumdrop is equally surrounded by six red gumdrops. Do the same thing to search for a position in which the white gumdrop is surrounded by four red gumdrops. How many of these positions are there in the cube?

    Look closely at the cube of red gumdrops. Notice that the eight gumdrops at corners would be shared with neighboring cubes. How many neighboring cubes share each corner atom?

52. Start with a bunch of gumdrops of two colors (let's say green and white) and some toothpicks. Build a cube with green gumdrops at the corners (representing chloride ions) and at the centers of each face. The cube should also have white gumdrops (representing sodium ions) at the center of each edge of the cube and at the very center of the cube.

53. What is the fraction of space occupied in a close-packed arrangement of equal-sized spheres that is actually filled with the spheres?

54. Consider how the chemical formulas of simple oxides and chlorides helped Mendeleev work out the structure of the periodic chart. How is this periodicity of properties related to the electronic nature of atoms? Particularly, how does the valence relate to those features that show periodicity?

55. Think of a good analogy to the continuum of electron behavior between equal sharing of electrons nonpolar to transfer of electrons (ionic behavior). Note how this behavior changes as the difference in electronegativity increases.

readings

1. Mallinson, J.C. *Foundations of Magnetic Recording,* 2nd ed., ch. 3. Boston: Academic Press, 1993.
2. Jacoby, M. Data Storage. *Chemical & Engineering News* 78 (June 12, 2000): 37–45.

websites

www.mhhe.com/kelter   The "World of Choices" website, contains activities and exercises including links to websites for: the Intel corporation; The University of Bristol's "Molecule of the Month" site; The International Union of Pure and Applied Chemistry, and much more!

Recycling and Chemical Mathematics

You will die but the carbon will not; its career does not end with you . . . it will return to the soil, and there a plant may take it up again in time, sending it once more on a cycle of plant and animal life.

—Jacob Bronowski, "Biography of an Atom—And the Universe"

"Better Things for Better Living Through Chemistry" was the slogan of a large chemical company not too long ago. The slogan rings true when we look at the myriad of products that have been made possible by people exploiting the chemical possibilities of their environment. Modern medicines, fertilizers, computers, TV sets, radios, fabrics, paints, cars, toys, and even books and newspapers are products, at least in part, of chemical technology. Much of what modern chemistry does for the world would be regarded by most people as good news. The bad news is that many of the items that we produce outlive their usefulness, then are mostly discarded as garbage. Every person in the United States generates an average of almost 2 kg of garbage each day. Multiply this by the 288 million people who live in the United States (year 2002) and you have half a billion kilograms of garbage per day!

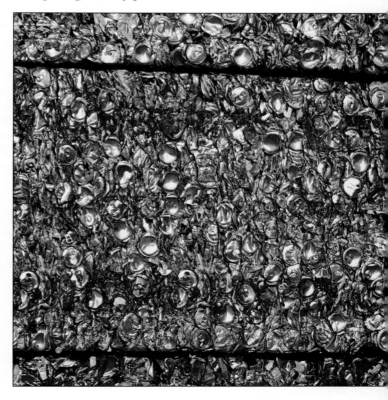

The traditional approaches to municipal trash disposal have been to bury it in huge holes in the ground, called **landfills,** or to **incinerate** the refuse at a very high temperature using an excess of oxygen for near complete combustion. It is becoming clear, however, that as we run out of space, clean air, and natural resources, traditional approaches are no longer good enough. **Recycling,** the recovery and reuse of the Earth's resources, is becoming an ever more important part of our product-laden lifestyle. Chemistry, as you will see, is the hub of the recycling wheel.

We have three goals in this chapter. First, we want to show you how chemistry is central to the recycling process. Next, we want to reinforce the notion, introduced in the Prelude, of the interrelationship between science, technology, and personal and social choices. Recycling offers outstanding examples of such choices. Finally, we want to introduce you to the key principles of "chemical mathematics," the quantitative aspects of chemistry, and to demonstrate how these will help you understand the issues involved in recycling.

We bury our excess products in tens of thousands of landfills across the country. As we begin to understand the finite amount of resources on Earth, society has increased the pace of recycling.

The principles of chemical calculation introduced here will be used (recycled) many times throughout this text. We will set the stage for our discussion by considering the Earth, our communal home, as a complex network of interlinked chemical cycles.

4.1   Nature's Recycling: The Earth as a Materially Closed but Energetically Open System

Everything on the Earth is made of atoms, mostly incorporated within molecules and ions. The vast majority of these atoms have existed as parts of our planet for billions of years. We can manipulate them physically and chemically to suit our needs, but what is already here is all that we can use. Materially, we have a virtually **closed system,** meaning one that does not receive matter from anywhere else and does not lose any matter either. The only significant amount of matter we gain from space arrives in the form of meteorites; and all the matter that is already here remains here, except for the few spacecraft that permanently leave Earth and the escape of some atmospheric gases to space.

Energetically, however, we do not have a closed system. The Earth constantly receives energy from the Sun and constantly releases it to space in the form of light and heat radiation. So energy continually *flows through* the Earth's atmosphere and crust (Figure 4.1). It is this constant throughput of energy that "stirs up" the materials of Earth and makes many interesting chemical reactions happen, including the reactions that sustain life.

Figure 4.1   The Earth is (effectively) a materially closed system, but it is an energetically open one. Energy from the Sun flows to the Earth's surface where roughly a third is reflected back directly, with the rest absorbed and then reradiated. This flow of energy in and out powers many interconnected material cycles, in which chemicals go through cyclical patterns of chemical change. These material cycles include all the chemical processes that sustain life.

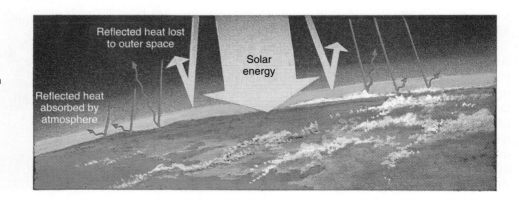

Life on Earth is possible because the flow of energy at the Earth's surface powers many **material cycles** within our materially closed but energetically open system. These cycles are transformations of matter and energy. The simplest involve only changes in physical state (that is, changes in form —solid, liquid, and gas) rather than chemical changes. The **water cycle** is a primary example (Figure 4.2). During the "turning" of the water cycle, the Sun's energy causes liquid water in oceans, rivers, and lakes to evaporate to form gaseous water in the atmosphere (vapor); the cooling of that gaseous water causes it to condense into clouds (composed of tiny droplets of liquid water), which can then release their liquid in the form of rain that falls back to Earth. The cycling of water between the solid state of water (in ice and snow) and the liquid and gaseous states is also part of the water cycle.

Many other material cycles involve complex cycles of chemical change in which specific types of atoms move through many different chemical forms. The **carbon cycle,** for example, involves the recycling of carbon atoms through different chemical forms (Figure 4.3). It operates in concert with other cycles, including the nitrogen cycle, the sulfur cycle, and the phosphorus cycle.

Many material cycles overlap, the components of one cycle feeding into other cycles. An example of one of the most fundamental of such chemical cycles is shown in Figure 4.4. Plants, using energy from the Sun, convert water and carbon dioxide into sugars and oxygen gas during the process of *photosynthesis*. These sugars are then indirectly recombined with oxygen to regenerate water and carbon dioxide while releasing energy during the process of *respiration* in plants and animals. This is the basic chemical cycle that allows plants and

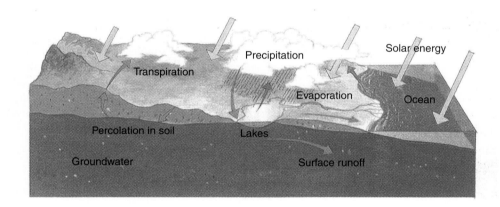

Figure 4.2   The water cycle (blue arrows), sustained by the flow of energy from the Sun (tan arrows).

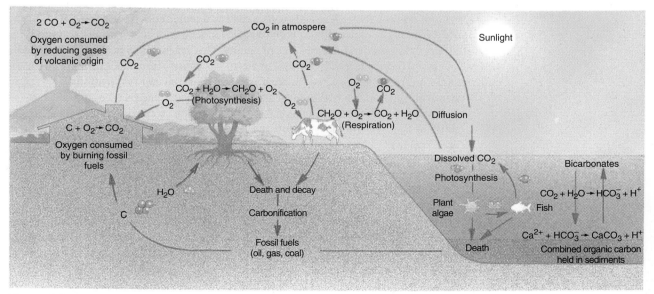

Figure 4.3   The carbon cycle.

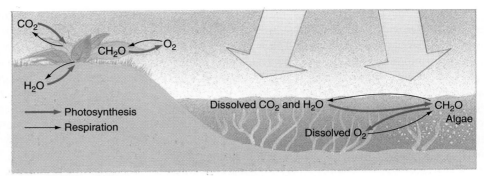

Figure 4.4   Photosynthesis and respiration. Together they form the most fundamental chemical cycle of life. CH_2O stands for biomass, and is the empirical formula for sugars.

animals to grow and to utilize the energy of the Sun to power all their activities. We will use aspects of the photosynthesis–respiration cycle to introduce you to the idea of chemical mathematics. (For an example of one ambitious attempt to study material cycles, see the Case in Point: Biosphere 2.)

4.2 | Introducing Chemical Equations

You have no doubt had some practice in dealing with mathematical equations. The quantity represented by one side of an equation must be equal to that represented by the other side, even though, on the face of it, they may differ in appearance. Now recall from Chapter 2 the fact that *atoms cannot be created or destroyed during a chemical reaction.* This is a direct consequence of the **Law of Conservation of Mass.** All of the atoms represented on one side of a chemical reaction or "equation" must also be present on the other side, even though they may appear rearranged into some different form. Let's consider the chemical reaction involved in respiration to see how this works.

Respiration

Glucose is a molecule containing 6 carbon atoms, 12 hydrogen atoms, and 6 oxygen atoms, hence its formula is $C_6H_{12}O_6$ (Figure 4.5). It is also known as "blood sugar" because it is the main form in which sugar travels through our body in the blood. Glucose is a **carbohydrate,** a compound of carbon, hydrogen, and oxygen, and is a major energy-providing fuel for the human body. When glucose combines with oxygen, a chemical reaction occurs that produces carbon dioxide and water and releases lots of energy. This reaction can occur directly if some glucose is burned (combined with oxygen) in air. In our bodies, the same overall reaction occurs in a very indirect manner when we breathe, the process we call **respiration.**

A **chemical reaction,** such as the reaction between glucose and oxygen, involves a *reorganization* of the atoms in one or more substances to form a different substance or substances. As shown in Figure 4.6, the observable result of a chemical reaction might be a change in color or perhaps *state* (that is, the generation of new solids, liquids, or gases). The chemicals involved might get hotter or colder, or give off forms of energy such as light or heat. Sometimes, however, there is no directly observable evidence that any reaction has occurred.

In chemistry, we use the shorthand notation of a **chemical equation** to describe what happens during chemical reactions. An equation uses the chemical formulas of elements or compounds rather than their names, and it lists the **reactants** (starting materials) to the left of a central arrow and the **products** (resulting materials) to the right. The arrow itself, often referred to as a *reaction arrow,* simply represents the progress of the reaction and can be taken to mean "to give" or "yielding." The overall equation that summarizes respiration is

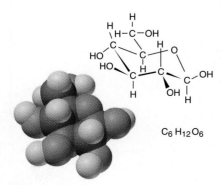

$C_6H_{12}O_6$

Figure 4.5   Glucose.

$$C_6H_{12}O_6(s) \quad + \quad 6O_2(g) \quad \rightarrow \quad 6CO_2(g) \quad + \quad 6H_2O(l)$$

glucose   oxygen   carbon dioxide   water

The "+" on the left side of the equation means "reacts with" and the "+" on the right side means "and." The number "6" in front of O_2, CO_2, and H_2O can be read to mean "6 molecules of." So we could read the equation for respiration as "one molecule of $C_6H_{12}O_6$ reacts with six molecules of O_2 to give six molecules of CO_2 and six molecules of H_2O." Also shown are the states of the reactants and products. Equations do not convey anything that could not be conveyed by words, but they do allow us to summarize all the details of a chemical reaction in a more concise (and numerically precise) form than writing them out in words (Figure 4.7).

Each chemical reaction is a small materially closed system, in the sense that every atom present in the reactants must also be present in the products. Atoms cannot be created or destroyed during chemical reactions, merely rearranged. To check that no atoms have been created or destroyed in a reaction we could first visualize the situation by drawing all the atoms in the reaction (Figure 4.8). A quicker way is to multiply the number of molecules of a particular compound ("6" in the case of O_2) by the number of each kind of atom in the compound, as given by its subscript ("2" for O_2). If a number is not shown, either in front of or within a formula, this indicates that the relevant number is "1," which need never actually be written.

A.   **B.**   **C.**

Figure 4.6   Chemical reactions involve chemical changes. **A.** Table sugar (a solid) treated with sulfuric acid (a liquid) turns to carbon (a different solid) and steam (a gas). During the process the reacting materials change color and phase. **B.** Two liquids react to form the solid Nylon 66. The nylon is formed at the interface of an aqueous solution of hexamethylenediamine (the bottom layer) with a solution of adipoyl chloride in hexane (the top layer). **C.** Energy is released from the Space Shuttle *Endeavor* in a chemical reaction that burns liquid hydrogen and oxygen. The two reactants (550,000 gallons' worth) are stored below the shuttle in a large external tank.

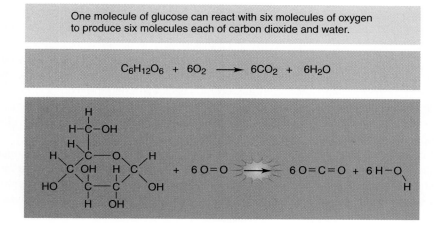

Figure 4.7   Three ways of writing a description of the process of respiration: in words, as a chemical equation, and as a chemical equation with structural diagrams.

Figure 4.8 An accounting of the atoms in the process of respiration.

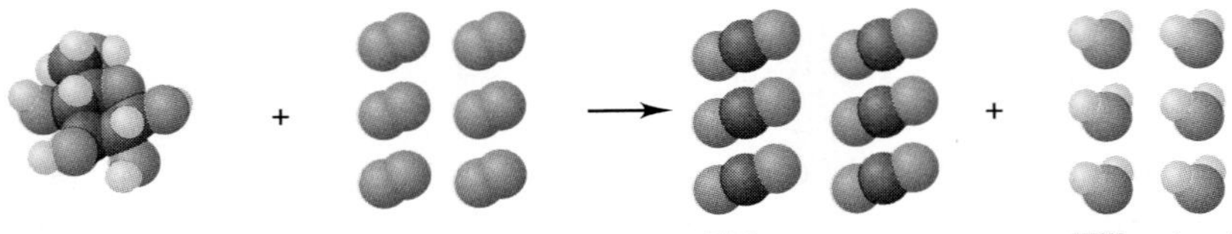

one molecule of glucose + six molecules of oxygen → six molecules of carbon dioxide + six molecules of water

$$1\ C_6H_{12}O_6 \quad + \quad 6\ O_2 \quad \longrightarrow \quad 6\ CO_2 \quad + \quad 6\ H_2O$$

$$1 \times \begin{cases} 6\ C = 6\ C \\ 12\ H = 12\ H \\ 6\ O = 6\ O \end{cases} + \quad 6 \times 2\ O = 12\ O \longrightarrow 6 \times \begin{cases} C = 6\ C \\ O = 12\ O \end{cases} + \quad 6 \times \begin{cases} H = 12\ H \\ O = 6\ O \end{cases}$$

Total	6 C	12 H	18 O	=	6 C	12 H	18 O

■ case in point

Biosphere 2: A World within a World

The Earth is the largest materially closed system that humans occupy. The second largest is a 3.15-acre greenhouse in Oracle, Arizona. It has been called Biosphere 2, the "2" being in recognition of the fact that the Earth is Biosphere 1. The original goal of the project was to determine whether a large-scale, closed ecosystem with human inhabitants could be sustained for 2 years, from September 1991 to September 1993. Once eight participating scientists entered the sealed enclave, no new air, water, or food would be allowed to enter the system; no electricity or other fuel would be supplied; and no waste material could be removed for the duration of the experiment. Sunlight would be the only source of external energy to enter the system, telephone and computer lines would provide communication with the outside, and radio and television signals could enter.

Earlier experiments with closed ecosystems were simpler in scale, including components like ocean water, sand, algae, microbes, and air. Some of these ecosystems died but others have been viable since as long ago as 1968. Several short-term mammal and human experiments have met with mixed results, but none of the experiments was as ambitious as Biosphere 2. In fact, one object of the study was to determine if a similar colony on Mars would be sustainable.

The total internal volume of the Biosphere is 200,000 cubic meters (m^3). This space and the soil beneath it had to be designed to sustain eight people for 2 years. Basic necessities are food, water, and air, and the ability to process waste materials. Food for people means growing plants, either for direct consumption or as food for the chickens, pigs, and goats, which supply eggs, meat, and milk, and as food for fish. The natural process that enables plant growth is photosynthesis. Using energy supplied by

sunlight, green plants take in atmospheric carbon dioxide and water to make the carbohydrates, vegetable proteins, and vegetable oils needed in the human diet. The plants release oxygen gas in the process. The animal life of Biosphere 2, through the process of breathing (respiration), takes in atmospheric oxygen and releases carbon dioxide. If everything could be arranged to come out even, a stable atmosphere with desirable levels of oxygen and carbon dioxide would be maintained. This steady state turned out to be very difficult to attain.

To get it all to come out even, around 4000 species were put into the Biosphere. This total included 1400 different animal species (no elephants or lions), 1100 plant species, 250 insect species, 42 fish species, 35 species of coral, 30 fungi, and the rest were microorganisms. Nobody really knows how many other species were present in the soil and the water before the whole experiment started.

Chemically, the biggest difficulty was keeping the oxygen and carbon dioxide concentration in the atmosphere under control. Initially, CO_2 concentrations went up from 350 ppm (that is 350 *parts per million,* which is the Earth's atmospheric concentration) to 3500 ppm in December 1991 and fell to about 1060 ppm in June 1992. As is true in Earth's atmosphere, carbon dioxide concentration goes down during the day when sunlight activates photosynthesis: the plants absorb CO_2 and release O_2. The concentration climbs back up during the night: plants and animals continue to respire, producing CO_2, but photosynthesis stops once darkness descends and so no CO_2 is absorbed.

The daily swings inside the Biosphere were much greater than in the regular atmosphere. The reason can be seen by considering the amounts, rather than concentrations, of carbon as it exists in various forms. A concentration of 1500 ppm atmospheric CO_2 inside the Biosphere corresponds to only 100 kg of carbon overall. This amount is much smaller than the amount of carbon contained in the living biomass and soils of the Biosphere. The ratio of organic biomass carbon to atmospheric carbon in Biosphere 2 is about 100:1, whereas Earth's ratio is about 1:1. A small change in growing conditions of the plants that make up a large part of the biomass, such as extended cloudy weather, will therefore have a huge effect on the concentration of CO_2 in the air. The concentration of CO_2 would fluctuate as much as 700–800 ppm during one day; this is in contrast to Earth's atmospheric concentration of about 350 ppm, which fluctuates only about 5 ppm during a day.

As the carbon dioxide levels increased, oxygen levels in the air gradually decreased after the Biosphere was sealed. By January 1993, the oxygen concentration was getting very low, 14.5% of the air compared to Earth's normal concentration of about 21%. Upon medical recommendation, pure oxygen was injected into Biosphere over a period of several weeks to bring the concentration up to 19%.

What was upsetting the balance of atmospheric gases? One potential culprit was the microbes in the soil. Their role was to compost animal and vegetable waste, enriching the soil and enhancing the intensive agriculture practiced in the experiment. They consume the oxygen produced by the plants and then form more CO_2, which enters the atmosphere and dissolves in the water. What was puzzling to the scientists studying the low oxygen levels was, if these microbes were eating up all the oxygen, why weren't the CO_2 levels even higher than they were? Aiding and abetting the microbes in the "cover-up" was none other than the concrete used to support the structure of the Biosphere and to form the artificial rocks and cliffs inside.

Concrete is a complicated mixture of substances but a major component is calcium hydroxide, $Ca(OH)_2$, which reacts to form calcium carbonate, $CaCO_3$:

$$Ca(OH)_2 + CO_2 \rightarrow CaCO_3 + H_2O$$

Engineers know this process as the carbonation of concrete, and it is one of the very slow steps that occurs as concrete gets harder and harder over a period of years.

A similar process was used intentionally to lower the concentration of CO_2 in the air:

$$CO_2 + 2NaOH \rightarrow Na_2CO_3 + H_2O$$
$$Na_2CO_3 + Ca(OH)_2 \rightarrow CaCO_3 + 2NaOH$$

The plan was to use a solar furnace to heat the resulting calcium carbonate to 950°C to regenerate CO_2 if it ever began to run short:

$$CaCO_3 \rightarrow CaO + CO_2$$

Biosphere 2 is now a research and public education facility. If you are ever traveling down Highway 77 (mile marker 96.5) in Arizona, you can stop in at the Visitor's Center and even take a tour of part of the facility. They even offer credit-bearing summer courses for undergraduates.

Some CO_2 was converted to $CaCO_3$ in this way, but none was ever reprocessed.

At the beginning of this textbook, we described science as the systematic study of the physical world—of nature. The Biosphere project was beset by mismanagement. It was begun with lofty goals that could not be met, due in large part to the complexity of the ecosystem. By 1994, it was ready to self-destruct. The science could not be studied systematically. Too much was going on. The place was even being overrun by millions of ants of the species *Paratrechina longicorpus*, commonly known as "crazy ants."

In January of 1996, Columbia University was contracted to take advantage of this magnificent research center. They have changed the focus from that of a self-contained ecosystem to the study of how changing atmospheric chemistry might affect global climate. Biosphere 2 is well-suited for this kind of study because of its small size and excellent climate control system. The main website (January 2002) is http://www.bio2.edu.

Anderson, Ian. Biosphere II: A World Apart. *New Scientist* (March 18, 1989): 34–35.
Lewin, Roger. Living in a Bubble. *New Scientist* (April 4, 1992): 12–13.
Nelson, Mark, et al. Using a Closed Ecological System to Study Earth's Biosphere. *Bioscience* 43 (no. 4): 225–236.
Turner, Mark Holman. Building an Ecosystem from Scratch. *Bioscience* 39 (March 1989).
Veggeberg, Scott. Escape from Biosphere 2. *New Scientist* (September 25, 1993): 22–24.
Vergano, Dan. Brave New World of Biosphere 2? *Science News.* (November 16, 1996): 312–313.
Broad, William J. Paradise Lost: Biosphere Retooled As Atmospheric Nightmare. *New York Times* (November 19, 1996).

We can illustrate how to check the total number of each atom on each side of the respiration equation as follows:

Reactant (left) side

- carbon (from one $C_6H_{12}O_6$) $= 1 \times 6\ C$ **= 6 C atoms**
- hydrogen (from one $C_6H_{12}O_6$) $= 1 \times 12\ H$ **= 12 H atoms**
  [oxygen (from one $C_6H_{12}O_6$) $= 1 \times 6\ O$ = 6 O atoms]
  [oxygen (from six O_2) $= 6 \times 2\ O$ = 12 O atoms]
- total oxygen $= 6 + 12$ **= 18 O atoms**

Product (right) side

- carbon (from six CO_2) $= 6 \times 1\ C$ **= 6 C atoms**
- hydrogen (from six H_2O) $= 6 \times 2\ H$ **= 12 H atoms**
  [oxygen (from six CO_2) $= 6 \times 2\ O$ = 12 O atoms]
  [oxygen (from six H_2O) $= 6 \times 1\ O$ = 6 O atoms]
- total oxygen $= 12 + 6$ **= 18 O atoms**

There are the same number of atoms of each element on both sides of the equation. We say that the equation is *balanced*. Only when it is balanced do we have a *quantitatively true chemical equation*. All of the reactions that occur can be represented by balanced equations—they are the only reactions that can take place. We sometimes use an unbalanced equation as a simple *qualitative* statement of what happens during a reaction (just listing the formulas, or even names of the reactants and products without worrying about the proportions in which they react or are formed).

The Law of Conservation of Mass states that atoms cannot be created or destroyed during a *chemical* reaction. During chemical reactions, all of the atoms (including all the electrons) of the reactants will also be present in the products. Therefore the mass of the products is always equal to the mass of the reactants. (See the Consider This box, "When Is a Law Really a Law" on page 60.)

Real Life versus Neat Summaries

We have summarized chemical reactions by saying that they reorganize the atoms in the reactants to form different substances (the products), accompanied by either the release or take-up of energy. That seems to be a fairly straightforward story, but, for the sake of simplicity, it actually omits much of the real story. Chemists write down the equations of the specific chemical reactions they are interested in, and the equations present the reactions as tidy processes in which the reactants are entirely converted into one set of products. Real life is a bit more complex.

In addition to the principal reaction of interest, most chemical reactions are accompanied by many *side reactions* that form different and often unwanted products. For example, in the manufacture of the herbicide 2,4,5-T (an ingredient in Agent Orange, a defoliant used in the Vietnam War), a side reaction led to the formation of small amounts of 2,3,7,8-tetrachlorodibenzodioxin, usually called dioxin. (See the Case in Point: Dioxin on the Side.)

Also, reactions may often fail to go to completion, meaning that some of the reactants may always remain (or be regenerated) in their original unreacted form. It is also important to realize that changes in the reaction conditions, such as variations in the concentrations of different reactants or in their temperature or pressure, can result in significant differences in the products that are formed. An important example is the reaction of hydrogen and nitrogen to form ammonia:

$$3H_2(g) + N_2(g) \rightarrow 2NH_3(g)$$

A total of 38 billion pounds of ammonia were manufactured worldwide during 1999, primarily for use in fertilizer. High pressure favors efficient conversion of the reactants to ammonia. High temperature favors rapid reaction but results in a less complete formation of ammonia. Hydrogen is much more expensive to obtain than nitrogen, so excess nitrogen is used to prevent wasting the expensive hydrogen. Since economics are important when so much material

■ case in point

Dioxin on the Side: When Chemical Reactions Produce Unexpected Results

Dioxin or TCDD—both are abbreviations for the compound 2,3,7,8-tetrachlorodibenzodioxin, $C_{12}H_4Cl_4O_2$. Dioxin has the distinction of having been called the most toxic compound ever made. Its presence in Agent Orange and in soil sprayed with a mixture of oil and dioxin in Times Beach, Missouri; in landfills in Love Canal, New York; and Sevaso, Italy has caused great concern. Over 265,000 tons of contaminated soil from the Times Beach area was incinerated in the mid 1990's to rid the area of dioxin.

Very little dioxin has ever been made intentionally, but considerable amounts have been made inadvertently as a by-product of manufacturing processes. Sadly, even extremely small amounts of the chemical can be toxic. The toxic levels for humans are still under debate but people are clearly less affected by it than are many small animals.

Dioxin first became infamous when it was detected in the herbicide 2,4,5-T (2,4,5-trichlorophenoxyacetic acid). The herbicide is made by reacting 2,4,5-trichlorophenol with chloroacetic acid. During the process, a very small fraction of the 2,4,5-trichlorophenol reacted instead, by a side reaction, to make dioxin, which ended up mixed in with the desired product (see diagram). Commercial 2,4,5-T was never more than about 10 parts per million (ppm) of dioxin, and once its presence and significance were realized, syntheses were improved so that the dioxin produced was below 50 parts per billion (ppb). It is a testimony to modern chemical analysis that this material could be detected in the commercial product, let alone in the environment after the herbicide had been sprayed over soil and vegetation. As a matter of fact, chemical methods of analysis in the 1970s were driven to lower and lower limits of detection precisely because people were so concerned

over accuracy and confidence in ultratrace detection of this compound.

The terms *parts per million, billion, trillion,* are used frequently in the press and will be used again in this book. To understand these terms it is useful to realize that the very familiar term "percent" really means "parts per hundred." A beverage which has 3.2% alcohol has 3.2 mL of alcohol per 100 mL of beverage.

- 1 part per million (ppm) means that there is 1 gram of the impurity per million grams of the major substance. That is the same as a millionth of a gram (1 microgram) per gram of substance. It is also 1 milligram per kilogram.
- 1 part per billion (ppb) means that there is 1 gram of the impurity per billion grams of the major substance. That is the same as a billionth of a gram (1 nanogram) per gram of substance or one microgram per kilogram.
- 1 part per trillion (ppt or pptr) means that there is 1 gram of the impurity per trillion grams of the major substance. That is the same as a trillionth of a gram (1 picogram) per gram of substance or one microgram per megagram (a metric ton). A few grains of salt in a swimming pool is about 1 part per trillion.

The herbicide 2,4,5-T was widely used in the United States to kill deciduous weeds along railroad tracks and powerlines and the underbrush in timberland. It was also used as a component of Agent Orange. Agent Orange was a material used by the military to defoliate jungle areas in Vietnam during the Vietnam War in order to make enemy troop movements more visible from the air. Many of those who handled Agent Orange have sued the U.S.

2, 4, 5 - trichlorophenol   chloroacetic acid   2, 4, 5 - T

dioxin

government for health problems they allege are the result of their wartime exposure to dioxin. Given that the exposure occurred over 30 years ago, direct proof of cause and effect is difficult to show. The controversy continues.

Until recently it was widely assumed that dioxin is entirely a product of recent human chemical activity, created in the manufacture of compounds related to trichlorophenol. The new analytical techniques available by the early 1980s showed that this assumption was wrong. Dioxin was found in parts per trillion levels nearly everywhere. We now know that dioxin is formed in small amounts whenever mixtures of fuels are burned as long as a source of chlorine atoms is present. Given the large number of consumer products that have chlorine in them, including most plastics and paper, municipal waste incineration was seen as an especially worrisome activity. Incinerators have since been redesigned to prevent detectable amounts of dioxin from being formed. The secret is very high temperatures, lots of excess oxygen, and a long residence time for the fuel in the combustion zone.

The dioxin molecule is simply so stable that whenever the right atoms are present at a high temperature, dioxin and other related compounds form. This represents yet another instance of a balanced reaction not telling the full story. Combustion is always written so that carbon always becomes carbon dioxide, and hydrogen always becomes

water. Tiny fractions of fuel do not behave this way; they react to form molecules that are not at all like carbon dioxide.

The numbers used in the names of these compounds indicate the positions of the chlorine atoms. In the compounds with one six-membered ring, the number 1 position is where the oxygen is connected and the numbers just go around the ring up to 6. In dioxin, the number 1 position is the top position on the righthand ring and then increases clockwise around the perimeter. See the accompanying structural formulas to compare the two reactions.

Bumb, R.R., et al. Trace Chemistries of Fire: A Source of Chlorinated Dioxins. *Science* 210 (October 24, 1980): 385–390.

Clapp, Richard, et al. Dioxin Risk: EPA on the Right Track. *Environmental Science & Technology* 29 (1995): 29A–30A.

Dioxin Risk: EPA Assessment Not Justified. Expert Panel, *Environmental Science & Technology* 29 (1995): 31A–32A.

EPA's Dioxin Reassessment. Editor, *Environmental Science & Technology* 29 (1995): 26A–28A.

Johnson, Jeff. Dioxin Risk: Are We Sure Yet? *Environmental Science & Technology* 29 (1995): 24A–25A.

Anon. Dioxin Risk: Incinerators Targeted by EPA. *Environmental Science & Technology* 29 (1995): 33A–35A.

Stehl, R.H. and Lamparski, L.L. Combustion of Several 2,4, 5-Trichlorophenoxy Compounds: Formation of 2,3,7, 8-Tetrachlorodibenzo-p-dioxin. *Science* 197 (September 2, 1977): 1008–1009.

is prepared, production is optimized by a balance among temperature, pressure, and mixture composition.

Another important complication is that a process which we may write out as a seemingly neat one-step reaction may in fact proceed by a large number of interlinked steps, each one associated with the formation of a variety of chemical **intermediates,** substances that are formed and then react before the final products result.

A well-known example is the decomposition of ozone, O_3, a different molecular form of the element oxygen. This decomposition proceeds in two steps. In step 1, an oxygen-to-oxygen bond in ozone breaks to give a molecule of ordinary oxygen (O_2) and a free oxygen atom:

$$O_3 \rightarrow O_2 + O \qquad \qquad \textbf{(step 1)}$$

The oxygen atom then reacts with another molecule of ozone, in step 2, to give two more molecules of ordinary oxygen:

$$O + O_3 \rightarrow 2O_2 \qquad \qquad \textbf{(step 2)}$$

The overall effect is the total of these two reactions:

$$\begin{aligned} O_3 &\rightarrow O_2 + O \\ + O + O_3 &\rightarrow 2O_2 \\ \hline 2O_3 &\rightarrow 3O_2 \end{aligned}$$

Note that the *free* oxygen atom [shown in red] is an **intermediate species,** formed but then reacted, but not a final product.

So chemical equations are a bit like the pictures in cookbooks, which show a neat pile of ingredients on one side and the final cooked dish on the other. The picture suggests the ingredients are converted into the dish in one neat step. In reality, there are many intermediate steps (washing and peeling vegetables, chopping up meat, trimming off fat, etc.), and many unwanted waste materials (side products) are discarded (such as vegetable peelings and fat, Figure 4.9).

exercise 4.1

Recognizing Balanced Equations

Problem

Which of these reactions are written as balanced chemical equations?

(a)   The *thermite* reaction used in welding:

$$Fe_2O_3 + 2Al \rightarrow 2Fe + Al_2O_3$$

(b)   The gas generator used in fireworks:

$$S + KClO_4 \rightarrow SO_2 + KCl$$

(c)   The overall process in the production of aluminum:

$$2Al_2O_3 + 3C \rightarrow 4Al + 3CO_2$$

(d)   The production of phosphoric acid for use in fertilizer:

$$Ca_3(PO_4)_2 + 5H_2SO_4 \rightarrow 3CaSO_4 + 2H_3PO_4$$

Note that when two or more elements are contained within parentheses, the quantity of each is to be multiplied by the subscript immediately after the parentheses. In (d) the expression $(PO_4)_2$ can be interpreted as "P_2O_8" for purposes of counting atoms. The actual structure is better represented by $(PO_4)_2$.

Solution

The key outcome of the Law of Conservation of Mass is that when a chemical reaction occurs, mass (matter) is neither created nor destroyed. *All atoms must be accounted for.* The number of atoms on the products (right) side of the reaction must equal that on the reactants (left) side of the reaction.

	Reactant Atoms	**Product Atoms**	
(a)	2 Fe, 3 O, 2 Al	2 Fe, 3 O, 2 Al	(balanced)
(b)	1 S, 1 K, 1 Cl, **4 O**	1 S, 1 K, 1 Cl, **2 O**	**(not balanced)**
(c)	4 Al, 6 O, 3 C	4 Al, 6 O, 3 C	(balanced)
(d)	3 Ca, 2 P, **28 O, 10 H, 5 S**	3 Ca, 2 P, **20 O, 6 H, 3 S**	**(not balanced)**

Figure 4.9   Chemical equations are like the "before" and "after" pictures in cookbooks. They indicate the starting materials and the end products, but do not give any impression of the messy reality in between.

Photosynthesis

Perhaps the most fundamental chemical process of life is **photosynthesis,** in which the energy of sunlight shining on living plants powers the conversion of carbon dioxide and water into sugars, such as glucose, with the accompanying release of oxygen gas. Overall, this reaction is the reverse of respiration. It is the process that actually forms the sugars that we use as a source of energy during respiration. It is also the source of the oxygen gas that we need to breathe to allow respiration to occur within our bodies (see Figure 4.4).

The overall reaction of photosynthesis can be summarized by the following chemical equation:

$$\underset{\text{carbon dioxide}}{6CO_2(g)} \quad + \quad \underset{\text{water}}{6H_2O(l)} \quad \xrightarrow{\overset{\textit{energy}}{\underset{}{\textit{from Sun}}}} \quad \underset{\text{glucose}}{C_6H_{12}O_6(aq)} \quad + \quad \underset{\text{oxygen}}{6O_2(g)}$$

The reaction requires energy (supplied by the Sun) to make it happen, but energy is not usually included as a "reactant" in chemical equations. You occasionally see it included as such, but the reactants and products are really only the atoms, molecules, and ions involved in the reaction. The size of any accompanying energy changes are conventionally summarized *after* the equation, in a form that will be introduced in Chapter 5. It is acceptable, however, to write "energy from Sun" *above* the reaction arrow, because particular conditions required for a reaction to happen are often indicated in this way.

Although the overall equation of photosynthesis is simple, the chemical change it summarizes actually occurs via an amazingly complex series of chemical reactions, involving over 100 individual chemical steps. The process of respiration,

$$C_6H_{12}O_6(aq) + 6O_2(g) \rightarrow 6CO_2(g) + 6H_2O(l)$$

which reverses photosynthesis and regenerates its starting materials, also occurs via a large number of *different* chemical steps. So although each process is the reverse of the other *overall,* photosynthesis and respiration do not actually proceed by the direct reversal of the chemical steps involved in one another. Our examples of respiration and photosynthesis reveal both the simplicity (overall) and the complexity (in the details) of the chemical processes that we can summarize using chemical equations.

The very existence of humans on the Earth depends on material cycles, such as the photosynthesis–respiration cycle, the carbon cycle, the nitrogen cycle, and so on. Yet we have begun to make use of materials in a way that drastically interferes with these natural cycles. We burn materials that would otherwise have decomposed naturally; we dig up materials from deep within the Earth that would otherwise have taken millennia to reemerge through slow geologic processes; and we bury materials in the ground, or dump them at the bottom of the sea, instead of reusing them or returning them to the Earth in the form in which they were found.

In order to live in chemical harmony with the Earth, we need to learn to use materials in ways that fit into the general pattern of natural material cycles. So we need to learn to return the materials we use to material cycles, or in other words, to **recycle** them. With the growth of recycling industries and technologies, we are beginning to meet that challenge, beginning to come to terms with the chemical realities of life.

In order to appreciate the scope of the recycling challenge, we will examine how we prepare a specific product from raw materials, how we use that product, and how it can be successfully recycled. As we do this, we will also come to appreciate the usefulness of chemical equations and learn more about the chemical mathematics involved in making best use of these equations. What product should we use to examine recycling in more depth? Many people associate recycling with aluminum cans, so we will look at the exploitation and reuse of aluminum (Figure 4.10).

Figure 4.10

A. Aluminum has many industrial uses. It is a favored building material because it is both strong and lightweight.

B. Aluminum finds its way into many commercial and household products, including the most notable: aluminum foil.

4.3 | Using and Recycling Aluminum

Just about any manufactured product can be recycled, given enough time and money. The materials of greatest concern, however, are those which we use most and are most practical with which to deal. These are commonly divided into four groups: aluminum, glass, plastics, and paper. Before any material can be recycled, it must first be processed from raw materials and used to create some product. As we look at the preparation of our product (the aluminum can), the use of chemical equations and mathematics to understand the processes will become ever more important.

Making the Can: An Introduction to Stoichiometry

Aluminum, a Group 3A element, is the third most abundant element in the Earth's crust, behind silicon and oxygen. It is used in such varied products as tuna fish cans, bicycle parts, and military tanks. Aluminum is never found naturally in its free elemental state because the metal is fairly reactive. Rather, it exists bound within compounds such as **bauxite,** which is a crystalline mixture of *hydrated aluminum oxide* (essentially Al_2O_3 combined with variable amounts of water), and other metal oxides including oxides of iron, silicon, and titanium.

The method by which aluminum is produced is named after two scientists, Charles Martin Hall of the United States and Paul Heroult of France, who in 1886 separately devised what we now call the **Hall-Heroult process.** Aluminum oxide, which has a melting point of 2030°C, is placed into a bath of molten cryolite (Na_3AlF_6). The resulting liquid has a melting point of roughly 1000°C. As you can see in Figure 4.11, carbon rods are placed into the molten mixture and electricity is passed through the rods, supplying the energy to power a chemical reaction between the rods themselves and the molten mixture. The equation for the overall reaction is

$$2Al_2O_3(l) + 3C(s) \rightarrow 4Al(l) + 3CO_2(g)$$

The molten aluminum settles at the bottom of the tank because it has a higher density than the molten reactants; the carbon dioxide bubbles off. You may infer from the equation that the carbon rods are eaten away as a result of the reaction, and, in fact, they must be replaced from time to time. The aluminum obtained from this process is more than 99% pure and can be further refined for special applications to greater than 99.9995% purity.

The advantages of using aluminum as a material are its abundance and its relative ease of preparation. It can be argued that because bauxite is a readily available substance, we should not be concerned with recycling aluminum. Yet the energy requirement to isolate aluminum is staggering, so recycling is an attractive option. In fact, obtaining 1000 kg (a metric ton or megagram) of aluminum from bauxite requires the energy equivalent of 120,000 kg of coal! In 1989, 86 billion aluminum cans were used in the United States alone, of which 60% (about

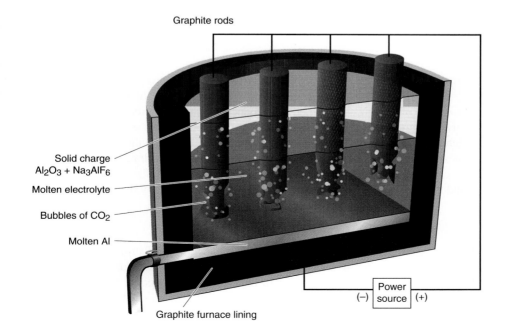

Graphite rods

Solid charge
Al_2O_3 + Na_3AlF_6

Molten electrolyte

Bubbles of CO_2

Molten Al

Graphite furnace lining

Power
source
(−)   (+)

Figure 4.11 This diagram shows the components of the Hall-Heroult process for the production of aluminum. Source: Silberberg, CHEMISTRY: THE MOLECULAR NATURE OF MATTER & CHANGE, 2nd edition. McGraw-Hill.

52 billion) were recycled. By 1998, over 102 billion aluminum cans were produced, with about 56 billion (56%) being recycled. The percent recycled has stayed fairly constant for the past decade, although the recycling rate in states that have a deposit on aluminum cans is significantly higher (80%) than those that do not (46%).

Let's put the numbers in a more recognizable framework, by asking:

- How much aluminum oxide is required to make one aluminum can?
- How much energy is needed to form that same aluminum can?

The answers to these questions can be obtained from the equation of the Hall-Heroult reaction. To extract the answers from the equation, however, requires some basic **stoichiometry,** a study of the quantities involved in chemical reactions—specifically the relationship between quantities of the reactants and products.

consider this:

Hall-Heroult Process

The Washington Monument, shown in Figure A, is capped with a small aluminum pyramid. The 2.85-kg pyramid, measuring 22.6 cm high and 13.9 cm wide, was attached in a "capping ceremony" on December 6, 1884. This was two years before the Hall-Heroult process, and aluminum had to be made via the reaction of aluminum chloride with sodium,

$$Al_2Cl_6(g) + 6Na(l) \rightarrow 2Al(s) + 6NaCl(s)$$

This was a difficult reaction to do, and so the total U.S. aluminum production in 1884 was 3.6 tons, about 1/10 that of silver production. Aluminum, even 97.5% pure as in the monument cap, was considered a precious metal, costing about $1 per ounce ($35/kg).

Two years later, the Hall-Heroult process allowed the mass manufacturing of nearly pure aluminum, and the price dropped substantially so that today refined aluminum costs about 75 cents per pound ($1.70/kg).

Figure A The Washington Monument capped with "precious" aluminum.

Source: National Archives.

Stoichiometry (pronounced stoik-ee-AH-metry) is a word created from the Greek "stoicheon" (meaning element) and "metron" (meaning measure). It is a means of determining the relative *amounts* of materials consumed and produced in chemical processes. Stoichiometry is not merely of interest to chemists. Materials cost money. Some processes produce by-products that pollute, requiring more money to be spent to avoid or deal with the pollution. So difficult social questions arise that involve issues of material supply, pollution, and money, and these questions often demand that difficult social choices be made. Dealing with such issues, which lie beyond the chemistry, requires some knowledge of the chemistry behind them.

Atomic Masses and Formula Masses

Stoichiometry is based on the Laws of Conservation of Mass and Definite Proportions. They enable us to use the number of atoms and their masses on the reactant side of a chemical equation to determine the numbers and masses of the atoms on the product side of that equation. Because chemical equations involve compounds as well as individual elements, we need a special unit of measure for these. A **formula unit** of a compound contains the atoms in the compound in the amounts indicated by the compound's formula. For covalently bonded compounds, it represents an individual molecule of the compound. Ionic substances, however, are not composed of molecules (so we cannot talk of a molecule of sodium chloride, NaCl, for example). We can, however, talk of a formula unit of NaCl (which is one sodium ion and one chloride ion). So the term formula unit is a general term applicable to any type of compound. To calculate the mass of a formula unit of Al_2O_3, we need to know the masses of aluminum and oxygen atoms.

We learned in Chapter 1 that the exceptionally small masses of individual atoms can be expressed in atomic mass units (amu). Since both protons and neutrons have a mass of about 1 amu, and electrons have a mass of only 0.00055 amu, the mass of an atom in amu is approximately equal to the number of protons plus neutrons in the nucleus. We can ignore the tiny mass of the electrons. This means atoms and the ions derived from them have the same masses, for the purposes of chemical calculations. Different isotopes of each atom exist, however, having different numbers of neutrons and therefore different masses. Fortunately, the proportions in which the different isotopes occur are virtually constant. This means that for each element, we can work out the *average mass* of one atom of the element. We call this value the **relative atomic mass** or **atomic mass** of the element. Table 2.2 lists the atomic masses of all the elements. You will need to refer to it frequently, when performing chemical calculations.

Twenty elements (Be, F, Na, Al, P, Sc, Mn, Co, As, Y, Nb, Rh, I, Cs, Pr, Tb, Ho, Tm, Au, and Bi) are monoisotopic; that is, only one isotope exists in natural samples of that element. For these, the molar mass is simply the mass of an individual atom expressed in grams. All other elements which are found in nature are a mixture of two or more isotopes, usually in a very constant isotopic ratio. For these elements, the atomic mass is really the average molar mass of all the atoms of that element. A good example is chlorine, which consists of two isotopes, ^{35}Cl and ^{37}Cl. In natural chlorine 75.77% of the atoms are ^{35}Cl having an isotopic mass of 34.968852 amu and 24.23% are ^{37}Cl having an isotopic mass of 36.965903. Therefore, the atomic mass equals the average of these atoms as shown,

$$atomic\ mass = 0.7577(34.968852) + 0.2423(36.965903)$$
$$= 26.496 + 8.957 = 35.453$$

Just as there are no families with 1.7 children, although that is reported as the average U.S. family, there are no chlorine atoms having a mass of 35.453 amu. Both of these values are average values, which are useful for calculations.

Just as each element has an atomic mass, each compound has a **formula mass,** which corresponds to the sum of the atomic masses of all the atoms in the formula. The formula mass is the mass of one formula unit, in other words. If the formula unit lists the number of atoms in a molecule of a covalently bonded compound, then the compound's formula mass equals its **molecular mass.** Remember, however, that the term molecular mass should not really be applied to ionic compounds (although it sometimes is) because ionic compounds do not contain any molecules.

Having set these definitions in place, we are ready to calculate the formula mass of Al_2O_3:

Al atoms have a mass of 26.98 amu
O atoms have an average mass of 16.00 amu*

Aluminum oxide, Al_2O_3, with two aluminum and three oxygen atoms in its formula has a total formula mass of

$$2 \times 26.98\ amu + 3 \times 16.00\ amu = 101.96\ amu\ per\ Al_2O_3\ formula\ unit$$
$$(so\ the\ formula\ mass = 101.96\ amu)$$

We can round this off to 102 amu, unless we need to be very accurate.

---

*Amu are intended as a *relative* measure of mass. They indicate *how much* more mass one type of atom has than another. Oxygen (16 amu) is four times as heavy as helium (4 amu). Amu are a lot easier to deal with than their actual mass equivalent of 1.66×10^{-27} kg.

The other reactant in the production of aluminum, namely, carbon, has an atomic mass of 12.01 amu.

Using the values for atomic masses in Table 2.2, you should always find that the total mass of the products of a chemical reaction equals the total mass of the reactants (although some minor discrepancies may arise due to rounding off when listing the values in the table of atomic masses).

The second-to-last line of the calculation in Exercise 4.2 tells us the mass of aluminum that can be derived from a given mass of aluminum oxide, provided the masses are quoted in atomic mass units. To be specific, 204 amu of Al_2O_3 will generate 108 amu of aluminum. Keep in mind that these masses represent the whole-number ratios of the different atoms and molecules involved in the reaction. "Half a molecule" of aluminum oxide isn't going to react with "three quarters of an atom" of carbon to give us an atom of aluminum. Laboratories and

exercise 4.2

Formula Masses

Problem

Use the atomic masses of aluminum, carbon, and oxygen to calculate the formula mass of carbon dioxide and calculate the total mass of the products of the Hall-Heroult reaction equation. Compare the masses on either side of the equation. What can you conclude about the Law of Conservation of Mass, at the level of accuracy at which masses are measured in chemistry?

Solution

We've already established that Al_2O_3 has a formula mass of 101.96 amu. The atomic mass of aluminum is 26.98, the atomic mass of carbon is 12.01 amu, and the atomic mass of oxygen is 16.00 amu. We can calculate the formula mass of CO_2 as:

$$\begin{array}{ll} 1 \text{ C atom} & = 1 \times 12.01 \text{ amu} = 12.01 \text{ amu} \\ 2 \text{ O atoms} & = 2 \times 16.00 \text{ amu} = \underline{32.00 \text{ amu}} \\ & \hspace{3.5cm} 44.01 \text{ amu} \end{array}$$

Therefore CO_2 has a formula mass of 44.01 amu.

The equation for the production of aluminum is

$$2Al_2O_3 + 3C \rightarrow 4Al + 3CO_2$$

Totaling the masses on both sides of the equation gives

$2Al_2O_3$	+	$3C$	$\rightarrow$	$4Al$	+	$3CO_2$
2(101.96 amu)	+	3(12.01 amu)	$\rightarrow$	4(26.98 amu)	+	3(44.01 amu)
203.92 amu	+	36.03 amu	$\rightarrow$	107.92 amu	+	132.03 amu

So overall:

$$\textbf{239.95 amu} \quad \rightarrow \quad \textbf{239.95 amu}$$

The Law of Conservation of Mass states that the total mass of substances does not change during a chemical reaction. So we can use this law to check that chemical equations are balanced.

Check

The stoichiometry of this equation shows it to be a balanced equation, with the mass of combined quantities of the reactants equal to the combined quantities of the products.

factories, however, do not have equipment that will measure mass in atomic mass units. Instead, they use machines that weigh materials, giving results in grams. Before we can give any meaningful indication of the amount of aluminum oxide needed to make an aluminum can, we need to consider realistic, practical units of mass that will effectively allow us to count out the right numbers of whole atoms, molecules, and ions. We do this using a fundamental unit of chemical mathematics known as the *mole.*

Counting in Moles

As we have said, the machines used by chemists to measure mass do not register masses in atomic mass units, they usually register mass in grams (or kilograms or milligrams). Once we have a balanced chemical equation for a particular reaction, like the Hall-Heroult reaction, how can we relate a mass of a chemical in grams to the actual number of atoms, or molecules, or ions shown in that equation? In order to do this, we need to *convert* the atomic masses and formula masses the equation represents into grams. Ultimately, we need to know how many atomic mass units there are in one gram.

In Chapter 1 we learned that there are 6.02×10^{23} amu in 1 g (that is, 602,000,000,000,000,000,000,000). We can now use this fact to enable us to count out a set number of atoms by using an element's atomic mass. For instance, we know that, on average, an atom of carbon has a mass of 12.01 amu. If we measure out 12.01g of carbon, we have effectively counted out 6.02×10^{23} atoms of carbon. Mathematically the calculation looks like this:

$$12.01 \text{ g} \times \frac{6.02 \times 10^{23} \text{ amu}}{1 \text{ g}} \times \frac{1 \text{ atom C}}{12.01 \text{ amu}} = 6.02 \times 10^{23} \text{ atoms C}$$

We have used the ratio of 6.02×10^{23} amu in 1 g as a **conversion factor:** the amounts involved do not change but the way they are measured do. Now you can see how we can use the atomic mass of any element to give us 6.02×10^{23} atoms of that element. Substitute 16.00 grams of oxygen into the calculation and you get this result:

$$16.00 \text{ g} \times \frac{6.02 \times 10^{23} \text{ amu}}{1 \text{ g}} \times \frac{1 \text{ atom O}}{16.00 \text{ amu}} = 6.02 \times 10^{23} \text{ atoms O}$$

Chemists find it useful to compute the amounts of material in units called **moles** (abbreviation *mol*). A mole is defined as being equal to the number of atoms in 12 grams of carbon-12, or 6.02×10^{23} (Figure 4.12). In practical terms, a mole of any element is the mass, in grams, equal to the atomic mass, in amu. If one average atom of chlorine weighs 35.45 amu, then 1 mole weighs 35.45 g.

This enormous number, 6.02×10^{23}, is known as *Avogadro's number* (after the Italian physicist Amedeo Avogadro). You should not be surprised to see a name being given to a number. The number 12, for example, is also known as a dozen. The mole is the "chemist's dozen," the basic reference quantity of atoms, or molecules, or ions. The reason that the mole is so useful to chemists is simple: if we know the mass of anything (an atom, an ion, a molecule) in atomic mass units, we automatically know that the mass in grams of 1 mol of these things must have the same value. This follows automatically from the fact that there is 1 mol of amu in 1 g.

One of the great uses of the mole is that it enables chemists to work out the masses of different chemicals that will contain *equal numbers of atoms, molecules, or ions.* It is easy to weigh out equal *masses* of different chemicals, but 10 g of carbon will not contain the same number of atoms as 10 g of sulfur because the carbon and sulfur atoms have different atomic masses (12.01 amu for carbon and 32.07 amu for sulfur). If we weigh out 12.01 g of carbon and 32.07 g of sulfur, however, we know we have *equal numbers of atoms* of each (6.02×10^{23} in each case).

Chemical reactions involve reactants participating in definite proportions of atoms, molecules, or ions. Using the mole makes working out the appropriate proportions very easy.

- 1 mol of anything $= 6.02 \times 10^{23}$ things
- 1 mol of carbon atoms $= 6.02 \times 10^{23}$ carbon atoms

Figure 4.12 Moles of several common elements and compounds, from left to right: calcium carbonate, oxygen (the capacity of the balloon), copper, and water.

exercise 4.3

Grams to Atoms

Problem

How many atoms of aluminum are there in a 16.0-g aluminum can?

Solution

The atomic mass of aluminum is 26.98 amu, therefore 26.98 g of aluminum contains 1 mol of aluminum atoms. Comparing one mass to the other establishes a ratio:

16.0 g of aluminum contains $\dfrac{16.0}{26.98} = 0.593$ mol of aluminum atoms and

$$0.593 \text{ mol} \times (6.02 \times 10^{23}) \text{ Al atoms} = \mathbf{3.57 \times 10^{23} \text{ Al atoms}}$$

In other words there are 357,000,000,000,000,000,000,000 aluminum atoms (approximately) in the soda can you hold in your hand, and that is a lot of atoms!

Check

The ratio of atoms to atoms should be equivalent to the original ratio of grams to grams.

$$\frac{16.0\text{g Al}}{27\text{g Al}} = \frac{3.57 \times 10^{23}}{6.02 \times 10^{23}}$$

$$0.59 = 0.59$$

- 1 mol of aluminum atoms = 6.02×10^{23} aluminum atoms
- 1 mol of aluminum oxide = 6.02×10^{23} formula units of aluminum oxide
- 1 mol of frosted flakes = 6.02×10^{23} frosted flakes

The next examples will help you get used to working with moles and grams. If you are a bit uncertain about working with the chemical units involved, see the Appendix at the end of this chapter for additional help.

The value 159.70 g is the formula mass of Fe_2O_3, specifically the **gram formula mass.** It is also referred to as the **molar mass:** the mass in grams of 1 mol of any element or compound. In solving Exercise 4.4, we have taken a quantity of iron(III) oxide measured in amu and changed the measure to its gram equivalent, which in this case is a convenient 1-to-1 ratio. Chemists use a method of canceling like units of measure (dimensions) to help streamline their calculations, which we shall use in Exercise 4.5. The method, called **dimensional analysis,** is more fully discussed in the Appendix at the end of this chapter.

We suggest that you estimate the answer in all problems that require numerical solutions. It really helps in problem solving! If you do not know how to perform calculations using scientific notation, like those in Exercise 4.5(c), see the Appendix at the end of this chapter.

We can interconvert between grams and moles of a given substance. But if we want to make the critical jump from grams or moles of one substance to grams or moles of another, we must return to our chemical equation for the necessary information. This is where stoichiometry comes in.

A useful strategy for solving stoichiometry problems is summarized in Figure 4.14, the simple "mole map." This shows that the mole is the essential quantity in chemical calculations. Using the map, we can get where we want to go by using conversion factors as a bridge. We can use the molar mass to go from mass to moles or back. We can use Avogadro's number to go from moles to number of particles and back again. Notice that there is no bridge directly from mass to number of particles, so we must use a more circuitous route. We'll use the mole map in Figure 4.14 to solve Exercise 4.7.

The mass (and therefore weight on Earth) of 1 mol of any elementary entity (such as a particular atom, molecule, or ion) has the same numerical value as the mass of one of the entities in atomic mass units, but is expressed in grams rather than atomic mass units.

- 1 atom of carbon-12 has a mass of 12 amu, so 1 mol of carbon-12 atoms has a mass of 12 grams
- 1 molecule of water has a mass of 18 amu, so 1 mol of water molecules has a mass of 18 grams and so on.

exercise 4.4

Practice with Formula Masses

Problem

Another important process involving aluminum is the thermite reaction. It is a powerful heat generator that produces molten iron from the reaction of iron(III) oxide with powdered aluminum. The balanced equation for the reaction is

$$Fe_2O_3 + 2Al \rightarrow 2Fe + Al_2O_3$$

This reaction is likely to be used in space as the heat source when welding together parts of a future space station. As Figure 4.13 shows, the reaction is violently hot; yet it will only work well if the proper amounts of iron oxide and aluminum are present. The first step toward getting the proper amounts is to calculate the formula masses of aluminum and iron oxide. The formula mass for aluminum is its atomic mass (26.98) since it exists as the uncombined element. Calculate the formula mass of the iron(III) oxide.

Solution

The formula mass of iron oxide is calculated as in Exercise 4.2, by adding together the atomic masses of the atoms present, all multiplied by the number of atoms present in the formula Fe_2O_3:

$$
\begin{aligned}
2 \text{ Fe atoms} &= 2 \times 55.85 \text{ amu} = 111.70 \text{ amu} \\
3 \text{ O atoms} &= 3 \times 16.00 \text{ amu} = \underline{\;48.00 \text{ amu}} \\
\text{Formula mass of } Fe_2O_3 &= \mathbf{159.7 \text{ amu}}
\end{aligned}
$$

This means that 1 mole of Fe_2O_3 will weigh 159.7 g.

Figure 4.13 The thermite reaction.

exercise 4.5

Converting Moles to Grams

Problem

Calculate the mass in grams of each of these:
(a)   35.2 mol of H_2O
(b)   0.0430 mol of $K_2Cr_2O_7$
(c) 9.18×10^{-8} mol of CCl_4

Solution

You must know (or calculate) the molar mass of each substance in order to solve the problem.

(a)   molar mass of H_2O is

$$2 \times \frac{1.01 \text{ g H}}{1 \text{ mol H}} + 1 \times \frac{16.00 \text{ g O}}{1 \text{ mol O}} = 18.02 \text{ g/mol } H_2O$$

So there are 18.02 g of H_2O per mole (the molar mass), which we can write as

$$\frac{18.02 \text{ g } H_2O}{1 \text{ mol } H_2O}$$

exercise 4.5 (continued)

To find the mass of 35.2 mol of H_2O, we use the molar mass as a conversion factor and multiply. In the process, we are able to cancel out the dimensional unit "mol H_2O":

$$35.2 \text{ mol } H_2O \times \frac{18.02 \text{ g } H_2O}{1 \text{ mol } H_2O} = \textbf{634 g } H_2O$$

(b)   The molar mass of $K_2Cr_2O_7$ is

$$(2 \times 39.10) + (2 \times 52.00) + (7 \times 16.00) = \frac{294.20 \text{ g } K_2Cr_2O_7}{1 \text{ mol } K_2Cr_2O_7}$$

therefore

$$0.0430 \text{ mol } K_2Cr_2O_7 \times \frac{294.20 \text{ g } K_2Cr_2O_7}{1 \text{ mol } K_2Cr_2O_7} = \textbf{12.6 g } K_2Cr_2O_7$$

Note that we completely label both the numerator and denominator of every molecule! This helps to neatly keep track of cancellations. Neatness counts—no kidding!

(c)   The molar mass of CCl_4 is $12.01 + (4 \times 35.45) = 153.8$ g/mol. Here we show how you can further simplify the conversion factor by using only units of measure and not the compound name; the end result is the same. (As a general rule, however, you should always include the substance name when you write conversion factors.)

$$9.18 \times 10^{-8} \text{ mol } CCl_4 \times \frac{153.8 \text{ g}}{1 \text{ mol}} = 1.41 \times 10^{-5} \text{ g } CCl_4$$

Check

Do the answers make sense? The key to determining if your answers are meaningful is to determine before you actually solve the problem arithmetically about what answers you expect. In part (a), for example, we were asked to find the mass of 35.2 mol of water. Roughly how much water is this? One mole, 18 g, of water is about 1 tablespoon. Therefore, 35 tablespoons of water is a little over a pint (there are 32 tablespoons in a pint). Our answer of 634 g (= 634 mL) is a little more than a pint (473 mL). Such estimations will not yield accurate answers. But they will tell us when we have made a serious blunder.

Figure 4.14 Crossing the mole bridge.

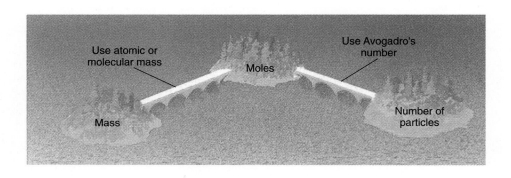

exercise 4.6

Converting Grams to Moles

Problem

How many moles of aluminum are there in

(a)   an aluminum can that weighs 15.4 g?
(b)   a sample of aluminum oxide weighing 3.30 nanograms (ng)?

Solution

The first step in this, and most stoichiometry problems, is to determine the atomic or molar mass of each substance. The atomic mass of aluminum can be read directly from the periodic table and, translated to grams, is 26.98 g/mol. You must calculate the molar mass of aluminum oxide:

$$\text{molar mass } Al_2O_3 = (2 \times 26.98) + (3 \times 16.00) = 101.96 \text{ g/mol}$$

Using dimensional analysis, we can get our answers as:

(a) $\text{mol Al} = \dfrac{1 \text{ mol Al}}{26.98 \text{ g Al}} \times 15.4 \text{ g Al} = \textbf{0.571 mol Al}$

(b) $\text{mol Al} = \dfrac{1 \text{ mol } Al_2O_3}{101.96 \text{ g } Al_2O_3} \times 3.30 \text{ ng } Al_2O_3 \times \dfrac{1 \times 10^{-9} \text{ g } Al_2O_3}{1 \text{ ng } Al_2O_3} \times \dfrac{2 \text{ mol Al}}{1 \text{ mol } Al_2O_3}$

$= \textbf{6.47} \times \textbf{10}^{-11} \textbf{ mol Al}$

exercise 4.7

Atoms to Mass

Problem

Find the mass, in grams, of 7.5×10^{27} atoms of aluminum.

Solution

A good problem-solving technique takes into account three key issues. Where are we going, where are we coming from, and how do we get there? That is what a map is for.

1. *Where are we going?* This is another way of asking, "What is it we want to learn?" In this problem, we are asked to determine the mass of aluminum.
2. *Where are we coming from?* We are coming from the knowledge of a particular number of atoms of aluminum.
3. *How do we get there?* "There" in this case is the mass of all those aluminum atoms. Start with the number of atoms (particles), use Avogadro's number to get to moles, and then use the molar mass to get to mass. Note that traveling this path (and back again!) is possible on our mole map.

$$\text{mass Al} = 7.5 \times 10^{27} \text{ atoms Al} \times \dfrac{1 \text{ mol Al}}{6.02 \times 10^{23} \text{ atoms Al}} \times \dfrac{26.98 \text{ g Al}}{1 \text{ mol Al}}$$

$= \textbf{3.36} \times \textbf{10}^5 \textbf{ g Al}$

exercise 4.7 (continued)

Check

Does the answer make sense? We see by the first step in the calculation that this is slightly larger than 1×10^4 mol of aluminum. (Can you prove this?) Each mole of aluminum weighs about 27 g, so the final answer should be slightly larger than 27×10^4 g. This is the same as 2.7×10^5 g, and our answer is slightly larger than this. Answer confirmed!

Oftentimes, we assess what the approximate answer to a problem should be before doing the problem. This way, we know right away if our answer "makes sense."

Stoichiometric Calculations

Remember our question of interest, "How much aluminum oxide is needed to make one aluminum can?" The answer lies in the Hall-Heroult equation.

Chemical equations indicate the *ratios* in which reactants react and in which products are produced. They can be interpreted as indicating the actual numbers of atoms, molecules, or ions that react together and are generated as products. More generally, however, they indicate the number of *moles* of reactants and products concerned. What they never do is directly indicate the masses of substances involved.

This means that the proper interpretation of the equation

$$2Al_2O_3 + 3C \rightarrow 4Al + 3CO_2$$

is that "2 mol of aluminum oxide are required to react with 3 mol of carbon to produce 4 mol of aluminum and 3 mol of carbon dioxide." We can also say that in this reaction, 2 mol of Al_2O_3 form 4 mol of Al. Also, 3 mol of C form 3 mol of CO_2. It is also correct to say that 3 mol of CO_2 are formed for every 4 mol of Al formed. These statements all quote valid *mole-mole ratios*. The simplest way to summarize all the relevant mole-mole ratios is simply to write in the mole numbers below the chemicals, as follows:

$$2Al_2O_3 \quad + \quad 3C \quad \rightarrow \quad 4Al \quad + \quad 3CO_2$$
$$\text{2 mol} \quad + \quad \text{3 mol} \quad \rightarrow \quad \text{4 mol} \quad + \quad \text{3 mol}$$

Another way of expressing the mole-mole ratios, which is very useful when solving stoichiometry problems, is as **mole-mole conversion factors.** One example of a mole-mole conversion factor for the Hall-Heroult reaction is

$$\frac{4 \text{ mol Al}}{2 \text{ mol Al}_2O_3}$$

This can be read as "there are 4 mol of Al per 2 mol of Al_2O_3 in the equation." Table 4.1 lists all the possible conversion factors for this equation.

We now have all the tools we need to answer this question: "How much aluminum oxide (in grams) would you need to produce one aluminum can?" A typical aluminum can has a mass of 16 g, which we will assume (for the sake of simplicity) is 100% aluminum. We have emphasized that chemical equations summarize the mole-mole ratios between the substances concerned, so we must first convert the figure of 16 g of aluminum into a number of moles of aluminum.

Remember that the balanced equation and corresponding mole-mole ratios can be summarized as

$$2Al_2O_3 \quad + \quad 3C \quad \rightarrow \quad 4Al \quad + \quad 3CO_2$$
$$\text{2 mol} \quad + \quad \text{3 mol} \quad \rightarrow \quad \text{4 mol} \quad + \quad \text{3 mol}$$

exercise 4.8

Practice with Mole-Mole Conversion Factors

Problem

During the commercial production of aluminum, a side reaction that occurs is the reaction of water with aluminum fluoride, AlF_3, which is present in the molten mixture. The balanced equation for this side reaction is

$$2AlF_3 + 3H_2O \rightarrow Al_2O_3 + 6HF$$

The hydrogen fluoride (HF) that is formed must be "neutralized" (made to lose its acid properties, in this case) so that this smokestack gas will not damage crops near the manufacturing plant.

Decide whether any of these conversion factors is correct for the side-reaction equation:

(a) $\dfrac{2 \text{ mol AlF}_3}{6 \text{ mol HF}}$
(b) $\dfrac{3 \text{ mol Al}_2O_3}{1 \text{ mol H}_2O}$
(c) $\dfrac{6 \text{ mol H}_2O}{3 \text{ mol HF}}$

(d) $\dfrac{1 \text{ mol Al}_2O_3}{2 \text{ mol AlF}_3}$
(e) $\dfrac{2 \text{ mol HF}}{6 \text{ mol AlF}_3}$

Solution

A cardinal rule when using conversion factors is that *the numbers in the equation travel with the atoms, molecules, or ions they refer to.* This means that the "2" stays with the AlF_3, the "3" stays with the H_2O, the "1" stays with the Al_2O_3, and the "6" stays with the HF. This is because changing any of these numbers would unbalance the equation, and the conversion factors must be the ones that refer to the balanced equation. This should enable you to appreciate that the correct conversion factors are **(a)** and **(d).**

Table 4.1

Conversion Factors for the Hall-Heroult Reaction

Aluminum Oxide	Carbon	Aluminum	Carbon Dioxide
$\dfrac{2 \text{ mol Al}_2O_3}{3 \text{ mol C}}$	$\dfrac{3 \text{ mol C}}{2 \text{ mol Al}_2O_3}$	$\dfrac{4 \text{ mol Al}}{3 \text{ mol CO}_2}$	$\dfrac{3 \text{ mol CO}_2}{4 \text{ mol Al}}$
$\dfrac{2 \text{ mol Al}_2O_3}{4 \text{ mol Al}}$	$\dfrac{3 \text{ mol C}}{4 \text{ mol Al}}$	$\dfrac{4 \text{ mol Al}}{2 \text{ mol Al}_2O_3}$	$\dfrac{3 \text{ mol CO}_2}{2 \text{ mol Al}_2O_3}$
$\dfrac{2 \text{ mol Al}_2O_3}{3 \text{ mol CO}_2}$	$\dfrac{3 \text{ mol C}}{3 \text{ mol CO}_2}$	$\dfrac{4 \text{ mol Al}}{3 \text{ mol C}}$	$\dfrac{3 \text{ mol CO}_2}{3 \text{ mol C}}$

One very useful general approach to solving such stoichiometry problems is to "map out" the conversions that must be done to reach your answer, and then use the appropriate conversion factors to work toward the answer. To determine the mass of aluminum oxide required for 16 g of aluminum, we need to multiply 16 g by the conversion factor that changes a mass in grams of aluminum into a number of moles, then multiply the result by the conversion factor that converts the number of moles of aluminum into a number of moles of Al_2O_3, then finally multiply that result by the conversion factor that converts the number of moles of aluminum oxide into grams. The appropriate strategy and actual calculation is shown next.

exercise 4.9

Satisfying the Demand

Problem

In 1994, 101 billion aluminum cans were produced, weighing 1.56×10^9 kg (more than 1 million metric tons). How much aluminum oxide is required to satisfy this annual demand for aluminum cans (assuming no recycling of used aluminum)? Express the answer in kg, and in metric tons of Al_2O_3.

Solution

The problem is an extension of the one we just finished. We know the number of grams of Al_2O_3 needed to form one can. We need only multiply the answer by 101 billion (1.01×10^{11}) to find the mass of a year's supply of Al_2O_3. We must remember, however, to convert the answer to kg and metric tons! Our strategy is:

$$\frac{30 \text{ g Al}_2\text{O}_3}{1 \text{ can}} \xrightarrow{\frac{1.01 \times 10^{11} \text{ can Al}}{1 \text{ year}}} \frac{\text{g Al}_2\text{O}_3}{\text{year}} \xrightarrow{\frac{1 \text{ kg Al}_2\text{O}_3}{1000 \text{ g Al}_2\text{O}_3}} \frac{\text{kg Al}_2\text{O}_3}{\text{year}}$$

This can be written as a single equation:

$$\text{kg Al}_2\text{O}_3 = \frac{30 \text{ g Al}_2\text{O}_3}{1 \text{ Al can}} \times \frac{1.01 \times 10^{11} \text{ Al can}}{1 \text{ year}} \times \frac{1 \text{ kg Al}_2\text{O}_3}{1000 \text{ g Al}_2\text{O}_3}$$

$$= 3.0 \times 10^9 \text{ kg Al}_2\text{O}_3/\text{year}$$

Now convert this to metric tons:

$$\text{metric tons Al}_2\text{O}_3 = 3.0 \times 10^9 \text{ kg Al}_2\text{O}_3 \times \frac{1 \text{ metric ton Al}_2\text{O}_3}{1000 \text{ kg Al}_2\text{O}_3}$$

$$= 3.0 \times 10^6 \text{ metric tons Al}_2\text{O}_3/\text{year}$$

$$16 \text{ grams Al} \xrightarrow{\frac{1 \text{ mol Al}}{26.98 \text{ g}}} \text{moles Al} \xrightarrow{\frac{1 \text{ mol Al}_2\text{O}_3}{2 \text{ mol Al}}} \text{moles Al}_2\text{O}_3 \xrightarrow{\frac{101.96 \text{ g Al}_2\text{O}_3}{1 \text{ mol Al}_2\text{O}_3}} 30 \text{ g Al}_2\text{O}_3$$

Setting out to use the correct conversion factors takes a bit of getting used to, but it is well worth the effort. Once you become comfortable with this technique, most stoichiometry problems become fairly simple. (Honest!) You will find more details of this approach to solving stoichiometry problems in the Appendix (starting on page 169); and you will get a lot more practice with it as you read through the text.

There are many ways to do stoichiometry problems, and you might or might not choose to adopt our approach. Whatever approach you do use, the last step in your problem-solving, after double-checking the math, is to decide if the answer makes sense. Does our answer of 30 g make sense here? Aluminum oxide is heavier than aluminum, so it makes sense that we would need to start with more than 16 g of aluminum oxide to make an aluminum can. If we obtained an answer that suggested less than 16 g of aluminum oxide could produce a 16-g aluminum can, we would know there must be something wrong.

Bauxite, the mineral that is used as a source of aluminum, contains well under 50% Al_2O_3. This means that the number of metric tons of bauxite needed to make aluminum cans would actually be much higher than the number you just calculated. Recycling would seem to be a sensible option! There are also other factors to consider, such as availability and renewability of resources (we can grow new trees, but we cannot grow aluminum). Another crucial factor is the amount of energy required to produce all the aluminum compared with the amount required to recycle aluminum already produced and used.

exercise 4.10

The Energy Needed to Manufacture One Aluminum Can

Problem

The joule (J) is the unit of energy defined by the Système Internationale (SI). (We introduced you to SI units in Chapter 1; the Appendix to this chapter includes a further discussion of the SI system.) Roughly 280,000 megajoules of energy are required to produce 1000 kg of aluminum (1 MJ = 1 × 10⁶ J). How much energy is needed per can, assuming 15.4 g Al per 16-g can?

Solution

This was one of the questions that was raised at the beginning of our discussion on stoichiometry. Again, you can use the strategy of mapping out the steps and then translating into a single equation in which the molecules are arranged so that you end up with what you want:

$$\frac{2.8 \times 10^5 \text{ MJ}}{1000 \text{ kg Al}} \xrightarrow{\frac{1 \text{ kg Al}}{1000 \text{ g Al}}} \frac{\text{MJ}}{\text{gram Al}} \xrightarrow{\frac{15.4 \text{ g Al}}{1 \text{ can Al}}} \frac{\text{MJ}}{\text{can Al}} \xrightarrow{\frac{1 \times 10^6 \text{ J}}{1 \text{ MJ}}} \frac{\text{J}}{\text{can Al}}$$

$$\text{J/can} = \frac{2.8 \times 10^5 \text{ MJ}}{1000 \text{ kg Al}} \times \frac{1 \text{ kg Al}}{1000 \text{ g Al}} \times \frac{15.4 \text{ g Al}}{1 \text{ can Al}} \times \frac{1 \times 10^6 \text{ J}}{1 \text{ MJ}}$$

$$= 4.3 \times 10^6 \text{ J/can}$$

Check

That seems like a large number, but we need to put it into perspective. A joule is a tiny unit of energy. The energy contained in 4.3 × 10⁶ J is about the same as you would get from digesting a large banana split or a double cheeseburger; it is the equivalent of burning the amount of gasoline that would half-fill an aluminum soda can.

We have not looked in detail at any calculations comparing the energy costs of the initial production of aluminum with the energy costs of recycling, but we have covered the key methods of calculation that allow these kinds of more detailed analyses to be performed.

4.4 | The Recycling Process

You have already seen that recycling is not a human invention; it is the process by which nature automatically cycles the atoms of the Earth through many different forms. It has also been applied by humans since long before the rise of interest in ecological issues. For centuries, farmers have given nature a helping hand by plowing unusable crop residues into the soil, where they serve as energy sources for microorganisms that break them down into compounds that can fertilize plants. For the same reason, farmers use animal manure as a natural fertilizer. If you have ever had a garden in the backyard or in the neighborhood, perhaps you have piled together grass clippings and leaves for the same purpose. This is called **composting** and is another way of recycling nutrients back to the soil (Figure 4.15). In all such cases the natural materials returned to the soil become decomposed by the action of bacteria and fungi. This process releases the chemicals in a form that can nourish new plant growth. So recycling is both nature's way and humanity's traditional way of maintaining a sustainable world. Only in the past two centuries did recycling temporarily go out of fashion as modern industrialized

exercise 4.11

The "Energy Cost" of Diapers

Problem

A 2000 Toyota Corolla can travel 32 miles per gallon of gasoline in combined city and highway driving. If the energy equivalent of 0.25 L of gasoline is required to manufacture one disposable diaper, how many miles could a Toyota travel using the energy equivalent of a 1-year supply of disposable diapers (let's say 1000 diapers)? Note that 1 gal = 3.785 L.

1000 diapers = ? miles

Solution

One of the possible methods is given next.

$$\frac{32 \text{ miles}}{1 \text{ gallon}} \xrightarrow{\frac{1 \text{ gallon}}{3.875 \text{ L}}} \frac{\text{miles}}{\text{L}} \xrightarrow{\frac{0.25 \text{ L}}{\text{diaper}}} \frac{\text{miles}}{\text{diaper}} \xrightarrow{\frac{1000 \text{ diapers}}{\text{year}}} \frac{\text{miles}}{\text{year}}$$

$$\text{miles/year} = \frac{32 \text{ miles}}{1 \text{ gallon}} \times \frac{1 \text{ gallon}}{3.785 \text{ L}} \times \frac{0.25 \text{ L}}{1 \text{ diaper}} \times \frac{1000 \text{ diapers}}{1 \text{ year}}$$

$$= \textbf{2114, which rounds to 2100 miles/year}$$

lifestyles developed. It never disappeared completely, even in industrialized nations, but its importance is slowly being recognized again.

How Is Recycling Done?

The goal of recycling is to recover as much of a particular material as possible for future use. The general steps involved are:

- Collecting trash, either directly from homes, offices, and factories, or from a municipal recycling center.
- Separating "recyclables" (material we can recycle) from the rest of the trash. This can be done at home, at municipal centers, or in certain cases, by recycling industries.

Figure 4.15 Composting.

Repeat layering as below

Kitchen scraps

Alfalfa meal (or alfalfa based cat liter)

2" to 3" manure or more grass clippings

Soil

6" to 10" layer green plant material and grass clippings

Figure 4.16 Many communities provide recycling centers where residents may bring sorted trash. Glass, newspaper, aluminum, and plastics are separated into bins to facilitate the process.

- Cleaning the recyclables to make sure that they are free of labels, food, and so forth.
- Processing the material so that it can be reused.

Figure 4.16 shows one aspect of the process.

Recycling Aluminum

So-called "aluminum" soda cans are not made of pure aluminum. They are really composed of a mixture of aluminum, magnesium, manganese, iron, silicon, and copper. A *homogeneous* (evenly mixed) mixture of metals is known as an **alloy.** The alloy used in the lid of the can contains more magnesium than the body. After the cans are manufactured, they are given a thin plastic coating on the inside so that the metal of the can does not react with the beverage it contains. The brand name and other labeling information is then painted on the outside. So there are many more chemicals than aluminum present in the cans that end up in the trash bin, ready for recycling.

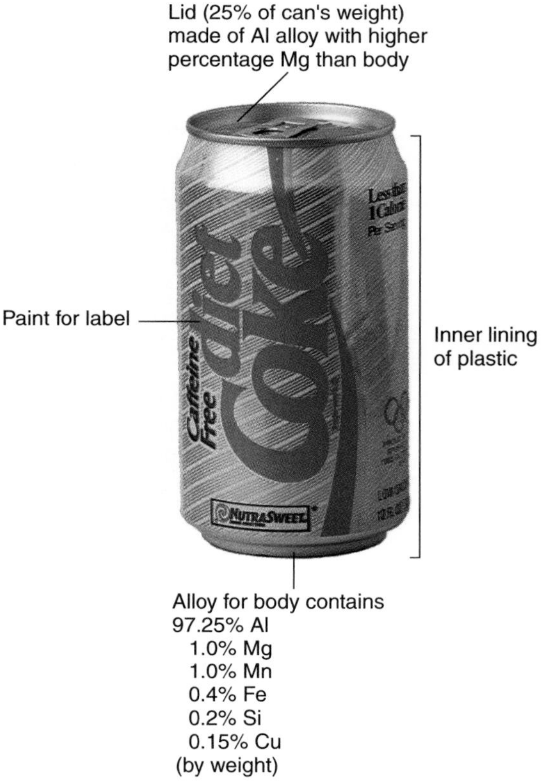

To recycle aluminum is not as simple as merely putting the aluminum cans in a furnace and cooling the molten metal in the shape of a new can, because all the different substances that make up the can must be dealt with separately. Despite this complication, recycling a can uses less than 10% of the energy that would be needed to manufacture aluminum cans from bauxite ore; and the basic material for the recycled can is available for free (although transport and other handling costs must be met).

The cans that arrive for recycling must be dried because moisture is explosive when in contact with molten aluminum, which is generated during the recycling process. The cans are melted in a "delaquering" furnace, which removes the paint and plastic. This generates a mixture with more magnesium than is appropriate for making the body of new cans. This is because the lid of the can (which accounts for 25% of the can's total weight) is made up of an alloy that has relatively more magnesium than the body. So a small amount of new aluminum must be

Lid (25% of can's weight) made of Al alloy with higher percentage Mg than body

Paint for label

Inner lining of plastic

Alloy for body contains
97.25% Al
1.0% Mg
1.0% Mn
0.4% Fe
0.2% Si
0.15% Cu
(by weight)

Figure 4.17 Aluminum recycling.

added. The molten aluminum/magnesium alloy can then be processed into new cans (Figure 4.17).

Recycling Silver—An Introduction to Oxidation and Reduction

For both economic and safety reasons, many industrial and academic chemistry laboratories recycle materials such as silver. The recycling saves money because silver, for example, can cost several hundred dollars per kilogram. Safe recycling prevents it from going into the "waste stream" where it can enter lakes and rivers and find its way into a variety of living organisms, for which it is a toxic substance.

Recycling silver converts silver ions in solution to silver metal. One way to do this is to put a piece of copper metal into a beaker containing the silver solution. Copper is a more *reactive* metal than silver, meaning that it has a greater tendency to lose its outer electrons than silver. This means that the outer electrons of the copper atoms will be transferred to the silver ions, creating copper ions and silver atoms in place of the copper atoms and silver ions that were present to begin with. Figure 4.18 illustrates this process. This equation summarizes how it occurs.

$$2Ag^+(aq) \quad + \quad Cu(s) \quad \rightarrow \quad 2Ag(s) \quad + \quad Cu^{2+}(aq)$$

silver ions in solution   solid copper   solid silver   copper ions in solution

This is a very simple reaction because it only involves the *transfer of electrons* from one chemical species to another. Reactions like this are called *reduction/oxidation reactions,* more commonly referred to as **redox reactions. Reduction** is a general term used for "the gain of electrons," while **oxidation** is a general term used for "the loss of electrons." What's confusing is that oxygen is not involved in many "oxidation" processes. They are called oxidations for historical reasons, because oxygen was the element that gained the electrons in the first such reactions to be intensively studied.

We can show the two distinct "half reactions" (the oxidation and the reduction halves), plus the overall redox reaction, as follows:

Cu	$\rightarrow$	$Cu^{2+} + 2e^-$	*Oxidation*
$2Ag^+ + 2e^-$	$\rightarrow$	$2Ag$	*Reduction*
$2Ag^+ + Cu$	$\rightarrow$	$2Ag + Cu^{2+}$	*Redox reaction*

The overall equation tells us that 2 moles of silver ions react with 1 mole of copper metal to produce 2 moles of silver metal and 1 mole of copper ion. The conversion factor is

Figure 4.18   We can recover silver metal from a solution containing silver ions by exposing the solution to copper metal. Within a few minutes a significant amount of silver metal forms, accompanied by the formation of blue-colored copper ions (Cu^{2+}), which enter the solution. The piece of solid silver on the right resulted from melting the tiny crystals of silver metal formed in the reaction.

$$\frac{2 \text{ mol Ag}}{1 \text{ mol Cu}}$$

Using the Reactivity Series

When we consider what is best for the environment, we can rarely find easy or clear answers. For example, recovering silver by the process just shown creates another dilemma—what to do with the copper ions? The copper ion is a hazardous "heavy metal" ion that cannot be released freely into the environment. Therefore, copper is not used to recover silver; rather, zinc or aluminum is used; both are cheaper and more environmentally friendly than copper. We have seen that we can recover silver from a solution of silver ions by adding atoms of a "more reactive" metal, such as copper. Similarly, it is possible to recover copper metal from a solution of copper ions by adding atoms of a metal that are more reactive than copper, such as zinc. The relative reactivities of metals are listed in the reactivity series in Table 4.2. The **reactivity series** ranks metals in their order of reactivity. When metals react, they lose their outer electrons to become ions. So the reactivity series essentially ranks metals according to how easily the metal atoms lose their outer electrons to form ions.

Table 4.2 shows zinc to be more reactive than copper, so zinc atoms react by losing their electrons more readily than copper atoms. This means that if we add zinc atoms to a solution of copper ions, the outer electrons of the zinc atoms transfer over to the copper ions, causing copper metal to be formed (Figure 4.19). The overall equation for this redox reaction is

$$Zn(s) + Cu^{2+}(aq) \rightarrow Zn^{2+}(aq) + Cu(s)$$

Think of the situation as a competition between the nuclei of the zinc and copper atoms to see which one keeps the outer electrons. The nuclei of the less reactive element will end up keeping the outer electrons, because the reactivity of a metal is a measure of the ease with which it loses its outer electrons (is oxidized) by transferring them to some other chemical species.

Since copper is listed above silver in Table 4.2, it will reduce silver ions, as we showed earlier. Likewise, zinc is listed above copper and will reduce copper ions. Therefore, what would you predict regarding the ability of zinc to reduce silver ions? This is why, as discussed earlier, zinc is used commercially to recover silver. Additionally, zinc is much cheaper than copper and it is not a severe environmental problem.

Recycling Plastics

There is not just one recycling process for plastics. The method depends on the type of plastic and the company doing the recycling. One well-established process recycles beverage bottles that are made of polyethylene terephthalate, or PET polymers. **Polymers** are chemicals made

Table 4.2

Reactivity Series of the Metals

	Ion	Atom, Molecule	
Ions Difficult to Displace	K^+	K	*Metals that react with water*
	Ca^{2+}	Ca	
	Na^+	Na	
	Mg^{2+}	Mg	
	Al^{3+}	Al	
	Zn^{2+}	Zn	*Metals that react with acid*
	Fe^{2+}	Fe	
	Ni^{2+}	Ni	
	Pb^{2+}	Pb	
Ions Easy to Displace	H_3O^+	H_2	*Metals that are highly unreactive*
	Cu^{2+}	Cu	
	Ag^+	Ag	
	Au^{3+}	Au	

A.   B.   C.   D.

Figure 4.19 A strip of zinc metal **(A)** is placed into a solution of copper ions **(B, C)**. Zinc is oxidized to Zn^{2+} and the copper ions are reduced to copper metal **(D)**. The metal settles at the bottom of the beaker, the zinc strip is partially dissolved, and much of the blue color is removed.

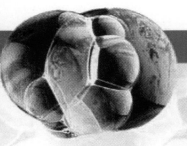

exercise 4.12

Silver Reduction

Problem

How many grams of copper are required to completely reduce 310 g of silver ion from a solution? Assume there are no side reactions.

Solution

Remember that the mole-mole relationship is the bridge that connects quantities of substances. Our strategy is therefore to convert grams of Ag to moles of Ag and use our mole-mole bridge to find the number of moles, and then grams of copper.

$$\text{g Cu} = 310 \text{ g Ag} \times \frac{1 \text{ mol Ag}}{107.9 \text{ g Ag}} \times \frac{1 \text{ mol Cu}}{2 \text{ mol Ag}} \times \frac{63.55 \text{ g Cu}}{1 \text{ mol Cu}}$$

$$= \textbf{91.3 g Cu}$$

Figure 4.20 shows molecular models of monomers with their structural formulas:

H H $C=C$ H Cl Vinyl chloride	H H $C=C$ H CH₃ Propene	F F $C=C$ F F Tetrafluorethylene	H H $C=C$ H [phenyl] Styrene

when a large number of smaller chemical units (the **monomers**, Figure 4.20) become linked together by chemical bonds. Figure 4.21 illustrates this general principle. All **plastics** are synthetic polymers of one kind or another.

The recycling process for PET bottles starts by separating the PET-containing bottle from its base cup, which is sometimes made of a different kind of plastic. The glue and the product label, along with the cap (usually made of polypropylene), are then removed. The remaining PET is melted down and processed into PET fiber that is sold to manufacturers for use in such things as blankets or sweaters (Figure 4.22A). (The Case in Point on Sorting Plastics shows the system of symbols used by the plastics industry to categorize recyclable plastics.)

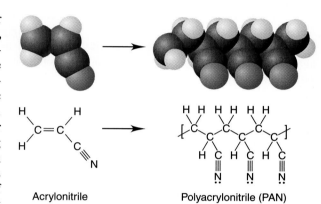

Acrylonitrile → Polyacrylonitrile (PAN)

Figure 4.21 The monomer acrylonitrile forms the polymer polyacrylonitrile, commercially known as orlon, which is used in carpeting and knitware. Here, and with the monomers shown in Figure 4.20, the double bond between the carbon atoms in the monomer breaks down into two single bonds, creating the link from one monomer to another, thus forming a chain.

A. Jackets

B. Bottles

C. Packaging

Figure 4.22 Recycled plastics are used for all kinds of consumer goods.
(A) Courtesy of Land's End.

consider this:

Plastic Beverage Containers

On average, over 2.5 million plastic beverage containers are thrown away *every hour!* As recycling hopefully becomes more popular, we can assume that many more uses for the recycled plastics will be found.

Uses Of Recycled Plastics

It takes about 480 2-L plastic bottles to make a high-quality 10- by 12-foot carpet for the bedroom. The plastic from 35 bottles can be converted into enough fiberfill for a sleeping bag. As Figure 4.22B and C illustrates, recycled plastics have many uses. The Federal Food and Drug Administration (FDA) began, in June 2001, to consider requests for using recycled plastics in food packaging. Their Office of Food Additive Safety lists their main safety concerns, and a number of proposed applications by companies, on their website (see the reading list at the end of this chapter).

4.5 | The Current Status of Recycling

"Paper or plastic?" You've had to answer that question at the checkout counter for at least the last few years. You might, on the face of it, assume that "paper" is for environmentally conscious consumers and "plastic" is for uncaring souls. Yet plastic does not necessarily have to be avoided by ecologically aware people. The bulk of landfill space is taken up by paper. The most critical problem in the United States is that we are simply running out of space to put our garbage, whether it is plastic, paper, or something else. The decision to pursue recycling is influenced, for better or worse, by more than just the availability of landfills and natural resources. Economics, technology, and public policy are also prime players in the recycling game.

The Status of Aluminum Recycling

Aluminum recycling is alive and well in the United States. Hundreds of millions of dollars each year are paid to "the average Joe and Mary"—people like you and me who recycle either for the money or out of a sense of what might loosely be called "environmental responsibility." Recycling of aluminum is also relatively easy no matter where you live, because there are over 10,000 facilities throughout the country.

The aluminum industry is committed to recycling because, as we mentioned before, the process uses less than 10% of the energy needed to manufacture aluminum cans from bauxite ore. Saving energy by recycling means saving money while still being able to produce quality products.

As more and more states mandate recycling, the aluminum recycling industry is expected to grow rapidly. As the recycling of other products, especially plastics, becomes more popular, aluminum recycling will level off. However, because the economics of aluminum recycling are so favorable, the process is becoming an accepted part of our lifestyles.

The Status of Plastics Recycling

The recycling of plastics presents a marked contrast to the aluminum process. There are many different types of polymers that need to be recycled. Plastics recycling also has a shorter history, only having come into its own within the last decade. Plastic containers are relatively light and take up a lot of room in the collection bin, so the collection process is not very efficient. Municipalities that have recycling programs generally deal only with polyethylene

■ case in point

The Landfill Question—Going, Going,…?

According to the Environmental Protection Agency, there were 5499 landfills in the United States in 1988. As shown in the accompanying table, that number is steadily being reduced. This tells us that as our population, and therefore the amount of trash, grows, there will be far *fewer* places to dump wastes. In addition, many of the chemicals in the "waste stream," especially plastics, are chemically unreactive. That is good if you want to use the product as a food container because the container will not react with the food. However, it also means that the same container will remain chemically and physically intact for hundreds of years.

One way to reduce the amount of trash that we toss away is to examine how we can use products more efficiently. To do this, we need to understand what we typically throw into landfills.

The composition of an average landfill, as determined by sampling landfills from throughout the United States, is given in the figure. It is interesting to note that disposable diapers, which have been portrayed as bloating up landfills, actually take up only about 0.8–3.3% of all landfill space.

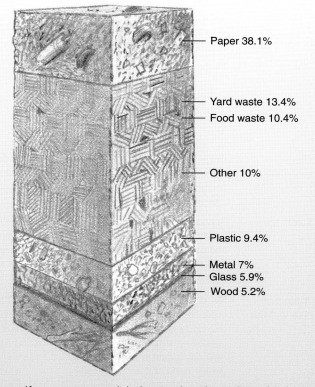

Paper 38.1%
Yard waste 13.4%
Food waste 10.4%
Other 10%
Plastic 9.4%
Metal 7%
Glass 5.9%
Wood 5.2%

Projected Number of Municipal Landfills Remaining in Operation over the Next Decade

Year	Number of landfills
1988	5499
1993	3332
1997	3091
2003	1594
2008	1234
2013	1003

Source: EPA Municipal Solid Waste Landfill Survey

If one way to minimize trash volume is to be less wasteful, with foods, for example, then the second way is to recycle. We do it with composting and we do it in the laboratory. It is now becoming feasible to do the same with many consumer goods. Recycling is becoming as much a part of our lifestyles as is throwing trash in the wastebasket instead of in the street. The table and the figure show how recycling of all kinds of materials and the types of things that are recycled, have increased during the last half-century. In 1996, over 57 million metric tons of potential waste was recycled, a 67% increase from the 1990 figure.

terephthalate ("#1," PET), and high-density polyethylene ("#2," HDPE). Over 345,000 metric tons of PET bottles were recycled in 2000, according to the National Association for PET Container Resources.

On balance, it seems that recycling is beginning to take hold as a necessary part of modern society. Table 4.3 summarizes the advantages and disadvantages of recycling materials currently being picked up curbside in recycling bins. Recycling can be economically feasible. It saves energy and natural resources. More intellectually interesting perhaps is that, as we pointed out earlier, recycling consumer goods is very much in harmony with the way the natural world works. We might have an insatiable appetite for things, but our resource supply is not infinite. We have begun to recognize this inescapable reality and to act upon it through a combination of chemistry, politics, and economics.

Table 4.3

Pros and Cons of Recycling

Material	Advantages to Recycling	Obstacles to Recycling
Paper	Saves more landfill space than recycling any other recycled material; reduces air and water pollution; abundant supply of newspaper and cardboard; low sorting cost; mills for recycled paper being developed	Weak market for mixed paper; recycled paper of lower quality; cannot be recycled indefinitely; hard to de-ink photocopy and laser-printed paper; de-inking plants costly to build
Plastic packaging	Reduces air pollution; conserves oil and gas	Only PET and HDPE recycled in quantity, nonpackaging plastic is rarely recycled; cannot be recycled indefinitely; generally cannot be used for food containers; difficult to sort; automatic sorting expensive; takes up a lot of space so expensive to pick up; virgin plastic can be cheaper; some plastics difficult to clean
Glass containers	Recyclable containers make up 90% of discarded glass; can be recycled indefinitely; can be recycled into food containers; labels and glues burn off in furnace; steady market for clear and brown glass	Containers break during sorting; broken glass hard to reuse; must be hand-sorted by color; poor market for green glass; often contaminated with unusable glass
Steel cans	Reduces pollution; conserves ore; can be recycled indefinitely; can be recycled into food containers; dirt and contaminants burn off in furnace; easy to separate with magnets; steel mills already set up to use scrap steel; strong market for recycled cans	None
Aluminum cans and foil	Recycling uses less than 10% of the energy of virgin production; reduces pollution; conserves ore; can be recycled indefinitely; can be recycled into food containers; dirt and contaminants burn off in furnace; well-developed system for collection and processing; strong market for recycled cans	Take up a lot of space so expensive to pick up

4.6 | Green Chemistry—A Philosophy to Protect the Global Commons

Recycling has been notable for its successes. Our society has changed from one that throws away the empties to one that most often asks, "can I recycle these?" However, as we take our first tentative steps into the new millennium, we wonder, "must we wait until the container is already made in order to save resources? Can we do more *in the production process* to help conserve and protect the global commons?"

The notion that pollution prevention is best done when the product is first prepared is the backbone of **Green Chemistry,** a philosophy in which we work toward **environmentally benign** chemistry and chemical manufacturing. Such chemistry ideally has no impact on the environment, including its air, water, land, or any living species. There is no need for cleanup, because there is no mess in the first place. This is a tall (truly impossible!) order. Yet we work toward it, because pollution has taken a toll worldwide, and chemical manufacturing has contributed to it. The Environmental Protection Agency states the mission of Green Chemistry:

> To promote innovative chemical technologies that reduce or eliminate the use or generation of hazardous substances in the design, manufacture and use of chemical products.

In the chapters to follow, we will explore the impact on the environment of human-made pollution, such as synthetic pesticides, smokestack emissions, and even nuclear waste. Environmental regulations of the type we will discuss have compelled the chemical industry,

pro con discussions

Baggin' It—Paper or Plastic?

pro

I choose paper bags at the grocery store rather than plastic. Paper is truly recyclable and is a renewable resource. Paper is made from wood, usually from trees that aren't suitable to be cut into lumber. Trees are renewable in the sense that as soon as an area is cleared of trees, another generation can be planted. This is in contrast to plastic bags, which are made from petroleum that cannot be renewed—once petroleum is taken from the Earth it is gone forever.

Proper management of woodlands allows a crop of trees to be harvested for paper production every 25–35 years. During all the years that these trees grow, they are removing carbon dioxide from the air, adding oxygen to the atmosphere, and providing large areas of land for wildlife to live and prosper.

Additionally, paper grocery bags are just the right size to fit into kitchen wastebaskets to dispose of kitchen waste. A paper bag filled with kitchen waste can degrade to environmentally benign substances in a landfill. Plastic bags, usually made of polyethylene, are undegradable and will stay in the environment unchanged for centuries. Paper bags are also just the right size to keep newspapers in for recycling. One can throw a paper bag filled with newspapers directly into recycling bins and they will be recycled right along with the newspapers.

Plastic bags are strong when lifting heavy loads but they tip over and spill their contents when the bag is put down. Paper, on the other hand, is both strong and stiff so that a bag will sit upright when being filled or emptied. Even worse than using plastic bags, though, is the practice of putting a paper bag into a plastic bag in order to take advantage of the stiffness of paper and the convenient handles on plastic bags.

con

This discussion really isn't about plastics, it's about people. It's about pitching in for the common good. It's about each of us knowing that social needs can be in harmony with individual convenience. Our family discards about six bags per week of "trash." Three of those bags are filled with plastics bound for the recycler. One of those bags is filled with aluminum, another with newspaper, and the last with what's left over. It doesn't take us any longer to sort plastics than it does to just dump the stuff in the garbage bound for the landfill.

Not too long ago, the grocery check-out clerk would say "paper or plastic?" and the expected response from anyone supposedly concerned about the environment would be "paper, please!" But that was before recycling technologies helped to develop alternative business markets for reused plastics. Also, a standard paper grocery bag weights 64 g. A plastic bag weights 10 g. There is a lot more materials used in a paper bag.

We have the carbon cycle, the nitrogen cycle, and the water cycle. Chemical technology has allowed the development of another cycle: the plastics cycle. Plastics can go from bottle to jacket to carpet and to who knows where, with a huge savings in natural resources. With a society that has its sights set on recycling, we can save huge amounts of natural resources. After all, plastic doesn't grow on trees!

further discussion

Try to develop your own arguments, first in support of PAPER and then PLASTIC. Then write down a few thoughts about what you think is a sensible and balanced view about the benefits and drawbacks of recycling.

government, and academic laboratories to work hard to reduce hazardous wastes. Green Chemistry is aimed at preventing the creation of these wastes.

One key to Green Chemistry is called **atom economy,** a term first used by Barry Trost, a professor of chemistry at Stanford University. Economical use of something means that there is no waste. You use all that you start with. In the chemical sense, atom economy means that all the atoms you start with end up in the product, with none left over as "waste." Atom economy is good for the environment and good for a company's economy, because leftover

■ case in point

Sorting It Out: Plastics and Your Recycle Bin

Ever wonder about those numbers in small triangles on the bottom of plastic soda bottles or liquid detergent containers? It seems that different packages have different numbers, depending on the properties of the package. Why are the numbers there? Who decides what number goes where? It is all part of the voluntary effort, begun in 1979, to recycle part of the millions of metric tons of plastics used each year. The number in the triangle designates a particular type of polymer that was used to make the product. The current coding system is shown in the accompanying table.

Many cities in the United States have plastics recycling programs, with the greatest successes coming in

recycling code 1 and 2 plastics. Higher code plastics have the potential to be recycled, but they tend to have similar physical properties and so are very difficult (or, more accurately "prohibitively expensive") to sort at this point.

Plastics recycling has not met with the same success as recycling aluminum. The recycling of PET beverage containers has shown a steady decline from 45% in 1994 to 22.3% in 2000. According to the Environmental Defense Fund between 1990 and 1996, new plastic packaging production outpaced recycled plastics by a 14 to 1 ratio.

Symbol/Plastic	Found in	Characteristics	Some Reuses
Polyethylene PET terephthalate	Soft drink bottles, peanut-butter-type jars	Most expensive; keeps oxygen out	Nonfood bottles, fiberfill, fibers, textiles, strapping, industrial paint, auto parts, insulation
High-density HDPE polyethylene	Containers for milk, water, and liquid detergents	Cheap; strong, good for handles; can be dyed many colors	Nonfood bottles, pipe, toys, trash cans, lumber substitute, flower pots, tubing
Vinyl or polyvinyl V chloride	Blister packs, containers for cooking oils and shampoos, food wrap	Very clear; resists degradation by oils	Piping, wire casing
Low-density LDPE polyethylene	Lids, squeeze bottles, bread bags	Flexible	
Polypropylene PP	Syrup and ketchup bottles, yogurt and margarine containers, bottle caps	Moisture-resistant; flexible; doesn't deform when filled	
Polystyrene PS	Coffee cups, meat trays, packing "peanuts," plastic utensils, videocassette boxes	Light but brittle; can be rigid	
Other	Ketchup bottles; handcream, toothpaste, and cosmetic containers	Plastics that may contain metals, glues, other contaminants mixed in	Wood substitute

Source: Society of the Plastics Industry

chemicals cost money to safely dispose of or recycle. Here are 6 guiding principles of Green Chemistry, including atom economy (see the reading by Hjeresen, Schutt, and Boese for the complete list):

- Prevention is better than cleanup.
- Maximizing atom economy is better than generating wastes.
- Making compounds using safer substances is better than preparation with more hazardous ones.
- Minimizing the use of hazardous solvents (to dissolve compounds) is desirable.
- Using as little energy as possible in preparing compounds is desirable.
- Products that are no longer useful should break down and not foul the environment.

Green Chemistry is still in its infancy. Still, there are examples of its successes. One such example is in some of the dry-cleaning stores of North Carolina and Nebraska, where liquid carbon dioxide has replaced compounds such as perchloroethylene, C_2Cl_4, as a dry-cleaning solvent. The CO_2 is more environmentally benign and can be recycled for further use. Two of us (PBK and JDC) have our dress clothes dry-cleaned in this way, and we notice the lack of that rather annoying "solvent smell" that is so prevalent with other solvents.

More Green Chemistry comes from DuPont, one of the largest chemical companies in the world. They have recently modified their process of manufacturing nylon, used in clothing, carpets, and other products, so that energy requirements have been reduced, water and air emissions are lowered, and all waste is recycled.

As we continue to take our tour through this chemical World of Choices, you will note some stories, exercises and data that will remind you of the principles of Green Chemistry. We will note these with the icon shown to the right. When you come across the Green Chemistry icon, think about, "How could this procedure be made a little more 'Green'?" Many chemical companies are thinking the same thing.

Green
Chemistry

main points

- The Earth is (virtually) a materially closed system but an energetically open one. In other words, energy flows in and out of the planet, but matter does not.

- The flow of energy through the Earth powers many *material cycles* in which matter is cycled through a variety of forms.

- Chemical reactions involve the *rearrangement* of atoms (although the atoms are often incorporated within molecules or ions).

- Chemical equations summarize chemical reactions, using chemical formulas to represent the reactants and products, rather than full names.

- All the atoms that appear on one side of a balanced equation must also appear on the other side. Chemical reactions, in other words, adhere to the Law of Conservation of Mass (matter).

- Real chemical reactions are more complex than the neat summaries depicted in equations. Some reactants usually remain unreacted, some side reactions generate additional minor products, and most reactions proceed through a variety of chemical intermediates which are not usually included in the equations.

- Chemical equations can be used to calculate the specific quantities of reactants and products involved in the reactions, either as masses or as moles. The general term for the process of performing such calculations is *stoichiometry*.

- In order to live in chemical harmony with the Earth, we need to learn to use materials in ways that fit into the general pattern of material cycles within our materially closed system.

- Recycling of materials is becoming an increasingly important and popular aspect of modern life and economic activity. The growth of recycling is being driven both by its economic benefits and by concern for the environment.

important terms

Alloy is a homogeneous mixture of metals, sometimes containing small amounts of nonmetals. (p. 153)

Atom economy involves having as much of the reactant as possible appear in the product. (p. 161)

Atomic mass is the mass of one atom in amu or 1 mol of an element in grams. (p. 141)

Bauxite is an ore of aluminum, consisting of aluminum oxide and water along with impurities. (p. 139)

Carbohydrate is a compound of C, H, and O in which there are twice as many H atoms as O atoms and the number of C atoms is similar to the number of O atoms; examples are sugars, starches, and cellulose. (p. 130)

Carbon cycle is the sequence of reactions characteristic of carbon atoms in the natural environment—the material cycle wherein carbon atoms are cycled through many different compounds. (p. 129)

A **chemical equation** is a formal method for describing reactants and products of a chemical change. (p. 130)

A **chemical reaction** is a process in which one set of chemicals is converted into different substances. (p. 130)

A **closed system** is a collection of matter shut off such that no other matter can enter or leave. (p. 128)

Composting is a process of allowing leaves, grass, etc. to react with water, air, and microorganisms to form a soil-like solid while giving off carbon dioxide. (p. 151)

Conversion factor is a ratio of quantity expressed in two different units, used to convert a quantity in one unit to the corresponding quantity in the other unit. (p. 143)

Dimensional analysis is a technique for solving numerical problems in which the units of the quantities guide the solution and serve as a check for the solution. (p. 144)

Environmentally benign means having no impact on the environment. (p. 160)

Formula mass is the mass of one formula unit or molecule in amu or the mass of 1 mol of a compound in grams. (p. 141)

Formula unit is the number of atoms of each kind in a molecule or the simplest formula of an ionic compound. (p. 141)

Gram formula mass is the mass of 1 mol of a compound or element in grams. (p. 144)

Green Chemistry is concerned with making the practice of chemistry as environmentally benign as possible by the reduction or elimination of hazardous wastes and efficient use of energy. (p. 160)

Hall-Heroult process is the process for the economical, commercial production of aluminum. (p. 139)

Incinerate is to burn to ashes in an excess of oxygen. (p. 127)

Intermediate is a substance formed in a chemical reaction which then reacts further so that it is not a final product of the reaction. (p. 136)

Landfill is a place where garbage and trash are buried in such a way that they will interact with the rest of the environment as little as possible. (p. 127)

Law of Conservation of Mass states that matter (mass) is neither created nor destroyed although matter is frequently transformed into different substances. (p. 130)

Material cycle is a description of the transformations of substances in the environment, usually implying that elements undergo continuous chemical change but often reappear in forms they have been in before. (p. 129)

Molar mass is the mass of one mole of a compound or element in grams. (p. 144)

Mole is Avogadro's number of atoms, molecules, or formula units; an amount of any substance equal to its molar mass expressed in grams. (p. 143)

Mole-mole conversion factor is the ratio of moles of two substances in a chemical reaction. (p. 148)

Molecular mass is the mass of a single molecule of a compound in amu. (p. 141)

Monomer is a compound of which many molecules will react together to form one much larger molecule. (p. 157)

Oxidation is the loss of electrons by a substance. (p. 154)

Photosynthesis is a sequence of reactions in green plants which results in water and carbon dioxide being converted into oxygen gas and carbohydrate under the influence of sunlight. (p. 138)

Plastic describes a high molecular-weight synthetic polymer. (p. 157)

Polymer is a molecule formed by a large number of identical smaller molecules (monomers) reacting together to form a much larger molecule. (p. 155)

Product is a substance formed in a chemical reaction. (p. 130)

Reactant is a substance consumed in a chemical reaction. (p. 130)

Reactivity series describes a ranking of chemical elements in order of their ability to oxidize or reduce each other. (p. 155)

Recycle is to utilize waste materials rather than throw them away. (p. 127)

Redox reaction is an abbreviation for "reduction and oxidation," a class of chemical reactions in which both reduction and oxidation occur. (p. 154)

Reduction is the gain of electrons by a substance. (p. 154)

Relative atomic mass is the average mass of one atom of an element. (p. 141)

Respiration is the reaction carried out in living organisms in which oxygen and carbohydrate are converted into water and carbon dioxide so that energy useful to the organism is obtained. (p. 130)

Stoichiometry is a means of comparative measuring that uses the fact that chemical reactions occur in ratios of moles and that calculations of amounts of reactants and products are possible. (p. 140)

Water cycle describes the sequence of forms characteristic of water in the environment. (p. 129)

exercises

1. Using your own words, define these words or terms:
   a. closed system
   b. respiration
   c. chemical reaction
   d. chemical equation
   e. photosynthesis
   f. bauxite
   g. mole
   h. plastic
   i. polymer

2. Do you agree that the Earth is a closed system both materially and energetically. Why or why not?

3. What are the recycling policies of the town where you live?

4. Is the human body an open or closed system? Explain.

5. You are drinking a glass of water. What is the source of the water (reservoir, well, etc.)? Trace its potential routes into your glass.

6. Indicate which of these is a chemical reaction:
   a. wood being cut by an ax (consider only the wood)
   b. rusting of a car
   c. combining carbon dioxide with lime to make limestone
   d. crushing a limestone sculpture

7. In our bodies glucose chemically reacts with oxygen in a multistep process that produces energy. However, when the oxygen supply is insufficient, glucose becomes lactate and the process ceases. This lactate (lactic acid) buildup causes the pain you may experience the morning after a hard workout. The overall equation representing the conversion of glucose to lactate is

$$\underset{\text{glucose}}{C_6H_{12}O_6} \rightarrow \underset{\text{lactic acid}}{2HC_3H_5O_3}$$

List the number of atoms of each element on both sides of the equation.

8. Which of these reactions are balanced chemical equations?
   a. $AgNO_3 + KBr \rightarrow AgBr + KNO_3$
   b. reaction of marble or limestone ($CaCO_3$) with sulfuric acid to form calcium sulfate ($CaSO_4$):

   $$CaCO_3 + H_2SO_4 \rightarrow Ca^{2+} + SO_4^{2-} + H_2O + CO_2$$

   c. exothermic reaction that produces the heat in heat packs:

   $$2Fe + 3O_2 \rightarrow 2Fe_2O_3$$

   d. reaction used in film processing to "fix" the image:

   $$AgBr + S_2O_3^{2-} \rightarrow Ag(S_2O_3)_2 + Br^-$$

   e. reaction used in the fermentation process:

   $$C_6H_{12}O_6 \rightarrow C_2H_5OH + CO_2$$

9. Which of these chemical reactions are *not* possible?
   a. $C_2H_2 + 2H_2 \rightarrow C_2H_2$
   b. $C_2H_2 + 2H_2 \rightarrow C_2H_6$
   c. $2C_2H_2 + 2H_2 \rightarrow 2C_2H_6$

10. Label the reactants and products of this chemical reaction:

$$C_6H_{12}O_6 + CH_3OH \xrightarrow{H^+} C_7H_{14}O_6 + H_2O$$

11. Using the reaction in question 10, how many oxygen atoms are present in the reactants? How many in the products? Does this make sense?

12. Balance this chemical equation: $C_{12}H_{22}O_{11} + O_2 \rightarrow CO_2 + H_2O$

13. If a plant leaf contains 100 molecules of CO_2, how many molecules of H_2O are needed to undergo photosynthesis? How many glucose molecules are produced? How many O_2 molecules are produced?

14. Calculate the formula mass (amu) of each of these compounds:
   a. K_2CrO_4
   b. NH_4OH
   c. $C_{10}H_{16}O$
   d. HCl
   e. $Cr_4(P_2O_7)_3$

15. What is the atomic mass of each of these elements? Sc, Pd, O, Na, Cl

16. Calculate the formula mass of these molecules:
   a. $Cu(OH)_2$
   b. H_3PO_4
   c. $NaCl$
   d. NH_3

17. Some catalytic converters in automobiles use such precious metals as platinum, palladium, and rhodium. These automotive catalysts remove hydrocarbons, carbon monoxide, and nitrogen oxides, which would otherwise be emitted into the air. As worldwide production increases, catalyst producers are now being pressured to recycle more of the precious metals used to make these catalysts. If in a typical year, 9200 kg of palladium were sold for automotive use in the United States, how many moles would this be?

18. Determine the mass in grams of each of these:
   a. 1.0 mol NH_3
   b. 1.0 mol $NaOH$
   c. 1.2×10^2 mol C
   d. 2.7×10^4 mol O_2
   e. 3.0×10^{-11} mol Ag

19. What is the mass, in grams, of 3.5×10^{22} molecules of water (H_2O)?

20. Given the number of grams, how many moles are there in each of the samples?
   a. 22 g $NaHCO_3$
   b. 220 g $HC_2H_3O_2$
   c. 2.22×10^3 g H_2O
   d. 2.22×10^4 g $CuCl_2$

21. A sample of rubidium peroxide (Rb_2O_2) has a mass of 0.85 g. Convert this to moles.

22. If the 6 billion people on Earth collectively produced 6.02×10^{23} pounds of trash, how much trash would each person contribute? (*Hint:* 6 billion can be written as 6.0×10^9.)

23. You eat a candy bar containing 23 g sugar. For this problem, assume the sugar is pure glucose. How many moles of O_2 are required to react with this much glucose? (refer to the combustion of glucose on p. 131.)

24. What is the mass of Cr in one formula unit of Cr_2O_3?

25. How many atoms of Cu are found in 10.0 g of copper pennies?
26. How many atoms of oxygen are found in 8.00 g water?
27. How many atoms of chlorine are found in 5.84 g $MgCl_2$?
28. Recently scientists have attached sections of DNA to gold "nanoparticles." These gold particles have a diameter of 13 nanometers and therefore a volume of 1.2×10^{-18} cubic centimeters. Gold has a density of 19.3 g/cm^3. How many gold atoms are in such a nanoparticle?
29. Roughly 124,300 kg of coal are needed to produce the energy for the extraction of 1000 kg of aluminum. If 16 g of aluminum are needed to produce 1 aluminum can, how many kg of coal would be needed to produce 12 aluminum cans?
30. How many moles are there in a sample of barium sulfate ($BaSO_4$) that has a mass of 9.90×10^7 ng?
31. How many grams of water vapor can be generated from the combustion of 118.0 g of ethanol according to the following balanced equation?

$$C_2H_6O(l) + 3O_2(g) \rightarrow 2CO_2(g) + 3H_2O(g)$$

32. How many grams of sodium hydroxide (NaOH) are required to form 61.4 g of lead hydroxide (Pb(OH)$_2$) according to the following balanced equation?

$$Pb(NO_3)_2 + 2NaOH \rightarrow Pb(OH)_2 + 2NaNO_3$$

33. Starkist Tuna projects a sale of 9.5 million cans of tuna for the coming year. Each can, minus the tuna, has a mass of 8.0 g. How much aluminum oxide (Al_2O_3) will be needed to make enough cans for all that tuna? Express the answer in kg and metric tons (1 metric ton = 1000 kg).

$$2Al_2O_3 + 3C \rightarrow 4Al + 3CO_2$$

34. You are preparing this reaction in the lab, starting with 1.07 g C_2H_2:

$$C_2H_2 + 2Br_2 \rightarrow C_2H_2Br_4$$

   a. How many moles of product are produced?
   b. How many grams of product are produced?
   c. How many molecules of product are produced?
35. Using H_2 and O_2 as reactants, write a balanced chemical equation for the formation of H_2O. What is the mole ratio of H_2 to O_2? H_2 to H_2O? O_2 to H_2O?
36. The synthesis of aspirin is a reaction of salicylic acid and acetic anhydride:

Salicylic	+	acetic	$\rightarrow$	acetylsalicylic	+	acetic
acid		anhydride		acid		acid
$C_7H_6O_3$		$C_4H_6O_3$		$C_9H_8O_4$		$C_2H_4O_2$

A chemist has a terrible headache, and no aspirin! She goes to the lab to synthesize 200 mg of aspirin. How much salicylic acid must you start with?

37. One serving of candy contains about 15 g sucrose. How many moles of sucrose, $C_{12}H_{22}O_{11}$, are you consuming with each serving of candy?

38. Based on the equation for the combustion of octane, how many grams of carbon dioxide will be released into the air from the consumption of 5 gallons of gas? (1 gallon = 3.8 liters)

$$2C_8H_{18} + 25O_2 \rightarrow 18H_2O + 16CO_2$$

39. How many recycled aluminum cans would be needed to make one screen door weighing 10 pounds?
40. An automobile contains about 5 quarts of oil in the crankcase. If this is typically changed once a year in each of the 50 million cars in the United States, how much used automobile oil must be disposed of each year?
41. In 1999 more than 35 million gallons of used motor oil was generated in Pennsylvania. If the used motor oil each year is burned to give heat that is used to generate electricity according to the following reaction, how many moles of carbon dioxide will be formed each year from this source? Density of oil = 0.78 g/cm^3.

$$C_{21}H_{44} + 32O_2 \rightarrow 21CO_2 + 22H_2O$$

42. The mineral portion of tooth enamel is hydroxyapatite, $Ca_5(PO_4)_3OH$. Hydroxyapatite reacts with tin(II) fluoride, an ingredient in some popular toothpastes, to form fluoroapatite, $Ca_5(PO_4)_3F$, which is more resistant to tooth decay than is hydroxyapatite. The reaction equation is:

$$2Ca_5(PO_4)_3OH + SnF_2 \rightarrow 2Ca_5(PO_4)_3F + SnO + H_2O$$

How many grams of hydroxyapatite will react with 0.22 grams of tin(II) fluoride? Tin(II) fluoride is also known as stannous fluoride.

43. How much sulfuric acid can be formed from the sulfur in a metric ton (megagram) of coal if the coal is 4.0% S?
44. Glass is a mixture of oxides melted together with silicon oxide. So-called "lead crystal" glass typically contains 22% PbO in with the SiO_2 and other oxides. How much lead is present in a lead crystal goblet which weighs 350 grams?
45. The white pigment, TiO_2, is made by the reaction of $TiCl_4$ with oxygen.

$$TiCl_4 + 2O_2 \rightarrow TiO_2 + 2Cl_2$$

The chlorine is recovered to make more $TiCl_4$. How much $TiCl_4$ is needed to make 500 kg of TiO_2?

46. Think up and supply the answer for a good example to explain how large Avogadro's number really is. For instance: "Calculate how long a clothesline must be to hang a mole of socks."
47. Sodium ion has a radius of 0.095 nm and chloride ion has a radius of 0.181 nm. Suppose that a mole of sodium chloride consisted of a line of alternating ions; how long would that line of ions be?
48. How many grams of oxygen are necessary to convert 5.0 moles of Fe_3O_4 into Fe_2O_3?
49. A lead tire weight has a mass of 50.0 grams. How many lead atoms does it contain?

50. Bromine has a density of 2.93 g/mL. How many grams of $PbBr_4$ can be made from 23.4 mL of bromine?

51. How many grams of oxygen are required to react with 14 grams of magnesium to form MgO?

52. What will be the volume of copper that can be formed by the reduction of 157 g of CuO to form pure Cu? You will need to look up the density of copper to solve this problem.

53. How many dioxin molecules are present in a 55-gram sample that contains 1.0 ppb of dioxin? (See the Case in Point on dioxin.)

54. What mass of H_2SO_4 can be made from the reaction of 88 g of SO_3 with an excess of water?

55. You were hired by a laboratory to recycle 6 mol of silver. You were given 150 g of copper metal (Cu). How many grams of silver metal (Ag) can you recover? Is this enough copper to recycle all the silver ions?

$$2Ag^+ + Cu \rightarrow 2Ag + Cu^{2+}$$

56. In this reaction, which reactant is oxidized? Which is reduced?

$$Mg(s) + Pb^{+2} \rightarrow Mg^{+2} + PB(s)$$

57. Calcium carbide, CaC_2, is used in manufacturing acetylene that is widely used in welding and cutting steel. How many grams of calcium are needed to react with 12 grams of carbon to form calcium carbide?

58. Carbon dioxide is formed in the reaction of $NaHCO_3$ with an acid. This reaction is also the basis of many antacid products. How many grams of carbon dioxide can be formed if 0.55 g of $NaHCO_3$ is reacted to form CO_2?

59. Complete and balance these reactions. Identify the physical state of products.
    a. __$Ca(s)$ + __$H_2O(l) \rightarrow$
    b. __$Zn(s)$ + __$HCl(aq) \rightarrow$
    c. __$Cl_2(aq)$ + __$NaI(aq) \rightarrow$

60. An Alka-Seltzer tablet contains sodium bicarbonate and citric acid. When dropped in water, carbon dioxide, water, and sodium citrate are produced:

$$3NaHCO_3 + C_6H_8O_7 \rightarrow 3CO_2 + 3H_2O + Na_3C_6H_5O_7$$

How much CO_2 is evolved if O·350g of $C_6H_8O_7$ react?

61. At your recycling plant, you have a copper shortage and only have 50.0 g left. How many grams of silver ions can you reduce with the copper?

62. What is being oxidized in the following reaction?

$$2Ag^+ + Zn \rightarrow Zn^{2+} + 2Ag$$

63. What is being oxidized in the following reaction?

$$2Fe^{3+} + Sn^{2+} \rightarrow Sn^{4+} + 2Fe^{2+}$$

64. Your "gold" ring has been oxidized! What metal could you use to help recover the gold?

65. What types of plastics can you find in the room in which you are sitting right now? Are they recyclable?

66. In addition to lawn clippings and leaves, what other "natural garbage" can be composted?

67. The Chambers Development Co. of Pittsburgh has recently followed the example of other companies that haul garbage by rail from big cities to county landfills. The Bergen County–Charles City County train leaves the station at North Arlington, New Jersey, every evening carrying 32 oversized "Trash Cans." Each of the 32 cans holds as much trash as an individual creates in 27 years. How many tons of New Jersey garbage is transported to the Charles City County landfill in Virginia on one train? (The average person generates 730 kg of garbage in 1 year, 907.185 kg = 1 ton.)

food for thought

68. Do you think that recycling should be mandatory? Why or why not?

69. Think of one creative recycling/reuse idea for any household item.

70. Look again at the list of polymers in the Case in Point on sorting plastics. How many of these items do you come across every day? List some more items that you think are polymers.

71. In the Prelude we discussed comparing risks versus benefits. List some risks and benefits to consider about recycling (for example, amount of money needed to sort and recycle plastic containers versus amount of money needed to build enough landfills to hold all the plastic waste).

72. Bottles and plastic food-packaging products are not the only culprits stealing space in America's landfills. Automobile manufacturers are now realizing the implications of discarding 9 million or so cars every year in landfill space. In fact a few auto makers, like BMW, have initiated auto disassembly plants. Approximately 85%, by weight, of BMW's cars are recyclable (2001 data). The company's goal is 90%. However, the disassembly of a car is quite expensive as foam and vinyl are more difficult to recycle than steel. Would you consider paying more for a car if you knew the manufacturer was including costs for recyclable parts and running disassembly plants?

73. How much CO_2 will be formed by driving all the automobiles in the United States each year? To answer this question you will have to make several estimates and look up some things before beginning the calculation.

74. The standard argument against the use of disposable diapers is that the 18 billion plastic-lined disposables that U.S. households use per year take up between 0.8 and 3.3% of landfill space and therefore exacerbate an already serious landfill problem. Both disposable and reusable diapers have environmental costs, as shown in the accompanying table, which uses data from a 1990 study

by consultants at Franklin Associates in Kansas. The data include all costs associated with diaper use, including packaging, disposal, cleaning, pins, and plastic pants. As you can see, there are all kinds of environmental issues involved in the diapering decision. Cost, convenience, and the comfort of the baby are three more things to think about. Which type of diaper, disposable or cloth, would you choose for your child? Justify your decision.

Environmental Costs of Diaper Use (per year per child)

Environmental cost	Cloth diaper risk	Disposable diaper risk
Energy use equivalent	400 L gasoline	200 L gasoline
Water use	40,000 L	10,000 L
Water pollution	10 kg	2 kg
Combustion products	15 kg	7 kg
Garbage to landfills	minimal	millions of diapers, contents

75. As discussed in the Case in Point on plastics recycling, most plastic products have been separated into seven broad groups. Each group is assigned a number that is found in the triangle label on most plastic containers. It is crucial that some kinds of plastics from different groups are not mixed. Some of these plastics are extremely difficult and costly to recycle, especially when they mix. What can we do to make plastics recycling more feasible?

76. What other processes besides photosynthesis, respiration, and composting occur on Earth to support the statement "Earth is a natural recycler"?

readings

1. Specter, M. Define "Recycled"? A Drive to Clarify Environmental Labels. *The New York Times,* December 16, 1991.
2. Downing, B. After Late '93, No More Lawn Waste to Landfills. *Akron Beacon Journal,* May 23, 1991.
3. Ditmer, J. Carpeting Takes Recycling a Step Up. *Denver Post,* August 8, 1991.
4. Bishop, G. Big Plastic Recycling Plant to Help Jersey Dispose of Its Polystyrene. *Star-Ledger,* August 4, 1991.
5. Wetzstein, C. High-tech Processes Let Recycled Plastics Safely Repackage Food. *Washington Times,* March 22, 1991.
6. Saltus, R. Drops In The (Plastic) Bucket? *Boston Globe,* August 27, 1990.
7. Jones, D. P. An Uncertain Future for Recyclers of Foam. *Hartford Courant,* December 23, 1990.
8. Coppertithe, K. G. *Rigid Container Recycling,* U.S. Department of Commerce, Washington, D.C., December 1989.
9. Thurman, M. High-tech System Boosts Recycling. *Marin Independent Journal,* September 8, 1991.
10. Rathje, W. L. Once and Future Landfills. *National Geographic,* May 1991.
11. Muir, F. R. L.A.'s New Recycling Plan Could Be Trashed by Meager Demand. *Los Angeles Times,* February 18, 1990.
12. Bernton, H. Recycling a Victim of Its Own Success. *Anchorage Daily News,* July 15, 1990.
13. Storck, W. J. Better Times Ahead for Chemical Industry. *Chemical & Engineering News,* December 12, 1991.
14. Austin, G. T. *Shreve's Chemical Processes,* 5th ed. New York: McGraw-Hill Book Company, 1984.
15. Which <u>WHAT</u> Are Best for the Environment? *Consumer Reports,* August 1991.
16. Koshland, Daniel E., Jr. The Dirty Air Act. *Science* 249, September 28, 1990.
17. Levin, Doron P. Imperatives of Recycling Are Gaining on Detroit. *The New York Times,* September 6, 1992.
18. Specter, Michael. New York, Amid Doubts, Broadens Recycling Plan. *The New York Times,* May 27, 1992.
19. Holusha, John. The 7 Levels of Plastics in Recycling- by-the-Numbers. *The New York Times,* October 11, 1992.
20. Thayer, Ann M. Catalyst Suppliers Face Changing Industry. *Chemical & Engineering News,* March 9, 1992.
21. Regan, Bob. '93 UBC Aluminum Recycling Off 6%. *American Metal Market,* March 28, 1992: 1–2.
22. Regan, Bob. 100 Billion Cans Likely '94 Total. *American Metal Market* (December 29, 1994): 1–2.
23. Binczewski, George J. The Point of a Monument: A History of the Aluminum Cap of the Washington Monument. *Journal of Metals* 47 (11), 1995: 20–25. http://www.tms.org/pubs/journals/JOM/9511/Binczewski-9511.html
24. Hjeresen, Dennis L., Schutt, David L., Boese, Janet M. Green Chemistry and Education. *Journal of Chemical Education,* 77 (12), December 2000: 1543–1547.
25. Cann, Michael C., Connelly, Marc. *Real-World Cases in Green Chemistry.* American Chemical Society, Washington, D.C., 2000.
26. Ritter, Stephen K. Green Chemistry. *Chemical & Engineering News* (July 26, 2001): 27–34.
27. U.S. Food and Drug Administration's Office of Food Additive Safety: Recycled Plastics in Food Packaging. Web site (January 2002) http://vm.cfsan.fda.gov/~dms/opa-recy.html

resources

Bronowski, Jacob. Biography of an Atom—And the Universe. *The New York Times,* October 13, 1968.

websites

www.mhhe.com/kelter   The "World of Choices" website contains activities and exercises including links to websites for: Biosphere 2 project; Environmental Protection Agency; the University of Nottingham (information about Green Chemistry), and much more!

Appendix to Chapter 4

Working with Exponents, SI Units, and Dimensional Analysis

This appendix is designed to introduce mathematical techniques and other skills necessary to "do science" that may be unfamiliar to you. The sections covered in this appendix (exponential notation, introduction to SI units, and more practice with stoichiometry) will be helpful in working problems not only in this chapter, but also in the rest of the text as well. Even if you feel you are a master of math and chemistry, it may be helpful to look over all the sections to clear out any remaining cobwebs.

A.1 Exponential Notation

Since the numbers used in scientific measurements are often very large or very small, a system has been devised to express these numbers using powers of 10. This system, known as **exponential notation,** may spare you from a painful writer's cramp. For example, 10,000,000 can be expressed as 1.0×10^7.

You may understand this shortcut expression as moving the decimal as many places as indicated by the exponent of 10. In which direction, you ask? In our example, the expression 1.0×10^7 indicates that the decimal point in the 1.0 must be moved seven places to the *right* to obtain the original number (10,000,000). Conversely, when the exponential expression contains a negative exponent, for example, 1.0×10^{-7}, the decimal must be moved seven places to the *left* to obtain the long form (0.0000001) of the number.

Mathematicians may look at this another way. They see 1.0×10^7 as 1.0 being multiplied by 10 seven times:

$$\mathbf{1.0} \times 10^7 = \mathbf{1.0} \times 10 \times 10 \times 10 \times 10 \times 10 \times 10 \times 10$$
$$= 10,000,000$$

Each time we increase the power to which 10 is raised, the decimal must be moved one more unit to the right. Or, in mathematical terms, we multiply by one additional factor of ten. For example:

$$1.0 \times 10^1 = 10$$
$$1.0 \times 10^2 = 10 \times 10 = 100$$
$$1.0 \times 10^3 = 10 \times 10 \times 10 = 1000$$
$$1.0 \times 10^4 = 10 \times 10 \times 10 \times 10 = 10,000$$
$$1.0 \times 10^5 = 10 \times 10 \times 10 \times 10 \times 10 = 100,000$$

However, for exponential notations containing a negative exponent, a mathematician would use the opposite operation. That's right—division! Each time we decrease the power of 10, the decimal must be moved one more unit to the left or divided by an extra power of 10. For example,

$$1.0 \times 10^{-1} = 1 \div 10 = 0.1$$
$$1.0 \times 10^{-2} = 1 \div 100 = 0.01$$
$$1.0 \times 10^{-3} = 1 \div 1000 = 0.001$$
$$1.0 \times 10^{-4} = 1 \div 10,000 = 0.0001$$
$$1.0 \times 10^{-5} = 1 \div 100,000 = 0.00001$$

What does 1.0×10^0 equal? Any integer raised to the zero power equals one. Thus, $1.0 \times 10^0 = 1.0$.

Just the Basics

Now that we have a general understanding of exponential notation, we can practice the method of converting very large or small numbers to the shorthand exponential notation form (or vice versa).

Let's consider the number 1992. The first problem we must address when writing 1992 in exponential notation is where to place the decimal point. The first objective is usually to produce a number between 1 and 10. By placing the decimal after the 1 in 1992 we create a number that is greater than 1 and less than 10.

$$1 < \mathbf{1.992} < 10$$

[Keep in mind that the number to the left of the decimal is usually written as a *single digit*.]

In order to get from 1992 to 1.992 we had to move the decimal three places. As discussed earlier, to get back to the original number we must multiply 1.992 by 1000 or 10^3. Putting our new decimal number together with the multiplication factor we get

$$1992 = 1.992 \times 10^3$$

Consider that the electric charge of one electron is 0.00000000000000000016208 coulombs (C). Since this number is rather cumbersome, we will write it in exponential notation. First we express the number as an integer between 1 and 10:

$$1 < 1.6208 < 10$$

Next, count the number of places the decimal point was moved to produce the new integer, 1.6208. The decimal point was moved 19 places to the *right,* thus, the correct exponential expression would be written with a *negative* number:

$$1.6208 \times 10^{-19}$$

Here are some further examples of writing in exponential notation:

| 1.6 | *would be* | 1.6×10^0 |

but you wouldn't bother with exponential notation;

379	*becomes*	3.79×10^2
134,000	*becomes*	1.34×10^5
0.0025	*becomes*	2.5×10^{-3}
0.0000783	*becomes*	7.83×10^{-5}
0.90	*becomes*	9.0×10^{-1}

A.2   Fiddling with Functions

Now that we have the conversion process mastered, let's talk about manipulating these numbers in mathematical computations. It would be pointless to convert the exponential notation form back to the original number when adding, subtracting, multiplying, and dividing. (It's almost as absurd as driving around in circles when you're lost instead of asking for directions.) Ergo, we have a few simple rules to follow when using exponents in mathematical operations.

Adding and Subtracting

When adding or subtracting numbers written in exponential notation, you may add or subtract only those *numbers with the same exponential degree.* That means you can add this set of numbers:

$$
\begin{array}{r}
1.25 \times \mathbf{10^3} \\
+\ 4.61 \times \mathbf{10^3} \\
\hline
5.86 \times \mathbf{10^3}
\end{array}
$$

but not this set:

$$
\begin{array}{r}
1.25 \times \mathbf{10^3} \\
+\ 28.6 \times \mathbf{10^2} \\
\hline
\textit{can't do}
\end{array}
$$

However, if we rewrite 28.6×10^2 as 2.86×10^3, since moving the decimal one place to the left can be compensated for by adding 1 to the exponent, we can now add the two numbers:

$$
\begin{array}{r}
1.25 \times \mathbf{10^3} \\
+\ 2.86 \times \mathbf{10^3} \\
\hline
4.11 \times \mathbf{10^3}
\end{array}
$$

As you can see, when performing addition or subtraction only the initial number is added or subtracted, while the exponent remains the same.

Example A.1
Worldwide Rubber Consumption

PROBLEM

In 1990, world rubber consumption was 1.42×10^7 tons. Due to a worldwide recession, rubber consumption deceased by 5%, or 7.1×10^5 tons in the next year. How much rubber was consumed in 1991?

SOLUTION

We must subtract 7.1×10^5 from 1.42×10^7, but first we must move the decimal point in one of the numbers so the exponents are equal. If we move the decimal two places to the right in 1.42×10^7 we get $142. \times 10^5$. Now, we are ready to subtract:

$$
\begin{array}{r}
142. \times \mathbf{10^5} \\
-\ 7. \times \mathbf{10^5} \\
\hline
135. \times \mathbf{10^5}
\end{array}
$$

Moving the decimal point back so that we have only a single digit to the left of the decimal (i.e., the answer expressed using a value less than 10), we obtain the correct amount of worldwide rubber consumption in 1991: $\mathbf{1.35 \times 10^7}$ **tons.**

Multiplication and Division

The process of multiplication and division is actually simplified when numbers are expressed in exponential notation. First, the initial decimal numbers are multiplied or divided as usual.

Next we deal with the exponents. When *multiplying—the exponents are added.* For example,

$$
\begin{aligned}
(9.8 \times 10^5)(3.2 \times 10^2) &= (9.8 \times 3.2) \times 10^{(5+2)} \\
&= 31.36 \times 10^7
\end{aligned}
$$

We must always check significant figures before recording the answers from mathematical computations. In our example, we see that the answer should have two significant figures so we must round 31.36 to 31. Are we there yet?—not quite. Because the initial number is greater than 10, we must move the decimal one place to the left and add 1 to the exponent of 10. Thus, our final answer is $\mathbf{3.1 \times 10^8}$.

In *division,* we divide the initial numbers as usual and then *subtract the exponent of the denominator from the numerator.* For example:

$$
\begin{aligned}
(1.4 \times 10^5) \div (2.5 \times 10^2) &= (1.4 \div 2.5) \times 10^{(5-2)} \\
&= 0.56 \times 10^3
\end{aligned}
$$

Again we must have an answer with the required number of significant figures (2 in this case). The answer seems correct. However, we now have an initial number which is less than 1. Thus, we must move the decimal one place to the right and subtract 1 from the power of 10 to reach our final answer of **5.6×10^2.**

We can summarize the steps for multiplication and division as follows:

1. Multiply or divide the initial numbers (those that appear before the factor of 10).
2. Add the exponents of 10 if multiplying; subtract the exponents if dividing.
3. Round your answer up or down to obtain the correct number of significant figures (this would be the least number of significant figures represented by one of your initial numbers).
4. Move the decimal point in the initial number of your answer to obtain a number between 1 and 10. To the exponent, add (if you move left) or subtract (if you move right) the number of places moved.

Example A.2

Industry Waste

PROBLEM

American Industry creates 7.6×10^9 tons of waste annually. If this amount were to increase by 8.0% in the next year, how many additional tons of waste would industry contribute?

SOLUTION

The number 8.0% (or 8/100) can be expressed as 0.08 or 8.0×10^{-2}. Now multiply:

$$(7.6 \times 10^9)(8.0 \times 10^{-2}) = (7.6 \times 8.0) \times 10^{[9 + (-2)]}$$
$$= 60.8 \times 10^7$$
$$= 6.08 \times 10^8$$

Again we must round and move the decimal point to obtain an answer to two significant figures: **6.1×10^8 tons** of industrial waste would be added to the 7.6×10^9 tons already produced.

A.3 An Introduction to SI Units

The United States is the last industrialized nation to use the **English system** of measurement, which is based on units such as quarts for volume, pounds for weight, and miles for distance. As a nation, we recognize the *international* nature of manufacturing and money exchange. We are making efforts to use the International System ("le Système International" in French, or **SI**) in the United States. The SI sys-

tem is based on the **metric system** of measurement, which we discussed somewhat in Chapter 1. According to the Omnibus Trade and Competitiveness Act (August 1988), federal agencies were required by Congress to implement the metric system in business-related activities (such as grants and procurements) by the end of the fiscal year 1992. However, as of February 1990, out of the 37 agencies surveyed by the GAO (Government Accounting Organization), only 6 had successfully fulfilled the guidelines (*Metric Conversion Plans, Progress, and Problems in the Federal Government,* GAO, March 1990). As you can see, it may be some time before we can truly call ourselves a global industrial nation with regard to our present system of measurement.

Although we primarily use the English system for measurement, we are still surrounded by the SI system. We see examples of the SI system when we go to the grocery store. For example, the label on a jar of peanut butter lists the weight as 12 ounces (English system) and the mass as 340 grams (SI). Both systems of measurement are also seen on soda bottles (with the volume measured in fluid ounces and milliliters) as well as just about any other product label. Since the SI system is unavoidable, we must try to understand its language. This section of the appendix will help you become "metric literate."

SI Prefixes and Base Units

Tables A.1 and A.2 show the seven fundamental SI base units and prefixes. You should make an effort to get to know (that means *memorize*) the units and prefixes listed in Tables A.1 and A.2. It is also important to relate SI units to English units. For example, in the Earth's gravitational field, if you weigh **220 pounds,** your mass is **100 kilograms (kg).** If you are **5.00 feet** tall, your height is **1.52 meters (m).** Table A.3 lists some additional English to SI relationships that will be used as conversion factors later in this appendix.

Fundamental Units

When using the metric system, we do not always stick to the fundamental units indicated in Table A.1. At times we must use derived units. There are two principal ways of arriving at a derived unit:

Table A.1

SI Units

Quantity	Name	Symbol
Length	meter	m
Mass	kilogram	kg
Time	second	s
Electric current	ampere	A
Temperature	kelvin	K
Luminous intensity	candela	cd
Amount of substance	mole	mol

Table A.2

SI Prefixes

Prefix	Symbol	Exponential Representation
pico	p	1×10^{-12}
nano	n	1×10^{-9}
micro	μ	1×10^{-6}
milli	m	1×10^{-3}
centi	c	1×10^{-2}
deci	d	1×10^{-1}
kilo	k	1×10^{3}
mega	M	1×10^{6}
giga	G	1×10^{9}

Table A.3

English to SI Conversions

Dimension	Conversion Factors
Length	1 mile (mi) = 1.6093 kilometers (km)
Mass	1 pound (lb) = 453.59 grams (g)
Volume	1 gallon (ga) = 3.7854 liters (L)

1. **Combine several fundamental units.** For example, distance (measured in meters) is considered a fundamental SI unit. But when we measure distance per unit time, we are actually measuring speed in meters per second, or m/s.

2. **Raise the fundamental unit to a power.** Volume of a solid object, which is measured in cubic meters (m^3), is derived from a fundamental unit—length.

Note: Fundamental units with prefixes attached to them, such as centimeter (cm), millisecond (ms), etc., are still considered fundamental.

TEST YOURSELF

1. Using Table A.2, put the following prefixes in order from smallest to largest:

   nano   milli   mega   micro
   pico   centi   deci   kilo

2. By attaching the prefix to the unit name, we form multiple SI units. Each can be represented by combining the prefix symbol with the unit symbol. Indicate the unit name for the following symbols:

   Example: ns
   Answer: nanosecond
   a. cm
   b. ng
   c. Mm
   d. ms
   e. kA

3. State whether the following units are *fundamental* or *derived:*
   a. density (g/mL)
   b. area (m^2)
   c. distance (cm)
   d. Newton (N = kg · m/s^2)
   e. volume (mL)
   f. weight (lbs)
   g. time (ms)

ANSWERS

1. pico < nano < micro < milli < centi < deci < kilo < mega
2. a. centimeter   b. nanogram   c. megameter
   d. microsecond   e. kiloampere
3. a. *derived,* volume is length cubed, and the ratio of mass to volume is also derived
   b. *derived,* area is length squared
   c. *fundamental*
   d. *derived* from mass, length, and time
   e. *derived,* volume is length cubed
   f. *derived,* English unit weight is derived from the SI unit of mass
   g. *fundamental*

A.4   Dimensional Analysis

Now that you have started to master the SI units and have prefixes mastered, we can focus on a very important skill—conversion. Whether it be converting between systems (i.e., metric to English or English to metric) or within a system (from one prefix or unit to another), the method employed involves using **conversion factors** (see Table A.3) to cancel units of measurement until you end up with only the unit you want. This technique is referred to as **dimensional analysis.**

When setting up problems using dimensional analysis, initially you are more concerned with *units* than with numbers. The units must cancel each other out. Let's illustrate this by making a pound-to-kilogram conversion.

Example A.3

Mass

PROBLEM
What is the mass, in kg, of a 275-lb box (1 kg = 2.2046 lb)?

SOLUTION
First you must clarify what the problem is asking for. In this case, we are looking for the number of *kilograms* this box weighs, therefore the answer should be in *kilograms.* (Fairly obvious, wouldn't you say!) Place the unit you want to end

up with at the left side of your solution followed by an equal sign:

$$kg = ?$$

Second, we set up our conversion factors. Usually, the necessary conversion factors are given in the problem (as in our problem). The conversion factor 1 kg = 2.2046 lb can be written as the following ratio:

$$\frac{1 \text{ kg}}{2.2046 \text{ lb}}$$

If the appropriate conversion factor is not stated, you can find a table of conversion factors on the **inside cover of this book** or in other reference sources. *When you choose a conversion factor, make sure that you pick one that will lead you to the unit asked for in the problem.* It may be necessary to use more than one conversion factor to get to the unit desired (we will demonstrate this in Example A.5).

To set up your equation, start with the desired unit followed by an equal sign:

$$kg =$$

Put the conversion factor to the right of the equal sign. Notice that we have placed the desired unit (kg) in the numerator:

$$kg = \frac{1 \text{ kg}}{2.2046 \text{ lb}}$$

Now we can introduce the amount we want to convert, in units of pounds. We multiply it by the conversion factor, which effectively makes it part of the numerator of the conversion factor:

$$kg = \frac{1 \text{ kg}}{2.2046 \text{ lb}} \times 275 \text{ lb} \left(\text{that is } \frac{275 \text{ lb}}{1} \right)$$

The initial dimensional unit (pounds) cancels out leaving us with the desired unit (kilograms).

Finally, after verifying that all units have been canceled (except for the one desired in our answer), we can proceed to *multiply the values in the numerator and divide by the values in the denominator* to arrive at the final answer. (Be sure to include the correct number of significant figures in your answer.)

$$kg = \frac{1 \times 275 \text{ kg}}{2.2046}$$
$$= \textbf{124.73918 kg, rounds to 125 kg}$$

Let's try another example!

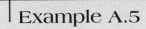

Example A.4
Miles in Kilometers

PROBLEM

How many miles are in 35.8 kilometers (km)?

SOLUTION

1. The problem is asking for miles:

$$\text{kilometers (km)} \rightarrow \text{miles (mi)}$$

2. Our conversion factor can be found in Table A.3.

$$1 \text{ mi} = 1.6093 \text{ km or } \frac{1 \text{ mi}}{1.6093 \text{ km}}$$

3. The problem should be set up with the desired unit in the numerator of the conversion factor and the initial unit in the denominator, as:

$$\text{mi} = \frac{1 \text{ mi}}{1.6093 \text{ km}} \times \frac{35.8 \text{ km}}{1}$$

Notice how the units cancel each other out!

4. Now we can multiply and divide as:

$$\text{mi} = \frac{1 \times 35.8 \text{ mi}}{1.6093 \times 1}$$
$$= \textbf{22.245697 mi, rounds to 22.2 mi}$$

5. After solving the problem, always ask yourself, "Does the answer make sense?" One mile is farther than one kilometer, therefore, our answer should be less than 35.8 km. We can see that our answer does indeed make sense!

Summary of Problem-Solving Steps

1. Decide what the problem is asking for and write down the unit of measurement desired in the answer.
2. List all necessary conversion factors.
3. Set up the problem making sure that the units are placed so that they cancel properly.
4. Multiply all the values in the numerator and divide by all those in the denominator.
5. Double-check that *all units cancel properly.* If they do, your numerical answer is probably correct. If they don't, your answer is certainly wrong.

This method seems quite tedious for such a simple problem. However, you will discover in future problems that the work involved is necessary for the sake of organization!

Let's extend this technique to a more cumbersome example.

Example A.5
Rate

PROBLEM

The World's Fair in Knoxville, Tennessee, held an apple-pie-eating contest. It takes a dozen apples to make 1 pie. The winner devoured 28 pies in 30 minutes. At this rate how many apples would the winner devour in 1 hour?

PROBLEM SOLVING PROCEDURE

Step 1. What is the problem asking for?

The problem is asking for the number of apples that will be devoured in 1 hour. Therefore, the unit that should appear in our answer is apples/hour.

Step 2. List conversion factors:
   a. 12 apples = 1 pie, which is *exactly the same* as 1 pie = 12 apples. Therefore you can use either

$$\frac{12 \text{ apples}}{1 \text{ pie}} \text{ or } \frac{1 \text{ pie}}{12 \text{ apples}}$$

   However, it is *not correct* to use

$$\frac{1 \text{ apple}}{12 \text{ pies}} \text{ or } \frac{12 \text{ pies}}{1 \text{ apple}}$$

   When you flip units, the numbers must be transferred with them.
   b. 28 pies per 30 minutes can be expressed as

$$\frac{28 \text{ pies}}{30 \text{ minutes}} \text{ or } \frac{30 \text{ minutes}}{28 \text{ pies}}$$

   c. Although it is not directly stated in the problem, you need a conversion factor from minutes to hours.

$$\frac{60 \text{ minutes}}{1 \text{ hour}} \text{ or } \frac{1 \text{ hour}}{60 \text{ minutes}}$$

Step 3. Set up the problem:
The units desired in the answer are written on the lefthand side of the equal sign. The information given in the problem and all conversion factors (written as factors) are placed to the right of the equal sign (so all "unwanted" units cancel on the right).

$$\frac{\text{apples}}{\text{hour}} = \frac{12 \text{ apples}}{1 \text{ pie}} \times \frac{28 \text{ pies}}{30 \text{ minutes}} \times \frac{60 \text{ minutes}}{1 \text{ hour}}$$

Step 4. Multiply and divide.
Consolidate (multiply) numerators and denominators; then divide the final numerator by the final denominator:

$$\frac{\text{apples}}{\text{hour}} = \frac{12 \times 28 \times 60 \text{ apples}}{30 \text{ hours}} = \frac{\textbf{672 apples}}{\textbf{1 hour}}$$

Step 5. Double-check the units!

$$\frac{\text{applies}}{\text{pies}} \times \frac{\text{pies}}{\text{minutes}} \frac{\text{minutes}}{\text{hour}} = \frac{\text{apples}}{\text{hour}}$$

Working with Prefixes

Dimensional analysis will often involve conversion between prefixes of the same unit. Let's try to incorporate prefixes

in our dimensional analysis scheme with the next sample problem. (It is especially important to pay close attention to the validity of your answer when using very large or small values.)

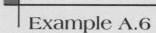

Example A.6

μm in 1 km

PROBLEM
How many μm (micrometers) are there in 1 km?

CRITICAL CONCEPT
Keep in mind that the prefix "micro" means 10^{-6}. It is a very small number. There are many μm in just a meter. You can write the conversion between μm and m as:

$$\text{(a)} \quad 1 \text{ μm} = 10^{-6} \text{ m}$$

or

$$\text{(b)} \quad 10^{6} \text{ μm} = 1 \text{ m}$$

It is generally easier to use (a) because this equation directly relates to the exponential representation for micrometers as indicated in Table A.2. Similarly,

$$1 \text{ km} = 10^{3} \text{ m because kilo} = 1 \times 10^{3}$$

Since both μm and km relate to m, we should use m as a *bridge* between μm and km.

SOLUTION
1 km (the value given in the problem) must be converted to μm. Therefore, the bridge between prefixes is km → m → μm. The conversion factors are

$$\frac{1 \text{ μm}}{1 \times 10^{-6} \text{ m}} \text{ and } \frac{1 \text{ km}}{1 \times 10^{3} \text{ m}}$$

The setup looks like this:

$$\underset{\text{want}}{\text{μm}} = \underset{\text{given}}{1 \text{ km}} \times \frac{1 \times 10^{3} \text{ m}}{1 \text{ km}} \times \underset{\text{conversion factors}}{\frac{1 \text{ μm}}{1 \times 10^{-6} \text{ m}}}$$

$$= 1 \times 10^{9} \text{ μm in 1 km}$$

CHECK
Does the answer make sense? A μm is very small. A km is very large. Therefore, we would expect that there would be a lot of μm in a km. Carelessly inverting prefix conversions (such as incorrectly stating that 1 m = 10^{3} km) is among the major sources of incorrect answers in general chemistry.

TEST YOURSELF

1. How many centimeters (cm) are in 1 megameter (Mm)?
2. How many kilometers (km) are in 2.5×10^8 millimeters (mm)?
3. How many μg are in these units?
   a. 1 centigram (cg)
   b. 35 nanograms (ng)
   c. 1.0×10^{-4} decigrams (dg)
   d. 4.89×10^3 milligrams (mg)
   e. 3.5 pounds (1 kilogram = 2.2046 pounds)
4. Convert 6.5 quarts to these units (1 quart = 0.94633 liters):
   a. kiloliters (kL)   b. milliliters (mL)
   c. centiliters (cL)   d. microliters (mL)
5. A slight increase in worldwide rubber production recently occurred, from 10,300 metric tons in 1999 to 10,820 metric tons in 2000. What would this increase be in pounds? Express your answer in exponential notation (1000 kg = 1 metric ton).
6. If a car weighs 3189 pounds, what is its mass in grams? Express your answer in exponential notation (1 kg = 2.2046 lb).
7. In the vacuum of space, light travels at a speed of 186,000 mi per second. How many kilometers can light travel in a year (1.6093 km = 1 mi)?
8. In July 1992, 14 cars from a Burlington Northern train derailed dumping 26,200 gallons of benzene solution into a Wisconsin river. How many liters would this be (1 gallon = 3.78 L)?

ANSWERS

1. 1×10^8 cm
2. 2.5×10^2 km
3. a. 1×10^4 μg   b. 3.5×10^{-2} μg   c. 10 μg
   d. 4.89×10^6 μg   e. 1.6×10^9 μg
4. a. 6.2×10^{-3} kL   b. 6.2×10^3 mL
   c. 6.2×10^2 cL   d. 6.2×10^6 μL
5. 1.14×10^6 pounds
6. 1.447×10^6 g
7. 9.44×10^{12} km/year
8. 9.90×10^4 L

The Role of Energy in Chemical Reactions

"All I want is a room somewhere
far away from the cold night air
and one enormous chair. Oh wouldn't it be loverly!
Lots of chocolate for me to eat,
Lots of coal making lots of heat—
Warm face, warm hands, warm feet—Oh wouldn't it be loverly!"

—Eliza Doolittle in *My Fair Lady*
by A. Lerner and F. Loewe

It's a cold day and you've been outside for too long. It feels good to go into the house and get warm. Maybe you sit in front of a fireplace with a gas or wood fire. Perhaps you stand beside a radiator filled with circulating steam or hot water, or find a chair next to a register releasing warm air from the furnace in the basement. You may go to the kitchen and pull up a chair by the stove, where a kettle of soup is simmering, and after enjoying some soup you might heat a cup of hot chocolate in the microwave oven.

All these ways of getting warm are examples of how we can harness energy for our comfort and survival. In our homes we use electrical energy to power most of our appliances, and the energy released by burning natural gas may heat our water. Energy gives us heat, light, and the power to make our machines work. Let's think about what energy is, where we get it, where it goes when we are finished with it, and how to use it efficiently.

Ultimately, most of our energy has come from the Sun (an exception is nuclear energy). Solar energy, which can be enjoyed while we sunbathe or harnessed in more sophisticated ways, is the energy that flows directly from the Sun. The release of energy by wood fires is possible because trees grow under the influence of sunlight, storing some of the Sun's energy for it to eventually be released by a fire. The fossil fuels (coal, oil, and natural gas) exist because plants grew in the sunlight many years ago and became transformed into the fossil fuels over millions of years. Even hydroelectric power results from the energy of sunlight evaporating water from the oceans, after which the water falls as rain in the

On a cold winter day, even bright sunshine is not enough to keep us warm. Vigorous exercise may warm us for awhile, but eventually we head indoors to a blazing fire (if possible).

mountains and delivers as available energy as it flows back to the sea.

The star we call the Sun is the source of energy that makes almost all of the interesting things on Earth happen, including the birth, growth, and achievements of every human being. Energy from the Sun *flows to the earth,* and powers the most significant chemistry on Earth as it is absorbed. Even those chemical processes that are not powered directly by the Sun are powered by the flow, or *dispersal,* of energy through the chemicals concerned.

Although particles of matter are the players in the drama of chemistry, energy makes these players move. The flow of energy makes things happen.

5.1 | What Is Energy?

We use the term **energy** in everyday life. We talk of feeling "full of energy," or of "not having the energy" to do some things. We talk of "energy sources," such as food (for our bodies), gasoline (for our cars), or electricity (for our homes). We talk of saving energy by driving our cars less or turning off the lights when we leave a room. We know that we use energy when we exercise. We have national debates about the sources and safety of our energy supplies. We worry that our overuse of energy may be damaging the environment, and countries even go to war over the rich black "source" of energy known as oil. Humanity's quest for and exploitation of energy has made countless fortunes, transformed countless lives, and caused countless deaths. But what is energy?

You need to know two things right at the start before we actually try to define energy. First, energy is never created or destroyed, it just moves from place to place. This is the **Law of Conservation of Energy.** It is also known as the **"First Law of Thermodynamics"** (**thermodynamics** is the study of energy changes). The Law of Conservation of Energy means that when we talk of some process "requiring energy," we mean that it requires energy to be *transferred* from one place to someplace else. For example, much of the energy in a candy bar is released as the bar is metabolized in your body and can then be transferred to your muscles to power the movements you make when you exercise, eventually leaving your body as heat. Second, energy is not a "thing" in the normal sense of that word, it is a *property* associated with physical systems. Think of energy as comparable with the "excitement" within the crowd at a football game or baseball match. You cannot point to it, pick it up, work out its size or shape or color; but it is there all the same, and it has powerful effects on any "system" that contains it.

The warmth of the fire comes from the chemical reaction we know as combustion.

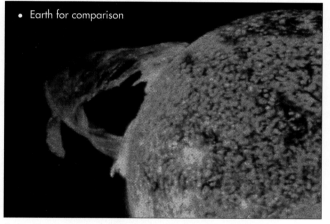

- Earth for comparison

Solar flares are spectacular demonstrations of the Sun's energy.

Solar energy is transferred to Earth by sunlight, which warms our atmosphere and is critical to plant and animal life.

Energy as the Capacity to do Work

If you consult a dictionary of science, you will probably find **energy** defined as "the capacity to do work," or perhaps as "the capacity to do work and transfer heat." That definition may simply make you wonder what is meant by **work.**

Flip through the pages to the back of the dictionary, and you will probably meet a definition of work, along the lines of "the product of force times the distance through which the force acts." This means that work is being done whenever some physical force is being used to move something. This idea can be made much clearer by referring to an example.

It takes energy to lift a rock up from the ground, and it certainly feels like "work" in the everyday sense of the word. It is also true that you need to use a physical force (the force exerted by your muscles) to raise the rock upward. Why should this be so? It is because you have to fight against the other physical force of gravity, which causes the rock and the ground to be attracted toward one another. Gravity is one of the "fundamental forces" of nature and this is the clue to really grasping what work and energy are all about. Doing "work" (in the strict scientific sense) involves making something move along a path it would otherwise not follow, such as moving it against the pull of a fundamental force. To do that, you have to use another fundamental force, so you are exploiting one force to fight against the other (Figure 5.1).

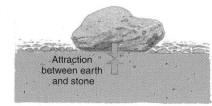

Attraction between earth and stone

Figure 5.1 All objects are attracted to each other by the force of gravity.

Let's separate the situation into two physical systems: (1) you and (2) the rock and the rest of the Earth. As you lift a rock (see Figure 5.2), we say that energy is being transferred from your body to the system composed of the rock and the rest of the planet. The energy embodied within the position of the rock increases, while the energy within your body decreases. Whenever any change within a system involves movement against a fundamental force, the energy of the system increases accordingly. As the rock rises upward, against the force of gravity, the energy embodied within its position is increasing. There is a specific name given to the energy embodied within the positional arrangement of things: it is called **potential energy.**

If you let go of the rock (Figure 5.2) it will begin to fall, pulled by the force of gravity. As it falls, it is losing the potential energy you just gave it, because its height above the Earth is decreasing. What is happening is that the energy is changing into a different form, known as **kinetic energy.** Kinetic energy is also known as "the energy of movement." Anything that is moving possesses a certain amount of kinetic energy, depending on how fast the thing is moving and its mass. As our rock falls, it is losing potential energy, but gaining a compensating amount of kinetic energy. In other words, the potential energy is steadily converted into kinetic energy. If we ignore the effects of air resistance, then just at the instant before the rock hits the ground, all of the potential energy you gave to the rock will be in the form of kinetic energy as the rock rushes downward. If the rock hits the ground it quickly comes to a halt, losing its kinetic energy. This energy does not disappear but is transferred into the ground causing the particles of the ground to move about a bit more quickly, or in other words, causing the ground to heat up a little.

We are making good progress toward understanding energy, but we have a way to go yet. For one thing, we said at the outset that energy is the capacity to do work (that is, to cause movement against a fundamental force); but we have now said that two fundamental forms of energy are potential energy (stored in objects' positions) and kinetic energy (associated with

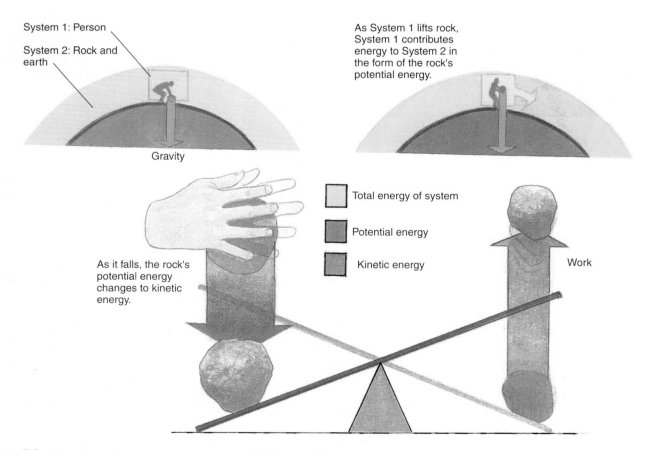

Figure 5.2 The nature of energy exchange as a rock is lifted and then dropped.

movement). To link these three facts together, we need to convince you that the position of the rock above the Earth, or its movement, can be used to do work.

Look at Figure 5.2, and consider what will happen if, instead of hitting the ground, the rock hits the plank of wood. The *other* rock in the figure will be propelled upward, so the movement of the falling rock will do work, because it will raise the second rock upward. Figure 5.2 demonstrates that things that are moving really do have "the capacity to do work." They really do have "energy." This should convince you that the kinetic energy of movement has the capacity to do work, but what about the potential energy stored in the rock when it was stationary, just as you released it after lifting it from the ground? Well, this is the potential energy that was converted into the kinetic energy that can do the work of propelling the second rock upward. So the positioning of things in such a way that a force (such as gravity) is "set up" ready to cause movement, can also, through the eventual effects of the movement, do work. "Potential" energy gets its name because of its capacity, or potential, to do work.

What has all this got to do with chemistry? Substitute the electric force for the force of gravity, and the answer becomes clear. Chemicals are composed of oppositely charged particles in shifting positions, and each position embodies a certain amount of potential energy. As an electron moves away from an atom's nucleus, the potential energy embodied in its position increases (because the movement is against the attraction between the opposite electrical charges of the electron and the nucleus). As an electron moves toward an atom's nucleus, the potential energy embodied in its position decreases. As any two electrons move closer together (against the force of repulsion between them), the potential energy embodied in their positions increases; and as they move apart it decreases. Also, as any two nuclei move closer together (against the force of repulsion between them) the potential energy embodied in their positions increases; and as they move apart it decreases. Also remember that electrons, atoms, molecules, and ions are also *moving*. So all the particles of chemistry possess kinetic energy as well.

Within chemicals, energy is constantly being converted between potential energy and kinetic energy as the particles move around and bump into one another; their electrons and nuclei also flit between varying arrangements embodying differing mixtures of potential and kinetic energy.

Chemical reactions are accompanied by energy changes, meaning that the chemicals either release or absorb energy as the reactions proceed. We discussed in Chapter 3 the reaction of Group 1A metals with water. Figure 5.3 shows the reaction of potassium (K) with water. This reaction,

$$2K(s) + 2H_2O(l) \rightarrow 2KOH(aq) + H_2(g)$$

clearly shows the release of energy. Such energy changes are due to changes in the movement and relative positioning of the electrons and nuclei within the chemicals concerned. So the energy released by chemical reactions, such as the burning of wood, is derived from changes in the motion and positional arrangements of the chemicals' electrons and nuclei. When energy is taken in by chemicals, it is held within the chemicals in the form of movement and rearranged positions of electrons and nuclei.

Let's look at a historically important example of the transfer of energy: the burning of compounds in a fire. Most natural fires are the result of the high-temperature reaction of oxygen with a compound containing carbon and hydrogen (and sometimes oxygen). This reaction is known as combustion. Table 5.1 gives a few examples of typical substances that are burned for the heat energy released during combustion.

Whichever fuel is burned, the products of complete combustion are carbon dioxide and water. The product molecules are much hotter than the reactant molecules were before the fire began. What this really means is that the gaseous product molecules are moving much faster than they would at lower temperatures. Also, the molecules are vibrating and spinning around in rapid chaotic motion.

As we will see in Chapter 6, wood is mainly cellulose, a complicated mixture of molecules having different numbers of carbon atoms. The number of hydrogen atoms in cellulose is *always double* the number of oxygen atoms present, and the number of oxygen atoms is always a bit smaller than the number of carbon atoms. Nonetheless, it falls into our pattern of fuels containing C, H, and O atoms.

Figure 5.3 When mixed with water (H_2O), potassium (K) produces a noticeable reaction.

Table 5.1

Typical Substances Burned for Heat Energy During Combustion

Substance	Chemical Formula
Natural gas	CH_4
Propane	C_3H_8
Butane	C_4H_{10}
Ethyl alcohol	C_2H_6O
Gasoline	C_8H_{18}
Paraffin	$C_{21}H_{44}$
Beeswax	$C_{30}H_{60}O_2$
Wood	$C_{6000}H_{10002}O_{5001}$

The glow of a wood fire is the result of a high temperature reaction between oxygen and a compound containing carbon and hydrogen.

A typical equation for the reaction of burning wood could be:

$$C_{6000}H_{10002}O_{5001} + 6000 O_2 \rightarrow 6000 CO_2 + 5001 H_2O$$

For the burning of paraffin candles (which contain only carbon and hydrogen atoms) the equation could be

$$C_{21}H_{44} + 32 O_2 \rightarrow 21 CO_2 + 22 H_2O$$

In all cases, energy is released by the burning reactions because the electrons and nuclei of the chemicals are shifting into lower energy arrangements as the reactions proceed. So when you warm yourself by a fire, you should be grateful to shifting electrons and nuclei for releasing the heat. When oil companies search for new supplies of oil for use as an energy source, they are really seeking out particularly energetic arrangements of electrons and nuclei. These energetic arrangements have been waiting beneath the Earth's surface for millennia to release some of their energy to warm us or power our machines.

5.2 | Energy Appears in Many Forms

Heat

Heat energy is in fact **kinetic** energy: it is the energy possessed by a substance due to the movement of its atoms, ions, and molecules. The **temperature** of a substance is a measure of the average kinetic energy of its particles. Heat and temperature are not the same thing. For example, a red-hot penny has a higher temperature than a tub of warm water; but there is more heat contained in the water. Although the average kinetic energy of the atoms in the red-hot penny is higher than the average kinetic energy of the molecules in the tub of water, there is more heat in the water because there are so many more water molecules in the tub than atoms in the penny. An everyday analogy might help. The average wealth of four executives in a car may be higher than the average wealth of the people in the crowd at a football match; but the audience may own much more money (in total) than the four executives in the car.

In common use, temperature represents the relative hotness or coldness of one thing *compared to* another. It is therefore fair to say that if it is 82°F (degrees Fahrenheit), you feel a lot warmer than if it were 41°F. But you are not *twice* as warm merely because 82 is twice 41. By the same token, if it were 3°C outside, that would not be three times as warm as 1°C. Neither Celsius nor Fahrenheit scales is useful when we want to compare average energies mathematically.

There is a temperature scale, called the **Kelvin** scale, that is directly related to the average kinetic energy of a system. This temperature scale is used primarily for scientific research applications, rather than more common situations such as measuring oven temperatures or

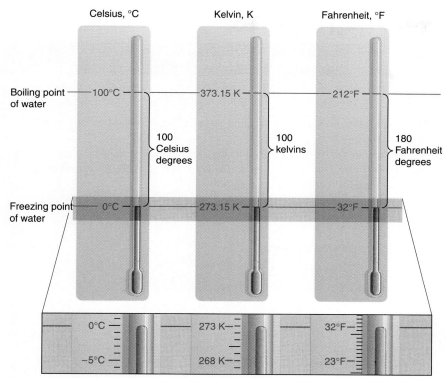

Figure 5.4   The Kelvin scale, also known as the absolute scale, is preferred for scientific work.

forecasting the weather. A substance whose atoms, molecules, or ions are completely stationary would have a temperature of 0 K (read as "zero kelvins") or **absolute zero.** A liquid at a temperature of 100 K ("one hundred kelvins") *does* have twice the energy of that same substance at 50 K. (Note that no "degree" symbol is used with the Kelvin scale.)

The **Celsius** (formerly called "centigrade") scale is in common use throughout most of the world. One degree Celsius has exactly the same magnitude as 1 K, but the two scales start at a different point; 0 K corresponds to $-273.15°C$. So to convert a temperature in Celsius to one in kelvins we need to add 273.15 degrees to the Celsius value (Figure 5.4).

The formulas we use to convert temperature among the three temperature scales are:

$$T_C = 5/9 \times (T_F - 32°)$$
$$T_K = T_C + 273.15$$

Can you derive the conversion formulas from Fahrenheit to Celsius and from Celsius to kelvin?

Light

Light is just the visible form of **electromagnetic radiation,** consisting of particular wavelengths in what is known as the electromagnetic spectrum (see Figure 5.5). When light is absorbed by chemicals, it causes electrons to jump into higher energy orbitals. So light can make electrons move away from the nucleus of an atom, as we would expect for a form of energy. Other electromagnetic radiations can energize chemicals in a similar way, causing more dramatic disturbances if they are more energetic (higher frequency) radiations such as X-rays. Such disturbances can be quite useful even at low energies, for example, microwave radiation used to cook food.

When electrons "fall" from a high-energy orbital into a lower-energy one, the energy difference between the two orbitals can come out in the form of light. This is what is happening when the chemicals within fireworks give out their brilliant colors of light. As we admire the beauty of the fireworks, we are admiring the light released by electrons shifting to lower-energy orbitals (having been temporarily pushed into high-energy orbitals by the energy released during the explosion).

exercise 5.1

Practice with Temperature Conversions

Problem

(a) The temperature on a pleasant autumn day is 17°C. What is that in degrees Fahrenheit (°F)?

(b) A frigid day in the Northern Plains of the United States can see the temperature drop to −40°C. **i.** Convert that to °F and K. **ii.** Is it possible to determine how much hotter than −40°C is a reading of 379°F?

Solution

(a) Using our conversion formula,

$$°F = (9/5 \times °C) + 32°$$
$$°F = (9/5 \times 17°) + 32° = 62.6° = 63°F$$

(b) **i.** $T_F = 9/5\, T_C + 32 = 1.8\,(-40) + 32 + (-72) + 32 = -40°F$
$\qquad T_K = 273.15 + T_C = 273.15 + (-40) = 233.15 = \mathbf{233\ K}$

ii. We have learned that measures of absolute hotness require us to use the Kelvin scale. It's easiest to get to kelvin via Celsius.

$$T_C = 5/9(T_F - 32) = 5/9(379 - 32) = 5/9(347) = 192.8 = \mathbf{193°C}$$
$$T_K = T_C + 273.2 = 192.8 + 273.2 = \mathbf{466\ K}$$

This final value, 466 K, is **twice as hot** as 233 K. One more thing—as a rule of thumb, we round off only when reporting final answers. We use unrounded numbers in calculations, as we did in the examples above.

Like it or not, the world thinks of temperature in degrees Celsius for most applications and in kelvins for some. To increase your familiarity to these two scales, and to be consistent with standard scientific usage, we will use only °C and K throughout the rest of this book.

Both Fahrenheit and Celsius temperature scales are defined in terms of the freezing point of water (32°F and 0°C). The difference between freezing and boiling points of water is 180°F and 100°C. In other words, a Fahrenheit degree represents 100/180 or 5/9 of a Celsius degree and a Celsius degree is 9/5 of a Fahrenheit degree. Since the freezing point is also 32° higher on the Fahrenheit scale than the Celsius scale, that difference must be added into the calculation. Thus, a temperature of 37.0 Celsius is converted to Fahrenheit by multiplying 37.0 by 9/5 (= 1.8) and adding the 32° difference the Fahrenheit scale has above zero on the Celsius scale.

$T_F = 9/5\, T_C + 32 = (1.8)(37.0) + 32.0 = 66.6 + 92 = 98.6°F$

This is the familiar normal body temperature of healthy people.

Figure 5.5 The electromagnetic spectrum lists the range of energies found in our universe. The visible region is only a small (but important) part of the energy range. It takes on such importance to us because our eyes respond to it! More powerful than the visible region are the ultraviolet, X-ray, and gamma-ray regions. Less powerful, but important nonetheless, are the infrared, microwave, and radio wave regions. Note the general relationship among the wavelength, frequency, and energy.

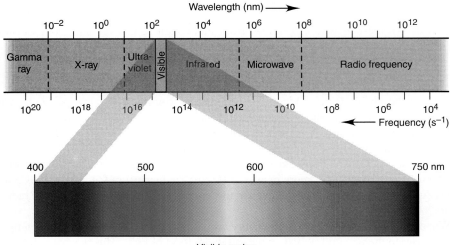

Fireworks are usually ignited by the rapid burning of gunpowder. This heats the atoms of metals added to the fireworks mixture up to very high temperatures. Metals like magnesium or iron burn in air at these high temperatures to release energy in the form of brilliant white light and also heat (Figure 5.6).

$$2Mg + O_2 \rightarrow 2MgO$$
$$3Fe + 2O_2 \rightarrow Fe_3O_4$$

The hotter the particles get, the more light they give off. This is called incandescence. The way fireworks manufacturers get colorful fireworks is by capitalizing on atomic emission: the ability of an electron to be excited to a higher energy orbital and then immediately release that energy in the form of light. Strontium is often added to fireworks to give a red color. In this case, an electron of the strontium atom accepts some of the heat energy and almost immediately gives the energy back as light. The symbol Sr* symbolizes a strontium atom having more energy than normal.

$$Sr \quad \rightarrow \quad Sr^* \quad \text{requires heat}$$
$$Sr^* \quad \rightarrow \quad Sr \quad \text{emits light energy, mostly red}$$

Other elements give off other characteristic colors. In fireworks, sodium is used for yellow light, barium is used for green, and copper is used for blue.

Figure 5.6 Burning magnesium, iron, or other metals is responsible for the brilliant displays in fireworks. Other elements are added to create a variety of colors.

Other Forms of Energy

Kinetic energy and potential energy are the two fundamental forms of energy. You may have heard about many other so-called forms of energy in your school science classes. We are thinking of such things as "wind energy," "wave energy," "electrical energy," "nuclear energy," "microwave energy," and so on. In fact, these are all found to be manifestations of kinetic energy and potential energy when they are examined closely. Wind energy, for example, is due to the overall motion (the kinetic energy) of the molecules in the air.

The motion of the air that we call wind is caused by the energy of the Sun, heating different parts of the Earth and the atmosphere to differing extents. Electrical energy is due to the motion of electrically charged particles as their potential energy and kinetic energy interconvert when they are subject to an electrical field. Nuclear energy is released when changes in potential energy occur within the nuclei of atoms (Figure 5.7).

Figure 5.7A A field of windmills generates electricity using the power of air currents.

Figure 5.7B Nuclear power plants, like this one, generate electricity by harnessing the power of the atomic reaction.

Figure 5.7C Electric current can be both the end product and the force.

The "energy debate" can sometimes seem very complex, demanding choices between "energy" in the form of coal, oil, gas, nuclear, wood, light, wind, waves, and so on. But in reality, it concerns choices over how the kinetic or potential energy available from such sources is released and used.

5.3 | Energy Coming Out and Energy Going In

In an exothermic reaction heat energy **is released.**
In an endothermic reaction heat energy **is absorbed.**

When a process (such as a chemical reaction) *gives off* energy, it is known as **exergonic** (i.e., energy-releasing.) If a process *takes in* (absorbs) energy, it is known as **endergonic** (energy-absorbing). Although these terms are technically correct, they are not used all that much. Instead, we usually use the words **exothermic** and **endothermic** which imply that only heat energy is given out or taken in by a chemical reaction. This is because the energy changes associated with chemical reactions usually do involve heat coming out or heat going in; and even when they do not (as in reactions that release light, for example), we can always arrange for the energy to be converted into the form of heat, to allow it to be quantified. We can perform a reaction within a closed, nontransparent chamber, for example, so that the only way for energy to get into it or come out of it is in the form of heat.

When chemical reactions occur, the chemicals either release energy or take in energy. If we could monitor the change in the chemicals' energy content as a reaction proceeds, we would find curves like the ones in Figure 5.8. A reaction that releases energy (Figure 5.8A) is called an exothermic reaction. A reaction that takes in energy (Figure 5.8B) is called an endothermic reaction.

Good examples of exothermic and endothermic reactions in action are provided by some of the chemical "hot packs" and "cold packs" that campers and hikers use to warm themselves on a cold day or cool an injury such as a twisted ankle (Figure 5.9).

One type of chemical hot pack is a small plastic bag loaded with powdered iron, salt, and a bit of water with other ingredients, all surrounded by a second plastic bag. The outer bag is impervious to oxygen gas but the inner bag allows oxygen gas from the air to move through easily. When the outer bag is removed, oxygen comes in contact with the iron, which oxidizes over a period of a few hours by the reaction:

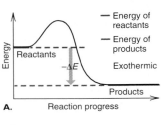
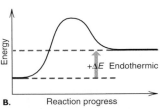

$$4Fe + 3O_2 \rightarrow 2Fe_2O_3$$

This is an exothermic reaction. Every mole of iron(III) oxide formed this way gives off 826 kJ of heat, which serves to warm whatever part of the body the pack is applied to.

One type of chemical cold pack is based on the endothermic dissolving of ammonium chloride. In this pack, a container of solid ammonium chloride is packed with a small amount of water in a separate pouch. Mixing the solid ammonium chloride with water initiates the endothermic process in which the ammonium chloride dissolves in the water. This draws heat energy in from the surroundings, such as whatever part of the body the pack is applied to, causing that part of the body to get colder.

Figure 5.8 Energy profile curves for **A.** an exothermic reaction and **B.** an endothermic reaction. There is a net loss of energy from the reaction system in an exothermic reaction. Energy is gained in an endothermic reaction. In any chemical reaction, the total energy of the universe is constant.

The Language of Energy Change

When energy goes in or comes out of a chemical reaction as heat, we can signify the energy change using a number following the symbol "ΔH," in which the "Δ" means "change in" and the "H" can be thought of as "heat content." The $\Delta H°$ values we will quote have been obtained under "standard conditions," set at 25°C and 1 atmosphere (atm) of pressure (the air pressure at sea level). The ΔH value associated with an exothermic reaction is always a negative number because the chemicals end up with *less* energy than they started with. For example, NASA's space shuttle is powered by an exothermic reaction in which hydrogen gas and oxygen gas combine to form water. This reaction releases around 16 million joules of energy per liter of water formed, and could be summarized using the equation and its energy change:

$$2H_2(g) + O_2(g) \rightarrow 2H_2O(l) \qquad \Delta H = -15.87 \times 10^6 \text{ J/L water formed}$$

Figure 5.9 Chemical hot packs and cold packs change temperature because of chemical processes that release or absorb heat.

A liter of water contains

$$1000 \text{ g H}_2\text{O} \times \frac{1 \text{ mol H}_2\text{O}}{18.02 \text{ g H}_2\text{O}} = 55.5 \text{ mol water}$$

$$\Delta H = \frac{-15.87 \times 10^6 \text{ J}}{\text{L}} \times \frac{1 \text{ L}}{55.5 \text{ mol}} \times \frac{1 \text{ kJ}}{10^3 \text{ J}}$$

$$= \frac{-286 \text{ kJ}}{\text{mol}}$$

so $\Delta H = -2.86 \times 10^5$ J/mol $= -286$ kJ/mol.

The ΔH value associated with an endothermic reaction is always positive, because the chemicals end up with *more* energy than they started with.

The Units of Energy

We use units of energy all the time. We talk about losing weight by burning calories. When we compare the efficiencies of air conditioners we talk about the number of British Thermal Units (Btus) they will consume. The internationally accepted unit of energy, however, is the joule (J). This is different from the more commonly used calorie. These units can be interconverted by knowing that

1 calorie (cal) = 4.184 joules (J)
1 Calorie = 1000 calories = 4184 joules (4.184 kJ)
1 Btu = 1054.5 joules

When converting among units, just remember that a calorie is bigger than a joule, and a Btu is much bigger than a calorie. We will generally use joules or **kilojoules** (one thousand joules) when discussing energy change. The abbreviation for the kilojoule is kJ.

exercise 5.2

Joules vs. Calories

Problem

In the United States, boxes of breakfast cereal list the available energy content (the energy released to the body when the cereal is digested) in "Calories." Confusingly, what they really mean by this unit is "kilocalories." That is, "one nutritional Calorie" is actually one kilocalorie (kcal), or 1000 calories of heat energy. A capital C in the word "Calorie" is often used to indicate nutritional Calories (each equal to 1000 calories). A small box containing a 30-gram (about 1-ounce) serving of frosted corn flakes may list an energy content of 120 Calories. The total energy available from the cereal is really 120,000 calories, or 120 kilocalories. In Britain, cereal boxes list the energy content in both kilocalories and kilojoules (kJ), so the confusion between calories and "Calories" no longer arises.

How many kilojoules of potential energy are contained in 30 grams of frosted corn flakes with an energy content of 120 Calories?

Solution

There are 4.184 joules in 1 calorie. Therefore, there are 4.184 kJ in 1 kcal. In 120 kcal there are (120 kcal × 4.184 kJ/kcal) = **502 kJ** of energy in 30 grams of frosted corn flakes. Recall our general rules for problem solving. The final step is to assess whether or not the answer makes sense. Thirty grams of frosted flakes is about 1 ounce, as pointed out above. If there are about 4 kilojoules in a kilocalorie, it makes sense that there are about 500 kJ (close to 4 × 120) in a serving.

Much of the study of chemistry is concerned with the energy changes that occur when chemicals react, because the generation of energy costs money and many industrial chemical processes require precise energy control. Additionally, many consumer goods are themselves based on energy exchange, such as the energy contained in a battery to power toys, calculators, and so forth. In essence, the energy changes accompanying chemical reactions depend on the difference in energy taken in when chemical bonds are broken and the energy given out when chemical bonds are formed. Section 5.4 takes a close look at this central concept of chemistry.

5.4 | Breaking and Making the Bonds

We have said that when any chemical change occurs in a system, energy is released or absorbed. In other words, energy is *transferred,* either into the chemicals from their surroundings, or out of the chemicals to their surroundings. With the exception of nuclear processes like fission and fusion, this energy transfer results from chemical reactions—the reorganization of atoms, molecules, and ions brought about by the rearrangement of their electrons. The *amount* of energy transferred depends on which chemicals are reacting together and the quantities of chemicals involved. To understand what determines whether a chemical reaction releases or absorbs energy, and how much energy is transferred, we need a meaningful and interesting reaction. The decomposition of hydrogen peroxide fits the bill nicely.

Meet Hydrogen Peroxide

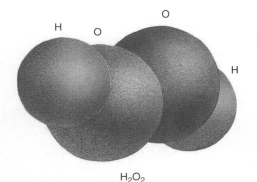

Figure 5.10 The molecular structure of hydrogen peroxide.

Hydrogen peroxide (H_2O_2) consists of molecules containing two oxygen atoms and two hydrogen atoms, bonded together as shown in Figure 5.10. Pure hydrogen peroxide is a thick, viscous liquid at room temperature. The clear watery "hydrogen peroxide" you can buy in drugstores is not pure hydrogen peroxide, but a 3% solution of hydrogen peroxide in water. The boiling point of hydrogen peroxide is 152.2°C, which is very high for a liquid with such a low molar mass (34 g/mol). Having such a high boiling point means that the attractions *among* hydrogen peroxide molecules are relatively strong, and a relatively high energy input is required to break them apart. The attractive forces between hydrogen peroxide molecules are the same kind of forces (known as hydrogen bonding) that lead to the unusually high boiling point, 100°C, of water (discussed on pages 248 and 251).

Hydrogen peroxide can be prepared in several ways, one of which is to react **peroxodisulfuric acid** ($H_2S_2O_8$), with water:

$$H_2S_2O_8(aq) \quad + \quad 2\,H_2O(l) \quad \rightarrow \quad H_2O_2(aq) \quad + \quad 2\,H_2SO_4(aq)$$

peroxodisulfuric acid   water   hydrogen peroxide   sulfuric acid

$$\Delta H = -99.1 \text{ kJ/mol } H_2O_2$$

The hydrogen peroxide product can be separated from the solution by boiling the solution and condensing (liquefying by cooling) the hydrogen peroxide as it boils off. This method of purification is an example of **distillation.**

$H_2S_2O_8$ + $2\,H_2O$ $\longrightarrow$ H_2O_2 + $2\,H_2SO_4$

The Importance of Hydrogen Peroxide

Hydrogen peroxide has a multitude of uses for two key reasons. First, it is very reactive. Second, it can react in a variety of ways. Dilute solutions of hydrogen peroxide (at a strength of about 3%) can be used as disinfectants and hair bleaches. This is the origin of the term "peroxide blonde." Hydrogen peroxide also chemically falls apart ("decomposes") with time. The products of the decomposition reaction are water and oxygen, and the equation is

$$2H_2O_2(aq) \rightarrow 2H_2O(l) + O_2(g)$$

$$\Delta H = -98 \text{ kJ/mol hydrogen peroxide decomposed}$$

The energy released per gram of H_2O_2 is

$$\frac{98 \text{ kJ}}{1 \text{ mol } H_2O_2} \times \frac{1 \text{ mol } H_2O_2}{34.0 \text{ g } H_2O_2} = 2.9 \text{ kJ/g } H_2O_2$$

This compares to the energy available to us from digesting the calories in a bowl of frosted corn flakes, which is 502 kJ/30 g = 16 kJ/g. Although the decomposition of hydrogen peroxide releases less energy per gram, it has a relatively low molecular mass and is therefore useful for rocket fuel.

This decomposition reaction of hydrogen peroxide is normally so slow it is not noticeable. However, the reaction is faster in the presence of light, so hydrogen peroxide is often sold in brown or opaque bottles. Under very specific conditions, however, the reaction can be made to happen explosively fast.

At the end of World War II, the Germans used the explosive reaction of hydrogen peroxide with hydrazine (N_2H_4) to power some Messerschmitt rocket fighter planes:

$$2H_2O_2(l) + N_2H_4(l) \rightarrow 4H_2O(l) + N_2(g)$$

$$\Delta H = -812 \text{ kJ/mol } N_2H_4$$

From the mid-1950s, hydrogen peroxide was used as torpedo propellant by the U.S. military. The two things that make hydrogen peroxide a useful fuel are that it reacts in ways that release a lot of energy and the reaction products include a gas.

How Can We Determine the Amount of Energy Change?

The energy change accompanying a reaction can be analyzed directly, using an instrument known as a **calorimeter** (see Figure 5.11). A calorimeter is essentially a closed chamber in which a reaction is made to occur, surrounded by some "heat reservoir" such as water, which can either receive energy from the reaction or supply energy to it. Careful measurement of the temperature change in the reservoir can allow us to calculate the energy change associated with the reaction.

The total energy change associated with a reaction depends on the number and type of chemical bonds that are broken and/or made during the reaction. The decomposition of hydrogen peroxide can be thought of as the net result of all the bond-breaking and bond-making processes summarized in Figure 5.12.

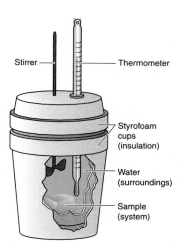

Figure 5.11 A calorimeter is used to measure heat exchange in chemical reactions.

Figure 5.12 The decomposition of hydrogen peroxide requires these bond-breaking and bond-making processes

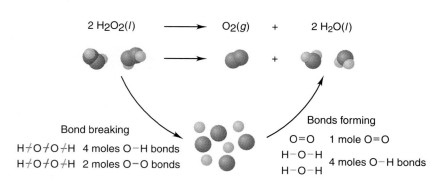

Chemists have already determined the energy changes associated with the breaking or making of most types of bonds, so we can now calculate the energy change accompanying a reaction by using these previously determined values.

Two fundamental facts summarize the energy changes involved in making or breaking bonds:

- Bond-breaking is an endothermic process—it *requires* energy.
- Bond-making is an exothermic process—it *releases* energy.

Table 5.2 lists the average **bond energies** of a variety of bonds. These values are in kilojoules *per mole* of the bonds concerned. If the bonds are broken, the stated amount of energy is absorbed by the chemicals. If the bonds are formed, the stated amount of energy is released. Using Table 5.2, the energy needed to break all the bonds in 2 mol of hydrogen peroxide can be calculated as follows:

$$2H_2O_2(aq) \rightarrow 2H_2O(l) + O_2(g)$$

$$\text{4 mol O—H bonds} \times \text{460 kJ/mol} = \textbf{+1840 kJ}$$
$$\text{2 mol O—O bonds} \times \text{145 kJ/mol} = \textbf{+ 290 kJ}$$
$$\textbf{Total} = \textbf{+2130 kJ}$$

The new bonds formed during the decomposition of hydrogen peroxide are those within two water molecules and one oxygen molecule. If 2 moles of hydrogen peroxide undergo decomposition, 2 moles of water molecules and 1 mole of oxygen molecules will be formed. We can calculate the energy released during the formation of these bonds as follows:

All chemical reactions begin with the chemicals taking in energy, the "activation energy" of the reaction, and end with the chemicals giving out energy. Whether energy is taken in or given out overall depends on the relative sizes of the energy changes in these two steps.

Table 5.2

Average Bond Energies*

Diatomic Molecules (Specific Values)

H—H	432.0	F—F	154.8	O=O	493.6
H—F	565	Cl—Cl	239.7	N=O	607
H—Cl	428.0	Br—Br	190.0	N≡N	942.7
H—Br	362.3	I—I	149.0	C≡O	1071
H—I	294.6				

Single Covalent Bonds (Average Values)

H—C	415	C—C	345	Si—Si	220	N—N	165	O—O	145
H—N	390	C—N	305	Si—F	565	N—O	200	O—Si	450
H—O	460	C—O	360	Si—Cl	380	N—F	285	O—P	335
H—Si	315	C—Si	290	Si—Br	310	N—Cl	200	O—F	190
H—P	320	C—P	265	Si—N	320	N—Br	243	O—Cl	220
H—S	365	C—S	270	Sn—Sn	145	N—I	159	O—Br	200
H—Te	240	C—F	485	Sn—Cl	325	P—P	200	S—S	250
		C—Cl	325			P—F	490	S—F	285
		C—Br	285			P—Cl	325	S—Cl	255
		C—I	215			P—Br	265	Se—Se	170
						P—I	185		
						As—Cl	320		

Multiple Covalent Bonds (Average Values)

C=C	615	N=N	420	C≡C	835
C=N	615	N=O	605	C≡N	890
C=O	750 805†	O=P	545		
C=S	575	O=S	515		

*All values in kilojoules per mole (kJ/mol).
†C=O bond energy in CO_2.

$$4 \text{ mol O—H bonds} \times 460 \text{ kJ/mol} = \mathbf{-1840\ kJ}$$
$$1 \text{ mol O==O bond} \times 494 \text{ kJ/mol} = \underline{\ \ \mathbf{-494\ kJ}}$$
$$\mathbf{Total} = \mathbf{-2334\ kJ}$$

The total energy change accompanying the decomposition of 2 moles of hydrogen peroxide is the sum of the two values just calculated:

Total energy absorbed   = + 2130 kJ (endothermic)
Total energy released   = − 2334 kJ (exothermic)
Net energy change (ΔH) =   − 204 kJ (exothermic)

This corresponds to −102 kJ/mol (we used 2 moles in this first example to ensure all the bond energy values were whole number multiples of the values in Table 5.2). Recall that the measured value of this ΔH is −98 kJ/mol and here we predict −102. These values are very close, considering that one is a measured number and the other is a predicted value based on average values in many compounds.

At this point in our discussion, we have a sense of *what energy is*. We also know how to determine the amount of energy change in a chemical reaction. However, after almost five chapters of chemistry, we still do not know why reactions occur. We turn to that central issue now.

exercise 5.3

Bond Energies

Problem

One of our favorite chemical demonstrations is the reaction of hydrogen with molecular chlorine to form hydrogen chloride. The equation for this reaction is:

$$H_2(g) + Cl_2(g) \rightarrow 2HCl(g)$$

Use the bond energy values in Table 5.2 to determine the amount of energy released during the formation of 2 moles of HCl.

Solution

The energy needed to *break the existing bonds* is

$$1 \text{ mol H—H bonds} \times 432 \text{ kJ/mol} = \ \ \ 432 \text{ kJ}$$
$$1 \text{ mol Cl—Cl bonds} \times 240 \text{ kJ/mol} = \underline{\ \ 240 \text{ kJ}}$$
$$\text{Total} = +672 \text{ kJ}$$

The energy *released* during formation of the new bonds is as follows:

$$2 \text{ mol H—Cl bonds} \times 428 \text{ kJ/mol} = 856 \text{ kJ}$$

The net energy change is the sum of the energy changes involved in bond-breaking and bond-making:

Bond-breaking: +672 kJ (endothermic)
Bond-making: −856 kJ (exothermic)
Net energy change = −184 kJ/2 moles of HCl = 92 kJ/mol HCl

So the reaction is exothermic overall.

5.5 | Why Do Chemical Reactions Happen?

The reason that chemical reactions occur is really fairly straightforward. You can even use a jar full of flies to understand it! Suppose we have trapped some flies within a jar, in the center of a large room (see Figure 5.13). In the absence of any chemical attractants or variations in light within the jar, the flies fly about, changing direction from time to time in a random exploration of the space available to them. If we lift off the lid of the jar the flies will inevitably *disperse* out of the jar, and once dispersed they will almost certainly never again become gathered together inside the jar. Their random exploration of space causes them to disperse, because there is more space available outside the jar than inside it. To put it another way, the flies disperse because there are *more ways* for dispersal to occur than for aggregation within the jar to happen. As we will now discover, this simple everyday situation embodies a guiding principle of all chemistry.

Spontaneity and Entropy

In general, there are many more possible ways for objects to be dispersed than to be concentrated together. Particles of matter are certainly objects, and therefore they follow this trend. Another way of expressing this natural tendency toward dispersal is to say that it is easier (and therefore more likely) for changes to make things more "disordered" rather than more ordered. Neatness (order) readily gives way to untidiness (disorder), a fact of which few students need to be persuaded.

So one reason why some chemical reactions occur naturally in the direction in which they do occur (such as $2H_2O_2 \rightarrow 2H_2O + O_2$) is that their particles become more dispersed (or more "disordered") as a result of the reaction.

What kinds of reactions result in increased dispersal of particles? Gas molecules are considerably more dispersed than molecules in the other states of matter. As a result, those reactions in which solids or liquids react to form a gas will (at least partially) be driven forward by the dispersal of the gas. By "driven forward" we really mean that the dispersal of the products makes the reverse reaction highly unlikely, because the dispersing products would need to aggregate together again, like our flies suddenly finding themselves back in the jar.

Reactions tend to proceed in the direction of greatest dispersal because the opposite process, involving concentration rather than dispersal, is so much less likely. The decomposition of ammonium dichromate, $(NH_4)_2Cr_2O_7$, is one example:

$$(NH_4)_2Cr_2O_7(s) \rightarrow N_2(g) + 4H_2O(g) + Cr_2O_3(s)$$

$$\Delta H = -315 \text{ kJ/mol } (NH_4)_2Cr_2O_7 \text{ reacted}$$

Figure 5.13 The random exploration of space causes dispersal, revealing the driving force of chemical reactions. Chemical reactions are driven forward by the tendency of energy and matter to disperse, due to the random exploration of possibilities. This is known as the Second Law of Thermodynamics.

$$(NH_4)_2Cr_2O_7(s) \longrightarrow N_2(g) + 4H_2O(g) + Cr_2O_3(s)$$

Ammonium dichromate decomposes to give three different substances.

$(NH_4)_2Cr_2O_7(s)$ → $N_2(g)$ · $H_2O(g)$ $H_2O(g)$ $H_2O(g)$ $H_2O(g)$ $Cr_2O_3(s)$

Each original mole of the reactant forms 5 moles of gas (1 of nitrogen and 4 of water vapor) and a total of 6 moles of substances overall. This means that the reaction is accompanied by a very significant dispersal of the matter that was initially present.

Something else happens when ammonium dichromate decomposes: *energy is released.* How does this affect the dispersal of particles? When heat is released into the surroundings, into the air in this case, the heated air molecules move about more quickly as a result of their gain in energy (remember that temperature is a reflection of the average kinetic energy of the particles). The energized air molecules become more dispersed, and they allow the energy to disperse outwards through the continual collisions between different molecules. The *dispersal of energy* is another key factor associated with reactions which occur spontaneously.

The tendency of matter and energy to disperse (or to move from order toward disorder) results from the tendency for the **entropy** of the universe to increase. Entropy can be defined in a strict mathematical way, but you need to think of it as a measure of the disorder (or degree of dispersal) within the system whose entropy is being discussed. Reactions that result in the dispersal of matter and energy in this way are **spontaneous.** This does not necessarily mean that they will take place immediately, or even *ever.* All it means is that they are chemically *prone* to occur.

The decomposition of ammonium dichromate is an especially favored reaction because it is accompanied by the physical dispersal of matter *and* the dispersal of heat energy. It is not necessary, however, for *both* matter and energy to become dispersed in a spontaneous reaction. It is sufficient for the overall tendency to be toward dispersal rather than aggregation (that is toward increased entropy in the universe as a whole, rather than decreased entropy). So a reaction in which matter becomes less dispersed can be spontaneous provided it is accompanied by sufficient dispersal of energy to overcome the effects on the aggregation of matter, and vice versa.

The reaction of ammonium nitrate, NH_4NO_3, with barium hydroxide octahydrate, $Ba(OH)_2 \cdot 8H_2O$, proceeds quite vigorously, even though it is endothermic. This is because matter is dispersed by the formation of ammonia gas

$$2NH_4NO_3(s) + Ba(OH)_2 \cdot 8H_2O(s) \rightarrow 2NH_3(g) + 10H_2O(l) + Ba(NO_3)_2(aq)$$

$$\Delta H = +63.3 \text{ kJ/mol Ba(OH)}_2 \cdot 8H_2O$$

The reaction gets cold enough that it can freeze the flask in which it is contained to the block of wood, as shown in Figure 5.14.

Forcing Nonspontaneous Things to Happen

Although a great many reactions are spontaneous, some reactions, such as the breaking down of water into hydrogen and oxygen, are **nonspontaneous.** Such reactions can be made to happen only by forcing them to do so, that is, by coupling them to other spontaneous processes that will drive the nonspontaneous processes forward. A football will not spontaneously go sailing into the air and between the goal posts. If we arrange for the football to be temporarily "coupled" to a swinging foot and leg, however, the football can be forced to behave in an otherwise nonspontaneous manner.

For a more rigorous chemical example, consider the way in which the release of energy from the chemical reactions within an electrical battery can be used to power the breakdown

Figure 5.14   The reaction of ammonium nitrate with barium hydroxide octahydrate gets cold enough to freeze the flask to a block of wood.

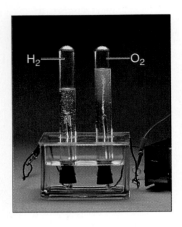

Figure 5.15 Water (H₂O) can be electrolyzed (split apart by electrical energy). Oxygen collects on the right side and hydrogen on the left in the lab set-up shown.

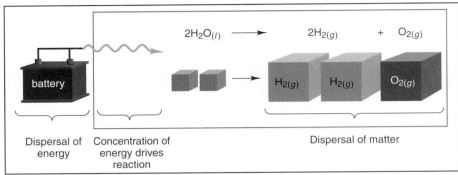

Overall dispersal of matter and energy

Figure 5.16 The overall dispersal of matter and energy.

of water, in the process known as the **electrolysis of water** (see Figure 5.15). The equation for the electrolysis of water is:

$$2H_2O(l) \rightarrow 2H_2(g) + O_2(g) \quad \Delta H = +286 \text{ kJ/mol water electrolyzed}$$

At first sight, something may seem amiss here, since this nonspontaneous reaction involves the dispersal of matter with two moles of a liquid generating 3 moles of gas (Figure 5.16). Remember, however, that there are *two* dispersal criteria. One concerns *matter* and the other *energy*. It is possible for there to be a concentration of energy (the opposite of dispersal) that makes up for a gain in the dispersal of matter, and vice versa.

It is also worth repeating that, even when we force a "nonspontaneous" process to occur, the way we do it is powered *overall* by the spontaneous dispersal of matter and energy (that is, by an increase in the entropy of the universe). Our electrolysis reaction, for example, is

exercise 5.4

Reaction Spontaneity

Problem

The reaction of solid barium hydroxide octahydrate, Ba(OH)₂ · 8H₂O, with solid ammonium nitrate, NH₄NO₃, is used in a "chemical cold pack" because the solution formed during the reaction feels very cold. This feeling of coldness is the result of an endothermic reaction—it takes in heat. The equation is shown below. The water that is formed acts as a **solvent** (a liquid in which other chemicals dissolve), and the ammonia dissolves in this water.

$$Ba(OH)_2 \cdot 8H_2O(s) + 2NH_4NO_3(s) \rightarrow Ba(NO_3)_2(aq) + 2NH_3(g) + 10H_2O(l)$$
$$\Delta H = +63.3 \text{ kJ/mol Ba(OH)}_2 \cdot 8H_2O \text{ reacted}$$

Can you rationalize the fact that this reaction is spontaneous?

Solution

The reaction is endothermic (when you touch the pack, your hand feels cold because it is losing heat to the cold pack). This means that energy dispersal acts *against* this reaction being spontaneous. *However,* the 3 moles of solid reactants are reacting to form 2 moles of gas, 10 moles of liquid and 3 moles of ions. So the reaction is accompanied by a substantial dispersal of *matter,* which allows it to be spontaneous.

We recognize that it is counterintuitive that a system which takes in energy feels cold rather than warm. The system is taking heat away from the rest of the universe which often includes our hand which touches the container. This causes our hand to feel cold.

powered by the dispersal of energy from the exothermic chemical reactions within the battery. If we consider the entire system of battery plus electrolysis reaction, rather than focusing only on the electrolysis, we will find that the entropy of the entire system is increasing overall. So otherwise nonspontaneous reactions can be made to become spontaneous by *coupling them* to other processes that allow the combination of a nonspontaneous reaction plus coupled spontaneous "driving" process to be spontaneous overall.

In summary, spontaneous chemical reactions are driven forward by the dispersal of matter and/or the dispersal of energy. The net effect of these two influences is the key factor. Reactions, and other physical processes, involving a mixture of dispersal and concentration of energy and matter can be spontaneous, provided the matter and energy of the universe ends up more dispersed *overall.*

◼ case in point

Entropy and Evolution

The very public debate on "evolution versus creationism" has been raging for almost 150 years. It is more than just a polite discussion of alternative views. It centers on issues such as "what is science?" and "who decides what will and will not be taught in the nation's public schools?" It is also an economic issue. Hundreds of millions of dollars are at stake when biology textbooks are chosen. In many states, the choice of such texts hinges on the acceptance of evolution or creationism. Teachers feel pressure from the community not to discuss "the wrong ideas." Money, freedom of speech, politics, and religion; all are involved in this highly-charged debate. It is a debate in which *energy dispersal* plays a central role.

One of the primary tenets of modern scientific ideas about biology is that species are **mutable**—that is, they have the capacity to change (evolve) and that these modifications affect their ability to survive in an ever-changing environment. Such changes have been occurring for over 2 billion years. Creationists assert that species are, except for very small changes, **immutable.** They claim that life on Earth is a fairly recent phenomenon and that the species that exist were placed here by a God.

How do you decide whether evolution or creationism is fact? There are several styles of evidence that can be considered. The collection of **direct evidence** involves seeing species actually evolve. A second kind of data, representing **indirect evidence,** can include such things as the fossil record and comparison of genetic material across species (to be discussed further in Chapter 16 which deals with biochemistry). You can also make intellectual arguments for or against specific assertions. The creationists rely on such arguments that say, in essence, that evolution just doesn't make sense. Foremost among these arguments is the concept that "evolution violates the Second Law of Thermodynamics."

We said in the text that entropy is the formal name for the increase in dispersal ("randomness") of matter and energy. The entropy of the *universe* increases as a result of all physical and chemical processes. This is called the Second Law of Thermodynamics. Those who assert that there is no spontaneous evolution of species argue that the highly organized chemical system that is a human is *less random* than loose clusters of atoms, or than simpler species (one-celled bacteria for example). Therefore, they argue, spontaneous evolution is impossible because such spontaneity produces a "system" (us) with *lower* entropy than the simpler species and/or chemicals that preceded them. This sounds reasonable enough, until we look more deeply at what is being said.

It is true that in any reaction, the entropy *of the universe* must increase. However, if we grant for the moment that humans have less entropy than previously evolved species or nonliving mineral substances, it is still possible for us to "spontaneously" evolve, because we are not the *entire universe*. We are merely individual *systems* within the larger universe.

It is perfectly permissible for parts of the universe (systems) to lose entropy as long as the rest of the universe gains enough to overcome that loss. Indeed, this happens all the time as an inevitable consequence of the increase in the entropy of the universe overall. For example, the chemistry of photosynthesis (which creationists are happy to accept) continually converts high-entropy starting materials (water and carbon dioxide gas) into a lower-entropy product (carbohydrates within plants). This happens spontaneously because it is powered by the dispersal of energy from the sun, so the combination of the dispersal of the sun's energy into the plant and the chemistry of photosynthesis increases the entropy of the universe overall.

Evolution and human existence do not violate the physical laws of nature, including the Second Law of Thermodynamics; but are direct *consequences* of these laws.

The application of force by the player results in the ball sailing into the air. This is a nonspontaneous reaction. There is no question that intentional force was applied in this case.

The continual and inevitable dispersal of matter and/or energy is known as the increase in the **entropy** of the universe. The entropy of a system can be thought of as a measure of its inherent "disorder," which is embodied within the dispersal of its matter and energy. The entropy of the universe increases because there are more ways for it to increase than to decrease. The random exploration of possibilities inevitably favors dispersal.

The unavoidable tendency of the entropy of the universe to increase is known as the **Second Law of Thermodynamics.**

We now have a sense of *why reactions occur* and that there exist two types of reactions—spontaneous and nonspontaneous. We also know that a nonspontaneous reaction can be made to happen by coupling it with spontaneous reactions, in such a way that the coupled process is spontaneous overall.

We presented two examples of spontaneous reactions—the decomposition of ammonium dichromate and a chemical cold pack. But there is an important difference between these two spontaneous reactions. The cold pack reaction occurs upon mixing. The ammonium dichromate decomposition will not occur until you heat it. So you have to first *add* some energy to the ammonium dichromate, perhaps provided by the heat of a flame, before the reaction that releases energy overall will occur. Why is the additional input of energy required? The reason lies in a branch of chemical theory known as **chemical kinetics.**

5.6 | An Introduction to Chemical Kinetics

We found in Section 5.5 that some reactions, such as the decomposition of hydrogen peroxide, proceed all on their own. Others, such as the decomposition of ammonium dichromate, do not proceed at all until they are given an extra jolt of energy, even though we call them "spontaneous." These differences result from differences in the size of an energy barrier associated with each reaction, known as the activation energy.

The Nature of Activation Energy

Recall that in our discussion dealing with the decomposition of hydrogen peroxide we said that even though the reaction is exothermic, some energy is required to break bonds before *more energy can be released in bond formation.* So (as happens when studying chemistry) you have to first put something in, in order to get more out.

Look at the diagrams illustrating the energy content of chemicals during exothermic and endothermic chemical reactions (Figure 5.17). We have seen these diagrams already, but we focused our attention on the initial and final energy levels. Look now at what happens as a reaction proceeds. To begin with, the energy content of the chemicals rises, regardless of whether the reaction is eventually going to lead to a release or absorption of energy overall. This initial rise in energy, taking the reactants up to the "top of the energy hill" in the diagrams, is known as the **activation energy** (E_a) of a reaction. It is the energy that needs to be put in to break existing bonds in the reactants, before new bonds can form.

All chemical reactions have an activation energy, so they all need to be supplied with some initial energy to break bonds and generally rearrange electrons. The energy content of the chemicals then falls as new bonds are formed. How *far* the energy level falls determines whether the reaction will be exothermic (giving off energy) or endothermic (taking in energy) overall. Also, as we shall see, the addition of an appropriate substance can lower the activation energy of some reactions, thus speeding the reaction but not changing the products or the energy change of the reaction.

Where does the energy needed to activate chemicals into reacting come from? It comes from the **collisions** between the atoms, molecules, and ions concerned. For some reactions, the collisions that occur at room temperature and pressure are energetic enough to provide the activation energy and initiate reaction. In other cases, the chemicals must be heated (causing their particles to move more quickly) until the collisions become energetic enough to supply the required activation energy.

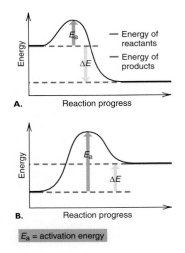

Figure 5.17 All reactions are initiated by an increase in the energy of the reactants, and end with an energy fall. Overall, the products of an exothermic reaction **A.** have less energy than the reactants; while the products of an endothermic reaction **B.** have more energy than the reactants.

A.

B.

C.

D.

Figure 5.18 The ammonium dichromate volcano. **A.** Solid ammonium dichromate (indefinitely stable). **B.** A burning match adds energy. **C.** The energy from the match initiates the reaction. **D.** The final volume of the product is much greater than the volume of the original ammonium dichromate.

You supply chemicals with the activation energy needed to start a reaction every time you light something with a match. The heat of the match jolts a few molecules of the substance into combining with the oxygen of the air. The resulting "combustion" or "burning" reaction initiated by the heat from the match then releases much more energy than your match put in and jolts many more molecules into reacting. Matches are devices we use to kick-start chemical reactions, by making a few molecules (or atoms, or ions) move about more quickly. An excellent example of the impact of activation energy is provided by the ammonium dichromate decomposition discussed earlier. The reaction is strongly exothermic, yet as is shown in Figure 5.18, substances will remain unreacted if there is insufficient energy to supply the needed activation energy to allow the reaction to occur. A match (or, more often, a lighted propane torch or Bunsen burner) supplies the necessary energy for the decomposition to occur.

The value of the activation energy differs for different reactions. As we shall see, however, there are ways in which an activation energy can be lowered to achieve the same overall reaction, but via a slightly different and faster route.

The formation of rust is an example of a *slow* reaction.

The Factors That Influence Reaction Rates

The speed of a chemical reaction is known as the **reaction rate.** This simply measures how quickly a reactant of the reaction is used up or a product of the reaction is produced. For example, consider again the decomposition of **hydrogen peroxide:**

$$2H_2O_2(aq) \rightarrow 2H_2O(l) + O_2(g)$$

$$\Delta H = -98 \text{ kJ/mol hydrogen peroxide}$$

The rate of this reaction could be measured in terms of moles of H_2O_2 reacted per minute (or per second etc.). But it could also be measured in terms of moles of H_2O produced per minute, or grams of H_2O produced per minute, or moles of O_2 produced per second, etc.

The study of the **rates** of chemical reactions, and the factors that influence the rates, is called **chemical kinetics.** There are many practical reasons for our interest in chemical kinetics. For example, the dosages in which medicines are prescribed are partly determined by how quickly they work in a particular part of the body and how quickly they are destroyed by the body. In plastic production, millions of dollars can be cut from the cost of a chemical manufacturing process if ways can be found to speed up the reactions concerned.

An explosion is an example of a *fast* reaction. The term reaction rate refers to the speed of a reaction.

Effect of Temperature on Reaction Rates

A substance can gain energy without an increase in temperature if the energy is going into breaking bonds, rather than speeding up particles. The barium hydroxide/ammonium chloride chemical cold pack, discussed previously, is an example of the temperature actually going

Green
Chemistry

exercise 5.5

Combustion Energy

Problem

The amount of heat released per mole when burning several carbon compounds is listed in this table. Graph these in such a way as to determine which gives the greatest amount of heat per mole of carbon dioxide released. What generalization can you make about the trends you see?

Compound	ΔH (kJ/mol)	Compound	ΔH (kJ/mol)
CH_4	882	$C_6H_{12}O_6$	2816
C_2H_6	1541	C_6H_6	3273
C_3H_8	2202	CH_3COOH	876
C_8H_{18}	5450	HCOOH	263
C_2H_5OH	1371		

Solution

Divide each ΔH by the number of carbon atoms in the molecule and graph this versus the oxidation number of the carbon in the compound. The conclusion is that the more negative the oxidation number of carbon in the fuel, the more energy is released upon combustion. This is a great advantage for methane (natural gas) since it has the lowest possible oxidation number for carbon.

Notice that the graph extrapolates well to zero energy released for carbon dioxide which has an oxidation number of +4.

Compound	ΔH (kJ/mol)	ΔH/C Atoms	Oxidation No. of Carbon
CH_4	882	882	−4.00
C_2H_6	1541	771	−3.00
C_3H_8	2202	734	−2.67
C_8H_{18}	5450	681	−2.25
C_2H_5OH	1371	685	−2.00
$C_6H_{12}O_6$	2816	469	0.00
C_6H_6	3273	545	−1.00
CH_3COOH	876	438	0.00
HCOOH	263	263	+2.00

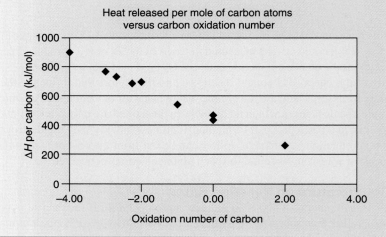

Heat released per mole of carbon atoms versus carbon oxidation number

down *with an input of energy.* However, the temperature of a substance does rise as it gains heat if that energy goes to speed up the particles in the substance. It falls as it loses heat (due to its particles slowing down).

It is a general rule that increasing the temperature of a reaction increases its rate, while lowering the temperature lowers its rate. A useful rule of thumb is that the rate of a reaction will approximately double for every 10°C rise in temperature. This effect of temperature should be easily understood: at higher temperatures, the reactants will be moving more quickly, and therefore the collisions between them will be more energetic. So as the temperature increases, a greater proportion of the collisions will occur with an energy that is equal to or greater than the activation energy of the reaction, so the reaction will proceed more quickly.

Effect of Concentration on Reaction Rates

The **concentration** of a chemical is a measure of how many particles of the chemical (whether atoms, molecules, or ions) are present in a given volume. It can be measured, for example, in moles per liter.

As the concentration of a chemical is increased, a given volume contains more particles of the chemical available to react by colliding with other particles, so the reaction becomes more likely. So, another general rule is that increasing the concentration of a reactant increases the rate of a reaction, while decreasing the concentration of a reactant decreases the rate of a reaction.

One of the most breathtaking examples of chemical kinetics in action is the **iodine clock** reaction, which depends on the effect of concentration on reaction rate. The reaction was popularized over 50 years ago by the late Hubert Alyea, known affectionately and accurately as the world's top chemical demonstrator. In this reaction, iodate ions (IO_3^-) combine with bisulfite ions (HSO_3^-), producing iodine that gives the solution a blue-black color in starch. There are several intermediate steps in the process which can be summarized by the following overall equation:

$$2IO_3^-(aq) + 5HSO_3^-(aq) \rightarrow I_2(aq) + 5SO_4^{2-}(aq) + 3H^+(aq) + H_2O(l)$$

Figure 5.19 shows what happens if one of the reactants of this reaction is present in different concentrations in different beakers. From left to right, the initial concentration of bisulfite ion is one-half that in the previous beaker. So the rate of reaction is also half as fast as in the previous beaker. So the color change runs through the series of beakers in a steady manner, hence the name *iodine clock.*

Effect of Catalysts on Reaction Rates

Many chemical reactions that proceed very slowly on their own may suddenly proceed very quickly (even explosively) if a tiny amount of some other chemical is added. If the additional ingredient is not actually *used up* during the reaction (so it can all be recovered unchanged at the end), it is known as a **catalyst** of the reaction. A catalyst is defined as a substance that speeds up a specific reaction by participating in the reaction but being recovered unchanged. Each atom, molecule, or ion of a catalyst can catalyze a specific reaction many thousands or millions of times. The catalyst may change temporarily during the reaction, but it is always returned to its original form by the end of the reaction, making it available to undergo repeated cycles of catalysis. This is the reason that an incredibly small amount of a catalyst can cause an amazingly large increase in the rate of a reaction.

Figure 5.19 Varied concentrations in the beakers causes the reaction to appear to race from beaker to beaker. The *iodine clock* is a perfect example of chemical kinetics. The top photo shows the reactants before mixing.

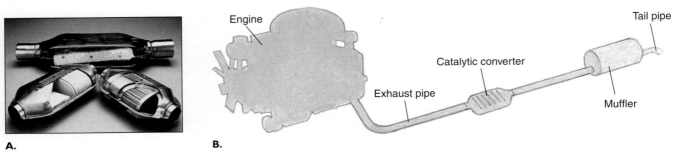

Figure 5.20 Catalytic converters speed up reactions in the exhaust gases from the engine. Combustion is speeded up so that more particles are consumed in the reaction. This results in fewer pollutants being released into the air.

Green Chemistry

Cars are fitted with **catalytic converters** (Figure 5.20), which contain catalysts able to speed up reactions in the exhaust gases. These reactions produce less-polluting products than would otherwise be released. Also, many of the most important chemical reactions in industry would not proceed at economical rates without the help of catalysts. (The details of how catalytic converters work will be discussed in Chapter 12.)

A catalyst works by *lowering the activation energy* of the reaction it catalyzes. So, in the presence of a catalyst, more of the collisions between reactants will be sufficiently energetic to initiate reaction than would be the case without the catalyst. The way in which catalysts achieve this effect is by opening up alternative "routes" or **mechanisms** by which reactions can proceed. The precise details of how the electrons become rearranged during the reaction will be different, but the overall effects of the rearrangement (in terms of bonds broken and bonds made) will be the same. The net effect of a catalyst can be to increase the rate of a reaction by as much as a billion or more, so catalysts are amazingly effective and important stimulators of chemical change.

Our bodies rely on natural biological catalysts all the time. We call these biological catalysts **enzymes.** They are molecules of **protein** that are able to accelerate specific chemical reactions required for our bodies to function. Saliva, for example, contains enzymes that catalyze the breakdown of food molecules into smaller molecules, beginning the process we call digestion. Our liver cells contain enzymes that can process the chemicals released from food and convert toxic chemicals into harmless (perhaps even useful) products. In general, the operation of every cell of the body depends on the activities of thousands of different enzyme molecules. We will have more to say about enzymes in Chapter 16.

5.7 | Why Bother Knowing All This?

We now have a sense of the *how* and *why* of energy change in chemical reactions. To round off this chapter, we must briefly emphasize the practical importance of what you have learned.

There are several reasons why it is important for you to understand what energy is and how chemical reactions are affected by the interplay of energy and matter. The first reason concerns **science literacy,** the understanding of scientific terms and processes. Many scientific terms and concepts are used in day-to-day decision making and discussions. "Energy" is one such term. Simply put, our desire for you to become scientifically literate means that we want you to learn about scientific terms such as energy and how we know about it and its uses, because it is important for you to know what you're talking about when you are involved in these decisions and discussions. The second reason is more practical. You have a stake in many energy-related choices, whether personal, social, or political. Understanding the concepts and the data involved in these choices helps you to make the appropriate choices for yourself and to understand the significance of your choices.

It is in this spirit of critical inquiry that we close this chapter by presenting some energy-related questions that you might be faced with in the future. You will not be able to answer them without further study. You will, however, *be able to make sense of the answers* should you come across them.

- I missed lunch. How much more should I eat at dinner to make up for it (or, how much weight will I lose if I just eat my normal amount)?
- Will gasohol burn as efficiently as gasoline? What if we change the ratio of ethanol in the gasohol?
- How much energy is provided to me when I digest 100 grams of popcorn and will I need more to run a 10 K road race? How does my body obtain energy after the popcorn is "used up"?
- How much energy was wasted when the Kuwaiti oil fields were destroyed in the 1991 Gulf War?
- Is it worth installing more insulation in the roof-space of my house?
- What's the best fuel for my space-heating: electricity or gas?
- Should I support the exploitation of sunlight, winds, and waves as energy sources, or are they just an impractical waste of time?

We learn about energy in order to increase our ability to analyze and answer these types of questions.

main points

- The Law of Conservation of Energy tells us that energy is never created or destroyed. It is only transformed between different forms and transferred from place to place.

- Energy is formally defined as "the capacity to do work."

- Potential energy is energy embodied in the *positional arrangement* of things.

- Kinetic energy is the energy of movement.

- Energy exchange is a part of all chemical reactions.

- An initial input of energy is required to initiate all chemical reactions. We call this the activation energy of a reaction.

- Reactions that release energy overall are known as exothermic reactions, while reactions that absorb energy overall are known as endothermic reactions.

- Chemical (and physical) processes occur spontaneously in the direction that leads to an increase in the entropy ("disorder") of the universe. This can be brought about by the net dispersal of energy and/or matter.

- A process that would be nonspontaneous on its own can be driven forward by being coupled to a spontaneous process, so that the combination of the two processes becomes spontaneous.

- The main factors influencing reaction rates are temperature, concentration, and the presence of catalysts.

- Learning about energy helps us to understand many of the choices faced in everyday life and helps us to make the choices most appropriate to our circumstances.

important terms

Absolute zero is the lowest temperature possible, at which atoms, molecules, and ions stop moving. (p. 183)

Activation energy is the energy needed to start a reaction going. (p. 196)

Bond energy is the energy required to break a given bond in a mole of gaseous molecules. (p. 190)

Calorimeter is a device used to measure the amount of heat released or consumed during a chemical reaction. (p. 189)

Catalyst is a substance added to lower the activation energy of a chemical reaction, thus causing a faster reaction. (p. 199)

Celsius is the name of a temperature scale, formerly called "centigrade," in which melting ice is defined as 0° and boiling water defined as 100° at sea level. (p. 183)

Chemical kinetics is the study of the speed at which chemical reactions occur. (p. 196)

Concentration is the measure of the amount of solute dissolved in a solution. (p. 199)

Direct evidence is the result of a measurement or observation. (p. 195)

Distillation is the process for purifying a liquid by evaporating it, condensing the vapors, and collecting the liquid in a clean container. (p. 188)

Electrolysis is the passage of electricity through a substance to cause chemical change. (p. 194)

Electromagnetic radiation is the collective term including visible light, ultraviolet, infrared, microwave, radio, X-rays, and gamma rays. (p. 183)

Endergonic is a process that takes in energy. (p. 186)

Endothermic is a process that takes in heat. (p. 186)

Energy is the capacity to do work or transfer heat. (p. 178)

Entropy is a measure of the disorder of a system. (p. 193)

Enzyme is a large molecule which acts as a catalyst in biological reactions. (p. 200)

Exergonic is a process that gives off energy. (p. 186)

Exothermic is a process that gives off heat. (p. 186)

First Law of Thermodynamics states that heat and work are interconvertible and together constitute energy of a system—in essence this is a restatement of the Law of Conservation of Energy. (p. 178)

Heat is the energy of a system caused by the continuous motion of its particles. (p. 182)

Immutable is unchangeable. (p. 195)

Indirect evidence is circumstantial evidence. (p. 195)

Kelvin is a temperature scale the zero of which is set to be absolute zero. (p. 182)

Kinetic energy is the energy of motion. (p. 180)

Law of Conservation of Energy is the scientific law that states that energy is neither created nor destroyed. (p. 178)

Mechanism is the sequence of molecular events that occur during a chemical reaction. (p. 200)

Mutable is capable of being changed. (p. 195)

Potential energy is energy possessed by virtue of location or situation. (p. 180)

Protein is a very large biological molecule containing C, H, N, O, and S. (p. 200)

Rate is the speed of a chemical reaction, sometimes expressed in moles per second. (p. 197)

Science literacy is the ability to read and comprehend scientific articles and arguments. (p. 200)

Second Law of Thermodynamics states that a process occurs spontaneously in the direction that increases the entropy of the universe. (p. 196)

Solvent is the substance, usually a liquid and often water, in which substances dissolve. (p. 194)

Spontaneous is occurring without continuous input of energy. (p. 193)

Temperature is the measure of the average speed of molecules or atoms of a substance. (p. 182)

Thermodynamics is the study of the interactions of energy with chemical or physical systems. (p. 178)

Work is the energy transferred when an object is moved by a force. (p. 179)

exercises

1. List five processes or activities that require energy (other than those listed in the text).
2. State and describe the importance of the First Law of Thermodynamics.
3. When you are lifting a barbell, in which position does the bar have a greater potential energy, in the low or raised position?
4. Convert these temperatures to the Celsius scale: 99 K, 325 K, 100 K, 461 K.
5. Convert these temperatures to the Kelvin scale: 32°C, 100°C, 9°C, 150°C.
6. Why are microwaves not visible to the human eye?
7. Why do the burners on an electric stove glow red when they are hot?
8. Our normal body temperature is at 98.6°F. What would it be in °C? What does this mean, with regard to energy exchange, about the chemical reactions that take place in us?
9. On reentry into the Earth's atmosphere, NASA's space shuttles are faced with temperatures as high as 2300°F. What would this temperature be in °C and K?
10. Potassium chlorate, which is used in household matches and smoke grenades, decomposes at 360°C. At what temperature (in °F) will potassium chlorate decompose?
11. The equation for the formation of water is shown. Using the bond energies listed in Table 5.2, how much energy is needed to break the bonds in the reactants (in kJ)? How much energy is given off during the formation of

the bonds in 2 moles of water molecules (in kJ)? How much total energy would be given off by the reaction forming 4 moles of H_2O (in kJ)?

$$2H_2(g) + O_2(g) \rightarrow 2H_2O(g)$$

12. Recently fast-food chains have started counting calories. On entering many McDonald's restaurants, you may find a calorie chart. A Big Mac contains 2092 kJ of available energy while the McLean with cheese contains 1548 kJ. What is the difference in **nutritional calories** (kilocalories) between a Big Mac and the McLean with cheese? (Which one will you order on your next McDonald's run?)

13. Label these processes as endothermic or exothermic:
    a. baking a cake
    b. making popsicles
    c. melting ice cubes
    d. burning wood in a campfire

14. One gram of chocolate chips contains about 9 Calories. How many grams of chocolate chips do you need to eat to obtain 600 kJ of energy?

15. Photosynthesis is the result of the reaction of carbon dioxide and water to make glucose and oxygen (see Section 4.2). A total of 2816 kJ of energy is required to form 1 mole of glucose. How much energy is required to produce one *molecule* of glucose? What is the source of this energy in nature?

16. Skiers, snowmobilers, ice fishermen, or people lost in a blizzard worry about frostbitten body parts. Stores now have available packets containing a mixture of powdered iron, activated carbon, sodium chloride, sawdust, and zeolite moistened with water which produce heat upon exposure to oxygen. This reaction is essentially the rusting of iron, as was discussed in the text.

$$4Fe(s) + 3O_2(g) \rightarrow 2Fe_2O_3(s)$$

Because the iron is crushed into a powder the reaction occurs fast enough (unlike the rusting of your car) to produce heat for up to 6 hours. Would you define this as an exothermic or endothermic reaction?

17. Prove that the combustion of methane is an exothermic reaction.

18. Use bond energy values to calculate the amount of energy change of 1 mol sulfur dioxide reacting with oxygen to give sulfur trioxide.

19. Use bond energy values to calculate the amount of energy released when 1 mole of acetylene (H—C≡C—H) is burned.

20. Photosynthesis is the result of the reaction of carbon dioxide and water to make glucose ($C_6H_{12}O_6$) and oxygen. Calculate the amount of energy change associated with the formation of 1 mole of glucose. What is the source of this energy in nature?

21. Nitrogen tribromide decomposes to form the elements via this reaction:

$$2NBr_3 \rightarrow N_2 + 3Br_2$$

Calculate the enthalpy change of this reaction for 1 mole of nitrogen tribromide.

22. In the reaction in Exercise 21, what mass of bromine is formed when 0.67 grams of nitrogen tribromide decomposes?

23. Use the bond energy table (Table 5.2) to determine whether this reaction is exothermic or endothermic:

$$N_2(g) + 2H_2(g) \rightarrow N_2H_4(g)$$

Reactants	Products
1 mol N≡N bond	1 mol N—N bond
2 mol H—H bonds	4 mol N—H bonds

24. Your professor has assigned a project in which you are to give a demonstration of entropy. What would you do for your presentation?

25. Is the reaction $CaCO_3(s)$ yields $CaO(s) + CO_2(g)$ endothermic or exothermic? Why?

26. Nitrogen triiodide decomposes explosively to form these elements via this reaction:

$$2NI_3(s) \rightarrow N_2(g) + 3I_2(g)$$

a. Is this reaction exothermic or endothermic? Explain your answer briefly.
b. Does this reaction have a positive or negative entropy change? Explain your answer briefly.

27. Balance these reactions and predict for each whether the entropy change is positive, negative, or near zero.
    a. ___ $KClO_3(s) \rightarrow$ ___ $KCl(s) +$ ___ $O_s(g)$
    b. ___ $N_2(g) + H_2(g) \rightarrow NH_3(g)$

28. List two reasons why chemical reactions occur.

29. Nonmetallic elements such as silicon and boron are just one type of fuel used in modern-day pyrotechnics. These elements release a sufficient amount of energy when oxidized to set off consecutive luminous explosions such as those seen on the 4th of July. The environmental and safety benefit in using these fuels is that no gas is given off as a result of the reaction. NASA has also utilized the energy released from the oxidation of solid fuels in their space shuttle booster rockets. However, these propellants consist of pulverized aluminum fuel, a special polymer fuel, a binder (PBAN), and ammonium perchlorate, which combine to produce many gaseous products. Why are each of these reactions suitable for their intended purpose? Would you expect these reactions to be exothermic or endothermic?

30. What is the difference between a spontaneous and a nonspontaneous reaction? Why is H_2O_2 useful? Knowing that light energy can speed up a reaction, why are the H_2O_2 solutions purchased in a store kept in a brown bottle?

31. What is the Second Law of Thermodynamics?

32. Chemical kinetics is the study of (fill in the blank). Why should we be concerned with chemical kinetics? Give examples.

33. The energy of combustion of coal is typically measured in units of Btu/pound. (A Btu is a British Thermal Unit and equals 1055 joules.) A high-quality coal (anthracite) is 84% carbon and releases 14,600 Btu/lb. Low-quality coal (lignite) is 37% carbon and releases 6000 Btu/lb. Use the data in Exercise 5.5 to estimate the apparent oxidation numbers of carbon in these coals.

34. Identify the catalyst in this reaction:

$$E + S \text{ yields } ES \text{ yields } E + P$$

35. Which graph represents a reaction with a catalyst? Explain.

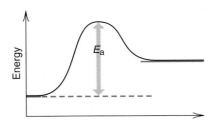

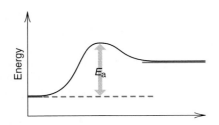

36. For the reaction $CO(g) + NO_2(g)$ yields $CO_2(g) + NO(g)$,
    a. Calculate ΔH.
    b. Draw a general graph of activation energy. Place the reactants and products on the graph, and indicate a measure of the ΔH you calculated.

37. Which reaction would you expect to occur faster, one with a low activation energy or high activation energy?

38. Answer each of these questions by supplying the correct word(s), number, symbol, or response
    a. When the products have a greater heat content than did the reactants, the reaction is said to be:
    b. Three features of a catalyst are:
    c. The energy necessary to get a reaction started is called:
    d. Is a measure of the change in orderliness during a reaction:

39. Explain why
    a. exothermic reactions feel warm to the touch
    b. endothermic reactions feel cold to the touch
    c. why the melting of ice is endothermic
    d. why breaking bonds is endothermic
    e. why condensation of steam into liquid water is exothermic

40. You are a food chemist working on an edible film to keep fresh fruits fresher. This edible film consists of a fatty acid and a protein which combine to form an even surface that prevents the penetration of moisture and oxygen. However, the chemical reaction takes too long to efficiently produce your product. Suggest several means by which the reaction rate can be increased.

41. Consider the decomposition reaction of NI_3. How many iodine molecules are formed when 5.3 grams of nitrogen triiodide decomposes?

food for thought

42. You are on the school board committee that is deciding which life science textbook to adopt. In general, what ideas in a textbook are acceptable and unacceptable to you, as a parent? How do you decide?

43. What type of energy concerns have you come across in your lifetime? What might be the energy debates of the future? Do you feel that you have enough background to make an "energy efficient" decision? If not, how can you find the answers to your questions?

readings

1. Clark, J.D. *Ignition.* Rahway, NJ: Rutgers University Press, Quinn and Boden, Inc., 1972.
2. Cotton, A. and Wilkinson, G. *Advanced Inorganic Chemistry.* 3rd ed., New York: John Wiley and Sons, 1972.
3. Shakhashiri, B. *Chemical Demonstrations.* Vol. 1. Madison: University of Wisconsin Press, 1983.
4. Wink, D.J. The conversion of chemical energy. *J. Chem. Ed.* 2:(69), 1992.
5. Conkling, John A. Pyrotechnics. *Scientific American.* July 1990: 96–102.
6. Fine, L.W. and Beall, Herbert. *Chemistry for Engineers and Scientists.* Philadelphia: Saunders College Publishing, 1990.
7. Kelter, P.B., Snyder, W.E. and Buchar, C.S. Using NASA and the space program to help high school and college students learn chemistry. *Journal of Chemical Education.* 64: 60–62, January 1987.
8. Pennisi, Elizabeth. Sealed in edible film. *Science News.* 141: 12–13, January 4, 1992.

websites

www.mhhe.com/kelter   The "World of Choices" website contains activities and exercises including links: to the Department of Energy; The University of Exeter in the United Kingdom, for conversion of energy units; The Gillette Company, which makes batteries; and much more!

Creating with Carbon
The Importance of Molecular Structure

'The people may eat grass': hasty words, which fly abroad irrevocable,—and will send back tidings.

—Thomas Carlyle, regarding a remark by Joseph-Francois Foulon, Finance Minister of France during the early part of the French Revolution. Foulon was hanged. (See the Case-in-Point on Lavoisier, Chapter 2.)

Many people start their day by eating a bowl of cereal, largely made of the chemical starch, composed of many glucose molecules linked together. Every cow starts its day by eating grass, largely made of the chemical cellulose, also composed of many glucose molecules linked together. So why can't we eat grass, like the cows, allowing the chore of mowing a lawn to yield a useful food harvest as a reward?

The answer lies in the subtle difference in the chemical structure of the two forms of glucose shown in Figure 6.1. The glucose molecules in starch are α-glucose, while those in cellulose are β-glucose. Our digestive systems contain enzymes that can break the linkages between α-glucose molecules in starch, releasing the glucose and allowing it to be absorbed from our intestines into our blood. These enzymes cannot break down cellulose, because of the different three-dimensional structure of the β-glucose molecules in cellulose. The guts of cows contain bacteria that have an enzyme called cellulase, which is able to break the cellulose down to glucose. So thanks to the assistance given to them by microorganisms in their guts, cows can live off cellulose or starch, while we are stuck with just starch.

Our inability to gain nourishment from grass, even though it is stuffed full of the glucose we need, is a great example of the importance of *molecular structure*. It is an example that comes from the world of **organic chemistry,** which is the chemistry of most carbon-containing compounds. The structural diversity of organic chemicals is both vast and hugely influential on our lives. We will use an exploration of organic chemistry to explore the more general importance of molecular structure.

Figure 6.1 Note the difference in α-glucose and β-glucose is at the carbon-1 position.

β-glucose

α-glucose

6.1 | Introducing Organic Chemistry

The word "organic" originally meant "derived from living things." As chemists analyzed the chemicals within living things, they found that the vast majority were built around rows or chains of carbon atoms. So **organic chemistry** became the chemistry of the carbon-containing compounds of life. Nowadays, organic chemistry also includes a wide range of synthetic carbon-containing compounds, including all of the plastics and synthetic fibers on which our modern lifestyles depend. You may sometimes see organic chemistry defined simply as "the chemistry of carbon-containing compounds." This is close to the truth, but somewhat inaccurate. Limestone, for example, is calcium carbonate ($CaCO_3$) and, although it does occur in living things (as a part of shells, for example), it is not usually regarded as an organic compound. Similarly, carbon dioxide (CO_2) is not usually regarded as an organic compound (although it is released from the body every time we breathe out). Such minor complications should not obscure the basic fact that when we talk of organic chemistry, we are referring to the chemistry of the vast majority of carbon-containing compounds.

The Organic Molecules of Life

Living things, including humans, are massive chemical machines largely composed of organic compounds interacting within a "sea" of water molecules. Water, the most abundant chemical of the body, is not itself organic; but most of the other chemicals that make up our bodies are carbon-based, organic chemicals. The molecules that form the outer membranes ("skins") of our cells are organic compounds. The genetic material within our cells that contains our genes, as parts of molecules of DNA, is composed of organic compounds. The enzymes that catalyze all of the chemical activities within cells are organic compounds. The key components of blood, which carry oxygen and nutrients to the body's cells and carry waste materials away from the cells, are mostly organic compounds. Muscles, fat, and skin are all based on organic compounds, and our bones are formed from a framework of organic compounds in which tough minerals become embedded.

Moving outside of the body, we find that the food we eat, the clothes we wear, the plastics we use so much in everyday life, and the fuel for our cars are all largely composed of organic compounds. Even the microbes that cause disease and the medicines we use to fight them are largely made of organic compounds. Life is sustained by organic chemistry, threatened by organic chemistry, and made interesting by organic chemistry.

Why can't we eat grass? The key is in the structure.

Principles of Organic Chemistry

Over 15 million different organic compounds are known, with many new ones being discovered and prepared each day. Over 90% of all known compounds are organic. There would seem to be so much to know in order to have even a rudimentary understanding of the subject. How can we structure our thinking to make some sense out of the vast expanse of organic chemistry? Two key ideas should always be kept in mind:

* *Structure dictates function.* This simply means that the three-dimensional structure of a molecule determines what it can do in a chemical sense. So the types of atoms present in the molecule and the overall shape in which they are arranged determine the chemical interactions and reactions in which the molecule can participate. Understand that, and you have understood the essence of organic chemistry. We have already seen how this principle explains why we can eat cereal for breakfast but not grass.
* *Organic molecules can be manipulated in systematic, repeatable, and predictable ways.* Organic chemicals can be classified into groups or "families" of related compounds that share key chemical characteristics. Learning about the reactivities of one member of such a family tells you much about the likely activities of all other members. Similarly, finding a way to modify one member in a particular way usually instructs you how to modify other members in the same way.

The idea that structure dictates function will be a main chemical theme of this chapter. In order to see how structure dictates function, we first need to know about structures. Specifically, how can we predict the structures of organic molecules?

6.2 | The Nature of Carbon and the 3-D Structures of Compounds

The simplest organic compound of all is **methane** (CH_4), the major component of "natural gas." It reacts exothermically with molecular oxygen in the air to warm up your house if you use gas heating. Methane is also produced by various natural processes. It is estimated, for example, that each year the world's cows release over 60 million tons of methane as a byproduct of their digestion of grass. Methane is also found in the atmosphere of Jupiter, Saturn, Uranus, and Neptune. Methane is the simplest member of a series of organic compounds known as the **hydrocarbons,** because they contain only hydrogen and carbon. Understanding why a methane molecule looks and behaves the way it does will provide a firm basis guiding our understanding of more complex organic compounds.

Molecules in 3-D

The central atom of a methane molecule is the carbon atom (see Figure 6.2). Note on the periodic table that carbon is in Group 4A, so it has four valence electrons. We pointed out in Chapter 3 that atoms often form compounds in a way that gives them a valence shell electron arrangement that is like (or **isoelectronic** to) that of the nearest noble gas. Carbon atoms can become isoelectronic with the noble gas neon by forming four covalent (electron-sharing) bonds with other atoms. In methane, these other atoms are all hydrogen atoms. At the same time, the hydrogen atoms attain the stable electron configuration of helium atoms. So the

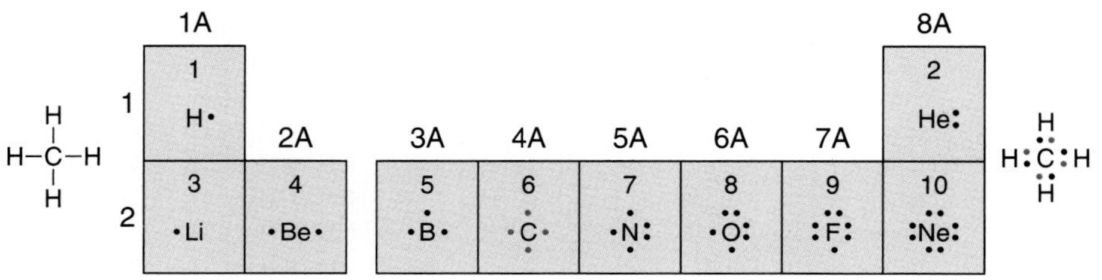

Figure 6.2 Carbon is in Group 4A of the periodic table.

A

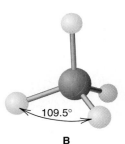

109.5°

B

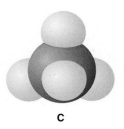

C

Figure 6.3 Balloons can help us to understand the 3-D shapes of molecules as with methane, CH_4. **A.** Four balloons tied together demonstrate the tetrahedral shape with carbon at the center. **B.** A "ball and stick" model with the carbon atom in black and the four hydrogen atoms in blue. The "sticks" connecting the atoms represent the chemical bonds holding the atoms together. **C.** A **space-filling** model shows the volume of space actually occupied by the electrons within the molecule.

structure of a methane molecule allows all of the atoms involved to attain stable outer electron arrangements, leaving no unpaired electrons on any atom.

We can represent any organic compound by its flat two-dimensional Lewis structure, using either dots or lines to represent the electron pairs of each bond as we showed in Section 3.3. To do so, however, ignores the fact that atoms and molecules are not flat, two-dimensional objects, but have more complex and interesting three-dimensional shapes. The three-dimensional shapes of molecules determine what chemical reactions and interactions they can participate in and, ultimately, give rise to much larger three-dimensional structures such as the human body. So chemistry happens in 3-D, and to understand it we must learn to think in 3-D and to represent chemicals as 3-D structures.

Three-Dimensional Structure of Methane

What does a methane molecule look like in 3-D? In order to get an impression for yourself you will need four equal-sized balloons. Each balloon represents one shared pair of electrons, forming a covalent bond between two atoms. Electrons are negatively charged, and like charges repel, so the four pairs of electrons around the central carbon atom of methane (represented by your four balloons) tend to be pushed as far apart from one another as possible in 3-D space. This idea or "model" for the behavior of electrons is called the Valence Shell Electron Pair Repulsion model, or **VSEPR** (pronounced "VESper.")

To visualize the VSEPR idea in action, blow up and tie off all four balloons. Now tie the balloons together so that they look like those in Figure 6.3A. Imagine the carbon atom at the center of the balloons and a hydrogen atom touching the top of each balloon. The hydrogen atoms are at the corners of a **tetrahedron,** with the bonds at an angle of 109.5°, as shown in Figure 6.3B. This angle between atoms is called the **bond angle.**

This is a central concept of organic chemistry: when a carbon atom is connected to four other atoms, the result is a tetrahedral shape with the carbon at the center.

The Alkanes

Methane is the simplest member of a series of hydrocarbons known as the **alkanes.** The first 10 members of the alkane series are shown in Table 6.1. The alkanes form a **homologous series** of organic chemicals, meaning a series that shares the same general formula and broadly similar chemical characteristics. The alkanes shown in Table 6.1 share the general formula C_nH_{2n+2}.

The bonds around each carbon atom in an alkane are arranged tetrahedrally, like those of the methane molecule we have just examined. This means that the carbon-carbon chain really follows a "zigzag" pattern in 3-D, and the molecules have 3-D structures, as depicted in Figure 6.4.

Table 6.1

The First Ten Unbranched Alkanes

Name	Number of Carbons	Molecular Formula	Structural Formula
Methane	1	CH_4	CH_4
Ethane	2	C_2H_6	CH_3CH_3
Propane	3	C_3H_8	$CH_3CH_2CH_3$
Butane	4	C_4H_{10}	$CH_3CH_2CH_2CH_3$
Pentane	5	C_5H_{12}	$CH_3(CH_2)_3CH_3$
Hexane	6	C_6H_{14}	$CH_3(CH_2)_4CH_3$
Heptane	7	C_7H_{16}	$CH_3(CH_2)_5CH_3$
Octane	8	C_8H_{18}	$CH_3(CH_2)_6CH_3$
Nonane	9	C_9H_{20}	$CH_3(CH_2)_7CH_3$
Decane	10	$C_{10}H_{22}$	$CH_3(CH_2)_8CH_3$

exercise 6.1

Molecular Shape

Problem

Can you predict the shape of the ammonia molecule?

Solution

The key to solving shape problems like this is to *count the number of atoms and lone pairs of electrons around the central atom.* The nitrogen atom in ammonia is bonded to three other atoms and has one lone pair, so a total of four balloons would be required to represent the shape of ammonia. Each balloon corresponds to an electron cloud, one containing the lone pair, and the other three containing the electrons that form the bonds with the hydrogen atoms. As with methane, the electron clouds will be pushed as far apart as possible by their mutual repulsion. This makes the electron pairs take up a tetrahedral orientation (see Figure 6.5). Because there are only three atoms around the central nitrogen atom, the molecular shape is trigonal pyramidal. The unshared lone pair of electrons is less restricted in its movement than the electron clouds forming the bonds between the nitrogen atom and the hydrogen atoms. Therefore, the lone pair pushes the other electron clouds together a little, as shown in Figure 6.6. The result is an ammonia molecule with an angle between the bonds (called the **bond angle**) of 107°.

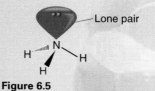

Figure 6.5

Ammonia

Figure 6.6   Ammonia has three bonding shared pairs and one unshared pair.

The alkanes are the major components of crude oil, which is used as the raw material to produce the vast range of organic materials made by the petrochemical industry.

Organic compounds are all based on the chemistry of carbon and its compounds. The most interesting and reactive parts of organic molecules, however, are often where atoms other than carbon and hydrogen are located. Oxygen and nitrogen are especially common in organic compounds. Therefore, we must say a few things about the compounds of oxygen and nitrogen in addition to carbon.

Ammonia, NH_3, is an **inorganic** (meaning "not organic") compound, because it does not contain carbon. The ammonia molecule, however, is involved in many organic processes when it reacts with organic chemicals. The Lewis structure of ammonia is:

$$H—\overset{..}{N}—H$$
$$|$$
$$H$$

Notice that the molecule includes a free electron pair, known as a **lone pair,** which is not involved in bonding between the atoms of the ammonia molecule. This lone pair of electrons plays a very important part, however, in the reactions between ammonia and other chemicals. During these reactions, the lone pair often forms a new bond between the nitrogen atom of the ammonia molecule and an atom of another chemical species.

Structure of Double Bonds

Although carbon atoms always react to become surrounded by four sets of shared electrons, this does not necessarily involve four separate single bonds to four different atoms. Carbon can also form **double bonds,** especially with atoms of oxygen and with other carbon atoms. In a double bond, four electrons (two distinct pairs) are shared between the atoms concerned.

Formaldehyde (CH_2O) is an example of a simple organic compound that includes a double bond. Formaldehyde is used as a starting material in making plastics and other polymers.

Methane

Butane

Hexane

Figure 6.4   The 3-D structure of the alkanes methane, butane, and hexane.

exercise 6.2

Structure of Water

Problem

The Lewis structure of water (another inorganic compound involved in many organic processes) is shown below.

H   H
:O:

What will water look like in 3-D?

Solution

How many atoms and lone pairs of electrons surround the central oxygen atom? There are two bonded hydrogen atoms and two lone pairs, making a total of four separate electron pairs (see Figure 6.7A). As before, the electron clouds will orient in a tetrahedral electronic shape. The unshared pairs will push the shared pairs together a bit. The three atoms of a water molecule will therefore be arranged in a bent molecular shape (see Figure 6.7B). We often do not show the lone pairs when drawing out the structures of molecules, but their hidden presence allows you to understand why water is drawn as a bent molecule, rather than a linear one.

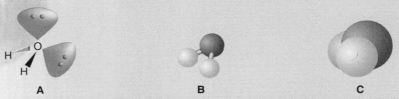

A   B   C

Figure 6.7   The VSEPR model of water shows that the unshared electron pairs of oxygen squeeze the shared pairs closer together, leading to the V-shape of the water molecule.

It is formed industrially from the reaction of methane (from petroleum) with steam or molecular oxygen, followed by air.

$$
\begin{bmatrix}
H & \\
 & 122° \\
116° & C = \ddot{O} \\
H &
\end{bmatrix}
$$

Molecular structure of formaldehyde

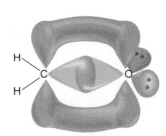

Figure 6.8 In double bonds, such as the one in formaldehyde, one of the electron pairs forming the double bond occupies an electron cloud with lobes above and below the carbon-oxygen plane.

The carbon atom of formaldehyde is surrounded by four pairs of shared electrons, but two of these pairs form the double bond with the oxygen atom. One of these pairs of electrons lies directly between the carbon and the oxygen atom, while the electrons of the other pair are found above and below the plane of the first pair (see Figure 6.8). Since there are only three atoms attached to the carbon atom of formaldehyde (and no lone pairs of electrons) you need only three balloons to represent the shape of formaldehyde. If you blow the balloons up and tie them off, what shape do you get? Figure 6.9 shows the formaldehyde atoms superimposed on the balloons. The molecule is flat (planar) since the four atoms all lie in the same plane. It looks like a trigonal planar triangle in 3-D, with the bonds separated by 122° (H—C=O bond angles) and 116° (H—C—H bond angle).

Table 6.2

Names and Formulas of the First Five Alkenes

Name	Molecular Formula	Structural Formula
Ethene	C_2H_4	$H_2C\!=\!CH_2$
Propene	C_3H_6	$H_2C\!=\!CHCH_3$
1-Butene	C_4H_8	$H_2C\!=\!CHCH_2CH_3$
1-Pentene	C_5H_{10}	$H_2C\!=\!CH(CH_2)_2CH_3$
1-Hexene	C_6H_{12}	$H_2C\!=\!CH(CH_2)_3CH_3$

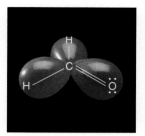

Figure 6.9 The electron pairs of the three separate bonds (one of them a double bond) of the formaldehyde molecule repel one another to form a flat planar molecule with a triangular shape.

Alkenes

When one of the C—C bonds of the alkanes shown in Table 6.1 is replaced with a carbon-carbon double bond, we generate a different homologous series known as the **alkenes.** The first five members of the alkene series with the double bond in-between the first and second carbons (called "*n*-alkenes") are given in Table 6.2. These alkenes share the general formula C_nH_{2n}.

Alkenes can readily be formed from the alkanes predominant in crude oil. The simplest alkene, ethene (also known as ethylene), is a particularly important raw material used to form a vast range of different organic compounds.

The alkenes are examples of **unsaturated** hydrocarbons, because they contain less than the maximum amount of hydrogen that can be bonded to the carbon framework. The corresponding alkanes are **saturated** hydrocarbons and can be produced by adding hydrogen to the alkenes. This "hydrogenation" reaction that changes an alkene into an alkane is an example of an **addition reaction.**

$$\text{>C=C<} + H_2 \xrightarrow[\text{catalyst}]{\text{Ni}} \begin{array}{c} | \quad | \\ -C-C- \\ | \quad | \\ H \quad H \end{array}$$

- Saturated hydrocarbons contain no C=C double bonds.
- Unsaturated hydrocarbons contain at least one C=C double bond.

Vegetable oils containing double bonds are hydrogenated to saturated fats for use in food products such as shortenings used in baking. We will revisit this chemistry in Chapter 17 (The Chemistry of Food).

Alkynes

Let's go one step further and look at another organic molecule, **ethyne** (acetylene) C_2H_2. Acetylene is the fuel for acetylene torches used to weld metals. Its structure includes a **triple bond.** Each carbon atom is surrounded by four pairs of shared electrons, but three of these pairs form the triple bond linking the two carbon atoms. The molecule has the following Lewis structure:

$$H\!-\!C\!\equiv\!C\!-\!H$$

As you can see in Figure 6.10, the 3-D shape of acetylene is linear. We would use a total of four balloons to show this shape; two balloons to represent the bonds around each carbon atom, one for the triple bond, and the other for the single bonds between each carbon atom and each hydrogen atom. How far apart can two balloons get? They can be at opposite ends of a straight line, or 180° apart. So when a carbon atom is bonded to only two other

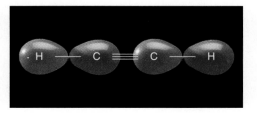

Figure 6.10 In triple bonds, such as the one in acetylene, two of the electron pairs forming the triple bond occupy electron clouds above and below, and on either side of the central electron pair.

atoms, a linear shape results. Since acetylene is a symmetrical molecule, each C can be looked at separately and considered a central atom. Acetylene is the first member of a homologous series known as the **alkynes,** which share the general formula C_nH_{2n-2}. Some other alkynes are shown in Table 6.3.

We have come a long way in our discussion of the shapes of molecules. We have seen that a VSEPR model is a way to predict the 3-D electronic and molecular shapes of compounds. We know that certain specific shapes result from the VSEPR model.

The chemistry of petroleum is an important part of organic chemistry (see also, Chapter 13). The dominant use of petroleum is as a fuel in motor vehicles. Although we will consider the importance of structure to chemical properties a bit later on, the Case in Point: Octane Rating will give you a sense of the different ways that molecules behave when there are even very slight structural differences between them.

Table 6.3

Names and Formulas of Some Alkynes

Name	Molecular Formula	Structural Formula
Ethyne (acetylene)	C_2H_2	$HC\equiv CH$
Propyne	C_3H_4	$HC\equiv CCH_3$
1-Butyne	C_4H_6	$HC\equiv CCH_2CH_3$
1-Pentyne	C_5H_8	$HC\equiv C(CH_2)_2CH_3$

■ case in point

Octane Rating

All of us have seen octane ratings listed on gasoline pumps with values such as 87 octane or 93 octane. We "sort-of" know that higher octane ratings correspond to better, more expensive gasoline, but how many of us know that the octane rating is named after an organic chemical compound?

Octane rating is a measure of how smoothly a fuel burns in an engine. A low octane gasoline tends to "knock" when it is burned in an engine. Knocking is a banging, chugging behavior and is very hard on engines. It happens if a fuel burns too rapidly, essentially instantaneously, when the spark is delivered to the mixture of fuel and air. Much more efficient burning occurs if the fuel burns slightly more slowly, pushing the piston rather than hitting it.

All compounds having a formula C_8H_{18} are octanes. If all the carbon atoms are connected in a single row, the compound is called simply octane. There are many possible ways to assemble C_8H_{18}, however, and even though the molecular formula of each of these is the same, the properties are different. Of particular interest is the speed with which each of these compounds burns. Table 6.4 shows the carbon arrangements of various **structural isomers** of octane. Structural isomers, also known as

constitutional isomers, are different compounds that have the same molecular formula but differ in the arrangement of the atoms.

The isomer, 2,2,4-trimethylpentane, informally called iso-octane, was thought to be the best possible automobile fuel, causing less knocking than any other compound known at the time the octane scale was formulated. Iso-octane was given a rating of 100. All the side branches of the molecule slowed its combustion so as to eliminate the objectionable knocking. Zero on the scale was given to the worst compound, heptane (C_7H_{16}), in which all seven carbons are attached in a single row. A fuel with an octane rating of 85 is therefore one that has combustion properties like a mixture of 85% iso-octane and 15% heptane.

A fuel's octane rating is based on how well it burns compared to the compound iso-octane. Since the octane scale was established, other compounds and blends of compounds have been manufactured which have anti-knock properties even better than iso-octane and so have octane ratings greater than 100.

There are still more isomers of C_8H_{18}; try to draw them but be sure they aren't just rearrangements of some listed in Table 6.4.

Table 6.4

Structural Isomers of Octane $(C_8H_{18})^{a,b}$

C—C—C—C—C—C—C—C

(*n*-octane)

C
|
C—C—C—C—C—C—C

(2-methylheptane)

C
|
C—C—C—C—C—C—C

(3-methylheptane)

C
|
C—C—C—C—C—C—C

(4-methylheptane)

C   C
| |
C—C—C—C—C—C

(2,3-dimethylhexane)

C   C
| |
C—C—C—C—C—C

(2,4-dimethylhexane)

C   C
| |
C—C—C—C—C—C

(2,5-dimethylhexane)

C   C
| |
C—C—C—C—C—C

(3,4-dimethylhexane)

C
|
C—C—C—C—C—C
|
C

(2,2-dimethylhexane)

C
|
C—C—C—C—C—C
|
C

(3,3-dimethylhexane)

C   C   C
| | |
C—C—C—C—C

(2,3,4-trimethylpentane)

C   C
| |
C—C—C—C—C
|
C

(2,2,3-trimethylpentane)

C   C
| |
C—C—C—C—C
|
C

(2,2,4-trimethylpentane)

C   C
| |
C—C—C—C—C
|
C

(2,3,3-trimethylpentane)

C—C
|
C—C—C—C—C—C

(3-ethylhexane)

[a]Remember, the structures in this table do not show hydrogen atoms. The real molecules have 18 hydrogens arranged so that each carbon atom has four bonds.

[b]Many of these compounds have straight chains with small carbon and hydrogen "branches." The names of the compounds are derived from the longest straight chain (5 carbons in 2,3,3-trimethyl pentane) and the branches (methyl groups, —CH_3, on the number 2 and number 3 carbons, counting from the left).

Introducing the Benzene Ring

We now know how to predict the structure of some fairly simple, yet important molecules. The structural basis of many not-so-simple, yet important, molecules is the benzene ring (C_6H_6). **Benzene,** shown in Figure 6.11, is an important industrial chemical, used in the syntheses of nylon and polystyrene. Benzene is also a constituent of gasoline and was once used as a solvent in many industrial processes. Its use has diminished in recent years because it has been shown to be carcinogenic.

The compound C_6H_6 was discovered by Michael Faraday in 1825 and its formula was determined in 1834. There was great puzzlement for years over its structure because a compound with six carbons would be predicted to have more than six hydrogens. Benzene didn't behave as expected based on its shortage of hydrogen atoms. It soon became clear that all six carbon-carbon bonds were chemically identical and nobody could figure out how that could happen until August Kekulé, in 1865, proposed that the molecule is a six-membered ring of carbons with alternating double and single bonds and one hydrogen on each carbon. Kekulé later said that he visualized the ring structure of benzene in a dream in which he saw snakes biting their tails and rolling like hoops.

Each carbon atom in the ring is connected to three other atoms (two carbon atoms and a hydrogen atom), and the atoms carry no lone pairs. Based on our previous discussion, you

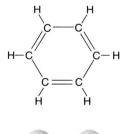

Figure 6.11   The VSEPR model leads to the correct conclusion that benzene is flat in 3-D.

should expect the resulting fragment of the molecule to have a triangular shape (trigonal planar) that could lie flat on a piece of paper.

We can make the same flat-paper argument for all of the carbon and hydrogen atoms in benzene, which means that benzene is a flat or "**planar**" molecule overall in 3-D (see Figure 6.12).

As with the formaldehyde molecule we discussed before, the double bonds that we draw between the carbon atoms involve electrons that are above and below the plane of the paper, so although the benzene molecule is flat with the nuclei in the same plane, there is a ring-like cloud of electrons above and below the plane. And since all the bonds are equivalent, benzene can be represented by either of the two equivalent structures shown in part **A** of Figure 6.12 or by the single structure in part **B** of that figure, with a ring representing the circulating electrons. The structure with the ring gives a more accurate impression, because we know that all of the carbon-to-carbon bonds in benzene are *identical* in length, strength, and all other characteristics.

Table 6.5 lists several important molecules that have at least one benzene ring in their structure.

Figure 6.12 Different ways of representing a benzene ring. The two structures at the left of the diagram (**A** and **B**) are chemically identical; but the benzene ring is often represented by the single structure in which a circle represents the ring-like clouds of electrons above and below the main plane of the molecule (**C**).

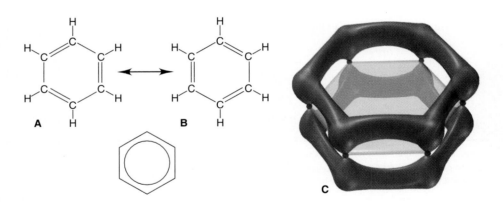

Table 6.5

Benzene-Related Compounds

Substance	Structure	Uses
Benzene		Manufacture of nylon and polystyrene
Toluene	CH_3	Paint solvent
Phenol	OH	Disinfectant and manufacture of adhesives and fibers

Table 6.5 (continued)

Substance	Structure	Uses
P-Dichlorobenzene		New mothballs
Aspirin		Pain reliever
Naphthalene		Old mothballs
Polystyrene		Plastic
TNT (trinitrotoluene)		Explosives
Saccharin		Artificial sweetener
2,4-D		Herbicide
PET (polyethyleneterephthalate)		Plastics

■ case in point

Benzene and the Price of Compact Disks

Many music afficionados would argue that there is no comparison between the quality of music that you hear when listening to a vinyl LP record and that from a compact disk (CD). The record, they argue, clearly wins because the sound is not broken down into many discrete bits, but rather contains the full range of sound that the artist puts forth. Nonetheless, CDs are now far more popular because the music imprinted on them will essentially never degrade, as will happen to every vinyl record.

CDs vary widely in cost from about $7 to $22, depending on the quality of the product, the recording artist, and where the CD is purchased. With prices varying so widely, one wonders what goes into the cost of the CDs. How do the prices of individual chemicals and processing impact the "the bottom line"?

One of the compounds found in crude oil is benzene, C_6H_6, which can be isolated in the refining process. The cost of obtaining enough benzene to be processed into a single CD is a fraction of a cent.

Benzene is then chemically converted to pellets of "Lexan," a superclear plastic used in, among other products, bulletproof windows. The conversion involves converting benzene to cumene, then to phenol, with acetone being produced as a by-product (and this is the way acetone is made commercially).

The phenol is converted to bisphenol A, which is combined with phosgene (a valuable synthesis agent, also used as a poison gas in past wars) to give the polycarbonate Lexan. This entire process costs about one cent per CD.

bisphenol A   phosgene

At this point, the Lexan pellets are pressed into a disk, the music is encoded onto the disk, a reflective surface of aluminum is added, and the product is coated with a lacquer. Total cost is now 10 to 15 cents. What drives the cost up?

The CD jacket and plastic box costs 30 to 35 cents. The artist gets up to $2.50 per CD sold. The companies that hold the patent on the CD technology each get a small royalty. Marketing, distribution, promotion, and profits for the recording company plus a markup that guesses what people will be willing to spend for a CD comprises the rest of the cost.

So many things go into determining the cost of a product. Yet in the earliest part of the 21st century, when advertising the product and paying the artist can cost millions of dollars, the cost of the chemicals that form the product is often insignificant.

benzene   cumene   phenol   acetone

Lexan

Functional Groups in Organic Chemistry

Most organic chemicals can be viewed as a hydrocarbon framework to which various other groups of atoms, often nitrogen or oxygen, have been added. We call the added groups **functional groups** because they bring to a molecule a specific set of chemical reactivities that can serve particular functions.

The incorporation of any functional group into a hydrocarbon framework creates a new series of related compounds, all possessing the characteristics bestowed by that functional group. For example, a molecule containing the hydroxyl functional group (—OH) is termed an alcohol. The best-known alcohol is ethanol, the one found in alcoholic beverages.

Table 6.6

The Key Functional Groups

Structure	Group	Example	Name	Selected Uses
$\diagup C{=}C\diagdown$	Alkene	$CH_2{=}CH_2$	Ethylene	Refrigerant; production of polyethylene
$-C{\equiv}C-$	Alkyne	$HC{\equiv}CH$	Acetylene	Welding; cutting; brazing
$-\overset{\mid}{\underset{\mid}{C}}-OH$	Alcohol	CH_3CH_2OH	Ethanol	Liquors; industrial solvent
$-\overset{\mid}{\underset{\mid}{C}}-O-\overset{\mid}{\underset{\mid}{C}}-$	Ether	$CH_3CH_2OCH_2CH_3$	Diethyl ether	Industrial solvent
$-\overset{O}{\overset{\|}{C}}-H$	Aldehyde	$CH_2{=}O$	Formaldehyde	Bactericide; fungicide; chemicals production
$-\overset{\mid}{\underset{\mid}{C}}-\overset{O}{\overset{\|}{C}}-\overset{\mid}{\underset{\mid}{C}}-$	Ketone	$CH_3\overset{O}{\overset{\|}{C}}CH_2CH_3$	Methyl ethyl ketone	Solvent in synthetic rubber industry
$-\overset{O}{\overset{\|}{C}}-OH$	Carboxylic acid	$H\overset{O}{\overset{\|}{C}}-OH$	Formic acid	Manufacture of textiles, pesticides; electroplating
$-\overset{O}{\overset{\|}{C}}-O-\overset{\mid}{\underset{\mid}{C}}-$	Ester	$CH_3\overset{O}{\overset{\|}{C}}-OCH_3$	Methyl acetate	Paint remover; pharmaceutical manufacture
$-\overset{\mid}{\underset{\mid}{C}}-NH_2$	Amine	$CH_3(CH_2)_3NH_2$	Butylamine	Manufacture of rubber, insecticides
$-C{\equiv}N$	Nitrile	$CH_3C{\equiv}N$	Acetonitrile	Industrial solvent; pharmaceutical manufacture
$-\overset{O}{\overset{\|}{C}}-NH_2$	Amide	$H-\overset{O}{\overset{\|}{C}}-NH_2$	Formamide	Manufacture of paper, glue; industrial solvent
$-\overset{\mid}{\underset{\mid}{C}}-SH$	Thiol	$CH_3CH_2CH_2SH$	Propanethiol	Herbicide; flavoring agent; additive to odorless poisonous gases (so they smell)

Source: New Jersey Dept. of Health & Senior Services Right to Know Program.

Table 6.6 shows the major functional groups found in organic chemicals. You should learn to recognize these functional groups. You will find they appear many times in the chemical structures and applications elsewhere in this book.

6.3   The Impact of Structure

In this section we examine a few interesting situations that reveal the way in which the structure of chemicals determines their chemical activities.

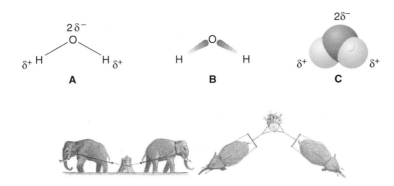

Figure 6.13 The structure of water.

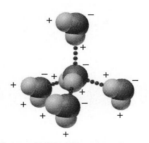

Figure 6.14 Water is a polar molecule with a partial positive charge (δ⁺) on each hydrogen atom and a partial negative charge (2δ⁻) on the oxygen atom. If water were a linear molecule it would be nonpolar overall, because, like two elephants pulling on either side of a tree, the effects of the two polar covalent bonds would cancel each other.

Figure 6.15 The polar nature of water allows the molecules to interact because the oppositely charged regions attract each other.

The Effect of Structure on the Properties of Water

We know from Exercise 6.2 that the atoms of water molecules are bonded in a bent shape. We also know from our discussions of periodic properties in Chapter 3 that oxygen tends to draw electrons toward itself, in this case away from the hydrogen atoms. So although the hydrogen and oxygen atoms share electrons, the sharing is not equal. In other words, the atoms are linked by a polar covalent bond that leaves a partial negative charge (δ^-) on the oxygen atom and a partial positive change (δ^+) on the hydrogen atoms. This is shown in Figure 6.13. A water molecule is **polar** overall, meaning that it has one partially positive end and one partially negative end due to the asymmetrical distribution of its polar covalent bonds.

What if water were linear? As shown in Figure 6.14, if water were linear, the two poles of charge would cancel out (like the effect of two identical elephants pulling on either side of a tree stump—the tree stump would not move).

So water is a polar molecule because of its electronic structure and its 3-D molecular shape. This is not a trivial fact—it has an enormous impact on the properties of water. The partial positive charge on each hydrogen atom is readily attracted to the partial negative charge on oxygen atoms of neighboring water molecules. For that matter, the partial positive charges on water's hydrogen atoms are attracted to partial negative charges on other molecules that contain oxygen or nitrogen. These interactions are absent between water and organic compounds which contain only carbon and hydrogen. The primary impact is that water molecules can interact with other water molecules to form an extended 3-D network of attracting **intermolecular forces** as shown in Figure 6.15. These specific intermolecular forces are known as **hydrogen bonds,** due to the central role of hydrogen atoms. Although labeled as "bonds," they are really temporary interactions that are much weaker than full covalent or ionic bonds. Nevertheless, a significant amount of energy is required to break these hydrogen bonds. They need to be broken when water either melts or boils, which explains why water has high melting and boiling points compared to similar-sized molecules with no hydrogen bonds between the molecules. We will spend an entire chapter (Chapter 7) discussing these and other implications of the structure of water in detail. For now, the purpose of this look at water is a reiteration of the point that structure dictates function. The extent of the interaction of water with selected organic molecules will provide many examples of this.

Water is a bent molecule that is polar because of its shape and the unequal sharing of bonding electrons between the oxygen and hydrogen atoms. Water molecules tend to stay with other water molecules because the molecules are all held to each other by the intermolecular forces we call hydrogen bonds.

Oils, on the other hand, are very different from water. Molecules in crude oil (from the Earth) and vegetable oil (from plants) are dominated by hydrocarbon regions. Crude oil contains thousands of different types of molecules. **Hexane,** a six-carbon chain, is one of the most abundant (accounting for about 2% of the molecules in crude oil, though the percentage varies widely depending upon the source of the oil). Mineral oil is a highly refined fraction of crude oil, with molecules typically containing 10- to 16-carbon atoms.

Carbon and hydrogen atoms attract bonding electrons to very similar extents, that is, they have very similar electronegativities. Moreover, bonds of one carbon to another are completely nonpolar since both atoms have the same electronegativity. This means that hy-

Figure 6.16 Some fatty acids found in nature.

Stearic acid

Oleic acid

Linoleic acid

Polar group

drocarbons such as hexane are not polar molecules, so they are not held together by strong intermolecular attractions such as hydrogen bonds.

So, to summarize, water molecules are actively attracted to other water molecules due to their polarity. They are *not* attracted to most oil molecules because oil molecules tend, as a group, to be nonpolar. This is why water and oil do not mix. If we try to force them to mix, the water molecules soon gather together again, pulled by their mutual attractions in a way that forces the oil molecules out of the network of interacting water molecules.

There are many types of oils containing compounds of varying polarities. Vegetable oils, for example, contain molecules called **fatty acids,** see Figure 6.16, which have a large nonpolar end and a small polar end (the shaded area in the figure). Gasoline, derived from crude oil, contains nonpolar molecules such as cyclohexane, C_6H_{12}, and more polar molecules, like ethanol, C_2H_5OH.

Like Dissolves Like

The inability of oil and water to mix illustrates a useful rule of thumb, namely, "like dissolves like." What this terse statement means is that molecules of one compound interact appreciably, and therefore mix easily, with molecules of a different compound only if the two are structurally similar in some important way. So water (a polar molecule) will mix with other polar molecules allowing polar substances to dissolve in water. Similarly, nonpolar substances such as oils will normally mix only with other nonpolar substances.

This rule has many practical consequences. Ducks and other waterfowl don't get their feathers wet because the feathers have a thin coating of a waxy, nonpolar substance that does not interact with water. This explains why water runs in droplets off a duck's back, rather than soaking into its feathers. On the other hand, oil is a particularly damaging pollutant to waterfowl because it is nonpolar and interacts very well with the waxy substance. Cormorants in regions of oil spills quickly become soaked with oil. A cormorant that has been washed with soap or detergent and water to remove this oil then becomes very vulnerable to getting soaked to the skin and not being able to cope (Figure 6.18), because the soap has washed away the protective waxy coating of its feathers as well as the oil. This raises an interesting issue: If oils and water do not mix, how is it that a watery solution of soap or detergent can so readily wash oil and grease away from ducks, dinner plates, or people? We will discuss this in Chapter 7.

Figure 6.18 The noninteraction of polar and nonpolar molecules protects healthy ducks and other waterfowl from the effects of the water. If anything interferes with this balance, the duck's natural weather proofing is damaged. When waterfowl are rescued from areas of oil spills, the cleaning process can cause as much harm to them as the petroleum. Just one more example of the wisdom of the saying "Don't fool with Mother Nature!"

hands-on

Figure 6.17 Why do the oil and water layers separate in the wave bottle?

Like Oil and Water

When two people just cannot "get along" we sometimes say that they are "like oil and water," meaning they just will not mix effectively.

Try this simple activity to explore the inability of oil and water to mix.

We use mineral oil, which contains hydrocarbons such as $C_{16}H_{34}$.

Collect:

- a small, colorless soda bottle (glass or plastic) with the cap
- enough mineral or baby oil to fill 1/2 of the bottle
- enough water to fill the other 1/2 of the bottle
- some food coloring
- a funnel (optional)

Put a couple of drops of food coloring into the water and swirl to mix. Add the water to the bottle, using the funnel to avoid making a mess. Add the oil to the bottle (again, using the funnel) and fill to the top. Seal the bottle tightly.

A photograph of what you should see is shown in Figure 6.17. Why are there two separate layers? The answer, and its implications for much more global issues, relates to our theme of *structure dictating function*. Before we explore the answer, however, give your bottle a very vigorous shake for about 30 seconds in an attempt to force the oil and water to mix (but do make sure the cap is on tightly!). Look at the bottle immediately after shaking. What happens? For a second or so you may think you have succeeded; but very soon the two layers separate once again. They just will not stay together.

Food for thought: Try the activity again, adding a squirt of liquid dishwashing detergent. What happens? Why? What does the observed behavior say about the structure of food coloring? We discuss the behavior of detergents in Section 7.3.

Use of Oxygenates in Gasoline

Green Chemistry

Most compounds in gasoline are hydrocarbons, that is, they consist only of hydrogen and carbon. **Oxygenates** are compounds which contain oxygen and are added to gasoline in large percentage (5–20%) in order to lower the amounts of pollutants given off by the automobile. The two classes of compounds usually considered are the alcohols, methyl alcohol (methanol)

Table 6.7

Oxygenates

Compound	Formula	% Oxygen	Octane Rating
Methanol	CH_3—OH	$(16/32)100 = 50.0$	110
Ethanol	C_2H_5—OH	$(16/46)100 = 34.8$	110
MTBE	CH_3—O—C_4H_9	$(16/88)100 = 18.2$	109
ETBE	C_2H_5—O—C_4H_9	$(16/102)100 = 15.7$	111

and ethyl alcohol (ethanol); and the ethers, methyl *t*-butyl ether (MTBE) and ethyl *t*-butyl ether (ETBE). These four compounds are listed in Table 6.7. All of these compounds burn well, supply lots of energy upon burning, and result in lower amounts of carbon monoxide being released in exhaust fumes.

Ethanol can be made either from petroleum by the reaction of ethylene with water

$$H_2C=CH_2(g) + H_2O(l) \rightarrow CH_3—CH_2—OH(l)$$

or by the fermentation by yeast of sugar or starch.

$$C_6H_{12}O_6(s) \rightarrow 2C_2H_5OH(l) + 2CO_2(g)$$

The technology for both of these means of producing alcohol is well worked out. Admittedly, it is a little tricky to remove the last bit of water from the alcohol and this is an important step if the ethanol is to be used as automobile fuel. An important feature is that ethanol can be produced from surplus grain or lower quality foods such as potato peels or orange rinds. Methods are also under development to manufacture fuel-grade ethanol from waste paper. This process requires that the cellulose in the paper first be converted into sugars by bacteria which can break the β-polysaccharides of cellulose into glucose which is readily fermentable.

Methanol is not approved for use as a fuel additive because it evaporates too easily, and would contribute severely to air pollution. Methanol is used, however, to make MTBE. Methanol is mostly made by reacting carbon monoxide with hydrogen over an appropriate catalyst. The carbon monoxide and hydrogen are both derived from coal that has been reacted with steam, again using a specialized catalyst.

$$C(s) + H_2O(g) \rightarrow CO(g) + H_2(g)$$

$$CO(g) + 2 H_2(g) \rightarrow CH_3OH(l)$$

Combining the densities of these compounds and of typical gasoline with the percent oxygen of the oxygenates, one calculates that 7.39 L of ethanol is necessary to bring the oxygen content of 100 L of gasoline to 2.7% oxygen. The corresponding numbers for MTBE and ETBE are 15.15 L and 17.32 L per 100 liters respectively, of oxygenated gasoline. In March 2000, the U.S. government issued a ban on the use of MTBE in automobile fuel. When MTBE gasoline leaks out of storage tanks and into ground water, the MTBE dissolves more completely in the water than hydrocarbons in gasoline. The terrible smell of MTBE makes for undesirable drinking water.

Gasoline must evaporate easily so that fuel vapors will be available for combustion to begin. However, if the fuel is too volatile, much of it will evaporate when the car is not in use. This results in air pollution. Since volatility is greater at high temperature than at low temperature, gasoline is formulated differently in winter than in summer. Regional differences also exist. Winter-time volatility is more of a problem in Minnesota than in Florida, for instance.

Current U.S. EPA regulations state that summer gasoline should have a vapor pressure of 9 psi (pounds per square inch) and winter gasoline 14 psi under standard conditions. We will have more to say about vapor pressure in Chapter 7, but in the meantime you should know that small molecules are more volatile than larger ones and that a high vapor pressure means that the compound evaporates more readily than one with a lower vapor pressure.

The implication of all this is that small molecules like butane (vapor pressure = about 50 psi) are suitable components of winter gasoline but not of summer gasoline. What should

exercise 6.3

The Impact of Structure on Vitamin Dosage

Problem

Vitamins are substances that are essential for our bodies to function but which our bodies do not produce for themselves. Vitamin A is important in vision, where it helps in the chemical transmission of images from the eye to the brain. It also helps keep the cornea moist. Vitamin C is required for the formation of blood and may also perform other positive functions, including boosting the immune system in ways that protect against illnesses ranging from cancer to the common cold.

vitamin A, a fat-soluble vitamin

vitamin C, a water-soluble vitamin

Excess vitamin A is toxic, but taking massive doses of vitamin C is apparently not dangerous and possibly helpful. Can you suggest an explanation for this difference using polarity and solubility as the key ideas? A crucial clue is that fat molecules have a long nonpolar hydrocarbon chain, with a small polar group of atoms attached at one end.

Solution

Vitamin C has a number of O—H groups that contribute to the polarity of the molecule. Therefore, like water, it is a polar molecule, making it able to mix with water by participating in intermolecular interactions with the water molecules. In other words, vitamin C is soluble in water. This means that excess vitamin C is readily excreted in urine and will not accumulate in the fatty tissues of the body. Vitamin A, on the other hand, is essentially nonpolar, making it soluble in fats. Excess vitamin A will therefore become concentrated in the body's fat cells, allowing it to build up in the body to levels that can have harmful effects.

the petroleum companies do with all the butane that they accumulate during production of summer gasoline? A little transforming makes them into the butyl part of ETBE which has a vapor pressure of 4 psi.

n-butane   isobutane   isobutene

pro con discussions

Should We Use Oxygenated Fuel Additives?

Green
Chemistry

pro

There are three important advantages to the use of oxygenated fuel additives.

- First, all three compounds that have been proposed have very high octane ratings.
- Second, the oxygen atoms in the molecules promote more complete combustion. More CO_2 is formed and less CO than is the case with nonoxygenated fuels.
- Third, these compounds are made largely from resources within the United States, ethanol being largely made from renewable resources.

People often argue that fuel ethanol results in using material such as corn that could be used for human food and converts it into automobile fuel. The corn that is used is not grown for human food; rather, it is ordinarily fed to cattle or hogs, which become human food. During the alcohol production step, only the carbohydrate becomes converted into alcohol; the yeast cells and corn protein are reclaimed after fermentation and fed to cattle.

While it once was true that some plastic parts in engines were damaged by gasoline/ethanol mixtures, that is no longer true. It was a pretty simple change for auto makers to use plastic components that are not harmed by alcohol in fuel.

The beauty of the production of both alcohols is that they come from sources readily available from within the United States, grain and coal. The use of ethanol reduces the U.S. dependence on foreign oil for our motor vehicle fuel. The grain and waste food origin of ethanol represents a truly renewable resource and the use of coal uses a resource which is plentiful within the United States.

con

Come on! The energy cost to produce any of these oxygenated fuels is more than the energy available from them when they are used as a fuel. As a result, a gallon of alcohol costs more than a gallon of gasoline. Putting ethanol into the fuel raises the price. It is only because of tax incentives on production of ethanol that it becomes even marginally cost allowable.

Besides the economic argument, it is unconscionable that we use corn and other grains to make automobile fuel when these products should be used to provide food for people or animals.

Besides that, the alcohols tend to form aldehydes when they are incompletely combusted.

$$CH_3CH_2\text{—}OH + [O] \rightarrow$$
ethyl alcohol
$$CH_3CH\text{=}O + H_2O$$
acetaldehyde

We always say that combustion leads to CO_2 and H_2O and most of the fuel does produce these products, but incomplete combustion frequently occurs. The resulting aldehydes are even more unpleasant in the atmosphere than are hydrocarbons or CO. Lots of people have severe allergic reactions to aldehydes and they contribute significantly to urban smog.

Another problem I've heard about is that ethanol in gasoline causes plastic parts of the fuel distribution system such as seals and gaskets to deteriorate rapidly. I don't want to use a fuel that damages my engine.

Since we are only talking about 10% of the fuel, the various concerns make it silly to go to the trouble of making such a change.

$$
\begin{array}{ccc}
\underset{\text{ethanol}}{\overset{\displaystyle \text{H} \quad \text{H}}{\text{H—C—C—O—H}}} & + & \underset{\text{isobutene}}{\overset{\displaystyle \text{H} \quad \text{CH}_3}{\text{H—C=C—CH}_3}} \longrightarrow \underset{\text{ETBE}}{\overset{\displaystyle \text{H} \quad \text{H} \quad \text{CH}_3}{\text{H—C—C—O—C—CH}_3}}
\end{array}
$$

The amounts of gasoline we use in a year are simply staggering. Annual consumption (1999) of gasoline in the United States is about 164 billion gallons. Of this, about 14 billion gallons are gasohol (10% ethanol/90% gasoline) so it amounts to 1.4 billion gallons of ethanol. Every gallon of ethanol we use in producing auto fuel is 1 gallon less fossil fuel that we must produce or import. Actually, various savings in modern production amount to another

0.25 gallons of fossil fuel saved for every gallon of ethanol blended into auto fuel. The energy balance is very tricky because energy needed to transport petroleum, alcohol, and corn must be figured in. We do not count the solar energy necessary to grow the grain, but the use of the protein fraction in animal feed, the drying, and transport of this material all must be included in the calculation.

You have seen that organic compounds are necessary for the continuation of life and as energy sources for both mankind and our machines. Some organic compounds, however, interact with living things in a very different fashion. Let's look at several classes of compounds that demonstrate the wide variety of ways organic compounds behave.

exercise 6.4

Growing Corn for Fuel-Grade Ethanol

Problem

Approximately 2.5 gallons of ethanol can be obtained from 1 bushel of corn. If the yield of corn is typically 150 bushels per acre, how many square miles of farmland are necessary to grow the corn to make all the ethanol used in the United States in a year, 1.4 billion gallons in 1999? (There are 640 acres per square mile.)

Solution

$$\text{Area} = 1.4 \times 10^9 \text{ gallons} \times \frac{1 \text{ bushel}}{2.5 \text{ gallons}} \times \frac{1 \text{ acre}}{150 \text{ bushels}} \times \frac{1 \text{ square mile}}{640 \text{ acres}}$$

$$= 5.8 \times 10^3 \text{ square miles}$$

This is a square area approximately 76 miles on a side.

exercise 6.5

More on Ethanol Production—You Make the Call

Problem

According to the U.S. Department of Agriculture about 80.2 million acres of corn were planted in 1998 with an average yield of 127 bushels per acre. Using information from Exercise 6.4, calculate the percentage of the area which is used for ethanol production. Do you think that major increases in ethanol production from corn are feasible?

Solution

$$\text{Area devoted to ethanol} = 1.4 \times 10^9 \text{ gallons} \times \frac{1 \text{ bushel}}{2.5 \text{ gallons}} \times \frac{1 \text{ acre}}{127 \text{ bushels}}$$

$$= 4.4 \times 10^6 \text{ acres}$$

Percentage of corn planting devoted to ethanol production

$$= \frac{4.4 \times 10^6 \text{ acres}}{80.2 \times 10^6 \text{ acres}} \times 100\% = 5.5\%$$

Chemical Warfare Agents

Soman ($C_7FH_{16}O_2P$) is a liquid "nerve gas" that was invented in Germany in the early 1940s, for use in World War II. Although a liquid at room temperature and pressure, it readily vaporizes and hence can also be described as a nerve gas. There is no evidence that Soman was ever actually used in the war, but it is one of a number of nerve gases that harm or kill humans by overstimulating the signaling between nerve cells that control, among other things, muscular activity. According to one account, when Soman is sprayed on a person

> the peripheral and central nervous systems stimulate themselves to their own destruction. This overstimulation causes violent muscle activity and malfunctioning of the various body organs. Most profoundly affected is the brain, which ceases to function before the body stops its *danse macabre*.

Many countries, including Russia, Germany, Britain, and the United States, have done research on these and similar chemical weapons for at least the past 70 years. One of the most recent uses of chemical weapons was in the infamous massacre of thousands of Kurds in Halabja, Iraqi Kurdistan, March 16, 1988. Another incident occurred in Tokyo, in 1995, when members of a religious cult put the nerve gas "Sarin" into a crowded commuter subway, causing both deaths and injuries. The threat of these types of "chemical weapons" remains with us today. Note the structural similarity of Sarin to Soman.

Soman

Sarin

How do chemical agents such as the so-called nerve gases actually cause their damage? Once again, we find that their chemical structure dictates their function. In our bodies, nerve cells interact with other nerve cells by releasing chemicals called **neurotransmitters.** The neurotransmitters diffuse across the tiny gap between two adjacent nerve cells known as the **synapse** and then bind to specific **receptor** molecules on the surface of the next nerve cell's membrane. The binding of the neurotransmitter initiates a series of chemical events that either stimulates or inhibits the transmission of a nerve impulse along the length of the nerve cell (see Figure 6.19). So the structure of the neurotransmitters dictates their function, by allowing them to fit into the receptor binding site on a nerve cell's surface.

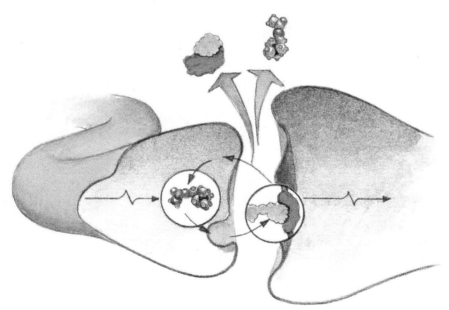

Figure 6.19   Nerve cells in our bodies interact across synapses by means of neurotransmitters.

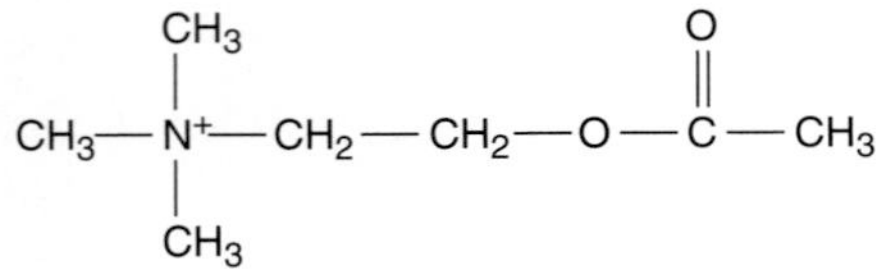

Acetylcholine

Figure 6.20

One of the most important neurotransmitters is **acetylcholine,** whose structure is shown in Figure 6.20. In order for nerve transmission to work properly, it is important that neurotransmitters such as acetylcholine are degraded to other products shortly after their release. This crucial degradation reaction, in the case of acetylcholine, is catalyzed by an enzyme called acetylcholinesterase.

Notice that acetylcholine has a positive charge (so it is technically a positive ion), while the site to which it binds on the enzyme has a negative charge. Many nerve gases work because they form positive ions which bind to the binding site on acetylcholinesterase more effectively than acetylcholine itself. This greatly reduces the availability of the binding sites to catalyze their normal reaction, leading to an accumulation of acetylcholine. This in turn causes the nerves concerned to be greatly overstimulated, leading to violent uncontrolled contractions of muscles and other life-threatening problems. All this happens because the chemical structure of the nerve agents allows them to bind to the acetylcholinesterase enzyme.

The nerve agents that work in this way (known as **acetylcholinesterase inhibitors**) include chemicals known as **organophosphates,** which are also used as insecticides. In small doses, they can be used to kill insect pests while causing minimal risks to humans, unless misused.

Pollutants That Mimic Hormones

What makes a woman a woman and a man a man? Some might say that the reasons are social, in addition to biological and chemical factors; but for this discussion we will stick with the biology and chemistry—the "science."

Our bodies contain chemicals known as **hormones** (from the Greek *horman,* meaning "urge on"). Each hormone is released by one or more organs of the body, such as the liver, kidneys, ovaries, or testes to travel through the blood and bind to and affect the cells of other organs. The hormones are chemical "messenger molecules," which are able to bind to the appropriate target cells and deliver the appropriate "message" due to their chemical structures.

Estradiol (an **estrogen**) and **testosterone** (an androgen) (see Figure 6.21) are sex hormones secreted by the gonads (the ovaries of women and the testes of men), the adrenal glands, and the placenta in pregnant women.

They control many physical sex characteristics such as the development of the sex organs, breasts, sperm, facial hair in men, and deposits of fatty tissues in the breasts, buttocks, and thighs of women. It is also becoming apparent that estrogens are even involved in brain function. Although estradiol is commonly referred to as a "female sex hormone" and testosterone as a "male sex hormone," both hormones are present in both sexes. The relative amounts present in each sex differ, however, with females containing more estrogens and males containing more androgens. As a matter of fact, the estrogens in a woman's body are made from male hormones such as testosterone.

Humanity's artificial manipulation of chemicals has now introduced new factors into the hormonal balance of nature. A range of chemicals released into the environment as pollutants is known to be able to mimic the effect of natural estrogen, although at first sight their

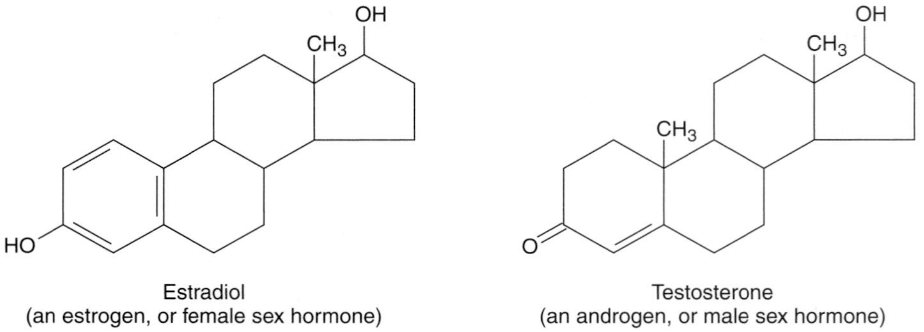

Estradiol
(an estrogen, or female sex hormone)

Testosterone
(an androgen, or male sex hormone)

Figure 6.21

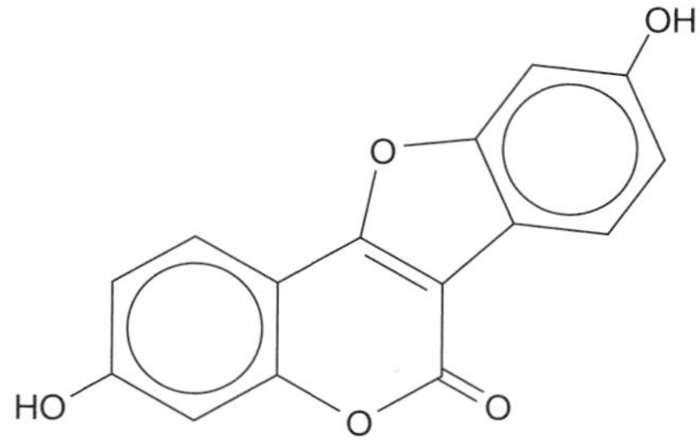

Coumestrol 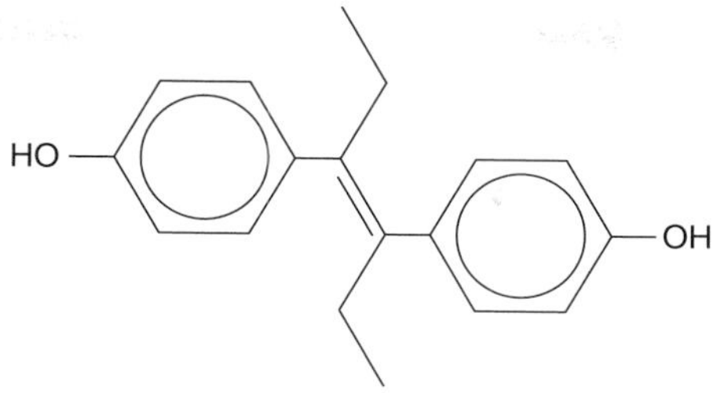   Diethystilbestrol

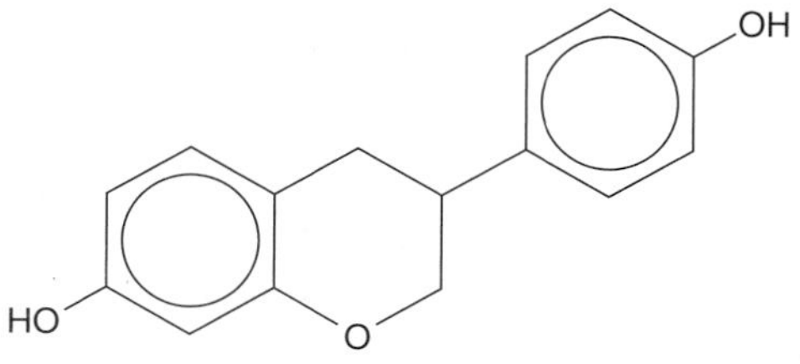

Equol 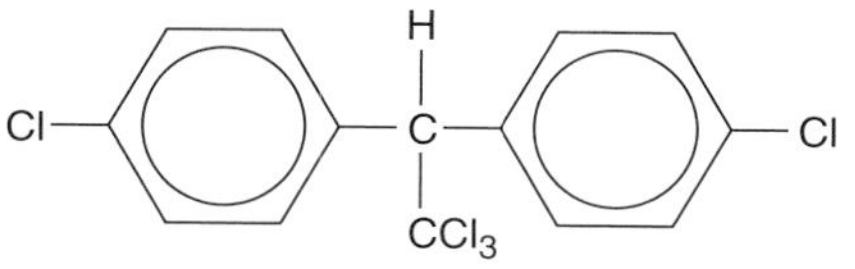   DDT

Figure 6.22   A variety of "gender-benders" have been found to mimic the behavior of estrogen. It is not yet understood why.

structures seem very different from that of estrogen (see Figure 6.22). These pollutants have become known as **environmental estrogens** or "gender-benders." The latter term emphasizes their ability to make male animals develop some of the characteristics of females and to over-stimulate normal female characteristics. For example, male alligators in one polluted Florida waterway have developed structures that look like ovaries, while the female alligators have begun to produce many more eggs than usual. We know of similar "gender-bending" effects in eagles, turtles, and panthers; there have been some indications that some humans may be affected.

What is it about the environmental estrogens that allows them to mimic estrogen? Nobody knows for certain, but it is believed to depend on structural features of the chemicals concerned, which allow them to share estrogen's ability to bind to the natural receptor molecules involved in estrogen's normal actions. These estrogen receptors are the molecules of the target cells for estrogen (often found on the cell surface), which initiate the chemical effects of estrogen once an estrogen molecule becomes bound to them.

Estradiol and testosterone are two examples of **steroids,** a broad class of molecules involved in controlling the characteristics and the development of living things. The next section considers another issue involving steroids.

The Use and Abuse of Steroids

The word "steroids" may make you think of athletes "popping" pills or secretly injecting themselves in darkened rooms. Such images are realistic, because, as we shall discuss, steroids are widely abused. Yet, as we have just discussed, many steroids are natural chemicals that are vital for the body to work properly. Estradiol, testosterone, cholesterol, vitamin D, and cortisone (an anti-inflammatory agent) are all naturally occurring steroids required, in appropriate amounts, for a healthy body.

Figure 6.23 reveals the structural link that allows us to group such compounds under the common name of "steroids." Steroids share a common structural skeleton, which is a multiringed system containing 17 carbon atoms. The (sometimes small) differences among the structures shown in Figure 6.24 can underlie major differences in their chemical function as dramatic as determining whether you are a male or a female.

Testosterone, the main so-called male sex hormone, and steroids able to mimic its effects are the main steroids abused by athletes. One effect of testosterone is to increase

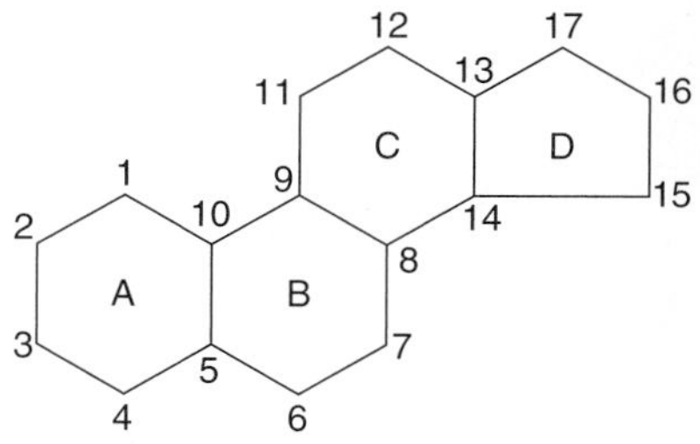

Figure 6.23   Steroids have a common skeleton, a 17-membered ring system. Many of the differences in steroid structure occur at carbon 17.

Testosterone

Vitamin D₃

Cholesterol

Cortisone

Figure 6.24   Compare the molecular structure of these compounds to the basic steroid structure of Figure 6.23.

the number of muscle cells in the body, accompanied by an increase in bone mass. Testosterone is known as an **anabolic steroid** because **anabolism** is the name given to the chemical processes in the body that build up new tissues. Many male and female athletes try to gain a competitive advantage by ingesting testosterone derivatives or additional testosterone. This definitely causes them to develop larger muscles and heavier bones, although whether it really improves athletic performance is not certain. Artificial testosterone derivatives are more commonly abused by athletes than natural testosterone. Exercise 6.6 will help you to understand why.

Effects of Steroid Abuse

You might think that, apart from the immorality of cheating in competition, taking anabolic steroids might offer a reasonable route to building up muscle mass. The abuse of these substances, however, is a great cause of medical concern. Anabolic steroids may well have many undesirable side effects, including increased aggressiveness and irritability, possible sterility in both men and women, and liver damage. There remains considerable uncertainty, however, about both the effects of anabolic steroids on athletic performance and any long-term harmful effects on the body.

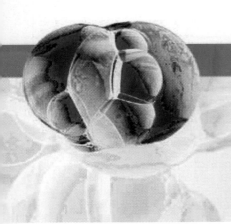

exercise 6.6

Variations On Testosterone

Problem

Table 6.8 gives the names and structures of some testosterone-like molecules that are used by doctors to treat people with genuine testosterone deficiencies.

How are these molecules different from testosterone? Are they likely to be more or less soluble in fat than testosterone? Why do you think some athletes prefer to use (or, we should say, abuse) one of the derivatives of testosterone shown here, rather than natural testosterone?

exercise 6.6 (continued)

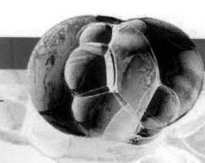

Table 6.8

Some Alkylated Steroids and Testosterone Esters Used in Hormone Therapy

Name	Chemical Structure	Usual Dosage for Androgen Deficiency
Methandrostenolone		2.5 to 5 mg daily for osteoporosis
Methyltestosterone		10 to 50 mg daily as a hormone replacement therapy in males
Oxymetholone		1 to 5 mg/kg daily for anemia
Testosterone		10 to 50 mg 3 times weekly
Testosterone enanthate		50 to 400 mg every 2 to 4 weeks for breast cancer in females and hormone replacement therapy in males
Sandrolone phenpropionate		50 to 100 mg weekly for breast carcinoma

exercise 6.6 (concluded)

Solution

The molecules shown differ from testosterone in the chemical group connected to the oxygen atom attached to carbon atom 17. Each of these groups is nonpolar, making these testosterone derivatives more fat-soluble than natural testosterone. This means that they will stay in the body longer than natural testosterone, allowing them to have a more lasting effect on the buildup of muscle mass. So athletes choose to use them, rather than natural testosterone, because their chemical structure produces greater effects.

Ultimately, all of the compounds are degraded by reaction with water (known as hydrolysis) to form natural testosterone, so they do not remain in the body indefinitely.

pro con discussions

Banning Steroids for Athletic Use

pro Performance-enhancing drugs should be forbidden for athletes at all levels from high school to the Olympics and professional sports. The nature of competition should be related to athletic ability, not the effect of a drug which an athlete might take. If steroids or other performance-enhancing drugs are allowed, there will be no stopping point as new products are developed that enhance muscle growth, improve oxygen utilization, or cause other physiological changes. It is a lot easier to make new compounds than it is to determine their long-term effects. We know a lot about how ethanol, caffeine, and tobacco affect the body so a user of these products can take them in full knowledge of the short- and long-term effects. This is clearly not so for synthetic steroids, which come out of the laboratory and are soon being tried out illicitly by athletes desperate to grow larger biceps, run farther and faster, or lift greater weights than would be possible with just a trained body.

con The ban on steroids in athletic competition has as much to do with personal choices as with chemistry. We do all kinds of things to our bodies every day to make ourselves look better, feel better, work and play better. Runners "carbo-load" with spaghetti the day before a marathon. They train at high altitudes in an attempt to generate more hemoglobin-containing red blood cells; then they race at sea level as if the high-altitude training is perfectly legitimate. Contestants in "beauty pageants" spend freely on cosmetic products; and how would many college students make it to that 8:30 chemistry class without several cups of coffee or a diet cola? Is the caffeine less of an external invasion to the body than steroids?

Steroids do seem to have a number of long-term side effects, yet even some of these are in dispute. For example, there is some evidence that aggressive behavior and testosterone level may be inversely, rather than directly, related. Are we prepared to limit steroid intake when the research results are not *absolutely* conclusive?

This is really an issue of personal choice. Many people abuse alcohol. We tried banning alcohol half a century ago, but it didn't work. Many people smoke, and smoking kills several hundred thousand people in the United States each year. How can we prohibit the use of steroids by athletes while still permitting so many lives to be lost due to smoking? At least athletes run and lift better because of their steroid intake. Cigarette smokers only cough louder.

A World of Choices

The case of testosterone derivatives that act as anabolic steroids raises questions that arise in many chemical issues. These chemicals can be used to do things that most people would perceive as being beneficial, but can also be used to do things most people would consider to

Figure 6.25 Some of the many uses made of synthetic polymers.

be immoral and/or harmful. They can be used to treat medical conditions such as cancer and Alzheimer's disease; but they can also be used to cheat in athletics and perhaps inflict lasting damage on the user's body. If the scientists who first developed these chemicals deserve praise because of their good uses, do they also deserve blame for the fact that they can be abused? Should chemistry, in general, be praised or criticized for giving these chemicals to the world? Who has the right to choose how to use them? Only scientists and/or doctors? Or only the people whose bodies are actually going to be affected by them? How can the discoveries of chemistry be channeled toward "good" and away from "bad;" and who decides on what is good or bad?

6.4 | Synthetic Polymers—Structures That Have Changed Our Lives

Human lifestyles have been changed profoundly by our ability to isolate pure substances from nature's mix of chemicals and use these substances in novel ways by linking them together into the long chemical structures we call polymers. Plastics are probably the most obvious example of the new polymers made by modern industrial chemistry. If you look around you right now, you are likely to find dozens of items that contain plastics. All of these plastics are organic chemicals. They are not the natural organic polymers (like starch and cellulose) we've examined so far, but are synthetic polymers that chemists have crafted out of simple starting materials available in crude oil. The world of synthetic polymers includes much more than just plastics. All of the synthetic fibers used in many modern clothes, automobile tires, guitar strings, astroturf, and movie film are polymers (see Figure 6.25).

Polymers—Building The Links On a Chain

As we explained in Chapter 4, polymers are formed by chemically linking together many smaller molecules called monomers. The term polymer comes from the Greek poly (= many) and meros (= part). Mono means "one." So polymers are composed of many parts, each single part being a monomer.

The first step in making a chemical polymer is to find a suitable monomer (or perhaps a variety of monomers) which can be linked together in a repetitive manner to form long chains or branching extended networks. The wide variety of linkable monomers has led to the vast array of polymers that we depend on in our day-to-day lives. Despite the diversity of the products however, most synthetic polymers are based on very few elements. Carbon is the main element, forming all or most of the basic framework of the polymers, with atoms of hydrogen, nitrogen, oxygen, chlorine, fluorine, and sulfur (all nonmetals) being incorporated within the frameworks in ways that provide the chemical variety of most polymers.

The living world is built around a few key naturally occurring polymers, including the DNA, proteins, cellulose, and starches that we have already mentioned. We will be looking at these naturally occurring polymers in much more detail in Chapters 16 and 17. In the past 100 years or so, however, a wide variety of synthetic polymers have been produced by industrial chemists.

The simplest synthetic polymers are made by allowing many identical small molecules to link together in a process known as **addition polymerization,** one of two types of **polymerization** we'll discuss. The process of addition polymerization involves the following steps (a) the addition of a chemical **initiator** to get the reaction started, (b) a **catalyst** to speed things up, and (c) an **inhibitor** to stop the reaction at the appropriate point. The simplest addition polymerization process uses ethylene (C_2H_4), also called ethene, as the monomer, and allows many ethylene units to link together to form long chains of **polyethylene** as shown in Figure 6.26.

As you can see from Figure 6.27, ethylene molecules contain a carbon-carbon double bond (a double bond between two carbon atoms) so are unsaturated. In the polymerization reaction that forms polyethylene, the double bond becomes a single bond, freeing two electrons per ethylene molecule to participate in the bonds that link the monomer units together. The resulting polymer contains no carbon-carbon double bonds, so it is saturated.

The polyethylene chain length can vary from several thousand monomer units to millions. The chains can have no branching, leading to tightly packed high-density polyethylene (HDPE). They can also be highly branched, leading to loosely packed low-density polyethylene (LDPE). Both HDPE and LDPE are strong solids that are used to make toys, bottles, etc.

Organic chemists have learned a great deal over the last 150 years about the reactions of organic compounds. The synthesis of new organic compounds has reached a high standard through our knowledge of how to carry out reactions to produce desired new compounds. We will mention only two types of reactions.

Alcohols react with carboxylic acids to make larger molecules called esters when the H of the alcohol and the OH of the acid combine to form water and the oxygen of the alcohol bonds to the carbon of the carboxylic acid functional group. Many esters have pleasant smells and are responsible for odors of some fruits and flowers.

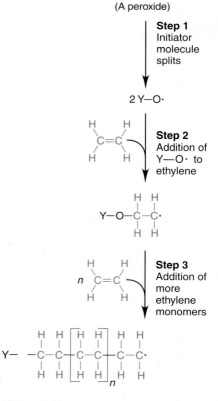

Figure 6.26 Polyethylene is a remarkably versatile polymer formed from linking together individual ethylene molecules.

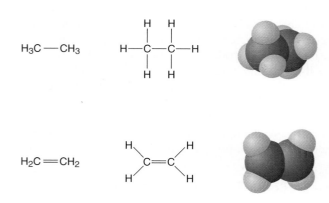

Figure 6.27 Three different representations of ethane and ethylene (ethene).

$$R—COOH + HO—R' \rightarrow R—COO—R' + H_2O$$

carboxylic acid   alcohol   ester

In a similar fashion, acids react with amines to make larger molecules called amides when one of the H atoms of the amine group combines with the OH of the carboxylic acid to form water. When this happens, the N of the amine bonds to the C of the carboxylic acid functional group.

$$R—COOH + H_2N—R' \rightarrow R—CO—NHR' + H_2O$$

carboxylic acid   amine   amide

We will soon see that the formation of esters and amides are the basis of much of the modern textile industry.

Expanding the Possibilities

A central principle of chemistry is that small changes in structure can lead to big changes in properties. If we substitute a chlorine atom for one of the hydrogen atoms of ethene, we get chloroethene, more commonly called vinyl chloride (see Table 6.9). This can be polymerized to give poly(vinyl chloride) or PVC, also shown in Table 6.9. PVC is more resilient than polyethylene and is used in such products as water pipes and vinyl house siding. Substituting other groups of atoms for one or more of the hydrogen atoms of ethylene yields monomers

Table 6.9

Polymerization of Some Common Monomers

Monomer	Polymer	Uses
Vinyl chloride	Poly(vinyl chloride) (PVC)	Indoor plumbing, toys, plastic wrap, vinyl siding
Styrene	Polystyrene	Outdoor furniture, insulation, packing peanuts
Propene	Polypropylene	Contained in indoor-outdoor carpeting
Acrylonitrile	Polyacrylonitrile	Orlon, Acrilon clothing, yarns, wigs

exercise 6.7

Saturated versus Unsaturated Fats

Problem

You may have heard of "saturated" and "unsaturated" fats. Doctors and health activists are currently advising us to increase the amount of unsaturated fats in our food and decrease the amount of saturated fats (see Section 6.5). What can you say about the basic structural difference between these two types of fats? Also, what do you think will be the difference between polyunsaturated fats and monounsaturated fats (two other terms familiar to us from food labels)?

Solution

Unsaturated fats are made from fatty acids with some carbon-carbon double bonds in their structures, while saturated fats only contain no carbon-carbon double bonds. A polyunsaturated fat contains at least one fatty acid with two or more double bonds per molecule, while monounsaturated fats contain only one carbon-carbon double bond per molecule. (Remember, poly = many, while mono = one.) A further structural feature of fats, called trans double bonds, also may affect health (see Section 6.5). We will explore the structures and behavior of fats in Chapter 17.

exercise 6.8

Styrofoam

Problem

Those "Styrofoam peanuts" that are used in packing are made of polystyrene; but they work so well because their volume is largely made up of gas, trapped within the polystyrene. The gas within the polymer protects the packed item in much the same way as an air bag would protect you in a head-on car crash—it absorbs and spreads the shock so you (or the packaged item) feel less of it. A typical styrofoam peanut has a mass of 0.040 g. A polystyrene (PS) monomer (C_8H_8) has a mass of 104 g/mol. To get a sense of how many bonds are formed in an addition polymerization reaction, calculate the number of styrene molecules (monomers) in a styrofoam peanut (remember that there are 6.02×10^{23} molecules per mole!).

Solution

This exercise is a reminder not to forget the skills in stoichiometry that you met in Chapter 4—you need them throughout your study of chemistry.

The strategy here is the same as in any other stoichiometry problem. We map out where we must begin, where we want to get to, and the "bridge" of conversions that will get us there:

$$\frac{0.040 \text{ g polystyrene}}{\text{peanut}} \times \frac{1 \text{ g styrene}}{1 \text{ g polystyrene}} \times \frac{1 \text{ mol styrene}}{104 \text{ g styrene}}$$

$$\times \frac{6.02 \times 10^{23} \text{ molecules styrene}}{1 \text{ mol styrene}}$$

$$= 2.3 \times 10^{20} \text{ molecules styrene/peanut}$$

that can be used to build many other useful polymers, such as those shown in the lower part of Table 6.9. Note that whatever the monomer, the process of addition polymerization, adding two carbons at a time, is still essentially the same. Variations of the ethylene-polyethylene theme can generate a rich diversity of chemical structures, each suited to its own range of uses.

There are many other synthetic polymers, including many that are made by linking monomers together in other ways. Nylon, for example, was designed and manufactured in an attempt to mimic the natural polymer silk. In its formation, two different monomers react to build the growing polymer. Each reaction of monomers leads to the increase in the polymer chain length and the release of a water molecule. This type of polymerization, in which a water or other small molecule is released, is called **condensation polymerization.** Glucose polymerizes to starch in plants via condensation polymerization. In our bodies, proteins are formed from appropriate monomers, called amino acids, by condensation polymerization. Figure 6.28 shows a sample of nylon polymer being pulled from the interface of the two liquids which react together to make it. This short introduction should have given you a good idea of the general chemical and structural principles involved in polymer formation.

Figure 6.28 Condensation polymerization, which results in the release of water as monomers chemically join, is the process that forms nylon.

Buckyballs—Structures of Beauty, Fun, and Perhaps Great Usefulness

You know by now that organic chemicals come in many different shapes and sizes, but even so, you may be surprised to learn that some come soccer-ball shaped. Figure 6.29 shows the structure of one member of a fascinating class of molecules known as **fullerenes,** including the soccer-ball shaped *buckminsterfullerene* or "*buckyball.*" These fascinating carbon-based structures are members of what the classic October 1991 issue of *Scientific American* labeled the "Fullerene Zoo," because they are some of the most exotic and interesting "creatures" of chemistry. The best-known of these molecules, C_{60}, was first identified in 1985 from the black soot formed by shooting a high-intensity laser at a piece of graphite (a pure form of solid carbon, used, for example, as your pencil "lead") in a helium atmosphere. Soon after that, C_{60} was identified in the soot of incomplete combustion (from such sources as smokestacks, wood fires, etc.).

The inhabitants of the Fullerene Zoo are molecules of the element carbon, derived from the C_{60} soccer-ball shaped buckminsterfullerene. This is named after the architect Buckminster Fuller, whose famous geodesic dome buildings (see Figure 6.30) look remarkably similar to the chemicals which are now known as fullerenes. We now know that buckminsterfullerene is quite common and has always existed in the soot of flames. Nevertheless, it is still a cause of great excitement because:

- Buckminsterfullerene is a recently discovered and very stable form of the element carbon (see Figure 6.31).

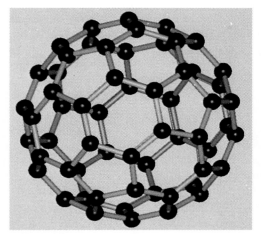

Figure 6.29   The C_{60} fullerene looks much like a soccer ball. It is easy to understand why it is nicknamed a "buckyball."

Figure 6.30   A geodesic dome designed by the architect Buckminster Fuller.

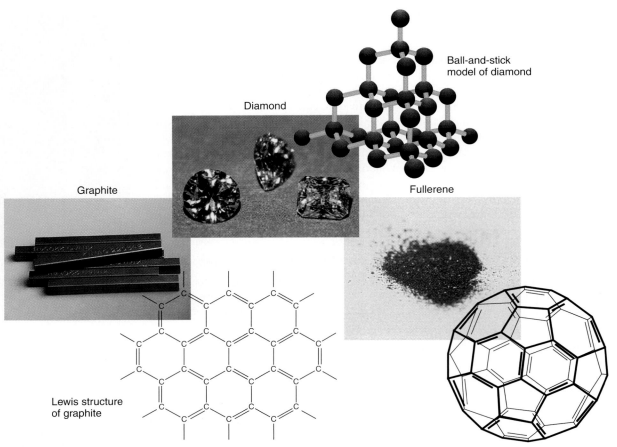

Figure 6.31   Diamond, graphite, and fullerene are all elemental forms of carbon. Differing arrangements of chemical bonds create the different properties (and values) for each. Just compare the cost of a one-carat diamond to the cost of a stick of graphite of the same weight.

- It, and the other fullerenes (perhaps after subtle modification) may possibly be exploited in many different ways, as catalysts, "cages" to deliver trapped drugs to specific sites in the body, miniature electrical components, etc.
- As a recent discovery, and is now being built upon by the discovery of a wide variety of related fullerenes, including tubular forms. Chemists are currently "playing around" with fullerenes like kids let loose in a toy store.
- Current work is looking at the application of buckyballs and related compounds, including nanotubes, to: deliver drugs throughout the body; convert and store electricity; and HIV-related research, as a way to stop the reproduction of the virus.

6.5 | Stereoisomers

Throughout most of this chapter we have emphasized the importance of chemical structure by comparing the structures of different compounds. We began, however, by examining how different forms of the same compound—glucose—had a great influence on the chemistry of these forms. The α-glucose molecules in the starch that we can eat differ from the β-glucose molecules in the cellulose we cannot eat simply in the way in which the bonds are arranged around a central carbon atom. The atoms involved are identical, and they were not structural isomers, such as we discussed in the Case-in-Point on Octane Rating (page 212). The 3-D structure in which α-glucose and β-glucose are arranged is different in a very subtle way.

These two forms of glucose are examples of **stereoisomers:** isomers that contain atoms bonded in the same order, but arranged differently in space.

Optical Isomers

How could a chemist tell you are a human, and not an extraterrestrial lifeform? A rude question perhaps! If it were a real challenge, however, one of the first things the chemist would likely do is examine the chemicals called amino acids that polymerize to form proteins in your body. Take a look at the two versions of the amino acid alanine shown here.

The difference between these forms of alanine is that they are mirror images of each other, as the illustration indicates. This means they are nonsuperimposable, in the same way as your right and left hands, also mirror images of each other, are nonsuperimposable.

If a chemist found that the alanine molecules in your proteins were the "D-" form on the right, they would probably look at you very strangely, because they would have clear evidence that you are not human, and not of this earth! All the alanine molecules in all proteins of all forms of life on earth are the "L-" version on the left. There are actually twenty different kinds of amino acids found in proteins, and almost without exception they are always in the L-form rather than the D-form.

Isomers that differ by being non-superimposable mirror images of each other are called **optical isomers.** The name comes from the fact that a beam of "plane-polarized light" will be rotated in different directions as it passes through a pure solution of each isomer (see Figure 6.32). Optical isomerism is not just of interest to chemists and extraterrestrial-hunters. Its real-life relevance became well known through the story of the drug thalidomide in the 1950s. In many countries, although not in the United States, thalidomide was used to treat the morning sickness that often accompanies early pregnancy. The preparation of the drug contained both D- and L-optical isomers, and while one was effective against morning sickness, the other proved to cause very severe abnormalities in the developing fetus. Many deformed "Thalidomide babies" were born, all because of this subtle difference in chemical structure. To guard against such tragedies, there is now a large effort to produce preparations of drugs that contain only the active stereoisomer—the one whose 3-D structure allows it to bind to some target chemical in the body and have the desired effect.

Geometric Isomers

Have you heard about the debate over *cis* and *trans* fatty acids? They are involved in an important nutritional controversy that is focused on a type of stereoisomerism. The oleic and

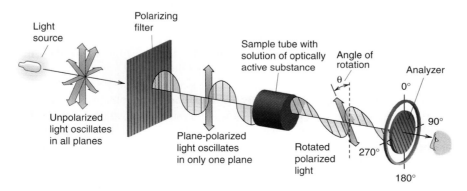

Figure 6.32 A polarimeter sends a beam of plane-polarized light through a solution. Different optical isomers rotate the light in different directions as it passes through.

linoleic acids shown in Figure 6.16 are examples of natural fatty acids, and are shown with both hydrogen atoms protruding from the same side of the double bond. No rotation of bonded groups can occur around a C=C double bond, so two hydrogens attached on the same side of such a bond must stay together on the same side. The alternative possibility is to have the hydrogen atoms on opposite sides of the double bond, where they would be equally stuck in that position. Two atoms or groups on the same side of a double bond are in the ***cis*** configuration. Those on opposite sides of the bond are in the ***trans*** configuration.

cis   trans

Molecules that differ only in this feature—one being *cis* and the other being *trans,* are called **geometric isomers.**

In the processing of many vegetable spreads used as butter substitutes, some unnatural *trans* double bonds are formed in the fatty acids of the vegetable oils concerned. In Exercise 6.7, we noted that some scientists claim to have evidence that consuming these *trans* fatty acids can be bad for our health. The debate is in progress, but the idea that such a subtle structural difference might actually make us ill is a yet another telling example of the importance of chemical structure.

main points

- Organic chemistry is the chemistry of carbon-containing compounds. Over 90% of all known compounds are organic.

- Living organisms are principally made of organic compounds. However, millions of nonnatural organic compounds have been synthesized in the last 100 years.

- The three-dimensional structure of a molecule determines its chemical behavior.

- Carbon always has four bonds, either to other carbon atoms or to other nonmetal atoms.

- The great variety of carbon-containing compounds arises from carbon's ability to form molecules containing long chains of carbon atoms.

- Functional groups are specific collections of atoms that give special properties and reactivities to molecules.

- Steroids, fuels, chemical warfare agents, and vitamins all behave as they do because of the composition and structure of their molecules.

- Polymers are an industrially and biologically important class of substances made by the reaction of appropriate small "monomers" bonded together.

- The "fullerene" compounds are a new form of chemically pure carbon whose properties and uses are still being explored.

- Stereoisomers include optical and geometric isomers and are important in many applications, including the preparation and activity of pharmaceuticals.

important terms

Acetylcholine is a compound that transmits nerve signals. (p. 226)

Acetylcholinesterase inhibitor is a compound that interferes with the functioning of acetylcholine in transmitting nerve signals. (p. 226)

Addition polymer is a polymer formed by adding together many monomers containing double bonds. (p. 232)

Addition reaction is a reaction in which a reagent X-Y adds to a multiple bond so X is attached to one of the carbons of the bond and Y to the other. (p. 211)

Alkanes are hydrocarbons that contain only single bonds. Alkanes not in a ring share the general formula C_nH_{2n+2}. (p. 208)

Alkenes are hydrocarbons that contain a carbon-carbon double bond. (p. 211)

Alkynes are hydrocarbons that contain a carbon-carbon triple bond. (p. 212)

Anabolic steroid is a steroid that aids in building body mass. (p. 228)

Anabolism refers to processes in the body associated with building new tissue (p. 228)

Benzene (C_6H_6) is a planar six-membered ring (hexagon) of carbon atoms. The carbon atoms are also bonded to hydrogen atoms. (p. 213)

Bond angle is the geometric angle at which two atoms are attached to a third atom. (p. 208)

Condensation polymer is a polymer formed by reacting molecules that split out other small molecules, usually water. (p. 235)

Constitutional isomers are the same as structural isomers. (p. 212)

Double bond is a covalent bond between two atoms that share four electrons. (p. 209)

Environmental estrogens are pollutants that make male animals develop some female characteristics. (p. 227)

Estrogens are female hormones. (p. 226)

Ethanol (C_2H_5OH) is ethyl alcohol, the alcohol of beverages and gasohol. (p. 221)

Ethyne (acetylene, C_2H_2) includes a triple bond. (p. 211)

Fatty acid is a constituent of a fat, an acid with a long hydrocarbon chain. (p. 219)

Functional group is a collections of atoms that give special properties and reactivities to molecules. (p. 216)

Geometric isomers are compounds differing only in that one is cis and one trans about a double bond. (p. 238)

Hexane (C_6H_{14}) is a saturated hydrocarbon, a constituent in gasoline. (p. 218)

Homologous series is a group of organic chemicals that share the same general formula and similar chemical characteristics. (p. 208)

Hormone is a compound made in our bodies to carry out a function at another location in the body. (p. 226)

Hydrocarbon is a compound made up of only carbon and hydrogen atoms. (p. 207)

Hydrogen bond is a moderately strong intermolecular attraction caused by the partial sharing of electrons between an atom of F, O, or N and the polar hydrogen atom in an F—H, O—H, or N—H bond. (p. 218)

Inhibitor is a substance that slows or prevents a chemical reaction. (p. 232)

Initiator is a substance that starts a reaction between two other molecules. (p. 232)

Inorganic refers to a compound made up primarily of atoms other than carbon. (p. 209)

Intermolecular forces cause the attraction of one molecule for another, but not such that the two react. (p. 218)

Isoelectronic means having the same electron arrangement. (p. 207)

Lone pair is a pair of electrons which are not shared in covalent bonding. (p. 209)

Methane (CH_4) is natural gas. (p. 207)

Neurotransmitter is a chemical that transmits a chemical message used in the nervous system. (p. 225)

Optical Isomers are nonsuperimposable mirror images of each other. (p. 237)

Organic chemistry is the study of most carbon-based compounds. (p. 205)

Organophosphate is an organic compound which includes a phosphate (PO_4) group. (p. 226)

Oxygenate is a gasoline additive that contains an oxygen atom. (p. 220)

Planar means all the atoms of the molecule or group are in a single plane. (p. 214)

Polar means the negative charge of a molecule is concentrated at one position. (p. 218)

Polyethylene is the addition polymer made from ethylene (C_2H_4). (p. 232)

Receptor is the location on a biological molecule that receives another molecule and transmits a chemical signal. (p. 225)

Saturated means containing as many hydrogen atoms as possible with no multiple bonds. (p. 211)

Soman ($C_7FH_{16}O_2P$) is a nerve gas originally created for use in World War II. (p. 225)

Space-filling diagram is a molecular diagram that shows how the electrons of all the atoms within a molecule occupy space. (p. 208)

Stereoisomers are isomers that contain atoms bonded in the same order, but arranged differently in space. (p. 237)

Steroid is a class of biologically active molecules that include a set of four rings fused together. (p. 227)

Structural isomers are compounds that have the same molecular formula but different arrangements of their atoms. (p. 212)

Synapse is a gap between adjacent nerve cells. (p. 225)

Testosterone is a male hormone. (p. 226)

Tetrahedron is a three-dimensional figure having four corners and four faces. (p. 208)

Triple bond is a covalent bond between two atoms that share six electrons. (p. 211)

Unsaturated means it can take on more hydrogen atoms and has one or more multiple bonds. (p. 211)

VSEPR is the Valence Shell Electron Pair Repulsion Theory that guides prediction of molecular structure. (p. 208)

exercises

1. Define "organic chemistry."
2. Classify these molecules as organic or inorganic: H_2O, $CH_3CH_2NH_2$, $CH_3CH_2CH_3$, NH_3, H_3PO_4
3. Do the starch molecules in pasta have α- or β-linkages? Why is this important?
4. Define "hydrocarbon."
5. Which *cannot* exist as a stable molecule? Explain your answer.

   $CH_2{=}CH_2$ $CH_3{=}C{=}CH_3$ $CH_3{=}C(CH_3)_2{-}CH_3$

6. Where are the electrons of the double bonds located in a benzene ring?
7. Draw Lewis dot structures of all possible structural isomers of C_5H_{12}.
8. Draw Lewis dot structures of all possible structural isomers of C_5H_{10}.
9. Are these following molecules saturated or unsaturated? CH_4, C_4H_{10}, C_4H_8, C_4H_6
10. Compounds called "halons" are used in fire extinguishers. One of the halons has the formula $C_2F_4Br_2$. Draw two different structural isomers of this compound.

11. Compounds called "freons" are used in fire extinguishers. One of the freons has the formula $C_2F_3Cl_3$. Draw two different structural isomers of this compound.
12. a. How many hydrogen atoms must there be in a hydrocarbon having 1–10 carbon atoms if there are no multiple bonds or rings in the structure?
    b. How many hydrogen atoms must there be if there is one double bond in the structure?
    c. How many hydrogen atoms must there be if there is one triple bond in the structure?
    d. How many hydrogen atoms must there be if there is one ring in the compound?
13. Work out algebraic formulas for the number of hydrogens in a saturated hydrocarbon, a hydrocarbon with one double bond, a triple bond, or a ring. Do the same for two double bonds.
14. How many carbon-carbon double bonds must there be in one molecule of each of these compounds?
    a. C_3H_8 b C_6H_8 c. $C_{10}H_8$
15. Name the functional group(s) in each of these molecules.
    a. $CH_3{-}CH{=}CH_2$
    b. $CH_3{-}CH_2{-}OH$

    $$c. \ CH_3{-}CH_2{-}CH_2{-}\overset{\overset{\textstyle O}{\|}}{C}{-}OH$$

16. Describe the shape of a molecule of formaldehyde (CH_2O).
17. Draw the Lewis structure of acetic acid, which is usually represented as CH_3COOH.
18. Describe what is meant by an unsaturated molecule.
19. Draw two reasonable structures for molecules having this formula: $C_3H_8O_2$.
20. Draw a structure of a three-carbon compound that is an amine.
21. Predict the structure of C_3H_4.
22. For use in home grills, propane (C_3H_8) is sold in tanks holding 20 lb of propane. How many moles of propane is this?
23. Demonstrate that organic compounds containing C, H, N, and O will always have even number molecular weights (to the nearest integer) unless the compound contains an odd number of nitrogens.
24. Draw the electron dot structures of each of the example compounds in your answer to Exercise 25.
25. Describe the molecular geometry about each atom bonded to more than one other atom in these compounds: CH_3COOH, C_2H_3Cl, C_4H_{10}, $CH_3{-}C{\equiv}C{-}H$
26. What are bond angles of each of the compounds in question 25?
27. Draw two isomers of C_2H_6O, one of which is an alcohol and one of which is an ether.
28. Draw a hydrogen bond between water and CH_3OH, connecting the correct atoms.
29. Explain why gasoline does not mix with water.
30. What is the structural difference between estradiol and testosterone?

31. Why is cholesterol classified as a steroid?

32. Assuming that carbon compounds like ETBE have an oxidation number of -2 (and give off 680 kJ per mole of carbon in combustion), calculate the heat released by the combustion of one mole of ETBE.

33. Write the formula for the ester formed by the reactions of:
    a. ethyl alcohol with acetic acid
    b. butyl alcohol (C_4H_9—OH) with benzoic acid (C_6H_5COOH)

34. Look up the formula and structure of aspirin and oil of wintergreen. What alcohol and acid reacted to make these two esters?

35. Esters can be converted back into their constituent acid and alcohol by reacting with water, especially in the presence of acids. Use this information to explain why old bottles of aspirin sometimes smell like vinegar.

36. Which of these molecules are polar and which are nonpolar?
    a. CH_3Cl
    b. C_2H_6
    c. cis-$C_2H_2Cl_2$
    d. trans-$C_2H_2Cl_2$
    e. CH_2Cl_2

37. Drugs can be administered in either polar or nonpolar forms. The body is primarily composed of the blood, which is largely water, and nonpolar lipids, which form cell membranes. How do you think buckminsterfullerene could be used to help transport a drug through the body to its receptor?

38. What geometry of carbon in buckminsterfullerene allows all of the carbons to connect, forming a sphere?

39. Draw the molecule that results from the reaction of

$$CH_3OH \text{ with } CH_3CH_2\overset{\overset{\displaystyle O}{\|}}{C}OH.$$ What class of compound is this substance?

40. Polyester, a material used to make yarn (such as Dacron) for clothing and many other uses is formed by the reaction of ethylene glycol (HO—CH_2—CH_2—OH) with terephthalic acid (HOOC—C_6H_4—COOH). Draw a molecule formed by the reaction of two of each of these molecules. Dacron actually has hundreds of each monomer in each polymer molecule.

41. How many grams of Dacron can be made from the reaction of 1.00 kg each of ethylene glycol and terephthalic acid? See Exercise 40 for formulas.

42. Nylon 6,6 is made from adipic acid (HOOC—$(CH_2)_4$—COOH) and hexamethylenediamine (H_2N—$(CH_2)_6$—NH_2). Predict what the two monomers must be for nylon 4,4.

43. Describe or draw what is meant by cis and trans.

44. Write the word (or phrase) for which each of these is a definition:
    a. Molecules that have the same molecular formula but different structures
    b. Molecules made from a large number of identical small molecules

c. Unsaturated fatty acid having hydrogen atoms on opposite sides of chain at the site of unsaturation

45. Answer each question with the correct word(s), number, symbol, or response.
    a. What is formed when an organic acid reacts with an alcohol?
    b. How many bonds do each of these following elements form in typical organic compounds?
       C _____   H_____   O_____   N _____
    c. What two functional groups react together to make amides?

46. Calculate the molecular weight of a compound in which three glucose molecules have condensed to form a trisaccharide.

47. Calculate the number of monomer molecules needed to make a polyethylene molecule having a molecular weight of 20,000.

48. Calculate the oxidation number of carbon in CH_3COOH, $C_6H_{12}O_6$, and C_4H_{10}.

49. What must the formula be of a hydrocarbon containing 79.9% carbon and 20.1% hydrogen?

50. What must the formula be of a hydrocarbon containing 83.2% carbon and 16.8% hydrogen?

51. What must the formula be of an ether consisting of 59.96% carbon, 26.63% oxygen, and 13.41% hydrogen?

52. Calculate the number of moles of carbon dioxide formed in the combustion of 60.0 g of diethyl ether.

53. Discuss the effect on the oxidation number of carbon by adding oxygen atoms or hydrogen atoms to organic compounds.

54. Calculate the percent by weight of carbon in each of these compounds:
    a. CH_3—O—H   methyl alcohol (methanol or wood alcohol)
    b. CH_3—CH_2—O—H   ethyl alcohol (ethanol or grain alcohol)
    c. CH_3—CH_2—O—CH_2—CH_3 diethyl ether

    d. CH_3—CH_2—O—$\overset{\overset{\displaystyle CH_3}{|}}{\underset{\underset{\displaystyle CH_3}{|}}{C}}$—$CH_3$ ethyl t-butyl ether (ETBE)

food for thought

55. Do you think steroids should be allowed in athletic competition? Why or why not?

56. Give additional arguments on one or both sides of the issue regarding whether the United States should make greater use of oxygenates, especially grain-based ethanol, in automobile fuel.

readings

1. Compton, James A.F. *Military Chemical and Biological Agents.* Caldwell, N.J.: Telford Press, 1987.

2. We wish to thank Prof. Dave Hage of the University of Nebraska's Chemistry Department for his advice and background in the steroid section.

3. Baum, Rudy. Fullerenes in Flames: Burning benzene yields C_{60} and C_{70}. *Chemical and Engineering News.* July 15, 1991, p. 6.

4. Baum, Rudy M. Flood of fullerene studies continues unabated. *Chemical and Engineering News.* June 1, 1992, pp. 25–34.

5. Baum, Rudy M. Systematic chemistry of C_{60} beginning to emerge. *Chemical and Engineering News.* December 16, 1991, pp. 17–21.

6. Browne, Malcolm W. Chemists' new toy emerges as superconductor. *New York Times.* September 3, 1991.

7. Browne, Malcolm W. Mirror image: chemistry yielding new products. *New York Times.* August 13, 1991.

8. Curl, Robert F. and Smalley, Richard E. Fullerenes. *Scientific American.* October 1991, pp. 54–63.

9. Stinson, Stephen C. Chiral drugs. *Chemical and Engineering News.* September 28, 1992, pp. 46–78.

10. Wood, Clair. Buckyballs. *Chem. Matters.* December 1992, pp. 7–9.

11. Angier, Natalie. How estrogen may work to protect against alzheimer's. *New York Times.* March 8, 1994.

12. Colborn, Theo; von Saal, Frederick S.; and Soto, Ana M. Developmental effects of endocrine-disrupting chemicals in wildlife and humans. *Environmental Health Perspectives.* October 1993, pp. 378–384.

13. McLachlan, John A. Functional toxicology: a new approach to detect biologically active xenobiotics. *Environmental Health Perspectives.* October 1993, pp. 386–387.

14. Raloof, Janet. Gender benders. *Science News.* January 8, 1994, pp. 25–27.

15. Raloff, Janet. That feminine touch. *Science News.* January 22, 1994, pp. 56–58.

16. Wallis, Claudia. The estrogen dilemma. *Time Magazine.* June 26, 1995, pp. 46–53.

17. Fussell, Betty. Crazy for corn. *Harper Perennial,* 1995.

18. Chang, Kenneth. A prodigious molecule and its growing pains. *New York Times,* October 10, 2000, pp. D1, D4.

19. Collins, Philip G., and Avouris, Phaedon. Nanotubes for electronics. *Scientific American.* December 2000, pp. 62–69.

websites

www.mhhe.com/kelter   The "World of Choices" website contains activities and exercises including links to websites for: Scientific American; The Royal Dutch/Sheil Group; and much more!

Properties of Water

. . . Water, water, everywhere, nor any drop to drink . . .
And every tongue, through utter drought
was withered at the root:
We could not speak, no more than if
We had been choked with soot . . .

—*Rime of the Ancient Mariner*
by Samuel Coleridge, 1798

Severe dehydration is likely to develop within a few days if no water is taken . . .
Symptoms include severe thirst, dry lips and tongue, an increase in heart rate and
breathing, dizziness, confusion, and eventually coma.

—*American Medical Association Encyclopedia of Medicine*

The ancient mariner's problem was that he was on a boat in the middle of the ocean but was dying of thirst because he was unable to drink the saltwater that was all around him. His dilemma is becoming a modern problem in that we have large amounts of undrinkable water due to the presence of other dissolved substances. Both quantity and quality of available water supplies are crucial for the continued health of the people of the world. In our day as in all of history, water plays a crucial role in our lives.

Life is entirely dependent on water: almost all the chemistry of life takes place in watery solutions within living cells. Our everyday lives revolve around the need for and uses of water. Think about a typical day: you awake and soon feel the need to take in water, perhaps in the form of a soft drink, or a cup of coffee or tea. You wash in water to which special chemicals called soaps and detergents have been added; then clean your teeth with a mixture of water and toothpaste. Each meal you take during the day contains lots of water, both within the food itself and as any drinks you consume with a meal. Many foods are cooked in boiling water; while others are kept cool with solid water, which we call ice. After a meal, the dishes are cleaned with water, again mixed with detergents; and we also use water to clean all sorts of other things, including our clothes. When you venture outside you may have to run through a shower of rain which takes water from the clouds back to the ground, from where it runs off to rivers, lakes, and the sea. The power of falling water can provide some of the electricity you depend on by turning the turbines of hydroelectric power stations.

The ancient mariner sailed through terrible trials on the ocean before reaching home with no living shipmates.

Water heated and turned to steam powers the turbines of gas, nuclear, and coal-fired power stations; and cool water is a versatile industrial coolant, used to keep all sorts of machinery from overheating. The seas, lakes, and rivers can be used as transport systems and recreational facilities. Water also comes to our aid every day, somewhere, to extinguish fires. These are just some of the most familiar everyday uses of water, whether in the solid, liquid, or gaseous phase, as it travels through the overall global water cycle in so many different ways.

As we look at the properties and uses of water in this chapter, please think of where in the water cycle the application fits—and how we try to deal with the pressures of increased population and increased water use by chemically keeping water fit to be a part of nature's cycles. Unfortunately, we have large amounts of undrinkable water due to the presence of other dissolved chemicals.

consider this:

Water Use

How is water used differently now than 200 years ago when *The Rime of the Ancient Mariner* was written?

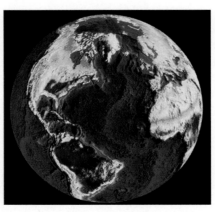

The oceans of the world contain over 95% of the water on Earth. The remainder is lakes, rivers, groundwater, snow and ice, and atmospheric vapor.

Water is absolutely required for all life on Earth. A person can only live for a few days without drinking water. People normally drink about 2 L per day counting the water which is part of food and beverages. Because all animals and plants need water, agriculture is a major user of water; most manufacturing needs water too. If we count all the uses of water, each of us in the United States uses 700 L of water every day.

Water seems like such a simple molecule—two atoms of hydrogen each bonded to a central oxygen atom. You are familiar with some of the other simple substances which have hydrogens bonded to a central atom, for example methane (CH_4) and ammonia (NH_3). But these substances are gases at room temperature. Water is a liquid at room temperature, even though water has about the same molecular mass as methane and ammonia. In Chapter 6, we discussed the reason for the structural difference—the shape—which leads to the physical difference in the properties of water from these other compounds. We used the VSEPR model to show that water has a "V-shape" with an angle between the bonds of about 105°. The elecronegativity difference between hydrogen and oxygen, combined with the V-shape, leads to a polar molecule that can hydrogen bond to other water molecules. One of the themes of the last chapter was that "structure dictates function," the behavior of this small molecule is dictated by its shape and polarity. Figure 7.1 shows three ways of depicting an individual water molecule. Collections of water molecules interact through hydrogen bonding. Liquid water is a collection of individual water molecules, each hydrogen-bonded to as many as four neighbors (Figure 7.2).

In this chapter we will discuss water in our world—first by considering its physical properties, and then its many uses, all of which are best understood in the context of water's structure, discussed in Section 7.3.

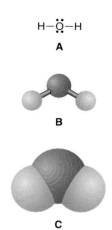

Figure 7.1 A water molecule can be represented in several ways. The Lewis structure **A** shows that electron pairs bond the two H atoms to a central O atom. A ball and stick model **B** gives an idea of the shape of the model but keeps the atoms an unrealistically long distance from each other. A space-filling model **C** shows more realistically how the atoms fit closely together in a bent shape.

7.1 | Water—Exceptional in Many Ways

On the evening of April 14, 1912, the transatlantic ship Titanic struck an iceberg off the coast of Newfoundland and sank, killing 1522 people. That ice floats in water, even in as mammoth a form as an iceberg, makes sense to us because the behavior of water is so familiar. Water, however, is nearly unique in that almost all other solid substances sink in the liquid from which they are frozen. This is because the density of a material in solid form is almost always greater than its density when liquid. Why is it, then, that ice is less dense than liquid water?

We are familiar with water in all three of its states: solid (ice), liquid water, and water vapor (steam). In ice, the structure is completely hydrogen-bonded. Since the oxygen electrons are arranged in a tetrahedral geometry, the four hydrogen bonds surrounding each water molecule are pointed toward the corners of a tetrahedron. This leaves a lot of unoccupied space in the ice structure. When heat is added, the ice begins to melt. This means that some of the hydrogen bonds break and the structure collapses. The same number of molecules of liquid water now take up less space than in solid ice, so the density of liquid water is greater than that of ice. Since a less dense solid will float on a denser liquid, ice floats on water. Occasionally, the importance of a commonplace observation like this is overlooked. It doesn't really matter that ice cubes float in lemonade but it is crucial that ice floats on top of frozen lakes. If ice were to sink in water, lakes would freeze from the bottom up and fish could not survive through winter. Instead, the layer of ice on top of the water acts as an insulator and prevents heat loss from the remaining liquid water so that liquid remains below the ice throughout a long cold northern winter.

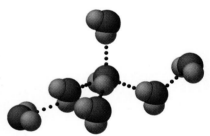

Figure 7.2 The hydrogen-bonding ability of the water molecule. Because it has two O—H bonds and two lone pairs, one H_2O molecule can engage in as many as four H bonds that are arranged tetrahedrally to surrounding H_2O molecules.

Freezing Point

There is one unique temperature at which both liquid and solid water can be present together in a stable situation. This temperature is 0°C, and is called both the **freezing point** and the **melting point** of water. The temperature at which ice melts is the same as the temperature at which water freezes. Using the chemist's shorthand,

$$H_2O(s) \rightarrow H_2O(l) \text{ occurs at the same temperature as}$$
$$H_2O(l) \rightarrow H_2O(s)$$

Actually, the melting point of ice changes a little bit depending on the atmospheric pressure. The change is so minor (melting point = 0.005°C at 0.5 atmospheres) that it is not worth worrying about until the pressure gets dozens of times higher than normal atmospheric pressure.

When we put ice cubes into a beverage what actually happens is that the ice warms up to its melting point (0°C) as it takes heat away from the warm beverage. That usually isn't

Ice melts when heat is added. The liquid water that is formed is less ordered than was the ice. When enough additional heat is added, water molecules can escape the liquid and become gaseous water vapor, which is even less ordered than was the liquid.

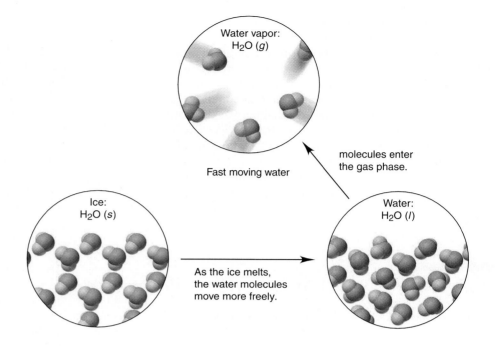

Water vapor: H_2O (g)

Fast moving water

molecules enter the gas phase.

Ice: H_2O (s)

Water: H_2O (l)

As the ice melts, the water molecules move more freely.

enough to get the drink cold enough so next the ice begins to melt. This process also requires heat and the heat is mostly drawn from the beverage, thus cooling it more. Some mixed drinks are simply shaken with a lot of ice and the ice is strained out before much of it has a chance to melt and dilute the drink. If enough cooling occurs just by the ice warming up to its melting point, the drink stays full strength.

All three states of water—solid, liquid, and vapor—are present during this eruption of Old Faithful in wintertime.

It is surprising to most people that heat is released when water freezes. It is never a surprise that one must add heat to ice to get it to melt. If we phrase the statement differently, though, it makes sense; heat must be removed in order to cause water to freeze. Isn't this saying the same thing as heat is released during freezing?

The amount of heat required to melt a quantity of ice at 0°C is called the **heat of fusion** and equals 334 J/g or 6.01 kJ/mole.

Boiling Point

If you add heat to a 0°C mixture of ice and water, the temperature of the mixture will not rise until all the ice has melted. The addition of still more heat causes the temperature of the liquid water to rise. While this is happening, a few more of the hydrogen bonds break and some of the molecules leave the liquid phase to become water vapor

$$H_2O(l) \rightarrow H_2O(g)$$

exercise 7.1

How Much Is There?

Problem

Use the data in Table 7.1 to calculate the number of water molecules on Earth.

Table 7.1

Sources of Water on Earth

Source	% of Total	Volume (L)
Oceans	97.33	1.348×10^{21}
Salt lakes	0.008	1.05×10^{17}
Polar ice/glaciers	2.04	2.8×10^{19}
Groundwater	0.61	8.45×10^{18}
Lakes	0.009	1.25×10^{17}
Soil moisture	0.005	6.9×10^{16}
Atmospheric water vapor	0.001	1.35×10^{16}
Rivers	0.0001	1.5×10^{15}
Total	100.0	1.385×10^{21}

Solution

The data in Table 7.1 indicate there are 1.385×10^{21} L of water on Earth. From there, we can convert, in order, from liters into milliliters, to mass of water (via density), to the number of moles (via molar mass), and finally to molecules, using Avogadro's number.

$$\text{Molecules } H_2O = 1.385 \times 10^{21} \text{ L} \times \frac{1000 \text{ g}}{1 \text{ L}} \times \frac{1 \text{ mol}}{18.0 \text{ g}} \times \frac{6.02 \times 10^{23} \text{ molecules}}{1 \text{ mol}}$$

$$= 4.63 \times 10^{46} \text{ molecules}$$

case in point

Ice Prevents Freezing

Fragile fruit such as oranges or strawberries is ruined if their internal temperature goes below the freezing point because the frozen water ruptures the fruit's cells causing them to burst. They are grown commercially in climates where this is unusual, but occasional cold snaps can do terrible damage to such a vulnerable crop. To prevent such damage during subfreezing weather, water is sprayed onto the crop during the cold weather. Some of the water freezes on the surface of the fruit but the continual drizzle of liquid water prevents the temperature from going below the freezing point. Remember, liquid and solid present together must be **at** the freezing point, not below. Remember too, freezing of liquid into solid releases heat to its surroundings; in this case, the temperature-sensitive fruit. As long as liquid water is present, the temperature can't go below freezing and the crop is protected.

Oranges coated with ice and freezing water are protected from internal freezing.

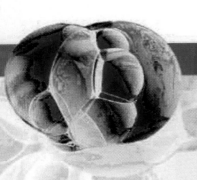

exercise 7.2

Breaking the Bottle with Ice

Problem

The density of liquid water at 0°C is 0.9998 g/cm³ and ice at this temperature is 0.9168 g/cm³. Use this information to explain why a tightly closed glass bottle filled with water will break when it is frozen.

Solution

When a bottle of water freezes, the density decreases but the mass stays the same so the volume must increase.

mass of the liquid = mass of the solid

since

$$density = \frac{mass}{volume}$$

$$mass = density \times volume$$

therefore,

density of the liquid × volume of the liquid = density of the solid × volume of the solid

cross multiplying to solve for volume of the solid,

$$volume\ of\ the\ solid = volume\ of\ the\ liquid \times \frac{density\ of\ the\ liquid}{density\ of\ the\ solid}$$

$$= volume\ of\ the\ liquid \times \frac{0.9998}{0.9168}$$

$$= volume\ of\ the\ liquid \times 1.090$$

In other words, the ice will occupy a volume about 9% greater than did the liquid from which it froze. That will cause a force great enough to break most rigid containers.

Figure 7.3 The boiling point of water changes dramatically as the applied pressure changes. This is a result of the vapor pressure being greater at higher temperatures. Our definition of boiling point is the temperature at which the vapor pressure equals the applied pressure. This graph shows vapor pressure at various temperatures. At 1 atm (760 torr) the boiling point of water is 100°C. At 0.5 atm it is only 81°C and at 2.0 and 3.0 atm, it is much higher. A table of vapor pressures at various temperatures is shown in Table 7.2.

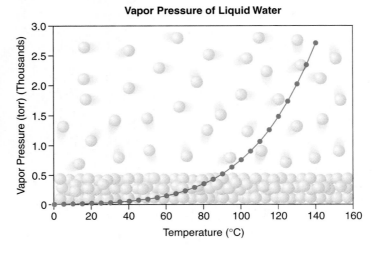

There is always water vapor in contact with solid or liquid water, and we can predict the amount of water vapor present. This so called "equilibrium vapor pressure" depends only on the temperature of the water and its immediate surroundings. The word **equilibrium** describes a dynamic situation in which water molecules are going from liquid to vapor and an equal number are simultaneously going from vapor to liquid. Table 7.2 and Figure 7.3 show that the higher the temperature, the higher the vapor pressure. When the temperature gets to the

Table 7.2

The Relationship between Water Temperature and Vapor Pressure

Temperature (°C)	Vapor Pressure of Water (torr)
Values for Solid Water (Ice)	
−40	0.097
−30	0.28
−20	0.78
−15	1.24
−10	1.95
−5	3.01
0	4.58
Values for Liquid Water	
−15	1.44
−10	2.15
−5	3.16
0	4.58
5	6.54
10	9.21
15	12.79
20	17.54
25	23.76
30	31.82
40	55.32
50	92.51
60	149.4
70	233.7
80	355.1
90	525.8
100	760.0
110	1,074.6
120	1,489.1
130	2,026.1
140	2,710.9
150	3,570.5
200	11,659.2
250	29,817.8
300	64,432.8
350	124,001.6
374	165,467.2

boiling temperature (boiling point), the vapor pressure of the water has reached the external air pressure of the atmosphere in an open container.

Notice something important in the last couple of sentences: we have defined the boiling point of a liquid in an open container as the temperature at which its vapor pressure becomes equal to the atmospheric pressure that surrounds it. That means that if the atmospheric pressure is lowered, the **boiling point** is also lowered; water doesn't have to be as hot to boil in lower pressure surroundings. This is significant when you boil water at high altitude, in the high mountains.

At sea level, water boils at 100°C. At high altitude the atmospheric pressure is lower than at sea level so water boils at a lower temperature than it does at sea level. Water boils at 87°C on the top of Pike's Peak in Colorado, where atmospheric pressure is only about 61% that at sea level. Because boiling water isn't as hot, it takes longer to cook a boiled egg in the Rocky Mountains than it does at a California beach front location.

Proving Fusion Values

Can you prove that the heat of fusion value for water, 334 J/g is equal to 6.01 kJ/mol? Can you think of some specific examples in which it would be useful to have values in J/g or kJ/mol?

The amount of heat needed to boil a given amount of water is called the **heat of vaporization,** which for water at 100°C is 2257 J/g or 40.67 kJ/mol.

Boiling water in a pressure cooker (a high-pressure environment) shows the opposite effect on water's vaporization compared to a low-pressure environment (Figure 7.4). When something boils in an open container, the surrounding pressure is that of the air in the room. In a tightly closed container, such as a pressure cooker, as water begins to vaporize the pressure goes up to about double that of the normal atmosphere. The temperature of boiling water is higher inside such a closed container than in a regular pan. Things cook much more quickly in a pressure cooker. A pressure cooker has a safety valve which releases steam when the

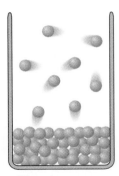  

Water in open
container at room
temperature
pressure = 1 atm
V.P. = 0.025 atm

Water in open
container at boiling
point, 100°C
pressure = 1 atm
V.P. = 1 atm

Water in closed
container at boiling
point, 120°C
pressure = 2 atm
V.P. = 2 atm

Figure 7.4 Each blue bubble represents a water molecule. The liquid water molecules lie in the bottom of the container packed closely together. Water vapor molecules are in the gas phase and are well separated from each other. The higher the temperature, the more water molecules there are in the gas phase. In an open container, the gaseous water molecules can escape into the surrounding air. In a closed container, the water molecules in the gas phase are trapped and so raise the pressure as more of them enter this phase and require more space. This allows the temperature of the boiling water to rise.

case in point

Distillation

If, as water boils away, the vapor is chilled enough to make it condense into a liquid in a clean container, the freshly condensed water will not contain impurities originally present in the water. Such water is called "distilled water" and is very pure. Recall from Chapter 6 that distillation is an important part of processing petroleum.

Much of the fresh water in desert regions such as Saudi Arabia is obtained by vacuum flash distillation of salty seawater. In this situation, oil is cheap but water is

precious. It makes good economic sense to use the abundant petroleum and abundant seawater to sustain life in the desert. The petroleum used in this process is what remains after the more valuable components such as gasoline and motor oil fractions have been removed from the crude oil. **Vacuum distillation** means that distillation is done at low air pressure. When this is done, the water boils at a lower temperature and less heat is needed to distill it.

pressure cooker gets up to about twice atmospheric pressure. Without this safety valve, the pressure would increase until the pan exploded. It is dangerous to heat liquids inside closed containers. In chemistry labs an apparatus designed to be heated while tightly sealed is called a "bomb"—for good reason.

Sublimation is the conversion of a solid directly to a gas. For water, this process is symbolized as

$$H_2O(s) \rightarrow H_2O(g)$$

This occurs at temperatures below the melting point and occurs because the vapor pressure of a solid is small but not zero. Is sublimation an exothermic or endothermic process? Why?

To this point, we have discussed several properties that you have heard of: freezing and boiling points along with sublimation. We also considered the energy exchange for freezing, melting, and boiling. If we take a closer look at energy exchange in water, we can begin to explain some common (and a few uncommon) changes that occur on Earth.

Specific Heat

Anyone who has gone swimming in a lake knows that lake water is a lot colder, even in the summer, than the air around it. Few people, at least in the northern part of the country, have gone lake swimming in the winter, but those who have know that the water is warmer than the surrounding air. It takes a lot of the Sun's heat to warm up a large lake and, conversely, one must remove a lot of heat to cool down such a body of water. This is why the lakes are cooler than the surrounding air during the day and are warmer at night—their temperature changes much more slowly than the air. Why does this happen? It has to do with the **specific heat,** the amount of heat necessary to raise the temperature of 1 g of a substance by one degree Celsius (or 1 K). The specific heat of water is equal to 4.184 joules per gram per degree of temperature change (J/gK). The amount of heat necessary to raise the temperature of one mole of a substance by one degree Celsius is termed its molar heat capacity. Table 7.3 gives the specific heat and molar heat capacity of water and of several other substances.

1.000 cal = 4.184 J, so the specific heat of water is 1.000 cal/g°C.

Let's consider a practical use of this physical property of elements and molecules—cooking an egg. An aluminum pot has a much lower specific heat than the water in the pan. This means that a relatively small amount of heat will cause a substantial increase in the temperature of the aluminum pot. The hot metal pot rapidly transfers heat to the water inside. Most of the heat from the stove ends up in the water even though the water and the pan are at the same

Table 7.3

Specific Heats and Molar Heat Capacities of Several Substances

Substance	Specific Heat J/g K	Molar Heat Capacity J/mol K[*]	Formula Mass
Water (liquid)	4.184	75.3	18.0
Ice (water)	2.05	36.9	18.0
Water vapor	1.84	33.1	18.0
Ethyl alcohol	2.4	112	46.0
Ethylene glycol	2.4	148	62.0
Lithium	3.6	24.9	6.94
Aluminum	0.88	23.7	27.0
Copper	0.38	24.1	63.5
Gold	0.13	26.0	197.0
Mercury (solid)	0.14	27.8	200.6
Argon (solid)	0.65	25.9	39.9
Nitrogen (solid)	1.63	45.7	28.0

[*]The number having these units is the heat needed to raise the temperature of 1 mol (6.02×10^{23} atoms or molecules) by 1 K.

exercise 7.3

Heating the Bath Water

Problem

Calculate the amount of heat necessary to heat 50 L of water from cold tap water temperature to the temperature comfortable for a bath. Please begin by selecting a suitable temperature for cold tap water and for a bath.

Solution

First we must figure the mass of 50 L of water. We do this knowing that 1 L of water has a mass of 1 kg or 1000 g. Fifty liters weighs 50,000 g, 5.00×10^4 g. Tap water temperature varies a lot from place to place and so does bath temperature so we'll use our best estimate and you can recalculate for your situation. Suppose that tap water is 10°C and that you prefer a hot bath of 45°C, so the temperature rise is (45 − 10) or 35°C. The change is equal to 35 K.

The heat necessary therefore =

$$\frac{4.184J}{gK} \times 5.00 \times 10^4 g \times 35K = 7.3 \times 10^6 J \text{ or } 7.3 \times 10^3 kJ$$

A good way to check your answers from these types of problems is to check to make sure that the units cancel to give you the final units you want. Do they here?

You might wish to look back at Chapter 5 and convert this amount to the energy equivalent in frosted flakes cereal. This is fun, but not especially practical. A bit later in the chapter we will relate it to something more meaningful: how much natural gas must be burned to get this much heat?

final temperature. So the net effect of cooking with metal pans, whether aluminum, iron or copper, is that water in them receives most of the heat supplied by the stove.

Remember that Δ°C = ΔK.

To figure the total amount of heat necessary to heat up a quantity of water, just multiply the specific heat by the mass of the water and by the number of Celsius degrees that the temperature rises. Note that this is a dimensional analysis problem.

Heat Capacity

The specific heat is higher for water than for nearly any other substance. This means that it requires much more of a heat input to raise the temperature of water than to raise the temperature of metals or compounds like alcohol. Water's specific heat multiplied by the mass of water involved is termed the heat capacity of the body of water. In the previous example, the **heat capacity** of the hot water in the bathtub is 209 kJ for every degree of temperature increase of that amount of water. For a cup of water, or about 250 g of water, the heat capacity is 4.184 J/gK × 250 g = 1050 J/K or 1.05 kJ/K. The specific heat doesn't change; the amount of heat needed to warm a big amount of water is simply more than for a small amount. A great amount of heat is needed to warm a sizable body of water such as a lake. For this reason, lake water stays cool throughout a long hot summer and oceans change their temperature even less.

This example helps show that, although heat and temperature are related quantities, they definitely are not the same. Both the specific heat and the mass of substance must be combined with temperature change to determine heat changes.

exercise 7.4

A Penny for Your Thoughts

Problem

As you saw in Table 7.3, every substance has a characteristic specific heat. The specific heat of copper metal is 0.38 J/gK. Consider a red-hot copper penny ($T = 600°C$, mass = 3.15 g) or the bathtub of water mentioned a bit ago; which will give off the most heat in cooling to room temperature (25°C)? Which contains more heat?

Solution

Since the question asks about the heat change of a piece of copper, the specific heat of copper, 0.38 J/gK must be used.

$$\text{Heat lost by the hot penny} = \frac{0.38 \text{ J}}{\text{gK}} \times 3.15 \text{ g} \times (600-25°C)$$

$$= 690 \text{ J}$$

This compares with 4.2×10^6 J for the bathtub ($4.184 \times 5.00 \times 10^4 \times 20$). Although the penny was much hotter (had a higher temperature), it contained far less heat.

exercise 7.5

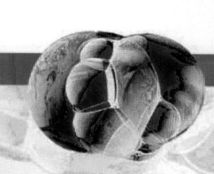

How Much Gas Is Needed to Heat That Bath?

Problem

When natural gas (methane, CH_4) burns, it releases 882 kJ per mole. A canister contains 120 g of methane. Is this enough to heat the bath water described in Exercise 7.3 to 45°C?

Solution

We know from Exercise 7.3 that 7.3×10^3 kJ is required to heat the bath water. If methane releases 882 kJ per mole, the number of moles of methane we need is

$$7.3 \times 10^3 \text{ kJ} \times \frac{1 \text{ mol methane}}{882 \text{ kJ}} = 8.3 \text{ mol methane}$$

The number of grams of methane =

$$8.3 \text{ mol methane} \times \frac{16.0 \text{ g}}{1 \text{ mol methane}} = 130 \text{ g methane}$$

We do not have enough methane to heat the bath water to 45°C. Alternatively, we can solve the entire problem in a one-dimensional analysis equation,

$$7.3 \times 10^3 \text{ kJ} \times \frac{1 \text{ mol methane}}{882 \text{ kJ}} \times \frac{16 \text{ g methane}}{1 \text{ mol methane}} = 130 \text{ g methane}$$

Food for thought: is it valid to assume that all the heat energy from burning the methane actually goes into heating the bath water? Where else might the heat be going?

exercise 7.6

Heat of Fusion

Problem

Let's calculate how much ice is needed to lower the temperature of 250 g of water from a temperature of 22°C to 0°C. Note that the ice, at 0°C, will melt as it absorbs heat from the liquid, which is cooling.

Solution

There are two parts to this problem—the first concerns how much heat needs to be removed from the water in order for its temperature to go down to freezing.

1. Heat released in cooling the water =

$$\frac{4.184\ \text{J}}{\text{gK}} \times 250\ \text{g} \times 22\ \text{K} = 2.3 \times 10^4\ \text{J}$$

Reminder Δ°C = ΔK

In the second part of the problem, we need to know how much ice must melt to absorb this much (2.3×10^4 J) heat. For this, we need to make use of the heat of fusion for water, in which 333 J/g of heat is absorbed as the ice melts to water at 0°C.

2. Each gram of ice needs 333 J to melt it. Divide the needed heat, 2.3×10^4 J by the 333 J/g to get the number of grams of ice needed.

$$\text{Mass of ice} = \frac{2.3 \times 10^4\ \text{J}}{33\ \text{J/g}} = 69\ \text{g}$$

This means that 69 g of ice will melt at 0°C on absorbing enough heat to bring the temperature of 250 g of water from 22°C down to 0°C.

Consider This

Which contains more heat: 100 g of water at 0°C or 100 g of ice at 0°C?

Heat of Fusion and Vaporization

When we run in warm or hot weather we perspire heavily. Perspiration is mostly water and its function is to cool down a person who is too warm. Perspiration cools us because it continually forms a thin layer of water that evaporates from the heated skin. The heat needed to evaporate this water comes from the body of the person who is exercising—resulting in a cooler person.

It takes a lot of heat to make water undergo **evaporation.** To convert a gram of liquid water to water vapor requires 2.44 kJ at 25°C; this is called the **heat of vaporization.** Thus, as water evaporates from heated skin, it uses large amounts of the skin's heat.

When gaseous water, or steam, condenses back into liquid water this same large heat of vaporization is released. Both boiling water and steam can be at 100°C, but a burn from steam is much worse than a burn from boiling water because so much heat is transferred to the skin when steam condenses into liquid water.

To this point in the chapter, our work with the water cycle has focused on the energy changes that accompany temperature changes and state changes in pure water. However, most of the water on Earth has substances dissolved in it. For example, sodium chloride and other compounds are dissolved in the saltwater in the oceans. Metal ions, such as calcium and magnesium, are found in water that comes from the ground and is consumed by families who do not have municipal water treatment systems or water softeners. In the next part of our discussion, we consider why so many substances dissolve readily in water, and why others, such as most substances in crude oil, do not.

7.2 | The Universal Solvent

Water is often called the "universal solvent." If a universal solvent is a liquid that will dissolve anything, the old puzzle is "in what container does one keep a universal solvent?" This is actually a real problem for storage of very pure water. Water is a good enough solvent that it dissolves small amounts of its container and thereby loses its purity. It is very difficult to store very pure water for very long. Let's define a few words used in this paragraph. To **dissolve** means that molecules or ions of one substance (the **solute**) are evenly mixed among the molecules of another (the **solvent**). The solvent is present in larger amounts in the final solution (a homogenous mixture).

Generally speaking, small molecules dissolve better than large molecules and polar molecules dissolve better (in water) than do nonpolar molecules. Therefore, methanol, CH_3OH, would dissolve better in water than hexadecane, $C_{16}H_{34}$, which is, in fact, insoluble in water. Hydrogen-bonding molecules, such as methanol, dissolve in water best of all. On the other hand, polar, hydrogen-bonding molecules like water dissolve very poorly in nonpolar solvents like carbon tetrachloride (CCl_4) but hydrocarbons dissolve very well in such solvents.

As you already know, water is a small, very polar molecule held in liquid form by hydrogen bonds. This same polarity and tendency to hydrogen bond make water very good at dissolving other polar, hydrogen-bonding molecular compounds like sugars and alcohols. The less polar and the less prone to hydrogen bonding a compound, the less well it dissolves in water. Remember "like dissolves like." Alcohols have an O—H group bonded to carbon, and sugars have several such O—H groups. As a result, alcohol and sugar both dissolve well in water.

Like Dissolves Like

When both solvent and solute are liquids, a few combinations of solvent and solute are mutually soluble in any proportion. Common alcohol (ethanol or ethyl alcohol, C_2H_5OH) is a good example. Any amount of alcohol will dissolve in any amount of water and any amount of water will dissolve in any amount of alcohol (Figure 7.5). The solubility of such pairs is said to be infinite.

As the length of the chain of carbon atoms increases in an alcohol molecule, its solubility in water decreases (see Table 7.4). As the carbon chain of the alcohol gets longer, the solubility gets lower because the hydrogen-bonding part of the molecule becomes relatively less important and the alcohol becomes less polar. Another way of thinking about this is to say that as the carbon chain gets longer, the molecule behaves more and more like a hydrocarbon and less like an alcohol. Molecules that are less polar than these long chain alcohols have even lower solubility. For instance, benzene (C_6H_6) has a solubility of only 1.8 g/L and many other compounds are even less soluble than this.

Solubility

In a solution of *n*-hexylalcohol (1-hexanol) in water, $C_6H_{13}OH$ is the solute and water is the solvent. When the maximum concentration of solute in water has been reached, called the **solubility** of the solute, there will be 5.9 g of the 1-hexanol in every liter of water. Any hexanol added above this amount will rise above the solution because its density is less, but it will not dissolve. In such a situation, the solution is said to be **saturated.** Since solutions are homogeneous, we can figure out how much solute is in any given amount of solution. When a solution is saturated, we need to know only the solubility and the volume of solution to calculate the amount of solute.

Saturation is related to temperature. To convince you of this, ask yourself the following question, "which dissolves more sugar: iced tea or hot tea?" The solubility of sodium chloride (table salt) is 0.357 g/mL in water at 0°C and 0.391 g/mL in water at 100°C—not much difference. Sodium carbonate decahydrate, $Na_2CO_3 \cdot 10H_2O$, also known as "washing soda," is soluble at the level of 0.22 g/mL in water at 0°C, but at 100°C, the solubility increases to 4.2 g/mL—a nearly 20-fold increase!

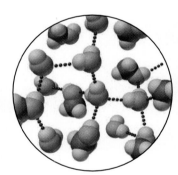

Figure 7.5 Like dissolves like: methyl alcohol (CH_3OH) bottom, is soluble in water (H_2O) top, because their OH bonds are so similar. The volume of the mixture is less than the sum of the volumes of water and alcohol before mixing.

Table 7.4

Solubility of Selected Alcohols in Water

Alcohol	Solubility (g/L)	Name of Alcohol
CH_3OH	Infinite	Methyl alcohol (methanol)
C_2H_5OH	Infinite	Ethyl alcohol (ethanol)
C_3H_7OH	Infinite	n-propyl alcohol (1-propanol)
C_4H_9OH	79	n-butyl alcohol (1-butanol)
$C_5H_{11}OH$	27	n-pentyl alcohol (1-pentanol)
$C_6H_{13}OH$	5.9	n-hexyl alcohol (1-hexanol)
$C_7H_{15}OH$	0.9	n-heptyl alcohol (1-heptanol)

exercise 7.7

Will It Dissolve?

Problem

Based on your understanding of polarity (if you are a little fuzzy on this, please review Chapter 6), predict which would be more soluble in water: ethylene glycol ($C_2H_6O_2$) or benzene (C_6H_6).

$$CH_2 - CH_2$$
$$|\qquad|$$
$$OH\qquad OH$$

Solution

The key is to assess the polarity and the hydrogen-bonding ability of the molecule as much as possible. Ethylene glycol is polar and takes part in hydrogen bonding, while benzene is not. The three-dimensional structures shown confirm our assessment on polarity. Ethylene glycol is mixed with water to make antifreeze. Benzene is an important laboratory and industrial-scale solvent for other nonpolar substances.

exercise 7.8

Saturated Solutions

Problem

How much $C_6H_{13}OH$ will form a saturated solution with 25 mL of water?

Solution

Think of it this way. The solubility of n-hexyl alcohol or 1-hexanol is 5.9 g/L so each mL will contain 0.0059 g. Multiply by 25 to see how much is in 25 mL.

$$\text{mass of hexanol} = 25 \text{ mL} \times \frac{1 \text{ L soln}}{1000 \text{ mL}} \times \frac{5.9 \text{ g 1-hexanol}}{\text{L soln}} = 0.15 \text{ g}$$

As a check, note that the units cancel to give the desired units; that is a very good indication that the problem is set up properly.

Gaseous Solutes

Luckily for the survival prospects of fish and aquatic plants, even gases like CO_2, oxygen, and nitrogen dissolve in water. When air is bubbled into water or water is sprayed into air, some of the air's nitrogen and oxygen molecules dissolve in the water. The solubility of these gases isn't very high and depends on the temperature of the water and on the pressure of the gas in the space above the liquid. Water in contact with air dissolves about two times as much nitrogen as oxygen because nitrogen is about half as soluble as oxygen but contains four times as much nitrogen as oxygen. Gases have a higher solubility in cold water than hot. This is part of the reason that many fish survive better in cold water than in warm; there can be more oxygen in cold water.

Ionic Solutes

Anyone who has spent any time in the kitchen knows that salt, baking soda, and cream of tartar all dissolve in water. These ionic compounds, all called salts, are a quite different type of substance from others we've mentioned as very soluble in water. Some ionic compounds dissolve well in water but others don't. Those that do dissolve in water do allow solutions to conduct electricity. We call such solutes **electrolytes.** We will have much more to say about electrolytes in Section 15.3. Sodium chloride (table salt) is the classic example of a water-soluble ionic compound. A crystal of solid sodium chloride, NaCl, is a vast collection of Na^+ ions and Cl^- ions mutually attracted to each other. When NaCl dissolves, its ions separate. This requires energy. The ions then become surrounded by water molecules (Figures 7.6, and 7.7). This later process makes the system more stable, thus releasing energy. On balance, more energy is released when the ions interact with water than is required to split the salt into ions. Thus, the process of dissolving NaCl is energetically favored.

Limestone ($CaCO_3$) and gypsum ($CaSO_4 \cdot 2H_2O$) are both ionic compounds but they do not dissolve appreciably in water because more energy is required to split up the compounds than is released when the ions interact with water. The dissolution is not energetically favored. Some salts are very water soluble and others are so slightly soluble that we usually call them insoluble. Just because a compound contains a particular metal ion does not necessarily mean that it will behave in a certain way. Even though calcium carbonate and calcium sulfate are insoluble, calcium bicarbonate is soluble. This solubility leads to calcium ion being present in some water supplies, which are said to be "hard water."

The ocean water that surrounded the ancient mariner contained such a high concentration of dissolved ionic compounds (especially NaCl) that it was (and is)

Electric power plants use enormous amounts of water to remove surplus heat from the power generation process. The water is then cooled by transferring the heat into the air via cooling towers. River water is used to carry off waste heat at power plants. Before the water can be returned to the river it must be cooled so it can dissolve enough oxygen from the air to sustain life in the river. When this is not done completely, the river gets too warm and is said to suffer thermal pollution.

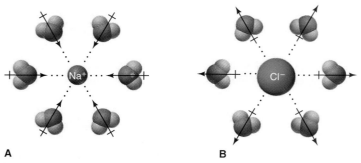

A   **B**

Figure 7.6 The electrons of water molecules are concentrated near the oxygen atom. This results in the oxygen having a partially negative charge and the hydrogens remaining partially positive. So when sodium chloride dissolves in water, the negative Cl^- ions will be attracted to the hydrogen end of the water molecule and positive Na^+ ions will be attracted to the negative oxygen end.

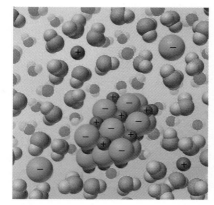

Figure 7.7 NaCl dissolving in water. The oxygen of a water molecule is partially negatively charged and the hydrogen remains partially positively charged. Therefore, negative ions (such as chloride) will be attracted to the positive end of the water molecules and positive ions (such as sodium) will be attracted to the negative ends of other water molecules.

Table 7.5

Major Elements Present as Ions in Seawater

Element	Parts per Million (ppm)	Molarity (M)
Cl (as Cl^-)	18,980	0.53
Na (as Na^+)	10,561	0.46
Mg (as Mg^{2+})	1,272	0.053
S (as SO_4^{2-})	884	0.028
Ca (as Ca^{2+})	400	0.010
K (as K^+)	380	0.010
Br (as Br^-)	65	0.0008
C (as HCO_3^-)	28	0.0023

harmful to drink. The oceans are salty because over huge durations of time rain has fallen on land, dissolved various soluble substances, and run them down to the sea. Ionic compounds do not evaporate so they remain behind in the ocean. After eons, the concentration of dissolved substances in the sea became what is shown in Table 7.5.

7.3 | Water Hardness

When soap is added to water from wells in some parts of the country, it forms a gray, slimy, insoluble product. This shows up as bathtub ring and when clothes are washed with soap rather than with synthetic detergents this goop causes dingy clothes even after washing. This same kind of water also causes rock-like deposits inside tea kettles and water pipes. Water that gives rise to these objectionable features is called **hard water.** Water that doesn't do any of these things is called **soft water.**

Slake means to add water. Therefore, **slaked lime** is calcium hydroxide that has been produced by adding water to calcium oxide (lime).

Hard water is water in which a lot of calcium (or occasionally magnesium or iron) is dissolved. It is called hard because rock-like material settles out of such water when it is heated. Water hardness occurs when it contains dissolved calcium bicarbonate ($Ca(HCO_3)_2$). The ions formed when calcium bicarbonate dissolves are Ca^{2+} and HCO_3^- (bicarbonate ion). Calcium bicarbonate usually leaches into water from limestone (calcium carbonate, $CaCO_3$) when some acid is also present.

$$CaCO_3 + H^+ \rightarrow Ca^{2+} + HCO_3^-$$

Or, water can act as an acid itself and dissolve limestone to a small extent. Notice calcium carbonate behaves as a base either way; in the first reaction calcium carbonate accepts a hydrogen ion and in the second case its dissolution in water releases hydroxide ion.

$$CaCO_3 + H_2O \rightarrow Ca^{2+} + HCO_3^- + OH^-$$

Water can be softened by removing its calcium ion. This eliminates the problem of insoluble "stuff" caused by the reaction of calcium ion with soap. One can also solve the hard water problem by using cleaning chemicals which do not interact with calcium in this way. We'll describe these two approaches, one at a time.

Soap and Detergents

First, let us discuss the use of different cleaning chemicals. Soap is a product of the reaction of sodium hydroxide (NaOH) or potassium hydroxide (KOH) (or a mixture of the two) with animal or vegetable fat. This reaction is called **saponification.** It results in three of the C—O bonds in a fat or vegetable oil molecule being broken and three fatty acid anions as well as one glycerine molecule being formed.

Fatty acid

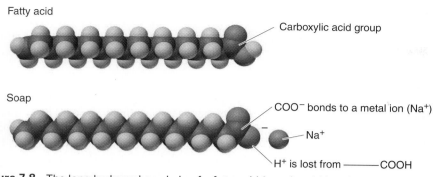

Carboxylic acid group

Soap

COO⁻ bonds to a metal ion (Na⁺)

Na⁺

H⁺ is lost from ——————COOH

Figure 7.8   The long hydrocarbon chain of a fatty acid (stearic acid here) remains the same in a soap. Only the carboxylic acid group on the end changes.

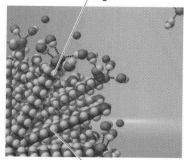

Soap molecules immersed in grease stain

Nonpolar molecule of grease

Figure 7.9   A micelle is a collection of dozens of soap or detergent molecules with their long hydrocarbon chains in the interior of a space and the ionic head groups on the outside.

To be a soap, something must be a salt, usually sodium or potassium, of a fatty acid. A fatty acid has a long chain of carbon atoms hooked together with mostly single, but occasional double bonds and with the rest of its valence sites occupied by hydrogen atoms. The feature that makes it an acid is the group of atoms:

$$\underset{\displaystyle -\overset{\textstyle \overset{O}{\|}}{C}-OH}{} \text{ (often written —COOH)}$$

called a **carboxylic acid group,** (see Table 6.6, page 217) at one end. In a soap, H⁺ is lost from the carboxylic acid group leaving behind a −1 charged anion (often written —COO⁻) which is ionically bonded to a metal ion such as sodium ion (Figure 7.8).

The reaction of calcium with soap is simply the trading of sodium ion for calcium ion. Since calcium has a +2 charge and the soap anion is only −1, two soap anions react with each calcium ion to form the insoluble calcium soap.

$$Ca^{2+} + 2NaC_{18}H_{35}O_2 \rightarrow Ca(C_{18}H_{35}O_2)_2 + 2Na^+ \tag{7.1}$$

The unusual way in which soap dissolves in water makes it an effective cleaner. The long carbon chain doesn't dissolve well in water but the anion at one end is highly attracted to water through ionic, polar, and hydrogen bonding. The soap molecule then is forced to do two things at once, both to dissolve in water and not dissolve in it. To do this split personality trick, soap molecules form clusters of molecules called **micelles** (Figure 7.9) in which the long carbon chains of several molecules get away from water by forming little spheres with the ionic group on the outside. The ionic groups are located out where the water is but the nonpolar parts are as far away from water as they can get.

Soap cleans grease off hands and clothes by causing the nonpolar grease to dissolve in the interior of the micelle while the entire micelle is suspended in water which then rinses down the drain.

The metals of Group 1A form soluble soaps but all the other groups of metals form insoluble soaps. Thus, when soap is added to hard water, the calcium ion forms the insoluble calcium soap which comes out of solution (Eq. 7.1) and sticks to everything it contacts.

Synthetic **detergents** form micelles and clean grease off things in the same way as soaps, but, unlike soaps, their calcium salts are soluble in water so no insoluble substances are formed when detergent is added to hard water. There are a variety of structures of synthetic detergents, but a very common one is the following:

$$CH_3-CH_2-CH_2-CH_2-CH_2-CH_2-CH_2-CH_2-CH_2-CH_2-CH_2-CH_2-SO_3^-\ Na^+$$
called sodium lauryl sulfonate

The similarities of detergent to soap are obvious and the difference is that soap molecules have a carbon atom bonded to the oxygen to which the sodium is bonded and detergents have a sulfur in that position. That small difference is enough to make calcium salts of soaps insoluble in water but the calcium salts of detergents to be soluble.

Water Softening

Water softening is the removal of calcium, magnesium, and iron ions, the so-called "hardness ions" from water and can be done by several methods.

Distillation

Rain water contains no metal ions and so is very soft. For years, to avoid the problems of hard water, people collected rain water in barrels to use when washing with soap. Rain water is actually a form of distilled water. **Distillation** is the oldest method of softening water and the one that also removes "nonhardness" ions like sodium. For centuries, distillation has been the method for preparing very pure water. The water is boiled and the steam carried away so that it condenses back to a liquid somewhere else where it is collected in a clean container. Rain water forms similarly when water evaporates from the Earth's surface, rises, and condenses back to liquid high in the colder air, at which point it falls to earth.

Ion Exchange

As we mentioned in Section 7.1, the amount of energy needed to boil large amounts of water makes distillation too costly except in unusual circumstances. Besides, there are not many common uses that require water this pure. Rather, water for household use is usually softened by replacing the calcium in the water by sodium from a water softener process called **ion exchange** (Figure 7.10). The hard water to be softened is passed through a cylinder packed full of plastic beads (called cation exchange resin) into which sodium ions are incorporated. Calcium replaces the sodium ions in the plastic beads because its greater positive charge is

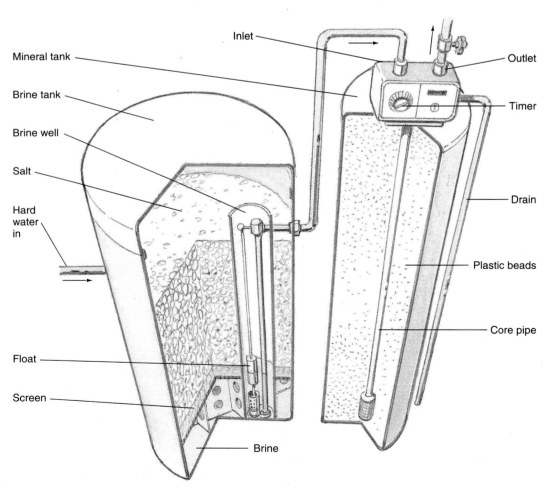

Figure 7.10   Ion exchange water softeners are recharged by passing strong saltwater through the exchanging material.

more strongly attracted to the negative charge of the resin than is the single positive charge of sodium ion. The result is that calcium comes out of solution and is replaced in the water by sodium ion. Since calcium has a charge of $+2$ and sodium only $+1$, two sodium ions must go into solution for every calcium that is removed.

$$2Na(resin) + Ca^{2+} \rightarrow Ca(resin)_2 + 2Na^+ \tag{7.2}$$

The resultant water is not pure in the sense that distilled water is pure, but the troublesome calcium ion is removed. After much of the sodium in the resin cartridge has been exchanged for calcium, the cartridge must be recharged with sodium for further use. A large excess of sodium chloride is added to the resin to drive off the calcium from the resin by the reverse of the reaction in Eq. (7.2), and the calcium ion is put down the drain along with considerable excess sodium chloride.

This softened water is great for washing, but not the best for drinking. People need both calcium and sodium for health, but the typical U.S. diet has more than enough sodium and not enough calcium. So, hard water is better than softened water for drinking.

Besides being inefficient for washing, hard water is also a problem because calcium carbonate will precipitate out of hard water, especially when that water is heated. (The word **precipitate** is both a noun and a verb. The noun means a solid that forms from solution and settles to the bottom. The verb means "to form an insoluble substance from a solution.") Pipes, tea kettles, water heaters, and other devices can become coated or clogged with rock-like deposits of $CaCO_3$. This can become dangerous when calcium carbonate is deposited in steam boilers. When this happens much more heat must be applied to the boiler to transfer the heat to the water since rock is poor at transferring heat. The steel boiler can get very hot. If a crack develops in the calcium carbonate layer, water can reach the hot steel suddenly. Red hot iron reacts chemically with the water to form iron oxide and hydrogen gas that can accumulate and later explode violently. The old river steamboats that occasionally had a boiler explode usually failed because of this. Modern steam boilers use very pure water to prevent hardness buildup. This saves fuel costs and eliminates the chance of disastrous explosions.

Whether your water is hard or soft, you expect it to be wholesome to drink. If you live in a large town or city, there will likely be a municipal water treatment plant. The next section describes how such plants typically guarantee safe water for our use.

Municipal Chemical Water Treatment

During the mammoth floods of the summer of 1993, the water treatment plant in Des Moines, Iowa, was put out of commission and the water system contaminated with untreated water (Figure 7.11). Water had to be trucked in from surrounding areas for several weeks. In the United States we have come to expect safe, clean water at the turn of a tap. When storms,

Figure 7.11 In the summer of 1993 the Mississippi River and many of its tributaries flooded due to heavy rains throughout the upper Mississippi Valley. The Des Moines River in Des Moines, Iowa, overflowed its banks and flooded the water treatment facility, leaving it inoperable. For more than two weeks, residents of Des Moines suffered the Ancient Mariner's plight of ". . . water everywhere nor any drop to drink. . . ." Photo by Bob Nandell. Copyright 1993 The Des Moines Register and Tribune Company. Reprinted with permission.

earthquakes, or accidents damage a city's water system, it is put right as quickly as possible. "You never miss the water till the well runs dry" is an old pioneer saying that still rings true.

Municipal water treatment varies from community to community depending on the quality of the water that is to be treated. Groundwater can be brought to the surface with wells; water can be pumped from rivers, lakes, or the ocean. Each of these sources presents different treatment problems. The most important treatment is the destruction of pathogenic microorganisms. For hundreds of years, people regularly died when pathogenic bacteria such as typhoid fever got into water supplies. These epidemics stopped abruptly in 1908, when the Jersey City, New Jersey, municipal water system began to add the disinfectant chlorine to the city's water to kill microorganisms. Other municipalities quickly followed suit. Millions of lives have been saved by water chlorination. Not only bacteria, but algae and viruses are killed by chlorine.

There are concerns about the safety of chlorine use because of the other things with which it reacts and the products of these reactions. Compounds such as chloroform ($CHCl_3$) and chlorophenol (ClC_6H_4OH) have been found in water after chlorine treatment. Chlorine has reacted with naturally occurring organic compounds in water to form these disagreeable compounds.

> **Allotropes** are different forms of the same element—for example, graphite and diamonds are both forms (allotropes) of carbon with very different chemical and physical properties.

Partly to avoid the disagreeable compounds formed when chlorine is used as a water oxidant, many modern water treatment plants use ozone as the primary oxidant. Ozone can't be stored for any appreciable time, so it is made from air at the location of its use. Air or pure oxygen is passed through an electric discharge and ordinary oxygen is converted into the **allotropic form** (see margin note to understand allotropes), ozone.

$$3O_2 \rightarrow 2O_3$$

Ozone is a terrific oxidizer and does not make chlorinated products. Chief among the substances it oxidizes are iron(II) and manganese(II), which are commonly found in well water. Low oxidation states of metals are frequently found in well water, where the water has been away from air for a long time. When there is no oxygen present, low oxidation states predominate. Ozone (or chlorine) oxidizes these ions to higher oxidation states.

$$6Fe^{2+} + O_3 + 6H_2O \rightarrow 3Fe_2O_3 + 12H^+$$

$$3Mn^{2+} + O_3 + 3H_2O \rightarrow 3MnO_2 + 6H^+$$

The oxides of ions with higher oxidation states are much less soluble than the lower oxidation state ions. Oxidation therefore rids water of objectionable metal ions by causing them to form insoluble solid oxides. Fe_2O_3 and MnO_2 are orange- and black-colored solids. They will stain clothes and anything else they may precipitate onto. Therefore, it is a good idea to eliminate them before they enter the water distribution system. Iron in water is not harmful (and even a source of a useful mineral) but manganese is not healthful.

Microorganisms are killed by ozone by mechanisms similar to their destruction by chlorine. Ozone, however, is much more reactive and so will not stay in water for more than a few hours. After ozone treatment, water is then treated with chlorine before flowing from the facility. This is done so the chlorine remains in the water for several days, keeping it free from contamination during distribution and use. Chlorine oxidation is even more extended if ammonia is added.

$$NH_3 + Cl_2 \rightarrow NH_2Cl + HCl$$

NH_2Cl (chloramine) is also a good oxidant and lasts even longer than chlorine in a water supply. Chlorine can leave water as a gas, but chloramine is a liquid at room temperature so is not so apt to escape in this way. Chloramine will form in dangerous amounts if bleach is mixed with household ammonia.

In most treatment facilities, an important step is filtration. This is usually done by allowing the water to trickle through several feet of sand and gravel to trap insoluble materials. Oxidation steps, either by chlorine, ozone, or even molecular oxygen, oxidize Mn(II) and Fe(II) found in well water to the insoluble compounds MnO_2 and Fe_2O_3. These are removed by settling, coagulation, and filtration. If they were not removed, these compounds would form in the water distribution system and clog pipes or stain plumbing and laundered clothes.

Many water treatment plants, especially those that rely on rivers or lakes for their water source, use sedimentation to clarify their water. In this process, aluminum sulfate ($Al_2(SO_4)_3$) and calcium hydroxide ($Ca(OH)_2$, also known as "slaked lime") are added sequentially to impure water.

$$Al^{3+} + 3\,OH^- \rightarrow Al(OH)_3$$

They react to form aluminum hydroxide, which is a gelatinous precipitate. In a process called **coagulation,** this solid slowly settles out of the water and entraps suspended material, much dissolved organic material, and many microorganisms, thereby removing them from the water.

Both distillation and resin softening are too expensive and usually unnecessary for treating huge amounts of water for municipal distribution. If raw water is very hard, some of the hardness is usually removed by chemical treatment at central water treatment plants. It seems surprising, but this hardness removal is done by putting in more calcium. Let's take a look at this.

Most natural hardness is due to calcium bicarbonate, a rather soluble calcium salt. It is made insoluble by addition of hydroxide ion which removes the hydrogen ion from the bicarbonate ion and forms carbonate ion.

$$Ca(HCO_3)_2 + 2OH^- \rightarrow CaCO_3 + CO_3^{2-} + 2H_2O$$

The best and cheapest source of hydroxide to use for this purpose is calcium hydroxide, $Ca(OH)_2$, so the overall reaction for removal of hardness results in adding one calcium ion but precipitating two for a net reduction of hardness.

$$Ca(HCO_3)_2 + Ca(OH)_2 \rightarrow 2CaCO_3 + 2H_2O$$

We started this section on water hardness with calcium and we ended it with calcium. We saw that calcium has very specific chemical behavior, different from other metals in many respects, and we can control that behavior chemically to meet our needs—whether to alleviate water hardness or to clean the grime from a bathtub by use of detergents. These chemical properties occurred because calcium itself was present. But there are properties of aqueous (water-based) solutions that do not depend on what is in solution—rather, the key is how much is in solution. We make use of these properties when keeping a car's engine running in the scorching heat of the summer or the brutal cold of winter. We also use them when making ice cream or even as a method of cleaning water. What unifies these different types of processes? They are all based on "colligative properties."

7.4 | Colligative Properties

When we talk about solutions of substances dissolved in water, we generally are emphasizing the behavior of the solute. The properties of the solution are also changed by the number of solute particles rather than their nature. These properties are called **colligative properties.** Four well-known colligative properties of solutions are their ability to lower vapor pressure and freezing point and to elevate boiling point and osmotic pressure.

Lowering Vapor Pressure

When you add sugar or some other solute to water, the vapor pressure is decreased. The vapor pressure of the solution decreases in proportion to the number of moles of sugar added compared to the number of moles of solvent. The vapor pressure of a sugar solution is less than that of pure water at the same temperature: it is less likely to evaporate or boil. (Recall the graph of vapor pressure of water versus temperature Figure 7.3.)

The vapor pressure of a solution looks like the lower curve in the graph (Figure 7.12). Since the boiling point at standard conditions is the temperature at which the vapor pressure of the solvent equals 1 atmosphere (atm), you can see that the boiling point of the solution occurs at a higher temperature when something is dissolved in the water.

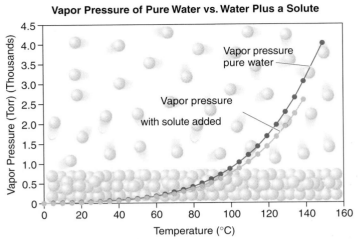

Figure 7.12 The vapor pressure of water increases greatly at higher temperatures but is always less when a solute is dissolved in the water than when the water is pure. When the water's vapor pressure equals the external pressure, the water is at its boiling point, 100°C for pure water at sea level.

Boiling Point Elevation

Arithmetically, the change in the boiling point, ΔT, is proportional to the concentration of the solution expressed as the number of moles of non-volatile solute per kilogram of solvent. This is a measure of concentration called **molality (m).** The proportionality constant, K_b, is called the **boiling point elevation** constant and is different for different solvents. See Table 7.6 for values for a few common solvents. The value for water is 0.512°C/m. The equation used to relate boiling point elevation to concentration actually gives the increase in the boiling point. This is designated with a Greek "delta" (Δ) in front of the T, which stands for change in temperature since the delta means "change in" the quantity that follows.

$$\Delta T = K_b \times m.$$

Ionic compounds and acids that dissociate into ions on dissolving in water have a greater effect on boiling point elevation (and other colligative properties) than their simple molality would predict. The effect on colligative properties is *approximately* proportional to their molality multiplied by the number of ions each formula unit dissociates into.

$$\Delta T \approx K_b \times m \times i$$

where i is the number of ions per formula unit. For example, $CoCl_2$, cobalt chloride, should have $i = 3$. In fact, i has been measured from anywhere between 2.80 to 4.58 for $m = 0.005$ to $m = 2.00$.

Table 7.6

Boiling Point and Freezing Point Modification of Selected Solvents

Solvent	Boiling Point (°C)*	K_b (°C/m)	Melting Point (°C)	K_f (°C/m)
Water	100.0	0.512	0.0	1.86
Ethanol	78.5	1.22	−117.3	1.99
Benzene	80.1	2.53	5.5	4.90
Acetic acid	117.9	3.07	16.6	3.90
Chloroform	61.7	3.63	−63.5	4.70

*At 1 atm.

exercise 7.9

Boiling Point Elevation

Problem

Calculate the boiling point of a solution of 2.5 mol of sugar ($C_{12}H_{22}O_{11}$) dissolved in 300 g of water.

Solution

$$\text{Molality} = \frac{2.5 \text{ mol sugar}}{0.300 \text{ kg water}} = 8.3m$$

$$\text{Boiling point elevation} = 0.512°C/m \times 8.3\ m$$

$$= 4.3°C$$

Since the boiling point of pure water is 100.0°C and the solution boiling point is elevated by 4.3°, the boiling point of the solution is 104.3°C.

Here's something to ponder. It is often said that we should add a pinch of salt to a pot when preparing to cook spaghetti. The thinking is that it will raise the boiling point of the liquid, thus cooking the spaghetti faster. Is this reasonable, given the rise in boiling point with the sugar solution above?

exercise 7.10

For a Greater Challenge, Try This . . .

Problem

Calculate the boiling point of an antifreeze solution in which 10 lb of ethylene glycol ($C_2H_6O_2$) are added to 2.0 gal of water.

Solution

First: **moles of ethylene glycol =**

$$10 \text{ lb } C_2H_6O_2 \times \frac{454 \text{ g } C_2H_6O_2}{1 \text{ lb } C_2H_6O_2} \times \frac{1 \text{ mol } C_2H_6O_2}{62 \text{ g } C_2H_6O_2} = \textbf{73 mol } C_2H_6O_2$$

Second: **kilograms of water =**

$$2.0 \text{ gal} \times \frac{4 \text{ quarts}}{1 \text{ gal}} \times \frac{946 \text{ mL}}{1 \text{ qt}} \times \frac{1.0 \text{ g}}{\text{mL}} \times \frac{1 \text{ kg}}{10^3 \text{ g}} = \textbf{7.6 kg } H_2O$$

Third: **molality** = 73 mol $C_2H_6O_2$/7.6 kg H_2O = **9.6 mol/kg**
Fourth: **boiling point elevation** = 0.512°C/m × 9.6 m
$$= \textbf{4.9°C}$$
Fifth: **boiling point** = 100.0° + 4.9° = **104.9°C**

This corresponds to 220.8°F. Ethylene glycol is antifreeze, but it is also important in preventing auto engines from boiling over when driving in hot weather.

exercise 7.11

How Does Antifreeze Prevent Freezing?

Problem

Calculate the freezing point depression of the ethylene glycol solution in Exercise 7.10. Does that solution really change the freezing point significantly, or, like the salt in the spaghetti pot, is it essentially meaningless in terms of temperature change?

Solution

The equation that describes freezing point depression is

$$\Delta T_f = K_f \times m$$

In the example about antifreeze,

$$\Delta T_f = K_f \times m = 1.86°C/m \times 9.6\ m = 17.9°C$$

Again, since pure water freezes at 0°C, the solution freezes at $0 - 17.9 = -17.9$ degrees Celsius (or $-0.2°F$). This is a significant change in freezing point!

People often assume that the more antifreeze they use, the lower the temperature their radiator will withstand. Up to a point that is true, but only up to a point. At some concentration the amount of ethylene glycol compared to water gets so large that it is more proper to talk about water as the solute that lowers the freezing point of the solvent ethylene glycol.

Water freezes at 0°C and ethylene glycol freezes at $-17°C$. Addition of the opposite compound lowers both of these values for intermediate concentrations of solutions so that they meet somewhere in the middle and the minimum freezing point of water-ethylene glycol mixtures is about $-40°C$ which is the same temperature as $-40°F$.

Freezing Point Depression

Those of us who live where it gets really cold in winter add antifreeze to our automobile coolant (water) to prevent it from freezing. The chemical used is usually ethylene glycol, $H—O—CH_2—CH_2—O—H$. Just as a solute raises the boiling point of water, it also lowers the freezing point formally called the **freezing point depression** (Figure 7.13). Antifreeze prevents freezing by lowering the freezing point to a temperature below what we expect will actually occur on winter's coldest day.

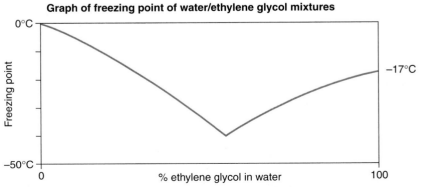

Figure 7.13 A mixture of water and ethylene glycol freezes at a lower temperature than does either compound when it is pure.

Depression of the freezing point is the third colligative property and is also related to vapor pressure. The argument concerning vapor pressure is trickier than with boiling point elevation and goes like this. For a solution to be at its freezing point, the vapor pressure of the solid must be the same as that of the solution. Since the presence of a solute forces the vapor pressure of the solvent down, the only way that it can equal the vapor pressure of the solid is for the temperature to be lower than the normal freezing point.

Addition of solute makes the solvent freeze at a lower temperature. The arithmetic that describes this is very similar to the arithmetic of boiling point elevation.

$$\Delta T_f = K_f \times m$$

For water, the freezing point depression constant, $K_f = 1.86°C/m$.

More and more, people are urged to buy a more environmentally benign anti-freeze made from propylene glycol, CH_3—$CH(OH)CH_2(OH)$. Propylene glycol is less toxic than ethylene glycol. A greater mass of propylene glycol than ethylene glycol is needed to reach the same molality and thus the same degree of protection.

Most people's favorite experience of freezing point depression occurs during the making of ice cream. To get the mixture of milk, cream, sugar and flavorings cold enough to turn into ice cream, it must be cooled well below the freezing point of water. This is done by surrounding the mixture of ingredients with ice mixed with salt. The little bit of liquid water on the ice surface dissolves enough salt to make a saturated solution of salt which has a very low freezing point. The temperature of the resulting slush is about $-17°C$ (about $0°F$). Note that although the salt melts some of the ice to make the slush, the temperature goes down.

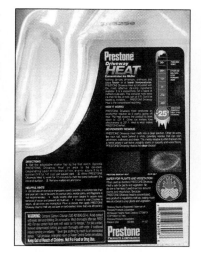

Melting ice on streets by applying calcium chloride or some other water-soluble compound is another practical application of freezing point depression. If salt can drop the freezing point of the ice to below the actual air temperature, the effect is to melt the ice, not a bad accomplishment on a cold winter day.

The zero point on the Fahrenheit temperature scale was actually defined originally as the temperature attained by mixing equal weights of ice and salt.

Osmotic Pressure

What do diabetes, water purification, and eating potato chips have in common? Important aspects of all three are explained by a colligative property called **osmotic pressure,** a property of solutions that is absolutely necessary to life as we know it. **Osmosis** is the passage of solvent through a semipermeable membrane. Water can diffuse through such a membrane, either a biological membrane or a thin sheet of plastic film, but solute particles cannot do so. A useful image is chicken wire. Mosquitoes can go right through chicken wire, but chickens can't because they are too large. Substances such as onion skin and cellophane behave as semipermeable membranes; they allow small molecules such as water to pass through, but ions or large molecules such as sugar cannot pass through.

Water can move in both directions through the membrane, but the greater flow is invariably from the pure solvent into the solution. This dilutes the solution, increases its volume, and pushes it up the tube at the right side (Figure 7.14). The final height of the column of solution is a measure of the pressure needed to stop the osmosis. This is called the osmotic pressure. The higher the concentration of solute in the solution, the greater will be the osmotic pressure difference and, therefore, the higher the column. The importance of this property shows up in many places.

Cell membranes of plants and animals behave as semipermeable membranes. If solutions are injected into a living organism, that solution must have a solute concentration that matches the osmotic pressure of the interior of the cells. If the injected solution is too dilute, water

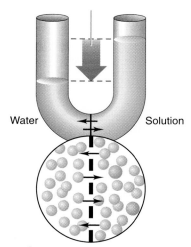

Osmotic

Water

Solution

Semipermeable membrane

Figure 7.14 Water can pass through the plastic sheet which serves as a semipermeable membrane but larger particles cannot pass through. Water will pass from the pure water into the solution to dilute the solution. When everything stabilizes, the difference in levels of the two vertical tubes is a measure of the osmotic pressure.

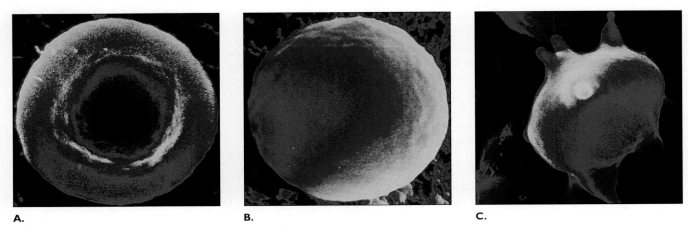

A.   **B.**   **C.**

Figure 7.15 Red blood cells in isotonic salt solution are plump, indented disks **A** as they are in whole blood. When the salt solution is too dilute (**hypotonic**) water flows into the cell until it is so plump **B** that is might burst. When the salt solution is too concentrated (**hypertonic**), the cells lose water until they shrivel up **C.** Hypertonic or hypotonic injections are dangerous, but if a salt solution is just right (**isotonic**), it can save lives.

will flow from the injected intercellular region into the more concentrated solution inside the cells and eventually cause them to fill up and burst. If, on the other hand, too concentrated a solution is injected, water will flow from the interior of the cells out into the surroundings and the cells will shrivel up and be badly damaged or killed.

Anything that increases the concentration of solutes in blood serum will have the same effect of draining red blood cells. In diabetes, the normal filtration of glucose (sugar) from the blood cells fails. Sugar levels rise and the excess sugar exceeds the kidneys' ability to absorb it. Rising glucose levels increase urine production as the body tries to eliminate the excess sugars. The excess glucose is actually acting like an osmotic diuretic pulling large quantities of water out of the body systems in an attempt to deal with the sugar overload. Dehydration results as one's blood sugar level increases and red blood cells lose water (Figure 7.15). There is not enough water available systemically and the viscosity of the blood makes it difficult for the heart to pump the thicker blood through small blood vessels.

 Green Chemistry

Reverse Osmosis

A completely different application of osmotic pressure can be used as a very practical means of purifying water. Imagine a system with a membrane separating pure water from an aqueous solution (see Figure 7.16). If a greater pressure than the osmotic pressure is applied to the solution side, water is pushed from the solution side to the pure water side with the result that the solution becomes more concentrated. This is a useful way to concentrate solutions without applying heat. More significant is the use of this process, called **reverse osmosis,** in purifying water. The solution gets more concentrated, but the amount of pure water increases. These devices are now easily installed in homes when water purification is desired. They are usually installed in kitchen sinks when water for human consumption is to be purified but water for washing clothes etc. need not be so purified.

And several municipalities throughout the world (Tampa Bay, Florida, is one important example) are installing large reverse osmosis facilities that will soon provide tens of millions of gallons of fresh water daily for their ever-increasing population.

Figure 7.16 Osmosis can be reversed by applying a pressure greater than the natural osmotic pressure. Applying pressure on the right side can cause water to go from the solution side to the pure water side. The final result is that more pure water is formed and the solution on the right side is more concentrated than it had been at the beginning.

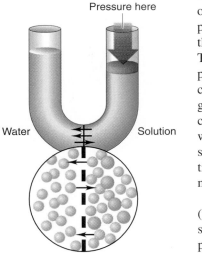

Pressure here

Water   Solution

Semipermeable membrane

■ case in point

Cholera

Cholera, a disease that even in 1991 killed thousands of people in a recent reappearance, does its damage through osmosis. The cholera bacteria causes the cells of the intestine to lose salt. This raises the salt concentration in the surrounding fluid and inside the intestine. Osmotic pressure then forces water through the cell walls of the intes-

tine where it is lost in perpetual diarrhea until the stricken person dies of dehydration or until antibiotics kill the bacteria responsible for the salt transfer. Treatment requires the victim to drink enough water, containing sodium and potassium chlorides and sugar, to at least equal that lost through diarrhea (several gallons per day) just to stay alive.

7.5 | Summing Up

There's much more that could be and will be said about water. Its properties allow life as we know it to exist on Earth and explain much of our weather and geologic changes. Water's importance will be readdressed in many of the remaining chapters of this book. Water seems so simple yet its bonding, structure, and polarity are the molecular features that make it behave in such wonderful ways.

main points

- Individual water molecules are composed of two hydrogen atoms covalently bonded to a single oxygen atom. The bonds are polar with the negative end toward the oxygen.

- Collections of water molecules in liquid or solid form are held to each other by hydrogen bonds.

- Small, polar, hydrogen-bonding molecules dissolve well in water, but large, nonpolar, nonhydrogen-bonding molecules dissolve only poorly in water.

- Water has an exceptionally high specific heat, heat of fusion, and heat of vaporization.

- Water is absolutely necessary for all earthly life.

- Water often must be purified prior to its use by humans.

- Some of water's properties change when substances are dissolved in it.

important terms

Allotropes are alternate forms of the same element that have significantly different chemical and physical properties. (p. 262)

Boiling point is the temperature at which a liquid boils and at which vapor pressure equals external applied pressure. (p. 249)

Boiling point elevation is the increase of the boiling point by dissolved solutes in a liquid. (p. 264)

Carboxyl acid group are the four atoms, COOH, carbon double bonded to one oxygen and single bonded to an —OH group resulting in an acidic H atom. (p. 259)

Coagulation is the process of making insoluble solids from dissolved materials. (p. 263)

Colligative property is a property of a solution that depends only on the concentration of a solute. (p. 263)

Detergent is a synthetically prepared cleaning agent. (p. 259)

Dissolve is to disperse one substance at the molecular level in another. (p. 255)

Distillation is the purification by evaporation and condensation at another location. (p. 260)

Electrolyte is a substance that on dissolving, allows current to flow through its solution. (p. 257)

Equilibrium is the state of a reaction at which there is no change in the concentration of the reactants and products (that state of a reaction in which the rates of the forward and reverse processes are equal). (p. 248)

Evaporation occurs when molecules of a liquid become gaseous. (p. 254)

Freezing point is the temperature at which a liquid changes to a crystalline solid (freezes). (p. 245)

Freezing point depression is the decrease of the freezing point by dissolved solutes in a liquid. (p. 266)

Hard water is water that contains significant concentrations of calcium ion. (p. 258)

Heat capacity is the amount of heat needed to raise the temperature of an object by one degree Celsius (or by one kelvin). (p. 252)

Heat of fusion is the amount of heat needed to melt 1 mol or 1 g of a solid. (p. 246)

Heat of vaporization is the amount of heat needed to vaporize 1 mol or 1 g of a liquid. (p. 254)

Hypertonic is having a salt concentration greater than blood serum. (p. 268)

Hypotonic is having a salt concentration less than blood serum. (p. 268)

Ion exchange is a process in which one kind of ion in solution is exchanged for another. (p. 260)

Isotonic is having a salt concentration equal to that of blood serum (p. 268)

Melting point is the temperature at which a solid melts and the same temperature as freezing point. (p. 245)

Micelle is a cluster of molecules of soap or similar substance in which a long nonpolar group of atoms intermingle to leave a small polar group extending into surrounding water. (p. 259)

Molality is the measure of concentration, number of moles of solute per kilogram of solvent. (p. 264)

Osmosis is the process in which solvent molecules move through a semipermeable membrane. (p. 267)

Osmotic pressure is the pressure developed as solvent undergoing osmosis expands. (p. 267)

Precipitate (**noun**) is a solid substance that comes out of solution to settle to the bottom of a solvent. (p. 261)

Precipitate (**verb**) is a process of coming out of solution. (p. 261)

Reverse osmosis is a process in which greater than osmotic pressure is applied to the solution side, so solvent (often water) is forced to the pure solvent side. The solution becomes more concentrated as a result. (p. 268)

Saponification is the process of making soap from fat. (p. 258)

Saturated is when the maximum possible concentration of solute is present in solution. (p. 255)

Slaked lime is calcium hydroxide. (p. 258)

Soft water is water having only very low concentrations of calcium ion. (p. 258)

Solubility is the amount of a solute that will dissolve in a standard amount of a given solvent. (p. 255)

Solute is a substance (solid, liquid, or gas) dissolved in a larger amount of another substance (the solvent). (p. 255)

Solvent is the substance (usually a liquid) in which another substance is dissolved. (p. 255)

Specific heat is the amount of heat needed to raise the temperature of 1 g of substance by one degree Celsius or kelvin (see Table 7.3). (p. 251)

Sublimation is the conversion of a solid directly into a gas. (p. 251)

Vacuum distillation is the distillation done under vacuum to lower the boiling point of a liquid. (p. 250)

exercises

1. What physical properties make H_2O polar and CO_2 nonpolar?
2. Explain why ice floats on water.
3. Explain why water expands when it freezes.
4. Is the making of ice cubes an endothermic or exothermic process?
5. If a pot of water boils at 120°C, what is the atmospheric pressure? (Use Table 7.2.)
6. Explain how the relative densities of liquid water and ice help to explain the effect of freezing on the weathering of rocks.
7. How much energy is necessary to heat 0.100 kg water at 40.0°C to 90.0°C?
8. What type of pot, weighing 300 grams, is better for boiling water, aluminum or copper? Why?

9. 0.248 grams of an organic compound is completely burned in an excess of oxygen and all the heat is transferred into 485 grams of water. The original temperature of the water is 17.8°C and its final temperature is 43.9°C. What is the enthalpy per gram of the organic compound?

10. 0.0248 moles of an organic compound having a molar mass of 120 grams per mole is completely burned in an excess of oxygen and all the heat is transferred into 785 grams of water. The original temperature of the water is 18.7°C and its final temperature is 34.9°C. What is the enthalpy change for the combustion of 1.00 mole of this compound?

11. How much heat must be removed to lower the temperature of 355 grams of water from 53.0°C to 18.0°C?

12. Imagine a container partially filled with water in which ice is floating. What would happen to the water level if all the ice melted? Does this more closely correspond to melting ice at the North Pole or the South Pole?

13. Table 7.2 shows that the vapor pressure of liquid water at −10°C is 2.15 torr and of ice at this same temperature is 1.95 torr. Water at this temperature is said to be "super cooled" and, sometimes (if everything is very clean), can be kept liquid at this temperature below its freezing point. What would happen if a container of ice and a container of supercooled liquid water were put side-by-side in an insulated container at −10°C?

14. A runner sweats 2.00 L in 1 hour. How much energy is needed to evaporate the sweat from the skin?

15. Which molecule is more water soluble, 1-decanol or methanol? Why?

16. Will 5 g n-heptyl alcohol completely dissolve in 5 L H₂O?

17. The solubility of n-heptylalcohol or 1-heptanol (C₇H₁₅OH) in water is 0.9 g/L. What is the molarity of a saturated solution of heptanol in water?

18. Draw a sketch of how water might interact via hydrogen bonds with:
    a. an ammonia molecule
    b. a methanol molecule
    c. a hexanol molecule
    d. a glucose molecule

19. Which of these compounds will form hydrogen bonds to water?
    a. methane
    b. ethylene glycol
    c. carbon tetrachloride
    d. ammonia

20. The solubility of AgCl is 1.3×10^{-5} M. How many grams of AgCl will dissolve in 1.0 L of water? Why do you suppose AgCl is so much less soluble than NaCl?

21. Caffeine (C₈H₁₀N₄O₂, molar mass = 194.19) has a solubility of 455 g/L at 65°C and of 13.5 g/L at 16°C. How much water will be needed to dissolve 155 g of caffeine at 65°C? How much of this will come out of solution when the hot, saturated, caffeine solution is cooled to 16°C?

22. What causes "bathtub ring" when the tub is emptied of water after you take a bath?

23. When you try to rinse vegetable oil off a spoon with water, the spoon does not come clean. Why does the addition of soap clean the spoon?

24. Define distillation, saponification, evaporation, sublimation, and precipitation.

25. Why can the calcium in hard water easily be replaced by sodium from a water softener?

26. List some good and bad attributes of both hard and soft water.

27. Why is ozone a better oxidant for water purification than chlorine?

28. Water hardness is measured in terms of the number of milligrams of calcium carbonate dissolved in 1 L of water. What would be the mass of the calcium carbonate that would remain if 500 mL of water having a hardness of 300 mg/L were allowed to evaporate? How many liters of this water would have to evaporate in order to deposit 1 g of calcium carbonate? (Assume mL H₂O = mL solution.)

29. When hard water is heated, calcium bicarbonate (Ca(HCO₃)₂), which is readily soluble in water, becomes calcium carbonate, which is very insoluble. Write the chemical equation for this process.

30. List the four colligative properties of solutions.

31. In a lab experiment, you are instructed to bring 0.0500 kg water to a boil. However, you notice that the water does not boil until 100.2°C. You then remember that you didn't completely clean your flask before beginning the experiment! How many moles of a nonelectrolyte impurity are dissolved in your water?

32. For a summer treat, you make popsicles by freezing Kool-Aid. If one popsicle contains 0.100 L water and 5.00 g glucose (assume the density of water is 1.00g/mL), what is the freezing point for the popsicle?

33. A solution of sugar in water freezes at −5.0°C. At what temperature will this solution boil?

34. Which temperature is the higher of each of these pairs? Select correct letter. If you believe both are the same temperature, write "equal." (The letter "m" means molal.)
    i. a. boiling point of pure water
       b. boiling point of 1.0 m sugar solution
    ii. a. freezing point of 3 m sugar
        b. freezing point of 3 m alcohol
    iii. a. boiling point of 1 m sugar
         b. boiling point of 2 m sugar

35. Which temperature is the lower of each of these pairs? Select correct letter. If you believe both are the same temperature, write "equal." (The letter "m" means molal.)
    i. a. boiling point of pure water
       b. boiling point of pure alcohol
    ii. a. freezing point of 3 m sugar
        b. freezing point of 3 m salt
    iii. a. boiling point of 1 m CaCl₂
         b. boiling point of 1 m NaCl

36. Which temperature is the higher of each of these pairs? Select correct letter. If you believe both are the same temperature, write "equal." (The letter "*m*" means molal.)
   - i. a. freezing point of pure water   b. freezing point of 3 *m* alcohol
   - ii. a. freezing point of 1.0 *m* sugar solution   b. boiling point of 1 *m* sugar
   - iii. a. freezing point of 3 *m* sugar   b. boiling point of 2 *m* sugar

37. A solution of ethylene glycol ($C_2H_6O_2$) in water is found to have a freezing point of $-10°C$. What is the concentration of the solution?

38. A solution of ethylene glycol ($C_2H_6O_2$) in water is found to have a boiling point of $115°C$. What is the concentration of the solution?

39. Calculate the fraction of molecules which are glycerine in an aqueous glycerine solution when the percent by mass of glycerine ($C_3H_8O_3$) is 50%.

40. The freezing point and density of various aqueous glycerine solutions are as follows:

Glycerine by Mass (%)	Density (g/cm³)	Freezing Point (°C)
10	1.024	−1.6
20	1.049	−4.8
30	1.075	−9.5
40	1.101	−15.4
50	1.128	−23.0
60	1.156	−34.7
70	1.183	−38.9
80	1.211	−20.3
90	1.238	−1.6
100	1.264	+17.0

Plot these freezing points versus the concentration of the glycerine. Compare this graph with the predicted line based on freezing point depression calculations. Explain the differences.

41. Use the data in Exercise 40 to calculate the freezing point of a solution in which half the molecules are water and half are glycerine ($C_3H_8O_3$).

42. What would be the freezing point of a saturated solution of 1-heptanol in water?

43. Calculate the molality of a solution made by adding 500 g of sucrose ($C_{12}H_{22}O_{11}$) to 500 g of water. The resulting solution has a density of 1.2296 g/cm³.

44. How much ice originally at $-20°C$ must be added to a glass of iced tea (600 mL) to bring its temperature from $15.0°C$ to $0°C$ while melting none of the ice?

45. What mass of ethylene glycol ($C_2H_6O_2$) is needed to keep 10.0 L of water from freezing at temperatures above $-20°C$?

46. Blood plasma is a 0.15 molal aqueous solution of NaCl and other solutes. How many grams of NaCl are needed to make 7.0 kilograms of blood plasma?

47. An aqueous sugar solution contains 1.00 L of water and 50 g of sugar ($C_{12}H_{22}O_{11}$). Calculate its boiling point. Suppose that this solution is distilled until 90% of the water has been removed. Calculate the boiling point of the new solution.

48. The solubility of NaCl is 35.7 g/100 mL of water. What is the molality of a saturated NaCl solution?

49. The solubility of atmospheric oxygen in water at $10°C$ and $20°C$ is 0.011 and 0.0090 g/L, respectively. How much oxygen will be removed from a lake having a volume of 60 billion L by raising its temperature from 10 to $20°C$? If the lake has an average depth of 10 meters, what is its surface area?

50. Seawater is often said to be a solution of NaCl in water. Table 7.5 shows that many other ions are there as well. Determine the amount in grams of salt (NaCl) dissolved in the oceans (refer to Table 7.1 for volume of ocean water).

51. Each drop of rain usually forms around a solid particle of dust. If each dust particle weighs a microgram, how much solid material comes to the earth in a 2.00-inch rainstorm over a 1000-km² area? It is usually estimated that there are 20 drops in a milliliter.

52. How much heat is necessary to evaporate the water which came down as rain in Exercise 51? (Assume ΔH_{vap} at $100°C$.)

53. The energy falling on one square centimeter (cm²) of the Earth's surface is 8.1 J/min. (Actually, this is the amount falling on the atmosphere where the Sun is directly overhead.) How much water can be evaporated in 1 hour in one square kilometer if all this energy reaches the ocean surface and is used to evaporate water? Assume an ocean temperature of $10°C$.

54. Some of the most accurate chemical measurements of all time are based on determining how much NaCl solution is needed to exactly precipitate the silver ion in another solution. Suppose that a silver alloy weighing 0.4785 g is dissolved in nitric acid and the resulting solution requires 31.74 mL of 0.02548 *M* NaCl to precipitate all the silver as AgCl. What is the percent silver in the original alloy?

55. A compound called ethylenediaminetetraacetic acid (EDTA) reacts with calcium ion to form what is called a "complex ion."

$$Ca^{2+} + EDTA^{4-} \rightarrow CaEDTA^{2-}$$

What is the hardness of a water sample if 200 mL of water require 32.7 mL of 0.0572 *M* EDTA to tie up the calcium ion completely? (Review Exercise 28 for information on how water hardness is measured.)

56. How deep would snow be if Avogadro's number of snowflakes fell on the city in which your university is located?

57. It has been estimated that 1×10^{25} snow crystals form in the Earth's atmosphere each year. How many moles of water would this be if each crystal has a mass of 0.1 mg? What fraction of the Earth's water does this represent?

58. How many grams of calcium hydroxide are necessary to precipitate aluminum hydroxide from a kilogram of aluminum sulfate in water purification by coagulation?

59. Calcium hydroxide is formed by heating calcium carbonate and adding water. How much CO_2 is formed in making the $Ca(OH)_2$ needed in Exercise 58? Remember to consider in your answer the CO_2 formed from burning fuel to heat the $CaCO_3$. Also, 177.9 kJ are needed to react one mole of $CaCO_3$ (see Exercise 5.5 for information about the energy of fuels).

food for thought

60. Read the poem *The Rime of the Ancient Mariner* by Coleridge, which was quoted in the introduction to this chapter. List several other features of water that are mentioned in the poem.

61. Examine Bartlett's *Familiar Quotations* under the index listing of water. Notice how many familiar quotations are based on water. Find one that you especially like and read the original source. Write a brief essay on the subject you have chosen.

62. How many grams of salt would be needed in an ion exchanger to remove all the calcium in enough water to take a bath if the hardness were 300 mg/L? First you will need to estimate volume of water needed to take a bath. Most bathtubs are about 48×22 inches and are filled to a depth of 5 inches. This is 5280 cubic inches of water. One cubic inch contains $(2.54 \text{ cm})^3 = 16.4 \text{ cm}^3$ so this is $8.6 \times 10^4 \text{ cm}^3$ of water. A liter is 1000 cm^3 of water so the bath holds 86 L of water. (Read Exercise 28 for a reminder of how water hardness is measured.)

63. We want to know what metal was used to make a metal statuette. The statuette has a mass of 5425 g. When it is put into water in a graduated cylinder, the water level goes up by 650 mL. The water is heated to boiling and is boiled for 30 minutes while the statue is immersed in the water. The statue is removed and placed into 800.0 mL of water originally at 24.8°C. The water temperature went up to 37.9°C. How can this information help to determine the metal from which the statue was crafted? What metal do you suspect?

64. Keep track of how much water you drink in one day. How does it compare to the per capita average for your country, which is available on the textbook's website? Why is it more or less than the per capita amount? How does this compare to water use in other countries? See Chapter 17 for additional information.

readings

1. Cholera. *Discover Magazine*. February 1992.
2. Freiser, H. 1972. Polywater and analytical chemistry— a lesson for the future. *J. Chem. Ed.,* 49: 445.
3. Powell, B. 1971. Anomalous water—fact or fiction. *J. Chem. Ed.* 48: 663–667.
4. Lide, D.R. *Handbook of Chemistry and Physics.* Boca Raton, Fla: CRC Press (any edition).
5. Water. *National Geographic.* November 1993.

websites

www.mhhe.com/kelter　　The World of Choices website contains activities and exercises including links to websites of: Honeywell International, for information on antifreeze; The United States Department of Labor; and much more!

Acids and Bases

CHAPTER

8

Plop Plop, Fizz, Fizz, Oh, what a relief it is.

—famous Alka-Seltzer® advertisement in the 1970s

Most of us are familiar with the type of stomach pain known as "acid indigestion." It might be brought on by a chemistry test that requires just a bit more study time. Perhaps it comes along with the deadline for a term paper, or maybe just before a first date. Unlike many other kinds of pain, we can often get instant relief by taking an indigestion remedy such as Alka-Seltzer®. The pain is a direct effect of a class of chemicals called acids. The instant relief is brought about by chemicals called bases that can counteract or "neutralize" acids. Acids and bases are two of the most significant categories of chemicals. They are crucial to the way our bodies work, the chemistry of the environment, the manufacturing processes used by industry, and many activities of everyday life. As an introductory example, we will explore the process of digestion to find out where acids fit in.

Digestion is the process by which food molecules are broken down into substances that can be absorbed into the blood and used as nutrients. We have already discussed two of the most important types of substance in food that must be digested, namely fats and carbohydrates (such as starch). The third main class of substances in food that need to be digested by the body is the **proteins.** Proteins are polymers that perform many important tasks in living things, including acting as enzymes. Muscle tissue (which we eat as meat and the flesh of fish) is mostly protein.

The digestion of food begins in the mouth where the enzyme α-amylase in saliva begins to break down starch into smaller carbohydrates and glucose. Chewing food helps in this process by providing a larger surface area for the saliva to work on. Digestion of starch continues in the esophagus (see Figure 8.1), stops in the stomach, and starts up once again in the small intestine where other enzymes finish converting starch into glucose, which is absorbed into the bloodstream. The digestion of proteins begins in the stomach, where enzymes chop the proteins up into smaller pieces, which are degraded further by enzymes in the small intestine. The digestion of fats does not begin until they reach the small intestine.

Why should the digestion of starch stop when the starch arrives in the stomach, the digestion of proteins begin only when they arrive in the stomach, and no digestion of fats occur in the stomach? The answer depends on the chemical substances found within the stomach. The

Most of us have used antacids at one time or another. They all employ the acid/base balance concept of chemistry to settle an upset stomach. The chemical composition of these products is based on the chemical principles discussed in this chapter. Next time you're in the drugstore, read the labels of some of these products and see if they make more sense to you once you know the underlying science.

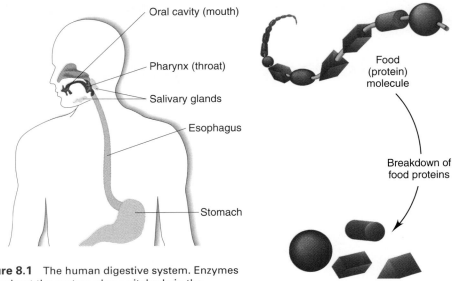

Oral cavity (mouth)

Pharynx (throat)

Salivary glands

Esophagus

Stomach

Food (protein) molecule

Breakdown of food proteins

Amino acid molecules

Figure 8.1 The human digestive system. Enzymes throughout the system play a vital role in the breakdown of proteins, carbohydrates, and fats.

$Zn(s) + 2HCl(aq) \longrightarrow ZnCl_2(aq) + H_2(g)$

Figure 8.2 Hydrochloric acid is a vital part of the gastric juice within the stomach, yet it is reactive enough to dissolve a strip of the metal zinc.

main component of the digestive fluid ("gastric juice") in the stomach is **hydrochloric acid** (HCl), with almost 2 L secreted each day by glands in the stomach. Hydrochloric acid is a necessary part of our delicate digestive system, yet it can dissolve a strip of zinc, as shown in Figure 8.2. Too much acid in the stomach gives us acid indigestion.

The presence of hydrochloric acid in the stomach causes the enzyme α-amylase to become inactive, stopping the digestion of starch until the food passes out of the stomach and into the small intestine. Hydrochloric acid in the stomach also activates enzymes that are able to break down protein molecules.

Hydrochloric acid is just one of countless chemicals known as "acids" that are involved in many of the most significant chemical reactions on Earth. Some acids are in our bodies where they allow us to live and stay healthy. Some are in the food and drinks we consume. Many acids are used in industry, in chemical reactions involved in producing all sorts of manufactured goods and materials. Some acids are such a significant threat to the environment that a form of pollution—"acid rain"—has been named after them.

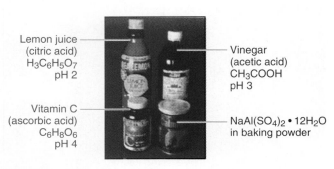

Lemon juice (citric acid) $H_3C_6H_5O_7$ pH 2

Vitamin C (ascorbic acid) $C_6H_8O_6$ pH 4

Vinegar (acetic acid) CH_3COOH pH 3

$NaAl(SO_4)_2 \cdot 12H_2O$ in baking powder

Common acids.

The sour taste of lemons and limes is a tip off to their acidic nature.

8.1 | What Is an Acid?

What comes to your mind when you hear the word "acid"? Some juices, such as lemon, orange, or grapefruit juice, are acids. These are things that make your mouth pucker. They, like other liquids that contain acids, taste sour. It is not safe to taste more dangerous acids, like sulfuric acid (H_2SO_4, found in many car batteries), but such acids are sour tasting as well. A list of the properties we often associate with acids and their chemical opposites, the bases, is given in Table 8.1.

Many common substances contain acids. Table 8.2 gives just a small sample. Note that the substances listed vary from very damaging (battery acid) to benign (such as cheese). As you look at the table, please recognize the depth of the involvement of acids in our lives.

"Fair enough," you may say, "but what actually *is* an acid? How are *strong* and *weak* acids different? How do we determine their strengths? How do we tell if something is an acid? How do acids react? And what about *acid* rain? Is it something to be concerned about?" Let's begin answering these questions by returning to the acid in our stomach, hydrochloric acid.

The Nature of Acids

What happens when hydrochloric acid and water mix? We can figure this out by remembering our discussions on bond polarity (Chapters 6 and 7). The HCl molecule is quite polar, so the partially positive hydrogen atom of the molecule is attracted to the partially negative oxygen atom of a water molecule with sufficient strength for the covalent bond between the H and Cl atoms to break. The resulting hydrogen ion (H^+) bonds to a water molecule to form H_3O^+, a **hydronium ion,** as shown in Figure 8.3. When the H—Cl bond breaks, both of the electrons that formed the bond remain with the chlorine atom, which becomes a chloride ion. This chloride ion can interact with the partially positive hydrogen atoms of other water molecules, as

Table 8.1

Properties of Acids and Bases

Acids	Bases
Taste sour	Taste bitter
Change the color of indicators (blue litmus paper turns red)	Change the color of indicators (red litmus paper turns blue)
Donate protons during acid-base reaction	Accept protons during acid-base reaction
React with some metals to produce hydrogen gas	Form insoluble hydroxides with many metal ions

Table 8.2

Common Acids

Substance	Contains This Acid
Battery acid	Sulfuric acid
Acid rain	Sulfuric, nitric acid
Stomach acid	Hydrochloric acid
Lemon juice	Citric, ascorbic acid
Wines	Tartaric acid
Vinegar	Acetic acid
Oranges	Citric, ascorbic acid
Underarm odor	3-methyl-2-hexenoic acid
Cheese	Lactic acid

Figure 8.3 Hydrochloric acid (HCl) interacts with water to form hydronium ions and chloride ions.

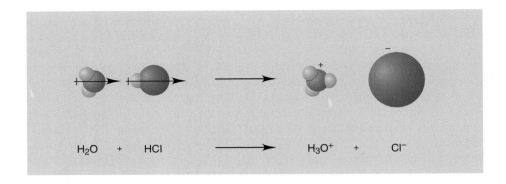

$$H_2O \quad + \quad HCl \quad \longrightarrow \quad H_3O^+ \quad + \quad Cl^-$$

shown in Figure 8.3. What happens to both the hydrogen and chlorine atoms of HCl when added to water is represented by the following equation:

$$H_2O(l) + HCl(aq) \rightarrow H_3O^+(aq) + Cl^-(aq)$$

hydrochloric acid   hydronium ion

Acids can be defined as:

- Hydrogen ion (H$^+$) donors
- Proton donors
- Substances that increase the hydronium ion (H$_3$O$^+$) concentration of an aqueous solution

By looking at what happens when hydrochloric acid interacts with water, we have uncovered the essential general chemical characteristic of acids. This is that *acids are substances that increase the hydronium ion (H$_3$O$^+$) concentration of an aqueous solution (a solution in which water is the solvent).* Since the acid molecule gives ("donates") a hydrogen ion to a water molecule to form the hydronium ion, another general definition of an acid is that *an acid is a hydrogen ion donor.* Hydrogen ions are simply protons, so yet another definition is that *an acid is a proton donor.*

Strong and Weak Acids

Hydrochloric acid is one example of a **strong acid.** Any time a strong acid comes in contact with water, it will *ionize completely.* So all HCl molecules entering the aqueous solution will immediately ionize to H$^+$ and Cl$^-$ ions. The chloride ions will be surrounded by water molecules and are not involved in the acid chemistry, so we will not consider them further. All of the hydrogen ions become linked to water molecules to form hydronium ions (H$_3$O$^+$), but chemists often continue to represent them simply as H$^+$. Remember that H$^+$, when used to represent hydrogen ions dissolved in water, really means H$_3$O$^+$. Many acids can release more than one hydrogen ion per acid molecule. Sulfuric acid (H$_2$SO$_4$) releases two hydrogen ions per molecule, 100% of the first H$^+$, and much less of the second H$^+$.

Actually, several water molecules, not just one, interact with the hydrogen ion. Therefore, the substances present in acid solutions are more like H$^+$(H$_2$O)$_4$ or H$_9$O$_4^+$. No one ever refers to such ions in introductory texts. We will speak of H$_3$O$^+$ or H$^+$.

$$H_2SO_4(aq) + 2H_2O(l) \rightarrow 2H_3O^+(aq) + SO_4^{2-}(aq)$$

Weak acids are those that only *partially ionize* in solution (for reasons considered later) so only a proportion (often a very small proportion) of the acid molecules dissociate into hydrogen ions and the corresponding negative ions. Acetic acid, CH$_3$COOH, is the acid that in diluted form is known as vinegar. Roughly 0.5% of the acetic acid molecules dissociate in solution. This means that for every 1000 acetic acid molecules added to water, on average, 995 remain as acetic acid molecules (CH$_3$COOH). Only 5 per 1000 will dissociate to give a hydronium ion and an acetate ion (see Figure 8.4).

We represent this partial dissociation by using a double arrow in the reaction equation, shown here.

$$CH_3COOH(aq) + H_2O(l) \rightleftharpoons CH_3COO^-(aq) + H_3O^+(aq)$$

acetic acid   acetate ion   hydronium ion

Calculating the amount of each ingredient in a solution is a critical step in the creation of many products. Just think of all the medicines, cleaning compounds, and cosmetics you rely on.

The arrows indicate what is known as a **chemical equilibrium.** The reaction shown is actually proceeding *in both directions at the same time, and at the same rate in each direction.* In other words, acetic acid molecules are continually dissociating as fast as free ions are re-associating to form acetic acid molecules (Figure 8.5). In a bottle of vinegar, for example, only about one-half percent of the acetic acid molecules are dissociated *at any one time,* but that will be a different one-half percent at different times.

Figure 8.4 Acetic acid is a weak acid. It dissociates very little, perhaps 1 to 5 molecules per 1000, depending on its initial concentration.

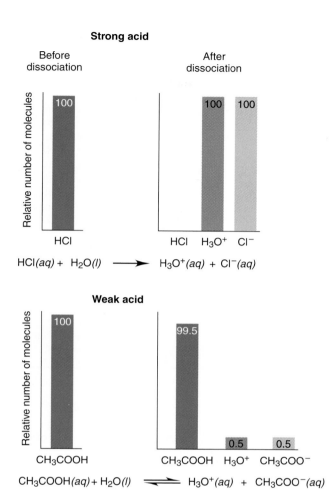

Strong acid

Before dissociation   After dissociation

Relative number of molecules

100

100   100

HCl   HCl H_3O^+ Cl^-

$$HCl(aq) + H_2O(l) \longrightarrow H_3O^+(aq) + Cl^-(aq)$$

Weak acid

Relative number of molecules

100   99.5

0.5   0.5

CH_3COOH CH_3COOH H_3O^+ CH_3COO^-

$$CH_3COOH(aq) + H_2O(l) \rightleftharpoons H_3O^+(aq) + CH_3COO^-(aq)$$

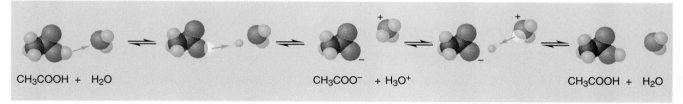

$CH_3COOH + H_2O$ $CH_3COO^- + H_3O^+$ $CH_3COOH + H_2O$

Figure 8.5 An individual acetic acid molecule can repeatedly lose and reacquire its hydrogen ion.

The actual proportion of acetic acid molecules which ionize (dissociate) depends on the concentration (molarity) of the acetic acid solution. The higher the concentration, the lower the percent ionized. Table 8.3 gives examples of some acids and their uses.

8.2   Molarity and the Acid Concentration

"Strong" and "weak" are terms that refer to the *extent of dissociation* of an acid. This is very different from "concentrated" and "dilute," which refer to the **molarity** of the acid, a measure of its concentration.

Molarity is one of the most important measures of concentration in the chemist's toolbox and we use it to describe the concentration of any solution, not just acids. It is defined as the number of moles of solute per liter of solution.

Table 8.3

Selected Acids

	Acid Name	Formula	Common Use
Strong	Sulfuric acid	H_2SO_4	Fertilizer manufacture
	Nitric acid	HNO_3	Fertilizer manufacture
	Hydrochloric acid	HCl	Manufacture of chloride-containing materials
Weak	Acetic acid	CH_3COOH	Vinegar
	Lauric acid	$CH_3(CH_2)_{10}COOH$	Typical fatty acid detergent when converted to sodium lauryl sulfate in toothpaste
	Tartaric acid		Soft drinks, baking powder
	Phosphoric acid	H_3PO_4	Cleaning products, fertilizer manufacture, plastics manufacture, food additive

$$\text{COOH}$$
$$|$$
$$\text{H—C—OH}$$
$$|$$
$$\text{HO—C—H}$$
$$|$$
$$\text{COOH}$$

$$\text{Molarity } (M) = \frac{\text{moles of solute}}{\text{liter of solution}} = \text{mol/L}$$

A 2×10^{-6} M HCl solution has 2×10^{-6} moles of HCl for every liter of solution. Three liters of this solution has 6×10^{-6} moles of HCl in it. Here's how:

$$\frac{2 \times 10^{-6} \text{ mol HCl}}{\text{liter of HCl solution}} \times 3 \text{ liters of HCl solution} = 6 \times 10^{-6} \text{ mol HCl}$$

Exercises 8.1 and 8.2 show different ways we calculate and use molarity of acids.

exercise 8.1

Molarity of Acids

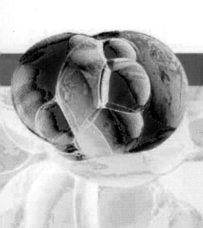

Problem

What is the molarity of a solution of 30.0 grams of acetic acid, CH_3COOH (molar mass = 60.0 g/mol), dissolved in enough water to make 250 mL of solution?

Solution:

We need to find moles of solute (acetic acid) and divide it by the solution volume in liters (0.250 liters) to get moles per liter.

$$30.0 \text{ g CH}_3\text{COOH} \times \frac{1 \text{ mol CH}_3\text{COOH}}{60.0 \text{ g CH}_3\text{COOH}} = 0.500 \text{ mol CH}_3\text{COOH}$$

$$M = \frac{0.500 \text{ mol CH}_3\text{COOH}}{0.250 \text{ L CH}_3\text{COOH solution}} = 2.00 \text{ } M \text{ CH}_3\text{COOH}$$

Food for thought: Is this solution more concentrated or dilute than the 2×10^{-6} M HCl solution?

exercise 8.2

Moles in the Solution

Problem

How many moles of HCl are present in 35.65 mL of 0.1723 M HCl solution?

Solution

The number of moles of HCl is the molarity multiplied by the volume in liters

$$\text{moles HCl} = \frac{\text{moles of HCl}}{\text{liter of HCl solution}} \times \text{volume HCl solution}$$

$$= \frac{0.1723 \text{ moles of HCl}}{1 \text{ liter of HCl solution}} \times 0.03565 \text{ liter HCl solution}$$

$$= 6.142 \times 10^{-3} \text{ mol HCl}$$

Given our understanding of molarity and of acid strength, we can say that a $1 \times 10^{-6}\ M$ HCl solution is a dilute solution of a strong acid, while a 2 M acetic acid solution is a relatively concentrated solution of a weak acid. So we need to use our words carefully when referring, for example, to the strength of acids, remembering not to talk of "strong" acid solutions when we mean "concentrated" acid solutions.

Since we know that in aqueous (water-based) chemistry, water is always present, it is general practice to *neglect* the presence of water by omitting it from acid dissociation equations. We know it is there. We just choose to ignore it in writing equations because it complicates equations that are already complicated enough. In this simplified form, the dissociations of hydrochloric acid (a strong acid) and acetic acid (a weak acid) can be written as:

$$HCl(aq) \rightarrow H^+(aq) + Cl^-(aq)$$
$$CH_3COOH(aq) \rightleftharpoons CH_3COO^-(aq) + H^+(aq)$$

We now know what an acid is. We know some of the characteristic properties of acids. We know some examples of acids. We know that some are strong, some are weak, and we know what these terms mean with respect to the extent of dissociation of the acid in water. Let's consider *why* some acids are strong and others weak.

What Determines Whether an Acid Will Be Strong or Weak?

A recurring theme in chemistry, and one we dealt with extensively in Chapter 5, is that spontaneous chemical reactions occur because, in the prevailing conditions, the products are more stable than the reactants. This statement holds true with acids. The chloride ion, Cl^-, and the hydrogen ion, H^+, are more stable in water than HCl, so hydrochloric acid dissociates completely when it is added to water. On the other hand, acetic acid, CH_3COOH, is more stable in water than its anion, CH_3COO^-, so acetic acid dissociates only a little. It is a weak acid. The relative stability of different chemical species in water is determined by their chemical structures in ways that are too complicated to discuss in detail here.

An Introduction to the Implications of Acidity

Try the following activity. Take a raw egg, still in the shell, and let it soak overnight in a cup of vinegar, which contains the moderately weak acetic acid. When you look at it the next day, the shell will be gone and the outer part of the egg white will, in fact, be white, as shown in Figure 8.6. The eggshell is primarily made up of calcium carbonate, $CaCO_3$, and some proteins which help hold the shell together. Acetic acid reacts with the calcium carbonate to release carbon dioxide gas (do you see evidence of that in the cup?)

Figure 8.6 The reaction of vinegar (acetic acid solution) with an eggshell (calcium carbonate) converts the eggshell into soluble products.

$$CaCO_3(s) + 2CH_3COOH(aq) \rightarrow CO_2(g) + H_2O(l) + Ca(CH_3COO)_2(aq)$$

$$CaCO_3(s) + 2CH_3COOH(aq) \rightarrow Ca^{2+}(aq) + CO_2(g) + H_2O(l) + 2CH_3COO^-(aq)$$

In Chapter 9, we will discuss a very similar reaction that dissolves marble and limestone structures such as buildings and statues, instead of eggshells.

Acids can have both intended and unintended effects on all kinds of things. The Case in Point on Rotting Paper shows just how careful we must be, even with books!

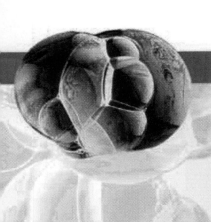

exercise 8.3

Acid Dissociation

Problem

Write the products of these dissociation processes—you may use the short form (neglecting water).

(a) Nitric acid, HNO_3, a strong acid.
(b) Malic acid, $C_4H_5O_3COOH$, a weak acid found in such fruits as watermelon and apples.
(c) Ascorbic acid (Vitamin C), $C_6H_7O_6H$, a weak acid.
(d) Hydrofluoric acid, HF, a weak acid that etches glass.
(e) Bicarbonate ion, HCO_3^-, an **amphiprotic** substance, which means that it can act as an acid (by donating a proton) *or* a base (by accepting a proton). Assume that it acts as a weak acid here.

Solution

(a) $HNO_3(aq) \rightarrow H^+(aq) + NO_3^-(aq)$
(b) $C_4H_5O_3COOH(aq) \rightleftharpoons C_4H_5O_3COO^-(aq) + H^+(aq)$
(c) $C_6H_7O_6H(aq) \rightleftharpoons C_6H_7O_6^-(aq) + H^+(aq)$
(d) $HF(aq) \rightleftharpoons H^+(aq) + F^-(aq)$
(e) $HCO_3^-(aq) \rightleftharpoons H^+(aq) + CO_3^{2-}(aq)$

We see that every acid here produces hydrogen ion, which increases the total hydrogen ion content of the solution. Keep in mind that acids can be strong or weak. The acid in part a, HNO_3, is strong. It dissociates completely. The hydrogen fluoride (part d) is a weak acid and does not dissociate very much.

8.3 | Bases—The Other Half of the Story

Each of the processes we have discussed so far has involved a hydrogen ion (H$^+$) and an anion of some sort. Hydrochloric acid (HCl) releases H$^+$ and Cl$^-$ ions. These are a matched set. The HCl is the acid. It gives up a hydrogen ion, leaving Cl$^-$. The Cl$^-$ ion is known as the **conjugate base** of hydrochloric acid. A **base** is a substance that can accept a hydrogen ion. HCl and Cl$^-$ are known as a **conjugate pair** or **conjugate acid-base pair** (Figure 8.7) whose members are an acid and the corresponding base generated when the acid loses a hydrogen ion. A wide variety of bases exist in nature and are exploited by industry. In fact, three of the top industrial chemicals are bases, as listed in Table 8.4.

Acids and bases are interconvertible. If the chloride ion we already discussed accepts a hydrogen ion, it will become HCl, an acid. If the HCl loses a hydrogen ion it generates Cl$^-$, a base. Acids and bases are a matched set. You don't have one without the other.

When an acid is relatively strong, its conjugate base is weak. When an acid is weak, its conjugate base is relatively strong. For example, HCl is a strong acid. In aqueous solution, it tends to ionize to form the more energetically stable H$^+$ and Cl$^-$ ions. The Cl$^-$ ion has essentially no tendency to automatically reassociate with H$^+$ ions to reform the less stable HCl in the presence of water. So Cl$^-$ is an extremely weak base, since the essential characteristic of a strong base is a strong tendency to combine with hydrogen ions.

The conjugate base of acetic acid (CH$_3$COOH) is the acetate ion (CH$_3$COO$^-$). The acetate ion is a much stronger base than the Cl$^-$ ion because it has a substantial tendency to recombine with hydrogen ions to form acetic acid molecules, as we have already discussed.

Lye is the common name for the solid form of the base, sodium hydroxide. This is used, among other things, for clearing blocked drains. It is a remarkably caustic substance. When added to water, NaOH completely dissociates into sodium and hydroxide ions:

$$NaOH(s) \rightarrow \underset{\text{sodium ion}}{Na^+(aq)} + \underset{\text{hydroxide ion}}{OH^-(aq)}$$

The hydroxide ion is the strongest possible base in water.

Just as HCl and Cl$^-$ make a conjugate acid-base pair, so water and the hydroxide ion make up another conjugate acid-base pair. Since the hydroxide ion is an extremely strong base, water is an extremely weak conjugate acid.

Water is also capable of acting as a base in the H$_3$O$^+$/H$_2$O conjugate acid-base pair, by accepting a hydrogen ion to form H$_3$O$^+$. So water molecules are capable of both accepting hydrogen ions (to form H$_3$O$^+$) and donating hydrogen ions (to form OH$^-$). Substances which can act as acids *or* bases in this way are called **amphiprotic**. (Take a look at Exercise 8.5 for an example of bicarbonate ions.)

The most common strong bases are hydroxides of Group 1A metals, such as sodium hydroxide (NaOH) and potassium hydroxide (KOH). Technically, the hydroxide ions of these compounds are the bases, since they are the chemical species that combine with hydrogen ions, but the compounds as a whole are also known as bases. We have defined bases as substances that will combine with hydrogen ions. Many bases do this by generating or releasing hydroxide ions (OH$^-$). Metal hydroxide bases that dissolve in water, releasing their hydroxide ions, are called **alkalis.**

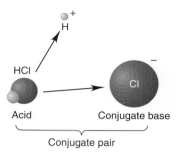

Figure 8.7 The HCl molecule and Cl$^-$ ion (the "before and after" of acid dissociation) are a conjugate acid-base pair. The stronger the acid, the weaker its conjugate base.

Bases can be defined as:

- Hydrogen ion (H$^+$) acceptors
- Proton acceptors
- Substances that increase the hydroxide ion (OH$^-$) concentration of an aqueous solution

Metal hydroxide bases that dissolve in water are known as **alkalis**. Alkalis cause an increase in the concentration of hydroxide ions in any solution to which they are added.

Table 8.4

Industrially Important Bases

Base (Formula)	Commercial Use	10^9 kg Produced (1999)
Calcium oxide (CaO)	Cement, paper	22.60
Ammonia (NH$_3$)	Fertilizers, plastics	18.96
Sodium hydroxide (NaOH)	Industrial synthesis	11.39

Table 8.5

Weak Bases

Base Name	Formula	Common Use
Ammonia	NH_3	Fertilizer, cleaning solutions
Hexamethylenediamine	$H_2N—(CH_2)_6—NH_2$	Nylon synthesis
Pyridine	C_5H_5N	Industrial solvent

Figure 8.8 Ammonia combines with water to form the ammonium and hydroxide ions. Ammonia is a weak base, so the equilibrium in this reaction lies far to the left.

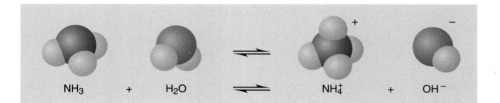

$$NH_3 \quad + \quad H_2O \quad \rightleftharpoons \quad NH_4^+ \quad + \quad OH^-$$

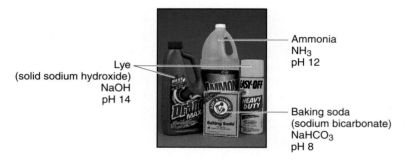

Lye
(solid sodium hydroxide)
NaOH
pH 14

Ammonia
NH_3
pH 12

Baking soda
(sodium bicarbonate)
$NaHCO_3$
pH 8

Although we are very familiar with foods as examples of acids, many common substances are bases.

Sodium hydroxide is a common strong base whose sodium and hydroxide ions become completely dissociated in water, as we have already discussed. Ammonia (NH_3) is a common weak base that generates hydroxide ions in a slightly more indirect way. The ammonia molecule does not contain any hydroxide ions, so it cannot release them into solution by simple dissociation. It can, however, react with water molecules to generate hydroxide ions:

$$\underset{\text{ammonia}}{NH_3(aq)} + H_2O(l) \rightleftharpoons \underset{\text{ammonium ion}}{NH_4^+(aq)} + OH^-(aq)$$

As with weak acids, the use of the double arrow tells you that the reaction settles into an equilibrium position, in which the reactants and products are all present at the same time, being interconverted at equal rates. In this particular equilibrium system, very little of the ammonia actually becomes converted into ammonium ions (Figure 8.8), which explains why ammonia is a weak base. Notice that the ammonia molecule clearly fits the definition of a base, because it can accept a proton and also generates hydroxide ions in solution. Table 8.5 gives some examples of weak bases.

8.4 | Oxides as Acids and Bases

Our working definition of an acid is anything that causes the formation of hydrogen ions when the substance is added to water. A base is anything that causes the formation of hydroxide ions when added to water.

Metal Oxides as Bases

Calcium oxide, CaO (see Table 8.4), reacts with water to form calcium hydroxide, $Ca(OH)_2$.

$$CaO(s) + H_2O(l) \rightarrow Ca(OH)_2(aq)$$

Calcium hydroxide then ionizes to form calcium and hydroxide ions.

$$Ca(OH)_2(aq) \rightarrow Ca^{2+} + 2OH^-(aq)$$

CaO is therefore properly regarded as a base because it produces OH^-. *Group 1A and 2A metal oxides are all strong bases.*

The other side of the coin is that in general, *oxides of nonmetals form acidic solutions in water.* For example, sulfur trioxide, SO_3, reacts with water to form sulfuric acid. As we discuss in Chapter 9, this is an important reaction that contributes to acid rain.

$$SO_3(g) + H_2O(l) \rightarrow H_2SO_4(aq)$$

Since nonmetal oxides (they form acids) and Group 1A and 2A metal oxides (they form bases) lack the water necessary to make them contain hydrogen or hydroxide ions, they are called, respectively, **acid anhydrides** and **basic anhydrides.**

exercise 8.4

Oxides in Water

Problem

What compound is formed when each of these oxides is dissolved in water?
(a) CO_2 (b) K_2O (c) P_4O_{10} (d) BaO

Solution

(a) $CO_2(g) + H_2O(l) \rightarrow H_2CO_3(aq)$ **an acid**
(b) $K_2O(s) + H_2O(l) \rightarrow 2KOH(aq)$ **a strong base**
(c) $P_4O_{10}(s) + H_2O(l) \rightarrow 4H_3PO_4(aq)$ **an acid**
(d) $BaO(s) + H_2O(l) \rightarrow Ba(OH)_2(aq)$ **a strong base**

exercise 8.5

Reactions of Bases with Water

Problem

Write the equations of the reactions that describe what happens when each of these is added to water:

(a) Potassium hydroxide, KOH
(b) Calcium hydroxide, $Ca(OH)_2$
(c) Fluoride ion, F^-
(d) Acetate ion, CH_3COO^-
(e) Bicarbonate ion, HCO_3^-, the amphiprotic substance acting as a weak base in this reaction.

Solution

(a) $KOH(s) \rightarrow K^+(aq) + OH^-(aq)$
(b) $Ca(OH)_2(s) \rightarrow Ca^{2+}(aq) + 2OH^-(aq)$
(c) $F^-(aq) + H_2O(l) \rightleftharpoons HF(aq) + OH^-(aq)$
(d) $CH_3COO^-(aq) + H_2O(l) \rightleftharpoons CH_3COOH(aq) + OH^-(aq)$
(e) $HCO_3^-(aq) + H_2O(l) \rightleftharpoons H_2CO_3(aq) + OH^-(aq)$

Note that each of the substances in these reactions is part of a conjugate acid-base pair. As a follow-up exercise, you might wish to consider which substance among these pairs is strong and which is weak.

8.5 | The pH Scale

We now know that both acids and bases react with water to release or generate ions, and that strong acids or bases become completely ionized in water, while weak ones do not. We have not yet mentioned a very specific measure of the "acidity" or "basicity" of a solution, known as the "pH" of a solution. You may be familiar with the term pH, because shampoos and cosmetic products are often described as being "pH balanced" or "pH controlled." If you are interested in farming or gardening, you may have heard of the importance of the pH of the soil. If you have a fish tank, you sometimes have to adjust the pH so that the fish won't die. On a larger scale, pH control in lakes and streams is equally vital for life to survive there. We will now investigate the true meaning of the term pH.

The **p** in pH can be thought of as representing the first letter in the word *power* (as in exponential power) and is mathematical shorthand for **the negative logarithm of, or −log.** The letter "p" is used in this way in several numerical relationships, and should always be written in lower case, even if at the start of a sentence. The **H** in pH stands for **the concentration of hydrogen ions (in moles per liter).** Another shorthand way to indicate that we are referring to the concentration of a chemical is to write its symbol within square brackets. [H^+], for example, means "the concentration of hydrogen ions (in moles per liter)." So we can summarize the meaning of the term pH by writing:

$$pH = -\log [H^+]$$

i.e., **pH = the negative logarithm of the hydrogen ion concentration (in moles per liter)** see Figure 8.9.

To clarify what we mean by pH let's work through this example. Vinegar is an aqueous solution of acetic acid (CH_3COOH) in which the concentration of hydrogen ions is typically 0.0010 M (i.e., 1.0×10^{-3} M). What is the pH of such a solution? The log of $1.0 \times 10^{-3} = -3.00$. The pH of our hydrogen ion concentration is the *negative* log, so we just take the negative of −3.00:

$$-(-3.00) = +3.00 = pH\ 3.00$$

Use of the "log" (for "logarithm") button on your calculator will make it easy to compute the pH if you know the hydrogen ion concentration. Suppose you are to figure out the pH of a solution with [H^+] = 4.64×10^{-5} M. On many nonprogrammable calculators, the following steps will work.

1. Enter 4.64×10^{-5} into the calculator.
2. Press the "log" button. The display should read -4.3335.

Since pH $= -\log [H^+]$, and [H^+] is a negative number, pH = 4.333. Why not 4.3335? We determine the number of significant figures in the pH by counting the significant figures in [H^+], three in our example, and reporting that number of digits *after* the decimal point in the value of pH.

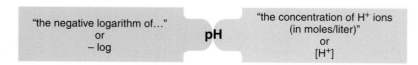

Figure 8.9 The p in pH means "the negative logarithm of." This method of describing hydrogen ion concentration conveniently allows dealing with an extremely great range of numbers from > 1 to < 0.00000000000001 moles per liter.

This means that $[H^+] = 4.3 \times 10^{-4}\,M$.

Just as we can find pH from the $[H^+]$, we can find the $[H^+]$ from pH. The pH $= -\log$ $[H^+]$. If we raise both sides of the equation to the power of 10,

$$10^{pH} = 10^{-\log[H^+]} \ or \ 10^{-pH} = [H^+]$$

This means that if the pH of a solution is 6.00, then the $[H^+]$ in that solution is $10^{-pH} = 10^{-6.00} = 1 \times 10^{-6}\,M$.

Calculators also help convert pH into $[H^+]$. Let's calculate the $[H^+]$ of a solution with a pH of 3.37 (two significant figures *after* the decimal point!). Do the following:

1. Enter 3.37 into your calculator.
2. Press the "+/−" key, so the value is −3.37.
3. Press the "2nd function" key, and then the "log" key. This is the same as pressing "10^x." Your display should read, 0.0004266.

exercise 8.6

Practice with pH

Problem

Calculate the pH of each of the following solutions, given their hydrogen ion concentrations:

(a) $1.0 \times 10^{-14}\,M$
(b) $1.0 \times 10^{-1}\,M$
(c) $1.0 \times 10^{-13}\,M$
(d) $7.53 \times 10^{-4}\,M$
(e) $6.24 \times 10^{-9}\,M$
(f) $0.200\,M$
(g) $2.45\,M$

Solution

(a) pH $= -\log (1.0 \times 10^{-14}) = -(-14.00) = 14.00$
(b) pH $= -\log (1.0 \times 10^{-1}) = -(-1.00) = 1.00$
(c) pH $= -\log (1.0 \times 10^{-13}) = -(-13.00) = 13.00$
(d) 3.123
(e) 8.205
(f) 0.699
(g) −0.389 (Note that a negative pH will result when $[H^+]$ is greater than 1.0 *M*.

Do the answers make sense? Note that the *lower* the hydrogen ion concentration, the *higher* the pH. As you work through the various pH problems, keep this relationship in mind. Also, note that the pH value is related to the exponent in the hydrogen ion concentration. So if you calculate the pH of a solution with a hydrogen ion concentration of $1 \times 10^{-9}\,M$, the pH should be 9, not 2, 3, or 0.009.

exercise 8.7

Draw the Conclusions

Problem

(a) Based on Exercise 8.6, how does the pH change as the hydrogen ion concentration *increases?*

(b) By what factor does the hydrogen ion concentration change for every pH change of *one unit?*

(c) If the pH of a lake is 5.0, what is the hydrogen ion concentration?

Solution

(a) As the $[H^+]$ *increases,* the pH *decreases.* This means, for example, that the more acidic a lake is, the lower will be its pH.

(b) For every one unit change in pH, the $[H^+]$ concentration changes by a factor of *10!* This means that small changes in pH, in a lake for example, indicate profound changes in the acidity of the lake.

(c) Working backward from our previous examples, take the negative of our pH and raise 10 to that power. That is,

$$[H^+] = 1 \times 10^{-pH}\, M = 1 \times 10^{-5}\, M$$

exercise 8.8

Hydrogen Ion Concentration from pH

Problem

Use the method discussed in the text to calculate the hydrogen ion concentration, given the pH of each solution:

(a) 12.00
(b) 4.00
(c) −1.00
(d) 10.00
(e) 6.95
(f) 6.35

Solution

In general, $[H^+] = 10^{-pH}$

(a) For pH = 12.00, $[H^+] = 10^{-12.00} = 1.0 \times 10^{-12}\, M.$
(b) For pH = 4.00, $[H^+] = 10^{-4.00} = 1.0 \times 10^{-4}\, M.$
(c) For pH = −1.00, $[H^+] = 10^{-(-1.00)} = 1.0 \times 10^{1}\, M.$
(d) For pH = 10.00, $[H^+] = 10^{-10.00} = 1.0 \times 10^{-10}\, M.$
(e) $1.12 \times 10^{-7}\, M.$
(f) $4.47 \times 10^{-7}\, M.$

exercise 8.9

Hydrogen—Ion Concentration and Relative Acidity

Problem

Is a solution of pH = 1.5 more or less acidic than a solution of pH = 1.0?

Solution

$$pH\ 1.0\ has\ [H^+] = 0.10\ mol/L\ ([H^+] = 10^{-pH} = 10^{-1.0} = 0.10\ M)$$
$$pH\ 2.0\ has\ [H^+] = 0.010\ mol/L$$

pH 1.5 is between pH 1 and pH 2, hence $[H^+]$ will be less than 0.10 so pH 1.5 is *less* acidic than pH 1.0. Remember the idea that the lower the hydrogen ion concentration, the higher the pH! Accurate calculation gives $[H^+] = 3.16 \times 10^{-2}\ M$, more acidic than pH = 2.0, but less acidic than pH = 1.0.

exercise 8.10

Acidity and pH Change

Problem

We know that a one-unit change in pH means a 10-fold change in the hydrogen ion concentration. By what factor will the $[H^+]$ change if the pH is changed by *two* units?

Solution

In Exercise 8.9, we had a two-unit pH change from part a to part d (pH 12.00 to 10.00). The $[H^+]$ concentration changed by a factor of *100* (10×10). If the pH change were four units, the acidity would change by $10 \times 10 \times 10 \times 10$, or 10,000 (or 10^4). Now you can really begin to see how significant pH changes can be!

8.6 | Water and pH

Pure water does not consist entirely of water molecules. A very small proportion of the molecules in pure water can dissociate to form hydrogen ions and hydroxide ions, as described by the following equilibrium equation:

$$H_2O(l) \rightleftharpoons H^+(aq) + OH^-(aq)$$

On average, so little water dissociates (5 molecules per 1 billion!) that the hydrogen ion concentration equals only about $1 \times 10^{-7}\ M$ at 25°C. This means that the pH of pure water must equal 7.00. This is usually referred to as **neutral pH.**

Now think about this: *for every H^+ ion that is formed from the dissociation of water, one OH^- ion is also formed.* This means that in pure water the concentration of OH^- ions must also equal $1 \times 10^{-7}\ M$, exactly the same as the concentration of H^+ ions.

Pure water has a pH value of 7 known as "neutral pH" at 25°C. At neutral pH, the concentration of hydrogen ions equals the concentration of hydroxide ions. In pure water at 25°C: $[H^+] = [OH^-] = 1 \times 10^{-7}\ M.$

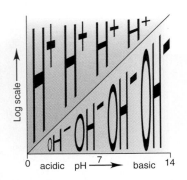

Figure 8.10 The pH value is a measure of the hydrogen ion concentration in a solution. As the $[H^+]$ decreases, $[OH^-]$ increases and pH rises.

What happens to the pH value if we add enough nitric acid (a strong acid) so that our $[H^+] = 0.1$ M? When we add lots more H^+ ions to the pure water equilibrium system, some of them will combine with OH^- ions to form water.

$$H^+ + OH^- \rightarrow H_2O$$

So when the H^+ concentration rises, the OH^- concentration automatically falls. The exact proportion in which this change occurs is described by the following very simple and useful fixed relationship between the pH and **pOH** values of an aqueous solution at 25°C:

$$[H^+] \times [OH^-] = 1.0 \times 10^{-14}$$

This allows us to calculate the $[OH^-]$ value of a solution if we know its pH value, and vice versa and at 25°C in water, pH + pOH = 14.00.

When we add sufficient nitric acid to pure water to produce a solution with $[H^+] = 0.1$ M, its pH will equal 1 (since 0.1 $M = 1 \times 10^{-1}$ M). The relationship between $[H^+]$ and $[OH^-]$ tells us that the $[OH^-]$ will equal 1×10^{-13} M.

Always remember that $[H^+]$ and $[OH^-]$ are complementary. As one goes up, the other goes down. As something becomes more acidic it also becomes less basic; and if it becomes more basic it must also become less acidic. This is illustrated in Figure 8.10.

The pH values of aqueous solutions are generally within the pH 0 to pH 14 range. Values outside of this range are possible, however. Substances that are neither acid nor base overall are called **neutral** and have a pH of 7.00 at 25°C. As we go toward the extremes of the pH scale, the acidic or basic behavior increases, so vinegar, at pH 2.5, is more acidic than milk, at pH 6.8. Similarly, ammonia, at pH 12 is more basic than baking soda, at pH 8.4. Table 8.6 and Figure 8.11 illustrate this and give the pH of some common substances.

Table 8.6

pH of Common Substances

Substance	Contains this Acid or Base	pH
Battery acid	Sulfuric acid	1.3
Stomach acid	Hydrochloric acid	1.5–3.0
Lemon juice	Citric, ascorbic acids	2.3
Vinegar	Acetic acid	2.5
Wines	Tartaric acid	2.8–3.8
Apples	Malic acid	2.9–3.3
Ant bites	Formic acid	3.0
Food preservative	Benzoic acid	3.1
Cheese	Lactic acid	4.8–6.4
Baking soda	Bicarbonate ion	8.3
Borax	Borate ion	9.2
Detergents	Carbonate, phosphate ions	10–11
Milk of magnesia	Magnesium hydroxide	10.5
Drain cleaner	Hydroxide ion	13+

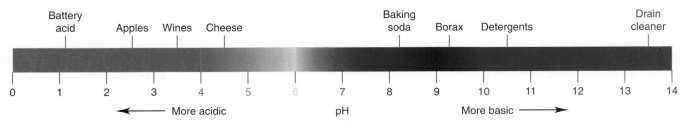

Figure 8.11   This is how some of our common substances from Table 8.6 compare on a graph of pH. The color represents that of "universal indicator," a mixture of substances that change color in much of the aqueous pH range.

exercise 8.11

pH and Concentration

Problem

Fill in the blanks.

pH	[H⁺]	[OH⁻]
5.00	1.0×10^{-5}	
12.00		
	1.0×10^{-3}	
		1.0×10^{-4}

Solution

The product of the values $[H^+] \times [OH^-] = 10^{-14}$. Therefore, when $[H^+] = 1 \times 10^{-5}\ M$, the question we need to raise is "what times $1 \times 10^{-5}\ M = 1 \times 10^{-14}\ M$?" The answer is $[OH^-] = 1 \times 10^{-9}\ M$.

pH	[H⁺]	[OH⁻]
5.00	1.0×10^{-5}	1.0×10^{-9}
12.00	1.0×10^{-12}	1.0×10^{-2}
3.00	1.0×10^{-3}	1.0×10^{-11}
10.00	1.0×10^{-10}	1.0×10^{-4}

This exercise reinforces the point that the lower the pH, the higher the pOH (and vice versa!).

We have grappled with some difficult ideas here. We now know that it is possible to numerically assess the concentration of an acid and a base and that small changes in pH or pOH accompany large changes in acid and base concentration. We also have reinforced the idea that many common substances are acids and bases. Now that we have considered quantitative (mathematical) issues, what about the qualitative side? Is there a way that we can *see* acid or base concentration?

8.7 | Seeing Acid/Base Concentration

Our bodies contain a variety of fluids, each of which has a fairly specific natural healthy pH. Blood, for example, has a normal pH range of 7.3 to 7.5. We began this chapter discussing stomach acid. This has a pH between 1.5 and 2.5. A rise in the pH of the stomach's contents can indicate the presence of a stomach cancer or anemia. The natural pH levels of the body's fluids must be maintained for the body to work properly. The protective shell of an egg will disintegrate in a low pH solution, such as vinegar. The proteins in egg white are also structurally affected by the vinegar. In the experiment we described earlier, they turn white as a result of this. From a broader environmental standpoint, the health of lakes and streams are closely linked to the pH of their water. Acids and bases are used a great deal in industry. Nitric acid, for example, is used in fertilizer manufacture and sodium hydroxide in the synthesis of many polymers. These examples should convince you that maintaining proper control over pH levels is vital in a wide range of activities, such as healthcare, catering, manufacturing, and managing the environment. To maintain proper control over pH values, we need efficient and reliable ways to determine the pH of solutions and substances in general. We can use a pH meter (see Figure 8.12), or we can use chemicals known as "acid-base indicators."

Figure 8.12 The read out on this pH meter tells us that this sample of HCl has a pH of 5.00.

Indicators

In chemistry, an **indicator** is any substance or process that signifies that an important chemical change has occurred. Color changes, sudden jumps in electrical current or the change in state from a liquid to a solid are among the ways that indicators reveal that changes have happened. **Acid-base indicators** are substances whose changes signify how acidic or basic the environment is to which they are exposed. They are also known as **pH indicators** because they are essentially indicating pH values.

The most commonly used acid-base indicators change color when pH changes. They are themselves conjugate acid-base pairs that just happen to undergo vivid color changes as they change between their acidic and basic forms. Recall that when we interconvert between an acid and a base we are really adding or removing a hydrogen ion. This changes the structure of the affected substance. For example, when acetic acid dissociates to form acetate ion (a weak base), its structure is changed (Figure 8.13):

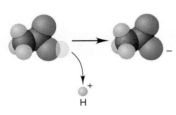

Figure 8.13 Acetic acid is a weak acid well-known for being the acid (3–5%) in vinegar. Acetic acid dissociates here, to form an acetate ion.

$$CH_3COOH \rightleftharpoons CH_3COO^- + H^+$$

This change in chemical structure is not accompanied by a color change, but many similar ones involving large organic molecules are accompanied by a color change. A change in chemical structure often causes a change in color because the way electrons interact with light depends on how those electrons are arranged in an atom, ion, or molecule. A very popular indicator, phenolphthalein, is orange below pH -1, colorless between pH -1 and pH 8, pink between pH 8 and pH 12, and then colorless again at pH values above 12 (see Figure 8.14). A very small amount of an indicator such as phenolphthalein can be added to a solution to

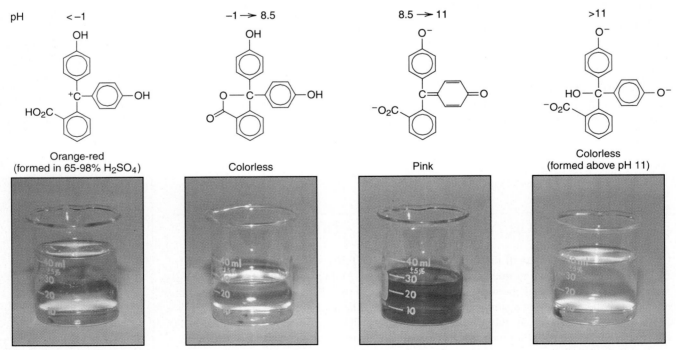

Figure 8.14 Phenolphthalein is a common acid-base indicator. The color changes shown are due to the corresponding changes in chemical structure.

Table 8.7

Indicators and Their pH Ranges

Indicator	pH Range	Acid Color	Base Color
Thymol blue	1.2–2.8	Red	Yellow
Methyl orange	3.1–4.4	Red	Yellow
Bromocresol green	4.0–5.6	Yellow	Blue
Methyl red	4.4–6.2	Red	Yellow
Bromocresol purple	5.2–6.8	Yellow	Purple
Phenol red	6.4–8.0	Yellow	Red
Thymol blue	8.0–9.6	Yellow	Blue
Phenolphthalein	8.0–10.0	Colorless	Red
Thymolphthalein	9.4–10.6	Colorless	Blue

The top row of flasks shows (left to right) methyl red, bromthymol blue, and phenolphthalein in acidic solutions. The bottom row of beakers shows the same indicators in basic solutions.

"report back to us" on the pH level of the solution. This does not indicate the precise value of the pH, but it does give valuable information about the range in which the pH must lie and can indicate the precise point at which a pH value crosses the boundary between two color ranges.

Table 8.7 lists some common acid-base indicators and the pH ranges in which they change form, and hence color. Many acid-base indicators are synthetic (i.e., purposely made) chemicals. Some, however, are natural products, and in fact many of the prettiest flowers and fruits owe their beauty to colorful acid-base indicators (see Figure 8.15 and the Case in Point on Anthocyanins).

One more critical piece of the acid-base puzzle remains: what do acids and bases do? How do they "behave chemically" in various situations? Why, for example, do Tums® help stomachs suffering from acid indigestion? The answers have to do with the typical chemical

Figure 8.15   Many flowers and fruits contain acid-base indicators which change color in response to a change in pH.

reactions of acids and bases. Although acids and bases participate in many kinds of chemical reactions, the most significant general classes of reactions are those in which an acid reacts *with* a base. Since the acid and base effectively counteract each others' acidity or basicity, or in other words "neutralize" each other, these are known as **neutralization** reactions.

case in point

Anthocyanins, Nature's Own Indicators

They are responsible for the red in radishes, the purple in red cabbages, and the melange of colors in autumn leaves. They can color a flower so that it is likely to be pollinated and then, via a change in pH, change that same flower color, indicating that pollination has occurred. "They" are chemicals called **anthocyanins** (*anthos* = flower, *cyan* = blue.) These remarkable compounds have the following three structures:

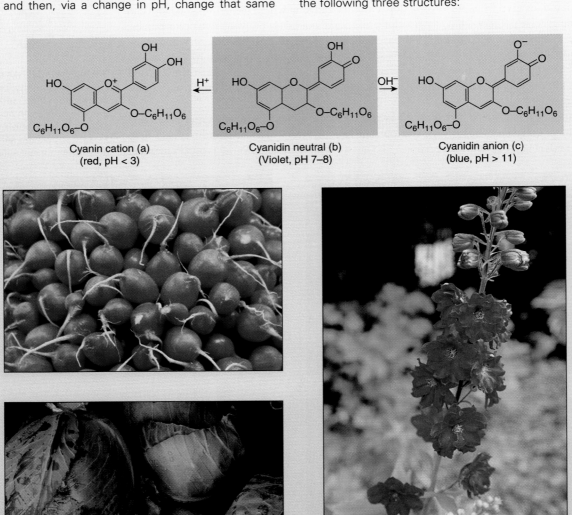

Cyanin cation (a)
(red, pH < 3)

Cyanidin neutral (b)
(Violet, pH 7–8)

Cyanidin anion (c)
(blue, pH > 11)

There are three important anthocyanins—pelargonidin, cyanidin, and delphinidin—that give many fruits and flowers their stunning color and acid-base behavior.

There are 150 known anthocyanins. They all have structures similar to cyanin and all have different colors. When mixed in different proportions, the result is a glorious array of flower and plant colors. As you might suspect, anthocyanins change color as they turn from acid to base form. An example is shown of three colors of cyanin. Plants often contain mixtures of several anthocyanins, which means that a variety of color changes are possible as the pH changes. Perhaps the most beautiful example involves red cabbage extract.

Try this experiment. Take several leaves from a red cabbage. After tearing them into smaller pieces (not too small) put them into a plastic zip-loc bag. Add 20 to 30 mL of hot water. After the water cools a bit, squeeze the bag to mash the leaves until the liquid turns deep blue or purple. Open the bag and decant (pour off) the liquid into a cup. You now have a usable supply of red cabbage extract, a very sensitive acid-base indicator.

Take some of this indicator, dilute it a bit with water and add some vinegar. Add a bit of baking soda to another diluted portion of the indicator. You can add the indicator to a variety of colorless (or nearly so) liquids (why can't you use colored ones?) and ground up solids such as aspirin, salt, or antacids. Compare your colors to those in the picture. Which of your substances were acids, bases, or neutral?

Red cabbage juice is a wonderful acid-base indicator because it changes color with relatively small changes in pH. We have added, from left to right: HCl, vinegar, vitamin C, nothing, baking soda, ammonia, and lye (sodium hydroxide).

exercise 8.12

You Be the Judge

Problem

Based on your tests with red cabbage extract (outlined in the Case in Point on Anthocyanins) classify each of the following substances as acidic, basic, or neutral; estimate its pH based on the figure in the Case in Point.

(a) ginger ale
(b) sugar
(c) dilute lemon juice
(d) Tums® (grind up first!)
(e) household ammonia (**CAUTION HERE:** Wear safety glasses. Ammonia has a very strong, offensive odor—do not inhale the vapor. It can be harmful or fatal if ingested!)

Solution

(a) ginger ale = acid, pH roughly 3
(b) sugar = neutral, pH roughly 7
(c) dilute lemon juice = acid, pH roughly 3
(d) Tums® = base, pH roughly 8
(e) household ammonia = base, pH roughly 11

8.8 | Neutralization

Combining acids with bases neutralizes the proton-donating and proton-accepting capacities of each. The simplest acid/base neutralization, shown in Figure 8.16, involves the hydronium ion of an acid reacting with the hydroxide ion of a base to form water, which has a neutral pH of 7.

$$H_3O^+ + OH^- \rightarrow 2H_2O$$

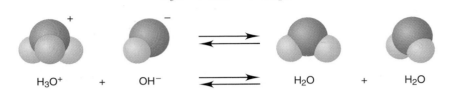

Figure 8.16 Hydronium ion and hydroxide combine to yield water.

$$H_3O^+ \quad + \quad OH^- \qquad \qquad H_2O \quad + \quad H_2O$$

If the entire formula (before ionization) of the acid and the base is included in the equation, it might look more like the following examples, where water and an aqueous salt are formed.

$$
\begin{array}{ccccccc}
\text{Acid} & + & \text{Base} & \rightarrow & \text{Salt} & + & \text{Water}
\end{array}
$$

$$\underset{\text{hydrochloric acid}}{HCl(aq)} + \underset{\text{sodium hydroxide}}{NaOH(aq)} \rightarrow \underset{\text{sodium chloride}}{NaCl(aq)} + \underset{\text{water}}{H_2O(l)}$$

$$\underset{\text{sulfuric acid}}{H_2SO_4(aq)} + \underset{\text{potassium hydroxide}}{2KOH(aq)} \rightarrow \underset{\text{potassium sulfate}}{K_2SO_4(aq)} + \underset{\text{water}}{2H_2O(l)}$$

How do we know when a neutralization is taking place? Most of these reactions show no obvious changes in color, physical state, and the like, although they all are exothermic. Most often, we use an indicator of some sort to tell us when neutralization has occurred. There is one important exception, in which the reaction itself indicates what is happening. **Carbonates** are a class of bases that react with acids to produce carbon dioxide gas upon neutralization (Figure 8.17). In effect, the reaction is its own indicator, because of the tell-tale bubbles of gas.

Remember:

$$CaCO_3(s) + 2CH_3COOH(aq) \rightarrow Ca^{2+}(aq) + CO_2(g) + H_2O(l) + 2CH_3COO^-(aq)$$

from earlier in the chapter (page 282).

Carbonates

Carbonic acid is formed when carbon dioxide (the gas in soda bubbles) dissolves in water.

$$CO_2(g) + H_2O(l) \rightarrow \underset{\text{carbonic acid}}{H_2CO_3(aq)}$$

Actually carbon dioxide, the oxide of a nonmetal, is itself spoken of as an acid even though it has no hydrogen atoms in its structure because *it forms* an acid when it is dissolved in water. The oxides of most nonmetals will form acids in water. These oxides, most of which are gases

Figure 8.17 Carbonates react with acids to form carbon dioxide gas. Many common products that relieve minor stomach distress contain carbonates. **A.** This mineral sample is treated with acid to test for the presence of a carbonate. **B.** Alka-Seltzer® signals the presence of carbonate with its fizzy bubbles when dropped in water.

A.   B.

case in point

Buffering the Blood

Buffers are chemical systems composed of conjugate acid-base pairs which are able to resist changes in pH. If the pH begins to rise, the acid of the acid-base pair releases hydrogen ions in a way that prevents or minimizes the rise in pH. If the pH begins to fall, the base of the acid-base pair combines with hydrogen ions in a way that prevents or minimizes the fall in pH.

Buffers are often used by chemists to keep the pH of solutions as steady as possible. Buffers are also very important in the body, where they serve to maintain the pH of different body fluids at appropriate values. A good example of the relevance of buffers to the body is given by the chemistry of blood.

For animals, blood is the fluid of life—it brings oxygen and nutrients to all our cells and takes waste materials away. In short, it keeps our cells, and therefore us, alive. A typical person has between 4 and 6 L of blood, containing water, proteins, carbohydrates, lipids, enzymes, hormones, and a host of other essential chemicals. As robust as a human life is, even small changes in the concentration of substances in blood can lead to health problems or death. A change in blood pH is one such example.

Human blood has a pH of between 7.3 and 7.5. Any change of more than 0.4 pH units can result in death by a

Carbonates and bicarbonates react with acids to form gaseous CO_2 and H_2O. The carbonic acid/bicarbonate system is a buffer system important to holding the pH of blood nearly constant.

series of complex processes that ruin enzyme function and oxygen transport by hemoglobin.

How does the blood maintain a nearly constant pH for an entire lifetime? The answer lies with buffers. The blood contains several buffers. Among the most important is the bicarbonate pair, H_2CO_3/HCO_3^-. Hydrogen ions released from acids that enter the blood can be removed by reaction with the bicarbonate ion to form carbonic acid:

$$HCO_3^- + H^+ \rightleftharpoons H_2CO_3$$

The reaction is "reversible," so if the blood later becomes threatened by the buildup of alkali resulting in removal of hydrogen ions, replacement hydrogen ions can be released from the carbonic acid:

$$H_2CO_3 \rightleftharpoons H^+ + HCO_3^-$$

Should there be an excess of carbonic acid built up, it can also be eliminated by the enzyme-catalyzed decomposition to water and carbon dioxide:

$$H_2CO_3(aq) \rightleftharpoons H_2O(l) + CO_2(g)$$

We breathe out the excess carbon dioxide.

If our breathing rate is slowed (known as **hypoventilation**) by, for example, asthma, pneumonia, or the inappropriate use of depressants, the body cannot get rid of carbon dioxide fast enough. Carbonic acid builds up in the blood, forcing the build-up of bicarbonate and hydrogen ions released from the carbonic acid. As a result, the pH of the blood drops, with associated health risks such as depression of the central nervous system or death.

If our breathing rate becomes too fast (known as **hyperventilation**), carbon dioxide is removed from the blood too quickly. This accelerates the degradation of carbonic acid into carbon dioxide and water. The resulting fall in

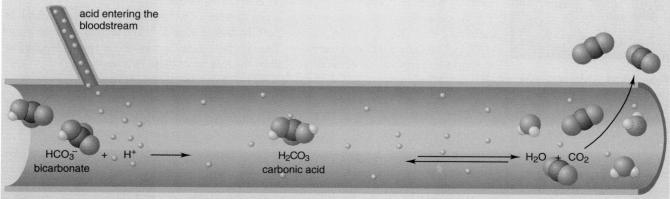

acid entering the bloodstream

HCO_3^- + H^+ $\longrightarrow$ H_2CO_3 H_2O + CO_2
bicarbonate   carbonic acid

$$HCO_3^- + H^+ \rightleftharpoons H_2CO_3 \rightleftharpoons H_2O + CO_2$$

Bicarbonate ion circulates in the bloodstream where it is in equilibrium with H^+ and OH^-. In the lungs bicarbonate ion adds a hydrogen ion and loses a water molecule to form carbon dioxide, which is exhaled.

carbonic acid levels encourages the combination of hydrogen ions and bicarbonate ions which reforms carbonic acid. The resulting fall in blood hydrogen ion levels leads to a rise in blood pH which can result in over-excitability or death.

The H_2CO_3/HCO_3^- buffer system is an equilibrium system that can be represented as:

$$H_2CO_3 \rightleftharpoons H^+ + HCO_3^-$$

All of the shifting concentrations discussed here can be understood in terms of something called **Le Châtelier's principle.** This is a very useful rule about chemical equilibrium systems that says that "if you impose a change in concentration, temperature, or pressure on a chemical system at equilibrium, the system responds in a way that opposes the change." The system tries to regain equilibrium. So if we increase the concentration of H^+ ions, for example, this increases the rate of formation of H_2CO_3, and therefore uses up at least some of the added H^+ ions. So, as predicted by Le Châtelier's principle, the system has responded to a change (addition of H^+ ions) by opposing the change (using up and therefore removing H^+ ions). The changes in reaction rates that make chemical systems adhere to Le Châtelier's principle occur automatically, due to the way in which reaction rates are affected by the concentrations of the starting materials and products of the reactions involved.

Le Châtelier's principle states that when change is introduced into a system at equilibrium, the system will react in a manner that will reduce the effect of the change. For example, in a chemical application, if heat is removed from a system at equilibrium, the reaction will be to produce heat to regain the state of equilibrium. However, we can apply this principle to many other situations. In nature, an environment can support a certain level of diversity. When this diversity, or balance of nature, is upset, circumstances change in the direction of restoring the balance. In the wilds, the balance of predator and prey is a very fine one. If the prey decrease sufficiently, the survival of the predator is at risk. If we eliminate the predator, the prey may overpopulate the capacity of the environment to support it and starvation and disease may be the result. Can you think of any other applications?

Figure 8.18 Carbon dioxide reacts with water to form carbonic acid. The acid is very weak, as is shown by the color that it turns red cabbage juice.

(carbon dioxide is a good example), are termed "acid anhydrides" (*an* = Greek for "without," *hydro* = "water") because they are acids from which water is absent. "Natural" (nondistilled) water, which contains dissolved CO_2 from the air, is almost always slightly acidic, at roughly pH 5 to 6.

You can demonstrate the acidity of carbon dioxide by using a cupful of your red cabbage extract (see the Case in Point on Anthocyanins). Put a straw into the liquid and blow for a minute or so (Figure 8.18). Your exhaled breath contains about 4% CO_2. Compare the color of the resultant liquid with some of your fresh extract. What is the pH of the extract and carbon dioxide? Keep this in mind when we study acid rain in Chapter 9.

As carbonic acid dissolves in water, it can lose one of its hydrogen ions leaving behind bicarbonate ions (HCO_3^-), or it can lose both hydrogen ions to leave behind carbonate ions (CO_3^{2-}). The carbonate ion is therefore a base because it can accept one or two hydrogen ions to form a bicarbonate ion or carbonic acid by the reverse of these two reactions:

$$H_2CO_3(aq) + H_2O(l) \rightleftharpoons H_3O^+(aq) + \underset{\text{bicarbonate ion}}{HCO_3^-(aq)}$$

$$HCO_3^-(aq) + H_2O(l) \rightleftharpoons H_3O^+(aq) + \underset{\text{carbonate ion}}{CO_3^{2-}(aq)}$$

The bicarbonate ion can either donate *or* accept H^+. So it can act as *either* an acid or a base. Do you remember that we have already said that substances that can act as acids or bases are called **amphiprotic?**

We now have a sense of what acids and bases are and some of the situations in which they are involved in our day-to-day lives. They have an impact on personal, ecological, and industrial decisions. One of the most significant aspects of acid/base behavior is the problem of "acid rain," the subject of Chapter 9.

case in point

Rotting Paper

The National Library of Congress is the largest library in the world, housing roughly 17 million books and 95 million maps, films, photographs, manuscripts, and recordings. Yet the library's collection is rotting away "at a rate of more than 200 volumes a day," according to *New York Times* science writer Malcolm W. Browne. Many of the library's written materials are in danger of decomposition.

The culprit is acidification of paper, and the problem is remarkably widespread, though it occurs primarily in books printed in the 19th and 20th centuries. Browne writes that, "some 300 million acidic books in North American research libraries and 400 million in Europe need treatment." How do you deal with rotting books? The problem and the cure have to do (naturally enough) with acid-base chemistry.

Very old manuscripts were made of cotton and similar vegetable fibers. Many of these works still exist and are in good condition. Around the mid-1800s, papermakers began using wood fibers, which were plentiful and inexpensive, to make paper. Wood and cotton are both largely composed of the same primary material, cellulose. However, words printed on wood-based paper tend to look blurred. To counteract the blurring, a "sizing agent" is added which prevents ink from spreading. Up until a few years ago, the sizing agent of choice was aluminum sulfate, $Al_2(SO_4)_3 \cdot 18H_2O$, commonly known as cake alum or patent alum. The alum ionizes in water as follows:

$$Al_2(SO_4)_3 \cdot 18H_2O(s) \rightarrow 2Al^{3+}(aq) + 3SO_4^{2-}(aq) + 18H_2O(l)$$

The aluminum ions (Al^{3+}) exist in water as $Al(H_2O)_6^{3+}$, formed when each ion combines with six water molecules. This complex then reacts further with water to produce hydronium ions, as follows:

$$H_2O + Al(H_2O)_6^{3+} \rightarrow Al(H_2O)_5OH^{2+} + H_3O^+$$

So, in effect, the aluminum ion acts as an acid, since it causes the generation of hydronium ions. Overall, the use of alum as a sizing agent acidifies the paper. Just because something contains acid does not mean it will begin spontaneously to rot. We all contain acids, and mostly do just fine. However, the cellulose that makes up the paper breaks apart due to reaction with water (hydrolysis) in the presence of acid. So alum plus cellulose, given time, leads to paper that Browne describes as becoming "as fragile as a dead leaf."

Two questions come to mind. First, can paper be made from wood that will not decompose? Second, is there anything that can be done to stop the rot in existing manuscripts? The answers are yes and yes.

To make stable new papers from wood, a sizing agent that is basic is added. Cellulose will not decompose readily in the presence of, for example, calcium carbonate. It is estimated that paper sized with this base will last three to four times as long as acid-sized paper.

Saving existing acidified books is more difficult, but possible. One method involves spraying books with a deeply penetrating solvent (such as Freon® or alcohol) containing a base such as methoxymagnesium methylcarbonate. The process is fairly expensive, however, so libraries must be selective about which books to save. So, in the short term, books were made from materials that were cheaper (wood-based). In the long term, we pay anyway, in money, time, and quality.

How important do you think it is to preserve books in their original form? Do you think photocopying or photographing the pages might be a reasonable alternative to preserving the actual books? Or should old books just be left to rot? Try to consider possible answers to these questions as part of a risk/benefit (or in this case more of a cost/benefit) analysis.

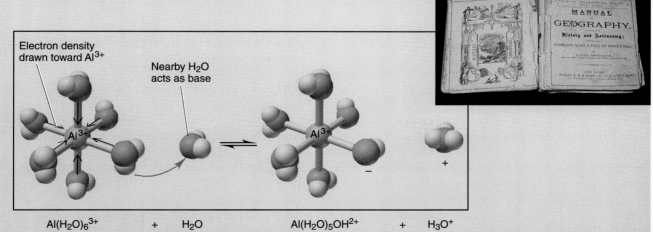

Electron density drawn toward Al^{3+}

Nearby H_2O acts as base

$Al(H_2O)_6^{3+}$ + H_2O $\rightleftharpoons$ $Al(H_2O)_5OH^{2+}$ + H_3O^+

Aluminum ion, surrounded by water molecules in solution, can act as an acid. This plays a key role in book pages becoming yellow and brittle.

main points

- Acids and bases have specific properties including taste and reactivity.

- Acid and base concentration is reflected in the pH scale.

- Every unit change in pH reflects a 10-fold change in acid and base concentration.

- There are many indicators of acid and base strength, including the change of color in the flowers or fruits of some plants.

- Hydrochloric acid in the human stomach is a critical component in digestion.

- Neutralization occurs when sufficient base is added to an acid to stop its proton-donating capability. Adding sufficient acid to a base will stop its hydroxide-donating ability (the base will be neutralized).

important terms

Acid is a substance that increases hydronium concentration in aqueous solution. (p. 278)

Acid anhydride is a nonmetal oxide that forms an acid when combined with water. (p. 285)

Acid-base indicator is a substance that has different appearance in acid and base. (p. 292)

Alkali is a substance that furnishes hydroxide ion in water by dissociation of a metal hydroxide. (p. 283)

Amphiprotic is a substance that can behave either as an acid or a base. (p. 282)

Anthocyanin is a class of substances responsible for the color of many flowers, fruits, and vegetables. (p. 294)

Base is a substance that furnishes hydroxide ion in water. (p. 283)

Basic anhydride is a group 1A or 2A metal oxide that forms a base when combined with water. (p. 285)

Buffer is a chemical system, typically composed of conjugate acid-base pairs that resist change in pH. (p. 297)

Carbonate, CO_3^{2-} ion, is an anion of limestone and hard water. (p. 296)

Chemical equilibrium is a state in which chemical reactions appear to have come to completion. (p. 278)

Conjugate base is that which remains when hydrogen ion is removed from an acid. (p. 283)

Conjugate pair is an acid and its conjugate base. (p. 283)

Digestion is a process that breaks down food into nutrients that can be absorbed into the bloodstream. (p. 275)

Hydrochloric acid is an aqueous solution of hydrogen chloride, HCl(*aq*). (p. 276)

Hydronium ion, H_3O^+, is a representation describing species found when water accepts a hydrogen ion from an acid. (p. 277)

Hyperventilation is breathing too rapidly, prefix "hyper" means "too much." (p. 297)

Hypoventilation is breathing too slowly, prefix "hypo" means "too little." (p. 297)

Indicator is a substance that has a different appearance in one set of circumstances than in another. (p. 292)

Le Châtelier's principle states that when a system at equilibrium is subjected to a stress, the system responds to partially alleviate that stress. (p. 298)

Lye is solid sodium hydroxide, NaOH. (p. 283)

Molarity is the number of moles solute per liter of solution. (p. 279)

Neutral pH means neither acid nor base. (p. 289)

Neutralization is the process of adding acid to a base or vice versa till neither is in excess. (p. 294)

pH is the negative logarithm (base 10) of hydrogen (hydronium) ion molarity. (p. 286)

pOH is the negative logarithm (base 10) of hydroxide ion molarity. (p. 290)

Protein is a large biological polymer of amino acids. (p. 275)

Strong acid is an acid that completely dissociates in water. (p. 278)

Weak acid is an acid that only partially dissociates in water (typically < 5%). (p. 278)

exercises

1. State four differences between an acid and a base.
2. List five substances that contain an acid and five which contain a base (try to include items *not* listed in Table 8.6).
3. What is the chemical formula for a hydrogen ion? Hydronium ion?
4. What is the difference between a weak acid and a strong acid? Can you give three examples of each?
5. What is the difference between a weak acid and a dilute acid: a strong acid and a concentrated acid? Can you give one example of each?
6. If you open a bottle of acetic acid in the lab, what common household item will it smell like?
7. What are the products of the addition of water to HBr?
8. Which 1.0 molar solution is more acidic, H_2SO_4/water or CH_3COOH/water? Why?
9. Write an equation showing the dissociation of each of these strong or weak acids. (You may neglect water in each equation.)
   a. Sulfuric acid, H_2SO_4, a strong acid used in automobile batteries
   b. Tartaric acid, $C_4H_5O_6H$, a weak acid used in the soft drink industry and in baked goods
   c. Carbonic acid, H_2CO_3, used in the carbonation of beverages
   d. Salicylic acid, $C_7H_5O_3H$, a weak acid used in aspirin production
10. Write an equation to show the reaction of each of these with water (if one occurs).
    a. sodium hydroxide, NaOH
    b. chloride ion, Cl^-
    c. cyanide ion, CN^-
    d. methylamine, $(CH_3)NH_2$
11. If you dissolve 5 mol HCl in 1 L water, how many hydronium ions are present?
12. What is the molarity of the solution in Exercise 11?
13. What is the pH of the solution in Exercise 11?
14. Phosphoric acid, H_3PO_4, is a triprotic weak acid. Predict the formulas of the compounds formed when:
    a. 1 mole of NaOH is reacted with one mole of H_3PO_4
    b. 2 moles of NaOH are reacted with one mole of H_3PO_4
    c. 3 moles of NaOH are reacted with one mole of H_3PO_4
15. Look at the ingredients on a can of regular soda. What acids are listed?
16. Which of the following solutions has the greatest $[H^+]$ concentration?
    a. 1×10^{-4} molar NaOH
    b. pH 4 solution of sulfuric acid
    c. 0.00001 molar HCl
17. Calculate the molarity of a solution of 43.0 grams of ascorbic acid (vitamin C, $C_6H_8O_6$) dissolved in enough water to make 250 mL of solution.

18. Calculate the molarity of a solution of 34.0 grams of ascorbic acid (vitamin C, $C_6H_8O_6$) dissolved in enough water to make 300 mL of solution.
19. Calculate the molarity of a solution of 64.0 grams of ethyl amine ($CH_3CH_2NH_2$) dissolved in enough water to make 250 mL of solution.
20. How many moles of HCl are needed to make 2.0 L of a 3.0 M HCl solution?
21. An acid is a proton _____ and a base is a proton _____.
22. What is the conjugate acid or base of each of these?
    a. conjugate base of HCl
    b. conjugate acid of HCO_3^-
    c. conjugate base of NH_4^+
23. Label the conjugate acid and base pairs of this reaction:

$$HCO_3^- + HF \rightarrow H_2CO_3 + F^-$$

24. What is the conjugate acid or base of each of these?
    a. conjugate acid of NH_3
    b. conjugate base of HCO_3^-
    c. conjugate base of H_2SO_4
    d. conjugate acid of HCO_3^-
25. Identify the acids and bases in this group: CH_3COOH, NH_3, $Ca(OH)_2$, HI, N_2H_4, CO_2.
26. Circle all the compounds that form basic solutions when added to water.

$$SO_3 \quad Na_2O \quad HNO_3 \quad NaCl$$
$$Ca(OH)_2 \quad NH_3 \quad CH_3CH_2CH_2OH$$

27. Circle all the compounds that form acid solutions when added to water.

$$SO_3 \quad Na_2O \quad HNO_3 \quad NaCl$$
$$Ca(OH)_2 \quad C_6H_{12}O_6 \quad CO_2 \quad CH_3CH_2CH_2OH$$

28. Circle all the compounds that form pH neutral solutions when dissolved in pure water.

$$SO_3 \quad MgO \quad C_{12}H_{22}O_{11} \quad NH_3$$
$$NaCl \quad CH_3COOH \quad C_3H_7OH$$

29. Why is CO_2 classified as an acid, even though it doesn't contain any hydrogen atoms?
30. If each of these were reacted with water, would the product be an acid or a base?
    a. NO_2, nitrogen dioxide
    b. K_2O, potassium oxide
    c. MgO, magnesium oxide
    d. SO_3, sulfur trioxide
31. Would you expect the pH of an orange to be close to 1 or 10?
32. The pH of a 0.02 *M* solution of a cola is 4.0. What is the hydronium ion concentration?

33. Calculate the pH of each of these hydrogen ion concentrations.
    a. $1.0 \times 10^{-3} M$ d. $1.0 \times 10^{-10} M$
    b. $1.0 \times 10^{-12} M$ e. $1.0 \times 10^{-7} M$
    c. $1.0 \times 10^{-5} M$

34. How many grams of HCl were added to a solution with a final volume of 1.50 L with a pH of 3.00?

35. A solution is prepared by dissolving 63 g of HNO_3 in enough water to make 10.0 L of solution. Calculate the pH of the solution.

36. 0.35 moles of a weak acid is dissolved in water and the volume is adjusted to 500 mL. The pH of the solution is 3.0. Calculate the percent of the acid which is ionized.

37. Given either the pH or pOH, find the other.
    a. pH 2.0   pOH = ?
    b. pOH 11.0   pH = ?
    c. pOH 6.0   pH = ?
    d. pH 4.25   pOH = ?
    e. pOH 8.98   pH = ?
    f. pH 3.67   pOH = ?

38. Answer each question with the correct word(s), number, symbol, or response.
    a. What is the pH of a solution having $[OH^-] = 0.00001$ mol/L?
    b. What is the pH of a solution if 0.010 mol of HNO_3 is dissolved in 100 mL of water?
    c. What is the pH of a solution having $[H^+] = 0.001$ mol/L?
    d. What is the pH of a solution having $[H^+] = 0.00001$ mol/L?
    e. What is the pH of a solution if 0.010 mol of NaOH is dissolved in 100 mL of water?
    f. What is the pH of a solution having $[H^+] = 0.01$ mol/L?
    g. What is the pH of a solution having pOH = 8.5?

39. In the summers and falls of 1986 and 1987, geographer Lee F. Klinger, who works at the National Center for Atmospheric Research in Boulder, Colorado, discovered extremely acidic rain water in southeast Alaska with a pH 3.6. (Normal rain water has a pH of about 5.6). The source of this acidity was not the expected sulfur compounds given off by oceanic organisms but rather organic acids originating from moss-filled peat lands. How many times more acidic is $[H^+]$ in the southeast Alaskan rain water compared to normal rain water? Could this be a significant enough change to cause the destruction of a large area of forests downwind from the peatlands?

40. Determine whether each of the following solutions is acidic or basic:
    a. $[H^+] = 1 \times 10^{-11} M$
    b. $[OH^-] = 1 \times 10^{-3} M$
    c. $[H^+] = 1 \times 10^{-5} M$

41. A solution has a hydroxide ion concentration of $8.1 \times 10^{-1} M$. Is this solution highly acidic or basic?

42. Calculate the hydrogen ion concentration given the following pH readings.
    a. 2.00
    b. 9.00
    c. 14.00
    d. -3.00 (Is this a reasonable question?)

43. Calculate the pOH for the following hydrogen ion concentrations.
    a. $1.0 \times 10^{-7} M$ c. $1.0 \times 10^{-11} M$
    b. $1.0 \times 10^{-4} M$ d. $1.0 \times 10^{-2} M$

44. How many moles of potassium hydroxide, KOH, would be necessary to prepare 2.0 L of a solution that is 0.50 M in KOH?

45. How many grams of calcium hydroxide, $Ca(OH)_2$, are required to make 750 mL of a solution initially 0.300 M in $Ca(OH)_2$?

46. Processed cheese must be at a pH of 5.5 to 5.9 to prevent the growth of unwanted microorganisms. You are the quality control supervisor who determines whether or not the cheese leaves the plant. If a report on a sample comes back with a $[OH^-]$ of 1.0×10^{-8} would you send the cheese to the market?

47. Draw reactions to illustrate how water acts as both an acid and a base.

48. What is the hydroxide ion concentration of the soda in Exercise 32?

49. Calculate the pH of a glass of wine if the hydronium ion concentration is $4.2 \times 10^{-5} M$.

50. Calculate the pH of a glass of wine if the hydroxide ion concentration is $2.4 \times 10^{-10} M$.

51. Are the results from Exercises 49 and 50 what you would expect? Use mathematical reasoning to determine your answer.

52. How can the pH of a cup of water be neutral when we know some of the water molecules dissociate to provide a hydronium ion? Why isn't water slightly acidic?

53. How do anthocyanins act as acid-base indicators?

54. Indicate whether these substances are acids, bases, or neutral. (Use the red cabbage indicator test as described in the Case in Point: Anthocyanins, Nature's Own Indicators if you are uncertain).
    a. crushed vitamin C
    b. hydrogen peroxide
    c. flour
    d. potato
    e. colorless mouthwash

55. Predict the products of the following reactions with water:
    a. $NO_2^-(aq) + H_2O(l)$ c. $MgO(s) + H_2O(l)$
    b. $H_3PO_4(aq) + H_2O(l)$ d. $HCN(aq) + H_2O(l)$

56. What are two major characteristics of a buffer solution?

57. When cider goes sour, some of the sugar has reacted to form acetic acid (CH_3COOH). What is the molarity of acetic acid in cider if 250 mL of cider requires 45.2 mL of 0.0045 molar NaOH to neutralize it?

58. Based on what you know about acids and bases, what do you think the term "acid rain" means? Why is it so harmful to the environment?

59. What is meant by an "indicator"?

60. If you add phenolpthalein to an ammonia-based household cleaner, about what color solution would you expect to see? If you added methyl red?

61. You find a beaker of a clear unknown substance sitting in the lab. In order to dispose of it, you must determine if it is neutral, basic, or acidic. Propose an experiment to test the substance.

62. A solution is prepared by dissolving 40.0 g of NaOH in enough water to make 1.00 L of solution. Then, 2.50 mL of this solution is diluted to 250 mL. What will be the pH of the final, diluted solution?

63. If an unknown substance found in the lab is determined to be a base, how would you neutralize the solution for disposal?

64. If an unknown substance fizzes and bubbles on neutralization by HCl, what type of compound might be present in the solution?

65. Write the products of this neutralization reaction:
$H_3PO_4 + 3NaOH \rightarrow$ _____.

66. Why do you take Pepto Bismol when you have an upset stomach?

67. Why do you think you should wear gloves when handling concentrated acids in the lab?

food for thought

68. You are asked to prepare a concentrated sodium hydroxide solution. What molarity would you prepare? You are then asked to prepare a dilute hydrochloric acid solution. What molarity would you prepare? In general, what are your cut-off molarities for concentrated and dilute solutions?

69. Earlier in the chapter, we said that pH is not exactly equal to $-\log[H^+]$ because there are many processes that occur in solution that affect how hydrogen ions act. Can you conjecture about what these processes might be? That is, what might H^+ do in solution that would affect its concentration?

readings

1. *Chemical and Engineering News,* April 10, 1995, p. 17.
2. Baum, S.J. 1989. *Introduction to Organic and Biological Chemistry.* 4th ed. New York: Macmillan.
3. Chiras, D.D. 1991. *Environmental Science: Action for a Sustainable Future.* Menlo Park, CA: Benjamin/Cummings.
4. Browne, M.W. Nation's library calls on chemistry to stop books from turning to dust. *New York Times,* May 22, 1990, pp. B5–B6.
5. Fieser and Fieser. *Organic Chemistry* 3rd ed. D.C. Heath, Boston, 1956.
6. Sugar maples sicken under acid rain's pall. *New York Times,* May 5, 1991.

websites

www.mhhe.com/kelter   The "World of Choices" website contains activities and exercises including links to websites for: J. T. Baker Chemicals, for information on the densities of acids; the Food and Drug Administration for information on cosmetics; and much more!

Acid Rain

Oh Shenandoah, I long to hear you,
Away, you rolling river,
Oh Shenandoah, I long to hear you,
Away, we're bound away, 'cross the wide Missouri

—Traditional American folk song

"They stop at Overlooks, idling their engines as they read signs describing Shenandoah's ruggedly scenic terrain: Hogback, Little Devil Stairs, Old Rag, Hawksbill, Panorama. Jammed along the narrow road, the cars and motor homes add to the miasmic summer haze that cloaks the hills."

That is how environmental writer Ned Burks describes summer in Virginia's 198,000-acre Shenandoah National Park, a monument to the natural beauty of the United States and to the problems that can occur when the needs of industry and automobile tourism conflict with the needs of the land.

The National Park is downwind from the Ohio Valley, a heavily industrialized area. This, combined with between 2 and 3 million park visitors, 90% of whom never leave their cars, generates a blanket of sulfur and nitrogen oxides that has helped reduce summer visibility by about 80% since 1948. The pH level in several monitored Shenandoah streams has become low enough to endanger fish such as brook and rainbow trout.

The title of this chapter is "acid rain," yet we have just mentioned both air pollution and the acidity of streams. Acid rain is about the *interactions* of nitrogen and sulfur oxides with water vapor in the air and with bodies of water.

Acid rain is really only one part of a more general process by which acids or acid anhydrides fall from the atmosphere to the Earth's surface. The technical name for the process is **acid deposition** because it can be found in rain, sleet, snow, and even as small solid particles that fall to the surface. Does acid rain have an important impact on the quality of the environment or on human health? There is fierce debate over the answer to that question, focused on a variety of separate issues, including these:

- Respiratory ailments possibly caused in part, or made more serious, by acid rain-laden smog.
- Acidification of thousands of lakes and streams, especially those near or downwind of industrial areas.
- Tree damage in forests, on a scale unprecedented in history.
- Accelerated weathering of buildings and monuments.

The view from the Blue Ridge Parkway in Shenandoah National Park is spectacular, if you can see it. On this day in late July 1997, visibility is greatly reduced by smog.

The interactions between the Earth's atmosphere and the chemicals on the Earth's surface are very complex and only partially understood. This makes it difficult for scientists to assess the specific environmental contribution of any one factor. So, as we begin our look at the chemistry of acid rain, keep in mind that there are many unanswered questions and points of sharp debate about its actual effects. Despite the disagreements and complexities, many convincing cases have been highlighted linking acid rain derived from specific sources to particular problems. For example, smokestack emissions from industrial plants in the northeast United States apparently cause acid rain in Canada.

9.1 | The Nature of Acid Rain

The world seems wonderfully fresh after heavy rain, especially in the countryside. It seems that the air has been washed clean, and rain does in fact wash impurities out of the air. This process has been going on for millions of years. Rain provides a mechanism for removing such things as dust, ash, and pollen from the atmosphere. Dust and ash enter the atmosphere naturally due to wind, fire, and volcanic activity, and the air is the natural medium for the transfer of many plant spores (pollen) and seeds.

It is perfectly natural for air to contain substances other than the gases that are traditionally regarded as parts of air. Our modern industrialized way of life, however, belches out many additional contaminants into the atmosphere, severely testing nature's ability to keep the air sufficiently clean for the health of plant and animal life. One major new source of atmospheric contamination is the burning of fossil fuels (coal, oil, and natural gas) to generate electricity, keep us warm, and power our automobiles and other machines. The varied and often large-scale processes of industry also release a bewildering array of chemicals into the skies (Figure 9.1). Some of the chemicals released in these ways can combine with water, in raindrops, to make the rain significantly more acidic than it would otherwise be.

The three major contributors to acid rain are recognized to be gases released during the burning (combustion) of fossil fuels in (1) industrial manufacturing processes, (2) power generation, and (3) personal automobile use. Hydrogen chloride (HCl) from coal combustion, and volatile organic compounds from various sources, play a small part in the formation of acid rain; but the most significant acid-forming gases fall into these two categories:

- The sulfur oxides SO_2 and SO_3 (known collectively as "SO_x").
- The nitrogen oxides NO and NO_2 (known collectively as "NO_x")

Figure 9.1 Industrial processes release a bewildering array of chemicals into the skies. Source: Fine, L. and Beall, H. "Chemistry for Engineers and Scientists," Saunders College Publishing, Orlando, FL. p. 325

Sulfur Oxides (SO$_x$)

Sulfur dioxide, SO$_2$, is a product of combustion in the boilers of factories, coal-fired power plants, and ore smelters. The smelters are designed to purify metals from compounds such as **sulfide ores,** which contain sulfur bound to other elements such as iron or copper. The first step in releasing the pure metal from its sulfide is "roasting," which simply means heating the ore in the presence of oxygen to convert the sulfide of the metal into the oxide.

Copper, for example, is often found in nature as the sulfide ores **chalcopyrite** (CuFeS$_2$) and **chalcocite** (Cu$_2$S). Although the industrial process for purifying copper from these ores is fairly complex, the overall chemistry is straightforward. In the case of chalcocite, for example, roasting of the ore converts some of it to copper(I) oxide and sulfur dioxide. The copper(I) oxide can then react with the remaining copper(I) sulfide to form copper metal and sulfur dioxide.

$$2Cu_2S(s) + 3O_2(g) \rightarrow 2Cu_2O(s) + 2SO_2(g)$$

$$2Cu_2O(s) + Cu_2S(s) \rightarrow 6Cu(s) + SO_2(g)$$

The sulfur dioxide gas is either released to the atmosphere or trapped by a "scrubbing" process designed to prevent its release. We will return to this topic in Chapter 13—The Earth as a Resource.

Some of the sulfur dioxide released when coal is burned comes from sulfide ores in coal such as **pyrite** (FeS$_2$) (Figure 9.2). Pyrite is known as "fool's gold" because it is golden colored and shiny and was often mistaken for gold by prospectors in the mid-19th century. It is particularly abundant in coal from the eastern United States. When the coal is burned, the sulfur atoms combine directly with oxygen, generating SO$_2$.

How Does SO$_2$ Contribute to Acid Rain?

Sulfur is a nonmetal element, and we learned in Chapter 8 that nonmetal oxides react with water to form acids. Sulfur dioxide will combine with water to give **sulfurous acid** (H$_2$SO$_3$):

$$SO_2(g) + H_2O(l) \rightarrow \underset{\text{sulfurous acid}}{H_2SO_3(aq)}$$

Dust in the air can catalyze a reaction between sulfur dioxide and molecular oxygen (O$_2$) to form sulfur trioxide (SO$_3$). The sulfur trioxide can then react with water to form sulfuric acid:

$$2SO_2(g) + O_2(g) \rightarrow 2SO_3(g)$$

$$SO_3(g) + H_2O(l) \rightarrow H_2SO_4(aq)$$

Figure 9.2 Pyrite (FeS$_2$) is known as "fools gold" because its glittering appearance led many 19th century prospectors to think they had found gold (Au). Sulfide ores such as pyrite are common in the eastern part of the United States.

Figure 9.3 The reaction of sulfur dioxide with water to form sulfuric acid.

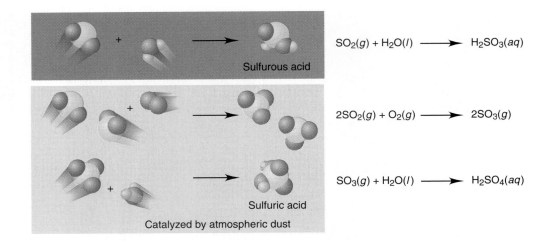

$$SO_2(g) + H_2O(l) \longrightarrow H_2SO_3(aq)$$

$$2SO_2(g) + O_2(g) \longrightarrow 2SO_3(g)$$

$$SO_3(g) + H_2O(l) \longrightarrow H_2SO_4(aq)$$

Sulfurous acid

Sulfuric acid

Catalyzed by atmospheric dust

This is the same overall chemical process that is exploited for the commercial manufacture of sulfuric acid. This process is summarized in Figure 9.3.

Although the mechanisms of acid rain–caused destruction are not well understood, the solubility of substances in acid is hypothesized to play a key role. For example, aluminum salts are soluble in acid and, therefore, tend to leach out of soils into acidified water. High concentrations of aluminum seem to interfere with the ability of plants and trees to gather nutrients from the soil. Acid rain also leaches sugars and amino acids that are necessary for growth in plant root systems. Temperature extremes, insect attack, acid rain, and other forms of air pollution have a "**synergistic**" impact on forest destruction, meaning that they combine to cause destruction greater than the sum of the damage caused by each factor individually.

Although many facilities which burn lots of sulfur-containing fuel have recently been redesigned to "scrub" (remove) the sulfur dioxide from the waste gases, over 20 million tons of SO_2 spew into the U.S. environment every year.

While sulfur dioxide emissions definitely cause significant problems due to acid rain, they do have one interesting and possibly positive side effect. Some of the gas forms particles in the air that reflect sunlight back to space. This may help to counteract the global warming which

exercise 9.1

Sulfuric Acid

Problem

How much sulfuric acid can result from the combustion of a train load (100 cars) of coal if the coal is 1.0% sulfur? A train car holds 100 tons of coal.

Solution

Think about this problem like any other stoichiometry problem. The overall chemical process, shown in previous equations, is that each mole of sulfur reacts with an excess of oxygen and water vapor to become a mole of sulfuric acid. The simplest meaning of 1% sulfur is that 100 units of coal contain 1 unit of sulfur. This can be used as a conversion factor in the solution.

mass $H_2SO_4 =$

(100 cars coal)(100 tons coal/car)(907 kg coal/ton coal)(1.0 kg S/100 kg coal)
(98.0 kg H_2SO_4/32.0 kg S)

$= 2.8 \times 10^5$ kg H_2SO_4

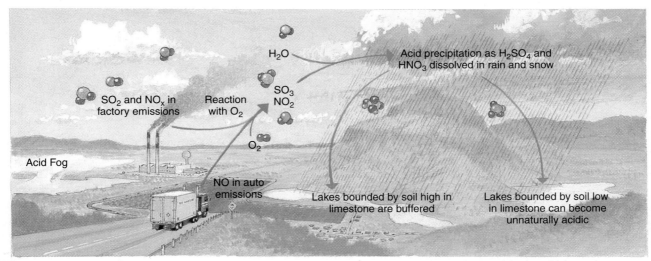

A complex cycle of interactions between industrial emissions and atmospheric chemistry results in acid rain, which can have dramatic environmental impact.

some researchers believe is occurring due to the "greenhouse effect" (discussed in Chapter 12).

Nitrogen Oxides (NO$_x$)

Modern life depends heavily on the automobile. Unfortunately, our ever-increasing dependence on cars causes increasing amounts of acid-forming nitrogen oxides to be released into the atmosphere. The relevant chemistry happens in the car's engine, where gasoline vapors are mixed with an excess of air and burned very rapidly. The energy released by this reaction causes the piston to move in the cylinder, and the movement of the piston can be used to move the car (see Figure 9.4). The reaction we want to occur in the cylinder is the combination of the hydrocarbons in gasoline with oxygen from the air. About four-fifths of the air, however, is nitrogen, and at the very high temperatures inside engine cylinders, nitrogen can combine with oxygen to form nitrogen monoxide (Figure 9.5):

The evening traffic "jam" in many cities dumps literally tons of pollution into the environment. The nitrogen dioxides are particularly high in automobile exhaust.

$$N_2(g) + O_2(g) \rightarrow 2NO(g)$$

When the nitrogen monoxide is released from the engine, it can react slowly with atmospheric oxygen to form nitrogen dioxide:

$$2NO(g) + O_2(g) \rightarrow 2NO_2(g)$$

Nitrogen dioxide can react with water to form nitrous acid, HNO$_2$ and nitric acid, HNO$_3$:

$$2NO_2(g) + H_2O(l) \rightarrow \underset{\text{nitrous acid}}{HNO_2(aq)} + \underset{\text{nitric acid}}{HNO_3(aq)}$$

We have now examined the most significant chemical processes that cause acid rain. We have discussed the gases that are the major culprits and the chemical reactions that lead to acidification. We have also identified some aspects of our lifestyle that are responsible for acid rain. However, there are several important questions that we have not dealt with:

- What damage is caused by acid rain?
- Who is responsible for acid rain?
- What solutions are there to the problem of acid rain, and what are their costs (economic, scientific, ecological, and personal)?

We will explore these issues throughout the rest of this chapter, after a little history.

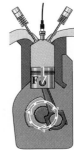

Figure 9.4 The piston of an automobile engine is moved by the expansion of the burning gasoline vapor. The movement of the piston causes the car to move. A by-product of this process is nitrogen monoxide which reacts with oxygen to form nitrogen dioxide.

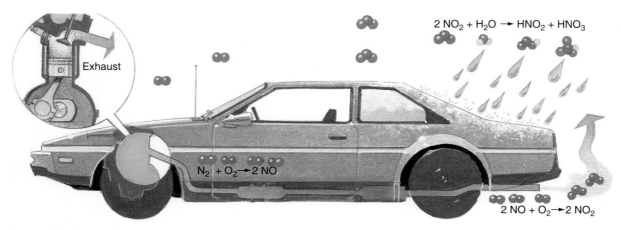

Figure 9.5 At the high temperatures of automobile combustion, nitrogen and oxygen can combine to form nitrogen monoxide, which forms nitrogen dioxide on reaction with oxygen in the air. Nitrous and nitric acids then form when the nitrogen dioxide reacts with water in the atmosphere.

9.2 | Historical Perspective

We think of acid rain as a relatively recent development, but according to Dr. Ellis B. Cowling, a plant pathologist and biochemist from North Carolina State University, scientists have been aware of the problem for over 300 years. Dr. Cowling gave a historical perspective of acid deposition in the February 1982 issue of *Environmental Science and Technology* in which he noted that:

> In 1661, J. Evelyn and J. Graunt of England noted the exchange of pollutants between England and France. They suggested dealing with the problem by relocating industries outside of towns as well as using taller chimneys.
>
> In 1727 an Englishman named Hales concluded that, "dew and rain contain salt, sulfur, etc. For the air is full of acid and sulphurous particles . . ."
>
> In 1734 Swedish biologist Linne (Carolus Linnaeus) described what we now know to be acid rain as, "corrode(ing) the earth so that no herbs can grow around it."

The understanding of acid rain increased with the general growth in chemical knowledge throughout the 18th and 19th centuries. Robert Angus Smith was a foremost 19th-century chemist. Rather than merely acknowledging the presence of foul air or stating a possible effect, he studied the phenomenon for many years. Smith's work, published in an 1852 report and an 1872 book, raised the idea that coal burning had an influence on the presence of acid rain and that "sulfuric acid in air" can damage plants and materials, including the corrosion of metals and fading color in textiles. It was Smith who first coined the term "acid rain" in his book *Air and Rain* published in 1872.

In the late 19th century, the United States was no stranger to acid deposition. The vegetation over several square miles at the junction of Georgia, North Carolina, and Tennessee, an area known as the Copper Basin, was destroyed by acid rain from a copper smelter (see Figure 9.6). A mining engineer wrote in 1896 that "No civilized community ought to be afflicted with the nuisance resulting from the continuous dissemination of scores of tons of sulfurous acid gas daily through that part of the atmosphere nearest the surface of the ground." This same engineer recommended that ore roasting be done in enclosed stalls so that gases could be directed into a tall chimney and delivered high into the atmosphere (shades of Evelyn in 1661!). Note that this would solve the problem in the nearby area but would spread the problem far away. We see such spreading nowadays, but often on a continental or global scale, as shown in Figure 9.7.

A. 1973

B. 1996

Figure 9.6   Photo **A,** taken in 1973, illustrates some of the worst industrial pollution known. This area in Ducktown, Tennessee, was known as part of Copper Basin, the site of copper mining and smelting for over 100 years. Large quantities of sulfur dioxide were dumped into the atmosphere and precipitated back onto the soil. No vegetation would grow in this contaminated soil.
B. In 1973 a reclamation effort was begun and has been so successful that it is hard to recognize the same site today.

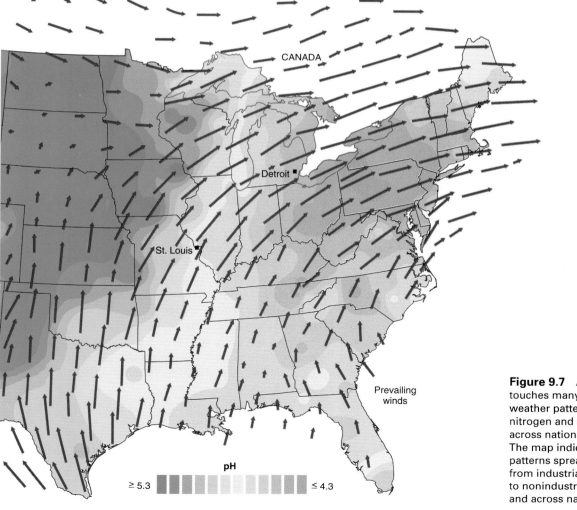

Figure 9.7   A. Acid rain touches many nations when weather patterns help spread nitrogen and sulfur oxides across national boundaries. The map indicates how wind patterns spread acid rain from industrialized areas to nonindustrialized areas and across national borders.

Figure 9.7 (continued)
B. These data support the notion that prevailing winds spread pollution (in this case, nitrogen oxides) across international boundaries.

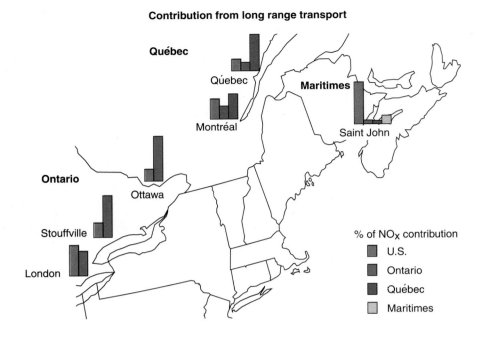

Contribution from long range transport

% of NO$_x$ contribution
- U.S.
- Ontario
- Québec
- Maritimes

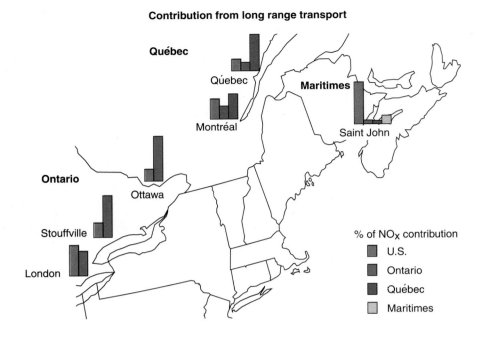

case in point

Closing the Sulfur Gap

We began this chapter by describing the acid rain problem at the Shenandoah National Park in Virginia. One of the scientists studying the ecosystem of Shenandoah, Dr. Art Bulger from the University of Virginia, has also studied acid rain in Norway and has seen what acid rain can lead to. Bulger says, and other reports have confirmed, that hundreds of lakes in southern Norway have no fish and two-thirds of its lakes have lost all their fish over the past century.

We do not normally think of Norway as an industrial giant. It is, however, downwind from the United Kingdom, Germany, and Poland, countries that are heavily industrialized and have a high density of automobile traffic. Norway has, in effect, suffered some of the environmental consequences of the economic vitality of its neighbors.

European countries recognize the problem of too much nitrogen and sulfur oxides and have taken steps to deal with the problem. The steps are based on the concept of "critical load." This is the lowest concentration of the pollutant of interest at which some aspect of the environment is harmed. Figures 1 and 2 show 1998 NO$_x$ and SO$_x$ emissions in Europe. Note areas of heaviest intensity.

The goal of the United Nations' Economic Commission for Europe (UNECE) Sulfur Protocol is to reduce the excess deposition of sulfur by at least 60% by the year 2010. Progress is already being made. In 1980, the United Kingdom generated 4.9 million metric tons of SO$_2$. By 1992, emissions had been reduced to 3.5 million tons. In 1998, the last year for which reliable data are available, SO$_2$ emissions had been cut down to 1.6 million tons.

A protocol also exists for nitrogen oxide emissions, but controlling these is problematic because such emissions largely come from automobiles. As we know from our experience in the United States, driving and emissions are much more difficult to regulate.

One interesting twist on the protocols is the understanding that ammonia and ammonium ions emitted from animal wastes can be oxidized to nitrates, acidifying the soil as follows:

$$NH_4^+(aq) + 2O_2(g) \rightarrow NO_3^-(aq) + H_2O(l) + 2H^+(aq)$$

So ammonia and ammonium ions are two more substances that must be considered when tackling the problems of acidification.

Emission of sulfur dioxide in 1998
(50 km × 50 km grid)

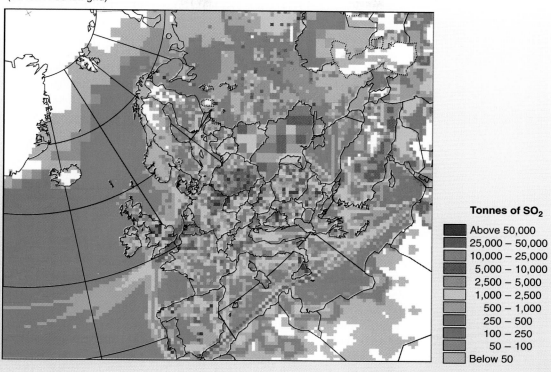

Tonnes of SO₂

Above 50,000
25,000 – 50,000
10,000 – 25,000
5,000 – 10,000
2,500 – 5,000
1,000 – 2,500
500 – 1,000
250 – 500
100 – 250
50 – 100
Below 50

Figure 1

Emission of nitrogen oxides in 1998
(50 km × 50 km grid)

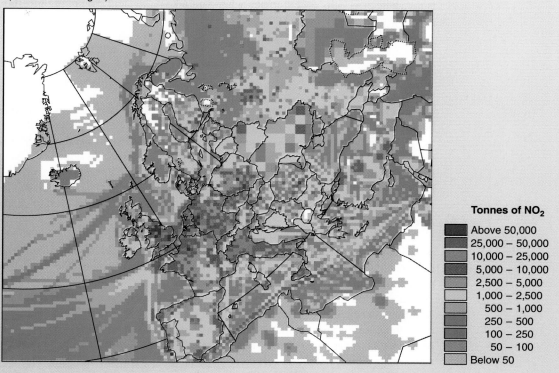

Tonnes of NO₂

Above 50,000
25,000 – 50,000
10,000 – 25,000
5,000 – 10,000
2,500 – 5,000
1,000 – 2,500
500 – 1,000
250 – 500
100 – 250
50 – 100
Below 50

Figure 2

9.3 | Chemical, Ecological, and Social Consequences of Acid Rain

As we enter the 21st century, the burning of fossil fuels by industry, in the home, and for transportation continues unabated. Worldwide, over 700 million motor vehicles were registered in 1999, including 216 million in the U.S. That year, we, as a nation, traveled over 2.5 *trillion* miles in our vehicles. Cars are only one part of the industrial engine that drives our economy and our way of living. What is the impact of our modern lifestyle on lakes and streams, forests and other plant life, urban structures and, ultimately, personal health? Let's consider the effect of acid rain on each of these areas in detail.

Lakes and Streams

The "natural" pH of unpolluted rainwater is between pH 5.5 and 6.0. Remember that rainwater is naturally slightly acidic because carbon dioxide is somewhat soluble in water and, as a nonmetal oxide, is acidic. Any rain with a pH value below about 5.5 is called "acid rain," because it is more acidic than would be expected naturally. Also remember that a change of one pH unit reflects a 10-fold change in hydrogen ion concentration. This means that the water in a lake with a pH of 3.6 is 100 times more acidic than water at pH 5.6. By "100 times more acidic," we mean that the concentration of H^+ ions is 100 times greater.

Since small changes in pH reflect substantial changes in acidity, even a small change in pH can have a noticeable impact on aquatic life. Observable changes in the plant and animal populations of the aquatic environment begin when the pH of a lake or stream falls below 6.0.

The land under and around some bodies of water is partly composed of rocks and soil containing carbonates and bicarbonates. These are compounds containing, respectively, the carbonate (CO_3^{2-}) and bicarbonate (HCO_3^-) ions. These ions can neutralize the acidity of acid rain by combining with hydrogen ions, so lakes and streams embedded in such rocks can, at least to some extent, resist the lowering of pH that acid rain would otherwise cause.

The reaction of the bicarbonate ion buffer with excess hydrogen ions in lakes and oceans is the same as it is in your stomach when you take an antacid containing bicarbonate ions to relieve a bout of acid indigestion:

$$H^+(aq) + HCO_3^-(aq) \rightarrow H_2CO_3(aq) \rightarrow H_2O(l) + CO_2(g)$$

With carbonate-based rocks in lakes, the buffering reaction is:

$$CaCO_3(s) + 2H^+(aq) \rightarrow Ca^{2+}(aq) + H_2O(l) + CO_2(g)$$

The end result of both buffering processes is that a lower concentration of H^+ ions is present in the water than would be the case without the neutralizing effect of the buffer.

Each body of water can be assigned an **acid-neutralizing capacity (ANC),** which measures its ability to neutralize incoming acidity. Lakes and streams with low ANC values tend to have beds of sediments made up of particles of broken-up granite or other rocks that do not have the capacity to neutralize acids. Most ANC is related to limestone dissolution. Parts of New England and huge parts of Canada have a surface dominated by granite rock, resulting in low ANC.

Many bodies of water in the United States have been significantly acidified. One report puts the figure at nearly 6000 lakes and streams. While the overwhelming majority of these have been acidified by acid rain, some, in mid-Atlantic coal mining regions, have been impacted by drainage from mining operations into streams.

In addition to affecting the quality and even the existence of plant and animal life, the acidification of lakes and streams can cause some metal ions normally bound within soil and rocks to dissolve in the acidified water, a process known as **leaching.** The metal ions involved are known as **trace metal ions,** because only very small amounts of them usually enter lakes and streams. Ions of aluminum, mercury, cadmium, and lead are common trace metal ions that can enter waterways in this way. The ions of many trace metals can harm a variety of aquatic organisms including algae, insects, fish, amphibians, and waterfowl. In particular, increased

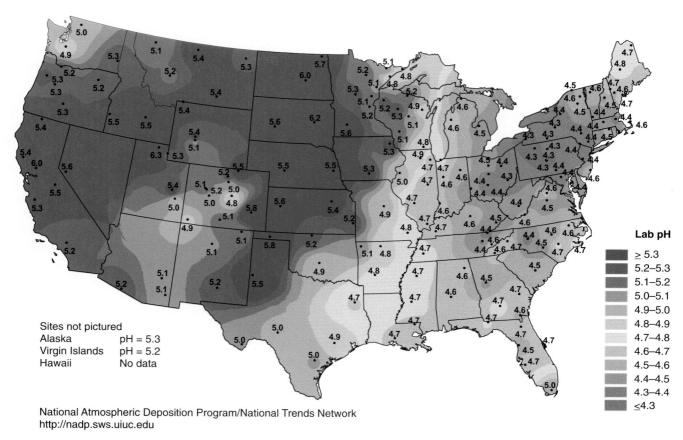

Lab pH

≥ 5.3
5.2–5.3
5.1–5.2
5.0–5.1
4.9–5.0
4.8–4.9
4.7–4.8
4.6–4.7
4.5–4.6
4.4–4.5
4.3–4.4
≤4.3

Sites not pictured
Alaska   pH = 5.3
Virgin Islands   pH = 5.2
Hawaii   No data

National Atmospheric Deposition Program/National Trends Network
http://nadp.sws.uiuc.edu

The northeastern United States suffers the greatest lowering of pH due to acid rain. The region is heavily industrialized, heavily populated, and contains waterways with low acid-neutralizing capacity. These 1999 data are from the National Atmospheric Deposition Program, a nationwide monitoring network involving dozens of government agencies, industrial companies and universities.

concentration of dissolved aluminum is toxic to fishes, although the mechanism is not well understood. Keep in mind that when the bottom rung of a "food chain," such as single-celled algae, is contaminated, the contamination finds its way up the other steps of the food chain, sometimes becoming concentrated at each stage. Thus, trace metal ions that were originally taken up by algae can eventually occur in fish that are caught for human consumption.

Forests

The decline in the ability of the world's forests to maintain healthy plant and animal life cycles is well documented and has been accelerating over the last 20 years. Areas in Europe seem to have suffered the most. According to one recent report, at least 11 tree species in Germany are showing signs of decline, including fir, spruce, pine, and larch. In 1984, 87% of white fir stands and 59% of Scots pine stands showed damage. By 1987, over 17 million acres in 15 European countries showed significant levels of forest damage. The effects continue today. The European Commission and the United Nations Economic Commission for Europe published an October 2000 report asserting that only 36.3% of trees throughout Europe are assessed to be "healthy." All others are at the "warning stage" or more severely damaged. The good news is that 27 European nations have signed the 2000 "Directive on National Emission Ceilings for Certain Atmospheric Pollutants," which calls for significant reductions in NO_x and SO_x emissions.

This forest has been severely damaged by acid rain. Dead and dying trees dominate the landscape and recovery will take decades.

The United States is by no means immune from the destruction. Forests from Maine to North Carolina contain areas of dead trees that look like European "waldsterben" (forest death). In the western United States, significant damage is occurring in the San Bernadino Mountains near Los Angeles, in southern Arizona, and around Mt. Rainier near Puget Sound in Washington State.

The damage is clear, but the causes of it may be complex. Acid rain appears to be involved, but is just one of many interrelated causes of forest destruction. Other suggested factors are:

- An increase in ground-level ozone (O_3) concentration: because it is a gas, ozone enters the pores (known as stomata) in leaves much more readily than do liquids. The ozone apparently causes leaf pores to open and the leaves to become brittle and crack. The ozone interrupts the photosynthetic pathway.
- A combined effect of acid rain and ozone in which the opening of pores caused by ozone allows acid rain to enter more readily.
- The presence of excess nitrogenous fertilizers: about 10% of all nitrogen-containing fertilizers sprayed on farm crops evaporates into the air. This excess can be carried away and eventually deposited by rain in forests. Nitrogen within compounds such as nitrates and ammonia is necessary for plant growth, but when trees get *too much* nitrogen, they seem to require many other nutrients in greater than normal amounts.
- Damage caused by toxic metals released from the ground by the effect of acid rain.
- Natural causes: there are cycles of life and death in forests. When we see some forests apparently dying, this may simply be due to natural cycles that are unrelated to acid rain or other pollutants.
- All of these: a widely accepted view says that there can be many causes of forest destruction. It is likely that each of the cited hazards acts synergistically, meaning that the overall effect of any combination of them is greater than the sum of the effects they would have individually. So the differing causes enhance one another's potential to cause damage.

The research into the specific causes of forest destruction is painstaking and time consuming, but it is driven by a great sense of urgency to more fully understand the impact of acid rain.

Urban Structures

If you live in or have visited a large city, you can see evidence of the decay of stone buildings and statues. Acid rain causes these structures to "weather," or in other words, wear away more quickly under the onslaught of rain, wind, and frost. This can blur fine architectural detail that was clearly visible not too long ago. For example, compare the photographs of Figure 9.8.

Mortar, **marble,** and limestone all contain calcium carbonate ($CaCO_3$) as a major or sole constituent. As we discussed on page 298, all carbonates will react with the hydrogen ions of acids to yield carbon dioxide and water:

$$2H_3O^+(aq) + CaCO_3(s) \rightarrow Ca^{2+}(aq) + CO_2(g) + 3H_2O(l)$$

Figure 9.8 This statue of George Washington was placed outside in 1944. As acid rain worked to dissolve the stone, details crumbled away. By 1994 there was significant damage to the entire statue.

This equation summarizes the chemistry behind the slow but steady erosion of many buildings by acid rain. If the acid involved is sulfuric (derived from emissions of SO_2), sulfate ions take the place of the original carbonate ions as summarized by the equation:

$$H_2SO_4(aq) + CaCO_3(s) \rightarrow CaSO_4(s) + CO_2(g) + H_2O(l)$$

The calcium sulfate formed is slightly soluble in water, so the insoluble calcium carbonate is replaced by the slightly soluble calcium sulfate, allowing some of the structure of an affected building to be literally "washed away" by rainfall.

Many stone buildings built from the late 18th through early 20th centuries were often beautifully ornate, showing great attention to detail. Acid rain threatens to make such detailed beauty a thing of the past.

Acid Rain and Human Health

In our industrialized society, there are hundreds of possible environmental causes of ill health including smoking, soot, street-level ozone, acid rain, and stress. Any one, or a combination of more than one, can contribute to illness such as respiratory problems. How do researchers assess the specific impact of acid rain and its precursor molecules on health? With so many variables in play, we can never say with absolute certainty that acid rain alone caused a particular set of symptoms, but we can look for suggestive evidence.

We do know that atmospheric sulfur dioxide levels in excess of 100 parts per billion (100 ppb) cause chest tightness and wheezing in the roughly 20% of Americans with asthma and other breathing problems. Young children with underdeveloped lungs are even more prone to breathing problems and sensitive to polluted air. Sulfur dioxide concentrations greater than 100 ppb are often found in big cities for hours on end. The air in many towns near smelters averages four to ten times that concentration. Figure 9.9 shows the estimated SO_2 emissions from electric utilities from 1980 through 1999. Strict government controls took effect in 1995, leading to a sharp decrease in SO_2 emissions (Figure 9.10). Electric utilities account for about 67% of all SO_2 emissions. Figure 9.11 lists sources of NO_x in the United States. Emissions have remained constant at about 23 million tons annually because only 30% of NO_x emissions are from electric utilities. The rest is from motor vehicle use, which is increasing.

There is some, as yet inconclusive, evidence that acid rain is responsible for high levels of lead leaching into some residential well-water samples in the Adirondack Mountain range of northern New York State. The relevant reaction is:

$$PbCO_3(s) + 2H^+(aq) \rightarrow Pb^{2+}(aq) + CO_2(g) + H_2O(l)$$

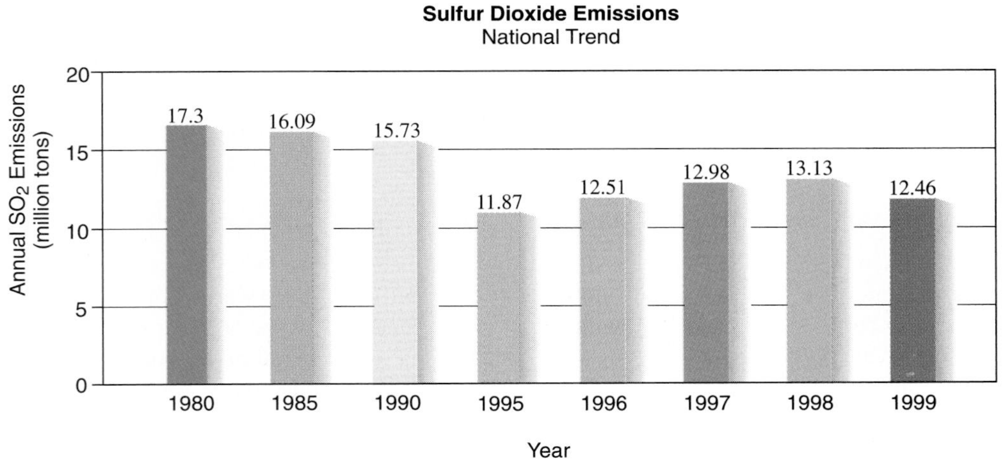

Sulfur Dioxide Emissions
National Trend

Figure 9.9 Title IV of the Clean Air Act Amendments of 1990 established the Acid Rain Program to help electric utility plants decrease the level of sulfur dioxide (SO_2) in their emissions. Economic incentives and not government penalties are the motivators in this program. The goal is to reduce the 1980 level of emissions by 10 million tons by the year 2010.

Figure 9.10 SO_2 emissions from electric utilities, by state. Note those regions of the country that have shown sharp decreases in SO_2 emissions. What might be the impact of these changes?

Lead has long been suspected of causing brain damage in children. The *direct* evidence, however, for a link between lead in drinking water and brain damage is not totally convincing.

Economic Considerations

Acid rain is such a pervasive and widespread problem that it is difficult to even begin to assess its financial impact. Is it in our economic interest to reduce the SO_x and NO_x emissions that lead to acid rain? How do we balance the economic and social health concerns? When put in those terms, economics seems to recede from its position as a primary consideration. But in the "real world" money is a critically important factor in industrial and governmental decisions. Another complicating aspect of the decision-making process is that the people who benefit economically from the polluting technology are often not the ones who suffer the resulting environmental consequences.

The key economic themes here are:

- Assuming levels of offending substances could and should be lowered, how technologically, can they be reduced?
- How much will it cost?
- What is the current and future medical price that we pay for inaction?

There are so many variables involved that global estimates of the costs are not very precise. Nonetheless, in the United States alone, over $800 billion dollars was spent on pollution abatement between 1980 and 1991 with $122 billion spent in 1994, the last year for which figures are available from the U.S. Department of Commerce's Bureau of Economic Analysis. Substantial efforts are being made to address the sulfur and nitrogen oxide emission levels in Europe.

Title IV of the Clean Air Act Amendment of 1990 mandated SO_2 emission reductions from 1980 levels of 10 million tons per year, and reductions in NO_x emissions of 2 million tons—both beginning in 2000. The Energy Information Agency of the U.S. Department of Energy estimates the annual cost of Title IV compliance reduction in SO_2 emissions at $836 million (1995 dollars) in the United States alone (see Case in Point: Closing the Sulfur Gap).

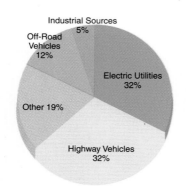

Figure 9.11 U.S. NO_x emissions in 1999 (23 million tons).

exercise 9.2

Emission Rate

Problem

In 1999, the United States emitted a little more than 12 million tons of SO_2. The total mass of the Earth's atmosphere is 5.2×10^{21} g. How many years would it take for this rate of emission to raise the concentration of SO_2 to 100 ppb if none of the SO_2 ever left the atmosphere?

Solution

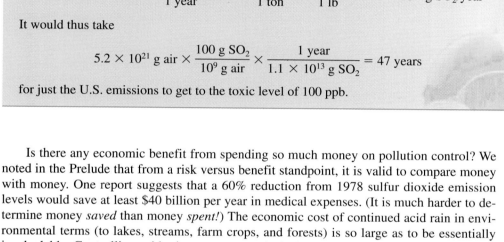

$$\text{mass } SO_2/\text{year} = \frac{1.2 \times 10^7 \text{ tons } SO_2}{1 \text{ year}} \times \frac{2000 \text{ lb}}{1 \text{ ton}} \times \frac{453.6 \text{ g}}{1 \text{ lb}} = 1.1 \times 10^{13} \text{ g } SO_2/\text{year}$$

It would thus take

$$5.2 \times 10^{21} \text{ g air} \times \frac{100 \text{ g } SO_2}{10^9 \text{ g air}} \times \frac{1 \text{ year}}{1.1 \times 10^{13} \text{ g } SO_2} = 47 \text{ years}$$

for just the U.S. emissions to get to the toxic level of 100 ppb.

Is there any economic benefit from spending so much money on pollution control? We noted in the Prelude that from a risk versus benefit standpoint, it is valid to compare money with money. One report suggests that a 60% reduction from 1978 sulfur dioxide emission levels would save at least $40 billion per year in medical expenses. (It is much harder to determine money *saved* than money *spent!*) The economic cost of continued acid rain in environmental terms (to lakes, streams, farm crops, and forests) is so large as to be essentially incalculable. Controlling acid rain costs a great deal of money, and yet it saves a great deal of money. We should also remember that the costs involved in controlling pollution can be beneficial to companies manufacturing pollution control equipment. Pollution control is a "growth industry," and the jobs and income generated by this industry must be fitted into the overall economic equation when considering the costs and benefits of adopting less polluting habits.

9.4 | Where Do We Go from Here?

Green Chemistry

Acid rain is an ever-increasing by-product of our continuing industrial development and of our reliance on cars as a means of personal transportation. As the developing countries become increasingly industrialized, they too are becoming major contributors to the global acid rain problem. In the past few years U.S. public policy has begun to reflect the need to control the emissions that cause acid rain.

Chemical treatment methods already exist that can reduce NO_x and SO_x emissions by up to 90%. Yet the economic, governmental, and individual choices needed to fully exploit these methods have yet to be made. Let's examine the chemistry behind some methods that can be used to reduce the acid rain problem.

The Chemistry of a Cleanup

Several methods are already in use to reduce SO_x emissions. An important one, called **flue-gas desulfurization (FGD)** has been used with success in the United States, Japan, and Germany.

pro con discussions

Should the Government Act Before It Is Sure of All the Facts?

 Green Chemistry

pro

When government is faced with a difficult decision, it is very easy to say, "let's study the problem some more; then, with fuller information, we can make a decision." This is sometimes a responsible plea from legislators for better information upon which to base a decision, but sometimes it is a stall to avoid doing what is necessary but that will be unpopular.

Installing flue gas desulfurizers on electricity-generating facilities, putting catalytic converters on automobiles, and cleaning coal before burning it are all ways to lower the amount of oxides of sulfur and nitrogen that enter our atmosphere. These measures would decrease the level of acid in rain. Each of these policies clearly has a cost to the consumer, but their benefits may be less obvious. How certain must government be of the importance of the long-term benefits before forcing citizens to spend money and effort to combat the chemical problem of acid rain?

Scientists have long recognized that it is easier to prove an hypothesis false than to prove one true. There-fore, opponents of such changes can point to the lack of absolute scientific proof that a policy change will result in a discernible improvement in environmental pollution. But the correlation between increased sulfur emissions and increased sulfuric acid in rain, and between the increased use of automobiles and the increased nitric acid in rain, are very indicative. The pattern of acid rain over the continent is consistent with known generation sites of SO_x and NO_x and well-known weather patterns. Can government not be reasonably confident of the link between these pollutants and acid rain?

We are not arguing that government should start banning things just because someone with some credentials comes in and starts quoting statistics. But when a number of scientists independently compile data indicating a problem, identify the principal causes, and suggest ways to diminish the problem, shouldn't the government listen to them? After all, the role of government should be "to promote the general welfare."

con

The argument that "the government should act when there is a reasonable degree of scientific evidence indicating a problem" sounds fine in political science classes. In the real world, the importance of economic growth is the key to a society where jobs are created and people are fed and housed. Governments act on this basis. The link between cigarette smoking and diseases of the lungs and circulatory system has been well-established for nearly 40 years; yet the U.S. government actually pays farmers millions of dollars to grow tobacco that will kill well over 400,000 people per year, because it is good for the economy.

The link between acid rain and forest destruction is a lot less well-established than the link between tobacco and fatal diseases. Peter Warneck of the Max Planck Institute in Germany said, "the catastrophic death of entire forests predicted for the 1980s did not take place. However, forest ecosystems react over long periods and in a complex manner to external forces such as pollutant deposition." Do we really want to affect the operation of vital industries in an effort to contain a problem that may not really be all that serious?

further discussion

Pro/Con Questions: Should governments regulate industrial processes when such regulation impacts on the economic vitality of the company? Do pollution controls negatively impact on the economy?

Quote Out of Context: We must be on our guard when quotes are taken selectively to support one side of an argument. The complete quote by Warneck is as follows: "the catastrophic death of entire forests predicted for the 1980s did not take place. However, forest ecosystems react over long periods and in a complex manner to external forces such as pollutant deposition. Reduction in the emissions of air pollutants is a vital prerequisite for the maintenance of healthy forests."

exercise 9.3

Acid Rain

Problem

Suppose rain is acidic due to sulfuric acid and its pH = 3.0. How many inches of rain would have to fall onto a limestone tombstone to make the inscription illegible if the letters were originally cut 0.5 cm into the stone? Limestone is nearly pure $CaCO_3$ and has a density of 2.7 g/cm³. Assume that half the acidity is consumed in this reaction.

Solution

The hydrogen ion concentration of this rain = 1×10^{-3} mol/L. If half of it reacts, that means that 5×10^{-4} mol of hydrogen ion reacts per liter of rain.

$$2H^+ + CaCO_3 \rightarrow CO_2 + H_2O + Ca^{2+}$$

The tombstone would have to be dissolved sufficiently that at least most of the top 0.5 cm is dissolved away. It doesn't matter what area of stone we consider, but for simplicity let's use a square area 10 cm on a side. A liter is a cube 10 cm on a side, so 1 L of rain on this stone would be 10 cm deep.

Calculate the number of moles of $CaCO_3$ in a piece of limestone $10 \times 10 \times 0.5$ cm = 50 cm³ of stone.

Moles of stone $(CaCO_3)$ = (50 cm³ stone)(2.7 g/cm³)(1 mol $CaCO_3$/100 g $CaCO_3$)
= 1.35 mol of stone

This would require 1.35 mol stone (2 mol H^+/mol $CaCO_3$) = 2.7 mol H^+.

Volume rain needed = (2.7 mol H^+) (1 L rain/5×10^{-4} mol H^+) = 5400 L.

Since each liter represents 10 cm of rain on our piece of stone, the depth of rain needed is 5.4×10^4 cm or 2.1×10^4 inches of rain. If an area gets 50 inches of rain per year, this is 425 years worth of rain.

In FGD, wet limestone $(CaCO_3)$ is sprayed into the hot gases escaping as a result of the burning of coal. This initiates a two-step process summarized by these equations:

Step 1: $CaCO_3(s) \rightarrow CaO(s) + CO_2(g)$

Step 2: $CaO(s) + SO_2(g) + 1/2O_2(g) \rightarrow CaSO_4(s)$

In step 1, the limestone decomposes when exposed to the heat of the flue gas generating calcium oxide and carbon dioxide. The carbon dioxide released may also cause another problem of current interest, the "greenhouse effect." We will have much more to say about this in Chapter 12—Air Quality. The calcium oxide solid is the substance that actually reacts with the sulfur dioxide gas to form the final product, calcium sulfate. The tiny lumps of solid calcium sulfate can be trapped and collected, preventing the sulfur locked within the compound from being released as SO_2 gas. Notice that this is the same overall process (the conversion of calcium carbonate to calcium sulfate) as occurs when marble is worn away by acid rain. In this situation, however, we use it to prevent acid rain.

Although FGD is very effective at reducing SO_2 emissions, it reduces the efficiency of power plants and it does nothing to reduce NO_x emissions. The product (calcium sulfate) also takes up more space than the $CaCO_3$ dug out of the ground to be used as the raw material for the process.

exercise 9.4

Power Plant Dilemma

Problem

A typical power plant burns one million metric tons (1 metric ton = 1000 kg = 1 Mg, or 2200 lb) of coal per year. The coal is 2.0% sulfur, by mass. If all of the sulfur is removed by the FGD process, how much $CaSO_4$ will the power plant generate in one year?

Solution

$$g\ CaSO_4 = 10^6\ Mg\ coal \times \frac{10^6\ g}{1\ Mg} \times \frac{2.0\ g\ S}{100\ g\ coal} \times \frac{1\ mol\ S}{32\ g\ S} \times \frac{1\ mol\ SO_2}{1\ mol\ S}$$

$$\times \frac{1\ mol\ CaSO_4}{1\ mol\ SO_2} \times \frac{136\ g\ CaSO_4}{1\ mol\ CaSO_4}$$

$$= 8.5 \times 10^{10} g!!$$

Another technology, called atmospheric fluidized-bed combustion, makes use of a turbulent "bed" of burning pulverized coal and limestone that is suspended in air. Thorough mixing of the coal and air in this fashion allows the mixture to burn at a lower temperature, thus producing fewer nitrogen oxides (normally formed from the reaction of nitrogen and oxygen in air at *high temperatures*). The presence of the limestone allows the removal of SO_x gases by the same chemistry as in the FGD process.

These and other technologies should certainly be promoted in order to reduce the amount of industrial pollutants released into the air. At the moment, however, almost half of the coal-fired power plants in the eastern United States have no pollution-control devices in spite of the relatively high sulfur levels in the coal they burn. Adding pollution controls to existing power plants is called "retrofitting," or "repowering." Installing them can cost millions of dollars per plant, and the costs are almost always passed on to the consumer in the form of higher electricity bills. Yet as we saw in the previous section, not dealing with the problem is often as expensive as enforcing a solution.

Government and industry are often at odds regarding pollution controls. Yet bridges can be built for the public good. For example, in 1990 alone, automobile manufacturers spent over $11 billion on pollution abatement and prevention technologies in cars. Devices such as catalytic converters are an important part of this. Have manufacturers gone as far and as fast as they could? Would electric cars eliminate automobile pollution? We consider this further in Chapter 15.

A Complicated Problem

We now know that many kinds of pollution, from many origins, interact to cause environmental damage. We must also realize that air pollution ignores national boundaries: pollution originating in one country can affect other countries (perhaps even all countries). It therefore seems prudent to develop an international policy to address pollution and antipollution law enforcement. The United States began attempts to improve air quality within its own continental borders through the 1970 Clean Air Act. The intent of the act was to have the Environmental Protection Agency (EPA) review and revise limits on air pollutant concentration by 1980.

There is an on-going pull-and-tug between the branches of government regarding the establishment of pollution-related regulations. The issues are political (state's rights versus

Would electric cars eliminate automobile-related pollution? On balance, would the creation and disposal of the power-pacs needed for these cars pose another pollution problem? Remember, there is a pro and con to every action.

federal control), economic (who pays, who benefits?), social (who is affected?), and chemical (what strategies are available?).

Debate about the appropriate governmental response to the acid rain problem will continue as will debate about its causes and the significance of its effects. Individuals must ask themselves about the lifestyle changes they are or are not prepared to make. Are you prepared to drive a smaller car, less often and more slowly, for example, as your personal contribution to reducing damage to the environment? Are you prepared to support public transportation by using it in preference to a car? Could you even imagine living your life without the use of a car at all? Or do you feel it is a problem for governments to fix, through legislation, rather than one that can possibly be affected by individual actions? The choices are yours, but ask yourself what the world might be like 50 years from now if emissions of SO_x and NO_x continue to increase. Then go outside for a breath of fresh air!

main points

- Acid rain is a result of automobile and industrial smokestack emissions.

- The chemistry of acid rain is fairly well understood.

- Acid rain has had important and measurable ecological consequences.

- As a result of weather patterns, acid rain–causing gases spread great distances from their sources, making acid rain a global rather than local issue of concern.

- There are several technologies available for reducing sulfur and nitrogen oxide emissions.

- There is a constant tension between the environmental consequences of our industrial economy and our modern lifestyle.

important terms

Acid deposition is the total effect of acid falling in rain, snow, etc. or dry solids. (p. 305)

Acid-neutralizing capacity (ANC) describes the ability of water or soil to neutralize falling acids. (p. 314)

Chalcocite is a copper ore composed of copper(I) sulfide. (p. 307)

Chalcopyrite is a copper ore composed of copper/iron mixed sulfide. (p. 307)

Flue-gas desulfurization (FGD) is a process for removing sulfur from combustion products of sulfur-containing fuel, usually coal. (p. 319)

Leaching is the slow dissolving of metals by passing solvent through granular solid. (p. 314)

Marble is calcium carbonate transformed geologically from limestone by high pressure and temperature. (p. 316)

NO_X is the combination of several oxides of nitrogen, especially NO and NO_2. (p. 306)

Pyrite is a shiny golden-colored iron ore, FeS_2. (p. 307)

SO_X is the combination of oxides of sulfur, especially SO_2 and SO_3. (p. 306)

Sulfide ore is a mineral in which the anion is a sulfide or disulfide ion. (p. 307)

Sulfurous acid (H_2SO_3) forms when sulfur dioxide (SO_2) comes in contact with water (H_2O). (p. 307)

Synergistic describes two (or more) processes that increase each other's effects such that the total effect is greater than the two effects taken separately. (p. 308)

Trace metal ions are metal ions present in very low concentrations in water or soil. (p. 314)

exercises

1. List the three main sources of acid rain. What do NO_2 and SO_2 react with to cause acid rain?
2. Why might lakes and streams have a lower pH in spring?
3. In your own words, define "acid rain" and discuss its environmental impact.
4. What are the two most significant acid-forming gases in the atmosphere? How do they interact with water to form acidic solutions?
5. Compare sulfur dioxide and carbon dioxide in terms of their acidic properties.

6. What is the primary human-made source of NO_2 in the atmosphere?

7. The evergreen population atop Camel's Hump in the Green Mountain Range in Vermont has diminished greatly. What could be some possible causes of this destruction?

8. Describe why nitrogen and oxygen do not react with each other in the atmosphere under normal conditions but do react in cylinders of internal combustion engines and in the atmosphere during lightning strikes.

9. True or False (explain your answer). Acid rain is a new phenomena, developing as a result of our industrialized society.

10. What is the pH range of acid rain? Why isn't acid rain defined as "precipitation with a pH less than 7"?

11. Explain what happens to a lake ecosystem when it is contaminated with acid rain.

12. Define acid-neutralizing capacity.

13. Is acid rain the only cause of forest destruction? Explain.

14. Why are building structures containing mortar, marble, and limestone susceptible to decay from acid rain? Include the relevant chemical equations.

15. Explain how sulfuric acid decays many statues and buildings.

16. List some potential health effects of acid rain.

17. What is the relationship between acid rain and potential lead poisoning?

18. Draw a qualitative graph of acid rain levels versus the continuing industrialization of society.

19. How much sodium hydroxide is necessary to neutralize the acid formed from 5.0 mol of NO reacting with an excess of oxygen and water.

20. How many moles of sulfuric acid are needed to decrease the pH from 5.5 to 3.0 of a 1.00-inch rain over a square area 100 km on a side?

21. How much nitric acid would be needed to dissolve 5.0g of lead carbonate?

22. If you dissolved 0.49 g of sulfuric acid in 100 L of water, what would be the pH of the water?

23. If 7.2 g of nitric acid were dissolved in 220 L of water, what would be the pH of the water?

24. Which would supply more hydrogen ions: 50.0 g of sulfuric acid or 50.0 g of nitric acid?

25. Which would supply more hydrogen ions: the acids formed from complete oxidation and hydration of 50.0 g of nitrogen monoxide or 50.0 g of sulfur dioxide?

26. There have been various attempts to remove sulfur from coal before it is burned. Why do you suppose these processes have not been widely adopted?

27. What is the role of calcium oxide in flue-gas desulfurization?

28. Compare the benefits and drawbacks to flue-gas desulfurization.

29. Why do you suppose that flue-gas desulfurization lowers SO_x emissions but not NO_x emissions?

30. If you are using flue-gas desulfurization to "clean up" 2 moles of sulfur dioxide, how many moles of calcium carbonate must you use?

31. If you drive up to the Blue Ridge Parkway in the mountains of North Carolina, you will see several trees without leaves and very dark bark. How might acid rain have affected these trees? Why do you think the damage is worse near the top of the mountain?

32. Dibenzothiophene ($C_{12}H_8S$) is a common sulfur-containing compound of coal. Write the balanced equation for its combustion.

33. What volume of air would be contaminated to a level of 1 ppm by combustion of a kilogram of dibenzothiophene? Air has a density of approximately 1.3 grams per liter.

34. What fraction of sulfur is present as pyrite and what fraction as dibenzothiophene if a kilogram of coal is shown to form 30.0 g of SO_2 and to contain 7.00 g of iron?

35. A rain event of 5.0 centimeters fell over an area of 400 square kilometers. The pH of the rain is measured to be 4.0 and the acidity was shown to result from nitric acid. What mass of NO in the atmosphere could give rise to this rainfall?

36. What color would red cabbage indicator (see Section 8.7) turn in the most acidic rainfall represented in the pH map on page 315?

37. Which indicator described in Table 8.7 would be most useful in determining the pH of suspected acid rain?

38. How much CO_2 is formed in making enough calcium oxide, CaO, to scrub (via flue-gas desulfurization) 1000 kilograms of sulfur from a coal-burning facility?

food for thought

39. The "white sands" of White Sands National Monument are composed of powdered gypsum. What would be the pros and cons of taking calcium sulfate from FGD processes and just adding it to the beautiful dunes of this area?

40. Do you think the government should play more of a role in pollution control by regulating industry emissions?

41. Propose some actions individuals can take to stem the impact of acid rain. How might these measures be different for people who live in the city compared to suburban or rural populations?

42. Using what you've learned about acid-base indicators, design a step-by-step experiment you could use to see if the rain falling in your back yard is acidic.

43. In the past 20 years the view of the Grand Canyon's north rim from the south rim has become foggy. In 1980 the EPA issued regulations that required control measurements from sources that hampered the visibility in the Grand Canyon. In 1987, one of the largest coal-

burning electric power plants in the United States (located within a few kilometers of the Grand Canyon) agreed to be tested for SO_2 emissions as they did not already have pollution controls. It was found that emissions from the power plant did contribute to the visibility impairment in the park, but the utility owners argued that other sources were more significant. In 1991, the EPA reached an agreement with the utility owners in which the SO_2 emissions would be reduced by 90% at an estimated capital cost of $430 million. Under the provisions of the 1990 Clear Air Act, the EPA was able to give the utility company "credit in the form of marketable allowances which can be sold or traded" because their emission reduction was better than the required levels. Do you think it is a fair compromise (marketable allowances) for industries that give off NO_x and SO_x emissions? Why do you think it took 10 years from the time the EPA issued regulations to the date of agreement for action to be taken by this company?

44. The U.S. National Acid Precipitation Assessment Program (NAPAP) spent $500 million on a 10-year study of acid deposition. The results of NAPAP's research were that acid deposition is in fact a cause of some surface water acidification; 4% of lakes and 8% of streams are chronically acidic, and 20% of the lakes and streams in the country have a low acid-neutralizing capacity. Thus, more than 90% of the U.S. surface waters are not considered acidic by NAPAP's standards. Do we need to be concerned with the effects of acid deposition on lakes and streams in the United States? Why or why not?

45. Pollution control in Eastern Europe is often nonexistent. Many cities are suffocated by blankets of sulfur-filled air; historical sculptures and buildings are crumbling to pieces; more than 600,000 acres of woodland have been damaged in Poland. The Danube River carries waste from eight nations. The life expectancy in Eastern Europe is actually lower now than in 1987, due to a combination of environmental and economic factors. Prior to the disintegration of the Iron Curtain, pollution emission levels were kept secret and confidential by the Communist party. During the political changes in the fall of 1989, the people of Eastern Europe complained about polluted air merely in protest of the communist government. However, today public complaints are few and environmental awareness is considered to be low. How can emission control plans be implemented in such unstable countries where concerns are focused on money, family, and employment rather than pollution? Should U.S. citizens be concerned with the pollution problems in countries as far away as Eastern Europe? (After all, we don't have to breathe their air, or do we?)

46. In light of our earlier discussion of costs and benefits, list costs and benefits associated with the copper smelting in the Copper Basin. How could benefits have been increased and costs minimized in the handling of this project? Why do you suppose that these things were not done at the time?

readings

1. Burks, Ned. Shenandoah Park on the brink. *American Forests,* Nov/Dec 1994, pp. 17–23.
2. Cowling, E. B. Acid precipitation in historical perspective. *Environmental Science and Technology,* February 1982, pp. 110A–123A.
3. Baedecker, P. A., and Reddy, M. M. The erosion of carbonate stone by acid rain. *J. Chem Ed.,* February 1993, pp. 104–108.
4. Baker, L. A., Herlihy, A. T., Kaufmann, P. R., and Eilers, J. M. Acidic lakes and streams in the United States: The role of acidic deposition. *Science,* May 24, 1991, pp. 1151–1154.
5. Chiras, D. D. 1991. *Environmental Science: Action for a Sustainable Future.* Menlo Park, CA: Benjamin/Cummings.
6. Mohnen, V. A. The challenge of acid rain. *Scientific American,* August 1988, pp. 30–38.
7. Schwartz, S. W. Acid deposition: Unraveling a regional phenomenon. *Science,* February 10, 1989, pp. 753–763.
8. Quinn, M. L. Early smelter sites: A neglected chapter in the history and geography of acid rain in the United States. *Atmospheric Environment,* 23: November 6, 1989, pp. 1281–1292.
9. Stevens, W. K. Study of acid rain uncovers a threat to far wider area. *New York Times,* January 16, 1990, p. 21.
10. Sugar maples sicken under acid rain's pall. *New York Times,* May 5, 1991.
11. Thompson, Jon. East Europe's dark dawn. *National Geographic,* June 1991, pp. 42–63.
12. Bliss-Guest, Patricia. United States of America National Report, Council on Environmental Quality, U.S. EPA, Washington, D.C., 1992.
13. EPA. Progress report on the EPA acid rain program. November 1999, http://www.epa.gov/acidrain.
14. DOE. The effects of Title IV of the Clean Air Act Amendments of 1990 on electric utilities: An update. 1997, http://www.eia.doe.gov/cneaf/electricity/clean_air_upd97/exec_sum.html.

websites

www.mhhe.com/kelter The "World of Choices" website contains activities and exercises including links to websites for the Environmental News Network, reporting on the status of sulfur dioxide emissions in China, Brazil and Argentina, and; the Norwegian Pollution Control Authority, and; and much more!

Water Quality: Chemical Concerns, Chemical Solutions

10

Out in the ocean they say the water's clear
But I . . . live right at Beacon here.
Halfway between the mountains and the sea,
Tacking to and fro, this thought returns to me:
Sailing up my dirty stream
Still . . . I love it and I'll dream
That some day, though maybe not this year,
My Hudson River and my country will run clear.

—"Sailing Up My Dirty Stream" by Pete Seeger (1961)

In the quotation we use to open this chapter, the folk singer Pete Seeger describes sailing on the Hudson River, near his home of Beacon, New York, in the early 1960s. In a later account he recalls, "lumps of toilet waste floating past me." In the United States and many other parts of the world, our waterways are still used as disposals for human and industrial waste, all composed of various chemicals. In many cases, such as with the Hudson River, the pollution is obvious, with mounds of clearly visible waste floating by (Figure 10.1). In other cases, the damage is more insidious: maybe a slight change in the color of the water, or a fish or plant species that is no longer where it was just a year ago, is all that indicates a problem. To understand water pollution we need to know about the causes, the effects, the risks, and, many would contend, the benefits of using worldwide waterways as dump sites for assorted natural and synthetic chemicals.

Understanding water pollution also means reinforcing one of the themes of this textbook, which is that the materially closed system of the Earth is built on chemical cycles, such as the water cycle, nitrogen cycle, carbon cycle, and so on. Pollutants in water do not simply stay there but become incorporated in some of the great cycles of the Earth, including the ones that can take pollutant chemicals into, and then out of, living things including ourselves.

Figure 10.1 Sometimes water pollution is obvious, as in this photo of debris on the Hudson River in New York State. Other times, the pollution is harder to detect, although it can be just as hazardous.

In this chapter we will look at the chemistry involved in the pollution of water, discuss the different forms of water pollution, and consider the availability and chemistry of clean-up procedures. To let you know that this is not a new problem, nor one that is localized in the United States, we begin by turning the clock back to 19th century London.

Growing urbanization and industrialization were responsible for the first obvious and widespread water pollution in Europe in the mid-19th century. As you can see from the Case In Point: On the Filth of the Thames, the Thames River in London was not the place to be near on a hot July day.

The pollution that existed in 1850s London was largely caused by the bodily waste of the people and horses of the city. Indoor plumbing did not exist, so chamber pots full of urine and feces were generally emptied from windows into the street. All vehicles were pulled by horses, so there were nearly as many horses as there were people in what was then the most populous city in the world (2,685,000 in 1850). Many of the streets were unpaved, there were few sewers, and those that did exist merely allowed rain water to wash the colossal amounts of waste into the river Thames.

It was about this time that, in Paris, Louis Pasteur and others were learning about the importance of microorganisms as causes of disease. The microorganisms responsible for many diseases, such as cholera and typhoid, were carried by water supplies. Water for household use was not piped into the house but was carried one bucket at a time from neighborhood wells. These were dug in the same ground on which all the waste was being dumped, so wells sometimes became badly contaminated.

We know how different life in Western cities is now compared to the mid-1800s. However, although the water-related problems are different, the concern that existed in 19th-century London is still with us. Nowadays, water pollution is also a cause for concern outside of cities, in farmland, on beaches, and anywhere humans come into contact with water.

case in point

On the Filth of the Thames

Note: Michael Faraday was one of the greatest chemists and physicists of the 19th century. He began his scientific career working in Sir Humphrey Davy's laboratory when Davy was doing some of the very early research on the relationship between electricity and chemistry. Although Davy isolated sodium, potassium, barium, strontium, calcium, and magnesium for the first time, his most important discovery is often said to be Michael Faraday!

Observations on the Filth of the Thames, contained in a July 7, 1855 letter addressed to the Editor of "The (London) Times" newspaper, by Professor Faraday.

SIR,

I traversed this day by steamboat the space between London and Hungerford Bridges between half-past one and two o'clock; it was low water, and I think the tide must have been near the turn. The appearance and the smell of the water forced themselves at once on my attention. The whole of the river was an opaque pale brown fluid. In order to test the degree of opacity, I tore up some white cards into pieces, moistened them so as to make them sink easily below the surface, and then dropped some of these pieces into the water at every pier the boat come to; before they had sunk an inch below the surface they were indistinguishable, though the sun shone brightly at the time; and when the pieces fell edgeways the lower part was hidden from sight before the upper part was under water. This happened at St. Paul's Wharf, Blackfriars Bridge, Temple Wharf, Southwark Bridge, and Hungerford; and I have no doubt would have occurred further up and down the river. Near the bridges the feculence rolled up in clouds so dense that they were visible at the surface, even in water of this kind.

The smell was very bad, and common to the whole of the water; it was the same as that which now comes up from the fully-holes in the streets; the whole river was for the time a real sewer. Having just returned from out of the country air, I was, perhaps, more affected by it than others; but I do not think I could have gone on to Lambeth or Chelsea, and I was glad to enter the streets for an atmosphere which, except near the sink-holes, I found much sweeter than that on the river.

I have thought it a duty to record these facts, that they may be brought to the attention of those who exercise power or have responsibility in relation to the condition of our river; there is nothing figurative in the words I have employed, or any approach to exaggeration; they are the simple truth. If there be sufficient authority to remove a putrescent pond from the neighborhood of a few simple dwellings, surely the river which flows for so many miles through London ought not to be allowed to become a fermenting sewer. The condition in which I saw the Thames may perhaps be considered as exceptional, but it ought to be an impossible state, instead of which I fear it is rapidly becoming the general condition. If we neglect this subject, we cannot expect to do so with impunity; nor ought we to be surprised if, ere many years are over, a hot season give us sad proof of the folly of our carelessness.

I am, Sir
Your obedient servant,
M. FARADAY.

The cartoon shown here, in response to Faraday's letter, was printed a few days later in the humor magazine *Punch*.

Something to think about: Is sticking white cards into the water, as described by Faraday, a good way to test for the cleanliness of the water? Why or why not?

FARADAY GIVING HIS CARD TO FATHER THAMES;
And we hope the Dirty Fellow will consult the learned Professor.

This cartoon from July 1855, appeared in the magazine *Punch*. Clearly, water pollution is not a new problem.

Dr. John Snow plotted cholera deaths in London for September 1854. Snow noted that most of those who died drank water from the Broad Street water pump (X on the map). The epidemic, which took more than 500 lives, stopped when the pump was disabled.

10.1 | What Is Water Pollution?

The Federal Clean Water Act (CWA) of 1972 defines water pollution as "the man-made or man-induced alteration of the chemical, physical, biological, and radiological integrity of water." This definition seems questionable because: (1) water can be polluted according to the official definition but still be safe for certain uses and (2) it is restricted to the effects of human activity. Waterways can be polluted from animal activity as well, but human activity produces far greater and more damaging pollution, so we will generally discuss only human causes of water pollution. We will also concentrate mainly on fairly simple chemical **pollutants;** but it is important to remember that more complex "biological" pollutants are just mixtures of many individual chemicals.

What Are the Causes of Water Pollution?

 Green Chemistry

Anytime some substance leading to a poorer quality of water is added to water, the water becomes polluted. "Poorer quality" is a very broad term that could refer to the cleanliness, taste, smell, or even appearance of water. Among the *sources* of pollution are:

- **Point Sources:** This means sources concentrated at a specific place ("point"). You may be familiar with scenes of waste spewing out of drain pipes directly into lakes and streams. Such direct inputs are called point sources (Figure 10.2A).
- **Diffuse Sources:** During a heavy rain, water can run off from farm fields into streams. The water can contain pesticides and herbicides (substances applied to kill insects and control weeds) and hazardous substances from fertilizers. The flow of pollution does not come from a single point, like a drainpipe, hence the source is termed diffuse (nonpoint) (Figure 10.2B).
- **Indirect Pathways:** For the past century, an enormous variety of wastes ranging from coal-burning residues to organic solvents such as acetone have been buried in metal drums. These containers begin to rust, typically within 20 years of their burial. The sludge in the drums leaks out and can travel through soils and groundwater until it percolates into a lake, stream, or river. The result is often pollution of the soil, the groundwater, and the waterway into which the sludge flowed. This is a major problem. For example, it is estimated that between 2 and 3 million drums are buried in Illinois (Figure 10.2C).
- **Atmospheric Sources:** Acid rain, which we discussed in Chapter 9, is a major cause of water pollution. This is dangerous to the species that live in waterways. Dangers are posed both directly, due to the acidity, and indirectly as a result of the leaching of metal ions from the soil. Winds can also carry sprayed pesticides, as well as particulate matter from industrial smokestacks, and deposit them on waterways.

How Are Water Pollutants Harmful?

Up to this point, we have considered pollutants in general terms. We have not limited our discussion to a few specific pollutants simply because there are so many thousands of substances that can pollute. Remember that "pollutant" is a very broad term covering chemicals that cause everything from bad smells to life-threatening illnesses. Table 10.1 lists some of the more common pollutants and their health effects on humans. As you look at this list, please keep in mind that the pollutants are not added to water specifically to poison it. Rather, there is usually some agricultural or industrial process that relies on the substances or their precursors.

Mercury(II) and lead(II) ions are of special concern because they are so dangerous and are found in relatively high amounts in water systems. These ions, along with other "heavy metal" ions like silver, form strong bonds with the carboxylate (—COO⁻) and the sulfhydryl (—SH) groups of certain amino acids (see Figure 10.3). Amino acids are essential parts (often called the "building blocks") of proteins in our bodies. The metal–amino acid bonds cause two things to happen. First, other bonds that are necessary to the proper functioning of proteins are broken. Second, the metal-protein complexes precipitate out as insoluble metal-protein salts.

A.

B.   C.

Figure 10.2   Water pollution originates at **A.** point sources that dump waste directly into bodies of water, **B.** diffuse sources that allow substances to enter the water supply from their point of use, and **C.** indirect pathways.

Table 10.1

Some Common Environmental Pollutants

Pollutant	Formula or Symbol	Contained In	Some Health Effects of Overexposure
Cadmium	Cd	Solder, batteries, lawn treatment chemicals	Causes high blood pressure, kidney damage, destroys red blood cells
Chlordane	$C_9H_6Cl_8$	Pesticides	Carcinogen
Lead	Pb	Plumbing, coal, gasoline	Anemia, kidney disease, blindness, mental retardation
Mercury	Hg	Industrial waste, pesticides	Central nervous system damage, mental retardation
Vinyl chloride	C_2H_3Cl	Polymer manufacture	Causes liver cancer
Octachlorostyrene	C_8Cl_8	Residues of semiconductor manufacture, aluminum production, and others	Suspected, but as yet unproven, damage to liver, kidneys, and reproductive systems

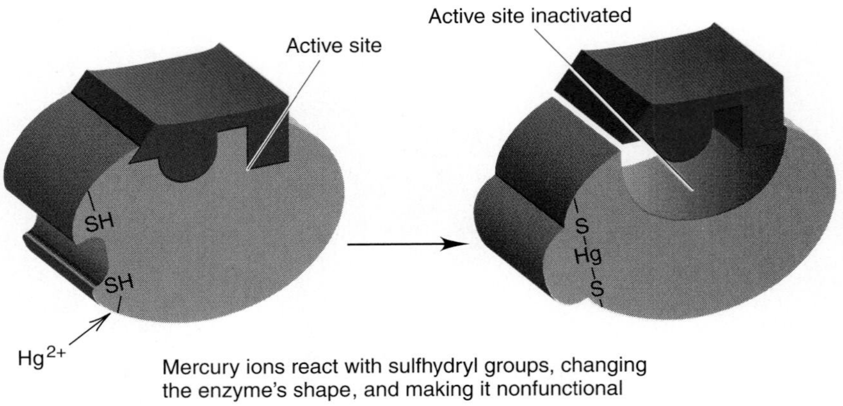

Figure 10.3 Mercury and lead poisoning can occur when the metals bond to the carboxylate or (as illustrated here) sulfhydryl (—SH) groups of amino acids in proteins or enzymes. These metal-protein complexes form insoluble (and biologically inactive) compounds.

Active site

Active site inactivated

Hg^{2+}

Mercury ions react with sulfhydryl groups, changing the enzyme's shape, and making it nonfunctional

Figure 10.4 Major pollution sites for the Great Lakes spotlight industrial centers for their detrimental effect on water quality for the entire system.

CANADA

Toronto

Milwaukee Detroit Buffalo

Chicago Cleveland

Toledo

Protein-containing substances, such as milk and egg whites, can be used to counteract heavy metal poisoning, because they will bind to the heavy metal ions and therefore prevent them from binding to the body's own proteins.

There have been several well-publicized cases of widespread water pollution in the United States. The Love Canal case is one that you might research on your own. Perhaps the most devastating long-term water pollution problem involves the Great Lakes area (see Figure 10.4) where a combination of factors nearly destroyed life in some of these vast waterways. The Case in Point: Pollution in the Great Lakes considers the causes and the extent of the damage.

Three major activities have contributed to the level of human-made water pollution that exists today:

- **Agriculture,** with its attendant use of pesticides and fertilizers.
- **Industry,** including many manufacturing processes.
- **The transport of oil,** which has led to some of the world's most famous pollution incidents.

The chemistry of these activities will serve as the subjects of Sections 10.2, 10.3, and 10.4. Section 10.5 will deal with ways of cleaning up the mess.

case in point

Pollution in the Great Lakes

"A vast ecological slum" is how one scientist described the condition of the southern Great Lakes in the 1960s. The rest of the lakes, which contain over 95% of the surface fresh water in the United States, had been used for 200 years as a 244,000-km² toilet bowl. Untreated sewage and industrial waste ran in from many different point sources. Farm and municipal run-off was added to the brew. Polychlorinated biphenyls, or PCBs (discussed later) were added by industries. Barrels containing toxic substances including crude oil, were dumped on the lake beds. Airborne pesticides like DDT, were being carried by the wind and deposited on the lakes.

Health Effects

According to science writer Bette Hileman, the health effects of such pollution can be divided into effects on three general populations: fish, birds and nonhuman mammals, and humans:

There is no *positively established* link between tumors that began appearing in Great Lakes fish in 1964 and the mass pollution in the lakes. However, the circumstantial evidence is strong. In laboratory environments, fish deliberately exposed to the same pollutants found in high concentrations in the Great Lakes developed skin and liver tumors. In the lakes, fish with liver cancer tended to be found in and around the most polluted sites.

DDT is a pesticide (see page 337) that was used in the U.S. from 1944 until being banned in 1977. Polychlorinated biphenyls (PCBs) were used in the production of electrical transformers until about 1976. Both substances are carcinogenic and PCBs cause a very painful and serious type of acne. Both are also chemically very stable. This is important because once they are released into a lake, they can remain for many years. These chemicals can also be **biomagnified,** meaning that the concentration of PCBs and DDT found in tissues of different species increases moving up the different levels of a food chain. The values in the Figure below graphically illustrate this effect.

The result of the biomagnification of PCBs around the Great Lakes area has been devastation of some animals high in the food chain. Gulls and terns have suffered high rates of "defective" (malformed, dead, etc.) embryos. Chicks are born with crossed bills, causing them to die within a few weeks because they are unable to catch enough fish to survive.

The effects of biomagnification all the way up the food chain to humans could be expected to produce a toll of human health problems. Several studies in Michigan, Wisconsin, and Ontario show that DDT and PCB levels in people who eat fish from the Great Lakes were, as of 1988, higher than levels in those who avoided eating the fish. These elevated concentrations were found in blood as well as in breast milk. In Sheboygan, Wisconsin, women who ate Great Lakes fish carried a higher level of PCBs in their breast milk (45 ppb vs. 39 ppb) than non-fish-eaters. Studies have found that smaller birth sizes, lower gestational age, and neonatal behavioral deficits are related to mothers' consumption of contaminated fish. Other contaminants however, in addition to PCBs, may have contributed to these effects.

Is Anything Being Done?

Fortunately, something *has been* done. In 1972, the United States and Canada signed the "Great Lakes Water Quality Agreement." The goal of this agreement was to deal with the most obvious pollution sources such as solid wastes and oil residues. Improvement was soon apparent. Fish populations began to increase in Lake Erie, which had lost virtually all of its fish to pollution. According to Hileman,[10] "The Cuyahoga River in Cleveland no longer caught fire, as it occasionally had in the sixties."

A 1978 United States/Canada water quality agreement pushed controls even further with the intent of banning some 350 hazardous substances from the waterways. Of particular concern were 11 especially toxic substances, including lead, mercury, and the insecticide DDT.

A subsequent strengthening of the 1978 agreement was signed in 1987. However, much remains to be done. Oil tankers still travel to the lakes via the St. Lawrence Seaway. Airborne pollutants that are deposited on the lakes represent a difficult long-term problem. Each year, millions of gallons of toxic compounds are still dumped into the lakes by industries and private homeowners. The fact that the Great Lakes have recently become a lot cleaner shows the benefits of cooperative action by the United States and Canada. It also showed the necessity for government intervention when the health of the lakes and its users was at stake.

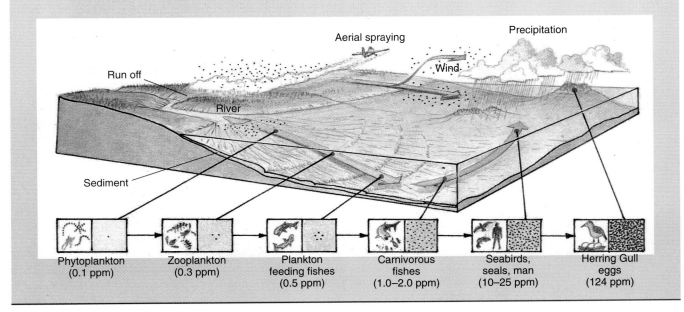

Phytoplankton
(0.1 ppm)

Zooplankton
(0.3 ppm)

Plankton feeding fishes
(0.5 ppm)

Carnivorous fishes
(1.0–2.0 ppm)

Seabirds, seals, man
(10–25 ppm)

Herring Gull eggs
(124 ppm)

Green Chemistry

10.2 | Agricultural Sources of Water Pollution

Crops need water to grow, so they are always raised in soil that receives a good water supply, either through natural rainfall or by irrigation. This means that chemicals sprayed on crops (Figure 10.5), or added to the soil around them, can find their way into the pathways that take rainfall through the soil and rocks and into rivers and lakes. Drinking water is largely drawn from rivers, lakes, or wells, so there are clear opportunities for agricultural chemicals to find their way into our water supplies (see Table 10.2).

Two main categories of agricultural chemicals are potential sources of water pollution: pesticides and fertilizers. **Pesticides** (meaning "pest-killers") are chemicals used to kill anything that farmers consider undesirable, such as insects, slugs, caterpillars, locusts, weeds, and fungi. Those pesticides used to kill insects are often referred to by the more specific term, **insecticides** ("insect-killers"). Pesticides used to kill unwanted plants (weeds) are known as **herbicides** ("plant-killers"). **Fertilizers** are chemicals that contain nutrients needed to promote the growth of plants. Insecticides, herbicides, and fertilizers are not all human-made substances. Examples of natural forms of each category exist and can be sources of water pollution. We will first examine some of the water pollution issues linked to pesticide use.

How Do Pesticides Kill?

This is not an easy question to tackle because there are so many complex processes that interact to sustain life. Insecticides, such as rotenone, destroy compounds necessary for respiration. Sodium fluoroacetate inhibits the breakdown of carbohydrates (such as sugars) in the pest.

Figure 10.5 Crop dusting orchards and fields is a standard practice of modern agriculture, but this general distribution of chemicals makes it easy for compounds to end up where they do not belong.

Table 10.2

Concentrations of Pesticides in Groundwater

Pesticides	Median Concentration in Groundwater in Several States (ppb)	Maximum Contaminant Level* (ppb)
Aldicarb	9.0	3.0
Atrazine	0.50	3.0
Chlordane	1.70	2.0
Trifluralin	0.40	5.0

*Maximum Contaminant Level (MCL) = maximum permissible level of a contaminant in water delivered to a user of a public water system.

exercise 10.1

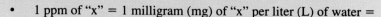

Parts per Million and Billion

Problem

Scientists often express concentrations of pollutants in parts per million (ppm) or parts per billion (ppb).

- 1 ppm of "x" = 1 milligram (mg) of "x" per liter (L) of water =

$$\frac{1 \times 10^{-3} \text{ g x}}{1 \times 10^{3} \text{ g H}_2\text{O}} = \frac{1 \text{ g x}}{10^{6} \text{ g H}_2\text{O}}$$

- 1 ppb of "x" = 1 microgram (μg) of "x" per liter of water =

$$\frac{1 \times 10^{-6} \text{ g x}}{1 \times 10^{3} \text{ g H}_2\text{O}} = \frac{1 \text{ g x}}{10^{9} \text{ g H}_2\text{O}}$$

Note that this is similar to the familiar percent calculation which is really parts per hundred or grams of x per hundred grams of water.

A drop of food coloring weighs roughly 0.05 g. If you add 1 drop of the food coloring to 8.0 L of water in a bucket, what is the concentration of food coloring in ppm and ppb? Is a ppm merely a "drop in a bucket?"

Solution

$$\textbf{ppm of food coloring (fc)} = \frac{\text{mg fc}}{\text{L water}} = \frac{50 \text{ mg fc}}{8.0 \text{ L water}} = \textbf{6.3 ppm}$$

$$\textbf{ppb} = \frac{\mu\text{g fc}}{\text{L water}} = \frac{50,000 \ \mu\text{g fc}}{8.0 \text{ L water}} = \textbf{6300 ppb}$$

$$\textbf{\% fc} = 0.05 \text{ g fc}/8000 \text{ g H}_2\text{O} \times 100 = \textbf{0.00063\%}$$

The terms ppm or ppb are usually more appropriate for describing trace concentrations of environmental contaminants. 1 ppm is *much less* than a drop in the bucket. It is more like a drop in a huge fish tank—1 ppb is akin to a drop in a swimming pool.

The largest group of pesticides are nerve poisons. Organophosphorus and carbamate compounds overstimulate nerves, causing convulsive death. Others, such as DDT and pyrethroids, prevent the transmission of nerve signals. Some pesticides on their own merely stun the victims, so other compounds are added to cause death. These additional compounds are known as **synergists.**

The Chemical Composition of Pesticides

There are two general types of pesticides—natural and synthetic. Natural means that the material is used in a form that is found in nature. Synthetic means that it is produced industrially. The chemical structures and chemical properties of natural and synthetic pesticides can be very similar. Knowing that a substance is "natural," however, often lessens public concern about its use. This is a very unreliable way to form judgments about the safety of chemicals.

The many widely used naturally occurring pesticides that are extracted from plants include:

- **Nicotine,** obtained from tobacco. This substance has been used since at least 1690 to kill the pear bug in France. By 1773, tobacco smoke was used to kill a variety of insects. Nicotine sulfate is still used today as a spray in small gardens.

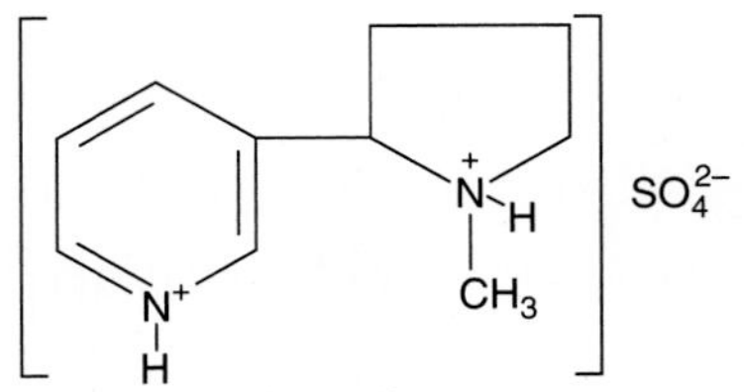

Nicotine sulfate

- **Pyrethrum,** extracted from several species of chrysanthemum. This insecticide, originally used in Iran (then called Persia) around 1800, is actually a combination of four different compounds, each having the following general form:

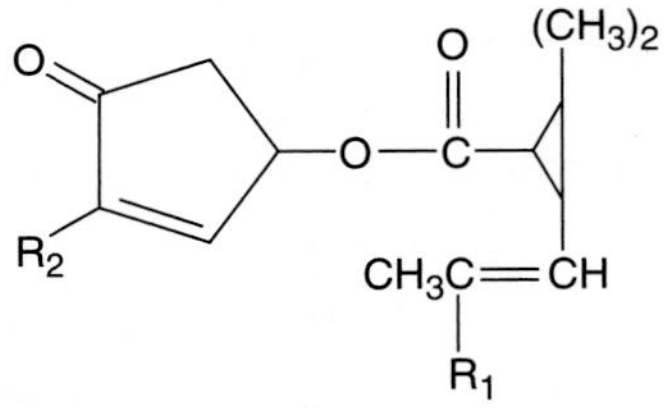

R$_1$ and R$_2$ are used in these structural formulas as general labels for a variety of more complex chemical groups that distinguish different but related compounds from one another. Pyrethrum and its related compounds are useful because they poison insects quickly, yet degrade rapidly in sunlight. This rapid degradation means they pose no long-term health hazards to humans.

- **Rotenone,** found in 68 different legumes (plants that contain fruit that grows in the shape of a pod that splits along its seams when mature to reveal its seeds). The first reported use of this compound was in 1649 in South America as a fish poison, and it is still used as a fish poison today. It has been used as an insecticide since 1848, although it only kills certain insects, and chemically degrades during storage.

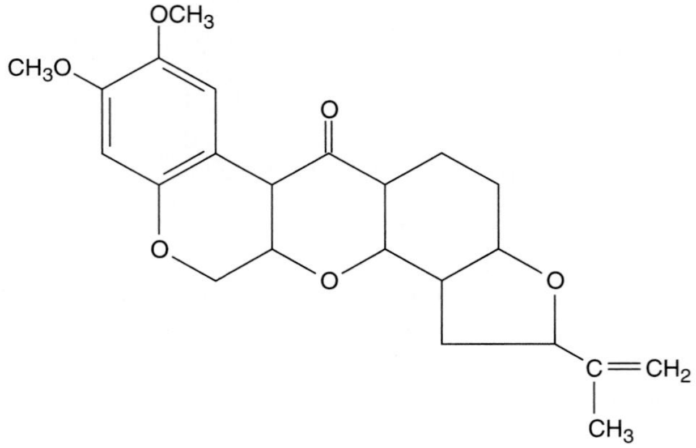

consider this:

Structure Similarities

As we work our way through the structures of the pesticides, can you note any similarities in structure within groups or among groups? Are there similarities in structure and function between pesticides and military nerve agents discussed in Chapter 6?

Most pesticides, like the three just examined, are organic compounds. Inorganic compounds, however, have been used as pesticides for millennia, dating back to 1000 B.C., when sulfur was reportedly burned to fumigate Greek homes. Lead hydrogen arsenate, $PbHAsO_4$, was used in fruit orchards but is very toxic to humans and other living organisms. Sodium fluoride, NaF, was used to kill cockroaches and grasshoppers. Cryolite, Na_3AlF_6, is currently used on plants that are especially sensitive to other chemicals. Certain mineral oils have also proven to be effective insecticides, while being relatively harmless to other animals. Most natural pesticides, however, come from plants.

Bruce Ames and Lois Swirsky Gold of the University of California at Berkeley have contended that the health risks associated with the small amounts of synthetic pesticides we consume are drastically overstated because we ingest greater amounts of so many natural ones. They point out, for example, that cabbage alone contains 49 natural pesticides, while lima beans contain 23. Also, according to Ames and Gold, 27 cancer-causing compounds (**carcinogens**) are naturally present at or above 10 ppm in foods ranging from apples to thyme. Among those which have been isolated are nicotine, rotenone, pyrethrins (discussed earlier), and certain pheromones. It is estimated that we eat 1.5 g of natural pesticides each day, perhaps 10,000 times the amount of synthetic pesticides that we consume.

Natural pesticides have been used by humankind for centuries. Our species has survived the effects of these with little difficulty. This peaceful co-existence, however, may not be so easy with synthetic pesticides.

The synthetic pesticide era began in the late 1930s with the mass production in Switzerland of an organic compound with the common name DDT.

$$Cl-\!\!\!\bigcirc\!\!\!-\!\!\underset{\underset{\displaystyle CH}{|}}{\overset{\displaystyle CCl_3}{|}}\!\!-\!\!\!\bigcirc\!\!\!-Cl$$

DDT

This substance, originally synthesized in the 1870s, is a very effective insecticide. Its first important use was in 1944 to kill lice on people during a typhus outbreak in Naples, Italy. Typhus is transmitted by the lice and DDT acts as a nerve poison to kill the lice. An uncontrolled outbreak of typhus in Serbia and Russia during the First World War (1914–1918) killed millions of people. The use of DDT in the 1944 Italian outbreak prevented such a catastrophe.

DDT was widely used in the United States until 1977. By then, the pesticide was found to be present in the blood and fat tissues of nearly all humans who were tested for it, and also in many other creatures, including predators living in aquatic environments such as ospreys, eagles, and cormorants.

The populations hardest hit by the biomagnification of DDT in waterways are probably birds. Predatory birds such as falcons and eagles (Figure 10.6), which feed on fish, are particularly susceptible. DDT and its eventual break-down product, DDE, are found in bird eggs, resulting in the production of thin shells that can break before the bird can hatch. A DDE level of 20 ppm will cause a falcon shell to fail due to a drop in the falcon's ability to produce calcium for the shell, not only terminating the baby's life but also that of the falcon family as these birds do not renest.

$$Cl-\!\!\!\bigcirc\!\!\!-\!\!\underset{\displaystyle C}{\overset{\displaystyle CCl_2}{\|}}\!\!-\!\!\!\bigcirc\!\!\!-Cl$$

DDE

As a result of the DDT ban, both the bald eagle (1994) and peregrine falcon (1999) populations have recovered sufficiently to remove both species from the U.S. Fish and Wildlife Endangered Species list. Although DDT is an important environmental hazard, for the past half century it and subsequent pesticides have helped reduce pest damage to crops and outbreaks of pest-borne diseases. On the other hand, it has required a substantial increase in pesticide use since 1945 to do it. In addition, many insect, fungus, and weed species have developed resistance to pesticides. This has resulted in a steady increase in crop loss due to pests despite the use of pesticides.

Figure 10.6
DDT and its breakdown product, DDE, can accumulate in birds who feed on fish, such as falcons and eagles. A DDE level of 20 ppm will cause a falcon egg shell to fail.

exercise 10.2

Why Was DDT Banned?

Problem

DDT is now banned because of the risk it poses to human health. Compare the structure of DDT (shown earlier) with the structure of **malathion,** a currently used insecticide (below):

(a) Which insecticide is likely to be more water soluble?

(b) Based on your answer to a, why might DDT have been banned?

Solution

(a) We learned in Chapter 6 that polar molecules will be more water soluble than those that are nonpolar. Malathion contains electronegative elements such as oxygen, nitrogen, and sulfur and is much more water soluble than DDT.

(b) DDT is fat soluble, and therefore can be expected to accumulate in fatty tissue within humans and other animals. It will not be metabolized and excreted in urine (note how this exercise is similar to Exercise 6.3 in Chapter 6 on organic chemistry, which compared the solubility of vitamins A and C). It can also become more concentrated (or **biomagnified**) at each level up the food chain, becoming more concentrated in the fat of fish-eating people, for example, than in the fat of the fish themselves. Water-soluble molecules like malathion are much more readily excreted from the body in urine.

Another important difference between malathion and DDT is that DDT is chemically much more persistent than malathion, meaning that malathion degrades into chemical break-down products much more quickly than DDT.

Types of Synthetic Pesticides

Synthetic pesticides fall into four structural classes: chlorinated hydrocarbons, organophosphates, carbamates, and pyrethroids. These classes differ from one another in their structures, dosages, and hazards.

- **Chlorinated hydrocarbons,** such as DDT, are the oldest synthetic pesticides. They tend to be persistent, so their use is waning. As their name indicates, they are essentially hydrocarbon molecules with chlorine atoms in place of some of the hydrogen atoms of the "parent" hydrocarbon. The examples **chlordane** and **heptachlor** are shown here. Note here, and with the other structures presented later, those features that the substances have in common and those that are different (structure dictates function!).

Chlordane   Heptachlor

- **Organophosphates** were developed in Germany during the Second World War as an outgrowth of the effort to make deadly nerve gases for use in the war. These phosphorus-containing compounds tend to chemically degrade within hours or days, so they do not undergo biomagnification. Two popular organophosphates, Counter® (terbufos) and Dyfonate® (fonofos), are shown here.

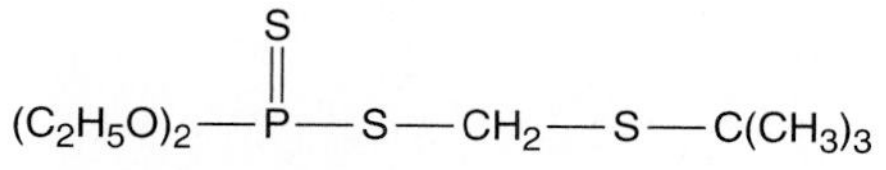

Terbufos 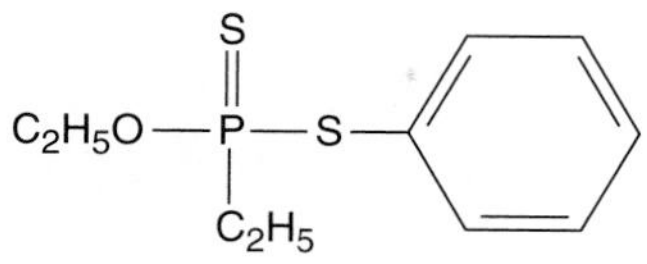   Fonofos

- **Carbamates** have been used commercially since 1956. They are called carbamates because they are formed from carbamic acid.

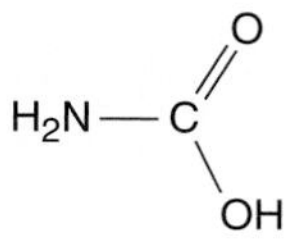

Carbamic Acid

Carbamates, like organophosphates, degrade rapidly. They vary in toxicity from the fairly nontoxic Sevin®, used in gardens, to the very hazardous Furadan®, which eradicates corn rootworms and other soil pests.

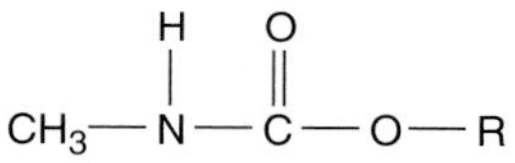

the general formula of a carbamate

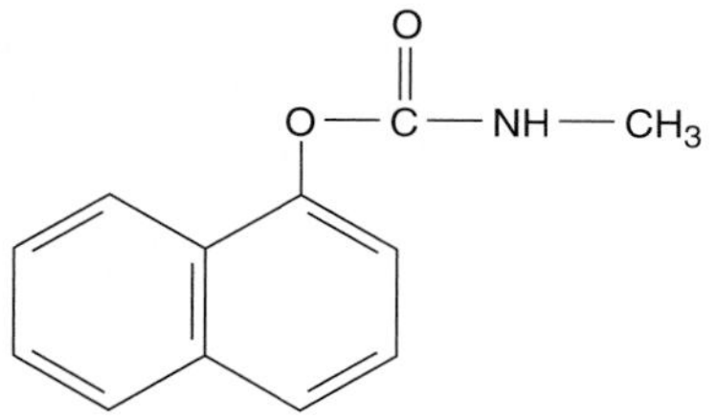

Carbaryl (Sevin) 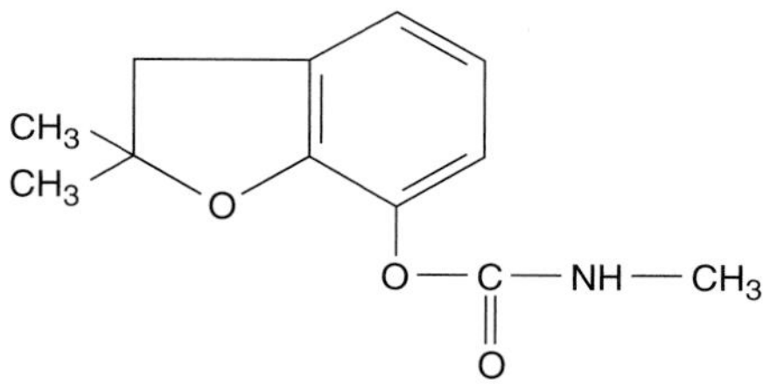   Carbofuran (Furadan)

- **Pyrethroids** are becoming very popular for insect control. As their name suggests, they bear a structural resemblance to pyrethrum. They are useful because only very small amounts are needed, and they chemically degrade within about a week. Two pyrethroids, allethrin and permethrin, are shown next.

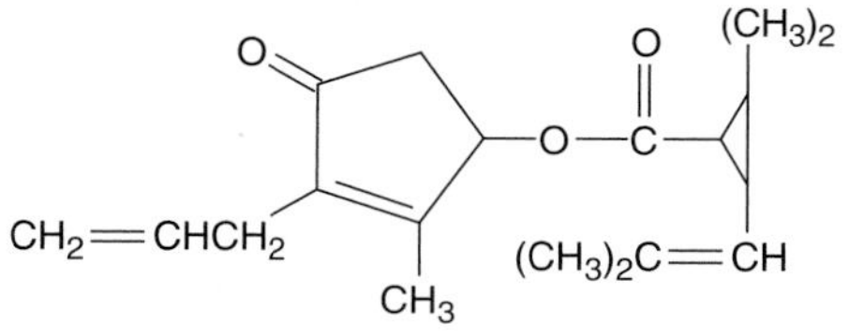

Allethrin (Pynamin®)

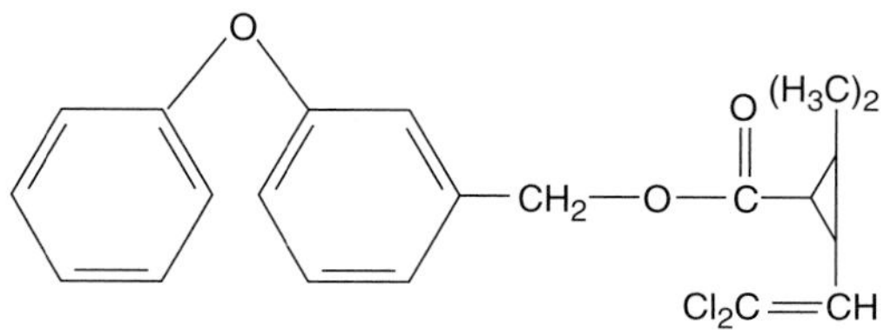

Permethrin (Ambush®, Pounce®)

Beautiful though it may be, purple loosestrife (*lythrum salicaria*) is an invasive weed that chokes out all other vegetation. The loss of plant diversity drives away the wildlife that feeds on it and upsets the balance of nature. Selective herbicides are used because they control the loosestrife but allow other plants to grow.

The overall goal of pesticide research is to create compounds that are effective as pesticides, relatively inexpensive, and decompose quickly after use into benign products.

Herbicides

Because plants grow in a wide variety of habitats, from arid to aquatic, many different herbicides have been developed, each suited to attack specific weeds in particular environments. Some herbicides are known as **nonselective** because they will kill all of the vegetation on the treated area. You would chose a nonselective herbicide to keep a path or driveway clear of weeds, because you want nothing to grow in the treated ground. Others are **selective** because they will kill only a specific type, or narrow range of plants. Selective herbicides are widely used in agriculture because farmers want to kill weeds without doing any harm to their crops. The exact way in which many herbicides work is not known. However, the ability of the herbicide to travel through the plant, and features of the metabolism of the specific target plant, are among the factors that influence the effectiveness of a herbicide.

Herbicides, whether selective or nonselective, can be divided into three categories:

- General contact
- Translocated
- Residual

As we discuss each type of herbicide, please pay special attention to their structures, looking for similarities and differences. Also, relate them structurally to the insecticides. Are there common features? Why might these be important?

As their name implies, **general contact herbicides** take effect wherever they come into contact with a plant. Sinox® (2-methyl-4,6-dinitrophenol), for example, is a selective general contact herbicide used to kill weeds among cereal, pea, and flax crops.

$$\text{structure of 2-methyl-4,6-dinitrophenol}$$

This herbicide is dusted over crops. It destroys the membranes within tissues of the weeds on which it is effective, causing sap to leak out and the plant to dry out and die.

case in point

Herbicide Concentration in the Environment

Most herbicides are applied to crops at a time when the weeds are most vulnerable. This means in the case of atrazine, a compound that kills many broadleaf weeds in crops such as corn, that it is applied at about the time of planting the corn. In midwestern states, this means that atrazine is applied typically at a rate of 0.9 kg/acre in the first week of May. The concentration of atrazine in rivers near the fields then varies during the season. The chart here shows the concentration of atrazine in the Platte River, near Ashland, Nebraska, during 1996. Notice that the concentration rises dramatically around May 10 after

having been at very low concentrations before application. The concentration then dies down later in the summer. Early rainfall events dissolve or suspend atrazine in the fields and then it flows into larger and larger rivers. Knowledge of the amount of water flowing in the river every day during the summer allows calculation of the mass of atrazine moving down the river at this point. In 1996, this amount was 10,000 kg. This seems like a lot, but approximately 3 million kg (7 million lb) was applied in the fields drained by this river.

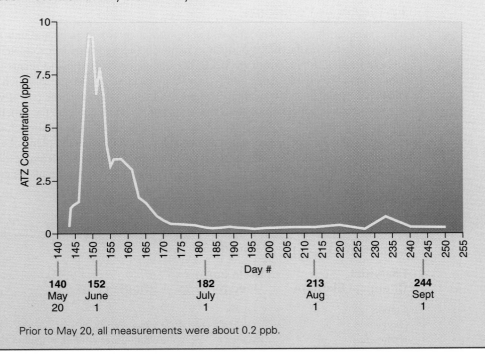

Prior to May 20, all measurements were about 0.2 ppb.

Translocated herbicides move through the **phloem** or **xylem** of plants, which are the tiny tubes that transport food and water within a plant. This allows the herbicides to be transported ("translocated") to all parts of a plant, thus killing the entire plant. Translocated herbicides can even kill new growth that begins to appear after spraying. One common translocated herbicide is 2,4-D (2,4-dichlorophenoxy acetic acid).

2,4-D

This is used to control broadleaf weeds in cereal crops, sugar cane, turf, pastures, and noncrop land. It can be absorbed through the foliage (leaves) or roots of a plant to cause an abnormal growth response which affects respiration and cell division.

Other translocated herbicides are combined with or contain oils which destroy the semi-permeability of living membranes (essential for the process of osmosis—see page 267).

Residual herbicides are ones that remain active in the soil to control weed seeds that germinate after application of the herbicides. One common example is Triallate®, which is applied to the soil to destroy wild oats.

Triallate® is absorbed by seedlings at an early stage of seed development and destroys the plant before it sprouts.

Fertilizers as Pollutants

Crops need nitrogen, phosphorus, and potassium in relatively large amounts to grow well. Nitrogen is part of many vital chemicals in plants including amino acids, nucleic acids, proteins, and chlorophyll. Phosphorus is part of the key chemicals allowing plants (and other living things) to make use of energy and is also found in important chemicals such as nucleic acids, lipids (fats), and attached to many proteins. Potassium regulates the rate at which plants lose water and take in carbon dioxide for photosynthesis. It also affects the functioning of at least 60 different enzymes. A variety of other elements are needed in smaller amounts. Table 10.3 shows all the elements required by plants and how they get to the plants. Fertilizers are used to augment these elements and boost production. Fertilizers differ from pesticides in that pesticides are applied to kill plants while fertilizers are intended to help plants grow. Yet both can cause water pollution.

Table 10.3

Elements Required by Plants

Essential Chemical Element	Main Form Used by Crop	Source
Carbon	CO_2	Air
Oxygen	CO_2, H_2O, O_2	Air
Hydrogen	H_2O	Soil water-rainfall, irrigation
Nitrogen	NO_3^-, NH_4^+, N_2	Soil organic matter, animal manures, fertilizers, air (legume crops)
Phosphorus	$H_2PO_4^-$, HPO_4^{2-}	Soil organic matter, soil minerals, fertilizers, animal manures
Potassium	K^+	Soil minerals, animal manures, plant residues, fertilizers
Calcium	Ca^{2+}	Soil minerals, limestone, fertilizers
Magnesium	Mg^{2+}	Soil minerals, dolomitic limestone, fertilizers
Sulfur	SO_4^{2-}, SO_2	Soil organic matter, animal manures, SO_2 in air, fertilizers
Boron	H_3BO_3	Soil minerals, fertilizers
Chlorine	Cl^-	Rainfall, soil, fertilizers, animal manures
Copper	Cu^{2+}	Soil minerals, animal manures, fertilizers
Iron	Fe^{2+}	Soil minerals, fertilizers
Manganese	Mn^{2+}	Soil minerals, fertilizers
Molybdenum	MoO_4^{2-}	Soil minerals, fertilizers
Zinc	Zn^{2+}	Soil minerals, animal manures, fertilizers

Source: Hileman, B. *Chem. & Eng. News,* March 5, 1990, p. 35.

Manure from livestock is a natural fertilizer. It is a major source of many nutrients in crops, and, when plowed into fields, helps the soil absorb and hold rain water. Both natural and synthetic fertilizers release nitrates (anything containing the NO_3^- ion) to plants at a rate that is close to the rate that plants can absorb it. However, nitrates are very water soluble and enter groundwater after rain or melted snow has washed them out of farm fields. This leads to a relatively high accumulation of nitrate in groundwater, creating a major groundwater pollution problem.

Excess phosphorus (P), present as the phosphate anion, PO_4^{3-}, can have a serious affect on fish because it serves as a nutrient for the growth of algae in waterways. When the algae die, they decompose using up oxygen in the water. This activity is illustrated by this equation:

$$biomass + O_2 \rightarrow CO_2 + H_2O$$

The more algae there are, the less oxygen will be available for fish, resulting in fetid waters with lots of dead fish.

consider this:

Are Synthetic Pesticides in Groundwater Harmful?

We have already seen that DDT accumulates via biomagnification which can lead to harmful results. Many pesticides are undoubtedly toxic and/or carcinogenic in *large doses,* and are therefore dangerous to farm workers, gardeners, and others who come in contact with high concentrations of the compounds. It is much less certain that exposure to very small amounts of these compounds within water is a threat to health. We are then talking of concentrations at levels of a few parts per billion or parts per trillion. Even experts in the field disagree on the level of hazard posed, or if one even exists. Therefore, our best answer to the question, "Are synthetic pesticides in groundwater harmful?" is "We don't know." Nevertheless, the concentrations of some pesticides in groundwater exceed "health advisory levels" in several U.S. states (see Table 10.2), which is cause for some concern.

Is Nitrate and Phosphate Pollution Widespread?

Figure 10.7 shows that nitrate concentrations of greater than 5 ppm in groundwater are widespread. It is not uncommon for groundwater in heavily fertilized areas to be well above the 10 ppm level federal limit in nitrates, occasionally as high as 100 ppm. In such places it is recommended that babies and pregnant women be given purified, bottled water.

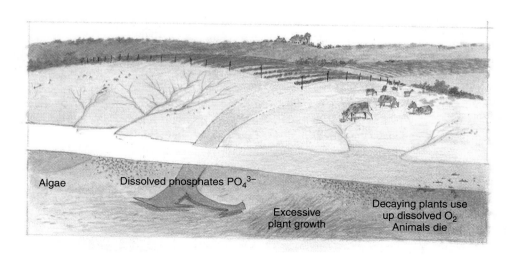

Algae   Dissolved phosphates PO_4^{3-}

Excessive plant growth

Decaying plants use up dissolved O_2 Animals die

In Chapter 4 we discussed the Earth as a materially closed system. We exist because we re-use and modify what nature has provided to us.

http://water.usgs.gov/nawqa/circ-1136/fig13.gig

Figure 10.7
High nitrate concentrations in groundwater are found in highly agricultural areas.

The biggest source of phosphate pollution used to be from laundry detergents. Since there are now strict controls on phosphates in detergents, agricultural runoff is the most significant source of phosphate in waterways.

Agricultural practices are changing. Less fertilizer is being used and it is being applied in such a manner as to minimize the amount that will leave the root zone of plants to work its way into groundwater. Irrigation using high-nitrate water actually removes nitrate from water by allowing crop plants to utilize the nitrate as fertilizer. Improved agricultural procedures should do a lot to minimize the groundwater nitrate problem.

Nitrate ions in water can definitely cause serious health problems in infants under 6 months of age. When nitrate-laden water is used to mix infant formula, the excess nitrate (once ingested) can interfere with the ability of the protein "**hemoglobin**" to carry oxygen in the babies' blood. The medical name of this syndrome is *methemoglobinemia*. This can cause babies to suffocate. Even though their lungs are receiving plenty of air, they are not receiving sufficient oxygen. The hemoglobin of infants, known as fetal hemoglobin, is most suited to carrying the oxygen supplied in the womb. This is a very different process from accepting oxygen obtained by breathing. By about 6 months of age most of a baby's fetal hemoglobin has been replaced by adult hemoglobin and, therefore, most of the danger from nitrate contamination is past.

case in point

Origin of Nitrate Ion in Groundwater

The concentration of nitrate ion in water is easy to measure and is significant in wells near agricultural land in the corn belt. The origin of this nitrate had to be known in order to change practices to lower the amount of nitrate entering our aquifers. Two reasonable sources were hypothesized: (a) excess ammonia fertilizer reacting to give nitrate and (b) cattle waste decomposing to give nitrate ion.

How can we distinguish these two sources of nitrate? After all, isn't one nitrate ion just like another? How can we learn the origin of nitrate ions if we can't trace them to their source? A very subtle difference exists in the fraction of the nitrogen that is present as the rare isotope with 8 neutrons, nitrogen-15. Ammonia fertilizer is made directly from air so has the same ratio of nitrogen-15/nitrogen-14 as does nitrogen in air. Nitrate ion from ammonia fertilizer will retain that isotope ratio. Nitrate that results from cattle manure has come from the fertilizer also, but by a less direct path. First the ammonia has been transformed into nitrate,

$$NH_3 + 2O_2 \rightarrow NO_3^- + H_2O + H^+$$

been taken up into grain plant, undergone dozens of complicated reactions in becoming protein in a grain of corn, then been through dozens more biochemical reactions on its way through the cow. Each of these many reactions has an **isotope effect.** This means that one of the isotopes, usually the lighter one, reacts preferentially to the heavier one so the ratio of the two isotopes of nitrogen in manure is very different from that in air.

Therefore, a measurement of the ratio of the two nitrogen isotopes can allow us to distinguish these two possibilities. If the nitrate has an isotope ratio similar to that of air, fertilizer is the culprit, but if the ratio is more similar to that in cattle manure, that source is the origin of the problem. In most cases, fertilizer is the source of the nitrogen atoms in the nitrate in groundwater.

No cases of methemoglobinemia in infants have been reported where water levels are below 30 ppm of nitrogen in the form of nitrate. To be on the safe side, safety limits of 10 ppm have been in place for a long time.

Excess phosphorus, present as the phosphate anion, PO_4^{3-}, can have a serious effect on fish, because it serves as a nutrient for the growth of algae in waterways. When the algae die, they decompose, using up the oxygen in the water. The more algae there are, the less oxygen will be available for fish, resulting in fetid waters with lots of dead fish.

What Can We Conclude About Agricultural Groundwater Pollution?

It is certain that a wide variety of pesticides, along with nitrate ion from fertilizer, are finding their way into groundwater supplies. What is not certain is how dangerous low concentrations of pesticides are. The 1988 amendment of the 1947 Federal Insecticide, Fungicide, and Rodenticide Act required pesticide manufacturers to submit complete hazard data on their products by 1997. A 1996 admendment extended the registration and fee collection process through the 2001 fiscal year. The problem is that between 25,000 and 50,000 products are affected! We are a long way from getting definitive answers.

10.3   Industrial Sources of Water Pollution

 Green Chemistry

As we have discussed, modern lifestyles can be maintained only in return for some deterioration in environmental quality. Almost all of the products that we come in contact with every minute of every day were manufactured using processes that create some pollution. As our case in point focused on the Great Lakes has shown, we have paid a heavy price for industrialization. However, we are also learning that, with new technologies, the pollution price does not have to be so high.

Figure 10.8
The Toxic Release Inventory (TRI) is a measure of hazardous substance releases by U.S. industries.

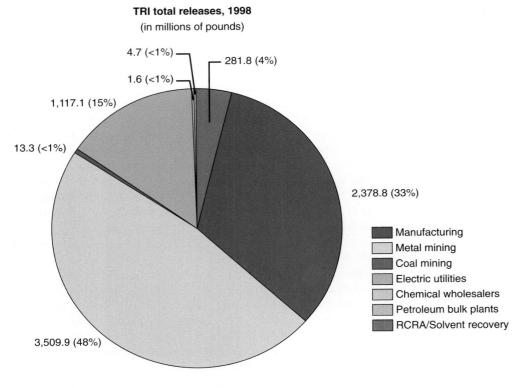

TRI total releases, 1998
(in millions of pounds)

4.7 (<1%)
1.6 (<1%)
281.8 (4%)
1,117.1 (15%)
13.3 (<1%)
2,378.8 (33%)
3,509.9 (48%)

- Manufacturing
- Metal mining
- Coal mining
- Electric utilities
- Chemical wholesalers
- Petroleum bulk plants
- RCRA/Solvent recovery

http://www.epa.gov/tri/tri98/data/new_rlm98at2_pie.pdf

The Scope of the Problem

The Environmental Protection Administration (EPA) monitors the level of industrial pollution throughout the United States in its Toxic Release Inventory (TRI). A total of about 23,500 companies representing many types of industries are required to report about hazardous substance releases into the air, water and ground. Until 1998, what the EPA calls "new industries" such as metal and coal mining, along with electric utilities, were not required to report this information. Therefore, over 60% of the toxic substances released went unreported. Even without these industries' data, it was clear that toxic releases were decreasing. Between 1987 and 1997, toxic substance releases of reporting industries in the United States went from 3.3 billion kg to 1.1 billion kg per year. Including the substantial contributions from the new industries, the 1998 figure is 3.32 billion kg (7.31 billion pounds). Although the overall trend is a good one, the EPA notes with caution that toxic releases into waterways increased from 82 million kg in 1995 to 102 million kg in 1998, a 24% increase (see www.epa.gov./tri/tri99/index.htm). The industrial distribution of the TRI is shown in Figure 10.8.

Table 10.4A gives you a sense of the kinds of pollution that can result from important industrial processes. Table 10.4B lists specific water pollutants from various sources and associated health risks.

What Can Be Done?

The extent of water pollution by industrial processes is very well understood. Many large companies have taken significant steps to reduce pollution. Nonetheless, as it did with the Great Lakes, the federal government has taken steps to mandate pollution control. The most far-reaching acts set minimum nationwide standards for water cleanliness and hazardous substance disposal. Table 10.5 describes these acts in more detail.

Would industries work to control their own emissions without government intervention? Some, including industry giant Dow Chemical, have voluntary programs in place. At least one environmental group has praised Dow for its 1986 Waste Reduction Always Pays (WRAP)

pro con discussions

Legislating Zero—Should We Pass Laws Stating That the Allowable Residue of Herbicides and Carcinogens in the Environment Is Zero?

pro

Political change is really a series of compromises. One member of congress wants to eliminate a program, another wants to increase the funding. In the end, they strike a balance that is acceptable to both combatants. The process of government moves forward. How do we get movement with regard to environmental safety? In 1995, many companies, and many members of congress pushed for rolling back hard-won environmental legislation. They argue, as in the "con" argument below, that safety is just too costly. We know that a zero detection limit is not scientifically reasonable. However, we fight for it, because when legislative compromises are made, we inch a little closer to the side of prevention and away from the side of too much permissiveness if we take a "zero-tolerance" stance. If we back off initially, then the compromise line moves ever closer to that in which pollution dominates and restraint is lost. It is all about positioning for the fight.

con

When questions are raised of allowable amounts of herbicides, insecticides, or carcinogens in our air, water, or soil, it is easy to say "don't allow there to be *any* of that stuff in my surroundings!" This response is highly unrealistic. It is impossible to prove that the amount or concentration of certain contaminants is zero. We can merely determine that the amount of it present is *too small* to detect. A chemist would say that the concentration is *below the detection limit*. The detection limit is the lowest concentration that can be reliably detected by the technique being used in the measurement. No one would try to weigh a grain of sand on a bathroom scale; the detection limit on such a scale is about one half pound so anything lighter than that would not show up. A finer balance is needed to weigh something so small.

A balance will be a useful analytical tool only if the sample can be separated from other components of the environment so that it alone can be placed on the balance. Usually this is not possible so other more sophisticated measurement techniques must be used. The most sophisticated of these techniques is capable of detecting amounts in the range of 1×10^{-15} g, usually from a sample that weighed about 10 g total. A few years ago, this detection limit was an impossible dream. The senior author of this text remembers when detection limits of a microgram (1×10^{-6} g) were very respectable. The point is, detection limits have been going down and down over the past 40 years and every indication is that this will continue to be the case. People are now seriously thinking about methods to detect single molecules.

Another observation that chemists have made is that the lower the detection limit, the more various substances are found in a sample. This should not be a surprise. When detection limits are micrograms, everything present at lower concentrations is not detectable so is judged to be absent. When detection limits are lowered to nanograms, many more substances show up.

Finally, we get to the issue of cost versus benefit. Cleaning water so herbicides are present below detection limits becomes very difficult, time-consuming, and costly. Is there any benefit to cleaning water to below detection limits for a wide number of objectionable compounds? Regulatory agencies have a very difficult time determining what concentrations of herbicides should be disallowed. They follow the philosophy that says "let's find the lowest concentration that causes any deleterious effect and then divide that concentration by ten or one hundred." The point is, they don't say zero allowance. To have said zero tolerance a few years ago when detection limits were ppm would have allowed water to be satisfactory that now would be disallowed with modern detection limits.

exercise pro/con 1

How Tiny Is Tiny?

Problem

Work out how much water would be needed to dissolve 1 mg of atrazine so that its concentration would be at the detection limit mentioned in the "con" discussion, 1×10^{-15} g in 10 g of water.

Solution

We can re-express this concentration as 1×10^{-13} g of atrazine in a liter (1000 g) of water. A milligram (1×10^{-3} g) of atrazine would require 10^{10} L of water to be this concentration. This would be a cube a bit more than 200 m (1/8 mile) on a side.

How much is too much? One mg (as shown in spoon) of atrazine is all that is needed to contaminate a small lake to detectable levels.

exercise pro/con 2

Atrazine in Water

Problem

Calculate the number of molecules of atrazine ($C_8H_{14}N_5Cl$) in a liter of water if its concentration is 0.010 ppb.

Solution

 # molecules =

$$1000 \text{ g H}_2\text{O} \times \frac{0.010 \text{g atrazine}}{10^9 \text{ g H}_2\text{O}} \times \frac{1 \text{ mol atrazine}}{215.5 \text{ g atrazine}} \times \frac{6.02 \times 10^{23} \text{ molecules}}{\text{mol}}$$

$= 2.8 \times 10^{13}$ molecules

Notice that even though this is a very low concentration, it still represents an enormous number of molecules.

10.4 | Oil Spills

At 9:30 P.M. on March 23, 1989, a single-hulled oil tanker named the *Exxon Valdez* left the Alaskan port of Valdez (see Figure 10.9) loaded with 1.26 million barrels (200 million L) of crude oil. Just before midnight, the tanker was given permission to change course to avoid large pieces of floating ice. What happened during the next few minutes is not entirely clear. But at 12:04 A.M. on March 24, the ship ran aground on Bligh Reef, a very shallow section of the sound. About 24 minutes later, the crew notified the Coast Guard that the tanker was leaking oil.

The tanker, which was gushing over 750,000 L of oil per minute, poured over 40 million L of its cargo into the sea. That's the bad news. The good news is that 160 million L did not leak out and were pumped into rescue ships. The grounding of the *Exxon Valdez* will always have a special place in environmental history because of the spate of questions about what *really* happened that night. Could the Exxon company have prevented the spill by using a double-hulled tanker? Why did it take so long (10 hours after the accident) for the cleanup to begin? Nonetheless, the *Valdez* was not the only oil tanker spill, and certainly not the largest, in recent years. In fact, it is number 34 on the all time list!

Hundreds of oil spills occur each year. There were 532 major tanker spills between 1974 and 1992, averaging 130,000 barrels per spill (a barrel holds 42 gallons of oil). The picture brightened between 1993 and 1999, when a total of "only" 43 spills of greater than 5000 barrels of oil occurred. The National Oceanic and Atmospheric Administration (NOAA) reported only 2 spills of greater than 5000 barrels in U.S. waters in 1998. For every 1 million metric tons of oil shipped, 1 metric ton is lost to spillage. Since the total effect of all these spills has been

Figure 10.9

Oil spilled from the *Exxon Valdez* spread over the surface of the water. As the more volatile components evaporated, droplets of less volatile compounds began to disperse within the water, forming a "mousse."

enormous environmental damage, it seems reasonable to wonder, "if this much oil is spilled each year, can't we switch to another energy source?" In part, the answer to that is "yes," as we will discuss in Chapters 14 and 15. However, the uses of oil extend far beyond energy production.

Crude oil is a complex mixture of organic compounds, mainly hydrocarbons, that has formed from the chemical decomposition of ancient plants and animals. It can be separated or **fractionated** into several fractions each containing compounds whose boiling points lie within a certain range (see Table 10.7). The fractionating process is known as **fractional distillation** because it involves heating the crude oil and collecting the different fractions from different parts of a large "fractionating column" (see Figure 10.10). Notice that the fractions with the lowest boiling points also have the smallest molecules as indicated by the number of carbon atoms they contain. Each fraction of crude oil can be exploited in several ways, most of which do not involve burning the oil as a source of energy.

Table 10.7

Petroleum Oil Fractionation

Fraction	Composition	Boiling Range, °C	Principal Use
Gas	C_1–C_4	Below 20	Heating fuel
Petroleum ether	C_5–C_6	20–70	Solvents, petrol additive for cold weather
Petrol (straight run)	C_6–C_{10}	70–200	Motor fuel, solvent
Kerosene	C_{10}–C_{18}	175–320	Jet and diesel fuel
Fuel oil	C_{12}–C_{18}	Above 275	Diesel fuel, heating fuel oil, cracking stock
Lubricating oils	Above C_{18}	Distill under vacuum	Lubrication of machinery
Residues	Above C_{18}	Nonvolatile liquid	Roofing and road materials

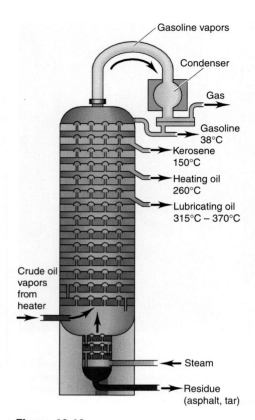

Figure 10.10
Fractional distillation results in the separation of crude oil into several groups of compounds based roughly on the number of carbons. Each fraction can then be further refined for use in everything from plastics to jet fuel.

What Happens to Spilled Crude Oil?

The site of the *Exxon Valdez* oil spill served as an excellent laboratory for scientists to learn more about what happens when crude oil mixes with water. Recall that water, being polar, does not mix with mineral oil, which is mostly nonpolar. Crude oil, however, is very different, because a few of its components include oxygen, nitrogen, and sulfur atoms, which can participate in polar interactions and possibly hydrogen bonding. We would therefore expect that a little crude oil could mix with seawater. Other components of crude oil, however, are nonpolar hydrocarbons that do not dissolve in water.

As oil spilled from the *Exxon Valdez,* it began to spread over the surface of the water. The most **volatile** substances in the water (meaning those with the lowest boiling points) evaporated into the atmosphere. These were mostly organic compounds containing from one to four carbon atoms. Droplets of the less volatile materials began to disperse into the water while some of the more soluble components dissolved in the water. Other components, including benzene, toluene, and xylenes, formed an emulsion.

Benzene   Toluene   Xylene

A **colloid** is a mixture of a material very finely dispersed in another in which it is not soluble. An **emulsion** is a colloidal dispersion in which tiny droplets of one substance (crude oil components, in this case) are dispersed throughout another substance. Emulsified crude oil can form a thick foamy "mousse." There were areas in Prince William Sound where this mousse was 4 feet thick.

Eventually, microorganisms, such as the *Pseudomonas* bacteria discussed later, began to break down some of the oil. The rest of the mousse was broken down by water currents into components including xylene and other compounds with higher molecular weights. These then washed to the shore as tar balls.

The Exxon Corporation spent over $1 billion cleaning up oil that had washed ashore. In a time of economic recession, scores of workers migrated from the continental United States literally to wash and scrub rocks for $16.69 per hour. One of the oft-debated issues of this incident is whether cleanup crews could have arrived sooner on the morning of March 24 when the oil starting spewing from the *Valdez.* That might have prevented much of the need for cleaning tar balls on the shore.

How Do We Clean Oil Spills?

Oil spills come in all shapes and sizes and happen under variable weather conditions. The strategy for cleaning up spills depends on where the spill is, how much oil is involved, and the weather at the time.

Containment

The preferred method for cleaning oil spills is to physically contain the oil using floating booms (see Figure 10.11). Once isolated, the oil can be skimmed off the surface. This works well in calm waters and before the spill has spread too far.

Combustion

Where the oil spill is continuous, it can be burned by igniting with lasers, bombs, or rockets. Burning is inexpensive and works well with large spills. On the downside, it releases pollutants into the air. It does not work well in cold water or where there are high winds, which make it difficult to sustain the burning.

Dispersal

When the waters are rough, chemical dispersants can be added. Dispersants break the oil into very tiny droplets that can be diluted in seawater and subsequently decomposed. High winds help keep the dispersants in contact with the oil. While dispersants work well on large spills, their environmental impact is not well understood.

Bacterial Degradation

Many components of oil can be naturally degraded by bacteria—single-celled organisms whose cell structure is considerably simpler than plant or animal cells. Bacteria use carbon-containing compounds as energy sources. As a result, oil slicks represent superb "food" for bacteria. The problem is that bacteria are fairly selective in their menu choices, so different bacteria are required to "eat" different components of oil. One especially useful kind of bacteria is *Pseudomonas,* which is used extensively as a spray on oil spills. The typical bacterial degradation of a carbon-containing compound begins with the oxidation of a "terminal" $-CH_3$ group (meaning a CH_3 group at the end of a chain) to form a $—COOH$ group:

$$-CH_2-CH_2-CH_2-CH_2-CH_3 \rightarrow -CH_2-CH_2-CH_2-CH_2-COOH$$

Sections of the long molecule are then sliced off and oxidized to carbon dioxide by repetition of the following reaction:

$$-CH_2-CH_2-CH_2-CH_2-COOH + 3O_2 \rightarrow -CH_2-CH_2-COOH + 2CO_2 + 2H_2O$$

Benzene-based toxic compounds can also be oxidized and degraded by bacteria. An intensive research effort is directed toward modifying bacteria using genetic engineering to produce new types able to oxidize chosen toxic substances, including some of the compounds found in oil.

What Are the Effects of Oil Spills?

The effects of an oil spill are as varied as the cleanup methods. Whether slight or great, there will be an economic impact on the company responsible for the spill. It is also likely that local industries that depend on clean water will be affected.

There will also be an environmental impact, although the extent of the environmental damage depends on many factors. The oil tanker *Amoco Cadiz* ran aground off the coast of France in 1978. More than 240 million L of oil were spilled, six times the amount released from the *Exxon Valdez.* Yet the environmental impact was not nearly as great because strong winds and warm seas off the French coast allowed the spill to disperse rapidly.

In general, birds and fish are the first to suffer from oil spills. Researchers investigating the *Exxon Valdez* accident detected contaminated clams and mussels. Oil-soaked seabird carcasses were scavenged by bald eagles, resulting in contamination of their young and their eggs. In the long term, these populations usually recover.

Figure 10.11
On July 2, 1997, the supertanker *Diamond Grace* gashed its hull on the reef in Tokyo Bay and proceeded to dump 4 million gallons of crude oil into the waters of the bay. As this photo shows, tugs positioned booms to contain the floating oil. Helicopters sprayed dissolving agents on the 3½-mile-wide oil slick and workers spread absorbent mats and skimmed the surface of the water to manually remove oil. While this spill was only about 1/3 the oil spilled by the *Exxon Valdez,* it was the worst in Japan's history.

If there is a silver lining to the grounding of the *Exxon Valdez,* it is that it (and other spills) have led to an unprecedented effort being made to learn more about the chemical, environmental, and social impacts of oil spills.

10.5 | Wastewater Treatment

London in the 1850s saw the beginning of governmental interest in large-scale water pollution. At that time, the health of millions of people was at risk because there were no effective ways of cleaning wastewater. The tenor of the times was described by Michael Faraday's letter in the Case in Point at the beginning of this chapter. A century and a half later, treatment to make wastewater safe to use again is performed in almost every town and city of the developed world. Our fresh water supply is limited, and in America we often reuse it countless times. In fact, water that enters the Mississippi River in Minnesota is used an estimated 17 times before it ends up in the Gulf of Mexico.

Treatment of municipal wastewater is designed principally to remove human waste, but it also removes dissolved metals, suspended materials, and ions such as phosphate. Normally this must be done very inexpensively with no added chemicals ending up in the treated water. Standard treatment is generally subdivided into three stages: primary, secondary, and tertiary (or advanced) (see Figure 10.12).

Primary Treatment

This step involves the simple removal of floating or suspended solids. This is mostly material that has gone down someone's garbage disposal or toilet and whatever goes into the sewer, such as leaves and rocks. Lots of suspended toilet paper fibers and floating grease are removed at this stage. The floating material is skimmed off the top and the remaining water is passed through ever finer screens to remove suspended solids. In some processes, iron ions or aluminum ions are added to the water. These precipitate as flocculent (fluffy and porous)

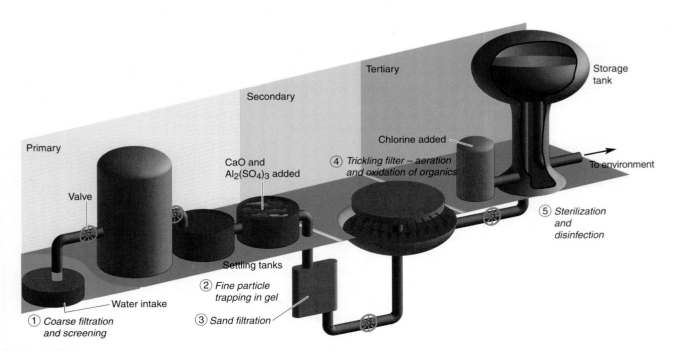

Figure 10.12
Standard wastewater treatment is divided into three stages: primary, secondary, and tertiary. Each step makes water successively less polluted. At the end of the process, involving both chemical and physical steps, the water is safe to introduce to the environment.

hydroxides that speed the settling of other solid materials. Other metal ions tend to coprecipitate with the iron or aluminum hydroxide, formed as follows.

$$Fe^{3+} + 3OH^- \rightarrow Fe(OH)_3$$

If the water has been in sewer pipes for some time, decomposition of various foods and other items is likely. Hydrogen sulfide gas, H_2S, is released, with the resultant smell of rotten eggs. Chlorine gas (Cl_2) can be added at this point to convert the hydrogen sulfide to (odorless) sulfate ion,

$$H_2S + 4Cl_2 + 4H_2O \rightarrow 8Cl^- + SO_4^{2-} + 10H^+$$

Secondary Treatment

The water that emerges from primary treatment looks a lot better than did the entering water. At this stage roughly 30% of all polluting substances have been removed. The remaining water is full of dissolved substances, most of them organic compounds. These are very difficult to remove chemically but most of them are readily metabolized by microorganisms, which are purposely allowed to do so. A "suite" of microorganisms (mostly bacteria) is brought into contact with the water in a variety of ways, but the most picturesque is the "trickling filter" (see Figure 10.13). A trickling filter uses huge circular tanks filled with rocks about the size of baseballs which are coated with huge numbers of microorganisms. The wastewater is trickled onto these rocks by what looks like a giant lawn sprinkler. As the water passes down through the bed of rocks, the bacteria convert about 70% of the dissolved organic compounds into carbon dioxide and living bacteria (biomass). Urea, $(NH_2)_2CO$, is a major component of urine which cannot be utilized by microorganisms. The urea reacts slowly with water to form carbon dioxide and ammonia.

$$(NH_2)_2CO(aq) + H_2O(l) \rightarrow CO_2(g) + 2NH_3(aq)$$

The ammonia formed in this reaction then is transformed by microorganisms, through a reaction with oxygen, to give nitrite ion which in turn can react with more oxygen to give nitrate.

$$2NH_3(aq) + 3O_2(g) \rightarrow 2NO_2^-(aq) + 2H_2O(l) + 2H^+(aq)$$
$$2NO_2^-(aq) + O_2(g) \rightarrow 2NO_3^-(aq)$$

At this point, 85 to 95% of all polluting substances have been removed.

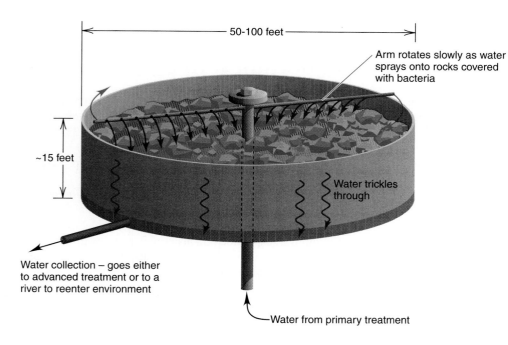

50-100 feet

Arm rotates slowly as water sprays onto rocks covered with bacteria

~15 feet

Water trickles through

Water collection – goes either to advanced treatment or to a river to reenter environment

Water from primary treatment

Figure 10.13
A "trickling filter" uses huge circular tanks filled with baseball-sized rocks coated with microorganisms. The wastewater is trickled onto these rocks, where the bacteria convert much of the dissolved organic compounds into carbon.

Tertiary Treatment

After the remaining solids (called sludge) settle out, the water looks clear and is often suitable for release to the environment. Many modern treatment plants go a step further, however, to tertiary treatment. Tertiary treatment removes nitrate and phosphate ions and other dissolved substances not previously removed or actually produced by the bacteria in secondary treatment. This removal of remaining organic substances is usually done by filtering the water through a bed of granular carbon that absorbs the dissolved organic substances. If phosphate is to be removed, the water is usually treated with calcium ions from $Ca(OH)_2$, forming the insoluble hydroxyapatite, $Ca_5(PO_4)_3(OH)$, which is allowed to settle out. Addition of calcium hydroxide also raises the pH and converts any ammonium ions to ammonia. The resulting water is sprayed into the air, allowing the dissolved ammonia to escape as ammonia gas, converting a water pollution problem into an air pollution problem. Removal of nitrogen from wastewater can also be done by rather specialized bacteria that convert nitrate ions into molecular nitrogen gas, a more suitable substance to put back into the air. Any residual microorganisms are killed by adding a carefully monitored dose of chlorine. At this point, the water can be safely returned to lakes, rivers, or streams.

What happens to all the sludge that has been removed from the water? In some instances it is spread onto agricultural fields as a soil-conditioning mulch, although in heavily industrialized cities this is discouraged because the sludge contains such a high concentration of heavy metals. Many municipalities use the sludge to feed a completely new set of bacteria that convert organic compounds into methane gas (CH_4), which is burned to make electricity or heat for use by the treatment facility.

In Conclusion

Our waterways have long been used as dumping grounds for human refuse, agricultural runoff, industrial waste products, and occasionally and unintentionally, crude oil. It is a testimony to human ingenuity that we have been able to create so much with what nature has given us, but this success also sets us the challenge of preserving our precious water supplies. Chemistry has the critical role to play in helping to preserve our water resources and governments working together with industries have begun to turn the tide toward cleaner waterways. It is our choice whether we pollute and whether or not we will clean up the mess. As always, chemistry is a world of choices.

main points

- Water pollution is the addition of some substance that causes the quality of water—taste, smell, appearance, cleanliness—to become poorer.

- Water pollution has been an important problem for over 150 years.

- Agriculture, industry, and oil transport are among the most important activities leading to water pollution.

- Chemical processes are important to pollution control efforts.

- Wastewater treatment involves primary, secondary, and tertiary treatment steps.

important terms

Biomagnification is the process by which the concentration of pollutant increases as predators eat plants and animals that originally have taken on pollutant substances. (p. 333)

Carcinogen is a substance which causes cancer. (p. 337)

Chlorinated hydrocarbon is a compound containing only carbon, hydrogen, and chlorine. (p. 338)

Colloid is a suspension of a material in a matrix in which it is not soluble but in which the suspended material is so finely dispersed that it does not settle out. (p. 352)

Emulsion is a colloidal dispersion of one liquid in another. (p. 352)

Fertilizer is a substance used to increase the growth rate of plants. (p. 334)

Fractional distillation is a procedure for separating mixtures by boiling and then condensing the compounds in order of increasing boiling point. (p. 351)

General contact herbicide is a substance that kills plants after contacting the plant's surface. (p. 340)

Hemoglobin is a protein (mol wt = 56,000) that carries oxygen through the blood system from the lungs to the rest of the body. (p. 344)

Herbicide is a compound used to kill selected plants. (p. 334)

Insecticide is a compound used to kill selected insects. (p. 334)

Isotope effect, different isotopes in reactants cause very minor but detectable changes in the reactivity of these substances. (p. 345)

On-demand generation is a strategy of making dangerous compounds just before they are needed. (p. 348)

Pesticide is a substance that kills a selected set of plants or animals. (p. 334)

Phloem is tissue that transports food materials in plants from where they are produced to where they are needed. (p. 341)

Pollutant is a substance present in the environment, usually resulting from human activity, that decreases the quality or value of its surroundings. (p. 330)

Residual herbicide is an herbicide that kills selected plants a considerable time after its application. (p. 342)

Selective, only harms certain organisms and not others. (p. 340)

Translocated herbicide kills plants only after being taken into the plant through the roots or leaves. (p. 341)

Volatile is easily evaporated. (p. 352)

Xylem is tissue which transports water and mineral nutrients in plants. (p. 341)

exercises

1. State the four general sources of water pollution and give an example of where each of these sources has previously caused water pollution (locally, nationally, or internationally). Look in recent newspapers and magazines for examples.

2. List four common water pollutants and where each may be found (see Table 10.1).
3. Give two causes for the rise in water pollution in Europe.
4. Can the causes of water pollution in Europe also be seen in the United States?
5. Which are easier to control, point sources or diffuse sources of pollution?
6. Explain why the pollutants mercury(II) and lead(II) are particularly harmful to the human body.
7. Describe the history of the pollution problem in the Great Lakes.
8. You are on the Great Lakes' Preservation Committee. Local industries are considering donating funds for a cleanup project of the lakes, but some members of the executive board are not convinced that there is a problem. Present some proof that the Great Lakes are in fact polluted? (See Case In Point: Pollution in the Great Lakes, if you are stumped). Devise a convincing plan of action to deal with the pollution sources which could be presented to the skeptical executives.
9. Surf the Internet to find four environmental pollutants not listed in Table 10.1. Why are the ones you chose dangerous? What is their "MCL"?
10. Pesticides and fertilizers are considered useful and even essential to the farming community. Why are they also classified as pollutants?
11. If 0.5 L of a 0.3 M solution of nicotine sulfate is sprayed on a flower garden, how many milligrams nicotine sulfate are released? Would this answer sound less dangerous if the answer had been expressed in pounds?
12. Name two organic and two inorganic pesticides. Describe how they interact with pests to control their populations.
13. Why is DDT no longer used as a pesticide?
14. What is the important structural feature for each class of synthetic pesticides?
    a. chlorinated hydrocarbons
    b. organophosphates
    c. carbamates
    d. pyrethroids
15. When developing a pesticide, do you want it to be long lasting or degrade quickly? Explain.
16. What is the difference between natural and synthetic pesticides? Which type of pesticide is better? Why?
17. What is the difference between a selective and a nonselective herbicide?
18. What are the three categories of herbicides and how do they work?
19. How do nitrates and phosphates from fertilizers contaminate ground water?
20. Why is nitrate-contaminated water especially harmful to infants?
21. As discussed in the chapter, on-demand generation appears to be an effective means of reducing hazardous waste. Why, then, is this process not used by every industry?

22. Find out more information about one of the Federal Water Pollution Control Acts listed in Table 10.5. Be sure to reference your information.

23. State five specific products made from crude oil.

24. State four strategies for cleaning up oil spills and indicate how each method works.

25. Why can crude oil be separated into different fractions? What physical property is used for the separations?

26. Suppose a 10-kilogram sample of crude oil is 8.5% kerosene. Explain how the kerosene could be isolated from the crude oil. Be specific about temperatures and yields.

27. In the *Exxon Valdez* oil spill, what two chemical or physical properties caused the oil to float on top of the water?

28. What is an emulsion? What components of crude oil form an emulsion with water?

29. What are the three stages of water treatment? What is the purpose of each of these steps?

food for thought

30. According to the 1992 U.S. National Report of the United Nations Conference on Environment and Development, a recent survey of drinking-water wells found that only 1 to 2% of the nation's wells are considered to be contaminated to a level that is harmful to humans. This survey also indicated that pesticide residues above harmful limits were found in less than 1% of drinking-water wells. Do these data indicate that we should no longer be concerned with contaminants in our groundwater? Why or why not?

31. As discussed in the chapter, bacteria are often used to clean up oil spills. Find out what happens to these bacteria after they finish their task of oil degradation. Could excess bacteria pose a threat to human health by themselves being a contaminant? (See the August 1993 *National Geographic* for a wonderful article on bacteria!)

32. In order to speed up the cleanup process of oil-stained beaches on the coast of Alaska, EPA scientists added nitrogen- and phosphorus-containing fertilizers to assist the growth of oil-eating microorganisms. Although difficult to determine, researchers concluded that "oil degraded up to five times faster and PAH levels dropped five times faster on beaches stimulated with sufficient fertilizer than on beaches left alone." We have discussed the ill effects of excess nitrates and phosphates in water in this chapter. Using a risk vs. benefit analysis, is there a significant risk in adding the fertilizer to the water which outweighs the benefits of cleaning up the oil spill?

33. In April of 1993, Exxon stated that "government scientists have mistakenly assumed that large numbers of biologic and sediment samples from Prince William Sound contained remnants of *Exxon Valdez* crude when, in fact, they did not." As a result of company-funded research, the area had been contaminated with oil long before the spill. Yet, chemist Jeffrey W. Short of the National Marine Fisheries Service in Juneau, Alaska, suggests that Exxon is putting the blame on other sources to conceal their liability. Whom are we to believe? How can we use science to determine the validity of scientific studies?

34. Milwaukee, Wisconsin, was the site of what is thought to be the largest *Cryptosporidium* (a microscopic parasite) outbreak in the nation. In March 1993, one of the city's water treatment plants experienced problems with turbidity (cloudiness) in the water due to the substitution of polyaluminum hydroxide for alum in the treatment process. According to Kim Fox, a federal environmental engineer, there was a high likelihood that the parasite passed through plant filters along with the particles causing the turbidity. There were 200,000 severe diarrheal illnesses reported at the time of the outbreak. What could have been done to prevent this water contamination?

35. The International Oil Pollution Compensation Funds (IOPC) compensates victims of oil spills. The London-based organization has 71 member states as of October 2001. When oil spills occur, those affected apply to the IOPC for funds. One recent case concerns a large oil spill from the tanker *Erika* off the coast of Brittany, France. About 19,800 tonnes of oil were spilled (one tonne = 1000 kilograms). The cleanup took place from June through September 2000, and included the collection of 200,000 tonnes of oil-laden wastes. IOPC set aside an account anticipating organizations and individuals applying for compensation. If you were part of the Secretariat (the governing body of IOPC), how much money would you set aside for claims? How do you use to decide? Additional information on the IOPC and the *Erika* spill is found on the textbook Website.

readings

Books

1. Baird, Colin. 1995. *Environmental Chemistry,* Salt Lake City, Utah: W. H. Freeman and Co.
2. Eby, Denise and Tatum, Roger. 1989. *The Chemistry of Petroleum,* Bativia, Ill: Flinn Scientific Inc.
3. Fletcher, W. W. and Kirkwood, R. C. 1982. *Herbicides and Plant Growth Regulators,* Great Britain: Granada Publishing Ltd.
4. Ashton, Floyd M. 1973. *Mode of Action of Herbicides,* New York: Wiley-Interscience.

Articles

5. Bock, C. Allen and Anaya, William J. Environmental laws, regulation and liability, *J. J. Keller & Associates, Inc.* November 1, 1991.
6. Hileman, Bette. Alternative agriculture. *Chemical and Engineering News,* March 5, 1990, pp. 26–40.
7. Ember, Lois R. Strategies for reducing pollution at the source are gaining ground. *Chemical and Engineering News,* July 18, 1991, pp. 7–16.

8. Haggin, Joseph. Hong Kong plans new generation chemical waste plant for 1993. *Chemical and Engineering News,* February 4, 1991, pp. 27–29.

9. Holloway, Marguerite. Soiled shores. *Scientific American,* October 1991, pp. 100–116.

10. Hileman, Bette. The Great Lakes cleanup effort. *Chemical and Engineering News,* February 8, 1988, pp. 22–39.

11. Lowey, Nita M. In '88, 6 oil spills every 7 days. *New York Times,* June 22, 1990.

12. Holusha, John. New techniques to turn an oil spill into a collectible. *New York Times,* April 2, 1991, p. F7.

13. Shabecoff, Philip. Largest U.S. tanker spill spews 270,000 barrels of oil off Alaska. *New York Times,* March 25, 1989 p. A1.

14. Degener, Richard. Deadly effects of insecticide reappearing in area wildlife. (Atlantic City) *The Press,* April 23, 1991.

15. The return of cholera. *Discover,* February 1992, p. 60.

16. Garber, Charri Lou. Wastewater. *Chem Matters,* April 1992, pp. 12–15.

17. Gibbs, W. Wyatt, The Arctic Oil & Wildlife Refuge. *Scientific American,* May 2001, pp. 63–70.

18. International Tanker Owners Pollution Federation, LTD. Tanker Oil Spill Statistics http://www.itopf.com/stats.html.

19. Grunwald, Michael. Monsanto Hid Decades of Pollution. *Washington Post,* January 1, 2002, p. A1.

websites

www.mhhe.com/kelter The "World of Choices" website contains activities and exercises including links to websites for Environment Canada; United Nations Pollution Release and Transfer Register; The Smithsonian Institution; and much more!

Behavior of Gases

The world's a nicer place in my beautiful balloon;
It wears a nicer face in my beautiful balloon.
We can stroll among the stars together, you and I,
For we can fly, we can fly!
Up, up and away in my beautiful, my beautiful, balloon

—"Up Up And Away" recorded by The Fifth Dimension

January 28, 1986, is etched in the collective memories of millions worldwide. That day, at 11:38 A.M. EST, the shuttle *Challenger* was launched from Kennedy Space Center in Florida. Just 73 seconds after launch an explosion, with causes as complex as the craft itself, occurred that tore the shuttle apart as its hydrogen fuel and oxygen combined to form gaseous water. The gas quickly cooled and condensed into thick white plumes of water droplets, spraying out at all angles like the legs of a spider (see Figure 11.1). Seven astronauts were killed in the accident. New terms like "O-rings" (seals that failed on the craft's solid rocket boosters) became permanent parts of our lexicon. And we learned about the extraordinary power of chemical reactions that produce gases.

The combination of hydrogen and oxygen gases releases energy, as described by this equation:

$$2H_2(g) + O_2(g) \rightarrow 2H_2O(g) \qquad \Delta H = -285.8 \text{ kJ/mol } H_2$$

Hydrogen is used as a fuel for the Space Shuttle precisely because its reaction with oxygen releases so much energy, which can be used to help power the craft into orbit provided the combustion reaction is carefully controlled.

The *Challenger* explosion was preceded by almost 50 years by what was perhaps the most famous aviation accident of the first-half of the 20th century. On May 6, 1937, the *Hindenburg,* a huge German passenger balloon known as a "dirigible," cruised in for a landing in Lakehurst, New Jersey. Suddenly, a spark ignited the hydrogen-filled ship and the ensuing explosion and fire sent the craft crashing to the ground. Thirty-five passengers died. History will record this as a tragedy, but also as a miracle, because 62 passengers survived. History will also note that for 60 years, the wrong explanation was generally accepted for the cause of the fire, as we discuss in the Case in Point New Light on the Hindenburg Disaster.

▇ case in point

New Light on the Hindenburg Disaster

The *Hindenburg* contained 7.2 million cubic feet of hydrogen gas encased within its 813-foot aluminum frame and outer covering. The dirigible had made nine successful transatlantic crossings before the ship was consumed by fire at the landing dock in New Jersey. Conventional thinking was that the highly exothermic reaction of hydrogen and oxygen gases caused the catastrophe.

In 1997, Addison Bain, the former manager of Hydrogen Programs at NASA's Kennedy Space Center, proposed an alternative explanation based on his own laboratory tests. He proposed that a static charge built up on the skin of the craft and set the highly flammable *skin itself* aflame. The disaster was not caused by hydrogen combustion. His evidence? Bain obtained and analyzed pieces of the *Hindenburg*'s covering. This skin was made of cotton and cellulose acetate, a chemical that protects the skin from deterioration, provides an aerodynamic surface, reflects sunlight to prevent the interior from overheating, and aids in weatherproofing the craft. On the skin was a coating of the highly flammable combination of aluminum powder and iron oxide that made it reflective to sunlight so the gas within would not unduly expand. The color of the *Hindenburg*'s flame was not that of the burning of hydrogen and oxygen, which combust without color. The colors *were* like that of the burning skin itself. Additional pieces of evidence are that hydrogen, being less dense than air, burns up. The flame on the *Hindenburg* burned down and sideways. Hydrogen diffused out in seconds, but the craft burned for about 10 hours.

The May 6, 1937, crash of the hydrogen-filled dirigible *Hindenburg* killed 35 passengers. The devastation is testimony to energy exchange that accompanies chemical reactions and to the impact that energy has on gases.

In a November 1997 *Popular Science* article (see reading #7), Dr. Bain remarks that "the moral of the story is, don't paint your airship with rocket fuel." An additional moral might be that hydrogen is a safe gas to use as a fuel—in dirigibles and, as we shall see in Chapter 15, in motor vehicles as well!

Figure 11.1
The *uncontrolled* reaction of hydrogen and oxygen led to the deaths of seven U.S. astronauts on board the Space Shuttle *Challenger*. In spite of this disaster, hydrogen and oxygen are still used extensively as rocket propellants because of their light weight, ready availability, and the ability to vary the amounts of the mixture burned during launch.

11.1 | Balloons and the Properties of Gases

Space shuttles rely on hydrogen fuel to help propel them into orbit. But the *Hindenburg* had no such need. Why did the dirigible contain hydrogen rather than helium, which is nonflammable and less dense than air?

Later in this chapter we will show how the density of gases is directly related to their molar masses. Knowing this, helium, which is nonflammable and less dense than air, would seem to be a good choice for a gas to fill dirigibles like the *Hindenburg*. Before the start of the Second World War helium was available only as a by-product of the Texas natural gas processing industry, and the United States refused to sell it to other countries. That is why other countries, most notably Germany, had to use hydrogen, which was produced industrially by the combination of silicon and aqueous sodium hydroxide:

$$Si(s) + 4NaOH(aq) \rightarrow 2H_2(g) + Na_4SiO_4(s)$$

The property of density has helped us to understand why balloons, even huge ones, can become airborne.

exercise 11.1

Calculation of Density

Problem

Calculate the density of air at 20°C, in grams per liter, if 500 mL is found to have a mass of 0.600 g.

Solution

First, convert 500 mL to L:

$$500 \text{ mL} \times \frac{1 \text{ L}}{1000 \text{ mL}} = 0.500 \text{ L}$$

Then divide the mass by the volume to get the density.

$$\text{density} = \frac{\text{mass}}{\text{volume}} = \frac{0.600 \text{ g}}{0.500 \text{ L}} = 1.20 \text{ g/L}$$

This means that 1 L, a bit more than a quart of such air has a mass of 1.2 g, 2 L has a mass of 2.4 g, etc. In general, gases are much less dense than liquids and solids. Water, you may recall, has a density of 1.00 g/mL or 1000 g/L.

In Figure 11.2 you can see a "hot air" balloon in flight. Based on the information in Table 11.1, can you explain how balloons filled with air rather than helium or molecular hydrogen, can be made to fly?

You now know that the density and temperature of gases are related properties, meaning that changes in one are associated with changes in the other.

On a molecular scale, we also note that gas molecules do not interact with one another, because the intermolecular attractions are too weak. This is why gases are gases rather than liquids or solids.

A Closer Look at an Explosion— Defining Pressure

We began this section by describing the explosion of the Space Shuttle *Challenger*. Let's end it by looking at the explosion process of a simpler system, a balloon filled to its largest possible volume with hydrogen and oxygen gases (Figure 11.3 inset). We provide sufficient initial energy for the reaction to proceed, forming water vapor and releasing energy (Figure 11.3). The reaction is fast, highly exothermic, and takes place as the hydrogen escapes from the ruptured balloon mixing with air. The temperature of the gas rises quickly and the flame expands to a much larger volume than was the original hydrogen due to its pressure suddenly dropping to atmospheric pressure. We define that *force per unit area* to be the **pressure** of the gas.

$$\text{pressure} = \frac{\text{force}}{\text{area}}$$

The sharp decrease in pressure of the gas and the sudden increase in the temperature of the gas from the reaction of H_2 and O_2, causes the flame to be much larger than the original balloon.

Figure 11.2
A hot air balloon flies because the warmed air is less dense than the surrounding cooler air.

Table 11.1

Density of Dry Air[a]

°C	Density
10	0.00125
15	0.00123
20	0.00120
25	0.00118
30	0.00117
40	0.00113

[a]g/mL at 1 atm.

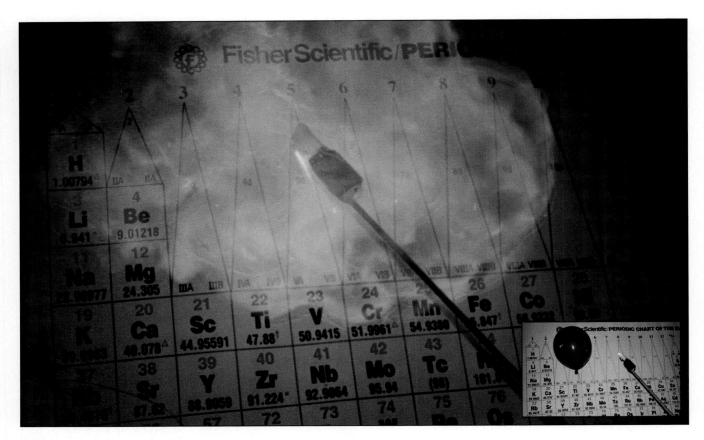

Figure 11.3
A balloon filled with hydrogen gas (inset). When the flame burns through the balloon, oxygen from the air reacts violently with the hydrogen.

consider this:

The old adage about staying as low as possible while leaving a room that is on fire does not necessarily hold true when certain chemicals are involved. For example, chlorine gas is heavier, and therefore more dense, than air. If chlorine gas is present in the blazing room, it will be near the floor. You should not stay low in this situation. What about ammonia?

When there is a fire in an area with many hazardous chemicals, we should go against the traditional thinking and avoid dangerous fumes by leaving the area in an upright stance. Most noxious chemical vapors are more dense than air, and will therefore collect close to the floor. Carbon monoxide, CO, has about the same molar mass as nitrogen gas, N_2, but carbon dioxide, CO_2 has a higher molar mass, and is therefore more dense. CO_2 is a much more common gas in a fire than Cl_2.

11.2 | A Closer Look at Pressure

To get a good idea of what pressure means we need to consider the terms **force** and **area,** used to define pressure, and look at some applications of pressure.

Recall that pressure is a measure of the amount of force applied to a given area,

$$\text{pressure} = \text{force per unit area}$$

$$\text{pressure} = \frac{\text{force}}{\text{area}}$$

■ hands-on

Why Wasn't Carbon Dioxide, a Readily Available and Nonflammable Gas, Used in Dirigibles?

To answer that question, let's generate the nonflammable gas, carbon dioxide (CO_2), in a balloon. To do this, you'll need:

- A balloon
- A plastic soda bottle
- A spoon, some baking soda, and about 50 mL of vinegar

With the help of a colleague, add about 2 tablespoonsful of baking soda into the balloon. *Do not* tie it off. Next, add the vinegar to the soda bottle. Carefully place the top of the balloon over the neck of the soda bottle, as shown in Figure 11.4A. When everything is in place, lift and tap the balloon so that the baking soda goes into the bottle and mixes with the vinegar. You should get a result like that in Figure 11.4B. Take the balloon off the bottle, being careful to keep the gas in the balloon from escaping. Tie the balloon off. Now throw the balloon up in the air. What happens and why?

Let's consider the reaction and answer the question just posed. Baking soda is the common name for sodium bicarbonate, $NaHCO_3$, which dissolves in water to form a weak base. Vinegar is a 3 to 5% solution of acetic acid, CH_3COOH. The acid and base react as follows (see Chapter 8 for a review of acid-base chemistry):

$$CH_3COOH(aq) + HCO_3^-(aq) \rightarrow CH_3COO^-(aq) + H_2O(l) + CO_2(g)$$

The result is that the balloon fills up with carbon dioxide gas and drops to the ground when you let it go.

The key to understanding why a balloon filled with molecular hydrogen rises and one filled with carbon dioxide sinks is density: the mass of a substance within a given volume. The force of gravity that pulls anything downward toward Earth is proportional to the mass of substance involved. Your experiment demonstrates that a balloon filled with carbon dioxide gas has a higher density than air, causing gravity to pull it towards the ground with greater force than the same volume of air. On the other hand, a balloon filled with hydrogen, which is less dense than air, will rise upward due to being displaced by the more dense air than is pulled more forcefully downward to take its place.

To be of any use in dirigibles, a gas must be less dense than air so that it will rise upward carrying the rest of the craft as it goes. This is why carbon dioxide could not be used in dirigibles, even though the gas is nonflammable.

Figure 11.4 A, B
After putting baking soda, sodium bicarbonate, into a balloon, place it over the neck of a soda bottle (photo **A**). After the balloon is lifted and the contents tapped out, they mix with vinegar forming carbon dioxide gas (photo **B**).

where **force** = mass × acceleration. Your body weight is a unit of force, because your body has mass, and gravity provides a constant acceleration. If you put all your weight on one foot instead of two, there is a lot of pressure (force per unit area) on that foot. If you then put on shoes with spiked heels, and put the same weight *just on the spike,* the pressure (force per unit area) on the spike is greater than on the foot before, because the area for the same force is so much smaller (see Figure 11.5). If this were done on a linoleum floor, how would you know whether or not there was more pressure on the floor from the spike or the bare foot?

Several different units for pressure are in use, but one of the most common is the **atmosphere** (atm), with 1 atm being approximately equal to the pressure of the Earth's atmosphere at sea level on a typical day. Since the precise value of atmospheric pressure varies, we need to be more specific in defining the standard unit of 1 atm. A **standard atmosphere** (meaning exactly 1.000 atm) is defined as the pressure which will support a 760-millimeter (almost 30 inch) column of mercury in a barometer (a device that measures pressure). See Figure 11.6 to visualize the measurement. Evangelista Torricelli first made such a device in 1643 and

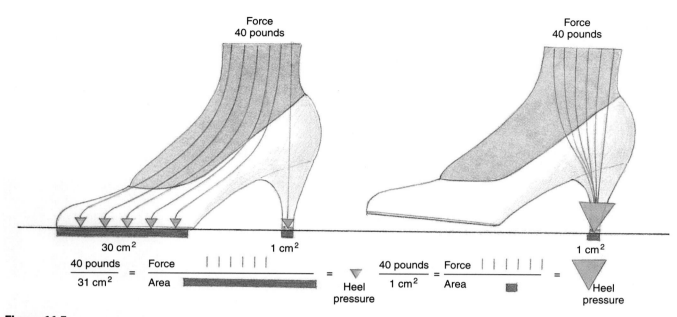

Figure 11.5
The pressure on the heel of a spiked shoe is greater than that on a shoe with the heel *and* sole touching the floor.

Figure 11.6
At sea level, the Earth's atmosphere exerts enough pressure to support a column of mercury to a height of 760 mm (29.92 inches). This is known as a "standard atmosphere."

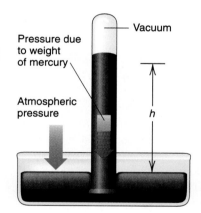

discovered that atmospheric pressure changed with altitude and with changes in the weather. In his honor, we also specify one atmosphere as being 760 **torr** (for Torricelli) of pressure. One torr is equal to the pressure necessary to support a column of mercury 1 millimeter (mm) high. In the United States, we express pressure in weather forecasts in inches of mercury in which one atmosphere equals 29.92 inches Hg. We also often express pressure as pounds per square inch, or **psi.** One atmosphere equals 14.7 lb per square inch. Pressure units called "pascal" and "bar" are becoming common. There are 1.013×10^5 pascals per atmosphere and a bar is 1×10^5 pascals. Therefore, 1.000 atmosphere = 1.013 bar.

consider this:

Hiking in high mountains can sometimes make hikers feel a bit lightheaded until they adjust to the lower air density (and resulting lower air pressure and lower oxygen concentration). Why is the air less dense at high altitude than at sea level? Based on your answer, can you justify why Jupiter and Saturn have atmospheres rich with hydrogen while the Earth's atmosphere is mostly nitrogen and oxygen and the atmosphere of Mars has a large concentration of carbon dioxide?

consider this:

Much of our discussion of gases has assumed a gas pressure of 1.0 atm. For many years, this was the so-called "standard pressure." Recently, **standard pressure** was redefined to mean exactly 1 bar (= 100,000 pascals). However, most chemistry textbooks and courses still use 1.0 atmospheres as standard pressure, so we will stick with this convention for now. There is also a **standard temperature** for gases of 0°C, or 273 K. Gases at these conditions are said to be at **standard temperature and pressure,** or **STP** conditions.

You may be familiar with units of pressure on inflatable balls or tires. An inscription on a basketball may read "inflate to 10 lb per square inch (10 psi)." You now know that atmospheric pressure is 14.7 psi, so how can inflating a ball to just 10 psi make the ball feel so hard? The answer is that the inscription really means we should pump the ball to a pressure that is 10 psi *greater than* atmospheric pressure. Pressure gauges, such as tire gauges, always measure pressure *in excess of 1 atm*. This is often written **psig,** meaning "pounds per square inch gauge." A tire gauge will read zero psi when the pressure inside the tire or ball is the same as outside, at 1 atm. Therefore, the actual pressure inside the properly inflated basketball will be 14.7 + 10 = about 25 lb per square inch or 25/14.7 = 1.7 atm.

In summary, pressure is a measure of force per unit area. In scientific usage it is expressed as atmospheres or millimeters of mercury. In everyday use, pounds per square inch is a more appropriate unit of pressure. Regardless of the units used, the same basic property of gases is being measured, namely, the force the gas exerts on a given area of any surface with which it is in contact.

11.3 | Dalton's Law of Partial Pressures Takes Flight

Let's shift our study of pressure away from the Earth's surface and prepare to return to the upper atmosphere as we, along with nearly 2 million people in the United States each day, take flight in a commercial jet. Onboard, the flight attendants make a variety of announcements including one about putting on oxygen masks if the cabin is "depressurized." Why would you need oxygen masks? How much pressure loss is enough to require the masks for life support? How high up can you be before requiring the masks? These questions relate to the nature of the air, the human body and the portion of the total pressure, called the **partial pressure,** exerted by oxygen at a given altitude.

When we breathe, air is taken into our lungs and some oxygen from it is transported to the tissues of the body by the protein hemoglobin. Dry air (we will consider the effect of water vapor in a few moments) is composed of about 78% nitrogen and 21% oxygen gases by volume, with minor amounts of other gases (see Table 12.2 in Chapter 12). If the oxygen gas were present alone without the nitrogen gas, the force per unit area, the pressure, exerted by the smaller amount of gas would be proportionately smaller—only 21% as large. If the nitrogen gas were present without the oxygen, the pressure would only be 78% of the original total. This suggests that the total pressure is the sum of the partial pressures of each gas. This fits the experimental work done almost two centuries ago by John Dalton, and is called **Dalton's Law of Partial Pressures** in honor of his work. We can write Dalton's Law in a more compact mathematical form,

$$P_{total} = P_1 + P_2 + \cdots + P_n$$

in which

P_{total} = the total pressure exerted by a mixture of gases
P_1 = the partial pressure exerted by gas 1
P_2 = the partial pressure exerted by gas 2
P_n = the partial pressure exerted by gas n

So if the pressure in the flight cabin is at 1.00 atmosphere, the partial pressures due to nitrogen and oxygen gases are, respectively, 0.78 and 0.21 atmospheres:

$$P_{total} = P_{N_2} + P_{O_2} + P_{minor\ components}$$
$$= 0.78 + 0.21 + 0.01 = 1.00\ atm$$

The human body at rest requires a partial pressure of O_2 of at least 0.040 atm to sustain life, so sea level pressure supplies us with more than enough oxygen (0.21 atm) to live. Is this true at, say, the top of Mt. Everest, 29,000 feet (8900 meters), where the total air pressure equals 0.286 atm? We can recast Dalton's Law in this way,

$$P_{O_2} = P_{total} \times (\text{fraction of the total that is } O_2)$$

We can calculate the partial pressure of O_2 at this altitude.

$$P_{O_2} = 0.376 \text{ atm} \times 0.21 = 0.079 \text{ atm}$$

The P_{O_2} of 0.079 atm is still somewhat larger than the bare minimum needed to sustain life of 0.040 atm. Why do so many who trek to Mt. Everest take along breathing equipment? The figure of 0.040 atm assumes no additional demands for oxygen. The hiker is at rest. The oxygen demand when moving is higher, hence the need for supplemental oxygen. Some mountaineers and guides have made it up Mt. Everest without breathing apparatus. The partial pressure of oxygen falls below 0.040 atm at an altitude of about 12,000 meters (39,000 feet). This means that passengers in a jet might barely survive depressurization as high as 39,000 feet, though the safety margin dictates the use of oxygen masks well below that altitude.

exercise 11.2

Dalton's Law of Partial Pressures

Problem

If the partial pressure of oxygen gas at 12,000 meters is 0.040, what is the total air pressure?

Solution

Oxygen gas makes up 21% of the atmosphere. The relationship between total pressure and the partial pressure of oxygen is

$$P_{O_2} = P_{total} \times (\text{fraction of the total that is } O_2)$$

We can switch that equation around to solve for P_{total}:

$$P_{total} = \frac{P_{O_2}}{\text{fraction of the total that is } O_2} = \frac{0.040 \text{ atm}}{0.21}$$

$$= 0.190 \text{ atm}$$

The air certainly gets thinner as you rise above the Earth's surface! It often takes people several days to adjust to the higher elevations (and lower partial pressures) in places such as the Rocky Mountains.

Our mention of 78% nitrogen and 21% oxygen is a description of dry air. Air often contains considerable amounts of water. This happens when we describe the weather as "humid." On such days, the partial pressure of water vapor in the air is noticeable. On page 251 in Table 7.2 we have shown the vapor pressure of water (partial pressure of $-40°C$ to $374°C$. On a warm day the temperature might be 30°C and the vapor pressure of water is 31.82 torr. This corresponds to $31.82/760 = 0.0419$ atmospheres. Therefore the highest possible partial pressure of water at this temperature is 0.0419 atmospheres and might occur on a foggy day when rain is falling or about to fall. Usually the air contains less water vapor than the maximum amount. Suppose that the amount of water vapor on a 30°C day is measured to be 12.0 torr or $12.0/760 = 0.0158$ atmospheres. Therefore, $(0.0158/0.0419) \times 100 = 37.7\%$ **relative humidity.** Usually these numbers are reported only to the nearest percent value so the weatherperson would report the relative humidity as being 38% on such a day. Relative humidity is simply the ratio of the actual partial pressure of water vapor in the air to the maximum partial pressure possible at ambient temperature, expressed in percent.

$$\text{relative humidity (\%)} = \frac{\text{actual } P_{H_2O}}{\text{maximum possible } P_{H_2O}} \times 100\%$$

exercise 11.3

Partial Pressure and Relative Humidity

Problem

What partial pressure of water would be present if the relative humidity were 38% at a temperature of 20°C?

Solution

The vapor pressure at 20°C is 7.54 torr (0.0231 atm) so the actual vapor pressure would be 0.38 × 17.54 = 6.7 torr or 0.0088 atm.

exercise 11.4

Relative Humidity and Temperature

Problem

What would happen to the relative humidity if 100% relative humidity air on a cold winter day (0°C) were brought into a warm house (25°C) while leaving its water content the same?

Solution

100% relative humidity at 0°C corresponds to a partial pressure of 4.58 torr (0.00602 atm). At 25°C, this would be 100 × (4.58/23.8) = 19.2%. Such air would be thought by most people to be too dry for comfort so humidifiers are often used to add water vapor to the air in our homes in cold weather.

We have discussed a variety of systems, including the Space Shuttle, balloons, and even lungs, filled with molecular hydrogen, air, and oxygen to show that gases have density, temperature, and exert pressure, and that these properties affect the behavior of a gas. Let's continue our exploration of gases by looking at one of the fashion trends of recent years as a model for the common behavior of gases.

11.4 | The Pump: The Behavior of Gases as a Fashion Statement

In the 1950s being fashionable meant slicked back hair for guys and ponytails for gals. The 1960s and 1970s ushered in hip-hugging bell-bottomed jeans with button fly fronts. By the 1980s, business suits with white shirts and short hair were all the rage. Nowadays we see all kinds of fashion statements from head to toe. One exception to that rule is on the basketball court, where $200 super shoes appear to have permanently replaced $15 "Converse All Stars" and "Keds" as *de rigueur*. What makes a basketball shoe cost $200? Leather costs quite a bit, and there are manufacturing, advertising, and sales costs; and then there is "the pump." This type of shoe gained fame during the 1992 National Basketball Association's Slam Dunk contest. Dee Brown, a guard with the Boston Celtics, squeezed the tongue of each of his sneakers several times to adjust their fit before doing a slam dunk with his forearm covering his eyes.

Figure 11.7
"The Pump" design is very popular among recreational basketball players. Air is introduced into an inflatable pocket intended to make the sneaker fit more comfortably. In 1995, Reebok typically included a CO_2 cartridge to inflate the pocket so that one did not have to pump the sneaker by hand. (A useful innovation? You judge!)

In this paragraph, when we say that "*T*" **or** "*n*" are constant, we mean that they do not change. Note the difference between the use of "constant" as an adjective ("the temperature is constant") and "constant" as a noun ("volume and moles are related by **a** constant").

He won the contest, and the pump mechanism on the tongue of his shoe became a nationwide phenomenon.

Preparing for the Game: Pumping the Pump

The purpose of the pump is to use air to make the shoe fit as comfortably as possible. The effect of pumping the pump depends on the design of the shoe you are pumping! Introduced in 1989, the Reebok Corporation, which marketed the shoes, had several designs (one is shown in Figure 11.7), each intended to adjust the fit in a different part of the shoe. When you pump the pump, air is forced into a pocket, which expands as a result, much as air blown into a balloon makes it larger. When the air-filled part of the shoe feels snug on your foot, you stop pumping. The pump was still available from Reebok in January 2002.

In a slightly more formal fashion, we can say that when we are pumping the shoe tongue, more moles of air (*n*) in the pocket are forcing an increase in the pocket volume (*V*) at constant pressure and temperature. This observation is true whether we are dealing with a $200 basketball shoe, a balloon, or an automobile tire: The volume of a gas is directly proportional to the number of moles of gas present (at constant temperature and pressure), as shown in Figure 11.8. We can express this mathematically as:

$$V = kn \qquad (k = \text{some constant number})$$

If we double the number of moles of a gas in a balloon, the volume will also double, provided the temperature and pressure do not change. Tripling the number of moles causes the volume to triple, and so on.

Playing the Game

You run up and down the basketball court, leaping like a gazelle fleeing from a predator. This is the stuff of which national television ad campaigns are made! What happens to that air-filled pocket under your foot each time you jump?

You finished pumping before you came on the court so the number of moles of gas will not change ("*n* is constant"). Let's assume that the temperature of the gas in the pocket does not change either ("*T* is constant"). When you step on the pocket, it is squeezed. The pocket presses harder against your foot. When you go into the air, the pocket relaxes and does not press as firmly against your foot (see Figure 11.9A and B).

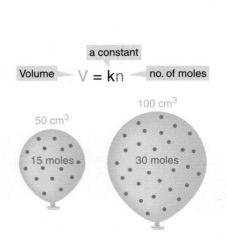

Figure 11.8
At constant temperature and pressure, the number of moles of a gas will be directly proportional to the volume it occupies (*V = kn*).

Figure 11.9
The volume of the gas inside the pocket is inversely related to the pressure exerted by the gas. In figure **A**, the gas pressure is higher and the volume of the pocket is smaller than in figure **B**.

Let's peer into the pocket of the shoe (Figure 11.10) and look at what happens to the gas molecules when you jump. As you run, the gas molecules move very quickly, constantly hitting the inside wall of the pocket. We have seen that the pressure of the gas is defined as the force exerted by the gas per unit of area with which it is in contact. When considering gas molecules bouncing off the inside of the pocket of a basketball shoe, this effectively means that pressure can be visualized as related to the number of collisions that the gas molecules make with the pocket wall in a given time.

When you squeeze the pocket by jumping, you are decreasing its volume so that the gas molecules inside will hit the walls more often. This means that the pressure will be higher. Smaller volume, larger pressure. Larger volume, smaller pressure. So volume and pressure are *inversely related* (at constant temperature); meaning that if the volume doubles, the pressure will be cut in half. If the volume is reduced to one-third of its former value, the pressure will be tripled. We can summarize this mathematically by saying that, at constant temperature:

$$P \times V = k \qquad (k = \text{some constant number}) \text{ or } P = \frac{k}{V}$$

This is a statement of Boyle's Law, named after the 17th-century scientist who first determined the relationship between temperature and pressure.

Traveling Home: The Celebration

You had a super game. Exhausted, you go to your locker. There, waiting for you is a mylar balloon left by a friend (just like the one on the left in Figure 11.11). You get cleaned up and walk slowly home in the dark, cold (−25°F) winter night. By the time you get home, you find that your balloon is not so fat anymore (as in the one on the right in Figure 11.11). What happened?

Recall from Chapter 5 that temperature is a measure of the kinetic energy of particles. Lowering the temperature of the gas in the balloon causes the following events:

- The kinetic energy of the gas in the balloon becomes lower.
- The particles of gas in the balloon move more slowly as a result of the lower temperature.
- The pressure (collisions per unit time) in the balloon becomes lower.
- The balloon becomes smaller until the pressure inside the balloon equals the pressure outside the balloon.

Ultimately, the pressure of the gas inside the balloon after the balloon has shrunk is the same as it was before. It is the *volume* that is reduced as a result of the drop in temperature. This is

Figure 11.10
"Pressure" can be thought of as the collisions per unit time the gas makes with the walls of its container—in this case, the walls of the bladder in the sneaker.

Constants are used to convert among different measures. Constants are therefore really conversion factors and they do not change (hence their Latin root *constare*—to remain steadfast). They are "constant." The speed of light constant is 3.00×10^{10} cm sec^{-1}. Yet this is also a conversion factor that says, "light travels 3.00×10^{10} cm in one second." It can be used to convert between time and distance. There are three teaspoons in a tablespoon. The conversion factor, or teaspoon/tablespoon constant is 3 teaspoons tablespoon^{-1}. Can you name some constants that we have used previously in this text?

Figure 11.11
This mylar balloon slightly shrinks when the gas inside is cooled. This illustrates the inverse relationship of gas volume to temperature (at constant pressure).

exercise 11.5

A Heavy Issue

Problem

A new *avant garde* dance troupe performs a ballet while wearing shoes with "the pump." The leading dancers in the group are Sadie and Arnold. Each pumps his or her dance shoes to an *initial volume* of 50.0 mL at an *initial pressure* of 1.0 atm in preparation for the performance. The performance begins. Sadie, known for being light on her feet, squeezes the shoe air pocket so that the internal gas pressure becomes 1.2 atm. Arnold, a dancer with a yen for pastrami sandwiches, squeezes his shoe air pocket so that the pressure becomes 1.8 atm. When they dance, they step on their shoe pockets, so that the pockets are made smaller. The gas molecules inside the pocket hit the walls more often, causing a higher pressure. What is the new pocket volume in each case?

Solution

One of the key steps in problem solving is to know clearly what your initial and final conditions are. Let's call the initial volume V_{start} and the starting pressure P_{start}. The ending volume will be V_{end}, with the ending pressure equal to P_{end}. If:

$$P_{start} \times V_{start} = \text{a constant, and}$$
$$P_{end} \times V_{end} = \text{that same constant, then}$$
$$P_{start} \times V_{start} = P_{end} \times V_{end}$$

In Sadie's case,

$$P_{start} = 1.0 \text{ atm} \qquad V_{start} = 50.0 \text{ mL}$$
$$P_{end} = 1.2 \text{ atm} \qquad V_{end} = ???$$

Using Boyle's Law, $P_{start} \times V_{start} = P_{end} \times V_{end}$, and rearranging, we get

$$V_{end} = \frac{P_{start} V_{start}}{P_{end}} = \frac{1.0 \text{ atm } (50.0 \text{ mL})}{1.2 \text{ atm}} = 42 \text{ mL}$$

In Arnold's case,

$$P_{start} = 1.0 \text{ atm} \qquad V_{start} = 50.0 \text{ mL}$$
$$P_{end} = 1.8 \text{ atm} \qquad V_{end} = ???$$

Again using Boyle's Law:

$$V_{end} = \frac{P_{start} V_{start}}{P_{end}} = \frac{1.0 \text{ atm } (50.0 \text{ mL})}{1.8 \text{ atm}} = 28 \text{ mL}$$

Do the Answers Make Sense?

We expect the final volume of the jumped-on air pocket in each case to be smaller than our initial volume because the pockets are being flattened. So far, so good. We also expect Arnold's pocket to be squeezed more than Sadie's, because Arnold is putting more pressure on it. So our answers do seem to make sense.

---

a result of the fact that *at constant pressure,* the volume of a gas is directly proportional to the temperature in kelvins (from the Kelvin scale known as the **absolute temperature** scale):

$$V = kT \qquad \text{(k is some constant)}$$

This equation is an expression of **Charles' Law,** named after the French physicist Jacques Charles (1746–1823) (who also is reported to have made the first solo balloon flight in a hydrogen-filled balloon).

exercise 11.6

How Much Did the Balloon Deflate?

Problem

The temperature in the locker room was 70°F. The balloon, with a volume of 3.0 L, was taken outside and cooled to a temperature of −25°F. What was the new volume of the balloon?

Solution

As we did with our pressure-volume problem, we can derive an equation to solve for the new volume. If we call our starting and ending volumes and temperatures V_{start}, V_{end}, T_{start}, and T_{end}, then if

$$\frac{V_{start}}{T_{start}} = \text{a constant, and}$$

$$\frac{V_{end}}{T_{end}} = \text{that same constant, then}$$

$$\frac{V_{start}}{T_{start}} = \frac{V_{end}}{T_{end}}$$

Temperature conversions were discussed in Chapter 5 on the role of energy in chemical reactions. Please refer to that chapter if you need a review!

Note that we must convert our temperature values into kelvins. Can you explain why? Is it possible to have a negative volume?

$$V_{start} = 3.0 \text{ liters} \qquad T_{start} = 70°F = 21°C = 294 \text{ K}$$

$$V_{end} = ??? \qquad T_{end} = -25°F = -32°C = 241 \text{ K}$$

$$V_{end} = \frac{V_{start}\, T_{end}}{T_{start}} = \frac{3.0 \text{ L } (241 \text{ K})}{294 \text{ K}} = 2.5 \text{ L}$$

The cold balloon will have a volume of 2.5 L (if P remains constant).

Does the Answer Make Sense?

We said that the volume of the cold balloon should be less than that of a warm balloon in order for the air molecules in the balloon to exert the same pressure. In this case, the volume has been reduced to 83% (2.5/3.0 × 100%) of its original volume. Note that we had to use temperature *in kelvins*. Calculations using Celsius or Fahrenheit temperature values directly would have given us a negative volume.

We can combine Boyle's Law and Charles' Law in order to relate P, V, and T in a single equation. The derivation goes like this.

Boyle's Law says that

$$P_{start}\, V_{start} = P_{end}\, V_{end}$$

In the same way, Charles' Law says that

$$\frac{V_{start}}{T_{start}} = \frac{V_{end}}{T_{end}}$$

Combining Boyle's and Charles' Laws, we get

$$\frac{P_{start}\, V_{start}}{T_{start}} = \frac{P_{end}\, V_{end}}{T_{end}}$$

This equation is often called the **combined gas law** and is useful when we are looking for P, T, or V when two of these change from the start to the end.

exercise 11.7

Combining to Change the Balloon's Volume

Problem

We have another balloon at a temperature of 21°C (294 K), a pressure of 1.97 atm and volume of 2.3 L. If we let go of the balloon it will rise into the sky. What must the pressure be in the balloon if the temperature is lowered to −53°C (220 K) and the volume is found to decrease to 1.9 L?

Solution

The combined gas law is designed for problems like this. We can rewrite the data in the following way.

$$V_{start} = 2.3\,L \qquad\qquad V_{final} = 1.9\,L$$
$$P_{start} = 1.97\,atm \qquad\quad P_{final} = \text{???}$$
$$T_{start} = 294\,K \qquad\qquad T_{final} = 220\,K$$

We can substitute our values into the combined gas law equation.

$$\frac{P_{start}\,V_{start}}{T_{start}} = \frac{P_{end}\,V_{end}}{T_{end}} = \frac{1.97\,atm\,(2.3\,L)}{294\,K} = \frac{P_{end}\,(1.9\,L)}{220K}$$

$$P_{end} = \frac{1.97\,atm\,(2.3\,L)(220\,K)}{294\,K\,(1.9\,L)} = 1.78\,atm$$

Does the Answer Make Sense?

The lowering of the volume from 2.3 L to 1.9 L would cause the *pressure to increase* by 2.3/1.9, or about 20%, if the temperature were constant (Boyle's Law). However, the reduction in temperature from 294 K to 220 K would *reduce the pressure* to 220/294 of its starting value, a 25% reduction, if the volume were constant. The net effect of the 20% increase and 25% decrease is a slight reduction in pressure, and this is what we observe.

Tying It All Together: The Pump, the Balloon, and Gas Laws

We have used "pump" shoes and a balloon to interrelate the number of moles, pressure, volume, and temperature of a gas. We have seen that

$$V/n = k, \text{ or } V = kn \qquad \text{(constant } P, T)$$
$$PV = k, \text{ or } V = k/P \qquad \text{(constant } n, T)$$
$$V/T = k, \text{ or } V = kT \qquad \text{(constant } n, P)$$

Or, putting everything together, and making a new constant, called **R**,

$$V = RnT/P$$

A more familiar form of this equation is

$$PV = nRT$$

This is called the **ideal gas equation** and describes how pressure, volume, number of moles of a gas, and temperature are related. It is called "ideal" because it is based on an assumption that gases behave ideally—that is, as if each particle of gas does *not* interact with any other particle. In all our calculations we will assume that we are dealing with ideal gases. The behavior of real gases deviates slightly from that predicted for ideal gases, but pretending that gases are "ideal" makes it possible to reach fairly accurate general approximations to the be-

▉ hands-on

How Temperature Affects the Density of a Gas

Let's turn, for a moment, to water balloons. A **fluid** can be thought of as *anything that flows*. In this sense, water is a fluid, but so are all gases. Let's use water balloons to show, in general terms, how the density of fluids (including gases) is affected by temperature changes.

To do this activity you'll need:

- One balloon filled with cold water
- One balloon filled with hot water
- One bucket roughly 2/3 filled with cold water
- One bucket roughly 2/3 filled with hot water

You may want to color code everything to indicate which balloons and buckets are hot and cold.

Take the "hot" balloon and carefully put it in the "cold" bucket. Take the "cold" balloon and put it in the "hot" bucket. What happens? What does this tell you about the relationship between temperature and density?

Figure 1 shows the results. The cold water balloon sinks, but the hot water balloon rises because cold water has a higher density than hot water. This demonstrates that *the density of a fluid is inversely related to its temperature.* For water, this applies above 4°C, as we discuss in Chapter 7. In other words, the higher the temperature, the lower the density of the fluid, and vice versa. Gases are fluids, so this relationship holds true for gases as well. Table 11.1 lists the density of air at various temperatures. Note how the density decreases as the temperature of air increases.

Figure 1
The blue balloon is filled with cold water and the red balloon is filled with hot water. The container is filled with water of an intermediate temperature.

havior of real gases in situations where the temperature is not too low and the pressure is not too high.

We have introduced every term in the ideal gas equation except for the "*R*." This is a constant number called the **ideal gas constant,** and it has value and units as:

$$R = 0.08206 \text{ L atm/K mol}$$

This constant is different from others we have worked with in that it has more complex units. Even though it would seem to be rather cumbersome, it is actually very practical because you can use it to relate so many measures—volume, pressure, temperature, and amount of substance (moles) to one another. So in that sense, this messy constant simplifies our problem solving. Section 11.5 considers a variety of examples.

11.5 │ Applications of the Ideal Gas Equation

You should now know what we mean by the "pressure" of a gas, and you have seen that the pressure, volume, temperature, and the number of moles of a gas can be related through the ideal gas equation. We will now consider some applications of this knowledge concerning such things as balloons, air pollution, and automobile safety. Remember that the conversion factor method (dimensional analysis, see page 172) can be a valuable tool in problem solving.

Figure 11.12 shows the ideal gas equation, $PV = nRT$, contains four "variables" (values that can vary), namely, *P*, *V*, *n*, and *T*. The value *R* is a nonvarying constant—it never changes. The variable *V* refers to the volume occupied by the gas. Provided we know the value of three of the four variables in any particular set of circumstances, we can calculate the value of the

case in point

When Gases Behave Nonideally: Air Conditioners, Refrigerators, and Heat Pumps

When we set a cup of warm coffee on a table we expect it to cool until it is the same temperature as the air of the room. If we put an ice cube on the same table, however, we expect it to become warmer until it melts to produce a pool of water at the same temperature as the air of the room. In general, we expect heat to flow from hot things to cold things until the two reach the same temperature, at which point we expect the temperature changes to stop. Air conditioners, refrigerators, and heat pumps are modern conveniences that we use to overcome these normal expectations by forcing heat to move in the unexpected direction: from cold things to hot things. How can we do this?

We know that in an ideal gas, the molecules are far apart from one another and that interactions between molecules are negligible. Suppose that this familiar situation changes substantially and that molecules are so closely packed that they interact substantially. Such a situation describes a liquid. Conversion of a gas to a liquid is called **condensation.** One way to cause condensation is to cool a gas sufficiently to allow the collisions between gas particles to become very gentle. Now weak forces of attraction between the particles *do* become significant. Another way to cause condensation is to increase the pressure so

much that the molecules are forcefully squeezed very close together. Not surprisingly, a combination of lowered temperature and increased pressure is usually the best way to condense a gas. Notice that we said that a gas must be cooled down in order to condense. That is, heat must be removed from the gas to the exterior of its container, which means that the process is *exothermic.* Condensation releases heat to the rest of the universe.

Evaporation is the reverse of condensation. A liquid evaporates into the form of a gas. When this happens the molecules get much farther apart, are no longer constrained to stay in the bottom of some container, and can go flying off into the space above and around the liquid's container. Heat must be taken in by a liquid as it evaporates so evaporation is an *endothermic* process.

Entire industries are built around the heat changes involved in the interconversion of gases and liquids. Air conditioners, refrigerators, and heat pumps are all devices that move heat from a cold place to a warm place. They achieve this by allowing a liquid (known as the **refrigerant**) to evaporate (endothermic process) in a cool place followed by compressing that same gas back into a liquid (exothermic process) in a warm place. Notice that the endothermic process, evaporation, occurs in the cool place

Figure 11.12
The *ideal gas equation* relates the pressure, temperature, number of moles, and volume of an ideal gas. "Ideal" means that each particle of the gas does not interact with any other particle. This assumption is reasonable unless the pressure is very high and the temperature is quite low ($PV = nRT$).

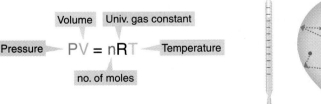
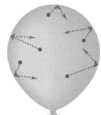

fourth "unknown" variable. This involves rearranging the equation accordingly. So we can use the ideal gas equation to calculate the pressure, volume, temperature, or number of moles of a gas, provided only one of these values is unknown. As we solve the problems together, note that we always begin by listing what we are given (what we already know) and what we want to calculate. This helps us focus on what we must do with the numbers available.

Application I—Density and the Floating Balloon

We explained earlier in this chapter that molecular hydrogen is less dense than air, and carbon dioxide is more dense than air. Let's compare the number of moles and, by extension, the masses of gases in three separate 5.0-L balloons. The first balloon contains H_2, the second contains air, the third contains CO_2.

The Question
Calculate the number of moles and then the mass of molecular hydrogen in a 5.0-L balloon at a temperature of 27°C and pressure of 1.0 atm.

and so cools that place even more, and that the exothermic process happens in a warm place and so warms it even more.

For a refrigerator, the cool place is inside the "fridge" and the warm place is the kitchen. For an air conditioner, the cool place is inside the house and the warm place is outside the house. Figure 1 illustrates the principles involved.

Only a few compounds have boiling points in the right range (within about 50°C of the temperature at which you want to refrigerate) to be successfully used as refrigerants, as shown in Table 11.2. The poisonous and foul-smelling gas, ammonia (NH_3), was a common refrigerant in the early days of refrigeration as was the equally objectionable sulfur dioxide (SO_2). The discovery that Freon®, CF_2Cl_2, acted as an effective refrigerant was regarded as a great advance because it was nonflammable, nontoxic, unreactive, and had no smell. Only many years later did we discover that Freon® can participate in reactions high in the atmosphere that contribute to the destruction of the "ozone layer." As we discuss in Chapter 12, the ozone layer shields the Earth

from some of the harmful ultraviolet radiation from the Sun. So the use of Freon® as a refrigerant became a serious ecological problem. We will consider some possible substitutes for Freon-12® in Chapter 12.

Table 11.2

Boiling Points of Common Refrigerants (at 1 atm)

Refrigerant	Formula	Boiling Point (°C)
Ammonia	NH_3	−33
Freon-21®	$CHCl_2F$	3.3
Freon-12®	CF_2Cl_2	−14
Freon-22®	$CHClF_2$	−41
Freon-113®	$C_2Cl_3F_3$	48
Sulfur dioxide	SO_2	−10

Source: From Lide, D. R. *Handbook of Chemistry & Physics,* any edition, Boca Raton, FL.: CRC Press.

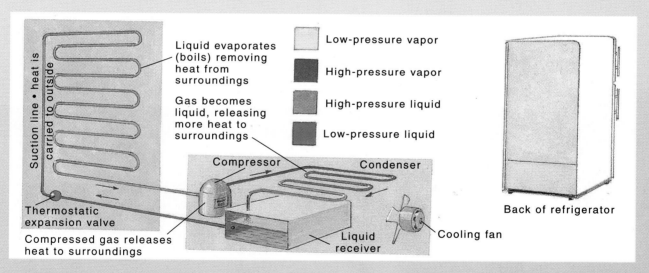

Figure 1
In a refrigerator, a substance such as Freon-12® (CF_2Cl_2) is allowed to evaporate (endothermic process), which cools the refrigerator. It is then compressed back to a liquid (exothermic process) giving heat off to the kitchen. This same strategy is used in an air conditioner, where the warm place is the outside of the house and the cool place is the house itself.

The Strategy

To find the correct answer to any question, we should first ask some other questions concerning the steps we need to take toward our answer. The first question is, "what do we want?" In this case, we want "moles, and then mass of molecular hydrogen." The next question is "what are we given?" In this case we are given the following values:

$$R = 0.08206 \text{ L atm/K mol} \qquad V = 5.0 \text{ L}$$
$$P = 1.0 \text{ atm} \qquad T = 27°C = 300 \text{ K}$$

We are *not* given the number of moles, *n*. Using the ideal gas equation to find *n* offers a direct path to the mass of molecular hydrogen by using the molar mass of H_2 to convert the number of moles into a mass.

The Mathematical Solution

The ideal gas equation, $PV = nRT$, can be rearranged to make n, whose value we wish to calculate, the subject of the equation, as:

$$n = \frac{PV}{RT}$$

We then solve this equation by substituting the known values of P, V, R, and T:

$$n = \frac{(1.0 \text{ atm}) (5.0 \text{ L})}{(0.08206 \text{ L atm/K mol}) (300 \text{ K})}$$
$$= 0.203 \text{ mol H}_2$$

We can convert to grams by multiplying moles of H_2 by the molar mass, 2.01 g/mol.

$$\text{g H}_2 = 0.203 \text{ mol H}_2 \times 2.01 \text{ g H}_2/\text{mol H}_2$$
$$= 0.41 \text{ g H}_2$$

In conclusion, there are 0.41 g of molecular hydrogen in the 5.0-L balloon.

We can also solve this problem mathematically by using the conversion factor method to create *one equation* to solve directly for grams of H_2:

$$\text{g H}_2 = \frac{2.01 \text{ g H}_2}{1 \text{ mol H}_2} \times \frac{1 \text{ K mol}}{0.08206 \text{ L atm}} \times \frac{1.0 \text{ atm}}{300 \text{ K}} \times 5.0 \text{ L} = 0.41 \text{ g H}_2$$

Whether you solved for grams of molecular hydrogen using the ideal gas equation or the conversion factor method, you still had to ask reasonable questions to come up with a meaningful answer.

exercise 11.8

The Mass of Air and CO_2 in the Balloons

Problem

Let's continue our comparison of the balloons by calculating the mass of air in a 5.0-L balloon and the mass of carbon dioxide in a different 5.0-L balloon. The average molar mass of air is 29.0 g/mol, and the temperature and pressure of the balloons are 27°C and 1.0 atm as before. Based on your answers to this exercise, discuss why a balloon filled with molecular hydrogen gas rises in air while one filled with carbon dioxide sinks.

Solution

We will calculate the mass of air using the ideal gas equation and solve for the mass of CO_2 (molar mass = 44.0 g/mol) using the conversion factor method (remember, the actual choice is up to you).

grams of air (using the ideal gas equation):

$$n = \frac{PV}{RT}$$

substituting numbers,

$$n = \frac{(1.0 \text{ atm}) (5.0 \text{ L})}{(0.08206 \text{ L atm/K mol}) (300 \text{ K})} = 0.203 \text{ mol air}$$

We can convert to grams by multiplying moles of air by the molecular mass, 29.0 g/mol,

$$\text{grams of air} = 0.203 \text{ mol air} \times 29.0 \text{ g air/mol air}$$
$$= 5.9 \text{ g air}$$

exercise 11.8

continued

grams of carbon dioxide (using conversion factors):

$$\text{g CO}_2 = \frac{44 \text{ g CO}_2}{1 \text{ mol CO}_2} \times \frac{1 \text{ K mol}}{0.08206 \text{ L atm}} \times \frac{1 \text{ atm}}{300 \text{ K}} \times 5.0 \text{ L} = 8.9 \text{ g CO}_2$$

We now have enough information to answer the final question posed in this problem, that is, why does a balloon filled with molecular hydrogen gas rise in air while one filled with carbon dioxide sinks? We know that a 5.0-L balloon filled with molecular hydrogen contains 0.41 g of the hydrogen, the same-sized balloon of air contains 5.9 g of air, and yet another 5.0-L balloon contains 8.9 g of carbon dioxide. We also know that substances will rise or sink based on their relative densities. We can calculate the densities (at 27°C and 1.0 atm) as follows:

$$\text{density of H}_2 = \frac{0.41 \text{ g}}{5.0 \text{ L}} = 0.082 \text{ g/L}$$

$$\text{density of air} = \frac{5.9 \text{ g}}{5.0 \text{ L}} = 1.2 \text{ g/L}$$

$$\text{density of CO}_2 = \frac{8.9 \text{ g}}{5.0 \text{ L}} = 1.8 \text{ g/L}$$

These values explain in a quantitative way why molecular hydrogen, rather than carbon dioxide, was used in the *Hindenburg* in 1937.

Application II—Air Pollution and the Ideal Gas Equation

As we will discuss in Chapter 12, a vigorous debate is proceeding about the effects of ever-increasing levels of carbon dioxide in the atmosphere. The main cause of the increase is the burning of fuels, especially oil and coal, as sources of energy. Exercise 11.9 will allow you to appreciate how burning even small amounts of fuel can lead to the addition of a substantial amount of carbon dioxide into the atmosphere.

exercise 11.9

Carbon Dioxide from Gasoline

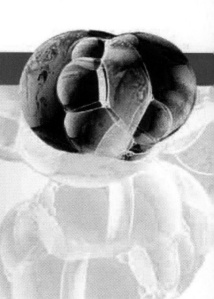

Problem

Gasoline can be represented as C_8H_{18} having a density of 0.704 g/mL. The equation for the combustion of gasoline is

$$2C_8H_{18} + 25O_2 \rightarrow 16CO_2 + 18H_2O$$

What volume of carbon dioxide at 0.96 atm of pressure and 25°C will be formed from the combustion of 5.0 L of gasoline?

Solution

First, rearrange the ideal gas equation to isolate volume on the left side of the equation:

$$V = nRT/P$$

continued

exercise 11.9

continued

The number of moles of carbon dioxide to use for "n" is the tricky part of this problem. There are 8 moles of carbon dioxide formed for every mole of C_8H_{18}, so we must calculate the number of moles from the volume, density, molar mass of gasoline, and mole-to-mole relationship in the equation.

$$n = 5.0 \text{ L C}_8\text{H}_{18} \times \frac{704 \text{ g C}_8\text{H}_{18}}{1 \text{ L C}_8\text{H}_{18}} \times \frac{1 \text{ mol C}_8\text{H}_{18}}{114 \text{ g C}_8\text{H}_{18}} \times \frac{16 \text{ mol CO}_2}{2 \text{ mol C}_8\text{H}_{18}}$$

$$= 247 \text{ moles carbon dioxide}$$

We then calculate the volume of this number of moles of carbon dioxide, using the rearranged ideal gas equation:

$$V = nRT/P$$

$$V = \frac{(247 \text{ mol}) (0.08206 \text{ L atm/mol K}) (298 \text{ K})}{0.96 \text{ atm}}$$

$$= 6.3 \times 10^3 \text{ L carbon dioxide}$$

The ideal gas equation has allowed us to determine that burning 5.0 L, a little over a gallon of gasoline, results in nearly *6300 liters* of carbon dioxide being released into the atmosphere! Multiply this by the hundreds of millions of automobiles and trucks burning gasoline, and add in all the coal- and oil-burning power plants, and you begin to see why carbon dioxide levels are rising due to human activities.

Application III—Automobile Air Bags

Forty-nine U.S. states, the District of Columbia, and U.S. Territories now require all occupants of an automobile to use a passive restraint system. The exception is New Hampshire, which only requires restraints on persons up to age 18. While this has historically meant seat belts, most new cars are now equipped with air bags. The purpose of an air bag is to restrain you from hitting a hard surface (steering wheel or dashboard) or flying through the windshield if a crash occurs. The air bag (Figure 11.13) is designed to distribute that *force* that is exerted on you over the maximum body *area* as the car decelerates. That is, to *minimize the pressure* on you resulting from the crash. As an example, let's say that you hit another car head on while going 48 kilometers (30 miles) per hour. There are three likely outcomes, depending on the restraint device that you use:

1. **No restraint:** Your head probably hits the steering wheel. About 5000 newtons (N) of force (over 1000 lb) is transferred from the steering wheel to your head, eyes, and nose. The area is somewhere around 20 cm² and the results are disastrous.
2. **Seat belt:** The 5000 N is spread over the area where the seat and shoulder belt touches your body (perhaps 200 cm²). The pressure is just 1/10 what it was with no restraint and only minor injuries may result.
3. **Air bag:** The force is spread over a large area (your entire head and upper torso—perhaps 2000 cm²). Injury is much less likely. Additionally, since the air bag is fairly soft, you decelerate more gradually than with seat belts or the steering wheel.

Knowing the relationship between force and area and pressure helps us understand *why* air bags are so effective. We need to look at a chemical reaction, though, to understand *how* the bags work.

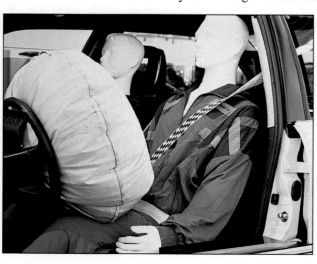

Figure 11.13
Close-up of inside of car with exploding air bag and crash test dummies being thrown onto it.

exercise 11.10

How Much Solid NaN₃ will React to Fill an Air Bag?

Problem

How many grams of sodium azide are necessary to produce the 74 L of nitrogen gas that fills an air bag in a crash (assume a temperature of 25°C and a pressure of 1.0 atm)?

Solution

We use the ideal gas equation to solve for the number of moles of nitrogen which will have a volume of 74 L:

$$n = \frac{PV}{RT}$$

$$n = \frac{(1.0 \text{ atm}) (74 \text{ L})}{(0.08206 \text{ L atm/mol K}) (298 \text{ K})}$$

$$= 3.0 \text{ mol N}_2$$

The chemical equation for the formation of the nitrogen gas is

$$2NaN_3(s) \rightarrow 2Na(s) + 3N_2(g)$$

which indicates that 3 moles of nitrogen come from 2 moles of sodium azide (molar mass = 65.0 g/mol).

So, the mass of sodium azide is

$$g \text{ NaN}_3 = 3.0 \text{ mol N}_2 \times \frac{2 \text{ mol NaN}_3}{3 \text{ mol N}_2} \times \frac{65.0 \text{ g NaN}_3}{\text{mol NaN}_3}$$

$$= 130 \text{ g NaN}_3$$

The N₂ is generated from the chemical decomposition of a relatively small amount (about 1/2 cup) of solid sodium azide. The ideal gas equation allows air bag designers to calculate an appropriate amount of sodium azide to use in air bags. We now see, however, that the term "air bags" is misleading, because the bags fill up with nitrogen, not air.

Inside the steering column of an air bag-equipped car is a deflated polymer-based bag, a capsule of sodium azide (NaN_3), some iron oxide (Fe_2O_3), and a mechanism that will set off a small detonator cap in the event of a head-on crash. The activation energy supplied by the detonator initiates the chemical decomposition of sodium azide as follows:

$$2NaN_3(s) \rightarrow 2Na(s) + 3N_2(g)$$

The sudden release of nitrogen gas fully inflates the bag (to a volume of about 74 L) within 40 milliseconds (0.04 s) of the crash. The sodium that is generated is mixed with iron(III) oxide, Fe_2O_3, to form sodium oxide, which is much safer than sodium metal,

$$6Na(s) + Fe_2O_3(s) \rightarrow 3Na_2O(s) + 2Fe(s)$$

Once inflated, the bag slowly deflates, helping to dissipate even more of the impact force as your head and body slam into it. A number of automobile manufacturers now put additional air bags in the driver's doors to help prevent side impact injuries.

11.6 | Solubility of Gases in Water

You unthinkingly open a can of soda and the contents spray out all over you. This spraying usually happens when the can (or bottle) is warm or has just been shaken (see Figure 11.14). What has happened?

Figure 11.14
Soda sprays from a can if it has been shaken because the bubbles of carbon dioxide (CO_2) expand rapidly when the pressure is released (by the shaking). This forces the liquid out of the opening.

Table 11.3

Temperature versus Solubility of Gas in Water (L gas/L water)

	CO_2	O_2
0°	1.71	0.049
10°	1.19	0.038
20°	0.88	0.031
30°	0.66	0.026
40°	0.53	0.023
50°	0.44	0.021

Note: Solubility is in liters of gas at standard temperature and pressure (STP) dissolved in 1 L of water when the gas (CO_2 or O_2) is at 1 atm pressure above the solution. When air (21% O_2) is in contact with water, the actual oxygen solubility values will be only 21% of those listed in the table. Source: From Lide, D. R. *Handbook of Chemistry & Physics,* any edition, Boca Raton, Fla.: CRC Press.

The principal gas present in containers of soda is carbon dioxide. Carbon dioxide is fairly soluble in water and solubility is greater at lower temperatures as you can easily show by letting a soda go "flat" when it becomes warm (see Table 11.3). This is due to the higher vapor pressure of the gas as it gets warmer. Shaking the bottle doesn't change the solubility of gas in the liquid, but it does distribute the gas into small bubbles spread throughout the liquid. Then, when the lid is removed, the internal pressure suddenly decreases until the pressure of CO_2 inside the bottle equals that in the atmosphere outside the bottle.

What Happens When the Pressure of the Gas Decreases?

The ideal gas equation tells us that the volume of a gas will increase when the pressure is decreased. This will occur whether the gas is all gathered together at the top of the bottle or whether it is distributed into tiny bubbles throughout the liquid. When the bottle has not been shaken there will be a rush ("whoosh!") of gas out of the bottle when it is opened but no liquid will spray out. *But,* a recently shaken bottle will have all those tiny bubbles throughout the liquid and each of them will grow in volume beneath the surface of the liquid and a froth of liquid and gas will be pushed out of the bottle as a result. When most of the carbon dioxide has left, the beverage tastes flat because the sour taste of the carbonic acid is gone. (Remember from Chapter 8 that CO_2 in water forms some carbonic acid, and that acids have a characteristic sour taste.)

Taking It One Step Further—Thermal Pollution

Water from rivers is frequently used to cool major industrial installations such as electric power plants. Power plants need to be cooled because of the enormous amounts of energy that build up in such a relatively small space. For example, if the cores of nuclear power plants were not cooled with water (or in some cases, liquid sodium—see Chapter 14 on nuclear energy), the nuclear core would suffer a "meltdown" in which the core melts and sinks into the earth, leading to a massive explosion and radiation release. One consequence of the resulting **thermal pollution** (raising of the normal water temperature) is a decreased solubility of gases in that water.

This heated water is then reintroduced into the river. Since it is now warmer, it can hold less dissolved oxygen, and oxygen is necessary for aquatic life in the river. A rise in temperature from 10°C to 30°C causes 31% of the dissolved oxygen to leave the water and enter the atmosphere. While small temperature changes do not seriously impact the aquatic ecosystem, the kind of change we have described can be disastrous to many species of fish and other forms of aquatic life. The aquatic food chain can be greatly affected by the removal of species that cannot cope with swift temperature changes. Therefore, the entire ecosystem feels the effects. Spawning and hatching are also temperature-dependent processes, which will feel the impact of thermal pollution.

One way to mitigate the impact of heating the water is to cool it before it is released back into the stream. This is done with a **cooling tower** (see Figure 11.15), in which the water cools.

Balloons, basketball shoes, air bags, air conditioning, thermal pollution are just of few of the wide range of things that are affected by gases and their behavior. We chose these few examples, but we could have picked such things as breathing or weather. In fact, we will discuss gases and cooking when we get to Chapter 17 which deals with the chemistry of food. Now that you have an understanding of gas behavior, you can confidently investigate the causes and implications of a most serious social and scientific issue—air pollution. That is the subject of Chapter 12.

Figure 11.15
River water is used to cool parts of power plants in which a great deal of energy builds up in a relatively small space. Before this water reenters the river, it can be cooled in a cooling tower. This helps prevent damage to the river's plant and fish life, which are sensitive to the water temperature.

main points

- Gases can be described by their properties such as density, temperature, and pressure.

- Pressure is a measure of force per unit area.

- The pressure and volume of a gas are inversely related (all other conditions constant).

- The volume and temperature of a gas are directly related (all other conditions constant).

- The volume and number of moles of a gas are directly related (all other conditions constant).

- These relationships lead to the ideal gas equation.

- The ideal gas equation can be used to determine a number of characteristics of a gas, including the density and molar mass.

- A number of "real-world" applications for these relationships were discussed, including automobile air bags, refrigeration, and thermal pollution.

important terms

Absolute temperature is the temperature measured with the coldest theoretically possible temperature set to be zero on Kelvin scale. (p. 372)

Area is the two-dimensional size of an object. (p. 364)

Atmosphere is the unit of pressure of air at the Earth's surface. (p. 365)

Charles' Law describes how the volume occupied by a gas is related to temperature. (p. 372)

Combined gas law $\dfrac{P_{start}\,V_{start}}{T_{start}} = \dfrac{P_{end}\,V_{end}}{T_{end}}$ (p. 373)

Condensation is the process of a gas becoming a liquid, a physical change. (p. 376)

Cooling tower is a large building in which warm water is allowed to cool. (p. 382)

Dalton's law of partial pressures says that the total pressure of gases is the sum of the partial pressures of each component gas of the mixture (p. 367)

Force is the mass of an object multiplied by its acceleration. (p. 364)

Ideal gas constant, $R = 0.08206$ L atm/K mol. (p. 375)

Ideal gas equation, $PV = nRT$, describes relationship of pressure, volume, and temperature of a given quantity of gas. (p. 374)

Partial pressure, the portion of the total pressure which is exerted by a gas. (p. 367)

Pressure is the force of gas pushing per area of surface. (p. 363)

psi, "pounds per square inch" is a unit of pressure. (p. 366)

psig, "pounds per square inch gauge" means the gas pressure in excess of the surrounding atmospheric pressure. (p. 367)

R is the ideal gas constant = 0.08206 L atm/K mol. (p. 374)

Relative humidity, the ratio of the actual partial pressure of water vapor in the air to the maximum possible pressure, expressed in percent. (p. 369)

Standard atmosphere is the atmospheric pressure at sea level on a nice day, equaling 760 torr. (p. 365)

Standard pressure, 1 bar = 100,000 pascals = 0.9869 atmospheres = 750.6 torr (p. 366)

Standard temperature, 0°C or 273.15 K. (p. 366)

STP is the "standard temperature and pressure." (p. 366)

Thermal pollution are the damaging effects of too much heat being added to a body of water. (p. 382)

torr is the unit of pressure equal to pressure needed to support a column of mercury 1 mm high; short for Torricelli. (p. 366)

exercises

1. Is the reaction of hydrogen and oxygen gases endothermic or exothermic? Explain.
2. Do you think a hot air balloon can fly higher on a hot day or a cold day? Why?
3. Calculate the density of 700 mL of a gas with a mass of 0.725 g at room temperature.
4. If you step on a nail protruding from a floorboard, you may need a Band-Aid (and a tetanus shot). Why, then, can people walk across a bed of nails without injuring themselves?
5. What is the difference between an atmosphere and a standard atmosphere?
6. If each tire on your car required 38 lb per square inch for maximum gas mileage, what would the actual pressure inside each tire be if the tire gauge read 38 psi? What would this be in atmospheres?

7. What is the actual pressure inside a football when a pressure gage reads 10.5 psig?

8. When more air is blown into a soap bubble, its size increases but its pressure stays the same. But, when more air is blown into a volleyball, its size stays the same, while its pressure increases. Think of a common example in which blowing more air into an object increases both its volume and its pressure. Explain all these observations.

9. A helium balloon is allowed to float to the ceiling of a room one day, and the next day it is found on the floor. Explain both of these circumstances.

10. You are holding two filled balloons in your hands. One contains helium and the other carbon dioxide. What would you do to tell which balloon is which?

11. Which gas is most dense at one atmosphere pressure and room temperature? N_2, Ar, SO_3, H_2, O_3, NH_3

12. What would be the relative humidity at 20°C if the actual partial pressure of water vapor were 12.0 torr?

13. At what temperature would the relative humidity become 100% if the actual partial pressure of water in the air is 12.0 torr?

14. If the number of moles of air in an inflatable flotation device is decreased by 1/3, how will the volume of the device be affected?

15. An air bag has a volume of 61.2 L and contains 3.0 mol of molecular nitrogen (N_2). How many moles of N_2 would be in an 80-L air bag at the same temperature and pressure?

16. Aircraft cabins are only partially pressurized so passengers are under a lower pressure in the air than they would experience on the ground. Flight crew are told not to eat foods that contribute to flatulence before flying. Flight menus are also constructed to minimize the amount of internal gas. Why are these precautions taken? Carbonated drinks that are distributed in flight do not pose any problems. Why is this so?

17. A gas has a pressure of 3.2 atmospheres (atm) and occupies a volume of 45 L. What will the pressure be if the volume is compressed to 27 L at a constant temperature?

18. A can of soda contains 30 mL of carbon dioxide gas at 40°C. What would be the volume of carbon dioxide in the can if the temperature were lowered to 10°C?

19. The term vital capacity refers to the maximum amount of air that can be forced from our lungs due to an increase in the internal lung pressure as we exhale. This volume varies from 3 to 7 L depending on the person. If an individual exhales 4.0 L of air at 37°C from her lungs into a bag, how much air will be collected if the external temperature is −10°C?

20. Many aerosol cans will explode if the internal pressure is greater than or equal to 3.0 atm. An aerosol can at 27°C will have an internal pressure of 2.0 atmospheres. Calculate the maximum temperature for storing this can before it will explode. (Assume that the volume and number of moles are constant.)

21. You may have been told to never inflate the tires on your car right after a long trip. Let's see why. Your car tires contain 32 lb per square inch (lb/in.²) at 24°C. After driving 240 miles, the air in the tires is at 44°C. Calculate the pressure of your car's tires assuming that the volume is constant. Why might it not be wise to inflate your tires immediately after a long trip?

22. A weather balloon is filled with 295 L of helium on the ground at 18°C and 750 torr. What will the volume of the balloon be on top of Mt. McKinley, Alaska at an altitude of 20,320 feet above sea level where the pressure is 340 torr and the temperature is −28°C?

23. The volume of gas in a container is 5.0 L. Keeping temperature and pressure constant, if the number of moles of the gas is increased by 50%, what will be the new volume?

24. You are blowing up balloons for a birthday party in a room where the temperature is kept constant at 69°F. If a balloon you are playing with is already inflated and tied, what can you do to decrease the volume of gas in the balloon?

25. If you are doing jumping jacks with shoes containing an air pocket, is the pressure greatest in the pocket when you are airborne or when you hit the ground? Explain.

26. At constant pressure, what is the relationship between volume and temperature?

27. In the winter in Wisconsin, will you have to put more or less air in your car tires to keep them properly inflated?

28. The volume of a playground ball is 14 L and has a pressure of 10 psig at a summertime temperature of 35°C. How many grams of air must be added to maintain this pressure and volume in winter when the outdoor temperature is 3.0°C?

29. In Aspen, Colorado, will you have to put more or less air in your car tires to keep them properly inflated? Compare the amounts of air needed at this altitude in summer and winter.

30. Why do instructors tell people not to hold their breath when they scuba dive?

31. Frequently, helium balloons are actually filled with a mixture of nitrogen and helium. This is done because nitrogen is much less expensive than helium. What is the density at 1.00 atmospheres and 25°C of a mixture of equal masses of He and N_2? Compare this with the density of air to determine whether a balloon filled with this mixture would rise.

32. The density of carbon dioxide, CO_2, is 1.976 g/L. If CO_2 takes up 24 mL of space in a can of soda, what is the mass in milligrams (mg) that this gas contributes to the mass of a can of soda?

33. Explain why large pleasure boats are required by law to cook with pressurized tanks of methane (CH_4) instead of the more common bottled propane (C_3H_8).

34. If 80 g of sodium azide (NaN_3) reacts in an airbag within 0.02 seconds after impact, what will be the volume of molecular nitrogen (N_2) in the airbag? Use the chemical

equation for the formation of molecular nitrogen gas as shown below. (Assume a temperature of 25°C and a pressure of 1.0 atm.)

$$2NaN_3(s) \rightarrow 2Na(s) + 3N_2(g)$$

35. What is the function of a detonator cap in an automobile airbag?
36. What is the best method to condense a gas? Explain why (at the molecular level).
37. If hot pudding is covered with plastic wrap and placed in the refrigerator while it is hot, water droplets will appear on the inner surface of the plastic wrap. However, if it is left uncovered, a "skin" will form on top of the pudding? Why does this occur? Is this an endothermic or exothermic process? (See the Case in Point: When Gases Behave Nonideally.)
38. Canisters that supply oxygen gas for airliners are filled with potassium chlorate and powdered iron. These generate oxygen by the iron-catalyzed decomposition reaction:

$$2KClO_3 \rightarrow 2KCl + 3O_2$$

How many liters of oxygen at 25°C and 0.90 atmospheres pressure can be generated by a canister which contains 50.0 grams of potassium chlorate?
39. The ozone concentration in the upper atmosphere averages about 30.0×10^{11} molecules per cm^3 in a region which is between 15 and 35 km altitude, at an average pressure of 0.001 atmospheres and temperature of $-25°C$. How thick would the ozone layer be if these same molecules were at 1.0 atmosphere pressure and 20°C?
40. If the density of air equals 0.00118 g/mL at 25°C and 1.0 atmosphere (see Table 11.1), calculate the apparent molar mass of air and rationalize your answer with the known composition of air.
41. What is the ideal gas equation? What does each variable represent?
42. Given these data, calculate the pressure of a sample of N_2 gas. Is this sample located at sea level?

Density = 0.00100 g/mL
Volume = 50.0 mL
Temperature = 20°C

43. Calculate the molar mass of a gas having a mass of 19.3 grams and a volume of 10 L at 1.00 atm and 5.00°C.
44. Why will a can of soda rupture when frozen?
45. Look at Exercise 11.10. If an airbag manufacturer accidentally used 230 g of sodium azide (instead of the calculated 130 g), what would happen when the airbag opens? Prove your thoughts mathematically.
46. An unknown liquid weighing 2.539 grams is vaporized to form a gas that has a pressure of 1.47 atmospheres in a volume of 0.575 L at 100°C. What is the molecular weight of this compound?
47. An unknown liquid weighing 0.769 grams is vaporized to form a gas that has a pressure of 1.25 atmospheres in

a volume of 0.375 L at 100°C. What is the molecular weight of this compound?
48. Which set of conditions is most likely to cause a gas to condense into a liquid?
    a. low pressure and high temperature
    b. low pressure and low temperature
    c. high pressure and high temperature
    d. high pressure and low temperature
49. As you will learn in Chapter 12 on air pollution, excess carbon dioxide, CO_2, adds to global warming by trapping heat in the atmosphere. Calculate the fraction of dissolved carbon dioxide which leaves a river when its temperature goes from 10°C to 30°C. Does thermal pollution only affect aquatic life?
50. You purchase a large bag of potato chips from a store at sea level. You then travel with the chips to Estes Park, Colorado, at an elevation of about 7800 feet. Will the bag be larger or smaller than at sea level? Explain.
51. Would you expect a lake at a constant temperature to contain more oxygen at an elevation of 1000 ft above sea level or 7000 ft above sea level if the temperature is constant? What affect could elevation have on plant and animal life in the lake?
52. Why is thermal pollution so dangerous to a lake ecosystem?
53. Can you think of an everyday item (other than those discussed in this chapter) that is influenced by the behavior of gasses?

food for thought

54. As was discussed in Section 11.1, the United States had a monopoly on the production of helium. The British airship R-100 used hydrogen on its voyage to Montreal and Toronto in August 1930. The hydrogen was produced by the reaction of sodium hydroxide solution with ferrosilicon (similar to the reaction shown on page 362 of the text). Why might this reaction be better for preparing hydrogen than the reaction of a **strong** acid with an **active** metal?
55. A key to surviving a front-end collision is the distance the driver is allowed to travel before impact with the windshield. Surprisingly, this distance includes not only the area between the driver's body and the windshield or dashboard but also the "crush depth"—the distance between the occupant and the barrier normally taken up by the front of the car, which is made available to the driver because the front end of the car gets shorter. Let us present three cases involving front-end collisions where the driver is traveling at 35 mph and hits a car traveling at 30 mph.
    a. Driver 1, who is not wearing a seat belt, will continue forward through the windshield and dashboard at a speed of 35 mph after the vehicle has completely

stopped, creating a force 80 times that of gravity or 80 Gs. (Astronauts experience 4 to 11 Gs during lift-off.) The available stopping distance is only 3 to 5 inches in this case.

b. A belted driver, driver 2, will be able to utilize the crush depth along with the compartment space as the driver slows down with the car, not after it has stopped. Although contact with the windshield will be prevented, the driver's head may hit the steering wheel.

c. Driver 3's car is equipped with an airbag but the driver is not wearing a seat belt. The inflation of an air bag will soften the blow to the upper torso only while its quick deflation may add a few inches to the driver's path. However, the speed and force experienced by the driver may still cause injuries, especially to the lower torso.

Which driver would you rather be? Do you consider crash ratings and what safety equipment is included when purchasing an automobile? Do you wear your seat belt at all times? Why or why not? Would a car made of steel protect you from injury in a head-on collision? Why or why not?

readings

1. Bell, W. L. Chemistry of air bags. *J. Chem. Ed.,* 1990, 67, 61.
2. Atkinson, G. F. Hydrogen and airships—a class project. *Chem 13 News,* October 1992.
3. Holden, Janet A. Safety belts work! Here's why. *Safety & Health,* February 1989, pp. 62–65.
4. Which cars are safest in a crash? *Consumer Reports,* April 1993, pp. 199–201.
5. Keebler, Jack. Demand propels argon airbag development. *Automobile News,* March 30, 1992, p. 4.
6. Information on the STS-51L/Challenger Accident (October 2001). http://history.nasa.gov/sts511.html.
7. Bain, Addison. What Really Downed the Hindenburg. *Popular Science,* November 1997, p. 71.

websites

www.mhhe.com/Kelter   The "World of Choices" website contains activities and exercises including links to websites for: weather information from the Unisys Corporation; the chemistry of airbags from the Smithsonian Institution; and much more!

Air Quality

This most excellent canopy, the air, look you this brave o'erhanging firmament, this majestical roof, fretted with golden fire, why, it appears no other thing to me than a foul and pestilent congregation of vapors.

Hamlet, Act 2, Scene 2, William Shakespeare

Think of Los Angeles, California, and you probably think of palm trees, Hollywood movie stars, fast cars, and trendy people. Los Angeles has Disneyland and skateboarding in the streets. It also used to have some of the foulest air in the United States. To understand why, you have to know the geography of the city. You also need to know how the choices that people make, such as where and how they live, shop, work, and commute contribute to **air pollution.** Public policy decisions have made Los Angeles a cleaner place to live (see Figure 12.1). Theirs is a success story. But their way of life presents a cautionary tale.

Los Angeles is located in the southwest corner of California. Its metropolitan area population of over 12 million ranks second only behind New York. The city never gets very cold in winter and is usually very warm during summer. It also has miles of coastline on the west and beautiful mountains to the north and east. That's the good news. The bad news is that the same features that make us want to vacation or live in Southern California are also responsible for concentrating pollution, especially during the summer, when many hours of sunlight interact with automotive and industrial emissions to cause "smog." Smog is a reactive mixture of airborne chemicals resulting from the interactions of volatile organic compounds and nitrogen oxides. It is the topic of Section 12.4.

As shown in Figure 12.2, the mountains around L.A. create a **basin environment,** meaning one in which polluted air can gather as if trapped in a shallow basin. In L.A., cool air from ocean breezes can enter the "basin" and then become trapped by the warm air above, which is heated by the sun. The mountains prevent the air from spreading horizontally, so, in effect, the pollution-containing air has a warm-air lid on it. The polluted cool air cannot escape through this "lid" because cool air is more dense than warm air. It cannot rise through the warm air above. This situation is known as **thermal inversion** because the normal situation of a decrease in air temperature with increasing altitude is turned upside down ("inverted"). It makes ground-level air pollution in Los Angeles stay in and around the city for quite some time. A similar thermal

Figure 12.1
A. The Los Angeles metropolitan area is home to over 12 million people. Many would consider its year-round warm, dry weather idyllic. For many years, it also had the worst air pollution problem of any large city in the United States. The brown haze on the horizon is called "smog," an unhealthy combination of volatile organic compounds, nitrogen oxides, and carbon monoxide cooked by the sun. **B.** Los Angeles has "turned it around" via municipal pollution control measures (see section 12.6).

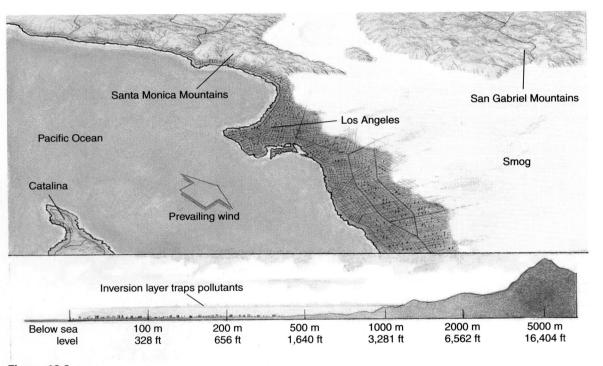

Figure 12.2
Pollutants that enter the air in Los Angeles often remain trapped because of the city's "basin environment"—the mountains on one side and the cool sea breezes on the other keep polluted air from dispersing. Reprinted from Chem. Eng. News, September 9, 1991, 69(36) pp. 27–28. Published 1991 American Chemical Society.

inversion effect can trap pollution in other cities such as Denver and Salt Lake City, ringed by mountains, with cool and polluted air at ground level.

According to one writer, "By the mid 1950s, Los Angeles' 'smog,' as the noxious vapor had been dubbed, was sufficiently thick and persistent to wilt crops, obstruct breathing and bring angry housewives into the street waving placards and wearing gas masks."

One major source of the air pollution is the city's automobile traffic. With over 6 million registered vehicles, Los Angeles is a city of cars and drivers. There is an

Table 12.1

Compounds Related to Automobile Emissions

Substance	Chemical Formula	Effects of Excess Dosage
Carbon monoxide	CO	Dizziness, headache, death by impairing blood's ability to carry oxygen
Carbon dioxide	CO_2	Greenhouse effect (long term)
Formaldehyde	$HCHO$	Irritates eyes, nose
Hydrocarbons	C_xH_y	Lead to smog formation
Nitrogen oxides	NO_x	Lead to smog formation
Lead	Pb	Brain damage, high blood pressure, impaired growth
1,3-Butadiene	C_4H_6	Cancer of heart, lungs, ovaries and other sites

average of only 1.1 riders per car, meaning that the driver is almost always the only occupant (Figure 12.3). The metropolitan area has an extensive system of freeways, yet the average freeway driving speed as of November 2000 was 17 miles per hour, forcing cars to use one out of every four gallons of gasoline idling in traffic.

Air pollution from personal automobiles takes a heavy toll on Los Angelenos. Table 12.1 shows a variety of compounds related to automobile emissions and their health effects. Respiratory ailments alone cost Los Angeles residents billions of dollars in medical expenses each year.

The experience of Los Angeles is by no means unique. In fact, Houston, Texas, has overtaken Los Angeles as the city with the dirtiest air in the United States. With its lax regulatory history, Houston has allowed businesses to have free reign with the air. Millions of U.S. residents live in areas that regularly exceed the 1970 Federal Clean Air Act standard for ozone (O_3). Millions also live under a blanket of other air pollutants.

If we are to truly understand the science of polluted air, we must consider these key questions:

- What do we mean by "clean air?"
- What are the sources of air pollution?
- What kinds of air pollution are of greatest concern?
- What are the short- and long-term consequences of air pollution?
- Is it possible to "turn the corner" on pollution, and is the cost (measured in lifestyle changes as well as money) worth the sacrifice? This is a "risk versus benefit" issue for individuals and governments to resolve.

Figure 12.3
Los Angeles is a city of drivers, most of whom travel alone. Over 6 million registered vehicles clog freeways and city streets so that the average road speed in the year 2000 was about 17 miles per hour.

case in point

The Dangerous but Exciting Life in the Biggest City in the World

This case in point was written by Dr. Carlos Mauricio Castro-Acuña, Professor of Chemistry at the Universidad Nacional Autonomo de Mexico—"UNAM" (the National Autonomous University of Mexico), a school with over 200,000 students. Dr. Castro-Acuña was born in Costa Rica and moved to Mexico City when he was a child. In addition to his research specialty in physical chemistry, Dr. Castro-Acuña is an accom- *plished chemical educator, whose teams have won many medals at international chemistry Olympiad competitions.*

Considering that I live in a small house surrounded by green areas and little traffic, a full 20-minute drive from my office at UNAM, it is easy to understand that sometimes I forget that I am living in the biggest city in the world.

However, I notice that I seem to get colds very often, my eyes occasionally itch, and headaches come from nowhere. It is at these times that I remember all too well the hazards related to being one of the nearly 20 million people who inhabit this valley at 2240 m (ca. 7340 ft.) above sea level.

Unfortunately, our city has many natural and artificial sources of pollution:

- Garbage, dust, and solid particles.
- Combustion products from millions of vehicles that consume leaded or unleaded gasoline.
- Millions of homes containing stoves, ovens, and heaters usually run on a mixture of propane and butane, which do not burn as cleanly as we might like.
- Emissions from chemical industries and even smoke from countless cigarettes.

From the earliest days of this ancient center of the Aztec empire to the vibrant and exciting Mexico City of today, many things have changed, but one thing remains the same: the geographical site. We are in a valley where the surrounding mountains do not allow winds strong enough to disperse the contaminants of our atmosphere so we need other mechanisms to clean the air. Occasionally, **natural convection,** moving the air in an up-down direction thus taking the polluted air to the upper levels of the atmosphere, brings us a healthier environment. Sadly, our city very often suffers **thermal inversion,** keeping toxic substances from going away. It is often as late as 11 A.M. daily before the sun heats the surface air sufficiently to break the inversion. For this reason, the city health department recommends that we do not exercise outdoors, especially running or jogging in the morning.

Another important factor is the city's altitude. At sea level, the concentration of oxygen in the air is about 275 g/m^3. At 2240 m, the air in Mexico City contains about 210 g/m^3 of oxygen. This causes the internal combustion motor to work less efficiently, and so our 4 million motorized vehicles contaminate the air as if there were many more vehicles.

Every day, more than 12 tons of airborne contaminants are generated in our city. There is no agreement on which is the main source of pollution—cars? industry? homes? In my opinion, the worst pollutant is *ignorance.* The government has tried many programs to reduce our levels of pollution. For instance, the "day without a car" program creates a mandatory rest day for every car to reduce the consumption of gasoline. However, many families get around the rules by purchasing a second car, resulting in ever-increasing levels of pollution.

To contribute to a better environment, the state-owned oil company dismantled its Mexico City refinery some years ago. A significant proportion of our electricity is produced in thermoelectric plants. These plants were consuming a fuel relatively high in sulfur. They now use the much cleaner natural gas, methane.

And the effort seems to be working. In the 15 months before January 2001, there have been no smog alerts. Levels of many pollutants, such as lead, are lower. New cars have pollution control devices.

Despite many programs and efforts, if our city continues to grow and we, the citizens, are not willing to change our bad habits, Mexico City will never again be "the most transparent region in the world," as it was called a century ago.

Student Questions: What cities in the world have the worst air quality? What do they have in common from the standpoint of population, location, economic system and so forth? How are they different? Can you draw any general conclusions about these cities?

12.1 | The Atmosphere

As far as we know, we live on a unique planet. Nowhere else in the solar system is there a planet with a protective blanket of air that can sustain life. The atmospheres on other planets vary from mostly carbon dioxide on Mars and methane and ammonia on Uranus, to nothing at all on Mercury (see Figure 12.4). Our atmosphere is a precious resource, although we don't often think of it in such terms. It is worth knowing about, and protecting.

How Did Earth's Atmosphere Develop?

The best evidence suggests that the Earth formed about 4.5 billion years ago. Volcanoes soon began to belch out enormous amounts of water vapor, hydrogen chloride, sulfur dioxide, and carbon dioxide gases that would kill any life form with which we are now familiar (Figure 12.5). Soon after, many of these gases reacted with each other, the rocks and metals at the Earth's surface, as well as other gases in the atmosphere, presumably including ammonia, NH_3, to leave behind a primordial atmosphere of nitrogen (N_2), methane (CH_4), hydrogen (H_2), and water (H_2O), but no oxygen.

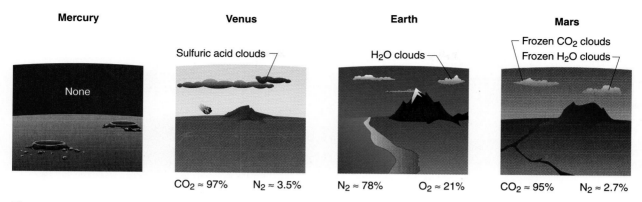

Mercury   **Venus**   **Earth**   **Mars**

Sulfuric acid clouds   H$_2$O clouds   Frozen CO$_2$ clouds / Frozen H$_2$O clouds

None

CO$_2$ ≈ 97%   N$_2$ ≈ 3.5%   N$_2$ ≈ 78%   O$_2$ ≈ 21%   CO$_2$ ≈ 95%   N$_2$ ≈ 2.7%

Figure 12.4
The atmosphere around the other planets in the solar system varies, depending to an extent on the size, and hence the gravitational attraction to keep in any gases. The atmospheres of Venus, Earth, and Mars contain gases largely in the 28 to 44 grams per mole range. The larger planets can retain lighter molecular weight gases such as hydrogen, helium, and ammonia.

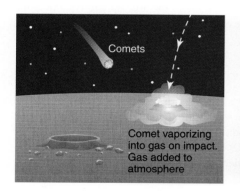

Comets

Comet vaporizing into gas on impact. Gas added to atmosphere

Figure 12.5
As the Earth was cooling, a complex mixture of gases such as hydrogen chloride, sulfur dioxide, and water vapor from volcanoes may have interacted to form the primordial atmosphere.

An atmosphere of this type probably persisted until about 2.0 billion years ago when, fairly suddenly, oxygen appeared in the atmosphere. Evidence of the arrival of oxygen can be found in the nature of the minerals in the rocks formed at that time. The red color of iron deposits laid down from that time onward is evidence of the oxidized form of iron, Fe$_2$O$_3$, presumably formed when iron reacted with oxygen gas.

The buildup of oxygen in the atmosphere is believed to have been caused by the chemical activities of living things, so oxygen accumulated in the atmosphere when living things evolved and began to release it as a waste gas. This began around 3.5 billion years ago, with primitive marine organisms that could use photosynthesis to sustain themselves, releasing oxygen as a by-product. This oxygen must have been a terrible toxic pollutant from the point of view of early **anaerobic** organisms (those that do not need and cannot tolerate the presence of oxygen). It wasn't until about 1.8 billion years ago that organisms arose for which oxygen was not toxic, and even later (1.3 billion years ago) that organisms evolved that actually required oxygen. The first organisms to require oxygen (known as **aerobic** organisms) utilized the oxygen much as we do now in respiration. Their energy source was the chemical reaction of oxygen with carbon-containing food molecules, the same process as we rely on today. Remember, oxygen will react with any carbon-hydrogen-oxygen compound to yield carbon dioxide and water-releasing energy.

$$C_xH_yO_z + nO_2 \rightarrow xCO_2 + y/_2H_2O$$

The oxygen-requiring organisms flourished *alongside* organisms capable of photosynthesis, setting up the continual cycling of oxygen through living things that occurs to this day. Plants participate in all parts of the cycle (Figure 12.6), because they release oxygen during photosynthesis but utilize it during respiration. Animals can only utilize oxygen, combining it into carbon dioxide and water which they excrete and thus make available to "feed" the photosynthesis of plants.

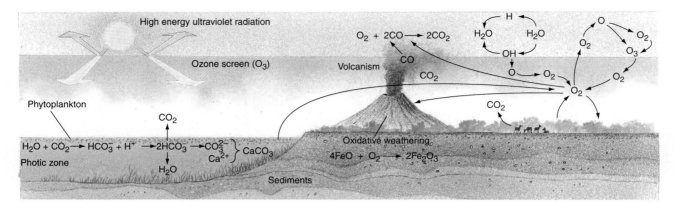

Figure 12.6
The oxygen cycle. Plants participate in all important parts of the cycle because they release oxygen during photosynthesis and utilize oxygen during respiration. Animals can utilize oxygen only during respiration.

Some time around 0.4 billion years ago the oxygen level in the atmosphere became high enough to allow ultraviolet (UV) radiation from the Sun to convert a little of it into the form of **ozone** (O_3) in the upper atmosphere,

$$O_2 \xrightarrow{\text{UV}} 2O$$
$$O + O_2 \longrightarrow O_3$$

The first step in the reaction can happen even at very low oxygen concentrations, but the second step can happen only when the oxygen concentration is high enough for the very reactive oxygen atom to have a reasonable chance of colliding with another O_2 molecule. Ozone is even more effective at absorbing a wider range of UV radiation than are ordinary oxygen molecules, but ozone can be formed only when a considerable amount of diatomic oxygen (O_2) is present.

This tenuous **ozone layer** then began to do what it still does for us today: absorb UV radiation and so reduce the amount of UV radiation that reaches the surface of Earth. More complex multicellular life forms, including humans, evolved underneath this protective ozone layer and would suffer considerable damage if the ozone layer were removed. With the ozone shield in place, a great diversity of complex life forms evolved and flourished. The atmosphere's oxygen level soared to about what it is today, accounting for about 20% of the volume of the air.

So the fact that our planet's atmosphere contains the oxygen gas needed to sustain most life on Earth is not a happy accident. The oxygen was put into the atmosphere by the chemical activities of evolving life forms. Life and the atmosphere evolved in harmony, each influencing the development of the other.

The Atmosphere Today

We have described the atmosphere as a "blanket of air." This common phrase implies that the atmosphere has a fairly uniform composition throughout the world. Yet, the air in urban centers like Mexico City, London, Calcutta, or Atlanta is often very irritating to our eyes, nose, and throat while the air in wild countryside areas such as Rocky Mountain National Park seems clean and refreshing. Is there an important chemical difference between the air in urban areas and that in more isolated locations?

Neglecting the variable amount of water vapor in the air, over 99.94% of the chemical composition of the air is the same no matter where you are on the surface of the Earth. The major components are listed in Table 12.2. The table is shown for dry air because the water content of the atmosphere changes from day to day depending on the weather. Nitrogen and oxygen make up more than 99% of the **troposphere,** the layer of atmosphere closest to the Earth's surface. The composition of the remaining fraction of a percent, containing mere traces of other substances, makes all the difference in the way the air looks and smells and affects your health. **Air quality,** in other words, depends on the precise mixture of chemicals

Table 12.2

Composition of Dry Air*

Component	Volume Fraction
N_2	0.7808
O_2	0.2095
Ar	0.00934
CO_2	0.00034
Ne	1.82×10^{-5}
He	5.24×10^{-6}
CH_4	2×10^{-6}
Kr	1.14×10^{-6}
H_2	5×10^{-7}
N_2O	5×10^{-7}
Xe	8.7×10^{-8}
SO_2	$<1 \times 10^{-6}$
O_3	$<1 \times 10^{-7}$
NO_2	$<2 \times 10^{-8}$
I_2	$<1 \times 10^{-8}$
NH_3	$<1 \times 10^{-8}$
CO	$<1 \times 10^{-8}$
NO	$<1 \times 10^{-8}$

*The fraction of water in the atmosphere varies from near zero on a cold dry day, to 0.05 on a warm humid day. It also varies with altitude and geography.

that comprise the variable 0.06% of the volume of dry air. Substances that, when added to air, contribute to a decline in air quality are called **air pollutants**. Among the important atmospheric trace gases that contribute to pollution are SO_2, O_3, CO, and NO_2. The concentration of these gases can change from day to day at any location. These gases will be the focus of much of our discussion on air pollution.

Nitrogen and Oxygen: The Main Gases of Air

Many of the trace substances in air, although they are only 0.06% of the volume of air, are the causes of the pollution that diminish air quality. Small increases in the concentrations of such trace gases as sulfur dioxide or ozone can cause big increases in pollution. Yet almost all of the air we breathe is composed of molecular nitrogen (N_2) and molecular oxygen (O_2).

Nitrogen

Nitrogen in the form of N_2 molecules is the main component of air, accounting for approximately 80% of the air's volume. The nitrogen molecules are held together by a triple bond involving six shared electrons overall (see Figure 12.7). This triple bond is very strong, meaning that a large amount of energy is required to break it. As a result, nitrogen molecules are very unreactive, being jolted into reactivity only by very high temperatures or the presence of very efficient catalysts. In fact, N_2 is used as an inert atmosphere in metallurgical, petrochemical, and other industrial processes in which oxygen or water vapor might react explosively with starting materials. The unreactivity of nitrogen means that it is easy to ignore the presence of nitrogen in the air, even though it is the main gas present.

Despite its chemical unreactivity, atmospheric nitrogen does contribute in important ways to air pollution. In automobile engines nitrogen is heated to very high temperatures in the presence of oxygen, with environmentally important results. In Chapter 9 we described the

Figure 12.7
Nitrogen, a gas at room temperature, is very unreactive because of the triple bond holding the two nitrogen atoms together. As a result, nitrogen is used as an inert atmosphere in industrial processes where oxygen or water vapor might react vigorously with starting materials.

The major industrial use of nitrogen is in the production of ammonia, NH_3, via the Haber process. Much of the ammonia is used in the preparation of fertilizer.

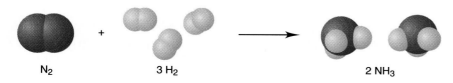

N_2 + $3 H_2$ $\longrightarrow$ $2 NH_3$

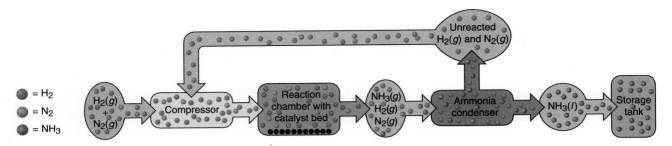

Figure 12.8
The formation of ammonia from nitrogen and hydrogen in the Haber process. Because nitrogen is such an unreactive substance, high pressure and fairly high temperature are needed to yield products.

specific reactions that lead from nitrogen and oxygen to nitrogen dioxide (NO_2) and nitric acid (HNO_3), a major cause of acid rain. We can actually see the yellowish color of NO_2 in air that is heavily polluted by motor vehicle exhausts. NO_2 is also the main chemical that absorbs sunlight during the complex process that leads to **photochemical smog** (discussed later).

The major use of nitrogen is in the production of ammonia (NH_3) by reaction of nitrogen with hydrogen (H_2) at 450°C, pressures of between 200 and 600 atm, and in the presence of a catalyst.

$$N_2(g) + 3H_2(g) \rightleftharpoons 2NH_3(g)$$

This is known as the **Haber process.** Le Chatelier's principle (see Chapter 8) tells us that high pressure will force the reaction to go toward the side with fewer gas molecules. In the Haber process, there are 2 moles of gaseous products and 4 moles of gaseous reactants. The high pressure used in the Haber process, therefore, favors the formation of ammonia.

This reaction is exothermic. This means that low temperatures also favor the formation of ammonia. However, low temperatures allow the reaction to go very slowly, so the temperature of 450°C that is actually used in the Haber process is a compromise between speed (kinetics) and completeness (thermodynamics).

The ammonia formed during the Haber process has a considerably higher boiling point (−33°C) than nitrogen or hydrogen. This allows pure ammonia to be obtained by passing the reacting gases through a cooling chamber in which the ammonia condenses into a liquid, which can be run off and collected, while unreacted nitrogen and hydrogen remain in the gaseous form (see Figure 12.8).

Enormous amounts of ammonia are made each year by the industrialized countries of the world. The total in 1999 in the United States alone was about 19 billion kg, making it second only to sulfuric acid in amount of inorganic compounds produced. Most of this ammonia is used to produce nitrogen-containing fertilizer, which is spread over the farmland and gardens of the world.

Oxygen

Oxygen accounts for just over 20% of the volume of air and is by far the most reactive major gas in the atmosphere. Its preparation, and the exploration of its properties, led to the advent of modern chemistry in the closing years of the 18th century.

Almost all of the oxygen in the atmosphere is in the form of diatomic molecules (O_2). A different and very rare form of oxygen known as **ozone** also occurs in the atmosphere and is composed of triatomic molecules (O_3). The O_2 and O_3 molecules are examples of **allotropes:** different molecular forms of the same chemical element.

Ozone is formed spontaneously when an oxygen atom reacts with an oxygen molecule (Figure 12.9):

$$O_2(g) + O(g) \rightarrow O_3(g)$$

The free oxygen atom can be formed by an electric spark occurring in air. The sharp smell associated with electric sparks and arcs is actually the smell of ozone. Ozone is good for us in

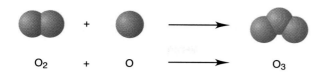

$$O_2 \;+\; O \longrightarrow O_3$$

Figure 12.9
Ozone, O_3, is formed from the reaction of molecular oxygen with atomic oxygen. Ozone is important in the stratosphere where it absorbs harmful UV radiation. It is, however, hazardous at the surface, where a concentration as low as 0.12 ppm of air can adversely affect human health.

the "right place" (high up in the atmosphere), but very bad for us in the wrong place (at ground level). The tiny concentration of ozone present high in the atmosphere forms the famous **ozone layer,** which absorbs some of the potentially harmful UV radiation from the Sun and prevents it from reaching the surface of the Earth. At ground level, ozone is a serious pollutant that contributes to the development of photochemical smog.

The combination of a substance with oxygen is an example of an oxidation reaction. The term oxidation was originally reserved for reactions in which oxygen combined with something else to form an **oxide.** As we explained in Chapter 4, the modern meaning of oxidation covers any process involving the complete or partial *loss of electrons*. The original meaning of oxidation, however, reflects the fact that oxygen is extremely reactive and will combine with a very wide variety of chemicals to form oxides. Most metals react with oxygen to form metal oxides. For example:

$$2Ca(s) + O_2(g) \rightarrow 2CaO(s) \text{ calcium oxide}$$
$$4Fe(s) + 3O_2(g) \rightarrow 2Fe_2O_3(s) \text{ iron(III) oxide}$$

Oxygen will also react with most of the nonmetal elements to yield oxides. Two examples of this process are:

$$S(s) + O_2(g) \rightarrow SO_2(g) \text{ sulfur dioxide}$$
$$C(s) + O_2(g) \rightarrow CO_2(g) \text{ carbon dioxide}$$

Recall from our discussion of acid rain that the nonmetal oxides (apart from hydrogen oxide, i.e., water) are called acid anhydrides because they react with water to form acids.

Oxygen will react with a very large number of compounds to form the oxides of each of the elements in the original compound. Of special importance are organic compounds, which are all based on carbon. Virtually all organic compounds will react with oxygen in this way. For instance, if supplied with enough energy (such as from a match) compounds of carbon and hydrogen, termed hydrocarbons, can react with oxygen to form carbon dioxide (an *oxide* of carbon) and water (an *oxide* of hydrogen). When this happens quickly it is called combustion or burning, and a great deal of energy is released. A very similar reaction occurs with all compounds containing carbon, hydrogen, and oxygen.

The following equations summarize the combustion of toluene (C_7H_8), a common ingredient in gasoline, and of ethyl alcohol (C_2H_5OH), the alcohol that it is safe to drink (suitably diluted and in moderation):

$$C_7H_8(l) + 9O_2(g) \rightarrow 7CO_2(g) + 4H_2O(g)$$
$$C_2H_5OH(l) + 3O_2(g) \rightarrow 2CO_2(g) + 3H_2O(g)$$

The carbon dioxide that is formed by such reactions is a "greenhouse gas," that may contribute to global warming (discussed later).

Air Pollution—What Is It, and Where Does It Come From?

It is difficult to write a short yet complete definition of **air pollution.** Most would agree that air pollution is caused by the addition of chemical substances to the air. Many people consider air pollution to be restricted to the human-made or human-induced addition of harmful chemical substances to the air. A significant amount of chemicals that can be regarded as air pollutants, however, enter the atmosphere even without the intervention of humans. For example, natural phenomena such as volcanoes, forest fires, and the activities of plants contribute plenty of pollution to the atmosphere (see Table 12.3). These sources of air pollution have been operating for billions of years, so living things including humans have evolved alongside them, and they are part of the natural chemical cycles of the planet.

Table 12.3

Natural Air Pollutants

Source	Pollutants
Volcanos	Sulfur oxides, particulate matter
Forest fires	Carbon monoxide, carbon dioxide, nitrogen oxides, soot
Windstorms	Dust
Plants (live)	Hydrocarbons, pollen
Plants (decaying)	Methane, hydrogen sulfide
Soil	Viruses, dust
Sea	Salt particles

Figure 12.10
Energy demands created by the exponential growth in the world's population during the last 200 years, along with the continuing industrial revolution, have put tremendous pressure on the Earth's environment.

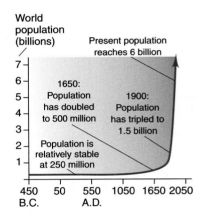

World population (billions)

Present population reaches 6 billion

1650: Population has doubled to 500 million

1900: Population has tripled to 1.5 billion

Population is relatively stable at 250 million

450 B.C.   50   550   1050   1650   2050   A.D.

In the last 200 years, human activities have led to an outpouring of new air pollutants, or additional production of existing ones, due to two main developments. The first development is dramatic population growth, portrayed in Figure 12.10. The energy needs of so many people have led to an ever-increasing use of polluting energy sources such as coal and oil. Second, industrialization, including automobile manufacture and use, has caused a very significant deterioration in air quality, especially in urban areas.

Although nonhuman activities contribute substantially to air pollution, these sources are essentially uncontrollable, and, as we have said, they are part of the natural chemical cycles of Earth. On the other hand, much of the air pollution caused by human activities could be significantly reduced, given the political and personal will to do so.

The Chemistry of Pollution

The remainder of this chapter considers the chemical interactions that lead to air pollution. Our focus will be on human causes. There are several approaches that we could take. For example, each section could deal with a specific pollutant or we could separate our sections into "causes" such as automobile driving and smokestack emissions. We have chosen instead to classify our discussion on the basis of three familiar phrases or "buzzwords": ozone depletion, the greenhouse effect, and smog and particulate matter.

12.2 | Ozone Depletion

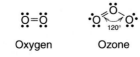

Oxygen   Ozone

Figure 12.11
Ozone is an allotrope of oxygen. Ozone's reactivity, described in the text, makes it useful in water purification. That same reactivity causes health problems in some people as well as damage to plants.

What is Ozone? Ozone is a pungent-smelling pale blue gas that irritates the mucous membranes of humans and other animals. It aggravates breathing problems by damaging the cells that line the trachea in the lungs, leading to swelling and inflammation. Ozone is an allotrope of oxygen composed of molecules each containing three oxygen atoms (O_3), rather than the normal two (O_2) (see Figure 12.11).

Ozone is even more reactive than diatomic oxygen. In other words, it is a very strong oxidizing agent. This means that a great many substances will become oxidized (by losing electrons) when ozone reacts with them. Among these are metals and organic compounds including amino acids. This powerful oxidizing ability makes ozone a useful water purifier because it reacts with and destroys microorganisms. Ozone's reactivity can also be exploited to remove odors from the air, by reacting with the odor-causing molecules. However, the oxidizing power of ozone also lies behind its many undesirable effects. In addition to being hazardous to health, ozone damages plants, causes rubber to crack, and leads to discoloration and pitting in painted surfaces because of its ability to react with compounds.

pro con discussions

On the Meaning of "Pollutant"

Since it is difficult to know just what we mean when using the word "pollutant," let's examine two points of view in favor of and opposed to the following definition: "a pollu-tant is something that has entered the environment as a result of human activity and that has a harmful effect on some aspect of the environment."

pro

The word "pollutant" has to mean something that is in the environment as a result of *human activity* and that has a harmful effect on some economic or ecological feature. It is true that many plants, including marigolds, for instance, release chemicals that harm insects; walnut trees have a substance in their roots that prevents many other plants from growing in their root area. These are substances that surely harm some of the organisms in the environment, but no one would reasonably call them pollutants because they are a natural part of the world. Pollutants are things like industrial wastes, herbicides, smokestack gases, and lead-containing paints. These things have a negative impact on the environment and they would not be in the environment if not for human activity.

con

The term "pollutant" means anything that is in the air, water, or soil that has a harmful effect on some part of the environment. Pollutants do not have to be generated by human activity. Trees are terrible polluters! Pine trees give off enormous amounts of volatile organic compounds such as terpenes, while oak trees emit isoprenes. These are, in part, responsible for the haze in the Blue Ridge and Great Smoky Mountains. Termites give off methane; vol-canoes emit sulfur dioxide, hydrogen chloride, and other noxious gases. These are all pollutants because they mess up the air, water, and soil; they can harm wildlife and can have a negative economic impact where their concentrations are too great. Aren't these outcomes exactly what we expect from "pollution"? In the sense that "chemicals are chemicals," regardless of their source, SO_2 from a volcano is as much a pollutant as SO_2 from a power plant.

Where Is Ozone Found?

Although we speak of "the atmosphere" as a single blanket of air, it is often useful to think of it as being divided into four layers. These four layers, from bottom to top, are the troposphere, the stratosphere, the mesosphere, and the thermosphere (Figure 12.12). Each layer has its own temperature profile and precise chemical composition (see Figure 12.13). Ozone exists in two of these layers: as a diffuse but more or less continuous layer in the stratosphere, and in specific areas (especially around urban centers) in the troposphere. The ozone in the troposphere is also called "surface ozone," because it is just above the Earth's surface.

The Importance of Stratospheric Ozone

As we discussed in Chapter 1, our Sun is a huge nuclear furnace. As it fuses hydrogen nuclei to form helium, the sun releases large amounts of energy into space. The Earth is continually bathed in some of that energy, in the form of electromagnetic radiation, including visible light and UV rays. UV radiation poses considerable hazards to life, because it can be absorbed by many molecules in a way that either destroys them or sometimes converts them into harmful substances. Fair skin tans when exposed to sunlight as a defensive mechanism that protects the body from UV radiation. The tanning is due to the accumulation of dark-colored molecules of **melanin,** which can absorb the UV rays and convert their energy to heat. Although some fair-skinned people like to encourage tanning, excessive exposure to UV rays can cause very severe burning in the short term and serious problems such as skin cancer in the long term. The cancers are initiated by damage to the molecules of DNA, which control the growth and maintenance of the body's cells (see Chapter 16).

Most of the UV radiation arriving from the sun is "screened" or "filtered" out by the atmosphere before the sunlight reaches the surface. The layer of stratospheric ozone (the **ozone layer**) is the main part of this natural "sunscreen" within the atmosphere. Without its

Figure 12.12
The layers of atmosphere closest to the Earth's surface.

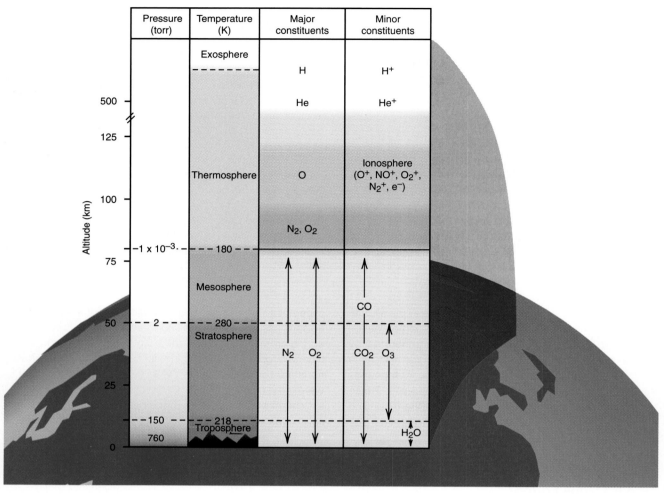

Figure 12.13
The atmosphere is composed of four main layers, characterized by their altitude and the direction in which the temperature changes in each region. The temperature decreases as we move up through the troposphere, but increases with height in the stratosphere, decreasing again in the mesosphere.

protective effects, sunlight would be much more hazardous for many species of animals (including humans), and it would cause considerable damage to plants.

The ozone molecules that make up the ozone layer protect us from UV radiation by actually absorbing much of it. This absorption of energy causes bond breakage so the ozone "photodissociates" (falls apart as a result of absorbing "light") to form an oxygen molecule and atomic oxygen:

$$\text{ozone destruction} \qquad O_3(g) \xrightarrow{\text{UV}} O_2(g) + O(g) \qquad \textbf{(12.1)}$$

If this were the end of the process, all of the ozone in the atmosphere would quickly be destroyed. Fortunately for life on Earth, a series of reactions continually regenerates ozone. First, molecular oxygen absorbs UV radiation, causing its double bond to split and releasing free oxygen atoms:

$$O_2(g) \xrightarrow{\text{UV}} 2O(g) \qquad \textbf{(12.2)}$$

Free oxygen atoms are very reactive, which is why we find virtually no free oxygen atoms on the Earth's surface—almost all of the oxygen atoms are bonded within molecules or ions. In the atmosphere, an oxygen atom can readily react with an oxygen molecule to form ozone thereby releasing heat.

$$O_2(g) + O(g) \rightarrow O_3(g) \qquad \textbf{(12.3)}$$

The formation of ozone is given by the sum of equations (12.2) and (12.3).

$$O_2(g) \xrightarrow{\text{UV}} 2O(g) \qquad (12.2)$$

$$\underline{2O_2(g) + 2O(g) \rightarrow 2O_3(g)} \qquad (12.3) \text{ (multiplied by 2, see below)}$$

$$3O_2(g) \xrightarrow{\text{UV}} 2O_3(g) \qquad (12.4)$$

Atomic oxygen is neither a reactant nor a product of the whole process, so we can eliminate it from the overall equation by multiplying both sides of equation (12.3) by 2. This yields equation (12.4), summarizing the overall formation of ozone.

Both the destruction and the reformation of ozone are powered by the absorption of UV radiation. In the absence of outside intervention the rates of ozone destruction and formation become equal, on average, allowing a relatively constant concentration of ozone to occur in the stratosphere, which continuously absorbs harmful UV radiation. As is often the case, it is only when human intervention disturbs this natural balance that problems arise.

How the Balance Is Upset by Human Intervention

Human activity releases several kinds of compounds into the atmosphere that threaten the stability of the ozone layer. Of greatest concern is a class of compounds known as **chlorofluorocarbons (CFCs).** Chlorofluorocarbons, as suggested by the name, contain chlorine, fluorine, and carbon.

The large-scale manufacture of CFCs began in the 1890s, when chemists discovered that compounds of antimony could serve as catalysts for CFC production. The production of CFCs is relatively simple. For example, CFC-11 ($CFCl_3$) can be made in one step by combining carbon tetrachloride, CCl_4, with hydrofluoric acid:

$$CCl_4(l) + HF(l) \rightarrow CFCl_3(l) + HCl(g)$$

It is used in air conditioning, refrigeration, and as a blowing agent (something that vaporizes to produce gas during polymer formation—as in forming polystyrene coffee cups, for example.) The most commonly used CFC is Freon-12,® CCl_2F_2, first synthesized by workers at General Motors in 1928. Other examples of CFCs and their uses are listed in Table 12.4. All these compounds are nonflammable, chemically inert, relatively nontoxic, and inexpensive. A related class of compounds, the **halons,** are bromine-containing compounds such as CF_3Br and CF_2BrCl. Notice there are no hydrogens in either the halons or the CFCs. The most important use of halons is in fire extinguishers. This type of fire extinguisher is especially valuable in areas where there is precision equipment such as computer hardware and jet airliners. Halons smother the fire by excluding oxygen, without leaving any damaging residue.

Table 12.4
CFCs and Halons: Uses and Atmospheric Lifetimes

Formula	Common Symbol	Atmospheric Lifetime (yr.)	Major Uses
CCl_3F	CFC-11	75	Polymer foams, refrigeration, air conditioning
CCl_2F_2	CFC-12	111	Polymer foams, refrigeration, air conditioning, aerosols, food-freezing
CCl_2FCClF_2	CFC-113	90	Solvent
$CClF_2CClF_2$	CFC-114	185	Polymer foams, refrigeration, air conditioning
$CClF_2CF_3$	CFC-115	380	Refrigeration, air conditioning
$CBrClF_2$	halon-1211	25	Portable fire extinguishers
$CBrF_3$	halon-1301	110	Total flooding fire extinguishing systems
$C_2Br_2F_4$	halon-2402	Unreported	Fire extinguishers

Source: Jacob, Anthony T. 1991. Chlorofluorocarbons and the Hole in the Ozone Layer. *Institute for Chemical Education.* University of Wisconsin, Madison.

In 1974, Sherwood Rowland and Mario Molina, working at the University of California at Irvine, theorized that CFCs might be accumulating high in the atmosphere and causing the destruction of the stratospheric ozone layer, particularly over the north and south polar regions. Many measurements of stratospheric ozone levels have been made since that time and the reality of ozone destruction is now perfectly clear. For their work, Rowland and Molina received a share of the 1995 Nobel prize in chemistry.

The events that allow CFCs to destroy ozone in the ozone layer can be summarized as: CFCs (from aerosol cans, air conditioning systems, refrigerators, etc.) escape into the atmosphere and mix with ground-level air. Since the CFCs are so inert, they do not react with any other chemicals, but slowly diffuse high into the atmosphere until they rise into the stratospheric ozone layer. CFC molecules do not react directly with the ozone in the stratosphere, but continue to diffuse upward until reaching the region where the level of UV radiation from the Sun has not been attenuated by the protective ozone shield. At this point, UV radiation can break the carbon-chlorine bonds within the CFCs releasing free chlorine atoms. The free chlorine atoms then react with ozone to form chlorine monoxide (ClO) and molecular oxygen (shown in equation (12.5)). Equation (12.6) shows that the ClO can react with atomic oxygen produced by the photodissociation of ozone.

$$Cl(g) + O_3(g) \rightarrow ClO(g) + O_2(g) \qquad \textbf{(12.5)}$$

$$\underline{ClO(g) + O(g) \rightarrow Cl(g) + O_2(g)} \qquad \textbf{(12.6)}$$

overall reaction $\quad O_3(g) + O(g) \rightarrow 2O_2(g) \qquad \textbf{(12.7)}$

The net result of these two reactions is the destruction of one ozone molecule and the reformation of a chlorine atom which can go through the process again. Note that this sequence of reactions is *catalytic* because each free Cl atom can cause the destruction of thousands or millions of ozone molecules.

As you may have suspected, the chlorine atoms can also participate in a great many other reactions with chemicals in the atmosphere. The overall chemistry of chlorine atoms in the atmosphere is actually very complex, but one simple consequence is that chlorine atoms destroy stratospheric ozone. The only mechanism known to remove the damaging chlorine atoms from the stratosphere is their diffusion downwards until they react with other chemicals and return to the Earth in raindrops.

Other compounds containing carbon-chlorine bonds would also release atomic chlorine on exposure to UV radiation, but most such compounds are more reactive than CFCs. They therefore react in the lower atmosphere to form such compounds as HCl, which dissolve in rainfall and are washed out of the atmosphere before they can get high enough to become involved in ozone depletion.

The oceans provide an interesting and, perhaps at first surprising, source of ozone destruction—bromine. The bromide ion, Br^-, comes from sea-salt spray in the form of an **aerosol** (see Section 12.5), as well as the frozen surfaces of the ocean and snow. The bromide ion is oxidized (loses an electron) in a way that is not yet understood, to form the very reactive bromine atom, Br. This can destroy O_3 to form O_2 in a series of reactions that are quite complex, but the outcome, as with chlorine, is the same—loss of ozone. The concentration of bromide ion in seawater is about 65 ppm, small when compared to nearly 19,000 ppm for chloride ion, so considering the concentration of bromine compared to chlorine from the oceans and industrial sources, bromine is far less important than chlorine in the destruction of ozone. It is interesting to note that this bromine-based ozone depletion has been around since the oceans and ozone layer have existed! It is part of the balance of nature that was present before the intervention of humankind.

The problem of ozone destruction has been most severe over Antarctica where ozone levels can fall nearly to zero in the Antarctic summer, only to reform a few months later. Figure 12.14 shows the sharp drop in the stratospheric ozone concentration in the Southern Hemisphere between 1980 and 2001. The pink color represents the region of greatest ozone depletion. The combination of months of uninterrupted sunshine and very cold clouds high in the atmosphere can result in almost total destruction of the ozone. A similar **"ozone hole"** has recently been detected over the Arctic Circle at times extending across heavily populated areas

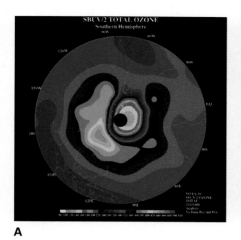
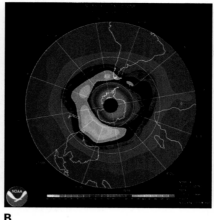
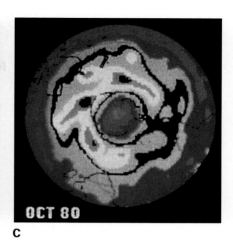

A   B   C

Figure 12.14
These images, taken by the National Oceanic and Atmospheric Administration's SBUV/2 polar orbiting satellite, show southern hemisphere ozone levels for **A,** October 16, 2001, **B,** October 1996, and **C,** October 1980. In each case, the area of greatest ozone depletion (ozone hole) is shown in pink and purple.

such as Northern Europe. The full environmental implications of these developments have not been firmly established. Significant depletion of the ozone layer over populated areas will certainly cause more skin cancers. No one knows the precise effects on plant or animal life, although many unwelcome possibilities have been proposed. Some scientists suggest that the yield of important crop plants may be significantly reduced by UV damage associated with destruction of the ozone layer.

Surface Ozone

The accumulation of ozone in the troposphere right next to the *surface* of the Earth is another kind of problem. In direct contact, ozone irritates human cell membranes and damages plants and some manufactured goods. A good proportion of surface ozone is produced by automobile use. Nitrogen oxides and hydrocarbons emitted by industrial processes and power plants also contribute to surface ozone. Automobiles without catalytic converters release nitric oxide (NO) which is converted to nitrogen dioxide by reaction with oxygen in the air:

$$N_2(g) + O_2(g) \rightarrow 2NO(g)$$
$$2NO(g) + O_2(g) \rightarrow 2NO_2(g)$$

The NO_2 can react on exposure to visible light to form NO and an oxygen atom:

$$NO_2(g) \xrightarrow{\text{light}} NO(g) + O(g)$$

As we will discuss in detail in Section 12.4 on smog, the use of catalytic converters significantly reduces the amount of NO_2 available for this reaction.

We know from our discussion on stratospheric ozone that oxygen atoms are very reactive and can combine with molecular oxygen to form ozone:

$$O_2(g) + O(g) \rightarrow O_3(g)$$

The problem of too much ozone at ground level and not enough in the stratosphere is troubling. Why can't we transfer by plane, or rocket, or whatever, ozone from down here to up there? Recognize first that there is a much higher ozone concentration in the stratosphere than at the Earth's surface. Even during the most polluted days in Los Angeles, when the ozone concentration can reach 0.5 ppm (compared to 0.02–0.03 ppm in rural areas), the surface ozone concentration in urban areas is significantly less than that at high altitude where the ozone density is about 5×10^{12} molecules/cm^3. This means that, assuming we could isolate pure ozone from polluted air, it would be impractical and ineffectual to transport the gas. Too

exercise 12.1

Grappling with Units to Understand the Literature

Problem

If the surface ozone concentration on a good day in Los Angeles is 0.10 ppm, how many molecules of ozone would there be per cm^3 of air at 25°C? The density of air is 0.0012 g/cm^3 at 25°C where a mole of air occupies 24.4 L at 1.00 atm pressure.

The quantity 0.10 ppm means that 0.10 L of O_3 is mixed into 10^6 L air or that 0.10 mol O_3 is mixed into 10^6 mol of air.

Solution

$$\frac{\text{molecules } O_3}{cm^3} = \frac{0.10 \text{ mol } O_3}{10^6 \text{ mol air}} \times \frac{6.02 \times 10^{23} \text{ molecules } O_3}{\text{mol } O_3} \times \frac{1 \text{ mol air}}{24.4 \text{ L air}} \times \frac{1 \text{ L}}{1000 \text{ cm}^3}$$

$$= 2.5 \times 10^{12} \text{ molecules/cm}^3$$

The concentration of ozone is often cited both ways, so interconverting them will prove useful when reading newspaper articles dealing with air pollution. We have two end-of-chapter problems (exercises 47 and 48) that take the issue one step further by comparing ozone levels at the Earth's surface with those in the stratosphere. Note that the density of air at 1 atm is 1.2×10^{-3} g per mL but that this value decreases at higher altitude.

much energy would be required. Also, ozone is highly explosive and toxic, making its purification and handling a risky business. Proposals have been seriously put forward, however, to fit commercial airliners with ozone-generators, in the hope that they might repair the ozone layer as they fly. So far nothing has come of such futuristic "atmospheric engineering" ideas.

As with so many environmental problems, establishing a safe ozone balance will require an appropriate combination of individual, social, economic, and scientific choices.

12.3 | The Greenhouse Effect

When gardeners and farmers want to grow plants in a climate that is really too cold for them, they raise their plants within the warmth of greenhouses—low buildings made almost entirely of glass. As long as the sun is shining, a greenhouse can become warm inside even when the air is cold outside. This is because the glass allows light energy from the sun to pass through it readily but acts as an insulating layer preventing heat from escaping outwards. Some of the entering light is converted into heat energy within the plants, soil, and air inside the greenhouse so even on a cold day sunlight can make the greenhouse warm. The famous global **greenhouse effect** is so named because the Earth's atmosphere acts in some ways like the glass of a greenhouse, allowing light energy in but restricting the flow of heat energy back out. The comparison is not perfect, but it has become so widely used that it is here to stay.

The notion of a greenhouse effect is not new. As early as the 1830s, the French mathematician Fourier proposed that specific gases in the atmosphere could trap heat as in a greenhouse. In 1896, the Swedish chemist Svante Arrhenius suggested that carbon dioxide could be the key gas. As we shall see, Arrhenius' proposition was right on the mark.

Causes of the Greenhouse Effect

When sunlight strikes the Earth, some of the radiation is reflected back into space and some is absorbed by chemicals in the atmosphere or on the surface. The radiation that is reflected has no net effect on the Earth. Much of the visible radiation, however, and most of the ultraviolet and infrared radiation is absorbed by water, soil, rocks, vegetation, or the air. Much of the energy of the visible and UV radiation which is absorbed upon striking the Earth is eventually re-emitted from the Earth as infrared (IR) radiation (see Figure 12.15). The key event in the greenhouse effect is that some of this IR radiation heading back toward space is blocked (absorbed) by the so-called "greenhouse gases" in the atmosphere. Carbon dioxide (CO_2) is the major greenhouse gas but certainly not the only one. As they absorb this energy, the molecules concerned move and vibrate faster, corresponding to an increase in air temperature. The higher the concentration of carbon dioxide (and other greenhouse gases) in the air, the greater this "greenhouse effect" temperature increase will be.

The greenhouse effect, as just described, is a natural part of the energy distribution mechanism of the Earth. A greenhouse effect has operated for billions of years, long before the arrival of humanity on the scene. The activities of humanity, however, particularly concerning the release of carbon dioxide gas from the combustion of coal, oil, gasoline, and natural gas, are threatening to cause a significant increase in the greenhouse effect, perhaps leading to undesirable warming of the planet. This warming caused by our influence on the greenhouse effect has become known as **global warming.**

You now know that carbon dioxide and methane are greenhouse gases, that CFCs may also contribute to global warming, and that CFCs are a cause of the depletion of stratospheric ozone. We will now look at the importance of each of these substances to the greenhouse effect.

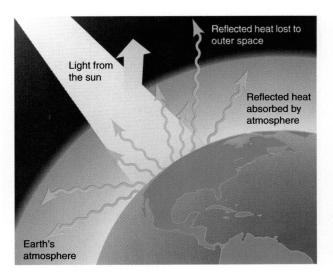

Figure 12.15
Much of the visible radiation that is emitted from the Sun passes through the Earth's atmosphere and reaches the surface. The ground warms and emits IR radiation, which is absorbed by "greenhouse gases" such as carbon dioxide, rather than being released back to space. This phenomenon is known as the "greenhouse" effect because a greenhouse absorbs light and retains the resultant heat.

Learning More About the Greenhouse Gases

Carbon Dioxide

Carbon dioxide enters the atmosphere as a product of the burning (combustion) of organic compounds and as an excretory product of the respiration of animals (including humans), plants, and many microorganisms. Fossil fuel combustion and industrial processes caused an estimated 1.3 billion metric tons of CO_2 to be released into the atmosphere in 1995. Carbon dioxide is taken up from the atmosphere largely by green plants and photosynthetic microorganisms during the process of photosynthesis. These are not the only routes by which carbon dioxide is released into and removed from the atmosphere.

It is now known that the ocean is an important part of the global "carbon dioxide cycle." Carbon dioxide is about 30 times as soluble in water as oxygen. More importantly, carbon dioxide from the air can combine with water as described in Chapter 9, to form bicarbonate and carbonate ions,

$$CO_2 + H_2O \rightarrow H_2CO_3$$

$$H_2CO_3 \rightarrow HCO_3^- + H^+$$

$$HCO_3^- \rightarrow CO_3^{2-} + H^+$$

Additionally, carbon dioxide is used in photosynthesis by organisms at or near the surface of the ocean. As these organisms die and sink to the deep parts of the ocean, bacteria can break them down in a process called remineralization. The overall equations for these complex processes are given below.

Photosynthesis:
$$106CO_2 + 16HNO_3 + H_3PO_4 + 122H_2O \rightarrow (CH_2O)_{42}(CH_2)_{64}(NH_3)_{16}H_3PO_4 + 170O_2$$

Remineralization:
$$(CH_2O)_{42}(CH_2)_{64}(NH_3)_{16}H_3PO_4 + 170O_2 \rightarrow 106CO_2 + 16HNO_3 + H_3PO_4 + 122H_2O$$

Figure 12.16
The destruction of the world's forests, especially the dense rain forests, is proceeding at a rate of about 1 acre per second. Although we often read about deforestation in developing countries, the process also happened in industrialized nations many years ago when they, too, were developing.

The increased release of carbon dioxide is a result of many of the things we consider most essential for civilized modern living: warm homes, offices, and shops, available electricity at the flick of a switch, fast and convenient transport, and an industrial system able to manufacture a bewildering array of consumer goods.

The other part of the equation, the destruction of forests, is proceeding at a rate of about one acre per second (Figure 12.16)! Most current attention is given to the destruction of the rain forests in South America and Southeast Asia, but we should not forget the countries that became developed earlier than these areas and destroyed vast areas of their own forest long ago. For example, most of Europe, the cradle of the industrial revolution, was once covered in forests. The continuing destruction of forests is steadily decreasing the Earth's capacity for converting atmospheric CO_2 into plant life (a form of **biomass**—matter held within living things).

As we have said before, carbon dioxide is not the only greenhouse gas. It was, however, the first to be related to global warming. Many other polyatomic molecules have similar effects. The impact of a certain gas depends upon two things; the gas's inherent ability to absorb infrared radiation and its concentration in the atmosphere. Carbon dioxide is not a particularly effective infrared absorber, but its concentration is very high compared to other greenhouse gases.

Methane

One other significant greenhouse gas is methane (CH_4), commonly known as **natural gas.** Methane is a more efficient IR absorber than carbon dioxide. The EPA estimates that CH_4 is 21 times more effective at trapping heat than CO_2! Methane is formed when vegetation decays in the absence of oxygen. Thus, dead plants in swamps generate methane (explaining why it is also known as **swamp gas**). As Table 12.5 shows, the flatulence of cows, buffalo, sheep and goats (called, collectively enteric fermentation) and the activities of termites also produce amazing amounts of methane.

CFCs

CFCs are excellent IR absorbers. Their documented increase in the atmosphere has led to concerns about their contribution to the greenhouse effect, as we discussed above. This, combined

Table 12.5

Major Sources of Methane*

Source	Low-High Estimates Millions of Metric Tons
Natural wetlands (bogs, swamps, tundras)	70–170
Rice paddies	20–150
Enteric fermentation (digestion-related releases from animals, especially cattle)	50–82
Biomass burning (forest fires, agricultural fires)	20–80
Gas drilling, venting, and pipeline transmission	33–68
Termites	10–50
Landfills	21–57
Coal mining	22–34
Oceans	5–20
Fresh waters	1–25
Decomposition of animal waste	10–20

*Annual global emissions of methane into the atmosphere, in millions of metric tons.
Source: EPA (www.epa.gov/ghginfo/, April 4, 2000).

with CFCs' long-term involvement in stratospheric ozone depletion, make it important to phase out CFC use as soon as possible.

HFCs, PFCs, and SF$_6$

Green Chemistry

Chemists have come up with alternative compounds to ozone-depleting substances such as CFCs. Three of the most important types are hydrofluorocarbons (HFCs, discussed in the Pro/Con box), perfluorocarbons such as CF_4 and C_2F_6, and sulfur hexafluoride, SF_6. These substances are used for applications including etching semiconductor chips, depositing metals on surfaces, and refrigeration. They are also by-products of smelting (see Section 9.1) and the production of aluminum. SF_6, used as an insulator in electric systems, is the most potent known greenhouse gas.

Even though these are potent greenhouse gases, they are present in relatively low concentrations in the atmosphere. CO_2 is still the main culprit in the struggle with greenhouse gases.

■ case in point

Structural Properties of Greenhouse Gases

How do we know which gases might contribute to the greenhouse effect? Greenhouse gases are those that both absorb IR radiation and allow visible light and UV radiation to be transmitted through them unhindered. The main structural factors that determine these properties of gases are as described next.

IR Absorbers

Simply put, a molecule will absorb IR radiation if the center of positive charge is different from the center of negative charge. Such molecules are said to have a dipole moment, by which we mean an overall uneven distribution of positive and negative charge. Also, molecules that contain three or more atoms can vibrate in ways that *create* dipole moments (see the figure), so all such molecules are of interest. Of special concern, however, is carbon dioxide, which is present in the atmosphere at 356 ppm, a far higher concentration than most other polyatomic molecules, with the exception of water. Methane, CH_4, is only present at 1.74 ppm, but is about 21 times more effective at absorbing IR radiation as CO_2, so its presence is also of some concern. Water is also a good absorber of IR radiation. It has been present in the atmosphere for the entire life of the Earth. We regard its contribution to surface conditions as normal. However, if the atmosphere and the oceans do warm up, more water will evaporate and enter the atmosphere, exacerbating the greenhouse effects.

IR Nonabsorbers (Transmitters)

These are

* Linear molecules containing only one element such as N_2 and O_2. These elements do not have a dipole moment.
* Single atoms such as Ar (argon).

The three major gases that make up our atmosphere, N_2, Ar, and O_2, are UV transmitters but they are not IR absorbers so they are not greenhouse gases. Carbon dioxide, water, and methane, however, are greenhouse gases (as discussed above). CFCs are very significant IR absorbers but their atmospheric concentration is low (0.47 ppb for CFC-12). CFCs themselves may contribute to global warming, but because of their interaction to destroy ozone, itself a potential greenhouse gas, the net effect may be negligible.

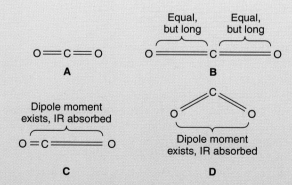

Incoming IR radiation can be absorbed by molecules, depending on the frequency of the radiation and the nature of the molecule. Molecules that have three or more atoms can move in ways that create a dipole moment (see text) and can therefore absorb IR radiation. Carbon dioxide, shown in **A** has 2 CO double bonds that lie along a line. If *both* CO bonds stretch so that the bond distances are longer than normal **B**, then there is still not a change in the dipole moment and IR radiation will not be absorbed. If one bond stretches while one contracts, the dipole moment is changed and IR radiation can be absorbed **C**. If bonds bend **D**, IR radiation can also be absorbed. CFCs contain carbon-fluorine bonds that, on stretching, also absorb IR radiation, contributing to the greenhouse effect.

Global Warming

What and how much does the evidence of continuing increases in the concentrations of greenhouse gases allow us to predict about global warming? In theory, the expectation is that the average temperature of the Earth will increase. The more debatable questions are:

- By how much will the temperature increase?
- How soon will this happen?
- How much polar ice will melt as a result?
- How much will the water of the ocean expand upon heating?
- How high will the waters of the ocean rise?
- What will be the effect of increased temperature on agriculture?
- What will be the general effect on weather and climate?
- What will be the effect of increased atmospheric carbon dioxide on growing plants?

These are all difficult questions because the answer to each will depend on a great many variable and often poorly understood factors. Scientists try to find answers by taking account of as many relevant facts and principles as possible. They use complex physical and mathematical models, combined with the power of modern computers, to make projections of the likely consequences of an increasing greenhouse effect. There is no single agreed-on projection, but many scientists make very worrisome predictions. Some predict that melting of polar ice and expansion of seawater will produce great floods devastating some of the most populated cities on Earth. Some warn of dramatic encroachment of the oceans across what is currently dry land and of major shifts in climate that will cause fertile areas to become deserts and perhaps turn existing deserts into fertile land. Some other projections, however, predict only very minor consequences, as they assume that natural systems of checks and balances will keep the climate relatively stable overall.

pro con discussions

Should CFCs Be Banned Worldwide?

As the old expression goes, *"better late than never!"* In 1978, the United States officially banned the use of CFCs in aerosol cans. Large quantities of CFCs continue to be released however (especially outside the United States) due to the destruction of old refrigerators and the many other uses of CFCs that are still not fully controlled by international treaties. Even if all releases of CFCs were stopped today, the catalytic destruction of ozone by chlorine would continue for years due to the long lifespan of CFCs already present in the stratosphere.

The 1990 Montreal Protocol on Substances That Deplete the Ozone Layer is an international agreement that called for the use of CFCs to be phased out by industrialized countries by the year 2000. Developing countries should complete the phase out by the year 2010. There are many who doubt that this timetable is being met.

However, industry is working hard to find alternatives to CFCs. In 1996, over 120 million kg of CFCs were produced worldwide. However, this is a significant drop from the 1988 level of 1 billion kg. This includes a drop in CFC-12 manufacture from 188 million kg in 1988 to 84 million kg in 1993.

Compounds that can be used as substitutes for CFCs usually contain hydrogen atoms in place of the chlorine atoms of the CFCs. These hydrogen atoms can react with hydroxyl groups (OH) found in many compounds in the troposphere. This means that CFC substitutes break down much faster than CFCs. Table 12.6 lists some CFC alternatives. It would appear then, that we must work as quickly as possible to ban CFCs and replace them with less harmful compounds.

Table 12.6

CFC Alternatives Under Development

Market	Current CFC	CFC Alternatives
Refrigerants	CFC-12 (CF_2Cl_2)	HFC-134a (CF_3CFH_2)
		HCFC-22 (CHF_2Cl)
		Blends/azeotropes
Blowing agents	CFC-13 ($CFCl_3$)	HCFC-141b (CH_3CFCl_2)
		HCFC-123 (CF_3CHCl_2)
		HCFC-22 (CHF_2Cl)

Source: *Chemical and Engineering News*, April 27, 1992, pp. 7–19.

con

The answer to the CFC problem might seem simple enough: merely synthesize an alternative compound such as HFC-134a (CF_3CFH_2). The synthesis of substitute compounds, however, is very difficult and involves some fairly new chemistry (see the figure). There is also evidence that these compounds, while not ozone-depletors, might well be potent greenhouse gases! Thus, although most but not all countries do want CFC use to be phased out, there are significant chemical and economic difficulties that make an immediate and complete phase out impractical.

While it is clear that the ozone layer is diminishing over various regions of the world, it is not clear that CFCs really are *principally* responsible for this phenomenon. Even though most experts in the field accept that CFCs are a major culprit, some highly respected scientists disagree. One of the main tasks of science is to challenge accepted explanations, and this process has led to many explanations being rethought in the past. We should not ban the use of CFCs so long as their role in ozone destruction is not certain.

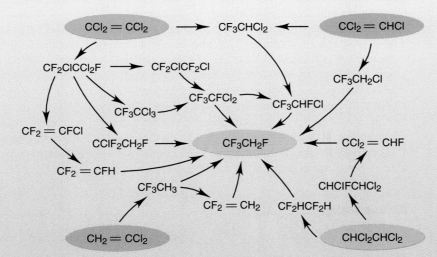

Although the production of CFCs is fairly straightforward, that of CFC substitutes is not. Many different routes are possible to synthesize the CFC substitute CF_3CH_2F, called HFC-134a.

further discussion

On September 20, 1995, the U.S. House Subcommittee on Energy and Environment grappled with this issue. Debate at that forum did not resolve the matter, however.

Table 12.7

Global Warming Potentials Compared to CO_2

Gas	GWP
Carbon dioxide (CO_2)	1
Methane (CH_4)*	21
Nitrous oxide (N_2O)	310
HFC-23	11,700
HFC-125	2,800
HFC-134a	1,300
CF_4	6,500
C_2F_6	9,200
SF_6	23,900

*Keep in mind that the concentrations of the most powerful greenhouse gases are still very low compared to CO_2, so their impact is not nearly as significant.
Source: http://www.epa.gov/globalwarming/emissions/national/gwp.html.

No one doubts that the concentration of carbon dioxide is increasing, but there is great uncertainty about whether this increase has begun to produce noticeable effects or how soon such effects will appear. One thing that is agreed is that once effects begin to be noticed, it will take a very long time for corrective action to take effect. As we discussed previously, other chemicals contribute to global warming. Table 12.7 lists the ability to trap heat, called **global warming potential** (GWP), compared to CO_2, which is said to have a GWP = 1.

One factor that complicates efforts to predict what changes are likely is that the Earth seems to undergo natural fluctuations in temperature caused by processes we don't understand. Unpredictable natural phenomena such as volcanoes, which release vast amounts of dust and gas into the air, also complicate the calculations. The most dramatic natural fluctuations in temperature are those that have been responsible for periodic Ice Ages. A few scientists suggest that the additional greenhouse effect caused by modern civilization is the only thing that has prevented the next Ice Age from arriving. Attempts to correct the "problem" of global warming may therefore tip much of the world into the opposite disaster. Despite such doubts, attempts are being made to reduce and eventually reverse the steady increase in the amounts of greenhouse gases in the atmosphere. The ways to achieve this are simple enough to state: burn less fossil fuel, cut down fewer trees, and plant more young trees. To actually achieve these aims on a large enough scale to make any difference will be very difficult and will certainly require the adoption and enforcement of strict international treaties. Individual actions alone cannot bring about this reversal.

12.4 | Smog

The word "smog" was coined in London in 1905 to describe a combination of **sm**oke and **fog** that was common there when individual coal-burning furnaces and fireplaces were widely used. The smog was made worse by the high sulfur content of the fuel resulting in a great deal of sulfur dioxide (SO_2) being mixed into the London smog.

In 1952, London experienced an infamous 4-day smog that was blamed for the deaths of about 4000 people (Figure 12.17). Because it is difficult to attribute specific individual deaths to smog, the 4000-death estimate was obtained by the method of examining the statistics for **excess deaths.** These are deaths in excess of those expected during a corresponding time period without smog. Visibility during this great smog was about 20 yards at noon, there was 100% relative humidity, and there were about 4000 μg of black sooty particles per cubic meter of air. To add to the misery, these soot particles were coated with a thin liquid layer of concentrated sulfuric acid. The sulfuric acid was produced when sulfur dioxide reacted with oxygen in the air to form sulfur trioxide, followed by reaction of the sulfur trioxide with water:

$$2SO_2 + O_2 \rightarrow 2SO_3$$
$$SO_3 + H_2O \rightarrow H_2SO_4$$

Chemically, the London smog of 1952 can be described as a **reducing smog,** because it contained the reducing agents SO_2 and CO. Note, for example, that when SO_2 is oxidized in air to SO_3, the oxygen in the air is reduced (see page 154 for a reminder of the meaning of oxidation and reduction). This type of smog was most intense during the winter and at night, when furnaces and fireplaces were most used.

Modern Smog

Today's big-city smog is chemically very different from the old-style London smog. It is a reactive mixture of substances produced largely from the interaction of automobile emissions with sunlight. This means that the highest incidence of this **photochemical smog** occurs during the summer when sunlight is most intense and during rush hours when car traffic is at its heaviest. Table 12.8 summarizes the important differences between reducing smog and photochemical smog.

The Role of the Automobile

Automobile engines burn mostly C_7 and C_8 hydrocarbons and gasoline contains nearly no sulfur or nitrogen-containing compounds. Automobile exhaust contains principally carbon dioxide and water, but the emissions of concern with respect to smog are unburned or partially

Figure 12.17
In 1952, London was the site of smog so intense that visibility was 20 yards at noon. The combination of sulfur oxides, water vapor, and soot resulted in the deaths of an estimated 4000 people. Reprinted with permission from C&EN. Jorge L. Sarmiento, *Ocean Carbon Cycle.* Copyright 1993 American Chemical Society.

exercise 12.2

How Much Soot Is Inhaled?

Problem

To get a sense of how quickly small amounts can accumulate, determine the amount of soot inhaled by one person in a day during the devastating London Fog of 1952 using these data:

- We inhale about 10 times per minute.
- We inhale 1 L of air in every breath.
- There were 4000 μg (4×10^{-3} g) of soot per cubic meter of air.
- There are 1000 L of air in a cubic meter.

Solution

We are now beginning to see how important dimensional analysis can be as a problem solving tool. In this case, the essential strategy (draw your own mole map) is to convert grams in one breath to grams in one day and cubic meters to liters!

$$\textbf{g soot} = \frac{4 \times 10^{-3}\ \text{g}}{\text{m}^3} \times \frac{1\ \text{m}^3}{10^3\ \text{L}} \times \frac{1\ \text{L}}{\text{breath}} \times \frac{10\ \text{breaths}}{\text{min}} \times \frac{60\ \text{min}}{\text{hour}} \times \frac{24\ \text{hr}}{1\ \text{day}}$$

$$= \textbf{58 mg soot} = \textbf{0.06 g of soot per day}$$

This seems like a very small amount and compared to a teaspoon of sugar, which weighs 8000 mg, it is. Figure 12.18 shows 58 mg of soot. Can you see why so much damage was done to the health of Londoners in 1952?

Figure 12.18
Fifty-eight milligrams of soot. Reprinted with permission from C&EN. Jorge L. Sarmiento, *Ocean Carbon Cycle.* Copyright 1993 American Chemical Society.

Table 12.8

Differences between Reducing and Photochemical Smog		
	Reducing Smog	**Photochemical Smog**
Major polluters:	SO_2, hydrocarbons, CO	O_3, NO_x, hydrocarbons
Time of heaviest pollution:	Winter (heating needed)	Summer when sunlight is most intense and rush hours when traffic volume is highest

burned hydrocarbons and nitrogen monoxide (NO). The NO is formed during the reaction between nitrogen and oxygen from the air at high temperatures. When NO reaches the atmosphere it reacts with the oxygen of the air to form the brown gas nitrogen dioxide (NO_2). Nitrogen dioxide is a particularly effective absorber of visible light from the Sun. This absorption of light not only gives smog its distinctive brown color but also powers the decomposition of nitrogen dioxide back to NO and free oxygen atoms:

$$NO_2(g) \xrightarrow{light} NO(g) + O(g)$$

The free oxygen atoms can then react with oxygen molecules to yield ozone (O_3):

$$O(g) + O_2(g) \rightarrow O_3(g)$$

Oxygen atoms and ozone can both react with hydrocarbons such as methane to produce hydroxyl radicals (OH). Note that the hydroxyl radical is not the same as hydroxide ion (OH⁻), which has a negative charge. Hydroxyl radicals are electrically neutral molecules that are highly reactive because they carry an unpaired electron. The word **radical** is used to describe many different chemical species that carry unpaired electrons. By way of a long complex series of reactions, unburned hydrocarbons, hydroxyl radicals, and nitrogen dioxide are converted into a set of compounds characteristic of photochemical smog. The most important of these compounds is called peroxyacetyl nitrate (PAN), and has the following structure:

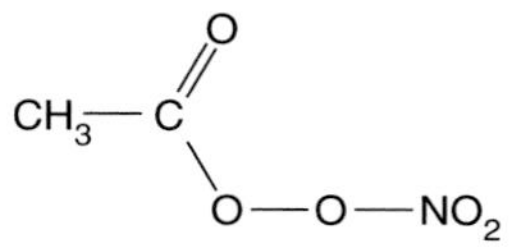

PAN

PAN and other similar compounds are powerful eye irritants, or **lachrymators** (meaning substances that cause tears). They also damage lung tissue and plants.

The Function of Catalytic Converters

The **catalytic converters** fitted to all new cars were designed to cut down on NO and CO emissions. A catalytic converter (Figure 12.19) is expected to perform a lot of chemistry in the very short time during which the exhaust is passing through it. In one part of the converter, air is added and carbon monoxide (CO) is oxidized with a palladium or platinum catalyst to carbon dioxide (CO_2):

$$2CO(g) + O_2(g) \rightarrow 2CO_2(g)$$

In another part of the converter, nitric oxide (NO) travels over a rhodium catalyst forming nitrogen (N_2) and oxygen (O_2):

$$2NO(g) \rightarrow N_2(g) + O_2(g)$$

The overall reaction, then, is summarized by this equation:

$$2NO(g) + 2CO(g) \rightarrow N_2(g) + 2CO_2(g)$$

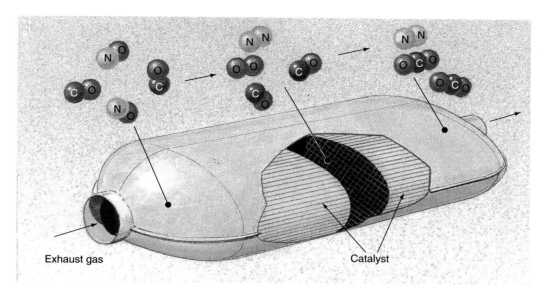

Figure 12.19
Catalytic converters are required on all new cars manufactured for sale in the United States. A converter transforms carbon monoxide to carbon dioxide, nitric oxide to nitrogen and oxygen, and unburned hydrocarbons to carbon dioxide and water by reaction with oxygen. Platinum, palladium, and rhodium catalysts permit these reactions to occur in less than a second.

Unburned hydrocarbons are oxidized to carbon dioxide and water. For all of this to happen in less than a second, all of the reactions must occur on the surface of a catalyst. The rhodium and platinum or palladium catalysts allow these reactions to occur in less than a second.

The Effects of Smog

Smog damages people, plants, and materials because some of the chemicals within smog are extremely reactive. The most notable effect on people is the irritation of the eyes, nose, and throat caused by ozone and PAN. Where industrial smog is severe, toxic chemicals such as SO_2 and H_2SO_4 can accumulate on soot, which delivers them deep into the respiratory tract. This can make people with existing respiratory problems seriously ill or even cause their death. Ozone and other strong oxidizing agents found in both kinds of smog can damage the DNA of plants. High levels of smog can reduce the rate of photosynthesis resulting in stunted crop growth and economic losses for farmers. Smog attacks the polymers found in rubber compounds, weakening the rubber and making it crack. Antioxidizing agents are added to rubber compounds before they are made into tires and other products to prevent the damage. The only possible positive effect of smog is that it may block some sunlight from reaching the Earth's surface, thus counteracting a small fraction of the greenhouse effect.

How can we lower smog levels? The most obvious answer is to use our cars less and turn to public transportation more. Using one engine to carry 50 people rather than 50 engines to accomplish the same task would greatly reduce the pollutants in the air. Some reduction in smog could also be achieved through the development of smaller more efficient automobile engines and improvements in catalytic converter technology. The issue of smog goes well beyond science and engineering and into politics, economics, and city planning.

12.5 | Particulate Matter

Our discussion up to this point would lead you to conclude that our atmosphere is composed of various gases whose composition is fairly constant over time and that human technological advances have caused a small but significant change in this composition. This conclusion is largely correct when applied to the atmosphere as a whole. However, the air close to the Earth's surface—the air that we breathe—has its own particular profile. This part of the atmosphere contains many small solid and liquid particles called **particulate matter.** This particulate matter can have a substantial impact on the health of millions, especially in urban areas. It is found in both urban and rural air and can cause us problems ranging from irritation of the eyes, nose, and throat to serious respiratory disease and possibly cancer.

Particulate matter, whether solid or liquid, is generally in the form of lumps or droplets measuring between 0.1 and 100 micrometers (μm). If the diameter of the particles is smaller than 2 μm they form what is called an **aerosol,** a mixture of tiny particles of solid or liquid suspended in a gas. Fog, smoke, and airborne volcanic dust are all natural aerosols. The effect of particulate matter on the health of animal (including human) and plant life seems to depend on the size of the particles.

Air pollution by particulate matter is not publicized as much as acid rain, gaseous emissions from automobiles, or the greenhouse effect. This may be because it is difficult to collect and study particulate matter.

Formation and Nature of Particulate Matter

Particulate matter reaches the air in three main ways:

- *Natural processes* such as the dispersal of sea salt from oceans, emissions from swamps, particles ejected from volcanoes, and dust blowing up from dry land.
- *High-temperature combustion* such as the burning of coal and other fuels. Cigarette smoke also contains particulate matter.
- *Reaction of substances already in the air to form particles* such as the reaction of sulfur trioxide with water to make sulfuric acid

$$SO_3(g) + H_2O(l) \rightarrow H_2SO_4(aq)$$

followed by the combination of sulfuric acid with ammonia gas to form ammonium sulfate solution which solidifies when water evaporates from it.

$$H_2SO_4(aq) + 2NH_3(g) \rightarrow (NH_4)_2SO_4(aq)$$

Once a particle forms, it acts as a surface upon which many different chemical reactions can occur. Figure 12.20 illustrates some of the many processes that can occur on an urban aerosol particle.

Samples of smog collected near Los Angeles in the late 1960s showed that particulate matter comes in two basic sizes—about 0.1 to 1 μm, which accounts for most of the particles present, and from 1 to 100 μm, accounting for the rest. Subsequent work has determined that the two kinds of particles have very different chemical characteristics. The small so-called "fine" particles tend to be acidic, reactive, and formed from combustion in automobile engines and factory smokestacks. Such particles are called "urban aerosols," because the concentration of automobiles and smokestacks is highest in urban areas.

The larger "coarse" particles result mainly from natural processes such as soil break-up from tillage. They account for a numerically small fraction of all the particles although they may be a significant fraction of the total mass. They are generally considered to be fairly harmless.

Urban Aerosols

Roughly half of the urban aerosol mass comes from car exhaust. Emissions from other forms of burning (smokestacks, fireplaces, etc.) are responsible for the rest. The composition of a typical urban aerosol is given in Table 12.9. Note that the aerosol contains large amounts of carbon, nitrates, and sulfates. Recall that automobile and smokestack emissions pro-

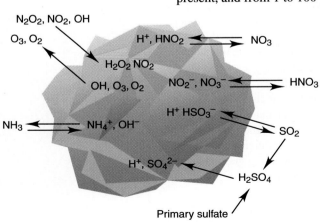

Figure 12.20
An urban aerosol surface can be a home for a variety of complex chemical processes, some of which are illustrated here. Reprinted with permission from *Science,* p. 749, February 10, 1989. Copyright 1989 American Association for the Advancement of Science.

duce these substances. Also note the wide variety of metals present, due mostly to the deterioration of metal and concrete items. The concentration of lead in modern urban air is much lower than it was 30 years ago when gasoline contained a lead-based additive to enhance its burning properties. Finally, note the total mass concentration of particles compared to that of air. Aerosols are truly "trace" components that nonetheless have real health consequences.

Health Concerns Resulting from Exposure to Particulate Matter

Fine hairs and tiny hair-like projections called cilia in our noses and lungs filter large (typically greater than 10 μm) particulate matter from the air we breathe. Large soot particles, for example, generally do not reach the lung surface to cause damage. On the other hand, smaller particles such as small soot particles and asbestos fibers can reach the sensitive lining of the lungs to cause a variety of respiratory ailments. Similarly, the small size of urban aerosols allows them to penetrate to nasal passages, the throat, and lungs causing irritation and inflammation and making any preexisting conditions such as emphysema or asthma worse. Recent studies have suggested that urban aerosols may be **mutagenic,** meaning they may cause the changes ("mutations") in DNA that can lead to cancer. The evidence for such a role, however, remains uncertain. What is clear is that increases in the concentration of urban aerosols bring a proportionate increase in their adverse health effects.

Again, we face difficult choices. Should we take action to decrease the concentration of particulate matter in the air? If so, what choices are we prepared to make? Stop driving? Scale-down factory production? Will better particulate control devices become available? Should there be economic incentives to encourage their use?

Table 12.9

Composition of a Typical Urban Aerosol*

Component	Concentration ng/m³
(Compared to air with a density of roughly 1.5×10^{12} ng/m³)	
C	16,000
Na	800
Al	200
Si	600
P	200
S	4,000
K	200
Ca	200
Ti	58
Mn	12
Fe	400
Ni	8
Cu	14
Zn	116
Br	90
Sr	49
Pb	558
NH_4^+	4,500
NO_3^-	5,300
SO_4^{2-}	10,000
Others	16,000
Total mass	60,000

*There is at least a 50% relative average deviation associated with each value.

Note: ng = nanogram = 1×10^{-9} gram; m³ = cubic meters.

12.6 | In Conclusion: Chemistry, A World of Choices

Air pollution reflects, as much as anything, the fact that "it's a small world." The atmosphere surrounds the entire world and, as we saw in our discussion on acid rain (Chapter 9), activities in one place by one group of people can soon become problems in other places for other groups of people.

We also know that, when scientific evidence is compelling, humankind will often act to try and correct the problem. We can act globally as with some of the worldwide initiatives discussed in the chapter, or act locally. In 1960, the ozone level in Los Angeles exceeded the federal government safe standard of 0.12 parts per million parts of air on 4 of every 5 days. In 2000, due to extremely stringent pollution controls on everything from industrial emissions to lawn mower emissions, the safe standard had been exceeded on "only" 1 out of every 8 days. This has occurred while the number of vehicles in the area has tripled and new industries have moved to Los Angeles.

Yet there is still much to do. While Los Angeles and other large metropolitan areas like Mexico City have made choices that have greatly improved air quality, other cities have not been as aggressive. Figure 12.21 shows the number of days, between 1990 and 1999, in which several large communities in the United States had an **air quality index (AQI)** greater than

Green Chemistry

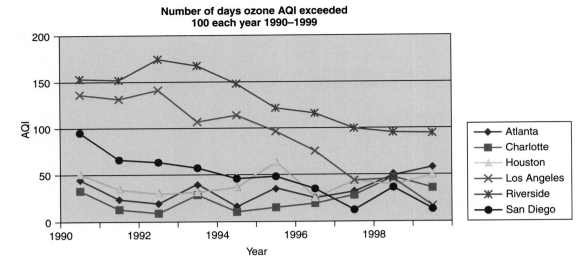

Figure 12.21
Pollution regulations enacted by individual cities can sharply change their air quality, as these data show.
Source: EPA Office of Air and Radiation

100. The AQI is an EPA measure related to surface ozone concentration, and a value greater than 100 is especially troublesome.

On balance, then, there is a rainbow's end that can finally be seen after the long period of dreary pollution data. The EPA's "Latest Findings on National Air Quality: 1999 Status and Trends" (http://www.epa.gov/oar/aqtrnd99/brochure/brochure.pdf) reaches several important conclusions about the state of our air:

- The average concentrations of every one of the six most important air pollutants, including carbon monoxide, lead, nitrogen dioxide, sulfur dioxide, surface ozone, and particulate matter, have decreased between 1979 and 1999.
- In 1999, over 150 tons of air pollutants were released nationwide.
- While air quality is improving in many cities, it is getting worse in some rural areas, occasionally exceeding the levels of nearby cities.
- The Clean Air Act has been an unqualified success because average pollution levels that increased between 1900 and 1970 have decreased since.
- Emissions of toxic substances such as benzene, C_6H_6, and perchloroethylene, C_2Cl_4 (used to make HFCs), have decreased by an average of 23% nationwide between 1990 and 1996.
- The concentrations of ozone-depleting substances measured in the stratosphere are beginning to decrease.

Will the good news continue? That depends on our choices—personal choices about how we get from place to place, public policy choices concerning the extent to which we are willing to regulate the activities of industries, and agreement among nations that air quality is a global concern.

main points

- Air pollution is a global issue meaning it affects nearly everyone on Earth.

- Air pollution has a larger impact in urban than rural areas.

- The atmosphere has changed over millions of years from one lacking oxygen to one that is oxygen-rich, due to the presence of life.

- Ozone is an important constituent of the stratosphere where it shields us from harmful amounts of UV radiation.

- CFCs and nitrogen oxides act to destroy ozone in the stratosphere.

- Nitrogen oxides, largely a result of automobile emissions, can lead to the formation of ozone at the Earth's surface. This can be harmful to the human respiratory system.

- Greenhouse gases are those that transmit visible radiation while absorbing IR radiation. This results in heat building up in the atmosphere like in a gardener's greenhouse.

- Carbon dioxide is the most important greenhouse gas. It is produced naturally by respiration and consumed by photosynthesis, among other processes. Nature's balance of carbon dioxide concentration has apparently been altered by the industrial revolution.

- Particulate matter is composed of tiny solid or liquid particles that result from a number of natural and industrial processes.

- Although there is some doubt among scientists about the specific mechanisms that give rise to various kinds of pollution, it is generally agreed that human technological activities have adversely affected the quality of the atmosphere.

important terms

Aerobic, requiring oxygen. (p. 391)

Aerosol, tiny particles suspended in air. (p. 400)

Air pollution is the result of something entering the atmosphere usually as a result of human activity, that has a harmful effect on some aspect of the environment. (p. 395)

Air quality is the description of deviation of air composition from that of pure air. (p. 392)

Allotropes are different molecular or crystalline forms of a pure element. (p. 394)

Anaerobic, not requiring oxygen. (p. 391)

Basin environment is a geographical region surrounded at least partially by hills. (p. 387)

Biomass is plant material such as wood, grasses, or leaves. (p. 404)

Catalytic converter is a device used on automobiles to react CO and NO into CO_2 and N_2. (p. 410)

CFC is the abbreviation for chlorofluorocarbon. (p. 399)

Chlorofluorocarbon is a small molecule containing only carbon, chlorine, and fluorine. (p. 399)

Excess deaths, statistically, more deaths than are expected in a given time period. (p. 408)

Global warming is the process by which the atmosphere and the surface of the earth show an increased temperature. (p. 403)

Haber process is the Nobel prize–winning process for making ammonia from hydrogen and nitrogen. (p. 394)

Halon is a small molecule containing carbon and bromine in addition to chlorine and/or fluorine but no hydrogen. (p. 399)

Lachrymator is a compound whose vapors cause copious tears to form. (p. 410)

Mutagenic is causing change in the DNA genetic material of an organism. (p. 413)

Natural gas, mostly methane (CH_4), is used as a fuel and raw material. (p. 404)

Oxide is a compound of oxygen with any other element. (p. 395)

Ozone is an allotrope of oxygen having three atoms per molecule, O_3. (p. 394)

Ozone hole is the dramatically lowered concentration of ozone in regions of the upper atmosphere. (p. 400)

Ozone layer is the region in the stratosphere in which O_3 is formed by the reaction of O_2 with sunlight and which diminishes the amount of UV radiation reaching the Earth's surface from the Sun. (p. 392)

Particulate matter are small pieces of solid or liquid suspended in the air. (p. 412)

Photochemical smog is a mixture of chemicals formed from interaction of sunlight with unburned gasoline and nitrogen monoxide formed mostly in automobile exhaust. (p. 409)

Radical is an atom or molecule having an unpaired electron. (p. 410)

Reducing smog is smog containing high concentrations of SO_2 and CO. (p. 409)

Thermal inversion is when cool air cannot rise through the warm air above (p. 387)

exercises

1. What is thermal inversion?
2. Define aerobic and anaerobic. Give an example of an organism of each type.

3. What evidence do we have to prove oxygen has been present in our atmosphere for only 3.5 billion years?

4. Why is ozone regarded as desirable in the stratosphere but a serious pollutant near the Earth's surface?

5. Why is depletion of the ozone layer so dangerous for the Earth? Does this depletion mean there is too much or too little oxygen in the atmosphere? What would happen if the ozone layer were completely removed?

6. Name four gases that contribute to air pollution. How does each gas contribute?

7. List four human activities that contribute to air pollution.

8. How was the "ozone shield" created? What is its function?

9. Air is comprised mainly of two main gases. What are they?

10. List three atmospheric trace gases.

11. What is the overall equation for ozone formation? Why do we not run out of ozone?

12. Why is N_2 such a stable molecule?

13. Are humans the only source of gaseous pollutants and particles found in our atmosphere?

14. Why do nitrogen and oxygen exist in the atmosphere as diatomic molecules (i.e., N_2 and O_2) rather than single atoms (i.e., N and O)?

15. The most reactive major gas in the atmosphere is _____.

16. How does the combination of N_2 and O_2 lead to air pollution? What is photochemical smog?

17. Why is CO_2 a greenhouse gas?

18. Would you categorize combustion as an exothermic or endothermic reaction?

19. List four sources of natural air pollutants (see Table 12.3).

20. What two factors have led to a decline in air quality?

21. How can the ocean be regarded as a source of air pollution?

22. Compare the Lewis structures of oxygen and ozone. Based on structure, why do you think ozone is more reactive than oxygen?

23. In which layers of the atmosphere is ozone found?

24. What volume of ozone at 740 torr could be formed by reaction of 50.0 L of pure oxygen in a tank at 20°C and 30 atmospheres pressure?

25. What is the difference between stratospheric ozone and surface ozone? What is the importance of stratospheric ozone?

26. Why is the series of reactions that cause the destruction of ozone by CFCs considered catalytic?

27. If CFCs are nontoxic and chemically inert, why are they so harmful to the atmosphere?

28. What is the main cause of surface ozone?

29. If there is much higher ozone concentration in the stratosphere than in urban surface areas even during times of high pollution, do we need to be concerned about tropospheric ozone? Why or why not?

30. Using a series of reactions, show how the Halon CF_3Br catalyzes the formation of oxygen from ozone. Can this transformation occur in the dark?

31. Why do communities declare health alerts when large amounts of ozone are found in the air on hot days?

32. Describe the greenhouse effect and its causes. Is the greenhouse effect itself necessarily bad? What are the problems associated with the greenhouse effect?

33. What is the definition of a greenhouse gas? Based on the properties of greenhouse gases given in the text, state two examples of molecules that act as greenhouse gases.

34. Name three important greenhouse gases. How is each related to global warming?

35. There is scientific evidence that CFCs contribute to the depletion of ozone. Why, then, is the use of CFCs not banned over the entire world?

36. What is the difference between smog and fog formed from water vapors?

37. State two differences between reducing smog and photochemical smog (see Table 12.8).

38. Design a laboratory experiment in which you make reducing smog.

39. Why is photochemical smog prevalent in the summer?

40. Draw the Lewis structure of NO_2. Why do you think nitrogen dioxide is a good absorber of visible light?

41. Explain the chemistry taking place in a catalytic converter.

42. Nitrogen and oxygen react at very high temperatures of lightning bolts to form NO, which then reacts with oxygen to form NO_2. Describe how this process is similar to what happens in an internal combustion engine to produce the air pollutants NO and NO_2.

43. What are the three primary routes for the formation of particulate matter? Give an example of a source for each.

44. What are the three main components of aerosols?

45. Why are aerosols so chemically active?

46. Using Table 12.9, give potential sources of the top three compounds found in urban aerosol.

47. Two different sources were used to compare the concentrations of ozone at ground level and in the upper atmosphere. One source quoted a highly polluted ground level concentration as being 0.5 ppm. Another source quoted the upper atmosphere concentration of ozone as 5×10^{12} molecules/cm^3. Convert 0.5 ppm to molecules/cm^3 to be able to compare these concentrations.

48. Which is the greater concentration of ozone: 5×10^{12} molecules/cm^3 in the stratosphere (20 km altitude, density of air = 1×10^{-5} g/cm^3) or 8×10^{12} molecules/cm^3 at the surface (density of air = 1.2×10^{-3} g/cm^3)?

49. Ozone is produced from oxygen at water treatment facilities for use as an oxidant in destroying pathogenic organisms and oxidizing iron(II) to iron(III) and manganese(II) to manganese(IV). Write a balanced equation for the oxidation by ozone of iron(II) while all the oxygen atoms form water.

50. Atmospheric pressure is related to altitude by this equation, where P = atmospheric pressure in atmospheres, h = altitude in kilometers, and T = temperature in kelvins.

$$\log P = -14.8 \, h/T$$

Find the pressure of the atmosphere at the altitude of many commercial jet planes, 30,000 feet where the temperature is approximately $-56°C$.

51. Why do you suppose that Freons cause ozone depletion through their chlorine atom rather than their fluorine atom?

52. Suppose that all the noble gas elements from 1000 L of air at 1.0 atmospheres and 25°C are separated from the rest of the air. What volume will the combined noble gases occupy at these conditions? Suppose further that each of the noble gases were separated from all the others. What volume of each gas would result?

53. Suppose that all the noble gas elements from 1000 L of air at 1.0 atmosphere and 25°C are separated from the rest of the air. What would be the experimental molar mass of this mixture of noble gases?

54. The temperature of the upper atmosphere changes dramatically. The highest region of the atmosphere, the thermosphere has a temperature of about 1200°C at an altitude of about 500 km. Explain why the temperature is so high at such a high altitude.

55. What could cause the termination of the catalysis by Cl atom of the destruction of stratospheric ozone?

56. What would be wrong with collecting ozone from the atmosphere of our cities and flying it up to the stratosphere where it would be off-loaded. If that doesn't work, why not use high-flying airplanes to make ozone high in the atmosphere?

57. Nitrogen and hydrogen, the reactants used to make ammonia via the Haber process, are only slightly soluble in water but ammonia is highly soluble in water. Explain.

58. Haber's original goal in making synthetic ammonia (1912) was for use as a fertilizer to increase food production but much of the ammonia made in early years was used to make gunpowder and other explosives for use in the First World War. Look up the reactions necessary to convert ammonia into explosives.

food for thought

59. As discussed in the text, hydrocarbons that cause smog are oxidized by oxygen in catalytic converters to produce carbon dioxide. However, we have learned that carbon dioxide contributes to global warming. Is the barter of hydrocarbon for carbon dioxide worthwhile? Use a risk versus benefit analysis to determine the value of each side (i.e., damage caused by the production of PAN if converters are not used versus the damage caused by addition of excess carbon dioxide to the atmosphere).

60. Did you know that in 1 hour a lawn mower will give off as much smog-causing hydrocarbon as a 1-hour drive in a car? Worse yet, 2 hours of chain saw use results in hydrocarbon emission equal to that given off during a 3000-mile trip in a new car. According to the California Air Resources Board, the annual hydrocarbon emissions of lawn mowers, chain saws, outboard engines, construction tractors, and other such equipment in the state equals the output of 3.5 million 1991-model cars, each driven 16,000 miles. Battery-operated electric mowers are now available. These battery-operated mowers are very heavy and pose another slight inconvenience for the customer—a full charge takes 20 hours, while an 80% charge takes only 4 hours. As stated in the text, establishing a safe ozone balance will require both individual and social, not merely scientific choices. What choices are you willing to make with regard to buying electric versus conventional gasoline-powered tools?

61. Following the record-breaking heat and drought during the summer of 1988, public concerns about global warming increased dramatically. The government also focused on this topic as numerous hearings and articles about "the greenhouse effect" appeared in newspapers and publications the following summer. Why did humans begin to notice global climate change only after their own lifestyle was threatened? Would threatening environmental issues exist if we could spark public attention before they became problems? What role do the media and politics play with regard to public perception of scientific data?

62. In general, the industrialized nations of this world agree that they must put ecology ahead of economic survival. The United States has not wholeheartedly adopted this policy. Although it contains less than 5% of the world's population, the United States emitted over 20% of the world's carbon dioxide in 1999. The average North American uses enough energy during the day to emit 4.8 metric tons of CO_2 compared to 3.0 metric tons per capita in Industrialized Asia, 2.7 metric tons in Western Europe, and 0.6 in China. The Japanese steel-making industry has adopted new electric-arc furnaces that cut carbon dioxide emissions in half and energy needs by two-thirds. In the United States, 70% of the steel is still made the old way using, on average, 22 million British thermal units (Btu's) of energy versus the 17 million used in Japan. Companies like Ford argue that the gap between savings in electricity bills due to energy-efficient devices and the payback covering the initial costs would take away from the development of competitive products. Should U.S. industry risk enhancing their worldwide competitiveness for ecological efficiency? Do you know of examples in which U.S. industries are putting "ecology first"?

63. We have proposed several questions at the end of this chapter regarding the economics of air pollution control. Find out what issues are pressing in the community

where you live (read the newspaper, watch the local news, write to agencies, etc.). Would you rather have your tax dollars spent on one of these issues or on new clean air technologies?

readings

1. Elmer-DeWitt, Phillip. A drastic plan to banish smog. *Time,* March 27, 1989, p. 65.
2. Carpenter, Betsy. The newest health hazard: Breathing. *U.S. News and World Report,* June 12, 1989, pp. 50–53.
3. Schneider, Stephen H. The changing climate. *Scientific American,* September 1989, pp. 70–79.
4. Jacob, Anthony T. Chlorofluorocarbons and the hole in the ozone layer. Institute for Chemical Education, University of Wisconsin, Madison, 1991.
5. Chiras, Daniel D. 1991. *Environmental Science,* Menlo Park, CA: Benjamin Cummings.
6. Tibbs, Hardin, B. C. Industrial ecology: An environmental agenda for industry. *Whole Earth Review,* Winter, 1992, pp. 4–19.
7. Wald, Matthew L. Lawn mower is new target in war against air pollution. *New York Times,* August 6, 1992.
8. Rensberger, Boyce. Taking Earth's temperature. *Washington Post, National Weekly Edition,* June 14, 1992.
9. The climate's wild mood shifts. *Washington Post, National Weekly Edition,* June 14, 1992.
10. Sarmiento, Jorge L. Ocean carbon cycle. *Chemical and Engineering News,* May 31, 1993, pp. 30–43.
11. Monastersky, Richard. Ozone on trial: Congress gives skeptics a day in the sun. *Science News,* October 7, 1995, p. 238.
12. Baird, Colin. 1995. *Environmental Chemistry,* New York: W. H. Freeman and Co.
13. Zurer, Pamela S. Complexities of ozone loss continue to challenge scientists. *Chemical and Engineering News,* June 12, 1995, pp. 20–23.
14. Hilleman, Bette. Web of interactions makes it difficult to untangle global warming data. *Chemical and Engineering News,* April 27, 1992, pp. 7–19.
15. Ayres Jr., B. Drummond. California smog cloud reveals a silver lining. *New York Times,* November 3, 1995, p. A7.
16. Cushman, Jr. John H. Why the U.S. fell short of ambitious goals for reducing greenhouse gases. *New York Times,* October 20, 1997.
17. Fisher, D. E., and Fisher, M. J. The nitrogen bomb. *Discover,* April 2001, pp. 50–57.
18. Intergovernmental Panel on Climate Change (IPCC) Summary for Policy makers (2001) http://www.meto.gov.uk/sec5/CR_div/ipcc/wg1

websites

www.mhhe.com/Kelter   The "World of Choices" website contains activities and exercises including links to websites for: a discussion by the Asia Pacific Space Center relating to Saturn's moon Titan; a simulation called "Smog City"; and much more!

The Earth as a Resource

There's gold in them thar hills!

—Countless old movies about the western U.S.

Lincoln, Montana, is in a beautiful mountainous region of forested hills drained by a marvelous trout stream. The bedrock beneath this beautiful area contains a fraction of an ounce of gold per ton of rock. Extraction of the gold requires that the forest be chopped down, the soil stripped away, and the rock ground to a powder and soaked in a cyanide bath to dissolve the gold. All the gold can be removed in about 15 years. Then all the mining-related activity will gradually shut down. However, for those 15 years, many people will earn a living in the gold mining industry. What are the risks and benefits of such an activity? How would your view change if you were a mine owner? A mine worker? The owner of land nearby?

Debate over this issue is going on right now in and around Lincoln, Montana, where plans are going forward to recover the gold in an area near the headwaters of the Blackfoot River. The benefits of going ahead are economic, at least in the short term. Lots of people will make good wages mining the gold and others will prosper from providing necessary services to the mine workers. The risks are both economic and environmental. Services such as schools, housing, and improved roads will be necessary during the boom phase of the project but will become unnecessary 20 years after the gold recovery has ceased and miners move away.

At the moment, lawsuits and the low price of gold appear to have stopped any plans to proceed with the mine. Later in this chapter we will discuss the methods for mining low-quality ore of this sort and the environmental impact that led the citizens of Montana to vote in opposition to expanding such mining in their state.

The dilemma facing the residents in and around Lincoln, Montana, is not all that unique in the United States, or in fact in many places throughout the world. Recent history includes "gold rushes" in California in the 1840s and in Alaska in the 1880s.

Lincoln, Montana, is the site for a proposed, new gold mine.

As you can see, the new gold discovery is located in the middle of "Big Sky" country. The conflict between preserving the natural beauty and reaping the economic rewards of a major gold strike faces the residents of the entire area.

Lincoln, Montana

Countries have made clear choices that the resources for which we mine are beneficial enough to take risks in getting them.

All the material substances that we use in all our societies come from the crust, the oceans, and the atmosphere of the Earth. We have already discussed the oceans (Chapter 7) and the atmosphere (Chapter 12) of the Earth. Let's now examine the Earth's crust (see Table 13.1).

13.1 | Introduction

The Earth, as was discussed in Chapter 1, has a spherical core made mostly of iron. This iron settled to the center when the entire Earth was molten early in its history. The central part of the core is a sphere of solid iron about 1200 km in radius, surrounded by a layer of molten iron about 2300 km thick (Figure 13.1). This iron accounts for about 35% of the mass of the Earth but it is inaccessible to us because between us and it is a layer of rock nearly 2800 km thick. This layer is the Earth's mantle. The crust of the Earth, like that of a loaf of bread, is a thin layer (about 40 km thick) at the surface of the planet. The crust and the top part of the mantle can be thought of as rafts floating on semi-liquid molten rock a bit deeper (50-100 km below

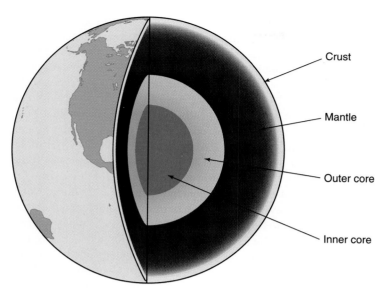

Figure 13.1
The Earth is a series of concentric layers with a very hot iron core in the center and a crust of rock at the surface.

Table 13.1

Average Composition of Earth's Crust

Elements	Percent
Oxygen	46.60
Silicon	27.72
Aluminum	8.13
Iron	5.00
Calcium	3.63
Sodium	2.83
Potassium	2.59
Magnesium	2.09
Titanium	0.44
Hydrogen	0.14
Phosphorus	0.12
Manganese	0.10
Total of all others	0.61
Including	
Copper	0.0070
Gold	0.0000005

Note: The elements listed here are the most common in the Earth's crust expressed in percent of weight. These numbers refer to rocks and soil, not oceans or atmosphere.

the surface) into the mantle. Big chunks of this floating mass, called plates, move very slowly in relation to each other. At the boundaries of these plates are volcanoes, earthquakes, and minerals coming up from the mantle to form new crust.

As we said back in Chapter 1, the Earth is a materially closed system. All the material that we can use from the Earth has been here since the formation of the Earth and will be here as long as the Earth exists. Wise management will allow us to continue to make constructive use of our resources well into our great-grandchildren's futures. The word **resource** is an interesting word. Since all the copper, for instance, present on Earth today will always be here, what can we consider to be a useful copper resource? A deposit of mineral that contains copper, even at very low concentrations, is said to be a copper resource. Only, however, if the copper can be won from the ore *economically* is the deposit said to be a **reserve** of copper. When we

exercise 13.1

The Impact of Metals

Problem

Here's one to "get you moving"—please name five things that you use every day that contain metal.

Solution

Here's our quick list:

*   Bicycles (the best ones contain titanium)
*   Cars (chromium, iron, palladium, rhodium)
*   Watches, soda cans, pens, computer disks, everything with batteries!

The list we have generated takes into account only a few of the uses of metals. The uses that you picked extend the list and begin to show the intimate relationship between our quality of life and our ability to acquire metals from the earth.

Figure 13.2
Manganese nodules litter regions of the bottom of the sea.
Though a valuable mineral, their location makes them
inaccessible.

speak of our reserves of a metal, we mean the amount that is still in the Earth in discovered places in circumstances that allow its economical utilization. If the price goes up or improved recovery technology is developed, deposits not previously designated reserves suddenly can be regarded as reserves. At the turn of the 20th century, the lowest grade of copper ore that could be mined economically was about 3% copper. Now, higher prices and improved technology allow mining of deposits with as little as 0.2% copper. Recall our discussion of the economics of recycling in Chapter 4; scrap wire is nearly 100% copper so it is a wonderful source of copper for new uses.

Mineral resources are nonrenewable (compare this to wood, which is renewable in the form of trees). Geologic processes are far too slow to keep up with human usage. If all known reserves of an element are used up, we must discover new deposits, find ways to convert resources into mined metal, substitute something else for its usage, recycle current products, or do without.

An example of how new reserves might come to be recognized in the future features the so-called "manganese nodules" (Figure 13.2), which are spherical lumps, typically baseball sized, known to litter regions of the ocean floor. They are rich in manganese, iron, copper, nickel, cobalt, and other metals. They are waiting to be picked up but are under more than a mile of water so the challenge to utilize this resource is very great.

Since these nodules are under open ocean, they are not in the territory of any country. International agreements require that the wealth from these nodules be shared if technology is ever developed to convert this potential resource into a reserve.

In this chapter we will be content to describe just four metals crucial to civilization: gold, copper, iron, and aluminum as well as the chemistry needed to *win* them from the Earth. Yet there are other important resources obtained from the Earth: petroleum, coal, and even soil itself. We will briefly visit the importance of each of these. The word "win" is actually used to describe the purification of a metal from its ore—does this mean that if society wins that the Earth loses?

About 75% of the elements of the periodic table are metal (Figure 13.3). A few metals, gold being a good example, are found uncombined in nature, but most are found in water-

Figure 13.3
Periodic chart with metals
highlighted in purple.

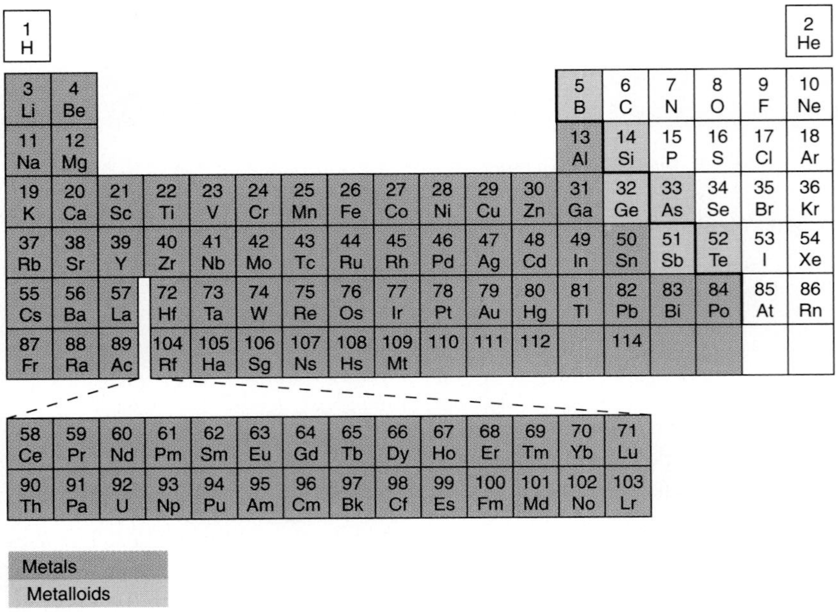

1 H																	2 He
3 Li	4 Be											5 B	6 C	7 N	8 O	9 F	10 Ne
11 Na	12 Mg											13 Al	14 Si	15 P	16 S	17 Cl	18 Ar
19 K	20 Ca	21 Sc	22 Ti	23 V	24 Cr	25 Mn	26 Fe	27 Co	28 Ni	29 Cu	30 Zn	31 Ga	32 Ge	33 As	34 Se	35 Br	36 Kr
37 Rb	38 Sr	39 Y	40 Zr	41 Nb	42 Mo	43 Tc	44 Ru	45 Rh	46 Pd	47 Ag	48 Cd	49 In	50 Sn	51 Sb	52 Te	53 I	54 Xe
55 Cs	56 Ba	57 La	72 Hf	73 Ta	74 W	75 Re	76 Os	77 Ir	78 Pt	79 Au	80 Hg	81 Tl	82 Pb	83 Bi	84 Po	85 At	86 Rn
87 Fr	88 Ra	89 Ac	104 Rf	105 Ha	106 Sg	107 Ns	108 Hs	109 Mt	110	111	112		114				

58 Ce	59 Pr	60 Nd	61 Pm	62 Sm	63 Eu	64 Gd	65 Tb	66 Dy	67 Ho	68 Er	69 Tm	70 Yb	71 Lu
90 Th	91 Pa	92 U	93 Np	94 Pu	95 Am	96 Cm	97 Bk	98 Cf	99 Es	100 Fm	101 Md	102 No	103 Lr

Metals
Metalloids

insoluble ionic compounds. Deposits of minerals from which metals can be recovered profitably are called **ores**. Mineral names often differ from chemical names for the same compound. For instance, Fe_2O_3 is called iron(III) oxide or ferric oxide by chemists but is **hematite** to geologists. We will try to use chemical names whenever possible in this text.

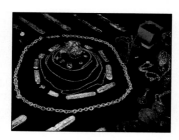

13.2 | Gold (Au)

Gold was one of the first metals to be won from the Earth. Before the Iron Age, before the Bronze Age, gold was prized, not for its utility, but for its enduring beauty (see Figure 13.4). Since most civilizations place a high value upon gold, it has been the basis of monetary systems for millennia.

Gold has chemical and physical properties that help to explain this fascination. Gold is chemically highly unreactive, it is extremely easy to work into any shape imagined by an artist, and it has a characteristic yellow (gold!) color. Since gold is so chemically unreactive, it often occurs in nature uncombined with other elements. That is, much gold occurs as nuggets of nearly pure gold. It is often found in small rivers where it has accumulated after washing out of nearby mountains. Water is necessary for human life, so people have always spent time near reliable sources of good water. The bright yellow luster of gold nuggets in a stream would certainly attract the eye of anyone going there to fetch water. Once a few pieces had been accumulated by a collector, it would be fairly easy to hammer them into shape. Gold ornaments have been found that date back to the Neolithic Age (late Stone Age). Gold's near immunity to chemical reactions allows gold artifacts to remain unchanged even if buried in the earth or on the bottom of the sea.

Figure 13.4
Gold objects have been prized by many cultures for centuries.

Modern Gold Mining

A lot of the gold from California and Colorado in the 19th century was found by miners who tore up the countryside to get it. Even today, streams in the gold-mining areas are completely denuded of vegetation because of the voracious search for gold.

We all have mental images of prospectors panning for gold in the gravel of western streams. Gold panning, in which ore is washed and repeatedly rinsed to separate out the gold, is successful because the density of gold (19.3 g/cm^3) is much greater than that of the gravel (typically 2.5 g/cm^3). The gold remains in the pan while the less dense gravel is swirled up and out. The various gold rushes of the late 1800s brought thousands of fortune seekers to California, Colorado, and other western states. Most of them spent long hours digging in the bottoms of streams.

■ case in point

The Gold Standard and Development of Coinage

For centuries, gold was the ultimate coinage metal but it wasn't until very late in the 19th century that the "gold standard" was made official in most western nations. The United States began using the gold standard in 1900. This meant that a paper U.S. dollar could, on demand, be exchanged for a specified amount of gold, usually in the form of gold coins, which were also in general circulation. The rate of exchange was set then to be $20.67 per troy ounce of gold. A troy ounce is similar to a regular avoirdupois ounce but not quite the same. There are 12 troy ounces in a troy pound whereas there are 16 avoirdupois ounces in an avoirdupois pound. A troy ounce = 31.103 grams and an avoirdupois ounce = 28.349 grams.

Other nations had their own gold exchange rates. This meant that, effectively, the entire western world used a single monetary system that remained unchanged for many years. Since any currency could be converted into gold, any currency could be exchanged for any other at a fixed rate of exchange. For instance, the British pound was worth U.S. $4.86.

(continued on next page)

This entire system fell apart during the economic collapse associated with the First World War. Attempts were made between 1919 and 1928 to reestablish much the same system as had been in place before the war, but the same stability was never obtained. The worldwide economic depression beginning in the late 1920s forced the permanent abandonment of the gold standard. Other nations' exchange rates also started changing so it became profitable to buy cheap U.S. gold at $20.67/oz. and sell it in other countries for currency that had more purchasing power than the $20.67 spent to buy the U.S. gold. One of Franklin Roosevelt's first acts as president (in 1933) was to change the value of the U.S. dollar to $35.00 per troy ounce of gold and to forbid the private holding of gold by U.S. citizens. All existing gold coins had to be turned in for paper money or silver coins. This action was taken to combat the loss of U.S. gold reserves.

Even at this new price, it soon was not economically reasonable to produce gold in the U.S. Gold simply cost more than its worth in dollars to mine and purify, so gold mining almost stopped. In 1975, the United States stopped fixing a price of gold and allowed gold to be sold like any other commodity for whatever price was agreeable to both buyer and seller. Today, at the time of writing (2002), the price of gold hovers around $270/troy ounce.

Figure A
The price of gold in the United States was determined by law for many years. From 1975, when government restrictions were lifted, the price rose dramatically to reflect the global trading environment.

Gold nuggets (called placer gold) settled in the stream beds, having been washed down from the high mountains when gold-containing rocks broke up. The so-called "hard-rock" gold miners went burrowing into mines blasted into mountainsides looking for the mother lode of gold from which the placer gold had come.

The richest deposits of gold in America have seemingly all been found and exploited by our ancestors. Most of the gold found in those years is still in circulation as coins, bullion, or objects of art. Gold mining continues, however, in towns such as Lincoln, Montana, and total production in recent years has been even greater than during the years of the great gold rushes.

Most of the gold mined nowadays is found either as tiny grains of gold immobilized in rock, usually quartz, SiO_2, or as gold telluride, $AuTe_2$. The rock, frequently containing as little as 0.0025% gold, is powdered and treated ("leached") with sodium cyanide solution in the presence of air. This causes one of the few reactions of gold to occur, as:

$$4Au(s) + O_2(g) + 8NaCN(aq) + 2H_2O(l) \rightarrow 4Au(CN)_2^-(aq) + 4OH^-(aq) + 8Na^+(aq)$$

Up until 1964, U.S. coins were made of **coin silver:** an alloy of 90% silver and 10% copper. In 1964, dimes, quarters, half-dollars, and silver dollars began to be minted in less expensive metals. Present-day coins of these denominations are a sandwich of metals, with the U.S. dime, quarter, and half-dollar composed of nickel clad on copper in a relative weight percent of 8.33% to 91.67%. An interesting problem arose in the selection of the new coin composition. Many coin-operated vending machines check the weight of the coin to assure that real money is being used. The new coins had to be the same size as the old and, if they had a density very different from the old silver coins, these vending machines would no longer function and huge costs would be incurred to make them work again.

The new Sacagawea dollar presented its own challenges. In addition to the golden color, the coin needed to have the same size, weight, and especially, electrical conductivity as other dollar coins so that vending machines would recognize it as they had the Susan B. Anthony dollar coins. The new Sacagawea coin meets these criteria with its core of pure copper to which is clad on each side an alloy of 77% copper, 12% zinc, 7% manganese, and 4% nickel.

Detractors of the policy of using less expensive, "base" metals in coinage used the term "debasement." For reasons related to the debasement of silver coins, pennies minted after 1983 are a sandwich with copper on the outside and zinc on the inside whereas earlier pennies are an alloy of about 90% copper. This was done when the value of the copper in a penny exceeded one cent and pennies were being taken out of circulation, melted down, and the copper used for other purposes (see Figure B).

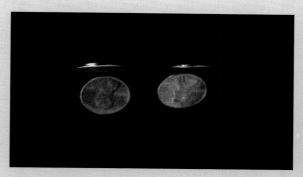

Figure B

Pennies minted after 1983 are a sandwich with copper on the outside and zinc on the inside. Pennies minted earlier are an alloy of about 90% copper. The change occurred when the value of the copper in the penny exceeded one cent. Pennies were being taken out of circulation, melted down, and the copper sold for other purposes. The value of zinc is much lower than the value of copper. When the edge of a post-1983 penny is filed away, the silver-colored zinc interior is revealed. When this filed penny is placed in aqueous HCl, the zinc dissolves and leaves behind the very thin copper "head" and "tail" of the coin.

The gold ore is allowed to sit for several days in contact with sodium cyanide solution in large ponds while air is bubbled through the solution (Figure 13.5). When enough time has elapsed, the solution which contains the gold cyanide ion, $Au(CN)_2^-$, is separated from the remaining powdered rock. Elemental gold is formed by reducing the gold cyanide complex, usually with zinc metal:

$$2Au(CN)_2^-(aq) + Zn(s) \rightarrow 2Au(s) + Zn^{2+}(aq) + 4CN^-(aq)$$

All this cyanide ion is an environmental problem because some of it is never recovered and cyanide is very poisonous. Its property of being poisonous is related to its ability to extract gold from powdered rock—cyanide in living things ties up (complexes) metal ions which are necessary for life.

The economics of winning the gold from this ore is marginal. Profitability depends strongly on the price of gold. The reserve in Lincoln is 205 million tons of ore at a price of $375/oz out of a total resource of 414 million tons. At the higher price of $600/oz more of the

The purity of gold used in jewelry is described as so many karats. Pure gold is 24 karats. Gold used in jewelry is typically either 18 karats (18/24 parts gold) or 14 karats (14/24 parts gold). This means 18 karat gold is 75% gold and 14 karat gold is about 58% gold. The remaining material is usually either silver or copper.

Figure 13.5
Gold is dissolved from its ore by dissolving in huge ponds filled with solutions of cyanide ion.

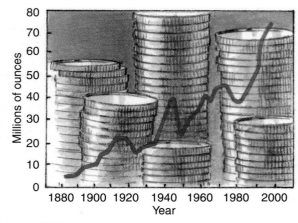

Figure 13.6
World production of gold has increased dramatically over the years. Economic forces can be detected in production trends.

resource would be economically recoverable. For all sorts of economic and political reasons, the price of gold moves up and down much more drastically than is true for other metals (see Figure 13.6).

Gold coins were routinely used for trade for many centuries. With the value of gold now at about $270 per ounce, gold coins are stockpiled as investments and a hedge against inflation. The belief that gold will always hold its value has been true for most of history. What circumstances might change that in the future?

Gold is amazingly malleable. It can be hammered into very thin sheets called gold leaf, so thin that a million sheets will form a stack less than 4 inches high. The melting point of gold, at 1063°C, allows gold to be melted on a hot charcoal fire so it could be melted and cast by craftsmen in very early societies. Gold's density, 19.3 g/cm^3, is among the highest of all the elements, so a one-dollar coin in 1910 would be only about a centimeter in diameter and a millimeter thick, about the size of a dime. This was too small for such a valuable coin, so gold coins were minted only in higher denominations. Current U.S. gold coins are minted in $50, $25, $10, and $5 face values but are not in general circulation.

exercise 13.2

How Much Must Be Moved?

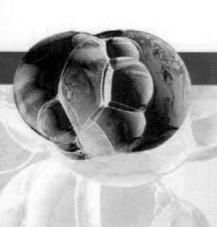

Problem

Suppose a troy ounce of gold (31.103 g) is worth about $375, although the value changes from day to day. If a quarry rock contains 0.0025% gold, how much rock must be processed to get 1.0 ounce (oz) of gold? (Assuming no gold is lost in processing—a hazardous assumption!)

Solution

$$\text{mass of rock} = (1.0 \text{ troy oz Au}) \times \frac{31.103 \text{ g}}{\text{troy oz}} \times \frac{100 \text{ g rock}}{0.0025 \text{ g Au}} \times \frac{1 \text{ kg rock}}{1000 \text{ g rock}}$$

$$= 1.2 \times 10^3 \text{ kg of rock}$$

Typical gold ore mined nowadays contains less than an ounce of gold per ton of ore. The ore near Lincoln, Montana, contains about 0.020 oz of gold per ton.

13.3 Copper (Cu)

Like gold, **copper** has also been used for millennia. Its use arose in Mesopotamia (modern Iraq) and in Egypt around 3500 B.C. Soon thereafter, copper was characteristic of the Mediterranean civilization on Crete and Cyprus. Indeed, the alchemical symbol for copper (♀) is also the modern symbol for female. The connection is that Aphrodite, the Greeks' goddess of love (renamed Venus by the Romans), was said to have sprung from the sea foam off the island of Cyprus (*Kupros* is the Greek word for Cyprus) carrying a hand mirror of polished copper.

Copper is the main ingredient in bronze, an alloy of copper and tin, which is harder than pure copper. The development of bronze led to the first real revolution in human technology when metal implements replaced stone ones. The resulting Bronze Age is usually dated as beginning around 3000 B.C. A related alloy, brass, is composed mostly of copper and zinc, but some modern bronze contains zinc and some modern brass contains tin.

Copper occasionally occurs in nature as the pure element but is more commonly found within chemical compounds. Most of these compounds are bright blue or green in color such as malachite and azurite and would be easily spotted and identified. Copper is easily won from many of its ores, essentially because copper metal is not very reactive. In general, the less reactive a metal, the easier it is to release it from its ores, and the more likely it is to occur in the unreacted elemental form. On the other hand, highly reactive metals such as sodium and potassium are never found in the elemental pure form because they react with so many things, and it is often very difficult to release them from their ores. This allows us to make sense of the fact that metals were found, exploited, and purified throughout history in roughly the reverse order of their reactivity. This is a fine example of the power of chemical reactivity to steer the development of civilization. It also reminds us of the power of the periodic table as a guide to chemical behavior.

The first copper ore to be exploited as a source of copper was probably malachite ($Cu_2(OH)_2CO_3$) (Figure 13.7), which can be found in the Sinai Peninsula. When malachite is heated in a charcoal fire (charcoal is nearly pure carbon) it decomposes into elemental copper:

$$Cu_2(OH)_2CO_3 + C \rightarrow 2Cu + H_2O + 2CO_2$$

A hot charcoal fire will also melt the copper, allowing it to be poured into molds and then, after some cooling, to be hammered into final shape if necessary.

Copper ores are usually blue or green, but cobalt and nickel also have brightly colored blue or green ores. In the past, ores of cobalt and nickel were sometimes mistaken for copper ores. The fact that no copper could ever be obtained from these ores was blamed on the Devil. They were therefore known as "Devil's copper" (*teufelkupfer* in German). When the cobalt and nickel were finally isolated many years later, the elements were given names of the Devil's helpers, Kobald and old "Nick" himself. Science, religion, and superstition can sometimes become strangely mixed as humanity stumbles toward an understanding of the world!

Figure 13.7
Malachite is the typical green color of copper compounds.

Properties of Copper

Copper is an orange-brown ("copper-colored") metal whose major importance to modern society is that it conducts electricity exceptionally well and is reasonably unreactive to oxygen and water. Conductivity of electricity can be measured by the electrical **resistance** that a material presents to the flow of electricity. A substance with a low resistance can transmit electricity efficiently. Only silver has a lower resistance than copper, but the rarity and high cost of silver make it uneconomical for use in electrical wiring. Copper can be drawn into fine wires that are strong and flexible and are used to carry electricity for power and for communication over thousands of miles.

Thousands of items are made of copper and its alloys.

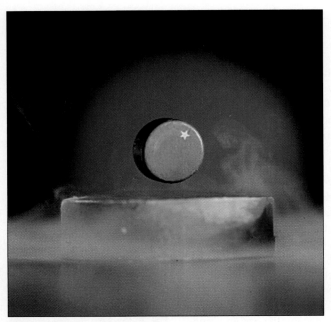

Figure 13.8
Superconductor materials are repelled by a magnet. This small magnet floats in midair above a piece of $YBa_2Cu_3O_8$ cooled to the temperature of liquid nitrogen.

Superconductors

In 1986 a compound of copper was found that conducts electricity with *zero resistance* at temperatures below about 90 K. A substance having zero electrical resistance is said to be a **superconductor** in contrast to ordinary **conductors,** which merely have low resistance. A temperature of 90 K ($-183°C$) may seem very cold but the compound's behavior is classified as "high temperature superconductivity," because such behavior had previously been observed only at temperatures below 20 K.

Since 1987, experiments with related compounds such as $YBa_2Cu_3O_8$ (Figure 13.8) have shown superconductivity at temperatures up to about 125 K and investigators are hoping eventually to achieve superconductivity at even higher temperatures.

Current high-temperature superconductors tend to be "ceramic oxides," with the feel and appearance of pottery. Unfortunately, nobody has been able to fashion them into wires suitable for electrical circuitry. Some electrical devices have been made using sheets of superconductors, however, and this field is bound to see many significant developments in the future.

Modern Copper Metallurgy

Most copper ore now mined is much lower in copper content than malachite and generally gives up its copper less readily than malachite. A typical modern ore is the black copper-iron mixed sulfide, chalcopyrite ($CuFeS_2$). Some usable ores contain only 1% copper, so much unwanted material must be separated from the copper during processing of the ore. A first step in the processing of chalcopyrite is the separation of the copper-containing mineral from the rock in which it is imbedded. This is done by "froth flotation" (Figure 13.9). The ore is powdered and suspended in water containing some detergent. The copper compound is taken up in the detergent micelles (see page 259) and floats to the top where the froth overflows into another container. The surrounding rock is not as attracted to the detergent so it stays behind to be discarded.

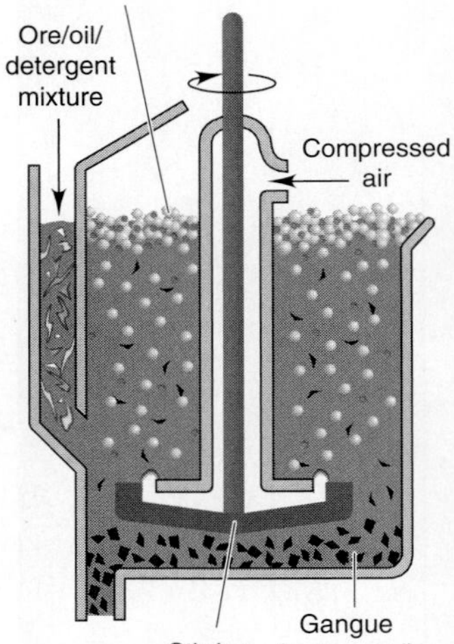

The purified $CuFeS_2$ is then heated to a high temperature ("roasted") in the presence of air. This converts some of the sulfur to sulfur dioxide, while the iron and copper form iron(II) sulfide, FeS, and copper(I) sulfide, Cu_2S. The FeS and Cu_2S are further heated with silicon dioxide, SiO_2, to make iron(II) silicate, $FeSiO_3$, sulfur dioxide, and copper metal. The iron silicate and copper metal are both liquids at the temperatures used. The iron silicate is much less dense than the copper so it floats to the top where it is removed as slag. The copper is usually cooled at this point and solidified into solid copper which is about 95% pure and filled with bubbles containing mostly sulfur dioxide. Copper at this stage of purification is called "blister copper" (Figure 13.10). The key equations summarizing its formation are:

$$2CuFeS_2 + O_2 \rightarrow Cu_2S + 2FeS + SO_2$$
$$2FeS + 3O_2 \rightarrow 2FeO + 2SO_2$$
$$FeO + SiO_2 \rightarrow FeSiO_3$$
$$Cu_2S + O_2 \rightarrow 2Cu + SO_2$$

Blister copper is purified by "fire-refining" in which air is blown into molten copper to remove

Froth of
detergent-
coated
ore

Ore/oil/
detergent
mixture

Compressed
air

Stirring
blade

Gangue
(rock, sand)

Figure 13.9
Copper sulfide attaches to the foam formed when detergent solution is bubbled with air in the presence of copper ore. The remaining rock settles to the bottom and is discarded.

Figure 13.10
Blister copper is the result of the first stage of the purification of copper ore.

■ hands-on

Preparation of Copper from Its Oxide

Start with a charcoal briquette, the kind which is not presoaked in a starter fluid. Cut one briquette in half so that there is a flat side. Hollow out a small hole in the flat side and save the powder removed from the hole. Put about 0.05 g of copper oxide or copper carbonate into the hole and mix it with the powdered charcoal. Heat this as hot as you can using a propane torch or an alcohol lamp and a blowpipe. You should be able to find a small bead of copper metal in the bottom of the hole after considerable heating. As with any hazardous chemical reaction, please use proper safety precautions!

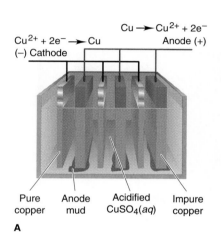

$$Cu^{2+} + 2e^- \longrightarrow Cu$$
(−) Cathode

$$Cu \longrightarrow Cu^{2+} + 2e^-$$
Anode (+)

Pure copper   Anode mud   Acidified $CuSO_4(aq)$ Impure copper

A

B

Figure 13.11
Diagram (**A**) shows how electrolysis is used for copper purification. Copper ions are reduced to copper metal at high temperature in the presence of carbon. Photo (**B**) shows slabs of copper being prepared for the electrolysis process.

the remaining sulfur as sulfur dioxide and allow existing sulfur dioxide to escape. This yields a product which is about 99% pure.

Final purification is done by electrolysis. The molten fire-refined copper is poured into thick slabs which are hooked up electrically (as shown in Figure 13.11) and immersed in dilute sulfuric acid. The flow of electricity causes the impure copper to be oxidized to copper(II) ions, which are then reduced to pure copper on sheets of existing pure copper. The impurities are of two sorts, those that are more easily oxidized than copper and those that are less readily oxidized than copper. The former, chiefly iron and zinc, are also more difficult to reduce than is copper so these go into solution and stay there. The metals that are more difficult to oxidize just fall to the bottom of the electrolysis chamber as a mud. This consists mostly of gold, silver, and the platinum metals. The mud is then processed to purify these valuable metals which often have a value great enough to pay for the processing.

If the ancients had needed to do all this to get reasonably pure copper, they probably would never have discovered copper and we might not have had the Bronze Age.

13.4 | Iron (Fe)

The use of **iron** developed much later than that of copper and bronze. Once developed, iron became the dominant metal of technology for nearly 30 centuries (see Figure 13.12). The exploitation of iron developed later than that of copper and bronze because it is much more difficult to win iron from its ores. It is also more difficult to work with iron than copper or bronze because iron melts at 1535°C, a temperature 452°C higher than the melting point of copper and not readily reached using a charcoal fire.

case in point

The Ice Man's Ax

In the summer of 1991, the remains were found of a man who had apparently frozen to death in the Alps of northern Italy. It soon became clear that the "Ice Man," as he was known to the press, had been there for a long long time (about 5300 years). He must have died when European civilization was just leaving the Stone Age and entering the Bronze Age. Beside the body there was an ax with a head originally thought to be made of bronze, but bronze implements from this date seemed out of place with respect to other historical evidence.

Chemical analyses on bronze, brass, and copper are very easy, accurate, and trustworthy, but most such analyses require some of the sample to be dissolved in acid. No one wanted to do this with such a precious object as this ancient ax head, so a nondestructive method of analysis was needed. Density measurements can be done to help identify the purity of metals, but copper and bronze have very similar densities, so this method would not be suitable to distinguish between the two. Besides, to do a density measurement would require the ax head to be taken off of its handle, doing unwanted damage to the artifact.

A technique known as **X-ray fluorescence** can very easily reveal the presence of even small amounts of tin or other elements in a piece of copper. When an atom is struck by an energetic X-ray, the X-ray can be absorbed, causing one of the innermost electrons of the atom to be ejected from the atom. An outer electron in the atom soon drops into the place left by the recently departed inner electron. In doing so, the electron that drops to the lower energy inner orbital gives off its surplus energy as another X-ray. The wavelength of this emitted X-ray indicates what type of atom is involved, all without ever harming the object in any discernible way.

Copper and tin emit characteristic X-rays of specific wavelengths when subjected to this procedure (see the graph). Analysis of the ax head in this way revealed that it was not made of bronze but was almost pure copper. X-ray fluorescence shows the actual composition to be 0.22% arsenic, 0.09% silver, and the rest copper. Copper without added tin had been known for hundreds of years before the improvement of bronze making, so the Ice Man's copper ax head is indeed consistent with other historical information.

This ax was made thousands of years ago and was recently found with the body of the "iceman." The head of the ax was shown to be nearly pure copper, including 0.22% arsenic and 0.09% silver.

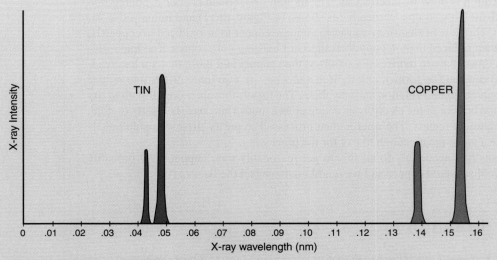

The X-ray fluorescence spectrum of copper shows peaks at 0.154 and 0.139 nm wavelength. The spectrum of tin shows peaks at 0.049 and 0.043 nm. Thus, the presence of minor amounts of tin are readily observed in metal which is mostly copper. These types of data are what Moseley used in recognizing the importance of an element's atomic number (see p. 73).

Figure 13.12
Iron is used in thousands of ways in our civilization. The Eiffel Tower is probably the most famous iron monument. The Forth Railway Bridge, built in the 1880s, is still able to carry travelers over the mile-and-a-half-wide Firth of Forth into Edinburgh, Scotland. Railway tracks, tools, and construction components are all produced with iron.

The dominant iron ore, Fe_2O_3, is a bright red-orange color, allowing it to be easily distinguished from most other rocks. The alchemical symbol for iron ($\male$) was originally the sign for the Roman god of war, Mars, and is now the international symbol for male. The connection among these three uses of this symbol is that iron implements were used in warfare as swords and spears and so were associated with the god of war and the war-like endeavors of males. The symbol actually represents a spear and a shield.

Iron was first produced for exploitation by heating its ore, iron(III) oxide, mixed with charcoal in the presence of air (Figure 13.13):

$$2Fe_2O_3 + 3C \xrightarrow{\text{air}} 4Fe + 3CO_2$$

It may seem strange that an oxide can be so readily reduced in the presence of air, but the real reducing agent is carbon monoxide formed by the reaction of charcoal with oxygen, rather than carbon itself:

$$2C + O_2 \rightarrow 2CO$$

Iron oxide reacts with carbon monoxide as:

$$Fe_2O_3 + 3CO \rightarrow 2Fe + 3CO_2$$

Iron ore often contains a significant amount of silicate impurities. These are removed by adding limestone, $CaCO_3$, to the ore being heated. The calcium carbonate decomposes to generate calcium oxide and carbon dioxide. Calcium oxide is basic and reacts with the more

Figure 13.13
Iron ore is reduced to metallic iron in a blast furnace. The diagram shows the reactions which occur at various levels within it.

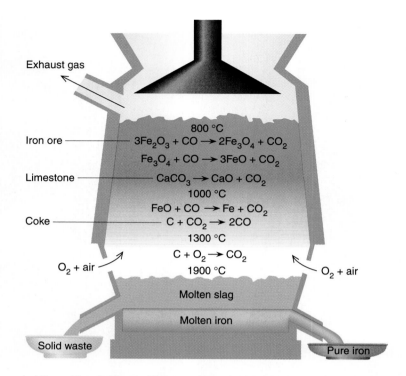

Exhaust gas

Iron ore
$$800\ °C$$
$$3Fe_2O_3 + CO \longrightarrow 2Fe_3O_4 + CO_2$$
$$Fe_3O_4 + CO \longrightarrow 3FeO + CO_2$$

Limestone
$$CaCO_3 \longrightarrow CaO + CO_2$$
$$1000\ °C$$
$$FeO + CO \longrightarrow Fe + CO_2$$

Coke
$$C + CO_2 \longrightarrow 2CO$$
$$1300\ °C$$
$$C + O_2 \longrightarrow CO_2$$

$O_2 +$ air
$$1900\ °C$$
$O_2 +$ air

Molten slag

Molten iron

Solid waste

Pure iron

acidic oxide of silicon, SiO_2, to give calcium silicate, which floats to the top of the molten reaction mixture as a waste slag:

$$CaCO_3 \rightarrow CaO + CO_2$$
$$CaO + SiO_2 \rightarrow CaSiO_3$$

In modern iron processing, the slag is frequently used in the manufacture of Portland cement.

Most of the iron made nowadays is immediately converted into **steel,** an alloy of iron and small amounts of carbon with the addition of such metals as chromium, nickel, vanadium, molybdenum, or cobalt. Small amounts of these various metals change the properties of the steel to make it stainless (resistant to rusting), very hard, or very tough.

13.5 | Aluminum (Al)

As we have moved from gold to copper and iron, we have been introduced to metals that are increasingly difficult to obtain. This is essentially because we are moving up the reactivity series (see Table 13.2). Aluminum is considerably more reactive than iron, and so aluminum did not become a routinely utilized metal until 3000 years after the widespread exploitation of iron had begun. Nowadays, aluminum is used to make even the most mundane objects of everyday life such as aluminum cans for beverages, but as recently as 130 years ago it was little more than a precious laboratory curiosity.

Aluminum is the most abundant metal in the Earth's crust, but the difficulty in isolating it from its ores meant that it was not obtained in pure form until 1827. It is so difficult to reduce aluminum (that is, to convert positively charged aluminum ions into aluminum atoms) that the ancients could not have carried this out even if they had known that this very useful metal was in every handful of clay. The reduction of aluminum had to await the harnessing of electricity because the first preparation of aluminum was achieved by reacting Al_2O_3 with molten potassium, which in turn is only produced by passing an electric current through potassium hydroxide (KOH):

$$Al_2O_3 + 6K \rightarrow 2Al + 3K_2O$$

This was a very awkward process, and aluminum remained a very expensive metal without practical uses until 1886 when Charles Hall in the United States and Paul Héroult in France

Table 13.2

Activities of Metals

Element	Reaction
Potassium	Reacts with cold water to form hydrogen
Calcium	Reacts with cold water to form hydrogen
Sodium	Reacts with cold water to form hydrogen
Magnesium	Reacts with steam to form hydrogen
Aluminum	Reacts with steam to form hydrogen
Zinc	Reacts with steam to form hydrogen
Iron	Reacts with steam to form hydrogen
Lead	Reacts with acids to form hydrogen
Hydrogen	Reacts with oxygen to form oxide
Copper	Reacts with oxygen to form oxide
Silver	Forms oxide only indirectly
Gold	Forms oxide only indirectly

Note: The elements listed here are ranked in decreasing order of their *activity*. Elements at the top of the list are very easily oxidized. Likewise, they are very difficult to isolate from their oxides because that reduction is not favored. Elements at the top of the list can reduce oxides or ions of elements listed below them.

independently discovered a process by which an electric current is used to form aluminum directly from aluminum oxide. (Review discussion of this in Chapter 4.) Both men dissolved Al_2O_3 in molten Na_3AlF_6. This compound melts at 1000°C. Once the aluminum oxide is dissolved, electricity is passed between carbon electrodes and molten aluminum is formed and sinks to the bottom to be drawn off. The essential reduction process is:

$$Al^{3+} + 3e^- \rightarrow Al$$

The enormous amount of electricity needed to produce aluminum (3 electrons at a high voltage for every atom) accounts for most of its cost. That cost is avoided when aluminum cans are melted for reuse, so recycling aluminum is very cost-effective. About 15.1 billion lb of aluminum are produced in the world each year, of which 1.9 billion lb are recycled. This recycling saves energy in terms of 7.4 million barrels of oil every year. Greater recycling efforts could save at least double or triple that amount of energy.

Properties of Aluminum

It is surprising that aluminum is as useful as it is because it is such a reactive metal. Pure aluminum can react with steam and with oxygen. It doesn't sound like a material to make good cooking pans, but thousands of meals are cooked in aluminum pans every day. This is possible because aluminum reacts with oxygen in a particular way. When pure aluminum comes in contact with air, it reacts to form aluminum oxide, Al_2O_3, but the aluminum oxide stays tightly bound on the surface of the piece of aluminum and prevents further air from reaching the remaining metal beneath it. So what we refer to as an aluminum cooking pan is really made up of internal aluminum coated with a thin protective layer of aluminum oxide. This is unlike the behavior of iron which oxidizes to form the familiar red-orange iron oxide "rust" that flakes off the iron and even seems to speed the advance of further rusting.

The actual aluminum of drink cans is also protected by a thin chemical coating, not of aluminum oxide, but of a varnish-like layer of plastic. If we scratch a thin circle of the varnish away from the inside of a can, the exposed aluminum will readily react with many solutions used to fill the can. Copper chloride solution can be used to demonstrate the reactivity of the exposed aluminum which reacts and dissolves due to the following reaction:

$$2Al(s) + 3Cu^{2+}(aq) + 8Cl^-(aq) \rightarrow 2AlCl_4^-(aq) + 3Cu(s)$$

This leaves the can held together at the scratched point only by the exterior paint. We could then very easily tear such a can in two.

Because aluminum is strong, lightweight, and durable it has been used in a huge variety of products.

case in point

The Statue of Liberty

A badly worn Statue of Liberty underwent a 2-year renovation in time for its centennial celebration, July 4, 1986. It was found that much of the iron structural work was badly rusted, especially where it was in contact with the copper skin of the statue. The copper was largely intact. The corrosion of the iron occurred where water had leaked in, of course in the presence of air. That the water was very salty from sea spray made things even worse.

Copper is more difficult to oxidize than iron. This is consistent with our statement that copper is more easily reduced from its oxide than is iron. Iron is thus said to be more *active* than is copper. See Table 13.2 for a list of such activities of metals. Iron in contact with more active metals such as zinc or magnesium is more resistant to oxidation.

The surface of the copper is green-colored instead of the orange-brown copper color because a layer of $Cu_2(OH)_2CO_3$ has resulted from surface oxidation. This green layer is called the *patina* of copper objects and is the same compound as the malachite ore from which our ancestors first won copper.

On the other hand, when iron comes in contact with water, it tends to oxidize:

$$Fe \rightarrow Fe^{2+} + 2e^-$$

The electrons released by this oxidation travel through the metallic iron to the edge of the water drop where oxygen is available. Oxygen in the presence of water picks up the available electrons and is reduced to hydroxide ions.

$$O_2 + 2H_2O + 4e^- \rightarrow 4OH^-$$

The hydroxide ions formed react with the iron(II) ions to form iron(II) hydroxide. This is further oxidized to form rust, an iron(III) compound which is mostly $Fe_2O_3 \cdot H_2O$.

$$Fe^{2+} + 2OH^- \rightarrow Fe(OH)_2$$
$$4Fe(OH)_2 + O_2 \rightarrow 2Fe_2O_3 \cdot H_2O + 3H_2O$$

Rusting of iron can be prevented by coating the iron with zinc, a more active metal. Any oxidation then occurs to the zinc coating and the iron itself is not harmed. This process is called **galvanizing** and is commonly used to prevent destruction of iron and steel objects. The photo to the right shows an old galvanized washtub standing on a steel frame which was painted many years ago but not galvanized. This washtub is in virtually new condition but the stand is badly rusted.

We have seen that the four important metals we have discussed have some similar "metallic" properties and some, such as reactivity, that are different. When we look at the structure of metals on an atomic level, we can see that there are some structural similarities that can be used to explain the similarities in physical and chemical behavior.

13.6 | The Structure of Metals

Metals form crystals of only a single kind of atom. The atoms are spherical and pack together as spheres would usually pack, fitting together in layers (Figure 13.14).

As discussed on p. 115, the valence electrons in metals are shared, but differently from the sharing in covalent compounds. The shared electrons in covalent bonds are shared between specific atoms, but the valence electrons in metals are shared among all the atoms in the crystal. If the bonding electrons of a covalent molecule go somewhere else, the bond is broken and a new molecule is formed. Since the atoms of metallic elements share their electrons widely, there are no specific neighbors which must be maintained. Any atom can be neighbor to any other, which explains why metals are very malleable. A piece of metal struck with a hammer just deforms; whereas a nonmetallic element, an ionic compound, or a covalent compound shatters into pieces. Likewise, since the electrons are free to move anywhere within the piece of metal, electricity conducts readily through metals. An electric current is just a flow of mobile electrons.

To summarize, the four metals we have discussed have a wide variety of uses. They have many properties in common and some unique characteristics as well. No matter how different they may seem, the metals are more like each other than they are like other resources of the Earth. Petroleum is mined from the Earth like metals, but that is where the similarity stops. Petroleum is so highly valued that countries that have large quantities of it wield enormous economic influence.

Figure 13.14
Atoms of metals fit together in an arrangement known as cubic closest packing (see the Hands-on box, page 118).

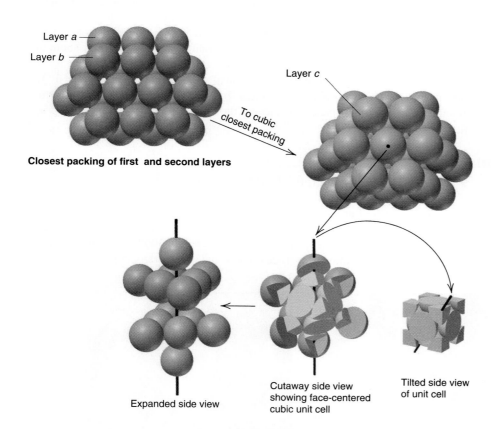

Layer *a*
Layer *b*

Closest packing of first and second layers

To cubic closest packing

Layer *c*

Expanded side view

Cutaway side view showing face-centered cubic unit cell

Tilted side view of unit cell

Cubic closest packing (*abcabc...*)

13.7 | Corrosion

Lay not up for yourself treasures on earth where moth and rust doth corrupt and where thieves break through and steal!

—Matthew 6:19

Rust has certainly been a problem for humanity for a long time. Objects made of iron often turn rusty and unusable. This is now known to be the reaction of iron with oxygen to form iron(III) oxide (Fe_2O_3) and iron hydroxy oxide (FeOOH), orange brown compounds which we recognize as rust. The annual economic cost of corrosion in the United States is estimated to be \$300 billion.

Of the four metals emphasized in this chapter, the oxidation of iron is by far the most serious. Gold does not react with oxygen; rather, gold objects stay shiny and new looking for millennia. Copper slowly turns from the shiny orange-brown color to a dark greenish-black color as copper oxide and copper carbonate form on its surface.

$$2Cu + O_2 \rightarrow 2CuO$$
$$2CuO + CO_2 + H_2O \rightarrow Cu_2(OH)_2CO_3$$

Aluminum reacts rapidly with oxygen, but the Al_2O_3 product sticks tightly to the surface of the aluminum object and protects the metal from further oxidation.

The extent to which metals react with oxygen is governed by the activity series of metals shown in Table 13.2. Metals at the top of the list react spontaneously with oxygen and those at the bottom do not. Those at the very top of the list will form oxides or hydroxides simply by reaction with water. Further, metals at the top of the list will reduce the oxidized form of metals lower on the list. For instance, copper ion will be reduced by zinc metal, as shown in Figure 4.20. This information can be put to use to prevent rust forming on large steel objects such as bridges or ships by attaching to it a piece of magnesium or zinc. This more active metal is oxidized rather than the iron object and is called a "sacrificial anode." Actual coating of steel with zinc (**galvanizing**) protects the steel as described in the Case In Point about the Statue of Liberty.

Rust can be removed from steel objects by chemical action. Phosphoric acid (H_3PO_4), gelled to make it thick, removes rust from iron objects by dissolving the iron oxide to form a soluble compound of iron and phosphate. This leaves the iron in a "passivated" state, less apt to react with further oxygen.

13.8 | Petroleum and Other Fossil Fuels

Nebuchadnezzar built much of ancient Babylon of "bitumen," which we would now call asphalt. That area of the world is still a dominant supplier of petroleum, which is much more important to modern life than it was in Nebuchadnezzar's time, nearly 3000 years ago. This is so because we have learned how to transform crude oil and coal into literally thousands of useful

These photos show "before and after" treatment of rusty steel by a commercial product containing phosphoric acid.

substances including motor fuels, lubricants, plastics, fiber, dyes, and so on. Chemistry leads the way in this technology.

Petroleum and coal are called **fossil fuels** because they are the residue left behind from decomposition of ancient plant life. Over millions of years, decayed plant material is slowly converted into coal and oil. Many of the molecules we now find in petroleum are recognizable as degraded versions of molecules found in the plants of today.

The overall oxidation state of carbon in plant material is much greater than in petroleum. In other words, the organic compounds in living plants contain more oxygen and less hydrogen than the organic compounds of petroleum. For example, cellulose, the most abundant organic compound in plants, is a polymer having the simplest ("empirical") formula of CH_2O. A typical petroleum molecule is octane, C_8H_{18}.

The great oil fields of Oklahoma, Texas, Saudi Arabia, and other OPEC countries produce 68 million barrels of crude oil every day. Crude oil is an incredibly complicated mixture of molecular compounds of carbon and hydrogen (hydrocarbons) plus small amounts of sulfur and oxygen and a few other elements. Recall from Chapter 9 that the sulfur content of coal and oil causes a problem when these fuels are burned or processed.

Natural Gas

Plant material decomposes in landfills to generate methane (CH_4) and carbon dioxide gas (CO_2). This can be summarized by the equation:

$$2CH_2O \rightarrow CH_4(g) + CO_2(g)$$

Remember, however, that CH_2O is the simplest approximate formula summarizing the ratio in which carbon, hydrogen, and oxygen are found in plant compounds having much bigger molecular formulas such as a typical cellulose molecule ($C_{600}H_{1002}O_{501}$), for example. Similar processes happened millions of years ago, often very slowly, to generate the supplies of buried natural gas that we can tap and exploit today. Not surprisingly, the reserves of natural gas are found in the same locations as supplies of the chemically more complex petroleum, also derived from ancient plants.

Hydrocarbons

As we mentioned in Chapter 6, hydrocarbons are organic compounds containing only hydrogen and carbon. The simplest hydrocarbon is methane, CH_4. More complicated compounds always have carbon atoms bonded to other carbon atoms in chains or rings with hydrogen atoms filling the bonding sites which are not utilized by carbon. The smaller molecules, those with 1, 2, 3, or 4 carbon atoms, are gases at room temperature; intermediate size molecules, with carbon numbers from 5 to about 16, are liquids at room temperature and bigger molecules are solids. Petroleum, a liquid mixture that is separated and purified into gases, liquids and solids, contains hydrocarbons with up to 50-60 carbons.

The complicated mixture of hydrocarbons in petroleum must be processed to separate individual compounds or classes of compounds for use in specific ways. The basic procedure involved is simply distillation. Crude oil is heated in huge **fractional distillation** columns, with systems of pipes collecting and taking away different **fractions** (parts) of the original oil at different heights up the column (see Figure 13.15). The lower molecular weight compounds are collected from the upper parts of the column because they condense back to liquid less readily than the higher molecular weight fractions, collected below them. Some material will not distill out at all and remains as blacktop asphalt. This can be used in road-building or is burned on-site to provide heat for the distillation of the more useful products.

Table 13.3 lists the various fractions of crude oil obtained by fractional distillation and some of their main uses. The original fractions contain mixtures of

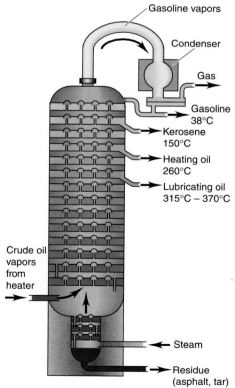

Figure 13.15
Fractional distillation is used to separate petroleum into many products. This simplified illustration shows how the 30-foot tower is used to separate the different "fractions" by differences in their boiling points.

Table 13.3

Products of Fractional Distillation and Their Uses

Fraction	Formulas	Uses	Boiling Point (°C)
Natural gas	CH_4 to C_4H_{10}	Fuel, cooking gas	<0
Petroleum ether	C_5H_{12} to C_6H_{14}	Solvent for organic compounds	30–70
Gasoline	C_6H_{14} to $C_{12}H_{26}$	Fuel, solvent	70–200
Kerosene	$C_{12}H_{26}$ to $C_{16}H_{34}$	Rocket and jet engine fuel, domestic heating	200–300
Heating oil	$C_{15}H_{32}$ to $C_{18}H_{38}$	Industrial heating, fuel for producing electricity	300–370
Lubricating oil	$C_{16}H_{34}$ to $C_{24}H_{50}$	Lubricants for automobiles and machines	>350
Residue	$C_{20}H_{42}$ and up	Asphalt, paraffin	Solid

compounds with similar boiling points but further purification can separate out individual compounds as required.

The mix of compounds present in crude oil never precisely matches our specific needs, so chemists have learned how to manipulate specific compounds obtained from crude oil in an enormous variety of ways. A single example is the conversion of the compound, ethane (C_2H_6), into ethene (commonly called ethylene), C_2H_4, and then polyethylene:

$$H_3C—CH_3 \rightarrow H_2C{=}CH_2 + H_2$$
$$nH_2C{=}CH_2 \rightarrow —CH_2—CH_2—CH_2—CH_2—CH_2—CH_2—CH_2—CH_2—CH_2—CH_2—$$

Coal

Coal fueled the Industrial Revolution in Europe and North America. Coal formed over many millions of years from the decayed remains of plant life in ancient times. It is often thought of as being made up of carbon, but the composition of coal is complex, containing thousands of simple and complex compounds. Primarily hydrocarbons, these compounds include many cyclic aromatic and other unsaturated molecules. According to the U.S. Energy Information Administration, in 1999, 1726 U.S. mines produced 1.1 billion tons of coal, about 24% of the world's output. Well over 75% of mined coal is burned to supply electricity. In the United States alone, the 474 billion tons of available coal represent an estimated 265-year supply.

Modern organic chemistry was first supported by government and industry in Germany in the late 19th century when the manufacture of synthetic dyes was perfected. Many of these were based on coal tar, the black gooey stuff obtained when coal is heated in steam in the absence of air. The products of this are coal gas, nearly pure carbon called "coke," and coal tar. A balanced equation cannot be written because none of these are pure compounds. Coal gas is a mixture of mostly hydrogen (H_2), methane (CH_4), and carbon monoxide (CO) and is used as a gaseous fuel. Its use in home heating and cooking almost stopped when supplies of methane (natural gas) were discovered and made available. Coke is the carbon source used instead of charcoal in the modern production of iron from iron ore.

$$4C + 3H_2O \rightarrow H_2 + 3CO + CH_4$$
product mixture called "coal gas"

In the equation, coal is represented as being just carbon, but coal actually is an enormously complicated mixture of carbon-based compounds (Figure 13.16).

Useful compounds isolated from coal tar include naphthalene and anthracene. These compounds had been in the coal all along but their presence wasn't noted when the coal was burned intact. Only when technologists began experimenting with coal did it become clear that many other interesting things came from coal.

We have spent most of this chapter discussing the chemistry below the Earth's surface. When considering the Earth's resources, we should not forget the precious soil covering much of the planet's surface.

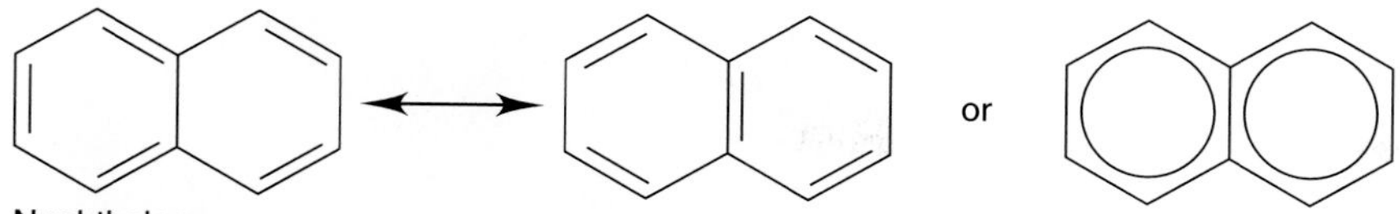

Naphthalene

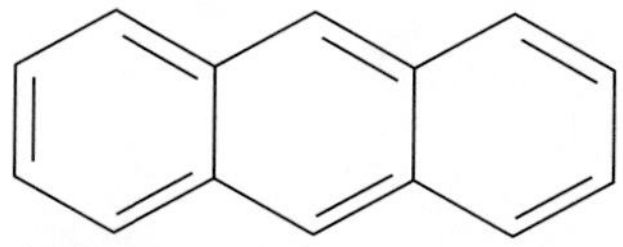

Anthracene

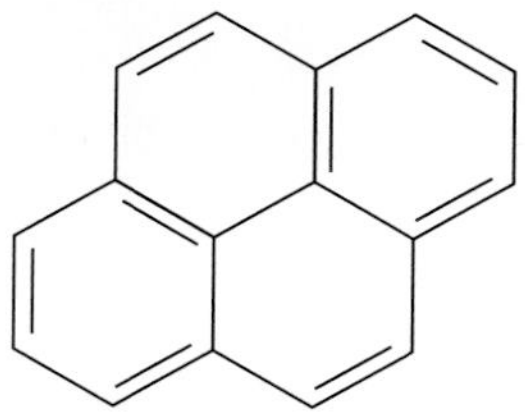

Pyrene

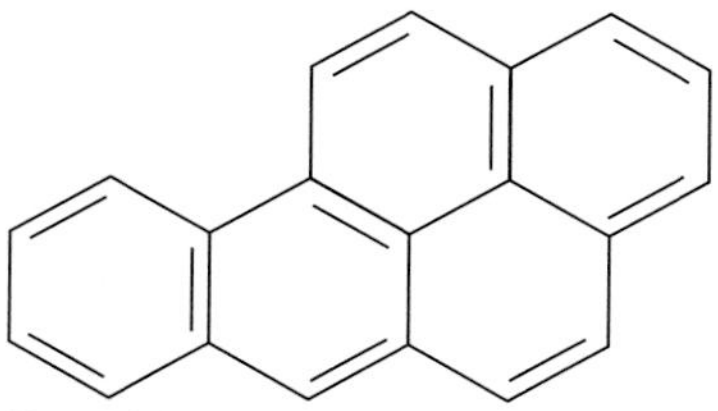

Benzo[a]pyrene

Figure 13.16
Naphthalene, anthracene, pyrene, and benzo[a]pyrene are examples of the simpler polyaromatic hydrocarbons present in coal. A compound is "aromatic" in this sense not because it has an aroma, but due to the delocalization of the double bonds in its structure. Naphthalene is shown with the double bonds in alternate positions. The actual bonding is more complicated, with some of the bonding electrons delocalized over the entire molecule as implied in the sketch on the right. The bonds are halfway between being double bonds and single bonds—circles are used to symbolize this.

13.9 | Soil and Clay

Most of the food that we eat is grown in the soil, or, in the case of farm animals, sustained by plants grown in the soil. In addition to food, the soil also provides the clay used to make bricks and pottery. The development of the systematic exploitation of soil to provide reliable food supplies and strong shelters was one of the key steps in the rise of civilization as we discussed in Chapter 2.

Most of the Earth's crust around our towns and cities consists of **soil.** Soil is just a general term for very complex and variable chemical mixtures including decomposed plant and animal life and tiny grains of rock. Most of us have noticed that soils are very different in color and appearance. The color and texture of a soil depends on its chemical makeup. Red or golden-orange soils are mostly clay but black soils have a large portion of humus (Figure 13.17). **Clay** is an inorganic substance composed mostly of calcium, aluminum, silicon, and oxygen. **Humus** is the decayed remains of plants and contains mostly carbon, hydrogen, nitrogen, and oxygen. The third major component of soils is sand (SiO_2) or powdered limestone ($CaCO_3$).

There are a variety of clays but they are all *aluminosilicates* (compounds of aluminum, silicon, and oxygen) and usually also include calcium or magnesium. The more intense the red color of a clay, the more iron is substituted for aluminum in its structure.

Bricks for construction and pottery for containing liquids have been made for millennia and are often the only clues we have to the lifestyle of some ancient peoples. Clay bricks and

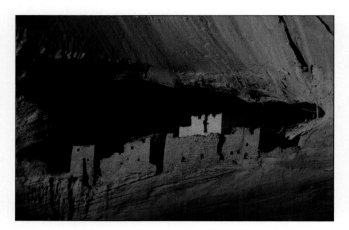

Figure 13.17
The red clay soil **A.** of the Southwest allowed the ancient ancestors of the Pueblo Indians to build these homes in the hills. The rich loam **B.** of the Midwest guaranteed excellent crops in that part of the country.

hands-on

Determining Soil Type

You can find out the proportions of sand, silt, clay, and organic matter in soil with a simple test. Fill a quart jar about one-third full of topsoil and add water to almost fill the jar. Fasten the lid tightly and shake vigorously until the soil is completely dispersed and all lumps are gone. Then set the jar down in good light and observe what settles out. The larger sand particles settle out first in just a few seconds. In a few hours the silt will have all settled out followed by the clay and eventually the organic matter (humus). These various layers are clearly visible after such an experiment. For lawn and gardens, the best mixture is about 40% sand, 40% silt, and 20% clay.

If the soil is more than 85% sand it is categorized as "sandy soil," if more than 40% silt it is categorized as "silt soil," and if more than 40% clay as "clay soil." Sandy and silt soils drain well but do not hold nutrients well; clay soils, on the other hand, drain poorly but hold nutrients well. Adding organic matter to soils nearly always improves it; it increases the ability of silty or sandy soils to hold water and allows better movement of water through clay soils. Increased incorporation of crop residue into soils also decreases the amount of carbon that enters the atmosphere as CO_2 and thereby works to counter the greenhouse effect.

pottery are easily made by mixing powdered clays with water and sometimes straw to make a mud that is formed to the desired shape. This is then allowed to dry and then heated to a high temperature, giving strength and rigidity to the mixture. (See Chapter 18 for more information about making pottery.)

13.10 | Spaceship Earth

All of the materials that we use or consume in our lives must be made from the chemicals of the Earth. We can utilize supplies of energy from the Sun, but the chemicals we use are the ones that were here from the beginning. The products we make from these chemicals must at some time and in some way be returned to the Earth. We are as self-contained as a solar-powered spaceship and must take care to keep our global "spaceship" well maintained and working properly. The main way to achieve this goal is to use the materials of the Earth wisely and recycle them in ways that prevent our supplies of essential raw materials from ever running out.

We began this chapter by looking at gold, which provides a good model for the careful management of a valuable resource. Over the centuries, society has recycled gold very efficiently because it is so highly valued. Most of the gold that was ever mined is still being used or carefully stored somewhere. Unfortunately, the same cannot be said about the other metals we have mentioned. Our junkyards and landfill sites are testimony to the fact that iron, aluminum, and other metals are generally used once and then thrown away. We have also been careless in our use of fossil fuels. Precious reserves of these fuels that formed over millions of years have been dug up and exploited with little thought for the future.

For the sake of our welfare and that of future generations, we need to learn to treat all of the chemicals of the Earth in the same way as we have always treated gold—as valuable resources to be cherished, used carefully, and recycled.

main points

- Civilization is based on soil, clay, and metals which have been utilized in agriculture, housing, tools, and weapons made from these materials of the Earth.

- Metals are seldom found uncombined in the Earth, but rather must be purified by chemical reactions from compounds of these elements. These reactions are all reduction reactions.

- Some metals are easily identified in nature and are easily reduced by technology available thousands of years ago, while others are difficult to identify and even more difficult to reduce from their compounds.

- Soil is a terrifically complex mixture. Different compositions of soil are good for different uses.

- Many compounds are classified as clays, but they are all based on silicon and oxygen.

- Coal and oil are called fossil fuels because they are the result of transformed plants buried millions of years ago. The most widely known use of these substances is as fuels, but they are also the basis of a great deal of organic chemistry.

important terms

Clay consists of fine particles of aluminum silicate in soil. (p. 439)

Conductor is a substance capable of conducting electric current. (p. 428)

Copper is an elemental metal, chemical symbol Cu. (p. 427)

Fossil fuels are residues such as coal or petroleum left from the decomposition of ancient plant life. (p. 437)

Fractional distillation is a process involving numerous vaporization-condensation steps, used to separate two or more volatile components. (p. 437)

Galvanizing is coating iron with zinc to prevent it from oxidizing. (p. 434)

Gold is an elemental metal, chemical symbol Au. (p. 423)

Humus is the decayed remains of plants in topsoil. (p. 439)

Iron is an elemental metal, chemical symbol Fe. (p. 429)

Reserve is a deposit of minerals that can be economically utilized. (p. 421)

Resource is a known deposit of a useful mineral. (p. 421)

Superconductor is a material that conducts electricity with zero resistance at temperatures typically below 90K. (p. 428)

exercises

1. Why is the Earth a closed system? What would happen if it were an open system?
2. Why are mineral resources nonrenewable?
3. Use Table 13.2 to predict which of the following reactions can actually occur.
   a. $6K + Al_2O_3 \rightarrow 3K_2O + 2Al$
   b. $Zn^{2+} + Cu \rightarrow Zn + Cu^{2+}$
   c. $H_2 + CuO \rightarrow Cu + H_2O$
4. The Earth's crust is 46.60% oxygen, 27.72% silicon, and 8.13% aluminum by mass. What is the ratio of moles of these three elements?
5. What is the difference between a gold resource and reserve?
6. What are the chemical symbols for gold, copper, iron, and aluminum?

7. Name and give values for three chemical or physical properties of gold.

8. Name and give values for three chemical or physical properties of iron.

9. Name and give values for three chemical or physical properties of aluminum.

10. Why is gold one of few metals to exist in its pure form in nature?

11. Weigh an aluminum can. Calculate the number of electrons necessary to reduce this much aluminum from Al_2O_3.

12. Will gold float or sink in water? In mercury?

13. "Cheap gold" jewelry will eventually turn green, whereas "real gold" jewelry will not tarnish. Why?

14. How large a piece of gold leaf can be formed from 1.00 troy oz of gold?

15. Which is heavier, a pound of gold or a pound of feathers? How about an ounce of gold or an ounce of feathers? (Check the Case in Point on The Gold Standard and Development of Coinage, p. 423.)

16. A typical wedding ring weighs 4.0 g and is 14 karat gold. How many kg of gold ore would be needed for one such ring if the ore contains 1.00 troy oz of gold per ton of ore? (Pure gold is 24 karat.)

17. The mass of the Earth is 5.983×10^{24} kg. Gold is 5×10^{-7} percent of this. How many moles of gold are there in the entire Earth? What volume would this occupy?

18. One day when you are swimming in a stream, you find a piece of quartz containing about 10 grams of gold. If you then powder and treat this rock to obtain the pure gold, will you actually end up with 10 g?

19. If a gold coin measured 2 cm³, and the price of gold were $300/troy ounce, how much is the coin worth?

20. If you had a gold double eagle ($20 coin) minted in 1910, what would it be worth right now based on its gold content? Check the business section of the newspaper for the current price of gold.

21. At the very end of the 19th century, William Jennings Bryan and others advocated the unlimited coinage of gold and silver coins with the value of 1 ounce of gold being equal to 16 ounces of silver. This policy was called "bimetallism" and was never adopted in the United States. It was used in many other countries. Look up the current price of gold and silver in the newspaper. How close to the ratio of 16:1 are these prices?

22. The price of gold is always listed to the nearest penny. How accurately must the purity of gold be known to justify this pricing?

23. How much would an error cost a seller of gold if the weight of a gold ingot truly weighing 1000.0 oz was off by just 0.05%?

24. The three elements, copper, silver, and gold (Group IB), are called the coinage metals. What features of these elements make them especially suitable for use in coins? Why is copper traditionally used in the lower-valued coins?

25. Calculate what the diameter would be of a gold coin worth one dollar at a gold price of $400/oz. Assume the coin to be 1.00 mm thick.

26. How much chalcopyrite is necessary to produce the copper in the power cord leading to an ordinary appliance? How much copper ore would this be if it is 1% copper?

27. Describe how you think our early ancestors might have discovered that copper metal could be obtained from malachite.

28. Why is it easier to win copper from the Earth than aluminum?

29. Draw a cartoon of how a copper ion is contained in a micelle during the processing of chalcopyrite.

30. How much copper can be obtained from a 100 g sample of chalcopyrite ($CuFeS_2$)?

31. Describe how electrolysis is used in the final purification step of copper.

32. How many moles of iron are found in one mole of each of the three types of iron oxide (see the Case in Point in Section 3.2)? How many grams of iron could be obtained from one mole of each type of oxide?

33. How many grams of iron could be obtained from 3.00 kilograms of iron(III) oxide, Fe_2O_3?

34. How many grams of iron could be obtained from 8.00 kilograms of magnetite, Fe_3O_4?

35. Why is so much energy needed to reduce aluminum so that it is available for use? Why is the recycling of aluminum so important?

36. Explain why many high-quality cooking utensils are made of stainless steel with a copper bottom?

37. Calculate the amount of carbon necessary to make a megagram of iron from Fe_2O_3.

38. A rectangular piece of aluminum foil 12 in. by 54 in. weighs 20 g. How thick is this piece of foil?

39. Look up the oxidation numbers of oxides of gold, copper, iron, and aluminum. Rationalize these numbers based on their position in the periodic table.

40. Look in an atlas to find town names which refer to the names of chemical elements. Try to find out the reason the names were given. For starters, consider Telluride, Colorado, and Iron Mountain, Michigan.

41. What is the difference between a conductor and a super-conductor?

42. List the metals studied in this chapter in order of increasing reactivity.

43. How is bonding in a metal different from in an organic compound?

44. Explain why metals are malleable and good conductors of electricity.

45. Look in the alloy section in any edition of the CRC *Handbook of Chemistry and Physics* for all the different recipes for brass and bronze. What characteristics do you see that would distinguish metals bearing these two names?

46. Use the CRC *Handbook of Chemistry and Physics* to identify the oxide with the highest oxidation number for

each of the transition metals from scandium to zinc. Which of these match the group number in the periodic table? Suggest a reason that the others do not.

47. Petroleum is derived from organic compounds in plants that originally contained considerable amounts of oxygen, but the petroleum contains very little oxygen. Explain what happened to the oxygen atoms originally in the plants.

48. For every mole of natural gas produced in the ground, how many moles of carbon dioxide are produced? What do you think happens to the carbon dioxide?

49. Would you expect these molecules to be solids, liquids, or gases at room temperature?
    a. C_3H_8
    b. $C_{17}H_{36}$
    c. $C_{33}H_{68}$

50. Calculate the oxidation number of carbon in anthracene. Use Exercise 5.5 to estimate the amount of heat expected from combustion of a mole of this compound.

51. Why does a sample of petroleum need to be purified by distillation?

52. Are the components of petroleum used only for fuels, solvents, and lubricants? Explain.

53. What are the differences between sand, clay, silt and humus?

54. A 40.0-g soil sample is acidified and the carbon dioxide that was liberated was recovered and found to weigh 0.73 g. What percent of this soil was limestone?

55. A second 40.0-g sample of the soil mentioned in Exercise 54 was heated to several hundred degrees in the presence of an excess of oxygen. The carbon dioxide that was liberated was trapped and found to weigh 3.55 g. What percent of the soil was humus (about 50% carbon)?

58. Why are there so many more ancient Greek weapons than Roman weapons in museums around the world?

59. Find several paintings of Aphrodite (Venus) and look to see whether she has a hand mirror. Read an artist's commentaries on the paintings to see whether the importance of copper and her mirror are mentioned.

60. Are we still in the Iron Age?

61. If your library system has the *Minerals Yearbook* published by the U.S. Department of Interior, Bureau of Mines, choose a metal or other mineral and research its history of production and uses over the last 60 years. Write a short essay on this substance and how it has impacted the U.S. and world economy.

62. Look up gold, silver, iron, or other chemical elements in Bartlett's *Familiar Quotations*. Try to understand how these elements have been selected by the original author to make certain points in their texts.

63. Read and analyze the meaning of W. J. Bryan's 1896 "Cross of Gold" speech.

readings

1. Matthews, Samuel W. Nevada's mountain of invisible gold. *National Geographic,* May 1968, p. 668.
2. Davidson, Keay, and Williams, A. R. Under our skin: Hot theories on the center of the Earth. *National Geographic,* January 1996, p. 100.
3. Jarnoff, Leon. Iceman. *Time,* October 26, 1992, pp. 62–66.
4. Hall, Alice J. Liberty lifts her lamp once more. *National Geographic,* July 1986, p. 2.
5. *Minerals Yearbook,* U.S. Department of Interior, Bureau of Mines, Washington, D.C., published annually since 1932.
6. Alternative Energy Institute, Inc. January 1999 Coal Fact Sheet http://www.altenergy.org/2/nonrenewables/fossil_fuel/facts/coal.html.

food for thought

56. When the Statue of Liberty was refurbished, there was discussion that she should be polished to look like shiny copper. It was decided not to do this. What do you think?

57. In the text, we listed the primary way of obtaining pure gold from rock, that of combining it with sodium cyanide. Look up some other ways that gold is isolated. What are the economic, social, and health advantages and disadvantages of each?

websites

www.mhhe.com/Kelter   The "World of Choices" website contains activities and exercises including links to websites for: The American Petroleum Institute; a discussion of corrosion from "The Corrosion Doctors"; the LaMotte Corporation, which makes soil test kits; and much more!

The Power of the Nucleus

I am become death, the shatterer of worlds!

—Hindu holy book, *Bhagavad Gita,* quoted by Robert Oppenheimer
on witnessing the first atomic bomb test in July 1945

August 1995 marked the 50th anniversary of the opening of the atomic age. A small group of scientists had been working since 1941 on the "Manhattan Project," devoted to building atomic bombs for use in the Second World War. The physics, chemistry, and engineering that led to the bomb were the result of the single-minded cooperation of hundreds of scientists and army personnel. More than fifty years of living with the bomb have led many to wonder whether it was right to develop and use such an awesome weapon.

Soon after the war ended, there were promises of abundant electrical energy derived from atomic sources. It was expected that such energy was to be so readily available that the standard of living in the world would soar. Sure enough, atomic energy was developed as a source of electricity but the controversy over its use continues to this day. Among the "hot-button issues" in the worldwide debate over atomic energy are nuclear power plant safety, nuclear waste disposal, and still, after six decades nuclear war. However, energy from the nucleus is used in many medical applications, space travel, and as a power source for millions of people throughout the world.

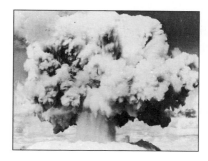

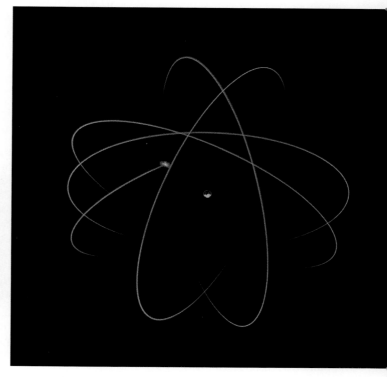

The mushroom cloud formed by the nuclear bomb dropped on the Japanese city of Hiroshima heralded the beginning of the Atomic Age. The devastation caused by the blast is only hinted at by the photo directly above this caption.

14.1 | The Radioactive Decay of Atomic Nuclei

In December 1984, Stanley Watras, an engineer at the Limerick, Pennsylvania, nuclear power plant set off warning alarms as he *entered* the plant at the start of the day. The warning system indicated that there was too much radiation in the area, Watras being "the area" in this case. You might expect that Watras had been contaminated with radiation due to his job in the nuclear power industry. In fact, the source of the radiation that triggered the alarm was the land that the Watras' home was built on. In December 1984, naturally occurring radon gas was percolating from the ground and into the Watras' home, where it was present at 2000 times the generally accepted safety level. Mr. Watras had inhaled radon gas, and once in his body it had decayed into other radioactive atoms. This radon gas is a naturally occurring source of "radiation" (defined later), and some of it is in the air that you breathe in your own home. Radiation and radioactivity are natural phenomena to which we are all exposed every second of the day, as shown in Table 14.1. But what is it? Why does it have a reputation for being very dangerous, and how can we exploit it? To begin to answer these questions, we will first refresh your memory about the atomic nucleus.

A Look at the Nucleus

As we have already discussed, atoms are composed of a central nucleus surrounded by moving electrons. The nucleus of an atom is made up of one or more protons, each with a charge of +1, and one or more neutrons, which have no charge. The only exception to that rule is the most abundant isotope of hydrogen, which contains no neutrons and has a nucleus composed of a single proton. The nucleus occupies a minuscule volume at the center of the atom, while the electrons (although much smaller than the nucleus in themselves) move about within much larger volumes of space known as the electron orbitals.

We have also explained that an element is identified by its atomic number (Z), which is the number of protons in the nucleus of each atom of that element. The mass number (A) is the total number of protons plus neutrons (Figure 14.1). The number of neutrons equals $A - Z$.

We have also briefly discussed isotopes, which are atoms of a particular element that have different numbers of neutrons, and therefore different mass numbers. For example, uranium-235 (^{235}U) and uranium-238 (^{238}U) are two isotopes of the element uranium. Atoms of both isotopes contain 92 protons (which defines the element as uranium), but atoms of ^{235}U contain 143 neutrons (from $235 - 92$), while atoms of ^{238}U contain 146 neutrons. As we will explore further in our discussion on nuclear applications, the uranium-235 isotope is used in nuclear

13			
Al			
26.98			

31	32		
Ga	**Ge**		
69.72	72.59		

49	50	51	
In	**Sn**	**Sb**	
114.8	118.7	121.8	

81	82	83	84
Tl	**Pb**	**Bi**	**Po**
204.4	207.2	209.0	(209)

Figure 14.1
Recall that an element is characterized by the number of protons it contains. The number of protons plus neutrons equals the mass number. Atoms of an element with differing numbers of neutrons are called isotopes. Aluminum is Al, its atomic number is 13, and its mass is 26.98.

Table 14.1

Typical Sources of Human Radiation Exposure

Natural Sources	Radon[a]	55%
	Inside human body	11%
	Rocks and soil	8%
	Cosmic rays from space	8%
Man-Made Sources	Medical X-rays	11%
	Nuclear medicine[b]	4%
	Consumer products[c]	3%
	Occupational uses[d]	0.3%
	Fallout from nuclear weapons tests	Less than 0.3%
	Nuclear power production	0.1%
	Miscellaneous[e]	0.1%

[a]Where significant amounts of radon are found in the home.
[b]Involves the use of radioactive materials in diagnosing and treating patients with cancer and other diseases.
[c]Building materials, tobacco, mining and agricultural products, water supplies, etc.
[d]Uranium mines, industrial and medical users, etc.
[e]Department of Energy facilities, smelters, transportation, etc.

power plants and atomic weapons. It is possible to separate these two uranium isotopes based on their differing atomic masses.

The Instability of Some Nuclei

The nuclei of some isotopes of some elements are inherently *unstable,* meaning they are liable to spontaneously undergo a change that results in the release of energy and of particles known collectively as **radiation.** The forces that account for nuclear stability are not completely understood. The key fact, however, is that only certain combinations of protons and neutrons will be found in energetically stable nuclei. For the lighter elements, the most stable ratio of protons to neutrons is one-to-one. Therefore, the nucleus of carbon-12 is stable (6 protons + 6 neutrons), as is that of calcium-40 (20 protons + 20 neutrons). The nucleus of an atom of the carbon-14 isotope (with 6 protons and 8 neutrons) is not stable. For heavier elements, more neutrons than protons are necessary to ensure nuclear stability. Iron-56 has the most stable nucleus, with 26 protons and 30 neutrons.

Ways of Achieving Nuclear Stability

Naturally unstable nuclei can spontaneously change or "decay" into stable nuclei by ejecting particles of matter and excess energy (Figure 14.2). As we have said, the ejected particles and ejected energy (which comes out as high-energy electromagnetic waves) are known collectively as radiation or **radioactivity.** Radioactive isotopes can become more stable by throwing off alpha particles, beta particles, gamma radiation, and in a few cases, by the process of **fission.** Fission occurs when a large nucleus fragments into smaller nuclei, releasing

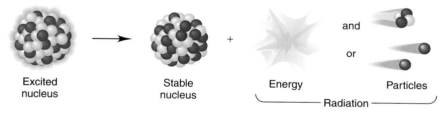

Excited nucleus → Stable nucleus + Energy and/or Particles — Radiation

Figure 14.2
The nuclei of some isotopes are energetically unstable. They can spontaneously release particles and energy known as radiation as they decay into more stable products. As we will discuss throughout the chapter, this radiation is used in industry, in medicine, to meet electrical energy needs, and for military purposes.

exercise 14.1

Calculate the Number of Neutrons

Problem

Calculate how many neutrons will be present in potassium atoms with a mass number of 39.

Solution

The atomic number (Z) of potassium is 19 and is the number of protons in its nucleus and the integer in the potassium block of the periodic table.

The mass number (A) = 39
The atomic number of potassium (Z) = 19 (see Figure 2.21)
The number of neutrons = $A - Z = 39 - 19 = 20$

Note that the number of neutrons, 20, added to the number of protons, 19, gives the mass number, 39.

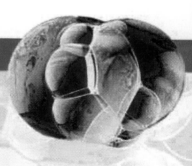

exercise 14.2

Review of Atomic Quantities

Problem

The mass number, atomic number, and number of neutrons each can be calculated if the other two are known. Show that you understand these relationships by filling in the blanks in this table.

Symbol	Protons	Neutrons	Electrons	Charge
$^{235}_{92}\text{U}$	_____	_____	_____	_____
$^{?}_{?}\text{?}^{-}$	17	20	18	_____
$^{55}_{26}\text{?}^{3+}$	_____	_____	_____	_____

Solution

Symbol	Protons	Neutrons	Electrons	Charge
$^{235}_{92}\text{U}$	92	143	92	0
$^{37}_{17}\text{Cl}^{-}$	17	20	18	−1
$^{55}_{26}\text{Fe}^{3+}$	26	29	23	+3

a burst of energy in the process. The opening scenario of the chapter depicts the enormous energy released when a fissionable element, uranium, undergoes decomposition rapidly.

Alpha Particles

● Proton
○ Neutron

α particle

Figure 14.3
An alpha particle (α) is composed of 2 protons and 2 neutrons. It is identical to a helium nucleus. Although relatively slow moving, alpha particles can cause severe damage if a source of them is ingested, as described in the text.

An **alpha particle** (α) is composed of 2 protons and 2 neutrons (see Figure 14.3). An alpha particle is identical to the nucleus of a helium atom, so it is often represented as ^4_2He (or $^4_2\text{He}^{2+}$, since it carries a double positive charge). Although alpha particles are ejected from radioactive isotopes, they are relatively slow moving compared to other forms of radiation and cannot even penetrate an object as thin as a piece of paper. Such particles can be very dangerous if a source of them is inhaled or ingested. Although relatively slow moving they contain a substantial amount of matter, and therefore can do a lot of damage when they collide with essential chemicals of living cells, such as DNA. (See the Case in Point: Effects of Radiation on the Body.)

We began this section with a reference to the dangers of naturally occurring radon gas. The particular isotope of radon that poses the danger is radon-222, which emits an alpha particle when it undergoes radioactive decay. Because radon is a gas it is easily absorbed by the body where the alpha particles can do damage. Let's look at this process in more detail, using a nuclear equation:

$$^{222}_{86}\text{Rn} \rightarrow {}^4_2\text{He} + {}^A_Z\text{X}$$

This equation reveals that the nucleus of the original radon-222 atom splits to form the alpha particle and the nucleus of a *different* atom. We can work out what nucleus is formed by calculating the values of the mass number (A) and atomic number (Z) of the new nucleus. In any nuclear equation the total of the mass numbers on each side of the equation must be equal, as must the total of the atomic numbers on each side. So the value of A in the equation must be 222 − 4 = 218; and the value of Z must be 86 − 2 = 84:

$$^{222}_{86}\text{Rn} \rightarrow {}^4_2\text{He} + {}^{218}_{84}\text{X}$$

So when the nucleus of radon-222 emits an alpha particle, a nucleus with 84 protons and a mass number of 218 is left behind. What element will this be? A glance at the periodic table (inside back cover) will reveal that it is the nucleus of a polonium atom (since the atomic number of polonium is 84). In fact, this polonium-218 nucleus is itself unstable, and so is liable to undergo a nuclear decay process of its own. Several such decay events can occur in sequence, until a stable atomic nucleus eventually results.

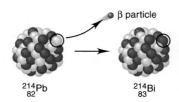

$^{214}_{82}Pb$ $^{214}_{83}Bi$

Figure 14.4
The emission of a beta particle (β) results in the transformation of a neutron to a proton. Because the proton count increases by one as a result of beta emission, a different element exists after the emission than before. In this figure, lead-214 emits a beta particle and is changed to bismuth-214.

Beta Particles

A **beta particle** (β⁻) is an electron that is ejected from the *nucleus* of an atom as it undergoes radioactive decay. How can an electron come from an atom's nucleus, when we know that the nucleus is composed only of protons and neutrons? The answer is that under some circumstances a neutron can change into a proton and a fast-moving electron (a beta particle) (see Figure 14.4). For example, the unstable isotope of lead with a mass number of 214 decays by emitting a beta particle:

$$^{214}_{82}Pb \rightarrow {}^{0}_{-1}e + {}^{214}_{83}Bi$$

Notice that this causes the original nucleus of the lead atom to become the nucleus of a bismuth atom, because the formation of the additional proton (from the original neutron) *increases* the atomic number of the nucleus by 1. The beta particle has almost no mass and a charge of −1, so it is represented in nuclear equations by ${}^{0}_{-1}e$ or β⁻.

 Beta particles are emitted with much higher energy than electrons in atomic orbitals. They are ejected outwards and can penetrate through several millimeters of skin. They are particularly hazardous when released from material ingested or inhaled into the body.

exercise 14.3

Alpha Emission

Problem

What is the product when a polonium-218 nucleus decays by emitting an alpha particle?

Solution

We can figure out the answer via the same strategy we used for radon-222 decay.

$$^{218}_{84}Po \rightarrow {}^{4}_{2}He + {}^{A}_{Z}X$$

The decay product is therefore lead-214 ($^{214}_{82}Pb$).

exercise 14.4

Beta Emission

Problem

The bismuth-214 nucleus that is formed from lead-214 by beta emission can itself emit a beta particle. What nucleus results from the emission of a beta particle from the bismuth-214 nucleus?

Solution

$$^{214}_{83}Bi \rightarrow {}^{0}_{-1}e + {}^{214}_{84}Po$$

Emission of a beta particle has resulted in a gain of one proton in the nucleus, which is now a polonium nucleus.

case in point

Effects of Radiation on the Body

"Irradiation may be compared to gunfire; for a living target there is no safety level for a striking bullet, but not every bullet that does hit will kill. An individual exposed to radiation is simply subjected to a greater risk than if he had not been irradiated. And the risk increases with the size of the dose."[4]

". . . there is no 'threshold' for the mutational effect of radiation. No matter how small a dosage of radiation the gonads receive, this will be reflected in a proportionately increased likelihood of mutated sex cells with effects that will show up in succeeding generations."[3]

These are two very strong, and somewhat frightening, quotations attesting to the effects of even the lowest levels of radiation exposure on the human body. That radiation from nuclear processes can cause damage makes sense when we recall that such radiation is exceptionally energetic. It is the large amount of energy released, especially in the case of gamma radiation, that makes exposure so very hazardous.

What happens when such energetic radiation enters the human body? About 50–60% of a woman's body is water. For men, the figure ranges from 60 to 65%, and 70% for infants. When assaulted by gamma rays, some of the water molecules absorb enough energy to break apart in an unusual way. Instead of sharing a pair of electrons in a covalent bond between a hydrogen and oxygen, the hydrogen and oxygen each take an electron, as summarized here:

$$H—OH \rightarrow H· + ·OH$$

The resulting species, H· and ·OH are not stable because they each have an unpaired electron, making them "radicals." The unpaired electrons of radicals make them highly reactive. Radicals can do a lot of damage in the human body, by reacting with vital chemicals of the body.

The hydroxyl radicals (·OH) produced as a result of gamma radiation can change the structure of enzymes, rendering them inactive. They can also combine with several of the monomers that make up DNA, leading to disruption of the operation of our genes, resulting in cell death or the onset of cancer.

The health effects of exposure to radiation depend on the intensity and length of the exposure and on the health of the person exposed. We are all exposed to continual low level "background radiation" almost entirely derived from natural sources every second of our lives. Very high radiation doses received over a short period of time, however, can result in a quick but agonizing death. Moderate short-term doses of radiation can lead to sterility and loss of vision. Long-term exposure to lower doses can lead to cancer. A more insidious but critical outcome of radiation exposure is the modification of DNA genetic material so that offspring, or the offspring of offspring, are affected. Table 14.2 summarizes the dose versus effect profile of radiation.

Table 14.2

Acute Effects of a Single Dose of Whole-Body Irradiation

rem[‡]	Effect
5–20	Possible late effect, possible chromosomal aberrations
20–100	Temporary reduction in white blood cells
50+	Temporary sterility in men (100+ rem = 1 yr. duration)
100–200	"Mild radiation sickness": vomiting, diarrhea, tiredness in a few hours
300+	Permanent sterility in women
300–400	"Serious radiation sickness," marrow/intestine destruction
400–1000	Acute illness, early deaths
3000+	Acute illness, death in hours to days

[‡]A **rem** is a measure of the biological impact of an absorbed amount of radiation.

Gamma Rays

After a nucleus emits an alpha or beta particle, it can still contain excess energy. This energy can be released in the form of **gamma rays** (γ), composed of highly energetic photons of electromagnetic energy (like light, only much more energetic, Figure 14.5). Gamma rays can be energetic enough to penetrate thick lead or concrete walls. Gamma radiation can travel rela-

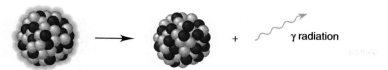

Figure 14.5
Gamma (γ) rays represent the release of excess energy from the nucleus on particle emission. This radiation is extremely high-energy and is associated with much of the health damage caused by radioactive substances.

tively long distances in air and can easily penetrate skin. (Much of the health damage associated with radioactive substances is caused by gamma radiation.)

Other Nuclear Transformations

Unstable nuclei can change to stable forms by undergoing transformations other than the emission of alpha particles, beta particles, or gamma rays. The other emissions, shown in Figure 14.6, are not of such widespread significance so we will introduce them very briefly:

- **Positron emission** involves the release of a **positron** (a particle like an electron but with a positive charge) when a proton is changed into a neutron. Sodium-22 can change into neon-22 in this way.

$$^{22}_{11}\text{Na} \rightarrow ^{22}_{10}\text{Na} + ^{0}_{+1}\text{e}$$

- **Electron capture** is a process in which an inner orbital electron is "captured" by (combines with) a proton in the nucleus to form a neutron. Tin-108 forms indium-108 by electron capture.

$$^{108}_{50}\text{Sn} + ^{0}_{-1}\text{e} \rightarrow ^{108}_{49}\text{In}$$

- **Spontaneous fission** results in a nucleus splitting into two smaller nuclei and some neutrons. It is an important decay mode for the heaviest human-made nuclei.
- **Induced fission** is a process similar to spontaneous fission, except it must be produced by a collision between the susceptible nucleus and another fast-moving particle such as a neutron. We will see in Section 14.5 that nuclear power plants as well as atomic bombs get their energy from induced fission.

Table 14.3 summarizes the important natural nuclear decay processes and their effects.

Radioactive Decay Series

Let's revisit the nuclear decay examples discussed in this section:

1. Radon-222 decayed to polonium-218 (α-particle emitted).
2. Polonium-218 decayed to lead-214 (α-particle emitted).
3. Lead-214 decayed to bismuth-214 (β-particle emitted).
4. Bismuth-214 decayed to polonium-214 (β-particle emitted).

If you get the sense that this is part of a long stepwise process, you are correct. This progression of decays is part of what is known as a **decay series.** Such series represent the sequence of reactions that an unstable isotope undergoes in its transformation toward stability. The four decays listed here are actually a small part of a much larger series in which uranium-238 decays to lead-206. This entire series is shown in Figure 14.7. Note that nuclei often have more than one possible decay route available, although one route tends to predominate over others. For example, polonium-218 can decay to lead-214 (α-emission) and astatine-218 (β-emission); however, 99.98% will decay via the α-emission pathway. There are other such series starting, for instance, with Thorium-232. All of these decay series form some isotope of lead, which has several extremely stable isotopes or bismuth.

One important aspect of decay series on health is that when a radioactive element (also know as a radioactive **nuclide,** since it is the nucleus that decays) enters the body, all of its future decay products and particle emissions may remain in the body to cause damage. When

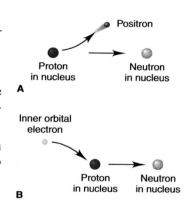

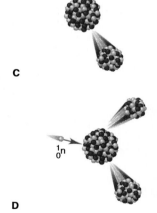

Figure 14.6
Other nuclear processes include **A.** positron emission, in which a proton is converted to a neutron (how is the element changed?); **B.** electron capture, in which an inner-orbital electron (1s electron) is captured by the nucleus (how is the element changed?); **C.** spontaneous fission, in which a nucleus splits into two smaller nuclei; and **D.** induced fission, in which a large nucleus is induced to split into two or more smaller nuclei by collision with a fast-moving particle, often a neutron. Induced fission is the basis for much of our nuclear power industry.

Table 14.3

Nuclear Processes

Process	Symbols	Results
Alpha emissions	α ($^4_2$He)	lowers proton and neutron count by 2
Beta emissions	β^- ($^0_{-1}$e)	raises proton count by 1
Gamma ray emissions	γ ($^0_0\gamma$)	energy burst only
Positron emission	β^+ ($^0_1$e)	lowers proton count by 1
Electron capture	E.C. ($^0_1$e)	lowers proton count by 1
Spontaneous fission	S.F.	splits nucleus into lighter fragments

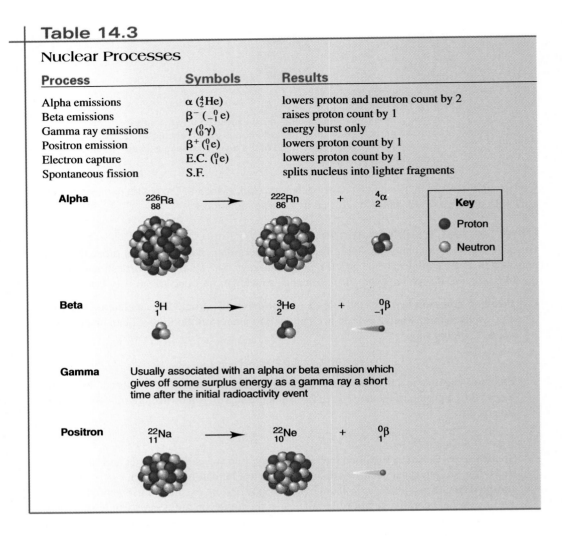

Alpha $^{226}_{88}$Ra $\longrightarrow$ $^{222}_{86}$Rn + $^4_2\alpha$

Key
● Proton
○ Neutron

Beta $^3_1$H $\longrightarrow$ $^3_2$He + $^0_{-1}\beta$

Gamma   Usually associated with an alpha or beta emission which gives off some surplus energy as a gamma ray a short time after the initial radioactivity event

Positron $^{22}_{11}$Na $\longrightarrow$ $^{22}_{10}$Ne + $^0_1\beta$

Figure 14.7
The radioactive decay series that converts uranium-238 into lead-206.

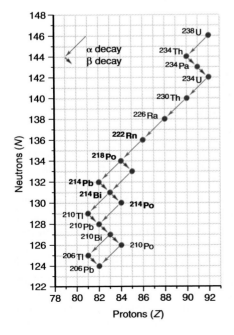

radon-222 enters the lungs, it decays to polonium-218. This nuclide is a solid which can attach itself to the lungs. It and another decay product in the series, polonium-214, are alpha particle emitters which may lead to health problems such as increased risk of lung cancer. So it is not the radon-222 itself which is the main problem. Rather, it is the *decay series products* that cause the long-term hazard, and which have been blamed for between 5,000 and 20,000 lung cancer deaths per year in the United States.

Radon-222 pollution is not a nationwide problem in the United Sates. It is found in significant amounts only in areas that have soil that is relatively rich in uranium-238. This nuclide will eventually decay into radon-222 (see the uranium-238 decay series) which, being a gas, can percolate out of the soil, into the basements of houses, and into the lungs of people like Stanley Watras.

How long will it be before all the radon-222 decays? If some goes in your lungs, how long does it remain radioactive? The answers to these questions and others concerning such issues as the burial of nuclear wastes all depend on the *rate of decay* of an unstable nuclide. In Chapter 5 we used the word *kinetics* to describe the study of rates of changes, in particular the rates of chemical reactions. The word kinetics is also applied to nuclear processes. We must analyze the kinetics of nuclear decay processes to properly assess both the hazards posed by radioactive nuclides and their usefulness.

14.2 | The Kinetics of Nuclear Decay

We cannot predict when a given nucleus will decay, but the overall rate of decay within a sample of any particular nuclide follows a remarkably predictable pattern. Each radioactive nuclide is associated with a particular **half-life;** the time taken for half of the atoms in the sample to undergo their decay process. The half-lives of several nuclides discussed in this chapter are given in Table 14.4.

Some half-lives are incredibly short, but others are incredibly long. The Earth is around 5 billion years old, yet almost half of the radioactive uranium-238 that existed when the Earth formed is still here. The osmium-186 nuclide has a half-life of 2×10^{15} years! It is possible that *any* particular osmium-186 nucleus on earth could decay by α-emission right now, but it will still take 2×10^{15} years for half the osmium-186 on Earth to decay. On the other hand, the half-life of nobelium-250, which decays by spontaneous fission is 250 μs. As we shall see later, the wide range of nuclide half-lives is one reason for their usefulness. It is also one reason for their danger.

Isotopes with short half-lives ("short" means anything between fractions of a second and a few years) can emit high levels of radiation, causing damage in spite of their short lifetimes. Isotopes with very long half-lives of thousands or millions of years are going to be with us into the future for longer than all of recorded human history, but their radiation levels are not too great because their decay is so very slow. Many people feel that the most dangerous isotopes are those with half-lives in the range of 10 to 100 years. These are short enough that the intensity of radioactive decay can be fairly great, but their half-lives are long enough that it becomes difficult to keep them isolated for the several half-lives necessary for them to "cool off."

On March 28, 1979, a nuclear reactor accident occurred at the Three Mile Island nuclear power station near Harrisburg, Pennsylvania (Figure 14.8). Operator error resulted in the release of substantial amounts of radioactive nuclides into the environment. The largest release was of xenon-133, which has a half-life of 5.2 days, decaying by beta emission to the stable cesium-133 nuclide. Given that *half* of the xenon-133 released decayed in 5.2 days, can we assume that the other half decayed 5.2 days later or that *all* the xenon-133 was gone in 10.4 days? The next section gives the answers.

Figure 14.8
The most well-known nuclear accident in the United States took place at the Three Mile Island nuclear power station near Harrisburg, Pennsylvania, on March 28, 1979. Substantial amounts of xenon-133, a nuclide with a half-life of 5.2 days, were released.

Table 14.4

The Half-Lives of Nuclides Discussed in the Chapter

Nuclide	Mode of Decay	Half-Life
^{250}No	Spontaneous fission	250 μs
^{131}I	Beta emission	8.04 days
^{192}Ir	Beta emission	73.83 days
^{222}Rn	Alpha emission	3.82 days
^{3}H	Beta emission	12.26 yr
^{235}U	Alpha emission	7.0×10^{8} yr
^{238}U	Alpha emission	4.51×10^{9} yr
^{239}Pu	Alpha emission	2.411×10^{4} yr
^{60}Co	Beta emission	5.272 yr
^{14}C	Beta emission	5730 yr
^{40}K	Beta emission	1.25×10^{9} yr

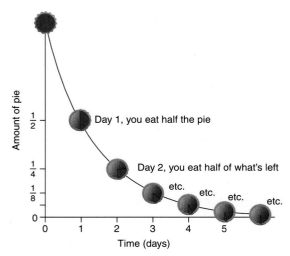

Figure 14.9
The "chocolate cream pie analogy" is an excellent way of visualizing the concept of half-lives and time. Imagine that the pie has a half-life of 1 day. That is, each day you eat one-half of the remaining amount of the chocolate cream pie. After the first day, 1/2 of the original pie is left. On the second day (2 half-lives), you eat half of the remaining pie (1/2 of 1/2). You now have 1/4 of the pie left. By the end of the third day (three half-lives), you have eaten 1/2 of the remaining pie, and you are left with $1/2 \times 1/2 \times 1/2 = 1/8$ of the original pie. It is therefore possible to predict how much of the pie you will have after any number of days. It is also possible to predict how many days it will take to finish a given amount of the pie. Food for thought: how many days will it take for the pie to be all gone?

Half-Life and the Chocolate Cream Pie Analogy

Suppose you are hungry and have a sugar craving. You buy a chocolate cream pie and eat half of it because you always like to leave half of any "treat" for later. The next day you again eat half of the pie, but eating half of what is available amounts to eating only one-quarter of the original pie. The day after that, you eat half of what is left of the pie, so you eat one-eighth of the original. The problem with following your "eat only half" policy is that you will never finish the pie, but will consume increasingly smaller amounts while always leaving half of what you began with at the start of each day (see Figure 14.9).

Radioactive nuclides decay in a way that rigorously follows the chocolate cream pie analogy. Each half-life results in less and less of the original remaining, but some will still remain (see Figure 14.10).

Let's pause and review where we are in our study of nuclear processes. At this point we know that:

- Many nuclei are energetically unstable and liable to decay until they attain a stable form.
- There are two important particles, alpha and beta, that can be emitted when nuclei decay. Gamma radiation is also given off during nuclear decay.
- The decay process often occurs in a series of steps.
- Particles emitted as nuclei decay are very energetic.
- The major negative consequence of nuclear decay for humans is the health hazard posed by radiation released during the decay processes.
- Nuclides decay at very different rates with half-lives ranging from microseconds to much more than the age of the universe.

You should now have sufficient basic understanding of nuclear processes to seriously begin to explore the applications and assess the hazards of radioactivity.

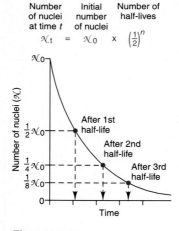

Figure 14.10
Radioactive nuclides decay as modeled by the chocolate cream pie analogy.

14.3 | Applications of Natural Radioactivity

Hundreds of radioactive nuclides are commercially available and many have been exploited for a very wide variety of commercial and medical applications. What follows is by no means an exhaustive survey. It is intended to give you an impression of the usefulness of radioactive nuclides by considering a few specific examples. As you read through each application, consider, "why is that particular nuclide used? Is it related to its half-life and/or the type of particle it emits, and/or any other reason?"

Commercial Applications

Finding the Fault

A metal pipe is being laid underground. The pipe sections are welded together. Are there cracks or gaps in the weld? **Gamma-ray photography** can supply the answer. As shown in Figure 14.11, a sample of iridium-192 (half-life = 73.8 days) is placed in the pipe and photographic film is wrapped around the weld. Gamma radiation emitted by the nuclide can penetrate the faults in the weld and will show up on the film when it is developed. This type of examination can be used to detect cracks in pipelines.

consider this:

When Is All of a Given Isotope Gone?

One way to answer this is to say "never"; we only go halfway to completeness in one half-life so we will never arrive at zero remaining. This isn't quite true, however, since we know that only a certain (very large) number of atoms are present in any sample and that eventually the last atom of a given isotope will undergo radioactive decay. Suppose that we begin with 6×10^{23} atoms of an isotope. After the first half-life, 3×10^{23} atoms would remain. After the second half-life, 1.5×10^{23} atoms of the isotope are still there. We can carry this forward many times (although, as with most such mathematical operations, there is a formula that allows us to solve explicitly for the answer). It turns out that after 79 half-lives, Avogadro's number of atoms of the radioactive isotope has been diminished to a single atom. When this atom decays, the sample has completely decayed. However, since half-lives are a statistical measure (that is, after a half-life, half the atoms are gone, on average), we cannot say for sure when that final atom will decay.

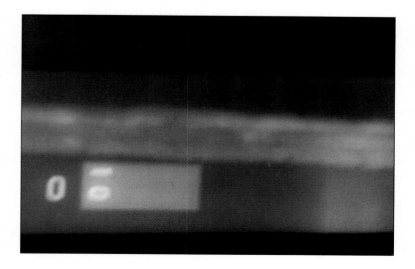

Figure 14.11
Iridium-192 can be used for detecting cracks or gaps in welds, as discussed in the text.

exercise 14.5

Half-Lives

Problem

We discussed the chocolate pie analogy in order to determine whether all of the xenon-133 (half-life = 5.2 days) decays after 10.4 days. Does it? How much of an original 20-gram sample remains?

Solution

The value 10.4 days represents two half-lives. We know from our analogy, and Figure 14.10, that after two half-lives $1/2 \times 1/2 = 1/4$ of the original sample remains; therefore:

$$1/4 \times 20 \text{ g} = 5 \text{ g } ^{133}\text{Xe remains}$$

Note: For additional clarity, you may wish to draw a descending curve of radioactive intensity (proportional amount of sample remaining) versus time as in Figure 1. Draw a horizontal line at 50% of the original height and then down to the time axis. This time is called the half-life of the isotope. At double, triple, and quadruple the half-life, draw lines across to the intensity axis. Notice that at each of these times the intensity has been cut in half. After two half-lives, half of one half ($1/2 \times 1/2 = 1/4$) of the original intensity (number of radioactive atoms) remains.

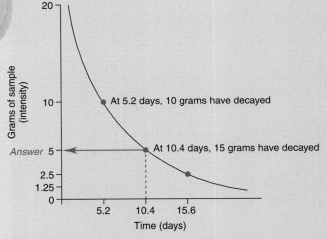

Figure 1
This type of curve can be used to determine how much of an original radioactive nuclide remains in a sample, as used in Exercise 14.5.

Detecting the Smoke

In a smoke detector (see Figure 14.12) a sample of americium-241 (half-life = 432.2 years) emits alpha particles that ionize the surrounding air, allowing current to flow in a circuit. Smoke from a fire blocks the flow of alpha particles (remember their poor penetrating ability), reducing the current flow to a level that causes the alarm signal to be triggered.

Controlling the Thickness of Paper

Paper products have varied thicknesses. Newsprint is very thin, while construction paper and poster board are relatively thick. In the paper manufacturing process, pulp is fed into rollers (see Figure 14.13) that press the newly formed paper to a desired thickness. A source

exercise 14.6

Practice with Half-Lives

Problem

Iodine-131 (half-life = 8.04 days) can be ingested primarily from cow's milk and from the air. It can accumulate in the thyroid gland and can be especially hazardous to children. A sample of air containing iodine-131 was collected and allowed to sit for 32 days before analysis. Why is this a problem? How might you make a mathematical correction to find out how much iodine-131 was originally in the sample?

Solutions

The delay before analysis is a problem because iodine-131 will be decaying to xenon-131 (via beta particle emission) as it sits. After 32 days, about four half-lives will have passed, meaning that $1/2 \times 1/2 \times 1/2 \times 1/2 = 1/16$ of the original sample remains. To make a correction, multiply the amount you determined to be in the sample after 32 days by 16 to get the amount in the original sample. Keep in mind that any error in the analysis gets magnified when you multiply. Therefore, it would have been best to do the analysis as soon as possible after collecting the samples.

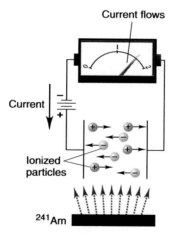

Figure 14.12
Smoke from a fire blocks the flow of alpha particles from a sample of americium-241. When these particles, which ionize the air around them to complete a circuit in the smoke detector, are blocked, the current flow is severely reduced and an alarm is triggered warning of danger.

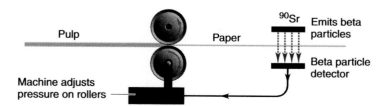

Figure 14.13
In the paper manufacturing process, pulp is fed into rollers that press the newly formed paper to a desired thickness. A source containing the beta particle emitter strontium-90 (half-life = 90 years) is placed beyond the rollers. Fewer beta particles can pass through thicker paper. In this way, the thickness of paper can be related to the amount of radiation detected and a control signal sent back to maintain appropriate setting of the rollers.

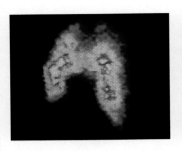
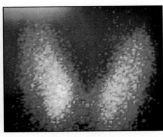

Figure 14.14
Iodine-131 is an important radionuclide that can be used in thyroid scans to assess nutritional deficiencies and tumor growth, among other health problems. Such a scan, made by beta emissions of absorbed iodine-131 exposing photographic film, is shown in the figure.

containing the beta particle emitter strontium-90 (half-life = 28.1 years) is placed beyond the rollers. Fewer beta particles can pass through thicker paper. In this way, the thickness of paper can be related to the amount of radiation detected and a control signal sent back to maintain appropriate setting of the rollers.

Medical Uses of Radioactivity

Radioactive nuclides are used extensively in medicine. Their primary applications are as cancer therapy agents and as tracers. If a tracer has sufficiently complete coverage of an area, it can be used in **imaging** in which a computer-generated picture of, for example, a brain, liver, or thyroid can be prepared. This is especially important in the detection of tumors (see Figure 14.14).

■ case in point

Carbon Dating

We emphasized in the Prelude that science is self-correcting, and from time-to-time there is a need to correct past errors. A good case in point deals with a technique known as **carbon-14 dating**. The method relies on the natural production of the carbon-14 (^{14}C) nuclide in the atmosphere. This unstable carbon isotope is produced by the bombardment of nitrogen-14 (^{14}N) with neutrons from cosmic radiation:

$$^{14}_{7}N + ^{1}_{0}n \rightarrow ^{14}_{6}C + ^{1}_{1}H$$

The ^{14}C produced decays via beta particle emission to re-form ^{14}N:

$$^{14}_{6}C \rightarrow _{-1}^{0}e + ^{14}_{7}N$$

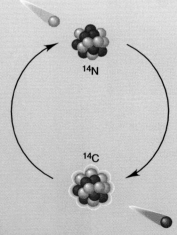

Figure A
Carbon-14 is produced naturally in the atmosphere by the bombardment of nitrogen-14 by neutrons from cosmic radiation with release of a proton. The carbon-14 then cycles back to nitrogen-14 via beta particle emission. Thus, it is assumed that the amount of carbon-14 in the atmosphere has remained constant over time. Recent measurements have shown that this is not quite true. This has forced scientists to make a correction when using carbon dating for objects in the range of 20,000 years old.

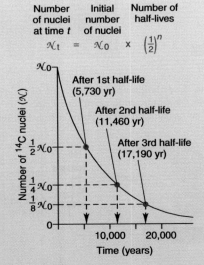

$$\underset{\substack{\text{Number} \\ \text{of nuclei} \\ \text{at time } t}}{\mathcal{N}_t} = \underset{\substack{\text{Initial} \\ \text{number} \\ \text{of nuclei}}}{\mathcal{N}_0} \times \underset{\substack{\text{Number of} \\ \text{half-lives}}}{\left(\frac{1}{2}\right)^n}$$

After 1st half-life (5,730 yr)
After 2nd half-life (11,460 yr)
After 3rd half-life (17,190 yr)

Figure B
Remember the "chocolate cream pie analogy"? This is an important understanding when determining the age of objects by carbon dating.

It had been assumed that since ^{14}C is produced from ^{14}N and decays back to ^{14}N, the total amount of ^{14}C in the atmosphere (less than 1000 kg) *remains constant* over time. Virtually all of the ^{14}C is present as carbon dioxide.

Living objects (people, animals, trees, etc.) exchange carbon, as CO_2, with the atmosphere during respiration and photosynthesis (see Chapter 5). So as long as they are alive, living things should contain the same ratio of ^{14}C:^{12}C as the atmosphere. When death occurs, ^{14}C is no longer taken into the living thing, so the rate of beta emissions from the decay of ^{14}C within the once-living material falls over time as the carbon-14 nuclide decays. Carbon-14 has a half-life of 5730 years, so determining the ^{14}C:^{12}C ratio of material that was once alive (including wood, paper, leather, etc.) has been used to determine the age of objects between about 500 and 20,000 years old.

Remember, however, that you need to compare the ^{14}C:^{12}C ratio of an object with the atmospheric ratio existing when *the object was alive!* Until recently, it has been assumed that the ^{14}C:^{12}C ratio in the atmosphere today is the same as the ^{14}C:^{12}C ratio in the atmosphere of thousands of years ago. We now know that this is not quite true (see Figure C). Recent research has shown that long-term changes in the Earth's magnetic field have modified the amount of cosmic radiation reaching the Earth's atmosphere. These changes are believed to have caused a significant increase in the amount of ^{14}C produced at certain times in the past. This means that objects dated using the ^{14}C technique are perceived to be somewhat younger than they actually are. The discrepancy is especially severe for samples roughly 20,000 years old, where there is a difference of about 3500 years between the age obtained by ^{14}C dating and more reliable methods. The error for younger objects is less. However, knowing the differences in ^{14}C concentration has allowed for correction of the ages of objects.

The most famous artifact that has been dated by ^{14}C dating is the shroud of Turin. This is a large piece of linen material said to be the shroud in which Christ was buried. The fabric bears an image of a man who was wounded and died in the manner described in the biblical account of Christ's crucifixion. The image was said to be somehow the result of the energy of the resurrection event. The history of the shroud is well documented back until the year 1300, but its authenticity has always been in question. If it were authentic, the linen of the cloth would have been grown and harvested about 2000 years ago but if it were a clever forgery, the linen would be much more recent. The Roman Catholic Church originally would not give permission to do carbon dating because it would involve burning a sizable piece of the shroud to convert the carbon of the fabric into CO_2, which would be examined for ^{14}C content. Only after the dating procedure had been improved, such that a much smaller amount of fabric was needed, did the Church give permission for the test to be done. Several very reliable laboratories were chosen. Small pieces of the shroud and other ancient linens of known age were submitted without the analysts knowing which sample was which. All the laboratories reported that the shroud linen was only about 800 years old so could not be the burial cloth of Christ. The current scientifically based understanding is that the shroud was woven and the image imprinted shortly before its first public appearance in 1356.

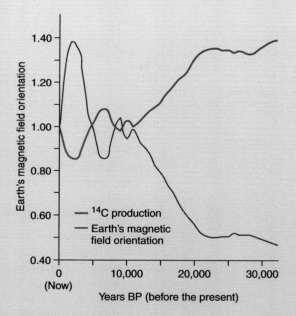

Figure C
Recent research has shown that long-term changes in the Earth's magnetic field mean that the amount of cosmic radiation reaching the Earth has not been constant. Specifically, the amount of carbon-14 produced at certain times in the past has been much higher than today, resulting in a ^{14}C to ^{12}C ratio that is artificially high. This means that fewer half-lives have passed than previously thought.
Source: Bard, E. et al. *Nature*, May 31, 1990, vol. 345 pp. 405–409.

Table 14.5 lists some of the radioactive nuclides used in diagnostic medicine. There are several reasons that these nuclides were chosen for use:

•   They are most often gamma-ray emitters. This is important because the radiation must be detectable outside the body.

Table 14.5

Some Nuclides Used in Diagnostic Medicine

Nuclide	Half-Life	Decay	Application
^{198}Au	2.69 d	β	Kidney function assessment
^{131}Ba	11.8 d	E.C.	Location of bone tissue
^{11}C	20.3 m	β$^+$, E.C.	Brain, pancreas imaging
^{51}Cr	27.7 d	E.C.	Assessment of kidney function
^{64}Cu	12.7 h	β, β$^+$, E.C.	Liver disease diagnosis
^{18}F	110 m	β$^+$, β$^-$, E.C.	Bone scanning, cerebral sugar metabolism, imaging
^{59}Fe	44.5 d	β$^-$	Determination of iron metabolism in blood
^{67}Ga	78.3 d	E.C.	Tumor localization
^{197}Hg	64.1 d	E.C.	Assessing spleen function, brain scanning
^{125}I	59.9 d	E.C.	Determination of blood hormone level
^{131}I	8.04 d	β$^-$	Location of kidney cysts, detection of blood clots, thyroid cancer
^{79}Kr	35.0 h	β$^+$, E.C.	Assessment of cardiovascular system
^{13}N	10.0 m	β$^+$	Lung function test, brain, heart, and liver imaging
^{24}Na	15.0 h	β$^-$	Detection of blood clots
^{15}O	122 s	β$^+$	Lung function test
^{75}Se	119 d	E.C.	Shape and size of pancreas
^{99m}Tc	6.0 h	I.T.	Imaging of many body parts including brain, lung, and kidneys
^{133}Xe	5.25 d	β$^-$	Lung capacity measurement

E.C. = electron capture   s = seconds
β$^-$ = beta emission   m = minutes
β$^+$ = positron emission   d = days
I.T. = internal transformation   y = years
"m" in the nuclide ^{99m}Tc means "metastable," or that the technetium nucleus is especially high in energy. It decays (γ-emission, $t_{1/2}$ = 6.0 hours) to ^{99}Tc.

- They have relatively short half-lives. This is important because longer radiation exposure means more cell damage.
- There must be an efficient route for them to get out of the body, such as by urination or sweating.
- They must have chemical properties that will not interfere with normal body processes.

Several nuclides are available whose gamma and/or beta emissions kill cancer cells. The large doses of radiation involved are aimed at the tumor or the nuclides are injected directly into the tumor. The treatment often results in healthy cells being destroyed along with the cancerous ones, so one of the main concerns during such treatment are to maximize the damage to cancer cells while minimizing the damage to healthy cells. Among the nuclides used in cancer therapy are iridium-192, radium-226, cobalt-60, gold-198, and phosphorus-32.

There is no single commercial or medical use of radioactive nuclides that is entirely risk-free, yet it is easy to argue that the benefit in human lives saved far outweighs the risk to life. Such an argument is possible but not as easy to make when the nuclide is used as part of an energy-generating power plant or a nuclear warhead. More people are affected and the stakes are higher by many orders of magnitude. The use of the power of the nucleus in war and for the production of energy are the two most dramatic and most debated aspects of nuclear issues. Let's consider them now.

14.4 | The Making of a Nuclear Bomb

Many countries have been successful at obtaining nuclear weapons technology. Others are trying hard, only to be thwarted by economic, technological, or military actions of other countries. Our focus in this section is on the atomic nature of the bomb. This requires us to understand nuclear fission.

case in point

Positron Emission Tomography

Wouldn't it be useful to physicians to be able to see exactly where certain molecules go in a living human body! The technique called positron emission tomography (PET) allows such measurements. PET involves measuring exactly the location in the body where radioactive decays occur and knowing exactly the molecule that has included the radioactive atom. Comparison of results from healthy and ill patients yields understanding of the causes of illness.

To understand PET, we must first talk about a fourth kind of radioactive decay. Of the first three kinds of radioactivity, alpha and beta particles are recognizable parts of atoms and gamma rays are electromagnetic radiation of very high energy. Positrons, the fourth kind of radioactive emission, are not part of ordinary matter. A positron is just like an electron but it has a positive charge. A positron is created and ejected from the nucleus when a proton in a nucleus is converted into a neutron.

$$\text{proton} \rightarrow \text{neutron} + \text{positron}$$

Since a positron is just like an electron except for its charge, positron emission is often called "beta-plus" and given the symbol β^+ in contrast to ordinary beta emission which then must be called "beta-minus" and be given the symbol β^-. Let's balance the equation for the formation of positrons.

$$^1_1H \rightarrow ^1_0n + ^0_1\beta^+$$

Atoms that undergo positron emission are those whose nuclei have fewer neutrons than the stable number. For instance, ^{11}C and ^{18}F have fewer neutrons than the stable isotopes of ^{12}C and ^{19}F.

Positron emission from these atoms proceeds as:

$$^{11}C \rightarrow ^{11}B + \beta^+$$

and

$$^{18}F \rightarrow ^{18}O + \beta^+$$

We are most interested in the positron and its fate. When a positron and an ordinary electron collide, they **mutually annihilate** each other and their mass is converted into energy in the form of two gamma rays of a characteristic energy. Since a positron leaving a nucleus must strike a swarm of electrons immediately outside the nucleus, it can't get far, so the gamma radiation originates very near the atom which gave off the positron. All we have to do is figure out where the gamma rays came from. This is done by sensing the directions in which the two gamma rays from each positron event are traveling and trace their paths back to an originating point, which is where the molecule containing the positron emitting atom was located.

The brain uses glucose as a source of energy. The photos in Figure A show the activity of glucose in the brain. Actually, the glucose used has a ^{18}F replacing one of the OH groups in the glucose molecule. This molecule has the same shape and size as a true glucose so goes to the same place in the brain. Because the half-life of ^{18}F is only 112 minutes everything must be done fast. That means that the fluorine-substituted glucose must be made very quickly, fed to the patient very quickly, and measured for several half-lives. The entire procedure is over in a few hours.

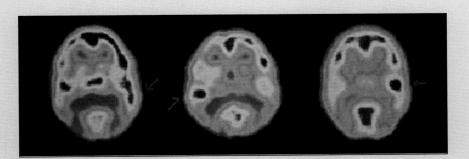

Figure A
This PET (positron emission tomography) scan shows brain activity with changing colors. The red areas indicate high activity levels as this individual listens to musical tones. The area of activity changes as the tones alter sequence and timbre. Changes in blood flow and glucose uptake accompany brain activity.

The Nature of Fission

As we have already mentioned, nuclear fission is the splitting of a heavy, energetically unstable nucleus into lighter, more stable ones. The nucleus of an atom of uranium-235, the isotope of uranium used in the Hiroshima bomb, can undergo induced fission when it is hit by a stray neutron (see Figure 14.15). This can form a variety of fission products, mostly nuclei of atoms with mass numbers between 85 and 160. An example of one possible fission reaction is summarized in this nuclear equation:

$$^{235}_{92}\text{U} + ^{1}_{0}\text{n} \rightarrow ^{87}_{35}\text{Br} + ^{146}_{57}\text{La} + 3^{1}_{0}\text{n}$$

The main point to notice here is that several neutrons are formed for each one that serves to initiate the reaction. The three neutrons formed could potentially start three other fissions that, depending on the products, could yield one, two, or three neutrons to initiate more fissions. This allows an accelerating nuclear "chain reaction" to occur, as illustrated in Figure 14.16. Each fission process releases a very large amount of energy, much more than the energy changes found in chemical reactions, because significant amounts of matter are converted into energy during these nuclear transformations. Remember the equation $E = mc^2$ introduced in Chapter 1. It reveals that large amounts of energy can be released from the destruction of small amounts of matter. In the Hiroshima bomb an amazingly fast fission chain reaction among uranium-235 atoms released in an instant enough energy to destroy a city. In nuclear power stations, similar chain reactions are carefully controlled, to allow them to be used as a source of heat energy that can be used to generate electricity (as we discuss later). In all cases where we try to employ the power of the nucleus, we are exploiting the vast amounts of energy that can be released when small quantities of matter are destroyed during nuclear transformations.

The key to making a huge and virtually instantaneous release of energy by nuclear fission (i.e., an atomic bomb) is to ensure that the neutrons that are released during fission actually cause more fissions rather than escaping into the atmosphere. Only in this way can you ensure that the rate of fission quickly rages out of control. In technical jargon, you want to make your packet of fissionable material (such as uranium-235) "*supercritical*," which is one of the three levels of fission found within a mass of fissionable material:

- **Subcritical** condition means the situation in which there are not enough free neutrons available to maintain fission. This is the "normal" situation.

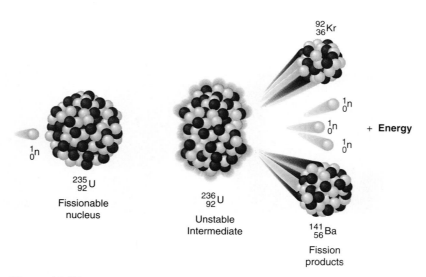

Figure 14.15
Induced fission of a heavy nuclide results in two lighter, more energetically stable, nuclides. This was the basis for the Hiroshima bomb and, when well-controlled, most nuclear reactors. The nucleus captures a neutron and becomes unstable in the process. It distorts and splits into two smaller fragment nuclei and free neutrons. The size of the fragmented nuclei and the number of free neutrons vary considerably.

- **Critical** condition means the situation in which there are just enough free neutrons to maintain a controlled fission. This occurs in nuclear power plants where the energy from the fission process can be tapped without losing control of the *fission rate*.
- **Supercritical** condition means the situation in which so many neutrons are available that the fission process goes out of control, creating a nuclear explosion (as used in atomic bombs).

How do we make a sample of fissionable material supercritical? In essence, we need to bring enough of it together to ensure that enough neutrons hit other uranium nuclei to cause further fission and release yet more neutrons. With a sufficient mass of material, an explosive chain reaction can be initiated. Of course, you want the explosion to occur where and when you choose, so careful control over the start of the process is essential. This control is achieved in a nuclear bomb by controlling when and where the critical amount of fissionable material comes together as we consider in the next section. When we move on to discuss nuclear reactors, we will look at other ways of controlling the fission process.

The Structure of an Atomic Bomb

Various uranium isotopes make up between 2 and 4 g of every 1000 kg of the Earth's crust. The uranium comes from such minerals as pitchblende (U_3O_8) and carnotite ($K_2(UO_2)_2(VO_4)_2 \cdot 8H_2O$). More concentrated samples can be recovered as a by-product of gold and copper mining. Only about 0.7% of all uranium is ^{235}U. This isotope can be separated from the rest of the uranium (^{238}U) by gaseous **diffusion,** a technique in which uranium

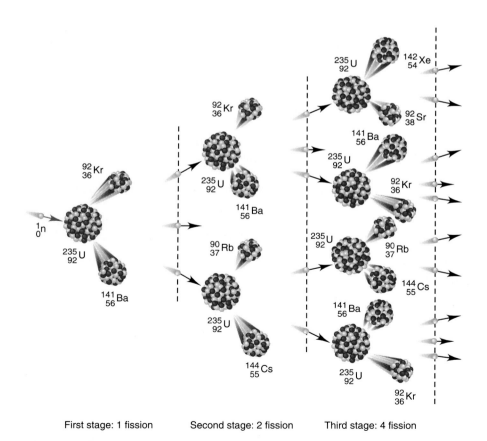

First stage: 1 fission   Second stage: 2 fission   Third stage: 4 fission

Figure 14.16
The release of neutrons from the fission of unstable nuclides can, if unchecked, cause an uncontrolled "chain-reaction" in which more and more available neutrons lead to more fission, leading to more neutrons, leading to a violent explosion that can occur in microseconds after the unchecked process starts.

exercise 14.7

Effect of a Nuclear Explosion on Gases

Problem

It is estimated that a nuclear explosion can result in temperatures in the region of $1 \times 10^8\,°C$. By what factor would a volume of gas, originally at $27\,°C$, expand as a result of such high temperatures assuming the pressure remains constant? What might be the effect of the expansion?

Solution

Recalling Charles' Law from Chapter 11, we know that the volume of a gas expands in direct proportion to temperature in kelvins (at constant pressure). V_1 is the original volume before the explosion and V_2 is the volume after the explosion.

$$V_1 = V \qquad T_1 = 27\,°C + 273.15 = 300\text{ K}$$
$$V_2 = ? \qquad T_2 = 1 \times 10^8\,°C + 273.15 \cong 1 \times 10^8\text{ K}$$

Rearranging,

$$\frac{T_2}{T_1} = \frac{V_2}{V_1} \text{ (ratio of volumes)}$$

$$\frac{1 \times 10^8}{300} = \frac{V_2}{V_1} = 3 \times 10^5$$

This tells us that the gas will expand by a factor of 300,000 in well under 1 second! This is, in large part, what leads to the immediate devastation from such bombs. The air exposed to the explosion undergoes a dramatic and almost instantaneous expansion, causing tremendous "blast" damage.

hexafluoride, UF_6, a gas, is passed through a series of porous membranes. The lighter $^{235}UF_6$ diffuses more rapidly than the $^{238}UF_6$, leading eventually to a high purity or **enriched** sample of uranium-235.

Plutonium-239, the material used in the Nagasaki bomb, is much more readily isolated than uranium-235, because the plutonium is prepared from uranium-238, which is far more abundant in the Earth's crust. The preparation involves the absorption of a neutron by uranium-238 and subsequent decay to plutonium-239 as follows:

$$^{238}_{92}U + \,^{1}_{0}n \rightarrow \,^{239}_{92}U$$
$$\downarrow$$
$$^{239}_{93}Np + \,^{0}_{-1}e$$
$$\downarrow$$
$$^{239}_{94}Pu + \,^{0}_{-1}e$$

This plutonium is easily separated from the unreacted uranium because they are different elements with different chemical properties. A rough sketch of a fission-based atomic bomb is shown in Figure 14.17. When the device is detonated, conventional chemical explosives surrounding the uranium or plutonium force the separate lumps of subcritical material inward toward each other so that they form a supercritical mass. A *tamper* of tungsten is used to prevent the mass from coming apart as long as possible, so that the initial energy of the explosion does not stop the

Figure 14.17
When a fission-based bomb is detonated, explosives surrounding the uranium or plutonium core push separate, subcritical amounts of the fissionable material together to form a supercritical mass of material. A tamper is used to prevent the mass from exploding apart as long as possible and it also reduces the number of neutrons that escape. The explosion takes place in less than a microsecond.

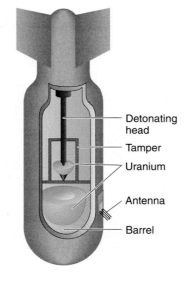

Detonating head

Tamper

Uranium

Antenna

Barrel

fission. The explosion takes place in less than a microsecond. Part of the energy goes into heat and the related expansion of gases. Most of the rest gets emitted as the fast-moving particles and energy known collectively as radioactivity.

14.5 | Using Nuclear Fission to Produce Electricity

Fortunately, very few people have directly experienced the power of the nucleus in the form of atomic bombs. Many more people, however, rely at least partially on nuclear power to generate the electricity they use at home and work. This electricity is generated by trapping some of the heat energy released by nuclear reactors. There are approximately 434 operating nuclear power plants worldwide. Currently the United States has 113 of these plants.

Conventional Reactors

In the case of a conventional nuclear reactor (see Figure 14.18), the fuel source is ^{235}U, which undergoes controlled fission (goes critical) generating heat energy. The heat energy is used to turn water into steam, which in turn drives the turbines that actually generate electricity. The fuel assembly is made up of 100 to 200 **fuel rods**—4 meters long—which contain the uranium-based pellets. The fuel rods are surrounded by a **moderator** such as water, which slows down neutrons (thus lowering their kinetic energy) so they can be captured by the fuel

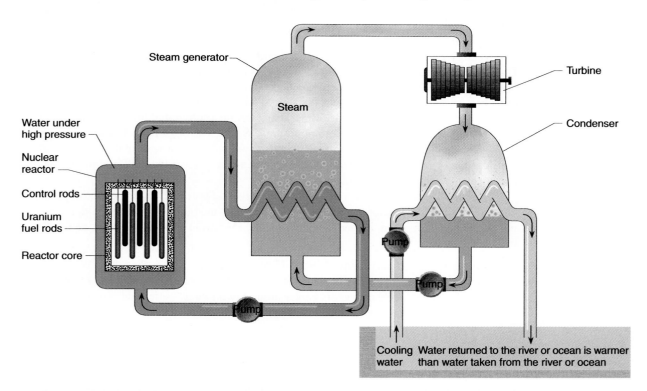

1. **Nuclear reactor:** Water under high pressure carries heat generated in the nuclear reactor core to the steam generator.

2. **Steam generator:** Heat from the reactor vaporizes water in the steam generator, creating steam.

3. **Turbine and condenser:** Steam from the steam generator powers a turbine, producing useable electricity. The condenser uses cooling water from a river or ocean to re-condense the steam from the turbine.

Figure 14.18
The essential parts of a conventional fission reactor are shown here. As discussed in the text, the uranium-based fuel rods undergo fission, which is controlled by neutron-absorbing boron or cadmium control rods. The moderator, typically water, slows neutrons down, thus allowing them to be better captured by the uranium in the fuel rods. Water surrounding the reactor core absorbs heat from the core and is converted to steam, which turns the power plant's turbines, generating electricity.

rods. Interspersed between the fuel rods are **control rods** of boron or cadmium. These are able to absorb neutrons (as shown for boron in the equation below) and, by moving them up and down, the power plant engineers can gain sensitive control over the rate at which the fission reactions proceed within the fuel rods.

$$^{10}_{5}\text{B} + ^{1}_{0}\text{n} \rightarrow ^{7}_{3}\text{Li} + ^{4}_{2}\text{He}$$

The systems that control the fission rate play an absolutely vital role because a typical nuclear reactor contains over 100,000 kg of ^{235}U. Compare this to the 0.45 kg needed for an atomic bomb! Without an appropriate number of control rods in place, the fuel could go supercritical and lead to a disaster.

Breeder Reactors

Just as with atomic bomb fuels, ^{239}Pu can be generated ("bred") from ^{238}U by neutron bombardment for use in nuclear reactors. The same rationale exists here as with bombs—the 238 isotope of uranium exists in nature in greater abundance than the 235 isotope. In a breeder reactor some ^{239}Pu is surrounded by a "blanket" of ^{238}U. As fission of the plutonium occurs, neutrons are absorbed by ^{238}U, which is converted to ^{239}Pu (see the process on page 464).

Therefore, the **breeding** of fissionable plutonium is occurring as heat is generated to turn the turbine. Breeder reactors do not have control rods. More heat is generated than with conventional reactors, so a more effective coolant than water, liquid sodium, is generally used to keep things under control. The reactors are thought to be more dangerous than conventional reactors because of the exceptional danger of plutonium and the reactivity of sodium with water should sodium escape from coolant coils. Just such an accident occurred on December 8, 1995, when several tons of sodium coolant leaked from the Monju reactor, 220 miles west of Tokyo in Japan.

Nuclear reactors are used widely throughout the industrialized world. By 2000, the United States obtained 20% of its power from a total of 104 conventional reactors. In contrast, France has 59 reactors, some of them breeders, and meets 75% of its electricity needs via nuclear power. As Figure 14.19 shows, the United Kingdom, Japan, Korea and many countries in Eastern Europe and the former U.S.S.R. also generate a substantial proportion of their electricity using nuclear reactors. According to the U.S. Nuclear Regulatory Commission, in the year 2000, there were 434 nuclear power plants in operation worldwide and another 62 under construction or on order. The international atomic agency places the 2000 figures at 438 operative plants with 31 under construction.

Nuclear reactors seem like a desirable alternative to fossil fuels. No greenhouse gases are emitted by a nuclear reactor, no acid rain is caused, and the oil politics of the Middle East do not play a part. Why then, is there so much public opposition to nuclear power, especially in the United States?

The Case Against Nuclear Power Plants

No new nuclear reactors for use in power plants have been ordered in the United States since 1978. Objections to nuclear reactors center around two issues—safety during operation and the disposal of radioactive waste such as depleted (used) fuel rods.

Safety During Operation

Even the most minor defects in design, workmanship, or judgment can have severe effects. Here are brief details of some serious reactor incidents:

- Chernobyl Nuclear Power Plant, Pripyat, U.S.S.R.: In early April 1986, several blasts blew the concrete lid off the core of the reactor number 4. An estimated 50 megacuries of radionuclides were released into the air. There were 31 deaths directly resulting from the blast and efforts to put out the fire, mostly firefighters exposed to massive doses of radiation. It is estimated that cancers attributable to this incident could result in the deaths of as many as 75,000 people. Many millions of dollars have been spent on the cleanup.
- Three Mile Island Reactors, Pennsylvania: In 1979, a combination of operator error and faulty equipment led to a loss of coolant, a resulting chemical explosion of built up hydrogen gas, and an escape of radioactive nuclides into the atmosphere. One nuclear

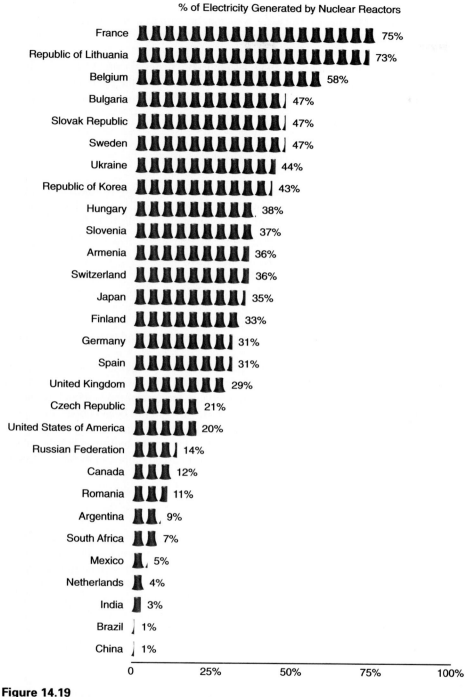

Figure 14.19
Many countries meet much of their energy needs via electricity from nuclear reactors. Can you suggest reasons why certain countries use nuclear energy while others seem to prefer different sources? Are the reasons primarily scientific, political, economic, or some combination of these?

engineer said that only multiple "strokes of luck" prevented a disaster across an area that a U.S. government report said, "might be equal to that of the state of Pennsylvania."

- Dresden Reactor, Illinois: This reactor went "out of control" in 1970, allowing the release of radioactive iodine for 2 hours.
- Lucens Reactor, Switzerland: In 1969, the reactor had a loss-of-coolant accident. The cavern inside a mountain holding the reactor had to be permanently sealed to prevent a massive release of radiation.

- U.S. government test grounds in Idaho Falls, Idaho: In 1961, a miniature test reactor at this facility underwent a nuclear runaway, a sudden buildup of fission rate caused by accidental withdrawal of control rods or earth movement, among other reasons. The resulting explosion killed three workers.
- Windscale Reactor, Great Britain: In 1957, graphite core material caught fire, releasing a cloud of radioactive iodine that traveled hundreds of miles from the plant. Dairy farming was banned for 60 days and cattle slaughtered within a 200 square mile area.

14.6 | The Exploitation of Fusion

As we discussed in Chapter 1, all of the atoms in the universe apart from hydrogen atoms were formed directly or indirectly by nuclear **fusion:** the coming together of two nuclei to produce a larger nucleus. In stars such as our own Sun, hydrogen nuclei fuse to form helium, liberating tremendous amounts of energy as heat and light. The fusion reactions are as follows:

$$2(^1_1H + ^1_1H \rightarrow ^2_1H + ^0_1\beta^+)$$
$$2(^1_1H + ^2_1H \rightarrow ^3_2He)$$
$$^3_2He + ^3_2He \rightarrow ^4_2He + 2^1_1H$$
$$\overline{4^1_1H \rightarrow ^4_2He + 2^0_1\beta^+}$$

The fusion reactions that occur on the Sun bathe us in heat and light and so provide the energy needed to sustain life on Earth.

Tremendously high temperatures in the region of 10^8 K are required to allow the positively charged nuclei to collide with enough energy to overcome their repulsive forces and participate in fusion. Sufficiently high temperatures exist in the core of the Sun but have never been sustained for significantly long periods on Earth. However, it has been known since 1951 that it is possible to generate high enough temperatures for the brief time needed to initiate *uncontrolled fusion*. Uncontrolled fusion is the process that powers the hydrogen bomb. Fusion releases much more energy per atom than fission, so fusion-based bombs can release up to 100 times the energy of fission-based devices.

Structure of the Hydrogen Bomb

The key to making fusion possible in the hydrogen bomb (sometimes called a **thermonuclear bomb** because of the heat required) is that uncontrolled *fission* can generate the temperatures necessary to initiate uncontrolled *fusion*. This principle was understood even before the atomic bombs were detonated over Japan in 1945. It was made practical in 1951 by the work of the mathematician Stanislaw Ulam and the physicist Edward Teller. Ulam proposed that mechanical shock from a fission bomb could be used to compress the hydrogen-based fuel that would initiate fusion. Teller added the key idea that *radiation pressure* from the fission could achieve the compression. The Teller-Ulam idea proved successful. The first hydrogen bomb was detonated on November 1, 1952.

Teller asserted that both fission and fusion bombs could be used for nonmilitary purposes such as blasting huge ditches to make it easier to explore for natural resources. Although nuclear bombs have yet to be used in such applications, the processes of producing fissionable material, storing warheads, and occasionally detonating them will have consequences far into the future.

Nuclides of Concern Released by Thermonuclear Explosions

Many radioactive nuclides are released from the detonation of a nuclear bomb. Some of the nuclides that are formed immediately decay into other, longer lasting nuclides. Some of the nuclides that are of greatest long-term concern are the following.

pro con discussions

Nuclear Power Plants

pro

Nuclear power plants produce electricity the same way that conventional plants do; they produce heat to boil water to make steam that expands and turns a shaft in a turbine that generates the electricity. Energy is neither created nor destroyed, merely converted from one form into another. In the case of nuclear power, the energy originates as the binding energy of heavy nuclei, specifically the 235 isotope of uranium. The electricity produced is indistinguishable from that produced by burning fossil fuels.

Here is the choice: which source of energy is more desirable? The coal-fired plant requires that huge amounts of coal be dug from the ground, transported to the power plant, burned, and the ash removed from the furnace and hauled someplace for disposal. Hundreds of people have died in coal-mining accidents. The most abundant element in coal is carbon, which forms carbon dioxide on burning. This, as we have said earlier, is a major contributor to atmospheric pollution, specifically, global warming. Coal also contains lots of material that isn't carbon. The most well-known impurity in coal is sulfur. Sulfur occurs in several forms: elemental sulfur, organic compounds which contain a sulfur atom, and metallic sulfides (often iron pyrite, FeS_2). This all ends up as sulfur dioxide and sulfur trioxide, which either goes up the smokestack or is trapped for later disposal. Many metals (often including small amounts of uranium) and clays are also part of what goes into a coal-fired power plant. The oxides of metals fused into the clay becomes the solid ash which must be disposed of.

This description of burning of coal in a section devoted to a description of nuclear power is necessary because those who say that we should not use nuclear power are therefore saying that we should rely more heavily on coal-fired power. As we will see in Chapter 15, there are other potential sources of energy, but nothing is on the horizon other than nuclear that will supplant coal as an available power source for electricity generation. A theme of this book has been that we must look at the big picture and count both the benefits and the risks of any decision. We cannot focus simply on the risks of nuclear power and the benefits of coal-fired power. We must include the risks and benefits of each mode in deciding whether or not to continue to use nuclear power.

Nuclear power does not emit smoke or particulate matter, does not contribute to either the greenhouse effect or acid rain, and results in far less waste material than coal-fired plants. Admittedly, the waste material is more hazardous than from coal-fired plants, but technology exists for safely storing such waste. At present, the United States is trying to decide what to do with the plutonium removed from thousands of warheads from decommissioned missiles and bombs. If it is used as a fuel for generation of electricity, it will become unavailable for further weapons or terrorist uses and we will have succeeded in beating our nuclear swords into plowshares.

con

24,400 years—that is the magic number; 24,400 years is the half-life of the plutonium-239 (^{239}Pu) nuclide. After 24,400 years, 50% of the tons of plutonium generated worldwide will still be here. After 48,800 years, 25% of the plutonium will still be here. It will take a quarter of a million years, 10 half-lives, for the plutonium generated in the past 60 years to be effectively gone. Can we really expect our society to deal responsibly with this? We live in a world in which "long-term" is measured in months and social decisions are only as permanent as the next election. The U.S. speed limit, instituted originally in 1973 to conserve oil, succeeded in conserving 200,000 barrels per day, or 3% of the total amount we use. Nearly 1.5 billion barrels of oil, the amount used in a 2- to 3-year period in

the United States, was saved between 1973 and 1995. Yet in December 1995, the national speed limit was eliminated: how quickly we forget.

Can a country whose policies are based on personal expediency be trusted to create the conditions to care for the world for 250,000 years? Can a world where engineers disable safety controls so that they can get home on time; where for over 40 years, we have dumped millions of gallons of radioactive wastes into a Washington river; where a nuclear power plant on Long Island is built so poorly that it can never open—can this kind of a world be trusted to properly deal with nuclides with half-lives of 24,400 years? Currently, we can hardly protect ourselves from ourselves one day at a time.

Strontium-90

This fission product decays via beta emission to yttrium-90.

$$^{90}_{38}\text{Sr} \rightarrow ^{90}_{39}\text{Y} + ^{0}_{-1}\text{e} \qquad (t_{1/2} = 28.6 \text{ y})$$

(where $t_{1/2}$ = half-life, y = years).

Strontium-90 that falls to the soil is taken up by plants and eventually by species that eat plant foods such as cows and humans. People take in additional strontium-90 by drinking cow's milk. Strontium is a group II (alkaline earth) metal that behaves chemically like another group II metal, calcium, found in bones and teeth. This is an important point: potentially damaging nuclides such as strontium-90 are incorporated into the body because they are chemically similar to substances already in the body. As is shown in Figure 14.20, strontium-90 concentrates in bone.

Cesium-137

Cesium is an alkali (group I) metal as is potassium, so Cs and K have similar chemical behavior. Isotopes of cesium can, like potassium, be found throughout the body, especially in the ovaries of women and in muscle tissue. Cesium-137, found in many fruits, vegetables, dairy, and grain products decays by beta emission as follows:

$$^{137}_{55}\text{Cs} \rightarrow ^{137}_{56}\text{Ba} + ^{0}_{-1}\text{e} \qquad (t_{1/2} = 30.17 \text{ y})$$

Iodine-131

Isotopes of iodine ranging from mass number 127 to 141 have been detected as fission products from nuclear explosions. Most of these isotopes have half-lives on the order of seconds or minutes, so are not of long-term significance. One exception is iodine-131, which undergoes beta decay with a half-life of 8.04 days:

$$^{131}_{53}\text{I} \rightarrow ^{131}_{54}\text{Xe} + ^{0}_{-1}\text{e}$$

This means that measurable amounts of the nuclide are present months after atmospheric nuclear tests. In the thyroid gland, iodine becomes incorporated into the hormone thyroxin,

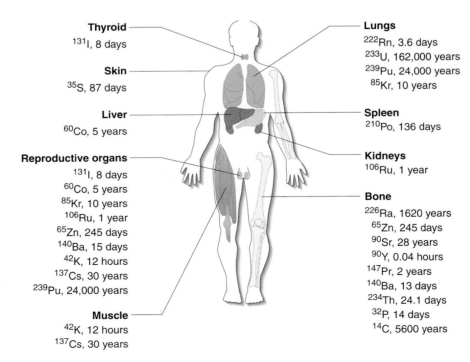

Figure 14.20
Strontium-90 is one of many radionuclides that can concentrate in different parts of the human body.

which stimulates oxygen consumption in the tissues. Therefore, iodine, whether radioactive or not, accumulates in the thyroid gland. Iodine-131 can be ingested in milk and other food items, resulting in unusually high beta and gamma radiation doses in the thyroid. Numerous studies (for example, see References 1, 4, and 13) have raised the possibility of a relationship between nuclear "fallout" (waste products) and growths called thyroid nodules, especially in children.

The Promise of Fusion

Fusion, the coming together of two nuclei to produce a larger nucleus, was discussed in the context of a thermonuclear weapon, using uncontrolled fusion. It is possible to have a short-term controlled fusion of deuterium and tritium nuclei which combine in a complex series of reactions to give helium nuclei and fabulous amounts of energy, owing to the transformation of mass to energy described earlier. The reactions can be summarized,

$$^2_1H + ^3_1H \rightarrow ^4_2He + ^1_0n$$

The key to controlled fusion is to force nuclei to overcome their repulsive forces (like charges repel!). To do this, the experimental goal is to press very energetic nuclei into a small volume for as long as possible. The benchmark value is 10^{14} particle-seconds per cubic centimeter. This means that if you can compress the fuel so tightly that its particle density is 10^{25} particles per cubic centimeter, but only for 10^{-10} seconds, the product is $10^{25} \times 10^{-10} = 10^{15}$ particle-seconds per cubic centimeter. This is greater than the 10^{14} threshold value, and fusion can be achieved.

When designing a machine to achieve fusion, the main point is to compress the fuel for as long as possible. The two important strategies are to use lasers (to generate heat) and magnetic fields (to contain the reaction). In the laser design, energetic beams work to compress the fuel. In the other major design, magnetic fields are used to confine the nuclear material along magnetic field lines (see Figure 14.21). In 1994, the magnetic field-based design at Princeton University generated 10 million watts of power for 0.5 seconds (compared to a commercial power plant that may generate 300 million watts), although more energy was used to fuse nuclei than was obtained from the fusion process. In the near future, an international team, including the United States, the European Union, Japan, and Russia expect to develop the plans for a magnetic field-based reactor that will accomplish controlled fusion for thousands of seconds.

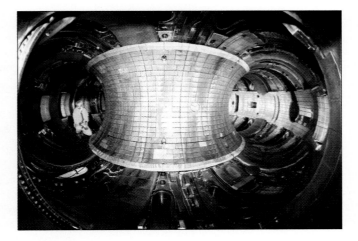

Figure 14.21
Prototype fusion reactors, such as the one shown here, use lasers and magnetic fields to force the fusing of light nuclei as they travel around these torus-shaped structures.

14.7 | Disposal of Radioactive Waste

One of the most pressing issues related to nuclear chemistry facing governments and citizens around the world is the problem of what to do with hazardous and long-lived radioactive waste. The two largest sources of radioactive waste are nuclear weapons and commercial nuclear reactors. Recently signed arms-control treaties, including the July 1991 START I and January 1993 START II agreements, will require well over 20,000 warheads (both United States and former U.S.S.R.) to be dismantled. Planners are not sure what to do with the plutonium-239 in the bombs. It is possible to use it in breeder reactors, but that is not the basis of U.S. nuclear reactors. Others worry about theft from reactors for subsequent reuse in bombs. Waste could be buried underground. The plutonium intended for burial would be **vitrified,** incorporated into borosilicate "glass logs." Each 3-meter long log would contain about 0.5 ton of a plutonium waste slurry mixed with 1.2 tons of the glass. The logs would be encased in steel and buried deep underground. This is the main option for the 40,000 metric tons of radioactive waste expected to be generated by the end of the century.

Another option includes dropping specially shaped waste canisters from ships toward the bottom of the oceans, where the force of their travel to the depths would work with their shapes to push them down several meters into the sea floor. This is currently viewed as a scientifically reasonable alternative. There are, however, political issues relating to international questions such as "who owns the sea floor?" Other options include converting the waste via neutron bombardment to nuclides with shorter half-lives or shooting the waste into space and diluting the impact of the waste by spreading it over the large surface of the oceans.

The biggest single question about most of the disposal proposals is how can anyone guarantee the safety of plutonium-239 (half-life = 24,400 years) for the 10 half-lives—a quarter of a million years—required for the waste to become relatively safe? As a result of such concerns, many states in the United States are vigorously resisting the burial of waste within their boundaries. The phrase "Not in my backyard!" comes to mind.

■ case in point

Hanford: A Disaster in the Making

What happens to the leftovers? Where do we bury the radioactive by-products of nuclear weapons manufacture? Roughly 75% is stored at the Hanford Nuclear Reservation in Washington state, 1 of at least 10 such storage sites across the United States. Among the waste nuclides are isotopes of iodine, cesium, and strontium: fission by-products of the neutron bombardment of ^{238}U, which makes ^{239}Pu, for atomic bombs.

The waste can exist as a liquid or a sludge. It can be acidic or basic and it is often hot due to the release of heat energy that accompanies the high levels of radioactive decay. Since the opening of Hanford in 1944 until its closing in the late 1980s, waste had been poured into trenches with concrete sides but no bottom.

Much of the information regarding storage safety at Hanford was not made available until 1986, when public pressure forced the U.S. Department of Energy to release 19,000 pages of data and history regarding conditions at Hanford. Among the revelations regarding the 1377 waste sites at Hanford are the burial of 6900 metric tons of nuclear materials including uranium, cesium, and strontium. The total radioactivity (as of July 1993) was 446 million **curies,** 10 times that released in the Chernobyl accident. About 20 metric tons of plutonium are at the site. There are 177 storage tanks that contain about 57 million liters of unidentified waste.

Some wastes are being cleaned. Other waste material will be vitrified, starting in 2007, if all is on schedule. According to the Oregon Department of Energy, the first 10% of Hanford's tank wastes will be in glass by 2018. The Department of Energy cites an overall cleanup date of 2070 at a cost of $147 billion—if all continues as planned.

14.8 | Where Do We Go from Here?

Humanity entered the atomic age in 1945. Some see atomic energy as an inhumane weapon of war that should not have been built or used. Others cite the rapid end of the war following the use of atomic weapons. All of us can see the plentiful energy of fission in use all around the world and some can visualize the even more plentiful supply of energy of fusion, which could become available sometime in the 21st century. Some see radioactive pollutants in our air, water, and soil and others see the life-saving medical diagnosis and treatment uses of selected radioisotopes.

All these attitudes about radiation, radioactivity, and atomic energy have a grounding in fact and are at least a part of the legacy of our exploration of the interior of the atom. Our uses of the power of the atomic nucleus can be for good or ill, but political and personal choices in this area should be made in the light of some understanding of the nature of radioactivity.

main points

- The nuclei of some isotopes of some elements are inherently unstable, resulting in the emission of particles and energy from these nuclei so that they decay into stable nuclei.

- Among the most significant ways that isotopes can become more stable is by releasing alpha particles, beta particles, gamma radiation, and, in a few cases, by fission.

- The energy released in nuclear transformations is much greater than that in chemical changes involving electron exchange. This intense nuclear energy has resulted in sharp debate about its problems and possibilities.

- A decay series is said to occur when a nuclide undergoes a series of emissions to reach stability.

- Each unstable nuclide has a rate of nuclear decay characterized as its half-life.

- Radioactive nuclides can be used for applications including dating of old objects, detection of tumors, meeting the energy needs of society, and as weapons of mass destruction.

- Nuclear fission is the splitting of a heavy, energetically unstable nucleus into lighter, more stable ones.

- Fission can occur at the subcritical, critical, or supercritical levels. Nuclear power generation relies on the controlled fission at the critical level. Supercritical fission is at work when nuclear warheads detonate.

- Fusion is the coming together of two nuclei to produce a larger nucleus. In this process, tremendous amounts of energy are liberated as heat and light.

important terms

Alpha particle is the helium nucleus emitted from certain heavy radioactive nuclei. (p. 448)

Beta particle is the electron emitted from nuclei having too many neutrons for stability. (p. 449)

Critical is the mass of fissionable nuclide that will just sustain a chain reaction. (p. 463)

Curie is a measure of the amount of radioactivity equal to 37 billion disintegrations (emissions) per second (DPS). This is the number of radium-226 nuclei in a 1-gram sample of the isotope undergoing radioactive decay in a second. (p. 472)

Decay series is the series of nuclides formed on successive radioactive decay events. (p. 451)

Enriched means having a greater fraction than normal of an unusual isotope. (p. 464)

Fission is a nuclear reaction in which one nucleus fragments into two smaller ones. (p. 447)

Fusion is a nuclear reaction in which two nuclei combine (fuse) into one. (p. 468)

Gamma rays, high-energy electromagnetic radiation emitted from many radioactive isotopes. (p. 450)

Half-life is the length of time necessary for half of an amount of a radioactive isotope to undergo decay. (p. 453)

Imaging is the use of radiation to gain a picture of hidden objects, often internal organs of living persons. (p. 458)

Induced fission is the nuclear fission triggered by a neutron striking an unstable nucleus. (p. 451)

Mutual annihilation, electron and positron collide and are both converted into two gamma rays. (p. 461)

Nuclide is a nucleus of a specific proton and neutron number. (p. 451)

Positron is a short-lived particle having the mass of an electron but a positive charge. (p. 451)

Radiation, energetic particles emitted from atoms undergoing radioactive decay. (pp. 447)

Radioactivity is the spontaneous change in the nucleus of an atom. (p. 447)

Rem is the unit of measure for the effect on humans of the energetic emissions from radioactive elements. An exposure of 1 rem results in 0.01 J absorbed per kilogram of body tissue. (p. 450)

Spontaneous fission is nuclear fission that occurs without the nucleus being struck by a neutron. (p. 451)

Subcritical is when not enough neutrons are available to maintain fission. (p. 462)

Supercritical is a mass of fissionable nuclide that is greater than a critical amount so a chain reaction occurs. (p. 463)

Vitrified is to change or make into glass, especially through heat. (p. 472)

exercises

1. Explain how radioactivity is related to the stability of atomic nuclei.
2. Use the Table of Isotopes in the *CRC Handbook of Chemistry and Physics* to identify the smallest isotope which is radioactive and the largest isotope which is not radioactive.
3. What are the three important types of radioactive particles? Which one is most harmful to humans? Why?
4. How is the ratio of protons to neutrons in a nucleus used to determine stability?
5. Label the stable isotopes in ^{31}P, ^{28}P, ^{37}P, ^{35}Cl, ^{37}Cl, ^{14}N, and ^{16}N.
6. Explain how the unstable nuclei of ^{20}F can become stable.
7. Why is there a difference between an alpha and a beta particle?
8. Indicate the difference between positron emission, electron capture, and spontaneous fission.
9. If ^{208}Po releases an alpha particle, what atom will be formed? Will this atom emit another alpha particle?

10. This reaction represents the emission of an alpha particle to form berkelium-248. Determine the particle present before decay ($^A_Z X$).
$$^A_Z X \rightarrow {}^{248}_{97}Bk + {}^4_2He$$
11. Cesium-137 can be found among the radioactive wastes at the Hanford nuclear research facility; $^{137}_{58}Cs$ emits a beta particle. Write the equation for the decay step.
12. Radon-222 has been identified as contributing to lung cancer. It is a radioactive isotope formed in the decay series of uranium 238. Radon-222 undergoes two alpha particle emissions, followed by two beta particle emissions, another alpha emission, and a final beta emission. What is the final isotope formed in this sequence?
13. Complete these radiochemical changes:
    a. $^{60}Co \rightarrow {}^{60}Ni +$ ____
    b. $^{11}C \rightarrow$ ____ $+ positron$
    c. $^{222}Rn \rightarrow$ ____ $+ alpha$
    d. $^{235}U + {}^1n \rightarrow {}^{93}Zr + {}^{130}Te +$ ____ ^1n
    e. $\beta^+ + \beta^- \rightarrow$ ____
14. Complete these radiochemical changes:
    a. $^{90}Sr \rightarrow {}^{90}Y +$ ____
    b. $^{18}F \rightarrow$ ____ $+ positron$
    c. $^{224}Rn \rightarrow$ ____ $+ alpha$
    d. $^{235}U + {}^1n \rightarrow {}^{97}Mo + {}^{125}Sn +$ ____ ^1n
    e. $^{237}Np \rightarrow {}^4He +$ ____
    f. $^{137m}Ba \rightarrow {}^{137}Ba +$ ____
    g. $^{235}U + n \rightarrow {}^{236}U \rightarrow {}^{60}Co +$ ____ $+ 4n$
15. Identify the missing product(s) in each of these nuclear transformations.
    a. $^{244}Pu \rightarrow {}^4He +$ ____
    b. $^{99m}Tc \rightarrow {}^{99}Tc +$ ____
    c. $^{235}U + n \rightarrow {}^{236}U \rightarrow {}^{90}Sr +$ ____ $+ 4n$
16. Which is uncharged, an alpha particle, a beta particle, or a gamma ray?
17. Which is heaviest, an alpha particle, a beta particle or a gamma ray?
18. Why are alpha, beta, and gamma particles so harmful to the body?
19. Radiochemical analysis has been used often in examination of precious works of art because the analyses can be done nondestructively. In 1968, the daughter products of uranium-238 (an impurity in lead-based paints) were analyzed to determine whether the painting *Christ and His Disciples at Emmaus* was really painted by Jan Vermeer in the 17th century. Modern pigments are purer than those of the 17th century and so have less uranium. Also, a much shorter time lapse would have allowed smaller amounts of daughter isotopes to accumulate. It turned out the painting was a modern forgery. When $^{238}_{92}U$ emits an alpha particle from its nucleus, what nuclide results?
20. Plutonium-239 ($^{239}_{94}Pu$), one of the primary radioactive materials used in nuclear bombs, decays to thorium-227 ($^{227}_{90}Th$) in the following sequence: $\alpha, \alpha, \beta, \alpha, \beta$. Write equations for all five steps.

21. True or False. Chemical reactivity is directly related to nuclear stability.

22. If you have a 0.1-g sample of ^{131}I, how long will it take for the sample size to decrease to 0.05 g? 0.025 g?

23. How much of a 20.0-g sample of a radioactive material is left after three half-lives?

24. A radioactive isotope shows an initial activity of 10,000 counts per minute. Forty-eight hours later its activity is 1250 counts per minute. What is the half-life of this isotope?

25. A radioactive isotope shows an initial activity of 900 counts per minute. Sixty minutes later its activity is 225 counts per minute. What is the half-life of this isotope?

26. How many half-lives have occurred if a radioisotope has lost 75% of its radioactivity?

27. What fraction of a radioactive nucleus remains after four half-lives?

28. A positron emission tomography (PET) scan is used in medicine to give a two-dimensional view of such organs as the brain, heart, etc. One of the isotopes used in a PET scan is fluorine-18 ($t_{1/2} = 109.7$ min). What fraction of fluorine-18 would be left in the brain 5½ hours after injection of a compound enriched with this fluorine isotope?

29. Various short- and long-term detection kits are used to determine the amount of radon in a home. The company you work for counts the gamma rays (which are emitted from lead-214 and bismuth-214) trapped in short-term detection kits. You are analyzing a sample 9 hours after it was collected which contains 5.2×10^{-6}g of lead-214 ($t_{1/2} = 26.8$ min). How many grams of lead-214 were in the original sample? Decide whether you think your answer is reasonable for a home measurement kit.

30. From July to September 1987, a foreign company deposited radioactive wastes at Koko Village, a small port in Nigeria. It wasn't until 1 year later that the wastes were discovered. They were found to contain such things as mercury, arsenic, paint and pigment residues, and bismuth-205 which gives off gamma rays. A study was conducted 6 months after the discovery to determine if the village outbreak of diarrhea and site workers' complaints of chest pains were caused by radiation injury or the villagers' diet. What fraction of bismuth-205 would be left in the blood of a port worker who had skin contact with the material during the cleanup 183 days before the test? The half-life of bismuth-205 is 15.3 days.

31. You are to determine the age of a piece of wood from an ancient campfire by carbon dating. If the carbon-14 content of the sample was determined to be one-fourth of that in living trees, how old would the campfire be ($t_{1/2} = 5730$ years for carbon-14)?

32. A bottle of wine is alleged to be 200 years old. How could radioactivity be used to determine the truth of this claim? Remember, if true, this wine is very valuable so it must not be damaged by the test, but it may be opened.

33. Use the Internet to find a positive use of radioactivity other than those listed in the text. Be sure to reference your source.

34. Several sets of products can result from the fission of uranium-235. Based on your own book or Internet search, propose one set of products not mentioned in the textbook.

35. What are the three mass-based levels of fission? Which level is necessary for use in atomic weapons? Explain your answer.

36. More energy can be produced from the fission of 1 pound of uranium than burning 3 million pounds of coal. Why hasn't nuclear energy replaced fossil fuels as our primary energy source?

37. Describe the synthesis of plutonium-239.

38. When an atomic bomb detonates, not all of the energy created is emitted as radioactivity. Remembering the Law of Conservation of Energy, what happens to the rest of the energy produced?

39. Nuclear reactions produce a large amount of heat. What types of environmental impact can the heat generated by a nuclear power plant have on surrounding ecosystems?

40. Explain the roles of plutonium and uranium in a breeder reactor.

41. How does a breeder reactor work? What method is used to keep the reaction under control?

42. Which nuclear process (fission or fusion) would be better suited as a source of energy for our society? Support your answer.

43. What is the function of a control rod in a nuclear reactor?

44. What key principle was used in the development of the thermonuclear bomb?

45. Why can the Sun be referred to as a natural nuclear reactor?

46. Compare fusion and fission. Which process releases more energy?

47. Compare a hydrogen bomb to an atom bomb.

48. What mineral in the body is similar to strontium-90? Why is strontium-90 so harmful to the human body?

49. What is the "key" to controlled fusion?

50. Describe the major problem associated with nuclear warhead disarmament.

food for thought

51. A sample of gin stored in a sealed glass bottle is purported to be a bottle used in the production of the movie *Casablanca* in 1942 and sealed immediately until this day. It was shown that the ^{14}C ($t_{1/2} = 5730$ years) level of the alcohol (C_2H_5OH) in this bottle is consistent with this age and that the ^3H ($t_{1/2} = 12.26$ years) level in the alcohol is 95% that of atmospheric water. What is your conclusion? Can the sample be authentic? Explain your reasoning.

52. How long would you consider it necessary to allow a sample of xenon-133 to decay before feeling safe in the same room with it?

53. On July 16, 1979, radioactive wastes were released from a collapsed reservoir of uranium mill tailings into the Puerco River near Church Rock, New Mexico. A flood wave then deposited sediments of radioactive thorium-230 (along with other elements) on the channel floor of various cross sections as the flood traveled 80 km downstream. Would this type of accident be as dangerous as the release of radioactive nuclides in the atmosphere from an explosion at a nuclear reaction center? Why or why not?

readings

Books

1. Dragani, Ivan G., Dragani, Zorica D., and Adloff, Jean-Pierre. 1990. *Radiation and Radioactivity on Earth and Beyond.* Boca Raton, FL: CRC Press.
2. Broad, William J. 1992. *Teller's War.* New York: Simon and Schuster.
3. Grossman, Karl. 1980. *Cover Up: What You Are Not Supposed to Know About Nuclear Power.* Sagaponack, NY: Permanent Press.
4. Kathren, Ronald L. *Radioactivity in the Environment: Sources, Distribution and Surveillance,* Harwood Academic Publishers, New York, 1984.

Articles

5. Gup, Ted. Up from ground zero: Hiroshima. *National Geographic,* August 1995, pp. 78–101.
6. Illman, Deborah L. Researchers take up environmental challenge at Hanford. *Chemical and Engineering News,* June 21, 1993, pp. 9–21.
7. Raymer, Steve. Chernobyl—one year after. *National Geographic,* May 1987, pp. 632–653.
8. McCrone, Walter. The Shroud of Turin: Blood or artist's pigment? *Accounts of Chemical Research,* 23, 1990, pp. 77–83.
9. Ahearne, John F. The future of nuclear power. *American Scientist,* January/February 1993, pp. 24–35.
10. Leary, Warren E. Studies raise doubts about the need to lower home radon levels. *New York Times,* September 6, 1994.
11. Sayle, Murray. Did the bomb end the war? *New Yorker,* July 31, 1995, pp. 40–64.
12. Hong, K. P. Residents protest near reactor. Associated Press dispatch in the *Lincoln Journal Star,* December 24, 1995.
13. Furth, Harold P. Fusion. *Scientific American,* September 1995, pp. 174–176.
14. Disposing of nuclear waste. *Scientific American,* September 1995, p. 177.
15. U.S. Department of Energy, Richland Operations Office Site, containing information on the status of cleanup efforts around and in Hanford, www.energy.state.or.us/nucsafe/HCleanup.htm (October 2001).
16. Oregon Office of Energy. Cleaning up Hanford's nuclear weapons wastes (2001). http://www.energy.state.or.us/nucsafe/hanclnup.htm.
17. Wald, Matthew L. Industry gives nuclear power a second look. *The New York Times,* April 24, 2001.
18. Lake, James A., Bennett, Ralph G. and Kotek, John F. Next Generation Nuclear Power. *Scientific American,* January 2002, pp. 72–81.

websites

www.mhhe.com/Kelter   The "World of Choices" website contains activities and exercises including links to websites for The Center for Strategic and International Studies; Lawrence Livermore National Laboratory for a discussion of fusion; Oregon's Office of Energy relating to the Hanford nuclear waste site; and much more!

Solar Power
The Chemical Energy Alternative

Little darlin', I see the ice is slowly melting,
Little darlin', it seems like years since you've been here,
Here comes the sun, here comes the sun, and I say-it's all right.

"Here Comes the Sun," The Beatles (1969)

January 2002 is typical of January winters in Nebraska. The green farm fields of summer are a memory now, covered with a blanket of snow. The temperature outside often hovers around $-10°C$. Yet some snow is melting over decaying vegetation, as chemical processes release the sun's energy stored during the growing season. It seems somehow contradictory to talk about heating from solar energy when it is so cold outside. But as we discuss below, energy released from nuclear fusion on the Sun provides abundant sources of energy on Earth.

The solar energy of long summer days makes possible, via photosynthesis (an endothermic process) the growth of trees whose logs we use in the fireplace. The energy "stored" in the cellulose of the wood is released via combustion (an exothermic process) when it is burned. When we burn coal or oil we are releasing energy that was trapped from the Sun many millions of years ago within the plants and animals whose remains gave rise to these fossil fuels. Scientists are working to use the Sun's energy economically to produce

Even during the frigid rural Nebraska winter, energy exchange drives physical and chemical processes. The snow absorbs energy from the Sun and melts, vegetation decays, and animals forage for food.

hydrogen that results from breaking apart water (an endothermic process). Hydrogen can be stored well and transported over long distances. It can even be used as a fuel in automobiles. The sunshine also gives rise to wind. According to one estimate, "80 percent of the electrical consumption in the U.S. could be met by the wind energy of North and South Dakota alone."[1] Photovoltaic cells, which convert sunlight to electricity, have been used for many years to provide power for small electronic devices. They may soon be used to meet the power needs of mid-sized cities.

Whether along the winter plains of Nebraska or in the scorching heat of a summer's day in the Middle East, the energy from the Sun that floods across the world is the most useful and most utilized form of energy we have. Many people believe that the day will come when we can trap enough of that energy directly to provide us with most of our heat and electricity, thus avoiding much of the cost and mess involved in our oil- and coal-based economies. Combine all the energy in all the known reserves of nuclear fuel, oil, natural gas, and coal and multiply that by 10. This is the amount of energy the Earth receives from the Sun every year. The other energy sources—wind, coal, oil, nuclear, and others—can be important, as discussed in the Case in Point: Alternative Energy Sources. Over the long term, however, solar power does seem to be the best alternative energy source and learning how to make best use of it will involve a lot of chemistry. The advantages of learning how to make efficient use of solar power are important to understand:

- The energy arriving on Earth from the Sun is available everywhere and is free, although some places receive more of it than others.
- The continual supply of energy from the Sun is assured at least for the next few billion years!
- It seems likely that we can develop ways of utilizing the Sun's energy that produce minimal pollution compared to today's heavily polluting energy technologies.

Solar energy is a very complex subject for discussion. To understand its possibilities we must examine each of the following wide-ranging subjects:

- The nature of the energy that the Sun releases
- How the earth processes that energy
- How we can chemically harness, store, and transport the energy for immediate and future use

■ case in point

Alternative Energy Sources

What kinds of things do we do each day that require an external energy source? We use computers and calculators. Our homes often need heat or air conditioning. Our cars, buses, trains, and planes need to get from here to there. Our industries need to run their manufacturing facilities. And we need lights to see it all.

Even with strong conservation measures, worldwide energy use will continue to increase as the world population rises. Scientists therefore search for sources that give the energy benefits of nuclear fission and the burning of fossil fuels with none of the attendant risks.

Is there such a thing as a perfect energy source? That depends on what is meant by "perfect." Such a source should:

- Be available forever and in infinite amount.
- Be free (remember, we are after perfection!).
- Produce energy and leave no dangerous residue.
- Be amenable to storage anywhere on Earth for any period of time.

Based on our criteria, there is no such thing as a perfect energy source because the conversion of any

[1]Hoagland, William. Solar energy. *Scientific American*, January 1996, pp. 170–173.

material source to useful energy requires equipment that costs money. There are, however, several energy sources that are closer to perfect than what we currently use. We call these **alternative energy sources.**

Just because something is an "alternative" does not mean that it is new. The potential and kinetic energy of moving water and wind and that contained in plants have been tapped for centuries and used in individual homes or farms and small communities. The serious push for *large-scale* alternative energy sources began in 1973 when major Middle Eastern oil producers withheld oil from the world market, causing shortages and steep price increases in oil and oil-based products. As in so many instances, world politics influences the direction of science and vice versa.

Although oil is for the moment plentiful, we recognize that oil and coal supplies are not infinite and that oil, in particular, continues to be used as a political weapon. Fossil and nuclear fuels are wrought with environmental consequences. We therefore seek alternative ways of generating energy for masses of people, many of whom live in huge urban areas. The following sources show promise as alternative energy sources.

Hydropower is literally **water power,** in which the kinetic energy from falling or flowing water is used to turn the turbine in an electricity-generating facility. In the 1930s, hydropower provided roughly 30% of U.S. energy needs. Building a hydropower facility is expensive as is upgrading aging structures. Additional research into new power plant components, including turbines, may allow the hydropower contribution to U.S. energy capacity to increase from the 2001 figure of 10%, according to the U.S. Department of Energy.

Biomass is the general name for plants and plant-derived materials. Among the sources of biomass are forest wood and municipal and industrial waste, especially wastepaper, which can be incinerated to harness the energy contained within. Biomass, in the form of dead plants, can return to the soil, preventing nutrient depletion. While biomass burning does produce covalent oxides (CO_2, NO_x, SO_x), it does burn cleaner than fossil fuels and is available as long as plants grow. Compost and manure can also decompose to produce methane gas which can be captured and burned. Biomass is readily available so it is much less expensive than oil and coal, and, as is true with all of our alternative fuel sources, cannot be used as a player in the theater of international politics.

Geothermal energy is heat harvested from underground sources. This can mean anything from using steam from geysers in northern California (supplying 7% of California's power as of 1998) to heat from underground rocks in New Mexico where steam energy to turn power plant turbines is harnessed by injecting water onto the rocks and directing the steam to the turbines.

Wind energy is the kinetic energy of moving air. Wind energy is harnessed by large windmills on which long blades turn slowly in response to the wind. The blades turn turbines producing electricity. Figure 1A shows that there is a "wind alley" throughout the mid-section of the United States in which, if the wind were harnessed, could supply a significant amount of U.S. energy needs.

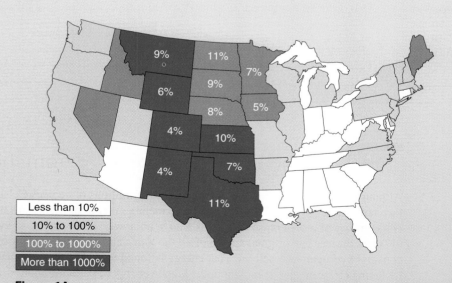

Figure 1A
The midsection of the United States is known as the wind alley. The colors show the potential that wind has to generate electricity as a percentage of the state's current electricity generation. The listed percentages reflect the portion of the total U.S. wind power possible from each state.

Sea energy is generated as a result of the differences in temperature between surface water and water several hundred meters below the surface. In a sense, sea energy is really solar energy because the 60-million-km² surface of tropical seas absorbs the energy equivalent of 250 billion barrels (over 40 trillion liters) of oil *each day!* The motion of tides is also a traditional energy source, powering such things as sawmills.

These sources have had a relatively minor impact on total U.S. energy generation. About 36% of all energy is converted into electricity in the United States and Figure 1B shows that alternative energy sources contribute less than 10% of this amount.

Exercise
Overview of the Alternatives

Table 15.1 lists a few of the advantages and disadvantages of the alternative energy sources discussed so far. Please complete the table.

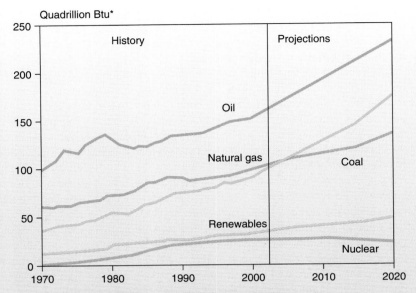

* 1 Btu (British thermal unit) is the amount of energy required to raise the temperature of 1 pound of water 1°F at or near 39.2°F.
A Btu ≈ 252 calories. A pound of coal can release about 13,000 Btu.

Figure 1B
Alternative energy sources have a small role to play compared to other sources that rely on either finite supplies or have significant hazards associated with their use.
Source: Energy Information Agency, Department of Energy

Table 15.1

Alternative Energy Sources

Source	Advantages	Disadvantages
Hydropower	No polluting emissions	
Geothermal		
Biomass		Some polluting emissions
Wind	Rapidly improving technology	
Sea		

15.1 | The Energy We Receive from the Sun

Although the core of the Sun has a temperature of millions of degrees Celsius, the Sun's surface temperature is "only" about 6000°C. Remember that temperature is a measure of the average energy of a system, so the Sun is a huge body with a wide range of average energies in different places. The original source of all the Sun's energy is the nuclear fusion process that converts hydrogen into helium. That energy is found within the Sun in many different forms including electromagnetic radiation, the kinetic energy of particles' motions, and the potential energy associated with particles' positions.

We know from our own experience that the energy from the Sun is not all the same. We can see some as light. We can feel some as heat. We know that too much exposure to the Sun's energy can damage our skin, so we apply protective chemicals as discussed in the Case in Point: Sunscreens. Green plants utilize some of the energy in photosynthesis. Why does it help for leaves to be green? Is there a special characteristic to having that color? The bottom line is that the most fundamental processes of life utilize energy from the Sun in different ways. In order to understand how this happens, we need to know more about how this energy reaches us, the forms it has, and what happens when it gets here.

The Sun Transmits

The **energy** that radiates outward from the Sun, solar radiation, is transmitted in all directions through space. A small part of this radiation reaches the surface of the Earth and enables the energy-storing reactions in plants that form the basis for life. We call this energy **electromagnetic** because of the changes in the electric and magnetic fields that are created by a traveling electromagnetic wave.

Figure 15.1 summarizes the essential features of waves. It shows that a wave has a length (**wavelength, λ**—pronounced lambda) and a height (amplitude), and implies a **frequency** (v, pronounced nu). The wavelength is the distance between the crests of two successive waves. The frequency is the number of waves that pass a point in a given time. Waves of **electromagnetic radiation** travel through space at the speed of light (c), which equals 3.00×10^8 meters per second. This makes good sense in the case of the visible electromagnetic waves that we call "light"—after all, light *should* be traveling at the speed of light! You might have been tempted to think, however, that high-frequency light moves faster than low-frequency light. This is not true. When traveling through essentially empty space all electromagnetic waves, whether visible or not, travel at the same speed (that of light) and the relationship between speed, frequency, and wavelength is:

$$c = v \times \lambda$$

The energy of light is directly related to the frequency and inversely related to the wavelength through the equation $E = hc/\lambda = hv$, where $h = 6.626 \times 10^{-34}$ J s (this known as Planck's constant, after German scientist Max Planck). This equation enables us to calculate the energy of a single bundle of energy known as a **photon,** the smallest quantity of light possible. Notice that long wavelength light has a lower energy than short wavelength light.

To make this mathematical relationship more meaningful, let's consider a wave of blue visible light from the Sun (or anywhere else, for that matter). Suppose that the wavelength of the blue light is 4.5×10^{-7} m, or, as more commonly expressed, 450 nanometers (nm). The energy, E, of the light wave, is

$$E = \frac{hc}{\lambda} = \frac{6.626 \times 10^{-34}\,\text{J s}(3.00 \times 10^8\,\text{ms}^{-1})}{4.5 \times 10^{-7}\,\text{m}}$$

$$\mathbf{E = 4.4 \times 10^{-19}\,J\ (per\ photon)}$$

What process gives rise to a photon? One possibility is that an electron within an atom gains energy from heat or light. The electron is said to be **excited,** much as we get excited and have too much energy (see Figure 15.2). The electron goes from its normal **ground state** to an

The dramatic solar display known as a prominence is a magnetic disturbance of the hot gases in the Sun's atmosphere. Scientists are particularly interested in this activity because it appears to alter the Earth's climate.

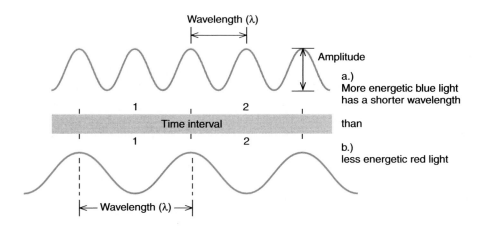

Figure 15.1
Waves have the properties of wavelength and amplitude. Since both waves here travel at the same speed, more waves of blue light than of red light arrive during the same amount of time. The **more energetic** light (blue) has the **greater frequency** and shorter wavelength.

Figure 15.2
An electron can gain energy and become "excited." This excess energy can subsequently be released. As this happens, the electron can emit a photon of a particular wavelength. If the wavelength is, for example, 450 nm, it is blue light.

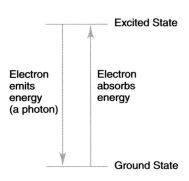

Excited State

Electron emits energy (a photon)

Electron absorbs energy

Ground State

excited state. The excess energy that was absorbed can be released as a photon with the properties of wavelength, frequency, and energy as we have discussed. If one atom releases a photon with an energy of 4.4×10^{-19} J, then a mole of atoms releases

$$E = \frac{4.4 \times 10^{-19} \text{ J}}{\text{photon}} \times \frac{6.022 \times 10^{23} \text{ photons}}{\text{mole}}$$
$$E = 2.6 \times 10^5 \text{ J/mol}$$

This is quite a bit of energy, compared for example, to our prototype bowl of frosted corn flakes that was discussed in Chapter 5. So you see, even visible light, which has a very low energy content compared to other electromagnetic waves, can pack quite an energy punch.

exercise 15.1

Converting Between Energy and Wavelength

Problem

What is the wavelength of electromagnetic radiation with an energy of 2.8×10^{-19} J? Is the wavelength longer or shorter than that of our 450 nm photon? Is the frequency higher or lower?

Solution

$$\lambda = \frac{hc}{E} = \frac{6.626 \times 10^{-34} \text{ J s}(3.00 \times 10^8 \text{ ms}^{-1})}{2.8 \times 10^{-19} \text{ J}} = \mathbf{7.1 \times 10^{-7} \text{ m}}$$

The wavelength of 7.1×10^{-7} m, or 710 nm, is longer than 450 nm, and the frequency would therefore be lower. Note that this can also be solved as a conversion factor problem in which the units must cancel if the problem is set up properly. Keep in mind the large energy difference between waves emitted from a single atom and those from a mole of atoms.

 case in point

Sunscreens: Protecting Us from What Gets Through

What do think of when you hear the word "sunscreen"? Undergraduates lying on the beach during spring break? Children whose parents lather the stuff on so heavily that bugs stick to their skin? Perhaps you think of folks who work outdoors such as roofers, farmers, or lifeguards. Such workers or beachcombers can be exposed to large amounts of UV radiation if they work outdoors in the southern United States or any other area that is relatively close to the equator. In more northern locations, UV radiation from sunlight is only a problem in the middle part of the day during the late spring and summer months when the Sun is high in the sky. At other times, UV radiation is scattered and absorbed by the atmosphere and does not reach the Earth's surface.

UV radiation is often classified into three types: UVA (400–320 nm), UVB (320–280 nm), and UVC (280–200 nm) (which of these types is most energetic?). UVC is absorbed by the ozone layer (see Section 12.2). The UVA and UVB radiation are only partially absorbed by the ozone layer and as that layer becomes more depleted less of this radiation is filtered out so more reaches the Earth's surface. With over 500,000 cases of skin cancer being reported each year in the United States, sunscreens become ever more important.

Why can certain types of UV radiation damage the skin? The reason is not all that different from that given in the last chapter when we discussed the effect of nuclear radiation on the body. Although UV radiation is not as powerful, UVB radiation which reaches the Earth's surface can cause damage because it contains enough energy to excite or ionize electrons in molecules. As soon as an electron has added energy, it will tend to get rid of that energy, possibly starting other reactions to do it, if necessary. The added energy is sufficient to break chemical bonds in molecules. Among the affected molecules can be proteins and DNA, which, as we shall see in Chapter 16 on biochemistry, is of primary importance to the body working properly.

In order to have true sunscreens, we need products that filter out wavelengths of about 200–400 nm, the entire UV range. The topical (applied to the skin) sunscreens are a convenient way to do this. Note, by the way, that sunscreens are different than suntan lotions, which merely allow the skin to become a slightly darker color and do not filter or block UV radiation.

How do sunscreens work? Many **chemical sunscreen agents** are organic (carbon-based) molecules that are applied as a cream, oil, or gel. Among the chemical sunscreen agents are oxybenzone, dioxybenzone, cinoxate, and para-aminobenzoic acid (PABA) whose structures are shown in Figure 1. (How are the structures similar?) The presence of so many alternating single and double bonds means that when a sunscreen molecule absorbs

energy its valence electrons, which absorb the energy, have the freedom to shift in certain ways among the orbitals that make up the molecules. These energetically excited electrons can dissipate their excess energy either by losing it as thermal energy or by losing some energy as thermal and the rest as light. This means that UV radiation can be absorbed and then dissipated as heat without initiating side reactions.

The bottom line then, is that UV radiation interacts with these compounds so that it does NOT interact with your skin. Another type of sunscreen is a **physical sunscreen agent,** an opaque substance that works by blocking or scattering UV light. Zinc oxide or titanium dioxide are physical sunscreen agents. Although effective, they are not clear when applied, so they are not popular. Clear chemical sunscreens are much more widely used.

What about the SPF rating? You may know about the **SPF,** or **sun protection factor,** the measure of a sunscreen's effectiveness. The SPF represents the factor by which you can increase your exposure to the sun and be affected the same as if you had no sunscreen. For example, if you get a mild burn after 30 unprotected minutes in sunlight, the SPF-6 lotion would allow you to stay "exposed" for 6 × 30 minutes = 180 minutes, before you would get the same mild burn. An SPF-15 lotion would allow exposure for 15 × 30 minutes = 450 minutes, or 9 hours. Why do many experts say that buying any product with an SPF rating of more than 15 is wasting your money?

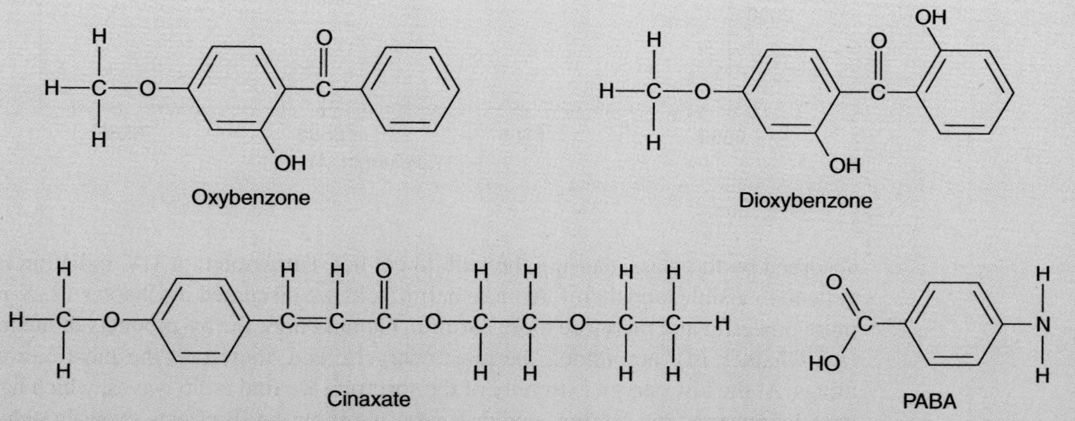

Figure 1
These are among the most popular ingredients in sunscreens. Are there structural similarities among them?

The Sun emits a wide range of wavelengths. This is typical of stars. Different wavelengths have different fates as we discuss later. Figure 15.3A shows the complete electromagnetic spectrum, a display of the range of wavelengths corresponding to *all* of the known energies in the universe. It is divided into regions that are used as benchmarks when discussing different wavelength groups. Note that the visible region takes up very little of the entire spectrum. About 7% of the energy used to power a 100-watt incandescent light bulb is reemitted as visible light. Most of the remainder is emitted as infrared (IR) radiation, some of which is

Figure 15.3
A. The electromagnetic spectrum displays the range of wavelengths corresponding to all the known energies in the universe. The relatively narrow visible region is expanded to show the component colors.
B. This is a small part of the solar spectrum—from 655 nm to 657 nm (often given in Angstroms (Å), 1 nm = 10 Å). Note the large absorption peak for hydrogen at 6563 Å (656.3 nm).

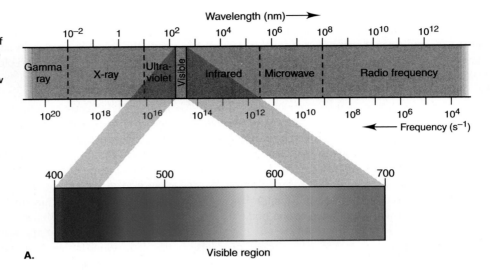

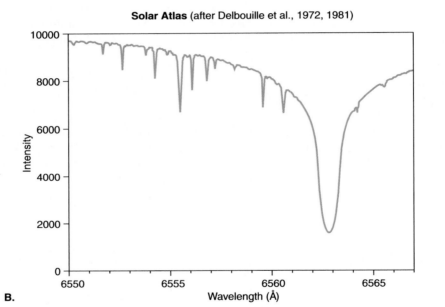

absorbed by the glass, causing the bulb to get hot. Ultraviolet, or UV, radiation is more energetic than visible, and therefore more harmful, as we discussed in Chapter 12. X-rays are even more powerful and therefore more harmful. Gamma rays, the by-products of nuclear processes (see Chapter 14), are among the most energetic and, therefore, the most harmful to living things. At the low energy extremity of the spectrum we find radio waves, which have very long wavelengths and are benign, which is why we show no ill effects even though we are constantly bombarded by the radio waves used for radio and television broadcasting.

The range of electromagnetic radiation transmitted by the Sun is very similar to the full electromagnetic spectrum. However, some atoms in the Sun absorb radiation before it leaves the surface, leading to dips in the spectrum such as the one for hydrogen shown in Figure 15.3B. The bulk of the emissions are between 2.5×10^{-7} m and 3×10^{-6} m (250 to 3000 nm) in the UV, visible, and IR ranges. Much of the Sun's energy is in the visible region, which corresponds to the wavelengths that human eyes recognize, as well as those that plants absorb to carry out photosynthesis. So, even though the visible range of the entire electromagnetic spectrum is rather small, we have evolved to take advantage of this particular set of wavelengths!

The overwhelming fraction of the energy that the Earth receives is what the Sun transmits to it. This is used in photosynthesis—storing some of the Sun's energy in plants. It is also what we have to work with if solar energy is to be exploited as a viable alternative to coal and oil.

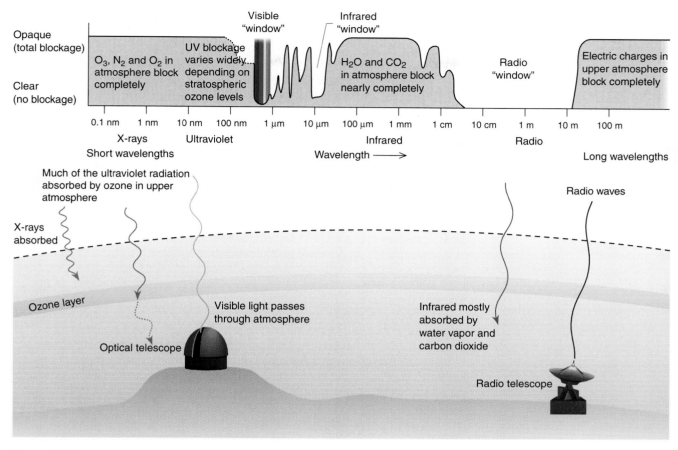

Figure 15.4
The atmosphere acts as a filter—it absorbs a wide range of wavelengths. Important among these are UV (by O_3) and IR (by CO_2 and H_2O).

The Earth Receives

And how! The Earth's upper atmosphere receives a total of about 1×10^{22} J of the Sun's energy *each day*. The actual amount of the Sun's energy that reaches the Earth's surface is far less than this amount, ranging from about 75% of the incoming radiation to nearly zero, depending upon weather conditions. A substantial amount of the energy is reflected by clouds, some is absorbed by molecules in the atmosphere, and still more is scattered by water vapor.

Figure 15.4 shows the range of wavelengths that are absorbed by the Earth's atmosphere and the major chemical species that absorb them (to review how molecules can absorb energy, please see the discussion on the greenhouse effect in Section 12.3). Note in particular the absorption of UV radiation by ozone, which we discussed in Chapter 12. Also, much of the IR radiation is absorbed by carbon dioxide and water molecules, which vibrate in response to this incoming energy.

What Happens to Energy That Strikes the Earth's Surface?

The Sun's energy can interact in many ways with objects on the Earth's surface. It can be absorbed; it can be converted directly or indirectly to heat; it can be reflected back toward the Sun; and some living things can use the Sun's energy to convert atmospheric carbon dioxide to organic molecules. It is through

Rows of solar reflectors at an experimental solar power station located near Albuquerque, New Mexico, are made up of heliostats, large mirrors mounted on movable frames that allow the mirror to be angled for maximum exposure to the Sun's rays.

this latter process, called photosynthesis, that life on Earth is possible. Most photosynthesis occurs when the Sun's energy becomes trapped and stored within the chemicals of the **chloroplasts** of leaves in plants. However, photosynthesis is also common in the oceans, where **phytoplankton** and algae-driven photosynthesis occurs. By examining the chemistry of photosynthesis, we should find clues that might reveal how we can mimic it to meet our domestic, industrial, and commercial need to trap and store energy.

exercise 15.2

Solar Energy and Soda Cans

Problem

We said in Chapter 5 that it takes 4.184 J of energy to heat 1 g of water so that the temperature rises by 1 degree Celsius (specific heat of water = 4.184 J/g°C).

(a) If the 60 million km^2 surface of tropical seas were heated by the Sun's energy so that the average temperature increase to a depth of 1 m (= 0.001 km, and the total heated volume = 0.060 million km^3 = 6.0×10^4 km^3) were 1°C , how many joules would be absorbed by the seawater (assume the density of seawater to be 1.00 g/cm^3)?

(b) Based on your answer to part a, and knowing from Chapter 4, Exercise 4.10, that it takes **4.3×10^6 J/can** to process the necessary materials to make an aluminum soda can, how many cans could be made if all the energy in part a could be harnessed?

Solution

(a) We look at this as a dimensional analysis problem. You want joules and you are given specific heat in J/g°C, density in g/cm^3, and volume in km^3. Standard factor-label technique has us organize units so that, after canceling, we end up with the desired units, J, in this case:

$$J = 6.0 \times 10^4 \text{ km}^3 \times \left(\frac{10^3 \text{ m}}{1 \text{ km}}\right)^3 \times \left(\frac{10^2 \text{ cm}}{1 \text{ m}}\right)^3 \times \frac{1.00\text{g}}{\text{cm}^3} \times \frac{4.184 \text{ J}}{\text{g}°\text{C}} \times 1°\text{C} = 2.5 \times 10^{20} \text{ J}$$

The message here is that the seas absorb massive amounts of energy transmitted by the Sun.

(b) # cans $= 2.5 \times 10^{20}$ J $\times \dfrac{1 \text{ can}}{4.3 \times 10^6 \text{ J}} = 5.8 \times 10^{13}$ cans (58 trillion cans)

15.2 | Photosynthesis—The Solar Energy Model

We discussed **photosynthesis** briefly in Chapter 4. At that time we said simply that the multistep process, powered by the energy of sunlight, converts carbon dioxide and water to glucose and oxygen,

$$6CO_2 + 6H_2O \rightarrow C_6H_{12}O_6 + 6O_2$$

Now we have the background to investigate the process in a bit more detail. In fact, photosynthesis does not necessarily require water. A more general equation for the process is

$$CO_2 + 2H_2A \rightarrow CH_2O + H_2O + 2A$$

where "A" can be oxygen, sulfur, or nothing, and CH_2O is the empirical formula for a carbohydrate. For example, *Chlorobium* is a member of a family of green and purple bacteria whose photosynthesis can include sulfur, with this reaction:

$$CO_2 + 2H_2S \rightarrow CH_2O + H_2O + 2S$$

Using this more general scheme, the equation representing the photosynthesis reaction in green plants is

$$6CO_2 + 12H_2O \rightarrow (CH_2O)_6 + 6H_2O + 6O_2$$

The complex process of photosynthesis occurs in two stages, with the first one, the "light stage," requiring primarily *visible light,* and the second one, the "dark stage," occurring whether or not light is present.

  Chlorophyll a, shown in Figure 15.5A, is a primary player in photosynthesis because it absorbs sunlight *with the subsequent release of an electron.* Although the process is very complex, involving many steps, the key point is that pigments, molecules that absorb visible light, often have many alternating single and double bonds in their structure. Compare the structure of retinal, the pigment in human vision, with chlorophyll a and chlorophyll b, shown in Figure 15.5A–C. The wavelengths of light that will be absorbed depend on the number of alternating single and double bonds as well as the atoms that are attached to the bonds.

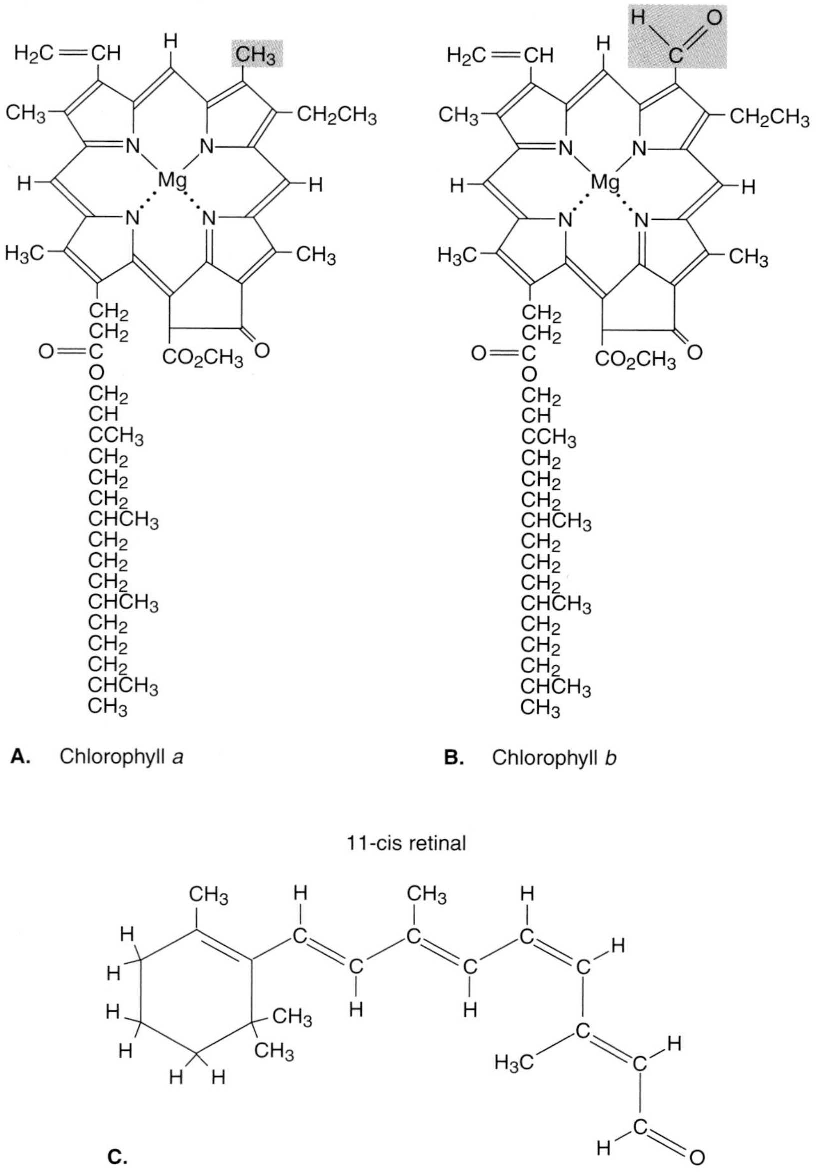

A. Chlorophyll *a*

B. Chlorophyll *b*

C. 11-cis retinal

Figure 15.5
Pigments, molecules that absorb visible light, have the common structural feature of many alternating single and double bonds. Note the structural similarities among chlorophyll *a* (**A**), chlorophyll *b* (**B**), and 11-*cis*-retinal (**C**), a pigment involved in our eyes' vision.

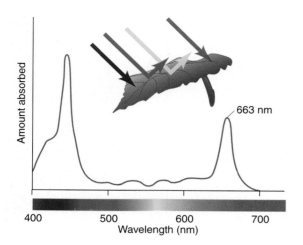

Figure 15.6

The absorption spectrum of chlorophyll *a*. Chlorophyll *a* is one of several leaf pigments. It absorbs red and blue wavelengths strongly but almost no green or yellow wavelengths. Thus, leaves containing large amounts of chlorophyll *a* appear green. The strong absorption at 663 nm can be used to quantify the substance in a plant extract.

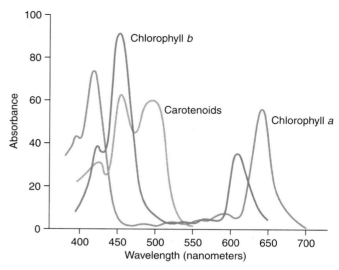

Figure 15.7

Chlorophyll *b* and carotenoids extend the absorbance range of chlorophyll *a*.

This solar-powered ejection of an electron from the molecule is the essential energy-capturing process of photosynthesis. All the subsequent events in photosynthesis participate in converting some of the energy captured in each carbohydrate molecule into forms that can be stored in living cells and utilized when and where required. Figure 15.6 shows the wavelengths of light that chlorophyll *a* absorbs. The absorbance is primarily in the blue and red regions. The molecule does not absorb much green light; therefore, we see the (reflected) green. The absorbance range of chlorophyll a is fairly limited, so plants contain other pigment or light-absorbing molecules, such as *chlorophyll b* and various *carotenoids,* which, as you can see in Figure 15.7, absorb different wavelengths of light. Energy captured by these other "accessory" pigments is then transferred to chlorophyll *a*. The net effect of the involvement of these additional pigments is to extend the **absorbance range** of chlorophyll *a,* allowing use of more of the Sun's energy.

By the end of the light stage of photosynthesis, water has been chemically *split* into hydrogen ions, molecular oxygen, and the released electrons:

$$H_2O \rightarrow 2H^+ + \tfrac{1}{2}O_2 + 2e^-$$

Since the electrons that perpetuate photosynthesis have already been released during the light stage, the dark stage does not require light. The outcome of this second stage is to convert carbon dioxide to carbohydrate.

These carbohydrates (as discussed in Chapter 17) become energy sources for animals (of which humans are but one example). We see then that the cycle of life on our materially closed Earth (Chapter 4) continues because we take in energy from the Sun.

Photosynthesis is the most important demonstration that the Sun's energy can be captured by chemicals in a way that allows the energy to be transferred and put to some controlled use. This is the conceptual basis of scientists' efforts to use solar energy to meet world energy needs.

15.3 | Spectroscopy

We can learn a lot about the behavior of matter by studying the way in which various materials interact with light. As mentioned in Chapter 5, atoms heated to a high temperature emit light that is of characteristic wavelengths for each element. This is a result of electrons

Figure 15.8
The beakers show copper(II) nitrate dissolved in water. The concentrations are 0.01, 0.05, 0.10, and 0.5 *M*. The beaker on the right is 0.01 *M* copper(II) nitrate to which ammonia is added so that its concentration is 0.20 *M*.

releasing energy when they fall from high-energy orbitals back to their expected orbitals. Look again at pages 76–77. The brightness (intensity) of such light can tell us how much of a certain element is present in a sample. Such studies are called **emission spectroscopy.** A complementary method of using light is called **absorption spectroscopy.** In such experiments, a light source is shined on a sample and the extent to which the light is absorbed is examined. Emission and absorption spectroscopy can be used to determine the concentration of elements at the parts per million level and below.

A spectrum, such as shown in Figures 15.6 and 15.7, shows the extent to which a substance absorbs various wavelengths of light. For light to be absorbed, the energy of a photon of the light must match the difference in two energy levels of the molecule or atom. Molecules having a series of alternating double and single bonds usually absorb radiation in the visible region (Figure 15.5). Chlorophyll and other plant pigments absorb sunlight and allow the plant to cause the light energy to force a series of endothermic reactions to occur.

Many of the transition metals form solutions that are brightly colored. The blue color of a solution of copper ion, for instance, means that blue light is transmitted through the solution while red and orange light, especially, is absorbed by the copper ions. The intensity of the blue color is a measure of the concentration of copper. Figure 15.8 shows several copper solutions of different concentrations. The intensity of the color can often be changed by adding compounds that react with the metal. Ammonia, when added to copper ion, causes the hue and intensity of the blue color to be greatly enhanced. Comparison of the color intensity with solutions of known concentration enable the determination of concentration of unknown samples.

15.4 | Energy Exchange Works Both Ways

A key idea in this chapter is that the Sun's energy can be stored in chemical systems. This occurs in photosynthesis. Whether or not the Sun is the immediate source, energy exchange is a two-way street and stored energy can also be released. The energy can take a number of forms, as discussed in Chapter 5. One form of energy release can accompany the exchange of electrons, which can, if harnessed properly, result in electron flow, or electric **current.** If the force of this flow, called the **voltage,** is great enough, useful work can occur, such as lighting a lamp or running a motor vehicle engine.

An excellent example of this is the Daimler-Benz "NeCar 5" (New Electric Car) shown in Figure 15.9. The vehicle, the fifth in a series of zero-emission cars introduced in 1994, has an electric motor propelled by the reaction of hydrogen (formed from liquid methanol in the vehicle's "reformer") and oxygen. The product is water and, therefore, is nonpolluting. As long as hydrogen and oxygen are available to the vehicle, it will run at speeds up to 90 miles

Figure 15.9
The Daimler-Benz NeCar 5 (New Electric Car) is propelled by the "clean" reaction of hydrogen and oxygen. Hydrogen is stored in a tank in the upper part of the car.

Electrolysis is the decomposition of a substance by electricity.

per hour with a range of 280 miles between refills of methanol. How is it that enough energy can be stored to run this car? How does this reaction produce a flow of electrons that allows the motor to operate? These two questions are the primary focus of this section.

Molecular hydrogen gas, H_2, is a desirable energy storage medium because it is relatively easy to produce from methanol which is less expensive to transport over long distances than electricity, and burns to form water vapor. No harmful emissions result. Disadvantages are that a molecule of H_2 doesn't hold much energy relative to other energy storage molecules such as methanol or gasoline. Also, much of the hydrogen produced worldwide is from reaction of methane (from natural gas) with steam and oxygen.

$$CH_4 + \tfrac{1}{2}O_2 \xrightarrow{(H_2O)} CO + 2H_2$$

This means that a nonrenewable resource (methane) is really being used when H_2 is part of the fuel cell. Nonetheless, research continues on storing energy derived from the sun by generating hydrogen gas.

The *reverse* of this use of H_2 as a fuel source is an important area of study called **photoelectrolysis.** Sunlight produces an electric current that provides the energy to split water into molecular hydrogen and molecular oxygen. These gases are then stored and recombined to reform water whenever the release of energy is required:

$$2H_2O(g) \rightarrow 2H_2(g) + O_2(g)$$

We have entire areas of study devoted to how we might split water. How different this is from the wonderful chemistry that goes on within living plants. It allows the effortless splitting of water as well as the storage of energy in carbohydrate bonds at ambient temperatures. It turns out that humans can split water too, but only either at very high temperatures or by using energy from a chemical reaction. This ability to generate electrical energy from a chemical reaction will serve as the theme for our next section.

Reduction and Oxidation—The Transfer of Electrons

How can chlorophyll *a* initiate the splitting of water at normal atmospheric temperatures while a high temperature or electrical energy is required for humans to split the same molecule? Try the hands-on activity on page 493.

The physical evidence from the hands-on activity tells us that some of the water has been split into gases, namely, H_2 and O_2. In order to understand how water can undergo electrolysis, we need to start with this representation.

We have previously discussed the polarity of water and how that polarity affects the way water interacts physically with other substances (like oil, for example). Because oxygen is much more electronegative than hydrogen (that is, it will attract the shared bonding electrons more than hydrogen), the electrons of the O—H bonds are far more likely to be found close to the oxygen nucleus than the hydrogen nucleus. We quantify this by saying that the oxygen has an **oxidation number** of -2 since the electrons derived from the hydrogen atoms spend most of their time around the oxygen atom. Oxidation number is, more than anything else, a clerical tool we use to keep track of electrons. If the oxidation number of the oxygen is -2 because it often acquires the hydrogen electrons, then the oxidation number of each hydrogen must be $+1$ because its electron spends most of its time around the oxygen atom. Adding the oxidation number of the oxygen (-2) to the total oxidation number of the hydrogens ($2 \times [+1] = +2$) demonstrates that the net effect is to have an electrically neutral water molecule (since $[-2] + [+2] = 0$).

The electrons in the molecular hydrogen are distributed *equally* between the two nuclei because the hydrogen atoms are of equal electronegativity. Therefore, the oxidation number of both atoms is 0. The same is true for the oxygen molecule.

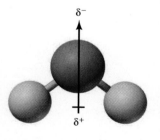

The polarity of water can be explained by its structure.

exercise 15.3

Energy versus Energy per Gram

Problem

We have said that hydrogen does not hold much energy relative to other molecules. When hydrogen is burned in oxygen it releases 242 kJ per mole of hydrogen burned.

$$2H_2(g) + O_2(g) \rightarrow 2H_2O(g) \quad \Delta H = -242 \text{ kJ/mol } H_2$$

The burning of methane results in 802 kJ/mol of CH_4

$$CH_4(g) + 2O_2(g) \rightarrow CO_2(g) + 2H_2O(g) \quad \Delta H = -802 \text{ kJ/mol } CH_4$$

This would seem to put hydrogen at a disadvantage as an energy source.

Let's take the calculation one step further. Please determine the amount of energy released (as heat) per gram of hydrogen and per gram of methane.

Solution

$$\text{Heat energy released per gram of } H_2 = \frac{242 \text{ kJ}}{1 \text{ mol } H_2} \times \frac{1 \text{ mol } H_2}{2.02 \text{ g } H_2}$$

$$= \textbf{120 kJ/g } \mathbf{H_2}$$

$$\text{Heat energy released per gram of } CH_4 = \frac{802 \text{ kJ}}{1 \text{ mol } CH_4} \times \frac{1 \text{ mol } CH_4}{16.0 \text{ g } CH_4}$$

$$= \textbf{50.1 kJ/g } \mathbf{CH_4}$$

Therefore, in spite of hydrogen's relatively low total energy output on combustion, it releases quite a bit of energy per gram. This makes it a very desirable energy source. In the gas phase, even at 100 atm pressure, hydrogen would require about 30 times the volume of liquid gasoline to contain the same amount of energy in the "gas tank." The new BMW 750 hL hydrogen-based electric car gets about 5.5 miles per gallon of liquid H_2, which is stored in the vehicle at its boiling point of $-253°C$.

In summary, as a result of the water-splitting reaction, each oxygen atom changes from an oxidation number of -2 to an oxidation number of 0, while each of the two hydrogens changes from an oxidation number of $+1$ to 0. In other words:

• Each oxygen atom has effectively *lost* two electrons.
• Each of two hydrogen atoms has effectively *gained* one electron.

As discussed in Chapter 4, we say that the oxygen atom has been **oxidized,** meaning it has *lost electrons*. The hydrogen atoms have been **reduced,** meaning that they have *gained electrons*. So the water-splitting process is accompanied by a movement of electrons "from oxygen to hydrogen." The result of this movement of electrons is the production of H_2 and O_2. Keep in mind that the shifting of electrons in this reaction does not happen unless forced to do so by the input of energy from a battery (or a flame or, desirably, the Sun.) On the other hand, the reaction of hydrogen and oxygen (the opposite reaction) to form water is a thermodynamically favorable reaction, although it requires a slight jolt of energy—an activation energy, as we discussed in Chapter 5, Section 5.6.

Equations "d" and "e" in Exercise 15.4 involve reduction and oxidation of water. Let's add each **half-reaction** (half of an entire reduction/oxidation reaction) together to get a whole reaction, being sure that the number of electrons gained in the reduction equals the number lost in the oxidation.

A good way to remember the changes in oxidation and reduction is with the acronym OIL RIG, *O*xidation *I*nvolves *L*oss (of electrons) *R*eduction *I*nvolves *G*ain (of electrons).

exercise 15.4

Reduction and Oxidation

Problem

State whether each electron movement represents reduction or oxidation.

(a) $Ag^+ + e^- \rightarrow Ag$

(b) $Cu^{2+} + 2e^- \rightarrow Cu$

(c) $Zn \rightarrow Zn^{2+} + 2e^-$ (the same as $Zn - 2e^- \rightarrow Zn^{2+}$)

(d) $4H_2O + 4e^- \rightarrow 4OH^- + 2H_2$ (or $2H_2O + 2e^- \rightarrow 2OH^- + H_2$)

(e) $2H_2O \rightarrow 4H^+ + O_2 + 4e^-$ (the same as $2H_2O - 4e^- \rightarrow 4H^+ + O_2$)

(f) $MnO_4^- + 8H^+ + 5e^- \rightarrow Mn^{2+} + 4H_2O$

Solution

(a) **Reduction.** The silver ion has been reduced by gaining an electron.

(b) **Reduction.** The copper(II) ion has been reduced to copper metal.

(c) **Oxidation.** The zinc metal has been oxidized by losing two electrons to form zinc(II) ion.

(d) **Reduction.** This seems more difficult to assess than the previous equations because there is a component, water, involved and two products. Nonetheless, the equation indicated that electrons are gained, so reduction occurred.

(e) **Oxidation.** Similarly, electrons are lost from the reactant (water) and gained on the product side.

(f) **Reduction.** Electrons are gained on the reactants side. The manganese is going from an oxidation number of $+7$ to $+2$. It has gained five electrons.

The gain of electrons by a substance (reduction) and the loss of electrons by a different substance (oxidation) always occur together as represented by the law of conservation of mass. Whether or not there will be a net energy gain or loss in the overall "redox" process will depend on the substances involved. Energy is required to split water. Energy is released from the battery that is forcing the splitting to occur.

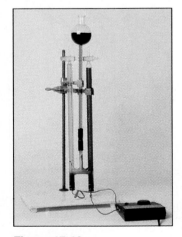

Figure 15.10
We can use bromthymol blue to indicate where acids and bases are present. The neutral (green) color changes to yellow in the presence of an acid and blue in the presence of a base.

a. $4H_2O + 4e^- \rightarrow 4OH^- + 2H_2$ (reduction)

b. $2H_2O \rightarrow 4H^+ + O_2 + 4e^-$ (oxidation)

overall: $6H_2O + 4e^- \rightarrow 4OH^- + 4H^+ + 2H_2 + O_2 + 4e^-$

The four electrons on each side cancel, leaving us with

$$6H_2O \rightarrow 4OH^- + 4H^+ + 2H_2 + O_2$$

This equation represents what happens when we electrolyze water. Note that in half reaction "a" the pH increases and in "b" it decreases. We have seen the gases in our tubes, but where are the acid (H^+) and base (OH^-)? If we add an acid/base indicator such as bromthymol blue to the solution, we get the evidence we need. Figure 15.10 shows that the indicator turns yellow in the presence of H^+ and blue in the presence of OH^-. If we had done the electrolysis *without separate* test tubes, just sticking the wires in the water, the H^+ and OH^- would recombine to water,

$$H^+ + OH^- \rightarrow H_2O$$

and the electrolysis would be

$$6H_2O \rightarrow 4H_2O + 2H_2 + O_2, \text{ or}$$
$$2H_2O \rightarrow 2H_2 + O_2$$

◼ hands-on

Splitting Water with Electricity

You Need

- A cup
- A baby food jar or butter tub
- Two metal 20-cm nonbraided *copper* wires, stripped at both ends
- 2 alligator clips
- 2 small test tubes (ask your instructor for these or buy them at the local science and hobby shop)
- A small 9-V (*rectangular*) battery
- Battery terminal clips
- Table salt, NaCl

The Activity

- Make a saltwater solution in the cup by stirring 2 teaspoons of table salt into a nearly full cup of water.
- Fill the baby food jar roughly 7/8 full with the saltwater.
- Place one end of a wire in each test tube. The other end of the wire should be above the tube.
- Fill the test tubes with the saltwater.
- Putting your pinky on the open end of a tube and wire and your thumb on the bottom, invert the tube and place it in the baby food jar with saltwater. Some saltwater may overflow.
- Do the same with the other test tube.
- Using the alligator clips, hook the exposed part of one wire to one of the battery clips. Do the same with the other wire.
- You are now ready to connect the battery. Do so.

What do you observe? If everything is connected properly, you should observe *bubbling* at each wire. After a time, the bubbles will begin to displace the water in each tube, and one tube will have a lot more gas than the other.

What's Going on Here?

To get at the answer, we need to realize that the only two gases that could be generated in water under these conditions are hydrogen (H_2) and oxygen (O_2). How? It is reasonable to assume that the battery supplied the necessary energy. It is also reasonable to assume that since we know that batteries produce electricity, and *electricity* is the *movement of electrons*, that a movement of electrons caused the water-splitting, or **electrolysis** of water.

   That should sound familiar because chlorophyll *a* initiates the splitting of water when sunlight causes an electron to move out of the chlorophyll *a* molecule. The bottom line is that the energy for this *endothermic* water-splitting reaction can come from inducing the *movement of electrons* in a way that avoids the need for very high temperatures. Whether the splitting of water occurs via *input* of energy from heat or from a battery, the outcome is the same—energy is stored in the hydrogen and oxygen products.

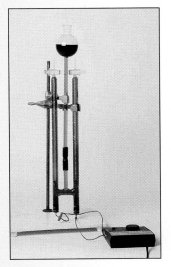

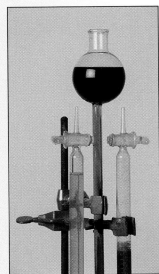

Water is "split" into hydrogen and oxygen in this hands-on activity (shown here using different apparatus). How do we know that the tube with greater gas volume contains hydrogen? Where does the energy come from to split the water?

Whether Sun-driven, battery-driven, or involving chlorophyll *a,* the splitting of water is really the reduction and oxidation, or, more commonly, the **redox reaction** of water.

As we discuss in the section on photovoltaics (Section 15.5), it is possible for us to use sunlight to drive many reactions useful to consumers in our daily lives. That we do not use it exclusively is partially because the technology is not sufficiently developed and partially because there are economic and political factors at work that impede its development. We therefore currently use other chemical processes to supply the electrical energy to meet many of our day-to-day needs. Once we decipher this, we can then proceed to look at the future in which the Sun is used more directly to meet individual and societal energy needs.

What Substances Can Supply the Energy?

How do we know whether a given combination of substances will supply the energy necessary to run a calculator, light a flashlight, or power a car? The photos in Figure 15.11 show what happens when a zinc strip is put into a solution containing copper(II) ion. Copper seems to be forming on the zinc. If, on the other hand, a copper strip is put into a solution containing zinc(II) ions, nothing happens. Summarizing:

$$Cu \text{ added to } Zn^{2+}: \quad Cu(s) + Zn^{2+}(aq) \rightarrow \text{no reaction}$$
$$Zn \text{ added to } Cu^{2+}: \quad Cu^{2+}(aq) + Zn(s) \rightarrow Cu(s) + Zn^{2+}(aq) \text{ (overall)}$$
$$Cu^{2+}(aq) + 2e^- \rightarrow Cu(s) \text{ (reduction)}$$
$$Zn(s) \rightarrow Zn^{2+}(aq) + 2e^- \text{ (oxidation)}$$

In Figure 15.11, copper is gaining two electrons from the zinc. The zinc is going into solution. After a while (about 24 hours) the zinc metal should no longer be visible, having been oxidized into solution as Zn(II). The blue copper ion should all be reduced to copper metal. This is in fact what happens.

Why does the reaction shown in Figure 15.11 spontaneously proceed? Remember from Chapter 5 that one criterion for spontaneous reactions is the energy of the products is lower

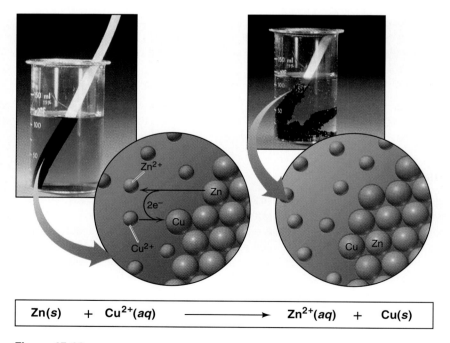

$$\textbf{Zn(s)} \quad + \quad \textbf{Cu}^{2+}\textbf{(aq)} \quad \longrightarrow \quad \textbf{Zn}^{2+}\textbf{(aq)} \quad + \quad \textbf{Cu(s)}$$

Figure 15.11
The spontaneous reaction between zinc and copper(II) ion. When a zinc metal strip is placed in a Cu^{2+} solution, a redox reaction begins (left) in which the zinc is oxidized to Zn^{2+} and the Cu^{2+} is reduced to copper metal. As the reaction proceeds (right), the Cu "plates out" on the zinc and then falls off in chunks. The Cu appears black because it is very finely divided. At the atomic level, Zn atoms each lose two electrons, which are gained by the Cu^{2+} ions.

than that of the reactants. We can say that the zinc metal and copper(II) ion in solution are less energetically stable than the copper metal and zinc(II) ion. The redox reaction goes, and energy is generated. What happens to the energy? It is dissipated in the solution in molecular motion, resulting in a small amount of heat.

The electrons are being forced from the zinc metal to the copper(II) ion in solution. This electron-moving force is known as **electromotive force,** or emf. We often refer to the emf by its more common name, **voltage (V),** with units of volts (V). The copper/zinc reaction generates an initial voltage of up to 1.10 V. The more positive the voltage (emf), the more force with which the electrons are moving. Also, the **power** of a system is equal to the **current** (electrons per unit time) multiplied by the voltage. So both quantities contribute to the power.

Let's do the same reaction again, but change the setup a bit, so that it looks like that in Figure 15.12. We'll use two beakers. The first contains zinc metal and zinc(II) ion. The second contains copper metal and copper(II) ion. An LED (light-emitting diode) is connected to each strip, and a **salt bridge,** a sodium chloride-filled glass tube that maintains neutrality without allowing the solutions to mix is also inserted. The half and overall reactions are the same as before. Zinc metal loses two electrons per atom and goes into solution as zinc(II). Copper(II) ions travel to the copper strip, pick up the two electrons, and deposit as copper metal on the copper metal strip. We call the zinc electrode the **anode,** the place where oxidation is occurring. We call the copper electrode the **cathode,** the place where reduction is occurring.

In this setup, the electrons travel through a wire to get from the anode (zinc strip) to the cathode (copper strip). The energy released in this spontaneous reaction is used, in part, to light the LED rather than merely being dissipated as heat, as before. The salt bridge contains ions, such as Na^+ and Cl^-, that migrate with positive (via Na^+) and negative (via Cl^-) charges so that charge balance is maintained in the isolated solutions. These can transfer enough positive charges from the beaker containing zinc to that containing copper to balance the electrons leaving the zinc for the copper beaker. The solutions themselves do not mix. In this way, we have used a spontaneous chemical reaction to generate electricity. In a less formal setup, all manner of salt-containing substances can be salt bridges, including a strip of bologna or a hot dog!

A battery is nothing more than one or more spontaneous reactions hooked together to harness enough energy to run something useful, such as a flashlight, car, or calculator.

Lighting the Flashlight

Most small appliances, like a flashlight or child's toy, require small, inexpensive batteries that put out enough current with a sufficient voltage to light a bulb or beep a horn. A very common design is the alkaline battery in which a graphite rod is an inert cathode, meaning it transfers electrons without itself being reduced. The graphite is surrounded by a starch-based paste that includes NaOH or KOH (hence the name "alkaline" for the battery) and MnO_2 (shown in Figure 15.13). The paste has the same function as the salt bridge in our Zn/Cu cell—it transfers charge. We call such an ion-containing material an **electrolyte.** This is surrounded by zinc, which is the anode, and finally by an outside cover. The half-reactions are

$$2MnO_2(s) + H_2O(l) + 2e^- \rightarrow Mn_2O_3(s) + 2OH^-(aq) \quad \text{(reduction)}$$
$$\underline{Zn(s) + 2OH^-(aq) \rightarrow Zn(OH)_2(s) + 2e^- \quad \text{(oxidation)}}$$
$$Zn(s) + 2MnO_2(s) + H_2O(l) \rightarrow Zn(OH)_2(s) + Mn_2O_3(s) \quad \text{(overall)}$$

The battery, which produces about 1.5 V when new, is not rechargeable. An example of a rechargeable battery in which the reaction can be reversed is a lead storage automobile battery shown in Figure 15.14. Here, Pb (anode) and PbO_2 (cathode) plates are stored in an H_2SO_4 solution (shown in Figure 15.15). The plates are separated by polymers such as rubber, polyvinyl chloride, or polypropylene. Typically, between 6 and 15 individual cells, yielding 2 V each, are hooked together to make one battery. The individual cell half-reactions are

Batteries come in all shapes and sizes. One of the most popular types is the alkaline battery, which is used to power many small appliances.

An LED, or light-emitting diode, is a device that will carry current in one direction only. Its light emission indicates that it is carrying electrons.

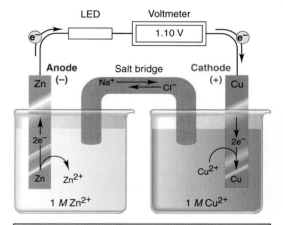

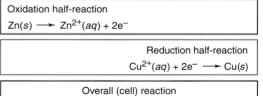

Oxidation half-reaction
$Zn(s) \longrightarrow Zn^{2+}(aq) + 2e^-$

Reduction half-reaction
$Cu^{2+}(aq) + 2e^- \longrightarrow Cu(s)$

Overall (cell) reaction
$Zn(s) + Cu^{2+}(aq) \longrightarrow Zn^{2+}(aq) + Cu(s)$

Figure 15.12
In this "redox reaction," the reduction and oxidation half-reactions occur in separate cells. The electrons generated in the oxidation of zinc move through the zinc strip and the wire and, with their energy, light the LED. The electrons then move through the copper strip to reduce Cu^{2+} ion. Na^+ and Cl^- ions move across the salt bridge completing the circuit and maintaining electrical neutrality.

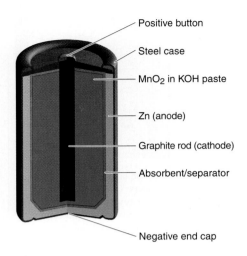

Figure 15.13
An alkaline battery used in small appliances such as a flashlight or toy.

Figure 15.14
An automobile battery is perhaps the most significant example of the interrelationship between chemical process and electrical energy exchange.

$$PbO_2(s) + 4H^+(aq) + SO_4^{2-}(aq) + 2e^- \rightarrow PbSO_4(s) + 2H_2O(l) \quad \text{(reduction)}$$
$$\underline{Pb(s) + SO_4^{2-}(aq) \rightarrow PbSO_4(s) + 2e^- \qquad\qquad\qquad \text{(oxidation)}}$$
$$Pb(s) + PbO_2(s) + 4H^+(aq) + 2SO_4^{2-}(aq) \rightarrow 2PbSO_4(s) + 2H_2O(l) \quad \text{(overall)}$$

Such batteries are used to power the starter of a car but not to cause the car to move. These batteries are continually recharged so they should, in concept, never run down. In practice, lead storage batteries can be recharged more than 1000 times before wearing away of the components forces you to replace the battery.

The Future for Electric Cars

A rechargeable battery that is suitable to use as the sole energy source in a car (to replace the gas-powered engine) needs to be able to last a long time and for many miles between charges. It also needs to provide the car with enough energy so the vehicle handles well. A survey done by scientists at the University of California, Davis, indicates that many U.S. households would buy electric cars as a second car for short trips, so the ability to reach highway speeds might not be an important issue in the development of electric vehicles.

Many redox combinations have been tried in electric vehicles including modifications of the conventional lead-storage battery. All have shortcomings in one or more areas. Nickel-cadmium batteries, shown in Figure 15.16, have been used for years dating back to their use powering the Telstar satellite. The reactions are

$$NiO_2(s) + 2H_2O + 2e^- \rightarrow Ni(OH)_2(s) + 2OH^- \quad \text{(reduction)}$$
$$\underline{Cd(s) + 2OH^-(aq) \rightarrow Cd(OH)_2(s) + 2e^- \qquad \text{(oxidation)}}$$
$$NiO_2(s) + Cd(s) + 2H_2O \rightarrow Ni(OH)_2(s) + Cd(OH)_2(s) \quad \text{(overall)}$$
$$\textbf{charged} \qquad\qquad\qquad\qquad \textbf{discharged}$$

Other designs incorporate sodium and sulfur, lithium and sulfide, nickel and iron, nickel and cadmium, and even zinc and air or aluminum and air in the redox reactions. All these batteries suffer from having a very limited range, typically less than 150 miles before they must be recharged.

Most automobile manufacturers now use a "nickel-metal hydride" (Ni-MH) battery that can store quite a bit of hydrogen (which combines with a metal such as $LaNi_5$, to form the metal hydride), and can be recharged longer than most recyclable batteries before failing. The reactions are

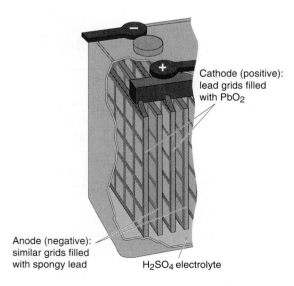

Cathode (positive): lead grids filled with PbO_2

Anode (negative): similar grids filled with spongy lead

H_2SO_4 electrolyte

Figure 15.15
The inside of a lead storage automobile battery.

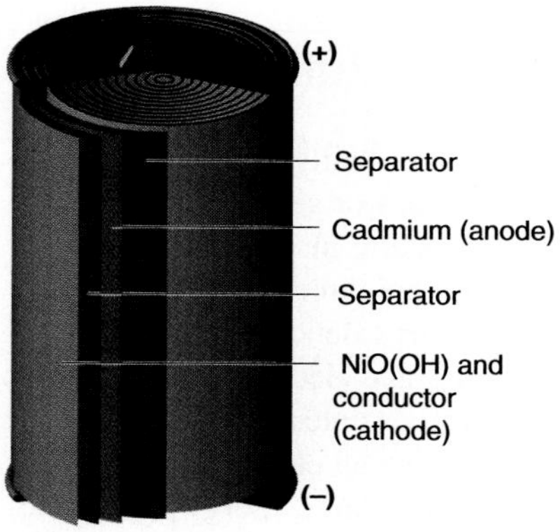

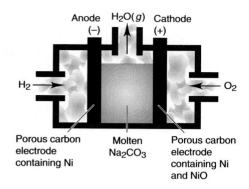

Figure 15.16
Nickel-cadmium ("Ni-cad") batteries are being tried in all-electric vehicles.

Figure 15.17
A schematic of an H_2/O_2 fuel cell. Hydrogen and oxygen (in air) are fed in and react to form water.

$$NiOOH(s) + H_2O(l) + e^- \rightarrow Ni(OH)_2(s) + OH^-(aq) \quad \text{(reduction)}$$
$$\underline{MH(s) + OH^-(aq) \rightarrow M(s) + H_2O(l) + e^-} \quad \text{(oxidation)}$$
$$MH + NiOOH \rightarrow M + Ni(OH)_2 \quad \text{(overall)}$$

All of the energy storage devices we have discussed in this section will eventually fail (as with rechargeable batteries) unless fuel is constantly fed in. The Daimler-Benz NeCar 5 vehicle has such a system, called a **fuel cell.** A fuel cell is like a battery, in that spontaneous chemical reactions produce electricity. The difference is that in fuel cells, the fuel (molecular hydrogen, in this case) is continuously fed in allowing the H_2/O_2 cell to produce energy as long as the fuel in the feed tanks holds out. The chemistry of the H_2/O_2 fuel cell is shown in Figure 15.17.

Electric cars may have a bright future, especially when it comes to the reduction of pollution. Public acceptance of the cars will depend upon cost, battery lifetime and availability. In 2000, only 759 electric cars were sold (or leased) in the United States. Table 15.2 shows the percentage of reduction in pollution that would occur as a result of worldwide use of electric vehicles. Can you explain the predicted increase in sulfur dioxide and particulate emissions?

We have seen that, whether powering an automobile or lighting a small battery-operated device, there are several kinds of chemical systems that can supply electrical energy.

Is it possible to have an energy source that does not require a continual supply of chemical reactants or that will almost never fail on its own? That is, can sunlight strike a substance and be reliably converted to electricity for many many years to come? The technology to do it on a relatively small scale has been available since 1953. In the past nearly half-century, the field of photovoltaics, which deals with converting sunlight into electricity, has begun to prove itself as an important way to meet global energy needs.

Table 15.2

Electric Vehicles Reduce Pollution

(Percentage Change in Emissions, if all vehicles were electrically powered)

	Hydrocarbons	Carbon Monoxide	Nitrogen Oxides	Sulfur Oxides	Particulates
France	−99	−99	−91	−58	−59
Germany	−98	−99	−66	+96	−96
Japan	−99	−99	−66	−40	+10
United Kingdom	−98	−99	−34	+407	+165
United States	−96	−99	−67	+203	+122
California	−96	−97	−75	−24	+15

Source: *Scientific American*, November, 1996.

15.5 | Photovoltaics—Converting Light Energy

Diamond, revered as a symbol of love, is made of carbon atoms covalently bonded together in three dimensions as shown in Figure 15.18. If you have ever worn or seen a diamond, you know that it keeps its brilliance and its shape. It also remains static in nature as diamond, rather than reacting to form, for example, a carbon oxide. These are three pieces of evidence that show diamond is *chemically unreactive*. In diamond, the four valence electrons of each carbon atom are involved in very stable bonding arrangements. The electrons are relatively low in energy. They are not likely to move out of their stable orbitals unless pushed by the addition of a substantial amount of energy. Therefore, diamond is also an electrical insulator; it does not conduct electricity. Figure 15.19 illustrates that there is a large energy gap, sometimes called a "band gap," between the energy level of the valence electrons in diamond (the "valence band") and the energy level they would need to be raised into, in order to conduct electricity (the "conduction band").

Figure 15.18
Diamond is very chemically unreactive. Carbons are covalently bonded in a stable arrangement which requires a substantial input of energy if electrons are to be moved out of their stable orbitals. Diamond is therefore an electrical insulator.

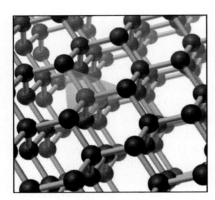

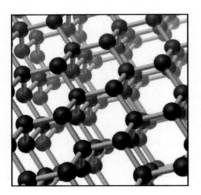

Silicon, which is just below carbon in group 4A of the periodic table (see Figure 15.20), forms crystals that have much the same structure as carbon and are also quite stable. However, the valence electrons in silicon (period 3) occupy higher energy levels than those in diamond's carbon (period 2) and the energy gap (or "band gap") between the valence band and conduction band is therefore smaller, shown in Figure 15.19. Silicon crystals will conduct some electricity if heated, so silicon is called an **intrinsic semiconductor,** meaning it can, without modification (i.e., "intrinsically"), conduct electricity (although poorly) under certain conditions.

Something interesting happens if we replace (or **dope**) some of the silicon atoms in the crystal with some arsenic (As) atoms as shown in Figure 15.21A. Arsenic is a group 5A metalloid (semimetal) and, as such, has one more electron per atom than silicon. This means that the part of the doped crystal around each arsenic atom is *electron rich* and is, consequently, less energetically stable than the pure silicon crystal. This makes the band gap between the valence and conduction bands smaller than before so less energy is needed to promote an electron to the conduction band. Electrical conduction, in other words, becomes easier.

Figure 15.19
The valence electrons of silicon occupy higher energy levels than those in carbon. Because higher energy levels have smaller gaps in energy among levels than lower levels, the energy gap between the valence band and conduction band is smaller in silicon than in carbon.

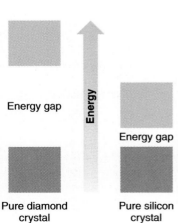

Energy gap

Energy

Energy gap

Pure diamond crystal

Pure silicon crystal

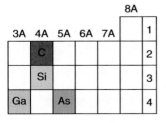

Figure 15.20
The positions of carbon, silicon, gallium, and arsenic are highlighted on this partial periodic table. Based on their positions, you should be able to explain whether each element is suitable for use as a semiconductor.

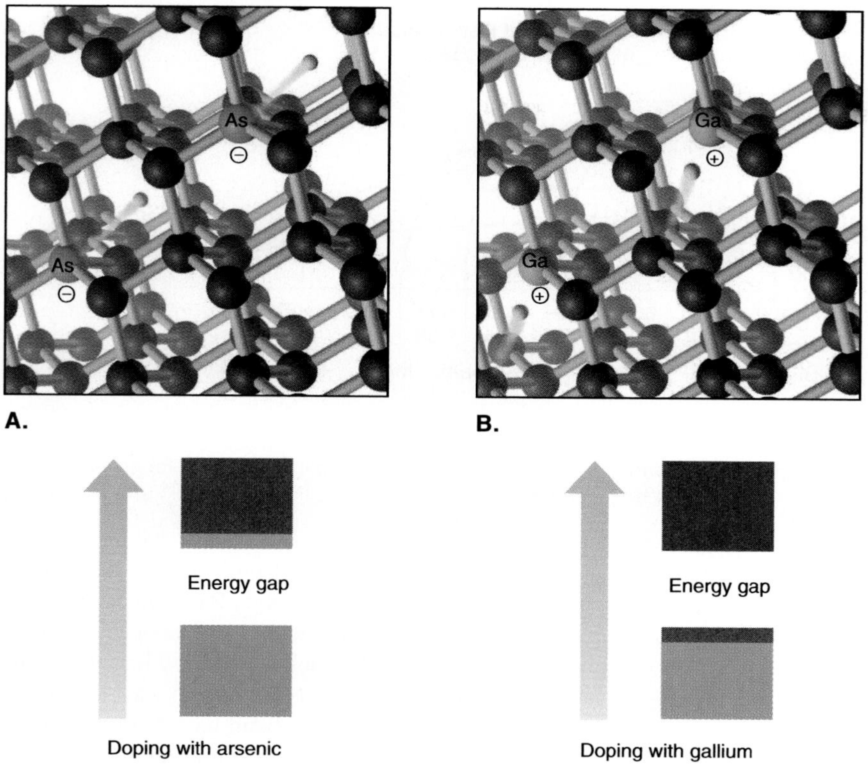

A.   **B.**

Energy gap   Energy gap

Doping with arsenic   Doping with gallium

Figure 15.21
Doping silicon with arsenic **A.** makes the crystal electron rich. Doping silicon with gallium **B.** makes the crystal electron poor. The band gaps in each case become smaller. When small amounts of both Ga and As are added to a crystal, the band gap becomes even smaller, allowing current to flow, as described in the text.

Suppose that, instead of doping a silicon crystal with arsenic, we use some gallium atoms. This causes an *electron deficit* to exist in the region of the gallium atoms because gallium (group 3A) has one fewer electron per atom than silicon. A positively charged "hole" is said to be formed. This less stable situation also reduces the band gap as shown in Figure 15.21B.

When small amounts of *both* arsenic and gallium are added to different parts of the silicon, the semiconductor has both electron-rich and electron-deficient areas and the band gap becomes even smaller. Sunlight throughout the visible and IR range (the particular range of wavelengths depends on the chemical composition of the semiconductor) hitting such semiconductor material can provide enough energy to allow electrons to move into the conduction band (called "jumping the band gap"), allowing both electrons and holes to move so that the flow of electrons can be harnessed as energy. This is the basis of a solar powered **photovoltaic device (PV device),** meaning one in which light energy is used to generate an electric current.

Many different materials can be combined to make semiconductors. Ones that are isoelectronic (of identical electronic structure) to silicon are popular. For example, a semiconductor of nearly pure gallium arsenide, GaAs, a so-called III–V device (named after their group numbers) works well as does cadmium telluride, CdTe, a II–VI semiconductor.

How Well Do They Work?

The first silicon-based photovoltaic systems were manufactured in 1954 at Bell Laboratories. These were able to convert about 6% of the sunlight that hit them into electricity (we say that their efficiency was 6%). While this does not seem very efficient, keep in mind that solar-powered calculators use silicon semiconductors that are only about 3% efficient. Why was the efficiency so low?

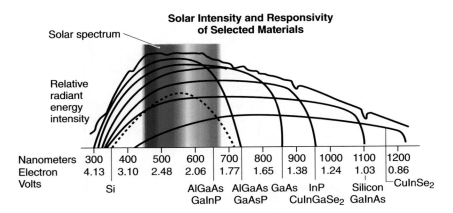

Figure 15.22
The wavelength ranges that will allow electrons to jump the band gap are shown for several semiconductors and compared to the solar spectrum.

Efficiency is affected by many things. Sunlight that is reflected from, rather than absorbed by, the PV device is lost. That means lower efficiency. To counter this, PV devices are often coated with a black surface to maximize absorption of sunlight. The best PV devices also have reflective interior rear surfaces so light that penetrates to the back is reflected into the semiconductor material. The most significant cause of low efficiency is that not all wavelengths of sunlight have enough energy to enable PV electrons to jump the band gap. For example, a GaAs semiconductor can use sunlight in the range of 850 nm, whereas amorphous (noncrystalline) silicon (called a-Si) requires wavelengths of about 360 nm or less. This means that wavelengths longer than 360 nm are not energetic enough to enable valence electrons to jump the band gap and conduct electricity. Figure 15.22 gives the wavelength data, compared to the solar spectrum, for several semiconductors.

Specially designed single gallium-arsenide photovoltaics have produced efficiencies of up to 26%. Multiple GaAs/Si cells have shown efficiencies just over 32%. While it is foreseeable that we may design devices with 35–40% efficiencies using AlGaAs/InGaAs (In = indium) PV devices, these may not be useful for anything other than specialized experiments because there is (not surprisingly) a trade-off between cost and efficiency, and high-efficiency PV devices are very costly.

Mass-produced silicon-based solar PV devices that have efficiencies of 10–15% offer promise for meeting current and future energy needs. Table 15.3 lists some potential uses of PV devices. Specialists in the semiconductor industry are not certain how much better semiconductors can get. There is a real trade-off between the cost of building new manufacturing facilities and the profit to be made from the manufacturing of PV devices. Although the price of a megabyte of computer memory has dropped to about 50 cents today from $550,000 in 1971, the cost of a manufacturing facility is now about $1.2 billion, vs. $4 million back then. Additionally, there are technological constraints that must be overcome. Primary among these is the ability to put transistors (switching and signal amplification devices that enable the computer to work) ever closer together on a semiconductor chip. In 1971, about 5000 transistors could be etched into a chip. In 1996, the figure was about 7 million. If the rate of increase holds, according to NASA, by the year 2007, 260 million transistors might fit on a chip. The ability to continue this improvement will be partly scientific, partly technological, and partly economic.

15.6 | Will It Happen?—The Future of Solar Power

For barely 10 years, initiated by the oil fuel crisis, there was genuine government sponsorship in developing alternative energy sources. During the Carter administration (1976–1980), the

Table 15.3

Potential Uses of Photovoltaic Devices

Home	Transportation	Business and Industry
Lawn mowers	Key ring security transmitter	Microwave sensors
Bug zappers	Radar detectors	Traffic control lights
Security lighting	Location transmitters	Hand-held radios
Wireless doorbells/speakers	Charging on-board	Remote cameras
"Invisible" fencing	Computers	Fans and in-line blowers
Watches	Emergency beacons	Outdoor lighting
Calculators	Storage trickle charging	Laptop computers
Tape recorders	Security alarms	Robots
Video camera recharging		Gate openers
AM-FM radios		Location transmitters
Power tools		Self-powered FAX machines
Flashlights		Timing devices
Bicycle lights and speedometers		

federal government and New York State gave tax breaks that reduced the cost of solar-powered hot water heaters by about 50%. About 10,000 solar units were sold on Long Island alone during the time when tax breaks were available. The program was repealed on December 31, 1985, as part of the government's budget cuts. Sales of these virtually maintenance-free solar-powered water heaters slowed dramatically.

The price of collecting and making use of solar energy is falling—from $5 per kilowatt-hour in 1970 to roughly 20-25 cents in 2001 according to the Oregon Office of Energy and the National Renewable Energy Laboratory. Still, sales of solar-powered sources is dwarfed by fossil fuel-powered sales—$800 billion (fossil fuels) to $1 billion (solar) in 1995. People in developed nations use about 10 times as much energy per person as those in nonindustrialized nations so the demand for power worldwide will continue to rise quickly. Does it make sense to pursue solar power even though it is not yet as inexpensive as other forms of power? What are some good alternatives? What is the role of government as an advocate for solar power? Do you have a role in this debate?

main points

- Energy from the Sun, so-called "solar power," is an excellent and underutilized energy source.

- There are a number of energy sources, such as wind and geothermal energy, that have been used successfully throughout the world. We call these "alternative energy sources" because they help meet energy needs without many of the disadvantages that conventional sources have.

- Electromagnetic energy travels in waves.

- Energy, wavelength, and frequency of electromagnetic radiation are mathematically related.

- The range of wavelengths that electromagnetic radiation can exhibit is described by the electromagnetic spectrum.

- Photosynthesis is a model process for using the Sun's energy to power chemical processes.

- Electron exchange occurs in many chemical reactions.

- The electron exchange that occurs in spontaneous chemical reactions can be used to power devices.

- Different types of electrochemical cells have a common structure; an anode part, a cathode part, and an electrolyte.

- "Photovoltaics" is concerned with the generation of an electric current from solar power interacting with a material.

- Photovoltaic materials are semiconductors, most often made by combining "electron-rich" and "electron-deficient" materials with silicon.

important terms

Absorbance range is the range of wavelengths that are absorbed by a sample. (p. 488)

Absorption spectroscopy examines the extent to which light from a known source is absorbed by a sample as a measure of its concentration. (p. 489)

Anode is an electrode at which oxidation occurs. (p. 495)

Cathode is an electrode at which reduction occurs. (p. 495)

Chloroplast is a substructure of a plant cell in which photosynthesis occurs. (p. 486)

Current is the flow of electrons through a medium, usually a wire. (p. 489)

Electrolyte is a substance that, on dissolving, allows current to flow through its solution. (p. 495)

Electromagnetic radiation is a general term that includes gamma rays, X-rays, ultraviolet, visible, infrared, microwave, and radio waves. (p. 481)

Electromotive force measured in volts, measures the tendency for a redox reaction to occur. (p. 495)

Emission spectroscopy studies the intensity of light given off by a sample as a measure of its amount. (p. 489)

Energy is the capacity for doing work or causing a change, either chemical or physical. (p. 481)

Excited state, an atom or a molecule possessing more than the minimum amount of energy is said to be in an excited state. (p. 482)

Frequency is the number of electromagnetic waves that pass a point in a second. (p. 481)

Ground state, an atom or molecule possessing the minimum amount of energy is said to be in a ground state. (p. 481)

Half-reaction is any reaction that is written with electrons as either reactants or products. (p. 491)

Intrinsic semiconductor is a material that, when very pure, conducts electricity less well than a conductor but better than an insulator. (p. 498)

Oxidation is the loss of electrons by an atom or group of atoms. (p. 491)

Photoelectrolysis is the decomposition of water into hydrogen and oxygen with light supplying the energy for the decomposition. (p. 490)

Photon is the smallest amount of light, a "particle" of light. (p. 481)

Photosynthesis is the use of light to power the endothermic conversion of carbon dioxide and water into biomass. (p. 486)

Photovoltaic device (PV) uses the conversion of light energy into electrical energy. (p. 499)

Phytoplankton are very tiny plants that live in water and convert sunlight into biomass. (p. 486)

Power is the rate at which work is done or energy is transferred; electrical power is the voltage multiplied by the current. The units of power are watts. (p. 495)

Redox reaction, a chemical reaction in which the oxidation state of at least one element is changed, consists of an oxidation and a reduction. (p. 494)

Reduction is the gain of electrons by an atom or group of atoms. (p. 491)

Voltage, a measure of the electromotive force, measures the extent to which a given redox reaction will occur. (p. 489)

Wavelength is a measure of the distance between crests in a wave of electromagnetic radiation. (p. 481)

exercises

1. List five possible alternative energy sources and describe how each creates energy.
2. True or false? As the energy of a wave is decreased its wavelength is increased.
3. What are the three measurable properties of an electromagnetic wave?
4. How is energy related to wavelength and frequency?
5. What is the frequency of radiation with wavelength of 500 nm? Would you be able to see this radiation?
6. The Sun is about 1.6×10^9 km from Earth. How long does it take a photon of the Sun's electromagnetic radiation to reach the Earth?
7. What is the energy of a photon of green light?
8. What is the wavelength of a beam of radiation with an energy of 5.72×10^{-19} J? Where does this fall in the electromagnetic spectrum?
9. It takes 6.72×10^{-18} J of energy to remove an electron from an unknown atom. What is the maximum wavelength of light that can remove an electron?

10. What types of electromagnetic waves do you use in your everyday life?

11. A friend with some chemistry background does not believe that UV radiation is harmful to human skin. Use your chemistry understanding to convince your friend to use sunscreen lotion (see the case in point on sunscreens).

12. Why is *chlorophyll a* important in photosynthesis? Name two other pigments in plants that absorb the wavelengths of light not included in the spectrum of *chlorophyll a*.

13. Based on its structure, why is *chlorophyll a* an important compound in photosynthesis?

14. Why is photosynthesis relevant to the theme of this chapter?

15. What chemicals in our environment absorb some of the energy we receive from the Sun? Explain the effect of high carbon dioxide levels in the atmosphere.

16. Why can't animals directly use energy from the Sun to supply their own energy? Can a person survive on the opposite of a vegetarian diet, eating only meat products? Explain.

17. The nutrient beta-carotene is found in carrots. Based on its structure, explain why carrots are orange.

18. You are a very creative research scientist! Can you propose the structure of a molecule that might make a good sunscreen? What are the important structural features of your sunscreen?

19. What nutritional macromolecule is the final product of photosynthesis?

20. Explain the difference between the role of water in photosynthesis and photoelectrolysis.

21. State whether each half-reaction represents a reduction or oxidation:
    a. $Cd \rightarrow Cd^{2+} + 2e^-$
    b. $Ni \rightarrow Ni^{2+} + 2e^-$
    c. $AuBr_4^- + 3e^- \rightarrow Au + 4Br^-$
    d. $PtCl_4 + 4e^- \rightarrow Pt + 4Cl^-$

22. Write the half-reactions for the following redox reaction:

$$Ca + Sn^{2+} \rightarrow Ca^{2+} + Sn$$

23. What is an electromotive force, and what units are used to express its value?

24. Calculate the power of a system with a current of 5.35 amperes and a voltage of 1.23 V.

25. In the sample activity described on page 493, the battery terminals were placed in separate test tubes. What would have happened if the electrolysis was conducted without separate test tubes? Write the chemical equation that expresses this situation.

26. A redox reaction occurs between aluminum metal and copper(II) chloride ($CuCl_2$). This effect can be seen if the protective coating inside an aluminum can is scratched off and a copper(II) chloride solution is poured in the can. The general overall reaction is expressed as:

$$\begin{array}{cccc} Al(s) & + \quad Cu^{2+}(aq) & \rightarrow \quad Al^{3+}(aq) & + \quad Cu(s) \\ \text{aluminum} & \text{copper(II) ion} & \text{aluminum(III) ion} & \text{copper} \end{array}$$

Can you balance the equation? Write the two half-reactions that make up this redox reaction and label each as an oxidation or reduction equation.

27. If aluminum foil is placed in a jar of copper(II) chloride solution, the same redox reaction as described in the previous problem will occur. What do you think will happen to the aluminum foil? What do you think would happen if a copper strip were placed in a solution containing aluminum ions?

28. A cell was created which involves these two half reactions:

$$Ce^{4+} + e^- \rightarrow Ce^{3+}$$
$$Sn \rightarrow Sn^{2+} + 2e^-$$

Indicate which equation represents the anode reaction and which represents the cathode reaction.

29. A series of reactions leads to the rusting of iron:

$$O_2(aq) + 2H_2O(l) + 4e^- \rightarrow 4OH^-(aq)$$
$$Fe(s) \rightarrow Fe^{2+}(aq) + 2e^-$$
$$Fe^{2+}(aq) + 2OH^-(aq) \rightarrow Fe(OH)_2(s)$$
$$Fe(OH)_2(s) + OH^-(aq) \rightarrow FeO(OH)(s) + H_2O(l) + e^-$$

Which of these reactions are oxidations and which are reductions? How would you classify any reaction that is neither an oxidation nor reduction? Combine these equations to give a balanced equation for the formation of one mole of FeO(OH) from one mole of Fe.

30. This following molecule is nitroglycerin, a well-known explosive:

Identify the "fuel" (hydrocarbon) portion of this molecule and sites for oxidation. Why do you think nitroglycerin is so explosive?

31. What is the purpose of the salt bridge in a redox reaction?

32. Explain the flow of electrons in the redox experiment shown in Figure 15.12.

33. Think of an analogy to help you remember the redox properties of the anode and cathode.

34. Why is a simple alkaline battery not rechargeable?

35. What is the paste-like material containing NaOH or KOH in an alkaline battery called? What is its role in the operation of a battery?

36. A hydrometer measures the density of solutions. How might this device be used to determine the condition of a lead storage battery?

37. What structural feature of diamond makes it an insulator?

38. What is a photovoltaic device? How can silicon be used in a photovoltaic device?

39. What is an intrinsic semiconductor? Can you think of a use for this type of conductor?
40. Explain the significance of band gap size in a photovoltaic device.
41. Does visible light have enough energy to activate most PV devices? Prove your assertion using example values.
42. Why do you think a watch battery is smaller than the battery used for a flashlight?
43. What are the problems associated with harnessing energy from the Sun? Why haven't we already replaced the use of fossil fuels with solar power?

food for thought

44. How would you define a perfect energy source?
45. The element helium was discovered on the Sun before being found on Earth. How was this done? Why was helium so long in being discovered on Earth?
46. In the beginning section, we mentioned that scientific development for large scale alternative energy sources occurred only after the Middle Eastern oil embargo. Can you cite other examples which support the phrase: "world politics influences the direction of science"?
47. Look at Table CIP 1 on the advantages and disadvantages of alternative energy sources. If you were on a federal budgeting committee that was presented with research proposals for each of the energy sources, for which source would you support funding and why?
48. The Itaipú hydroelectric powerplant located on the Brazilian-Paraguay border generates 79 billion kWh per year (far above the 20 billion kWh produced by Grand Coulee, the largest U.S. plant). A treaty between Brazil and Paraguay made possible the building of this dam that uses the waters of the Paraná River separating the two countries. By 1995, the Itaipú was expected to supply Paraguay with about 64% and Brazil with about 22% of all of the electricity consumed in each country. Can you think of any other examples where a scientific undertaking has brought people of different nations together?
49. Many of the theoretical approaches to solar energy presented in this chapter seem viable. Why, then, is solar energy not our main source of energy?

readings

Books

1. McCrea, S. and Minner, R. Editors, 1992. *Why Wait for Detroit?* Ft. Lauderdale, FL: South Florida Electric Auto Association.
2. Raven, Peter H. and Johnson, George B. 1986. *Biology.* WCB/McGraw-Hill Higher Education.

Articles

3. Borman, Stu. Energy. *Chemical and Engineering News,* June 17, 1991, pp. 42–46.
4. Hazelton, Lesley. Really cool cars. *New York Times Magazine,* March 29, 1992.
5. Kelter, Paul B., Snyder, William E., and Buchar, Constance S. Using NASA and the space program to help high school and college students learn chemistry. *Journal of Chemical Education,* March 1987, pp. 228–231.
6. Davis, Ged R. Energy for planet Earth. *Scientific American,* September 1990, pp. 55–62.
7. Dostrovsky, Israel. Chemical fuels from the Sun. *Scientific American,* December 1991, pp. 102–107.
8. Mallouk, Thomas E. Bettering nature's solar cells. *Nature,* October 24, 1991, pp. 698–699.
9. Haggin, Joseph. Current directions of research on solar energy look promising. *Chemical and Engineering News,* November 2, 1992, pp. 25–27.
10. Wald, Matthew L. Putting windmills where it's windy. *New York Times,* November 14, 1991.
11. Lewis, Ronald M., and Fischer, Richard G. Pediatric drug information. *Pediatric Nursing,* May-June 1987, pp. 200–201.
12. Wells, C. H. J. Suntans and sunscreens. *Education in Chemistry,* July 1993, pp. 95–100.
13. Illman, Deborah L. Automakers move toward new generation of 'greener' vehicles. *Chemical and Engineering News,* August 1, 1994, pp. 8–16.
14. Hoagland, William. Solar energy. *Scientific American,* January 1996, pp. 170–173.
15. Hutcheson, G. D., and Hutcheson, J. D. Technology and economics in the semiconductor industry. *Scientific American,* January 1996, pp. 54–62.
16. Sperling, Daniel. The case for electric vehicles. *Scientific American,* November 1996, pp. 54–59.
17. Greenman, Catherine. Fuel cells: Clean, reliable (and pricey) electricity. *New York Times,* May 10, 2001, p. D8.
18. Editor's Introduction to the Series of Articles in the Edition. The future of fuel cells. *Scientific American,* July 1999, pp. 72–94.
19. Energy Information Administration. Annual Energy Outlook 2001, with Projections to 2020. December 2000, 262 pages. http://www.eia.doc.gov/oiaf/aeo/.
20. Hanisch, Carola. Powering tomorrow's cars. *Environmental Science & Technology,* November 1, 1999, pp. 458A–462A, also available at http://pubs.acs.org/hotartcl/est/99/nov/hanis.html.

websites

www.mhhe.com/kelter   The "World of Choices" website contains web-based activities and exercises linking to: The University of Sydney in Australia's photosynthesis simulator; Varian Scientific Instruments; the Solarbuzz Information Center; and much more!

The Chemistry of Life

And God said, Let the earth bring forth the living creature after his kind, cattle, and creeping thing, and beast of the earth after his kind: and it was so.

Genesis 1:24

You are lying on soft grass at the base of a giant redwood tree, looking up at the massive trunk and the canopy of branches and greenery above you. Your attention is caught by a squirrel moving gracefully between the high branches. The squirrel, the tree, the grass, you, and all the unseen microorganisms around you are very different living things. Amazingly, all living things, including bacteria, grasses, trees, squirrels and people, are mainly built by the chemical actions of just one class of molecule— the proteins.

Life is put together and maintained by enzymes—the protein molecules that catalyze most of the chemical reactions of life. Many of the key structural and functional parts of living things are also made largely of protein, such as our muscles, the hairs on our heads and on the body of a squirrel, the delicate "cytoskeleton" of protein that forms a scaffolding within living cells, and the proteins that are embedded within cell membranes and control what enters and leaves each cell. To understand the chemistry of life we need to understand proteins: what they are, what they can do, how they are made, and how they sustain the endless march of life from generation through generation.

You may be wondering why we are not telling you that our genes, made of the famous double-helix of DNA, are the most fundamental chemicals of life. It is true that the genes of living things are vitally important and without them nothing could live at all. We will look at these genes and how they work later in the chapter. But genes are important because they determine which proteins a living thing contains. Although genes can be thought of as chemicals that store a master copy of the "instructions" for making life, these instructions control the manufacture of specific protein molecules. The proteins are the chemicals that actually do the work of sustaining life. So let's consider what proteins are and what they do.

16.1 | Proteins

Protein molecules are long chain-like polymers with many different side groups strung out along a repeating backbone. The protein chains are normally folded into complex specific shapes, allowing them to perform specialized functions within living things. The monomer building blocks of proteins are known as **amino acids**.

Amino Acids

The simplest amino acid is named glycine and has the structure shown in Figure 16.1. It has a "central" carbon atom bonded to a —COOH (carboxyl) group, an NH_2 (amine) group, and two H atoms. All amino acids have a —COOH group, an —NH_2 group (except proline, see Table 16.1) and one H atom on the central carbon (more properly called the alpha carbon atom, since it is often not actually in the center of the molecule). The differences among amino acids depend on the different "side groups" bonded to the variable site around the alpha car-

Table 16.1

The Amino Acids

The names, structures and molecular masses of the 20 amino acids found in proteins

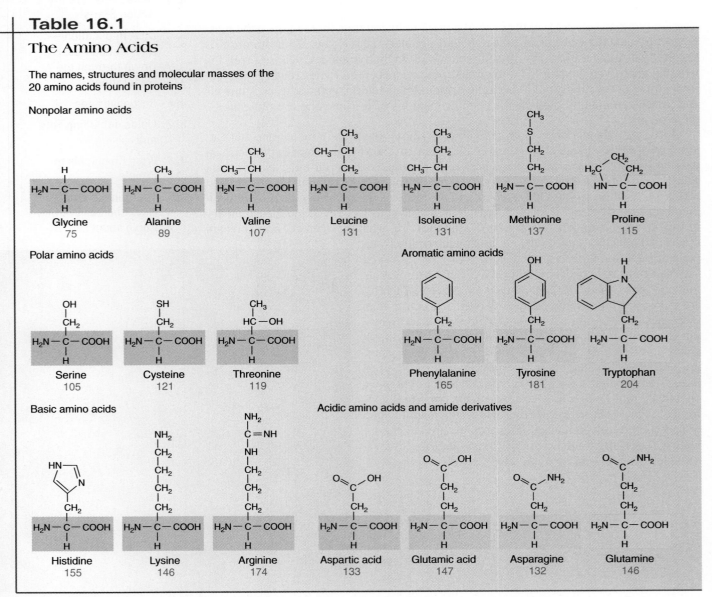

Nonpolar amino acids

Glycine 75   Alanine 89   Valine 107   Leucine 131   Isoleucine 131   Methionine 137   Proline 115

Polar amino acids   Aromatic amino acids

Serine 105   Cysteine 121   Threonine 119   Phenylalanine 165   Tyrosine 181   Tryptophan 204

Basic amino acids   Acidic amino acids and amide derivatives

Histidine 155   Lysine 146   Arginine 174   Aspartic acid 133   Glutamic acid 147   Asparagine 132   Glutamine 146

Table from P. M. Gamon and K. B. Sherrington. 1990. *The Science of Food*, 3rd Edition, Oxford: Pergamon Press, p. 112.

bon atom. Table 16.1 shows the structures of all the amino acids used to make proteins. Twenty different amino acids are found within proteins, arranged in long protein chains in an enormous variety of specific amino acid sequences. The side group of an amino acid, the part that varies among amino acids, is often symbolized as R. The full variety of side chains or R-groups can be seen in the table of amino acids (Table 16.1).

In aqueous solution, the amino groups of amino acids typically exist in their charged form, as NH_3^+, while the carboxyl groups typically exist in their charged form as COO^-. This can be considered as due to the transfer of a hydrogen ion from the acidic carboxyl group onto the basic amino group. Thus, although called amino *acids*, these molecules actually have both acidic and basic groups within their structure.

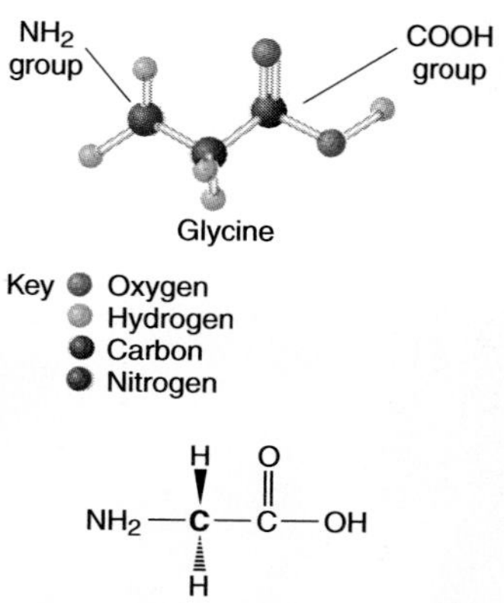

Peptide Bonds

The amino acids within protein molecules are linked together by peptide bonds. A peptide bond forms when a carboxyl group from one amino acid molecule reacts with the amino group of another amino acid in a condensation reaction that releases a molecule of water (see Figures 16.2 and 16.3). When two amino acids become linked by a peptide bond a **dipeptide** is formed. It still has both an amine group and a carboxyl group just like the original amino

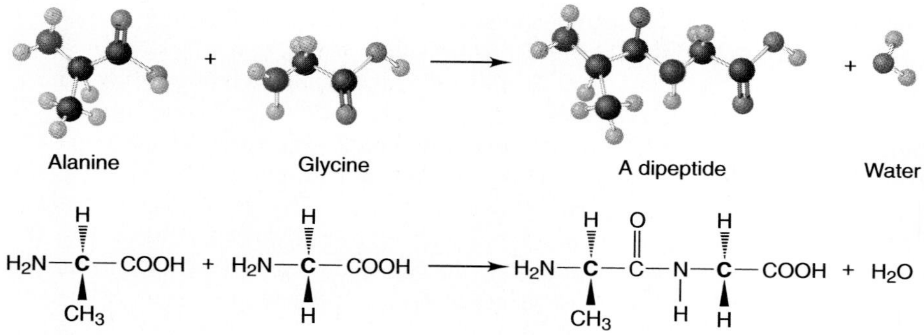

NH₂ group   COOH group

Glycine

Key Oxygen
     Hydrogen
     Carbon
     Nitrogen

$$NH_2 - C - C - OH$$

Figure 16.1
Structure of glycine.

Alanine   Glycine   A dipeptide   Water

$$H_2N - \overset{H}{\underset{CH_3}{C}} - COOH \; + \; H_2N - \overset{H}{\underset{H}{C}} - COOH \; \longrightarrow \; H_2N - \overset{H}{\underset{CH_3}{C}} - \overset{O}{C} - \overset{}{\underset{H}{N}} - \overset{H}{\underset{H}{C}} - COOH \; + \; H_2O$$

Figure 16.2
Formation of a dipeptide. A dipeptide is formed when the carboxylic acid of one amino acid reacts to form an amide with the amine group of another amino acid. In this example, alanine reacts with glycine.

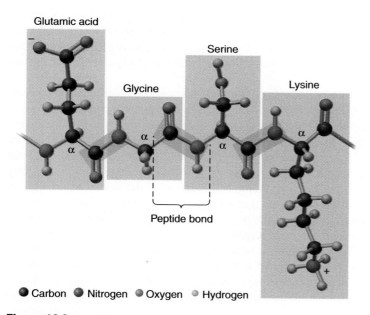

Glutamic acid

Serine

Glycine

Lysine

Peptide bond

● Carbon ● Nitrogen ● Oxygen ○ Hydrogen

Figure 16.3
In this sequence of four amino acids connected by peptide bonds within a larger protein, match the four different side groups of the four amino acids with those shown in Table 16.1. Note that the alpha (α) marks the central carbon of each amino acid.

acids, so other peptide bonds can be formed at either end, to form a tripeptide, tetrapeptide, and so on. The sequential addition of amino acids on such peptides (which occurs at one end only in nature) eventually results in protein molecules with dozens, hundreds, or even thousands of amino acids in a long **polypeptide** chain.

exercise 16.1

Counting Polypeptide Possibilities

Problem

An enormous number of protein structures can be made from the 20 amino acids that serve as the building blocks for protein synthesis. Calculate the number of dipeptides possible, and then the number of tripeptides. What does this suggest about the possible number of polypeptides containing a hundred amino acids each?

Solution

Each of the 20 amino acids can be the first amino acid and each can be the second. This is equivalent to the question "How many 2-letter words can be made from a 20-letter alphabet?" The answer is $20 \times 20 = 20^2 = 400$. So 400 different dipeptides could be made from the amino acids used in protein synthesis. The number of possible tripeptides is $20^3 = 8000$. Things get out of hand when we ask this question about real proteins. A small protein with 100 amino acids will have just one of 20^{100} possible amino acid sequences. $20^{100} = 1.3 \times 10^{130}$, a number too big to evaluate on many calculators. To give you some idea of how large this number is, suppose we put one molecule representing each sequence into a box. The size of the box would need to be bigger than the known volume of the universe! This should convince you that a specific sequence of amino acids within a protein is a very unique arrangement of matter indeed.

Protein Structure

The amino acid sequence of a protein, which simply tells us which amino acids are present and the order in which they are connected, is known as the protein's **primary structure**. All molecules of any particular protein share the same amino acid sequence, and therefore the same primary structure. Different proteins have different amino acid sequences and therefore different primary structures.

As soon as a protein is formed, it begins to fold into a variety of specific structures, particularly helices and sheets (see Figure 16.4). These distinctive structural motifs are known as a protein's **secondary structure**. The main forces holding secondary structure in place are hydrogen bonds, which we first discussed on page 218. The hydrogen atom bonded to nitrogen in the peptide bond can form a hydrogen bond to carbonyl oxygen atoms ($C{=}O$). It cannot do so to its immediate neighbor because the protein cannot twist that tightly, but it can readily form a hydrogen bond to an oxygen in an amino acid three units down the chain. This can happen if the chain of amino acids coils up into a helix (like a spiral staircase). Hydrogen bonds can also form between parallel strands of a protein chain to form sheets. Hydrogen-bonded helices and sheets are the two main contributors to protein secondary structure.

The way in which the different regions of secondary structure become folded together to form the final structure of a protein is known as the protein's **tertiary structure** (see Figure 16.4C). In some proteins, the tertiary structure is strengthened by the formation of covalent bonds that form between the —SH groups on the side chain of molecules of the amino acid cysteine within the protein chain. The —SH groups are known as sulfhydryl (thiol) groups, and the —S—S— linkages formed when two sulfhydryl (thiol) groups react together are called **disulfide bridges**.

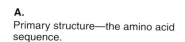

A.
Primary structure—the amino acid
sequence.

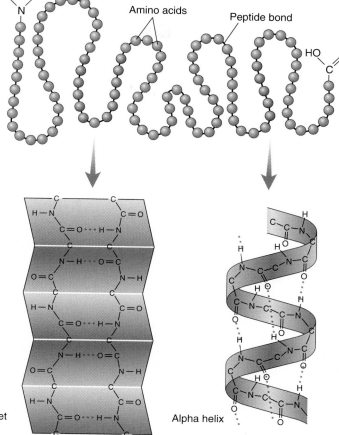

Amino acids

Peptide bond

B.
Secondary structure with folding as a result
of hydrogen bonding (dotted red lines).

Pleated sheet

Alpha helix

C.
Tertiary structure with secondary folding
caused by interactions within the polypeptide
and its immediate environment.

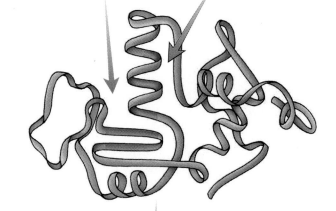

plus 3 other subunits

D.
Quaternary structure refers to the
relationships between individual subunits.

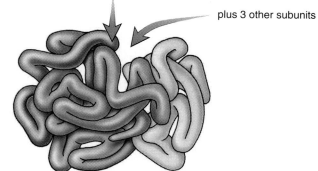

Figure 16.4
The primary structure of a protein **A.** with its amino acid sequence
begins to fold (forming the secondary structure) as a result of the
formation of hydrogen bonds **B.** The tertiary structure exhibits
secondary folding caused by the interaction of polypeptides and
their environment **C.** The quaternary structure **D.** refers to the
relationship between individual subunits.

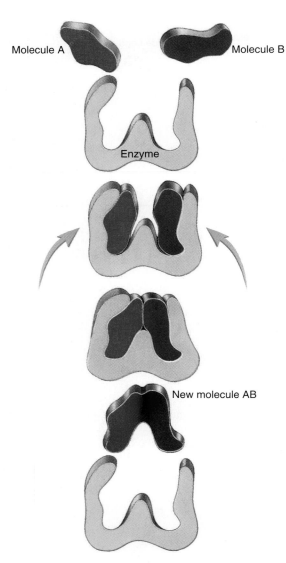

Figure 16.5
Enzymes catalyze reactions by binding to specific chemicals, the substrates, then releasing the products after the catalyzed reaction has taken place.

The amino acid sequence determines what the protein can do. It controls the crucial process of folding up of the protein chain to form a specific three-dimensional structure. The pattern of folding determines the way in which the various chemical groups that are part of a protein are arranged in space, and this, in turn, determines the protein's reactivity, or in other words what the protein can do. This folding process occurs automatically as the protein chain is formed. It is controlled by forces of attraction and repulsion between surrounding water molecules and the amino acids of the protein, and also between the different amino acids themselves. Remember, all of the movements involved in the folding process and the formation of any disulfide bridges are controlled by the chemical interactions between the amino acids of the primary structure. The way in which these amino acids interact with surrounding water molecules and other chemicals in the environment in which the protein forms are also controlled by these same chemical interactions.

After folding, some proteins are immediately ready to perform their specific tasks as enzymes or structural materials. Many proteins, however, must be chemically modified before they become functional. The modifications include the addition of sugar groups (discussed on page 522) to form glycoproteins, the addition of phosphate (PO_4^{3-}) groups to form phosphoproteins, the combination with metal ions to form metalloproteins, the addition of lipids (discussed on pages 524–526) to form lipoproteins, and combination with a wide variety of small chemical groups known as prosthetic groups.

If several protein chains become loosely bonded together, by noncovalent interactions, a protein with **quaternary structure** is formed (see Figure 16.4D). Such proteins are known as multisubunit proteins. The protein hemoglobin, discussed in the Case in Point: Powerful Proteins, is one such multisubunit protein.

We have seen that a lot of things can happen to a newly formed protein before it can perform its proper role in living cells, but all of these further modifications of proteins are catalyzed or otherwise assisted by preexisting proteins. One important role of some proteins is to mediate the chemical processing of other proteins. Let's now review all the main things that proteins can do—it's an impressive list.

The Functions of Proteins

The main functions of proteins can be summarized as:

- **Proteins can act as enzymes**, which are proteins able to bind to specific substances and catalyze particular reactions involving these substances (see Figure 16.5). Enzymes in your saliva, for example, catalyze the breakdown of some chemicals in food, starting the process we know as digestion.
- **Proteins can act as structural materials**, forming much of the physical framework of living tissues. The structure of your skin and bones, for example, is supported by a protein called collagen.
- **Proteins can form contractile tissue**, such as the contractile fibers of muscles, which allow us to move about and lift things.
- **Proteins can transport other materials** by binding to them in certain locations of the body, for example, and releasing them in other locations. For example, the protein called hemoglobin transports oxygen around the bloodstream.
- **Proteins can act as messenger molecules**, such as hormones, which are released from one site in the body and travel through the blood to another site where they can bind to specific cells and initiate specific changes in the cells.
- **Proteins can act as "receptors,"** able to bind to "messenger molecules" such as hormones, and initiate some chemical change in a living cell as a result of this binding process.

- **Proteins can act as "gates" and "pumps,"** which are able to control the passage of other chemicals through some barrier, such as the membrane of a cell.
- **Proteins can activate or inhibit other molecules,** such as other proteins or the DNA of genes, by selectively binding to the molecules that they control.
- **Proteins can act as defensive chemicals,** able to bind to disease-causing microorganisms or diseased cells to trigger a series of events leading to the destruction or neutralization of the microorganisms or diseased cells. The "antibodies" that help us to fight off disease are large protein molecules.

 # case in point

Powerful Proteins

Proteins are involved in just about every chemical reaction important to life, and they are also used in lots of medical and commercial applications. Here are a few examples to help you appreciate how powerful and versatile proteins are.

Hemoglobin

Take a breath, feel the air fill your lungs. As you do this, many of the oxygen molecules from the air you breathed in are becoming attached to the protein called hemoglobin in your red blood cells and whisked around the body in your blood. Every cell of your body needs a continual supply of oxygen, and hemoglobin is the molecular vehicle that transports the oxygen from where it is plentiful (the lining of your lungs) to where it is required. Each oxygen molecule actually becomes attached to an iron ion held at the center of a nonprotein chemical group known as a heme group, which is itself bound to the basic protein structure. Each hemoglobin unit is actually made up of four protein subunits, each with its heme group and iron

ion, as shown in Figure 16.6. So hemoglobin is a "multi-subunit" protein that carries an essential nonprotein component as part of its structure. The heme group absorbs light outside of the red region of the spectrum, so the hemoglobin molecule appears red. It is the hemoglobin packed into our red blood cells that makes them red, and makes blood red. Without our hemoglobin to continually transport oxygen around the body we would die within seconds.

Proteases

The time has come when you really must get your clothes washing done! You trek off to the laundromat, add plenty of "enzyme-laced" laundry detergent (advertised as capable of removing even the toughest stains), and sit down to munch on a bag of peanuts washed down with cola while the detergent does its work. As you sit there, you are relying on a group of enzymes known as the "proteases" both to clean your clothes and to digest your peanuts. Proteases are enzymes that break down proteins by catalyzing a

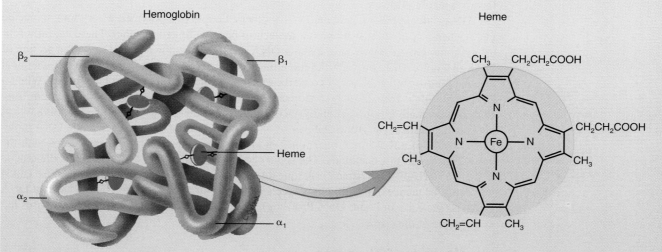

Figure 16.6
The structure of hemoglobin, a multi-subunit protein composed of four protein molecules each carrying one heme group. Oxygen molecules can become bound to the central iron ion within each heme group.

reversal of the condensation reaction that formed the proteins' peptide bonds. Proteases from various sources, particularly from some bacteria, are purified and added to biological laundry detergents, allowing them to break down the proteins that hold together food and grass stains on clothing. Your digestive system contains a variety of protease enzymes, each one able to break the peptide links between specific types of amino acids or, in some cases, to chop off amino acids one at a time from a specific end of a protein chain. So these enzymes busily break the proteins in your peanuts down into amino acids that can be absorbed into your blood and used to make the proteins you need, rather than the proteins peanuts need.

Insulin

Many people have "insulin-dependent diabetes," a condition in which cells within the pancreas do not make enough of a small protein called insulin. One crucial normal role of insulin is to bind to special receptor proteins on the surface of some body cells and promote the entry of glucose (and some other sugars and amino acids) into the cells. Glucose is the main energy source for cells, so an inability to take up glucose can have very serious consequences indeed. People with severe insulin-dependent diabetes must take regular injections of purified insulin in order to stay alive. If they miss an injection, or even inject too little insulin, they can fall into a life-threatening coma when their brain cells become deprived of glucose. The ability to manufacture and/or purify supplies of insulin allows our modern understanding of chemistry to be used to permit people with this type of diabetes to enjoy all the activities of normal life, despite having a condition that, untreated, can very quickly lead to death. For such people, insulin is a protein that wields the ultimate power.

Keratin

When you want to open a ring-pull cola can, separate some sheets of paper that are stapled together, scrape a sticky label off a window, pluck the strings of a guitar, or just scratch an itch, you can use the strength of your fingernails to do the job. The strong material you are making use of is keratin, a structural protein. It forms fingernails (and toenails), the powerful claws of birds of prey, the tough hooves of horses, and many more structures in the animal world that require a strong material to be grown to the shape required. Keratin is also the main component of animal hair (including your own), which can be wound into remarkably strong ropes or cables. If you have long hair, twist a handful of it a few times and pull on either end of the cable you have formed. This should help you appreciate the strength of keratin.

16.2 | The Nature of DNA

It has been said of the chemicals of life that "DNA gives the instructions—proteins do the work." Having looked at the structure of proteins and the kinds of chemical "work" they are able to do, we must now consider DNA and how it "gives the instructions" required to manufacture specific proteins.

DNA—deoxyribonucleic acid—is another example of a polymer, a very long molecule made when lots of smaller monomer molecules become linked together. There are only four different monomers involved in making DNA and they are shown linked to form a section of DNA in Figure 16.7. Each of the monomers is itself composed of three distinct chemical groups: a nitrogenous base, a **deoxyribose** sugar group, and a triphosphate group. When a DNA molecule is formed in a reaction catalyzed by enzymes, only one phosphate atom from each triphosphate group is actually incorporated into the DNA chain. The combination of phosphate group, sugar group and base is called a **nucleotide**. You can see that a DNA molecule consists of a repeating "backbone" of alternating sugar and phosphate groups with the bases attached to this backbone. The sequence in which the bases are arranged is the crucial thing. The DNA of each person is unique due to differences in the base sequences (also called nucleotide sequences) of different people's DNA. A murderer convicted on the basis of DNA analysis (see Case in Point: DNA Profiling) is betrayed by the specific sequence of bases left at the scene of the crime.

The Scope for Differences

The information stored or "encoded" in DNA takes the form of variable ways in which just four chemical groups, the "bases of DNA," are arranged on an unchanging "backbone." You

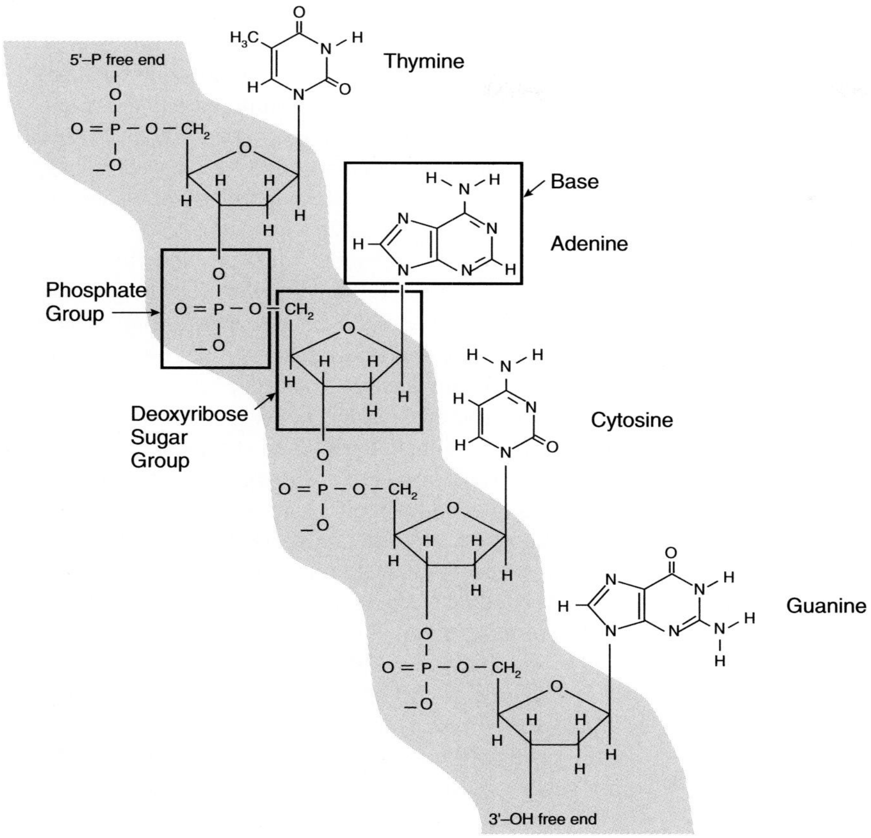

Figure 16.7
A section of single-stranded DNA containing the four different bases found in DNA.

might be tempted to think that such a simple code could not possibly allow sufficient variation to be responsible for the differences among men, women, fish, worms, plants, and bacteria. A few moments considering the possibilities, however, can persuade you otherwise. Given four bases, there are 16 different base sequences that are just 2 bases long. Sixteen equals 4^2, and in general, the number of possible base sequences for a DNA molecule "n" bases long equals 4^n. Suppose we consider DNA molecules just 17 bases in length. There will be 4^{17} possible sequences for such molecules amounting to about 3 unique molecules for every person on Earth. Real DNA molecules are many thousands of bases long. The DNA in a human cell is estimated to be about 3 billion bases in length (shared out between 46 distinct molecules within our 46 "chromosomes"). This means that there are around $4^{3 \text{billion}}$ different possible DNA sequences for a human being. This is literally an unimaginably large number—no wonder we are all different—and also very different from fish, worms, plants, and bacteria!

■ case in point

DNA Profiling

In recent years a major new technique has become available to investigators trying to identify the perpetrators of serious crimes. That technique is DNA profiling or DNA fingerprinting and is based on the fact that no two people (apart from genetically identical twins) share the same DNA sequence. DNA profiling has become an everyday feature of news reports of murder, assault, and rape cases worldwide. Whenever any biological fluids or tissues are left at the scene of a crime or on the victim, it is possible to analyze these materials to gain a DNA profile that should match the DNA profile of the culprit. Blood, skin scrapings, or semen are all suitable samples for analysis, even when quite old and dried. Under suitable conditions tiny quantities of these samples can be enough to gain overwhelming evidence of a suspect's guilt. Furious debate proceeds between lawyers and scientists regarding how convincing DNA profiling evidence really is. The answer to this issue depends on the circumstances involved, the quality of the sample, the professionalism of the people performing the analysis, the exact type of analysis involved, and the closeness of the genetic relationship between various suspects. Under the best conditions and circumstances, however, there is little doubt that a properly conducted DNA profiling procedure can provide a foolproof link between a suspect and the scene of a crime. The U.S. National Academy of Sciences has endorsed DNA evidence in court as fully reliable if the sample is properly handled leading up to, and during, the analysis. Let's look at the basic principles involved.

There are actually several different ways to undertake a DNA profiling procedure, but the crucial step involves using an enzyme (called a restriction endonuclease) to cut samples of the DNA at specific sites. The sites that the enzyme recognizes and cuts are found at different locations in different people's DNA, so each individual's DNA is chopped up into a specific pattern of fragments of characteristic lengths. These fragments are applied to a gel and pulled through the gel by an electric current. This process is known as gel electrophoresis, and it works because the DNA fragments are electrically charged due to the nega-tive charges carried by the phosphate groups on the DNA. The different-sized DNA fragments are pulled at different speeds through the gel so they end up spread out into a pattern that looks a bit like a supermarket bar code (see Figure 16.8). This is not the bar code for a can of beans, however, but the "bar code" that identifies a particular human being! In order to visualize the final pattern of a DNA profile test, radioactive "probes" must be added to bind to DNA fragments and make them show by exposing photographic film. In a full analysis, several different restriction enzymes and several different probes must be used and the collective results analyzed carefully with a computer. Despite these and other complexities, however, we have described the essential principles of the technique. DNA profiling is also useful in paternity disputes to prove who is the father of a child, in medical analysis of genetic relationships between different people and populations, and in the identification of decayed remains. The identity of the bones long thought to be the remains of Tsar Nicholas II of Russia (murdered during the Revolution of 1918) was confirmed by comparing DNA extracted from the bones with DNA from some of the Tsar's known living relatives. Differences in the base sequence of DNA also underpin all the physical differences between different people and different species. DNA can be thought of as an "information store," and the information it contains profoundly influences what each living thing looks like and can do. The base sequence of your DNA not only determines many of the differences between yourself and your best friend, it also is the reason you grew up to be a human rather than a fish, a worm, a plant, or a bacterium.

The last of the Romanovs—Tsar Nicholas II and his family—were killed in 1918 and their bodies hidden. DNA testing was used to identify the remains of the royal family.

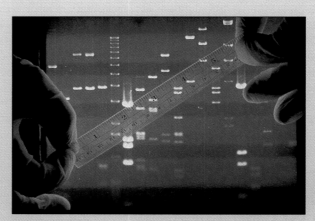

Figure 16.8
Each individual's genetic material, contained in the nucleus of each cell in the form of deoxyribonucleic acid (DNA), is unique. The analysis of DNA by the complex process of DNA fingerprinting is used to prove conclusively whether people are related, to establish paternity, or to identify individuals from evidence left at a crime scene, such as blood or hair. The process of gel electrophoresis is used to obtain the band pattern.

16.3 | The 3-D Structure of DNA

So far we have described DNA as a single-stranded polymer chain with specific sequences of the four bases connected to the unchanging sugar-phosphate "backbone." This is true, but it is not the form in which DNA occurs in nature. As we shall see, the ability of DNA to act as the foundation of life depends on noncovalent interactions between distinct strands of DNA. One of the most significant events in the history of science was the 1953 discovery, by James D. Watson and Francis Crick working at Cambridge University in England, that DNA normally exists in the form of two strands intertwined in a spiral formation, known as a **double helix** (see Figure 16.9). The strands are held together by hydrogen bonds between the bases, but a very strict rule controls which bases can "pair" up in this way. Can you identify the "rules of base-pairing" by looking at Figure 16.9? Hopefully, you will soon notice that A is always paired with T, and G is always paired with C. So there are really only two possible types of base-pairs: the A-T pair and the G-C pair, although these can occur either way around. Adenine and thymine (A-T) form a pair of **"complementary" bases**, meaning bases that can pair up within the regular structure of DNA. Guanine and cytosine (G-T) form the other possible pair of complementary bases. Two strands of DNA whose sequences match the rules of base-pairing are called complementary strands of DNA, and stable double-helical DNA can form only between complementary strands.

Note in Figure 16.9, that the A-T base-pair is held together by two hydrogen bonds, while the G-C base-pair is held together by three hydrogen bonds. This makes the G-C link somewhat stronger than the A-T link, meaning more energy is required to separate a G-C pair than an A-T pair. Although hydrogen bonds are much weaker than covalent bonds, the overall effect of many hydrogen bonds allows the double-helical structure of DNA to be relatively stable. It can, however, become "unzipped" when specific enzymes and other proteins bind to it—a feature that is vital for DNA's function, as we shall see. A word about terminology. Sections of double-helical DNA are routinely described as DNA molecules, although strictly speaking they are composed of two distinct molecules held together by hydrogen bonds. To be even more strict, DNA is actually an ion rather than a molecule, due to negative charges carried by oxygen atoms of the sugar-phosphate backbones. Such subtleties are usually ignored, allowing us to talk of molecules of single-stranded DNA and molecules of double-stranded DNA, as appropriate.

There are roughly 3 billion base-pairs in the double-helical DNA strands in each human cell, although only a small percentage of the DNA is known to perform specific functions. To begin to understand the functions, in other words to begin to understand exactly what it is that DNA does that can be so important, we need to look at how DNA is delineated into individual sections known as genes.

In 1963, Francis Crick, James Watson, and Maurice Wilkins shared the Nobel Prize in medicine and physiology for their conclusion on the true structure of DNA. Rosalind Franklin's work was also crucial to the success but she died before the Nobel Prizes were awarded.

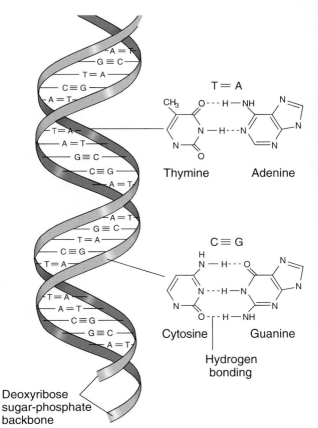

Deoxyribose sugar-phosphate backbone

Figure 16.9
The DNA double helix.

exercise 16.2

Pairing Up

Problem

Given this base sequence from one of the DNA strands of a gene, what would be the sequence of the complementary DNA strand?

TTAGCCAATGATC

Solution

Original DNA: T T A G C C A A T G A T C
                        | | | | | | | | | | | | |
Complementary DNA: A A T C G G T T A C T A G

16.4 | Genes—Where the Action Starts

The DNA of human cells is found within 46 distinct structures in the cell nucleus known as chromosomes (see Figure 16.10). Chromosomes are not naked DNA, but instead contain DNA coiled around a mixture of protein molecules with the combination of DNA and protein being known as **chromatin**. As shown in Figure 16.10, the DNA strands within chromosomes are wrapped around protein much like the dough around hot dogs in "pig-in-a-blanket" hors d'oeuvres. More protein surrounds the structure like a scaffold around a building.

Figure 16.10
DNA is folded together with protein molecules to form chromosomes.

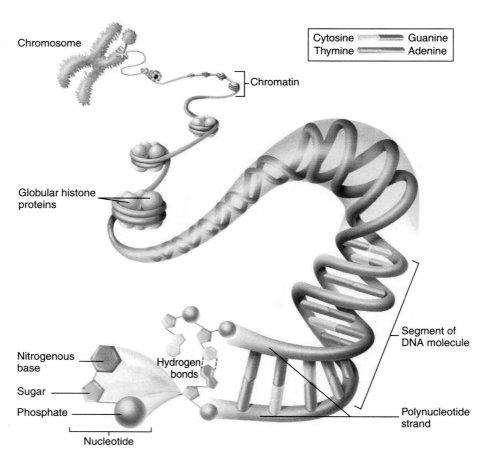

Let's begin with a simple first definition of a **gene,** then add details and exceptions later. In essence, a gene is a long section of a DNA strand that contains a particular sequence of bases able to direct the manufacture of a particular protein, or in other words "encode" a specific protein.

How Genes Make Proteins

Since the base-pairs of DNA are held together by weak hydrogen bonds, they can be broken rather easily when an enzyme called RNA polymerase binds to the double helix near the start of a gene. This enzyme moves along one strand of the DNA catalyzing the manufacture of a strand of **ribonucleic acid (RNA)** that is complementary to one strand of the DNA. As its name implies, ribonucleic acid is virtually identical to deoxyribonucleic acid (DNA). The two differences are that each sugar group in RNA is derived from a molecule of **ribose,** rather than deoxyribose; and a base called uracil (U) is found in RNA in place of the thymine of DNA. So the A-T base-pair of DNA is replaced by an A-U base-pair between a strand of DNA and RNA, or within double-stranded RNA.

As shown in Figure 16.11, the enzyme RNA polymerase manufactures a single-stranded RNA copy of a gene. This copying process is possible because the rules of base-pairing allow the new RNA molecule to be built upon the "template" of an existing DNA strand. The RNA polymerase enzyme can only link nucleotides into the RNA molecule if they can form base-pairs with the existing bases of the DNA template. So the RNA strand is perfectly complementary to the DNA strand that acts as the template controlling its formation. The manufacturing of an RNA copy of a gene is called **transcription** because the base sequence stored in the gene's DNA is being "transcribed" or copied into the slightly different form of a base sequence in RNA.

The form of RNA made by RNA polymerase is called **messenger RNA (mRNA)** because once it is formed it leaves the cell nucleus and takes the chemical "message" stored within a gene out into the cell cytoplasm where the synthesis of new proteins occurs. The messenger RNA molecule is able to direct the synthesis of a new protein by determining the amino acid sequence of the protein. This occurs because each set of three sequential bases in mRNA (known as a **codon**) can direct the incorporation of one specific amino acid into a growing protein chain (see Table 16.2). This is achieved during a process known as **translation** because the base sequence of the mRNA (and, indirectly, of the original gene) is now being "translated" into the quite different form of the amino acid sequence of a protein (see Figure 16.11C).

exercise 16.3

Forming an RNA Strand

Problem

Draw a complementary RNA strand below the uncoiled section of DNA.

Use the abbreviations U, A, G, and C for uracil, adenine, guanine, and cytosine respectively.

Uncoiled DNA: T G A C A C T T A G C A A G G T

Solution

DNA: Template: T G A C A C T T A G C A A G G T
                         | | | | | | | | | | | | | | | |

Complementary RNA: A C U G U G A A U C G U U C C A

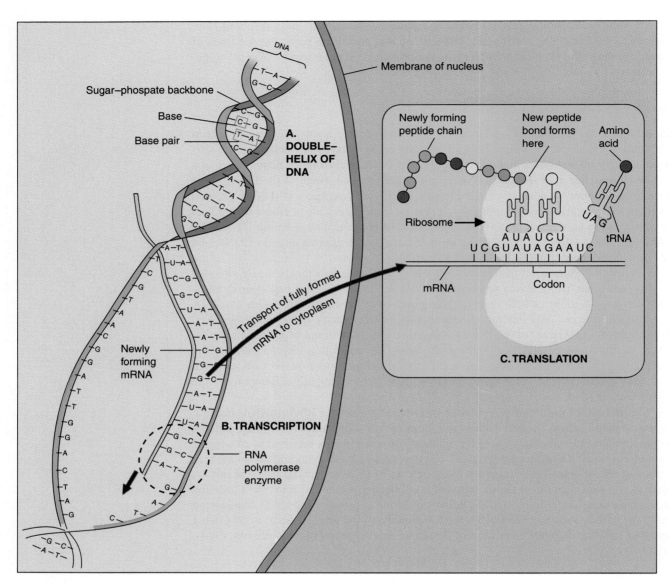

Figure 16.11
Transcription and translation.

The synthesis of a protein takes place on complex structures called **ribosomes**, structures composed of several distinct protein molecules and specialized RNA molecules called **ribosomal RNA (rRNA)**. Although ribosomes are the sites of protein synthesis, they cannot catalyze protein synthesis without the assistance of many other enzymes that form temporary associations with the ribosome as protein synthesis occurs.

A ribosome becomes attached to the start of an mRNA molecule and then moves along it. Each time a codon of the mRNA is exposed at a special site on the ribosome, other RNA molecules known as **transfer RNA (tRNA)** molecules can bind to the codon by base-pairing (see Figure 16.11C). This is possible because each tRNA has a matching set of three bases, an **anticodon**, that is able to form base-pairs with one particular codon. Each tRNA molecule has also been previously attached to a specific amino acid molecule, which, when it arrives at the ribosome, is linked by enzymes into a growing protein chain.

So as a ribosome moves along an mRNA molecule, the base sequence of the mRNA determines which tRNA molecules become temporarily bound to the mRNA codons, and so

Table 16.2

The Genetic Code Showing Which Amino Acids (Represented by Their Abbreviations) Are Encoded by Each Codon

	U	C	A	G	
U	UUU ⎫ Phe UUC ⎬ UUA ⎫ Leu UUG ⎬	UCU ⎫ UCC ⎬ Ser UCA ⎬ UCG ⎭	UAU ⎫ Tyr UAC ⎬ UAA ⎫ Stop UAG ⎬	UGU ⎫ Cys UGC ⎬ UGA Stop UGG Trp	U C A G
C	CUU ⎫ CUC ⎬ Leu CUA ⎬ CUG ⎭	CCU ⎫ CCC ⎬ Pro CCA ⎬ CCG ⎭	CAU ⎫ His CAC ⎬ CAA ⎫ Gln CAG ⎬	CGU ⎫ CGC ⎬ Arg CGA ⎬ CGG ⎭	U C A G
A	AUU ⎫ AUC ⎬ Ile AUA ⎭ AUG Met	ACU ⎫ ACC ⎬ Thr ACA ⎬ ACG ⎭	AAU ⎫ Asn AAC ⎬ AAA ⎫ Lys AAG ⎬	AGU ⎫ Ser AGC ⎬ AGA ⎫ Arg AGG ⎬	U C A G
G	GUU ⎫ GUC ⎬ Val GUA ⎬ GUG ⎭	GCU ⎫ GCC ⎬ Ala GCA ⎬ GCG ⎭	GAU ⎫ Asp GAC ⎬ GAA ⎫ Glu GAG ⎬	GGU ⎫ GGC ⎬ Gly GGA ⎬ GGG ⎭	U C A G

determines the order in which amino acids become linked into a protein chain. This is the essential outline (which is embellished by many details) of the way in which the base sequence of a gene determines the amino acid sequence of the protein encoded by the gene.

Each cell of the human body contains around 50,000 genes. The exact number is not yet known. Most of these genes encode specific protein molecules, in the manner just described. We say "most," however, because some genes simply code for functional RNAs such as the rRNAs and tRNAs. Another complication is that the proteins encoded by more than one gene must sometimes combine to produce a larger "multisubunit" protein able to perform a particular function. At least 30,000 different proteins, however, can be made by human cells when and where appropriate. The point about genes only becoming active where and when appropriate is crucial: genes can be switched "on" and "off" in various ways, usually due to the binding of specific proteins to the DNA close to a gene. This means that although each human cell contains the same genes, only the genes needed to make liver cells are active in liver cells, only the genes needed to make brain cells are active in brain cells, and so on. Genes can also be switched on and off at different stages in the life cycle of any one type of cell, ensuring that the proteins they encode are only made when required and in the quantities required.

exercise 16.4

Using the Genetic Code.

Problem

Use the genetic code table (Table 16.2) to work out the amino acid sequence of a segment of protein encoded by the following sequence of bases on mRNA:

GGUGGGUGGCCAAAUGCUGCCGGAGGCGUGGUUUCA

Solution

gly-gly-trp-pro-asn-ala-ala-gly-gly-val-val-ser

This is the basic procedure scientists must use whenever they wish to work out the amino acid sequence of the protein produced by any gene.

16.5 | DNA Replication—How Genes Pass Down through the Generations

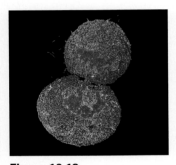

Figure 16.12
Cell division.

You began life as a single cell, which then divided into two, then four, and so on until that first cell had generated the many billions of cells that you contain today. Even once we have stopped growing, many of the cells in our body continue to multiply by cell division (see Figure 16.12). Cells in the skin, for example, are constantly growing then dividing to generate new cells to replace those that constantly die and rub away on the skin's surface. A similar process of constant cell division keeps the lining of our stomach and intestines in good condition, by replacing cells that are continually shed and lost in our feces. The repair of wounds is also made possible by cell division, allowing new cells to be produced to replace and repair wounded tissue. Cell division also allows our bodies to generate the specialized egg cells (in women) or sperm cells (in men) that carry copies of our genes into our children, and thus sustain the process of inheritance that keeps our species (and all species) going. Every time a cell divides in two, one copy of the original cell's genes must go into each of the two new cells. So the growth, maintenance, and reproduction of living things depends on a mechanism of gene copying.

Since it is really all of the DNA that is being copied, this essential chemical copying process is known as DNA replication.

Making Copies of DNA

How is it possible that all of our genes, containing billions of base-pairs along with proteins, can be reproduced nearly flawlessly as cells divide, throughout our entire lifetime? The key lies in the double-helical structure of DNA, particularly the rules of base-pairing. Recall from the beginning of this section that DNA in the cell nucleus is arranged in a double helix with complementary base-pairs involved in hydrogen bonding that keep the two DNA strands together (see Figure 16.9) As shown in Figure 16.13, when cell replication occurs, the DNA double helix (parent strand) unzips, leaving two individual strands called templates. These template strands are the "pattern pieces" used to construct new DNA (daughter) strands. This occurs in much the same way as a single DNA strand acts as a template for the production of a complementary mRNA during transcription. Various enzymes catalyze the linking up of nucleotides in a sequence complementary to the template strand, as shown in Figure 16.13. When the process is complete, two identical DNA double helices have been created in place of the single original one. Before a cell divides, each chromosome is duplicated by DNA repli-

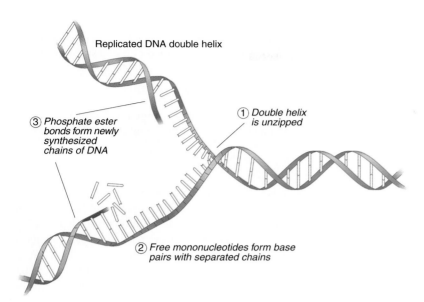

Replicated DNA double helix

③ Phosphate ester bonds form newly synthesized chains of DNA

① Double helix is unzipped

② Free mononucleotides form base pairs with separated chains

Figure 16.13
DNA replication in which the original double helix unwinds or "unzips" allowing the individual strands to act as the templates on which new complementary strands are assembled. In this way, one original double helix generates two identical copies to replace it.

cation, and then when cell division occurs one copy of each chromosome is packaged into each of the resulting cells.

The chemical process of DNA replication underpins the ability of humans to have children, of plants to release seeds that can grow into other plants, and of all cells to multiply by splitting in two. Many subtleties and complications can be built upon that simple chemical foundation. It takes two humans, for example, to create a new human because we reproduce by sexual reproduction, in which genes from two individuals are mixed together to create the **genome** of the child. It is not appropriate to explore such complications in this book; but it is important to emphasize that the chemical copying process of DNA replication is the fundamental reason why living cells can make more living cells, thus ensuring that life continues even though each living individual must die.

The genetic material (DNA) of the cell is found within a single large membrane-bound organelle known as the **nucleus**. The all-important genes embodied in the DNA never leave the nucleus (apart from when the nucleus itself breaks down during cell division), but they are constantly used to direct the manufacture of specific proteins. The actual process of protein synthesis itself occurs in the cell cytoplasm, on small granules of protein and RNA known as ribosomes. It is here that amino acids within the cell are linked into new protein chains whose amino acid sequences are determined by the chemical structure of the genes within the nucleus.

16.6 | Carbohydrates

Proteins and the nucleic acids (DNA and RNA) are the most fundamental of the chemicals of life, but many other classes of chemicals also play vital roles. We now turn to a diverse range of chemicals known as **carbohydrates**. These serve as an energy source to power living things, and are also used as a basic raw material for building cells, and people. Sugars and starch, familiar to us as food ingredients, are carbohydrates.

Carbohydrates consist of only three elements, carbon, hydrogen and oxygen. The most common formula for a carbohydrate is $C_6H_{12}O_6$. Carbohydrates (carbon + hydrate) are so named because their formulas can be written as $C_x(H_2O)_y$. In other words, the ratio of hydrogen to oxygen atoms in a carbohydrate is always 2:1 just as it is in water. The structures of most carbohydrates feature a ring containing carbon atoms (usually five) and one oxygen atom. The carbons of such a ring are bonded to O—H groups and H atoms. Sometimes two or more of these rings are hooked together to make bigger molecules. We will discuss these structures shortly. The O—H groups are what make sugar and starch so soluble in water.

The best-known function of carbohydrates is as a source of energy in our food. The energy available to the body from carbohydrates is the result of a complex series of chemical reactions that has the overall effect of the complete oxidation of glucose (blood sugar), $C_6H_{12}O_6$, by oxygen.

$$C_6H_{12}O_6 + 6O_2 \rightarrow 6CO_2 + 6H_2O$$

The glucose will either have been eaten directly or, more commonly, formed from other carbohydrates in the diet. The oxygen is inhaled in every breath we take.

Monosaccharides

The simplest carbohydrates are the **monosaccharides**, which means "single sugars." The three monosaccharides we will consider, glucose, galactose, and fructose, all have the same familiar formula, $C_6H_{12}O_6$, but different structures and properties. Recall from Chapter 6 that compounds that have different structures but the same formula are termed isomers. Glucose, galactose, and fructose are three monosaccharide isomers (see Figure 16.14). They are all white crystalline, sweet-tasting compounds. Of the three, glucose is most important in human nutrition because sugar in this form is used by our bodies as a source of energy and is the sugar found in human blood. For this reason, glucose is often called "blood sugar." When carbohydrates are eaten, our body transforms them into glucose for utilization by the body's cells.

Disaccharides

Two monosaccharides can react to join together to form a larger **disaccharide** molecule and release one water molecule.

$$2C_6H_{12}O_6 \rightarrow C_{12}H_{22}O_{11} + H_2O$$

Because water is released, this is known as a condensation reaction (see Figure 16.15). The most important disaccharides in our diet are sucrose, composed of one glucose and one fructose; and lactose, composed of one galactose and one glucose (both shown in Figure 16.15). The bond holding the two monosaccharides together through an oxygen atom is called a glycosidic bond.

Sucrose is the sugar that most of us think of when we hear the word sugar. It is the sugar in the sugar bowl. Lactose is the sugar in milk. Fructose is the sugar in honey and fruits. The reactions causing the formation of disaccharides look straightforward, but simply mixing glucose with fructose will not cause sucrose to be formed. Rather, the coupling of one sugar with another to make a more complex carbohydrate happens only under the influence of an enzyme.

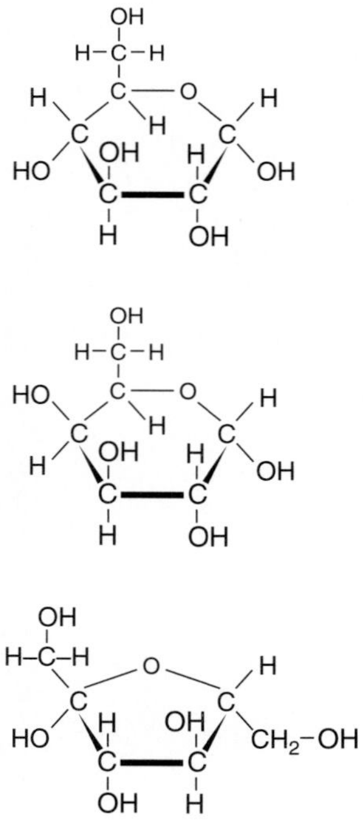

Complex Carbohydrates

The carbohydrates we call complex are composed of many glucose molecules joined together into a large polymer. The joining reaction is the same reaction as the one that forms disaccharides, just repeated many more times. For example, the complex carbohydrate known as starch, found in large quantities in many foods, is formed when many thousands of glucose molecules become linked. Our bodies can digest the starch in foods into individual glucose molecules, which are then used as a source of energy. Another component of food, called fiber, is made of indigestible carbohydrate molecules, which our digestive enzymes cannot break down. The main component of fiber is cellulose.

Starch and cellulose are both composed of many glucose molecules linked together. So why can we utilize one as a source of energy but not the other? There are two forms of glucose, called α-glucose and β-glucose, which differ in the three dimensional arrangement of their H and OH groups (see Figure 16.16). Starch is a polymer composed of α-glucose molecules (see Figure 16.17), while cellulose contains β-glucose (see Figure 16.18). That subtle

Figure 16.14
Structures of glucose, galactose, and fructose. All three compounds are monosaccharides with the formula $C_6H_{12}O_6$. They differ in their structures and properties.

Alpha-glucose

Galactose

Fructose

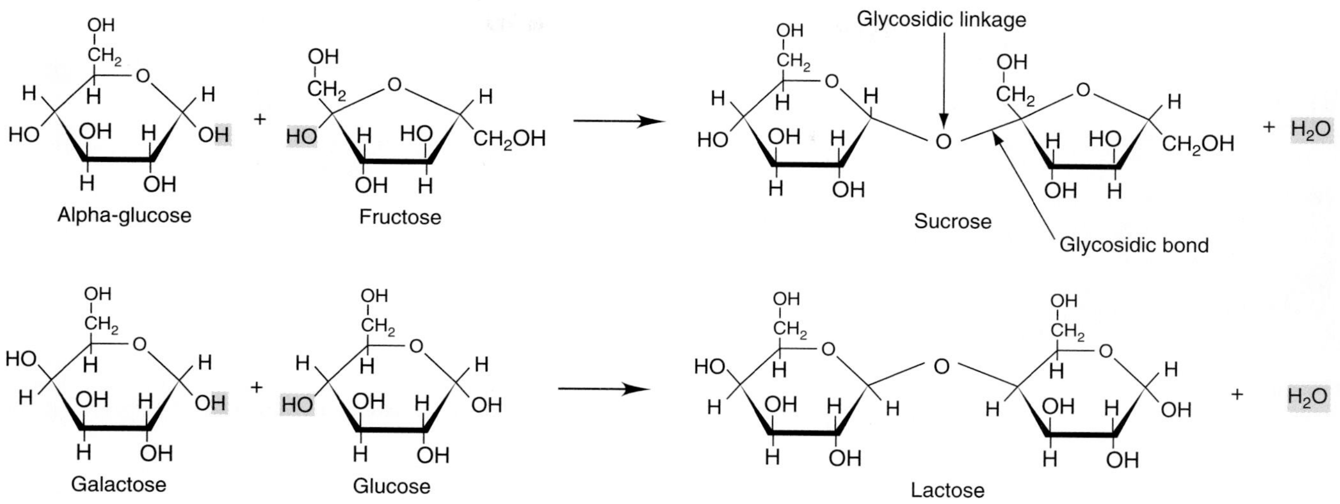

Figure 16.15
Formation of sucrose from glucose and fructose; and lactose from glucose and galactose. The linkage between the two rings through the oxygen atom is called a glycosidic linkage. The highlighted atoms in the two monosaccharides become a water molecule in this reaction.

β-glucose

α-glucose

Figure 16.16
Note the difference in α-glucose and β-glucose is at the carbon-1 position.

Figure 16.17
Amylopectin, a component of starch, formed by the α-linkage of many glucose molecules.

structural distinction means that the digestive enzymes able to bind to and break the "α-linkages" between glucose molecules in starch cannot attack the β-linkages in cellulose.

Cellulose in plants is principally a structural chemical responsible for the rigidity of plant cells walls. Starch, on the other hand, serves largely as an energy storage molecule. It is found in large amounts in seeds and tubers (such as potatoes) and can be degraded to energy-yielding glucose molecules when energy is required to power new plant growth. Some animals, such as cattle, can use cellulose as a food energy source because bacteria in their gut do possess enzymes that will degrade the cellulose into a usable form. This explains why cows can survive by eating grass, but we cannot.

When we eat amounts of food that supply more than our immediate energy needs, some of the extra carbohydrate is converted in our livers into a complex carbohydrate called

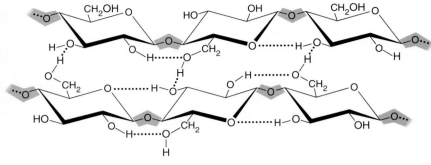

Figure 16.18
β-Linkages of glucose lead to cellulose, a linear polymer capable of packing into fibers in, for example, cotton or wood.

Figure 16.19
The structure of glycogen is composed of branched chains of bonded glucose molecules.

glycogen, or animal starch. Glycogen is a lot like starch but it is highly branched rather than linear as shown in Figure 16.19. Glycogen is stored in the liver and the muscles. Between meals or during intense physical exertion, the glycogen can react very quickly with water and separate into glucose. This glucose is almost immediately available to be used as a source of energy.

16.7 | Lipids, Including Fats and Oils

Lipids are a rather diverse range of the chemicals of life that share the property of being insoluble in water. Some of them, the fats and oils, are familiar to us as components of food. One of the most important roles for the lipids of life is to form the thin cell membranes that make each living cell distinct (see Figure 16.20). They also serve as important energy storage molecules, as discussed in Chapter 17.

The fats and oils familiar to us as key food components belong to a specific class of lipids known as triglycerides. Technically oils are triglycerides that are liquid at room temperature while fats are triglycerides that are solid at room temperature. A triglyceride molecule is made when four other molecules come together. Three of these molecules are so-called "fatty acids" and the fourth is called glycerol (glycerine). A typical fatty acid structure is shown in Figure 16.21. Notice the —COOH group at one end of the fatty acid and recall from Chapter 6 that this is characteristic of organic acids. The R in the general formula of a fatty acid (Figure 16.21D), represents any of several possible hydrocarbon groups, typically with 11–17 carbon atoms and about twice that number of hydrogen atoms. A triglyceride is formed when the —COOH groups of three fatty acids combine with the H of the three —OH groups of glycerol to make a single triglyceride molecule and release three water molecules (see Figure 16.22). Since the product of a carboxylic acid reacting with an alcohol is called an ester, fats and oils are also often called triesters. As you might guess, molecules having one or two of the glycerol OH groups substituted by fatty acids are termed monoglycerides and diglycerides, and are monoesters and diesters respectively.

The term lipids is an unusually general one, since it includes a variety of compounds that are not triglycerides but share the characteristic of being insoluble in water. Some are derived from triglycerides in a simple way, such as some "phospholipids" in which one of the fatty acid chains of a triglyceride is replaced by a small phosphate-containing group of atoms. Phospholipids are vital components of the lipid bilayer membranes, illustrated in Figure 16.20, that surround all cells. They spontaneously form bilayers when in contact with water due to the

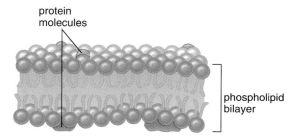

Figure 16.20
Structure of a lipid bilayer membrane.

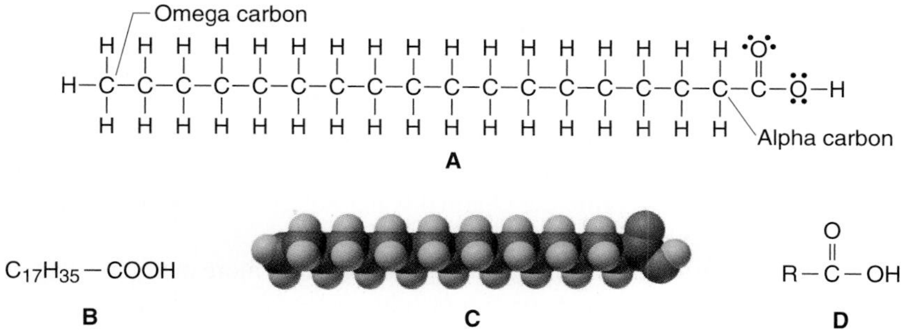

Figure 16.21
A fatty acid can be represented in several different ways.

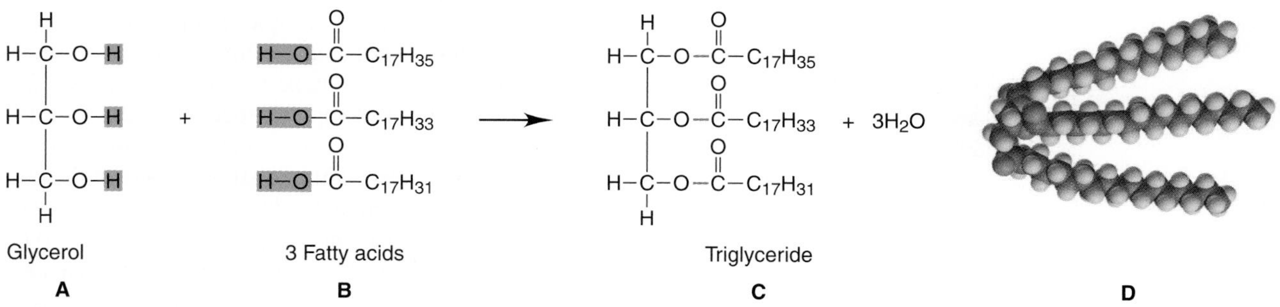

$C_{17}H_{35}$ — COOH

B

D

Figure 16.22
A triglyceride is formed from the reaction of one molecule of glycerol with three fatty acid molecules. The three fatty acids are usually not all identical to each other. They may differ in carbon number or number of double bonds.

Glycerol

A

3 Fatty acids

B

Triglyceride

C

D

$+ 3H_2O$

Figure 16.23
Chemical structure of a phospholipid.

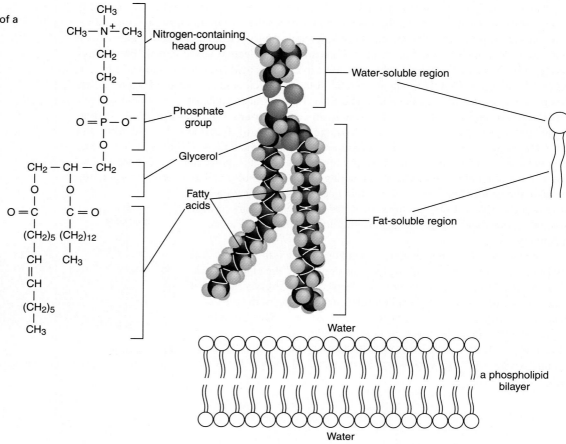

differing affinities for water of the different parts of the molecules. This is illustrated by the phospholipid in Figure 16.23. In the presence of water, the uncharged hydrocarbon chains cluster together away from the water, while the charged water soluble head groups interact with the water.

Others lipids, such as the well-known "cholesterol," considered in more detail in Chapter 17, have completely different chemical structures from triglycerides (see Figure 16.24).

16.8 | From Chemicals to People

Figure 16.25 illustrates the different levels of organization within a person, from chemicals upward. We know that everything within us is composed of atoms or ions made up of protons, neutrons, and electrons. These atoms combine together to make molecules such as amino acids, sugars, proteins, DNA and so on. The chemicals of life operate together within cells (see Figure 16.26), and clusters of similar cells form tissues such as muscle tissue, skin tissue, and so on. Different specialized tissues are found within distinct organs of the body such as the brain, heart, lungs, stomach, liver, and kidneys. Various organs work together to form systems of the body such as the digestive system, the nervous system, and the circulatory system; and all the systems operating in harmony sustain a healthy human being. At all levels, however, everything that happens is dependent on chemistry, indeed it *is* chemistry, since all we have within us are chemicals interacting in ways that depend on their structures and which achieve the overall amazing function of producing a human being. Each one of us is a wonderful example of how chemical structure dictates function.

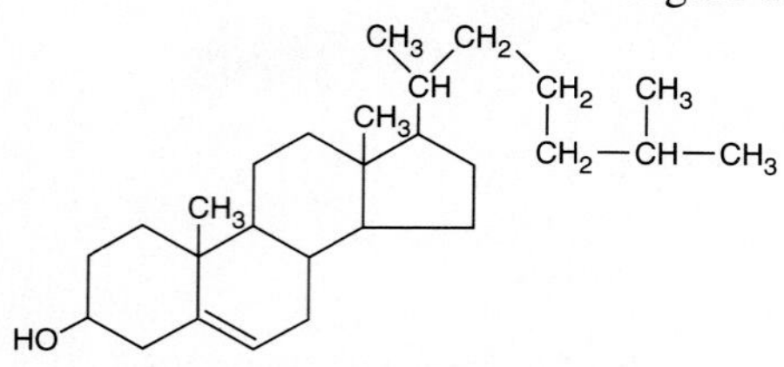

Figure 16.24
The molecular structure of cholesterol.

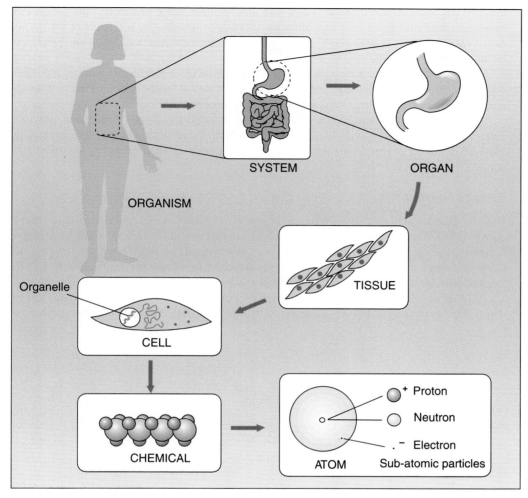

Figure 16.25
Deconstruction of the body into systems, organs, tissues, cells, chemicals, atoms, protons, neutrons, and electrons.

There is a lot more to understanding the chemistry of life than just knowing how DNA encodes protein, the functions of proteins, and the structures of carbohydrates and lipids. That knowledge just introduces the key chemical characters in the wonderfully complicated chemical drama being played out in all of us every second of the day.

We cannot examine the complexities of biochemistry in detail in this course, but we should complete this review of the chemistry of life with a brief look at three important issues—metabolism, energy supply, and mutation.

Metabolism

We are made from the food we eat, which provides us with chemical raw materials used to build our cells and to provide a source of energy. As we discuss in Chapter 17, our food is composed of proteins, lipids, carbohydrates, minerals, vitamins, and water. The proteins, lipids, and carbohydrates must be broken down into simple building blocks, such as glucose and amino acids, before we can use them as raw materials. The food that is combined with oxygen to release energy for us is also broken down in

Figure 16.26
Generalized cell structure.

that process, into carbon dioxide, water, and other simple waste products. This process of breaking chemicals down into simpler ones, often accompanied by a release of energy, is called **catabolism**. It is achieved by many individual chemical reactions, each one catalyzed by an appropriate enzyme.

Building the raw materials back up into our own proteins, lipids, carbohydrate, and other essential chemicals is called **anabolism**, and is also achieved by the action of many individual enzymes, each catalyzing some small step in the process.

The combination of catabolism and anabolism, comprising all the thousands of enzyme-catalyzed reactions that sustain life, is our **metabolism**. Generally speaking, the reactions of catabolism release energy, while those of anabolism require an input of energy to make them happen.

Energy Supply

When chemicals such as carbohydrates and lipids are broken down in catabolism, some of the energy released must be trapped in a chemical form that can to be used to power the energy-requiring processes of anabolism. The main way in which this energy is trapped is by the coupling of energy-releasing reactions to the manufacture of **adenosine triphosphate (ATP)** from **adenosine diphosphate (ADP)** and phosphate ions (see Figure 16.27).

ATP acts as an "energy currency" within living things. Energy-requiring processes can be coupled to the hydrolysis of ATP back into ADP plus phosphate. Provided the energy released

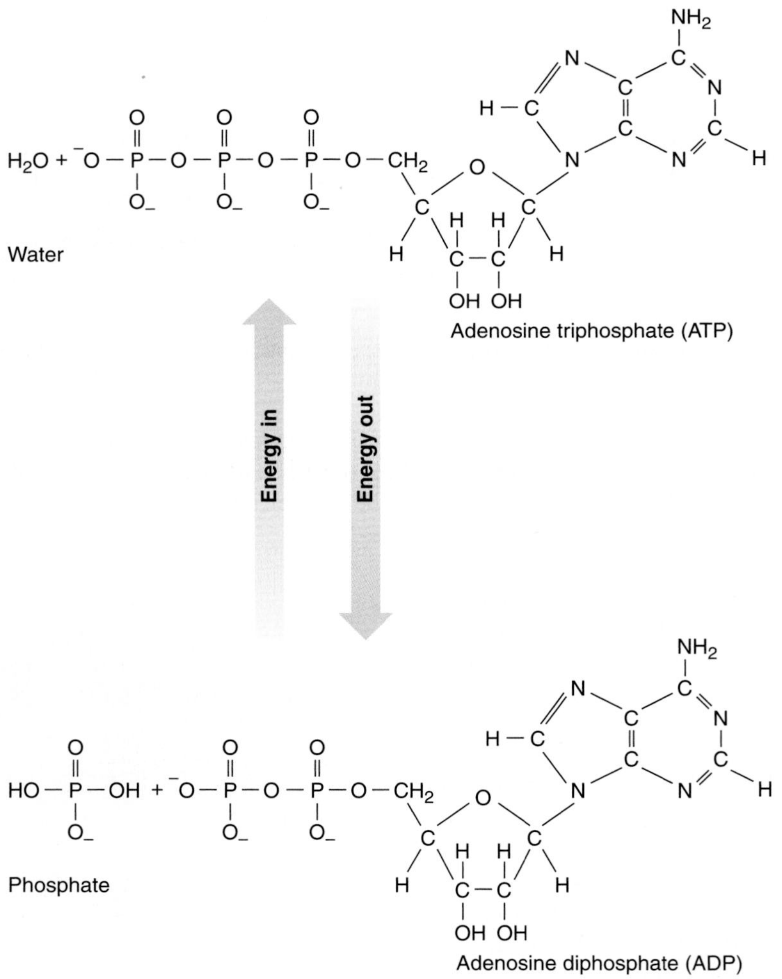

Figure 16.27
The formation and hydrolysis of adenine triphosphate (ATP).

by the hydrolysis of ATP is greater than the energy needed to make the coupled reaction proceed, ATP hydrolysis can be used to drive forward energy-requiring reactions. The reactions that build our proteins, and replicate our genes, for example, must be powered in one way or another by the hydrolysis of ATP

Mutation

The DNA of genes is not an unchanging information store. From time to time accidents happen, causing one base to be substituted for another, or one base to be deleted, or a new base added, or causing much larger rearrangements of the base sequence of genes. These changes are called **mutations**. They can occur spontaneously, or may be initiated by the effect of radiation or the reaction of "mutagenic" (mutation-generating) chemicals with DNA.

These changes in DNA are responsible both for disease and for evolution. They can cause disease when their effect on the proteins encoded by the affected genes is damaging. They power evolution when their effect opens up new possibilities, allowing the affected organisms to develop and function in new ways.

Our understanding of molecular genetics—the science of our genes—is currently undergoing explosive development. We have learned how to manipulate genes for ourselves, in the process known as "genetic engineering," and we are learning all there is to know about all the genes that make up a human, and all the genes that make up many other organisms as well. We will close this chapter with a look at some key developments at this "genetic frontier."

16.9 | The Genetic Frontier—New Choices We Can Make about Life

Scientific progress has recently opened up a whole new set of chemical, medical, and social choices that have been unavailable to previous generations. We are now able to alter genes, and to "mix and match" the genes of different species using technology known as **genetic engineering**. We can use the cells of adult animals to create genetically identical copies, or clones, of the donor animal. We can control and modify the chemical that directs the activities of all living things—DNA.

This is enticing but dangerous territory. On the one hand, it offers the opportunity to understand and cure many serious diseases and perhaps to grow crops more efficiently and more safely. On the other hand, it raises such specters as "designer babies" modified to suit the possibly misguided preferences of their parents, of "human cloning" being used to create many genetically identical copies of chosen people, and all manner of unpleasant and unpredictable consequences of tampering with the central chemistry of life and evolution.

Recombinant DNA Technology

The basic techniques of genetic engineering are quite simple to describe, even if they are more complex to achieve in practice. To alter DNA we need tools that will cut DNA at selected sites and then "paste" different sections of DNA together. The necessary tools are a set of naturally occurring enzymes, which can be purified and added to samples of DNA to modify them in the ways we wish.

Enzymes called "restriction endonucleases" will cut DNA at selected short sequences, as shown in Figure 16.28. Some of these enzymes produce staggered cuts, creating sections of DNA with "sticky-ends" that will "stick" through base pairing to the matching ends of other DNA molecules cut with the same enzyme. So if a chosen sequence of DNA cut from the DNA of one species is mixed with chosen sequences cut from a different species, the sticky ends can hold the different DNAs together. The join can then be made permanent using another enzyme, "DNA ligase," to repair the cut in the sugar-phosphate backbone by forming appropriate new covalent bonds.

In simplest outline form, this explains how chemistry can be used to cut genes from one species and insert them into another species. Why would we want to do this? There are many

Figure 16.28
Using enzymes to transfer genes between species.

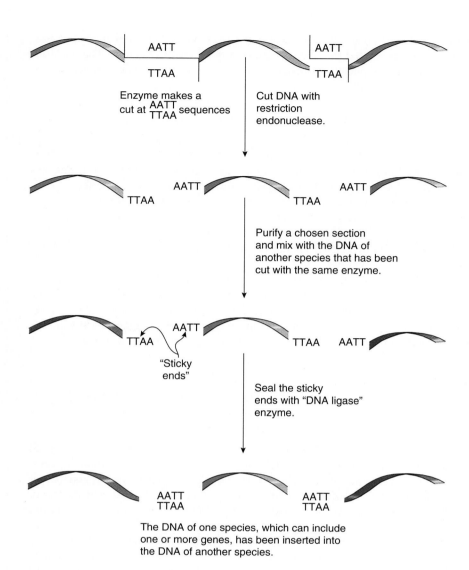

Enzyme makes a cut at $\frac{AATT}{TTAA}$ sequences

Cut DNA with restriction endonuclease.

Purify a chosen section and mix with the DNA of another species that has been cut with the same enzyme.

"Sticky ends"

Seal the sticky ends with "DNA ligase" enzyme.

The DNA of one species, which can include one or more genes, has been inserted into the DNA of another species.

reasons. For example, the gene coding for a medically useful protein, such as the insulin needed to treat diabetes, can be cut from human DNA, and inserted into bacterial DNA. The genetically engineered bacteria can them be grown in huge numbers and used as a source of human insulin, which can be purified as required. Other examples come from agriculture, with genes that confer resistance to specific pests and diseases able to be transferred into genetically engineered versions of the crops. The technique is also of great use to medical researchers who wish to isolate and study the effects of particular genes, and their corresponding proteins, that may cause such serious diseases as cancer.

The Polymerase Chain Reaction

In order to study and manipulate DNA molecules, it is often necessary to make a great many copies of particular pieces of DNA. This is done using another natural enzyme, called "DNA polymerase," whose natural task is to replicate DNA strands in the manner shown in Figure 16.13. The technique is known as the "polymerase chain reaction" (PCR). You may have heard of it in the news media. It can be applied to a tiny sample of DNA obtained from the scene of a murder, assault, rape, or other crime in order to produce sufficient copies of the DNA for forensic analysis. The practical process is quite complicated, so we will avoid the details, but it all depends on the ability of the enzyme to replicate DNA, as directed by the rules of base-pairing. Almost every scientist working with DNA must at some time use the polymerase chain reaction to make many copies of segments of DNA they wish to study or manipulate.

Gene Therapy

One of the main reasons scientists want to manipulate genes, and be able to transfer genes between different bits of DNA, is to hopefully cure some of the very serious diseases caused by defective genes. Trying to cure a genetic disease by adding new genes to the DNA of a cell is known as gene therapy. Attempts at gene therapy have met with mixed results so far, although it is bound to become more successful and widespread in the future.

Cloning

A Scottish sheep called "Dolly," pictured in Figure 16.29, has attained a celebrity status equal to that of presidents and film stars. She is the first mammal ever to have been "cloned" from the body cells of an adult. She was created at the Roslin Institute, near Edinburgh. Cells were cut from the udder of a six-year-old ewe and then fused with unfertilized eggs from which the genetic material had been removed. Twenty-nine of the eggs that appeared to have developed normally were implanted into the uterus of "surrogate mother" ewes. One of these eggs developed through a full "pregnancy" and gave rise to a live lamb, which was named Dolly. This lamb is genetically identical to the ewe whose cells gave rise to it, so it is a clone, produced artificially, without sexual reproduction and with no father being involved.

Figure 16.29
Dolly the sheep, with her "creator."

Scientists want to use the cloning technique to generate herds of cloned sheep that have had medically important genes added to their DNA. The hope is that the proteins encoded by these genes can be produced in the ewe's milk, and easily purified for use as medicines by humans. The intention is commendable, but many people react to the cloning of mammals in horror, because it raises the possibility of humans being cloned in a similar way. "Herds" of genetically identical humans, specially equipped with genes selected by their creators to make them do and act in ways their creators desire. Some of the worst speculations of science fiction seem about to become feasible.

The Human Genome Project

As we learn how to purify, isolate, and manipulate DNA, it is natural (and good science) to want to know everything there is to know about the DNA of humans. That is a massive task. The entire complement of DNA within a human's cells is known as their genome. The genome of a human contains around 50,000 genes, stored within a much larger total volume of DNA. There are also huge stretches of DNA separating many of our genes, some of it performing important regulatory functions, some of it probably without function—we just don't know for sure. Each person has a slightly different genome from every other person. Genes come in different variant forms, and none of us has exactly the same mixture. Unraveling all the complexity of human DNA is being undertaken by The Human Genome Project, a consortium of research teams from around the world, most notably in the United States, Europe, and Japan. The project's goal is to determine the base sequence of at least one version of every human gene, then study all the variants, and search for function within all the bits of DNA between the genes. The first major milestone was reached in January 2000, with the announcement that a first "rough draft" of a complete set of human genes had been obtained. It will take many more years to get a complete draft, then analyze all the variants. Already, however, the knowledge gained is revealing details of the genes involved in major diseases, such as cancer, and is being used to assist in the search for new treatments and cures.

The basic *chemical* challenge facing The Human Genome Project has been to work out the exact base sequence of all our genes (also known as the "nucleotide sequence," since the bases are held within nucleotides). Various slightly different methods are used. In one of the

pro con discussions

Genetic Engineering

pro

Genetic engineering offers great benefits to medicine, agriculture, industry, and basic research. We can mass-produce medically useful proteins very efficiently by putting the human genes for these proteins into bacterial cells, which will then pump out reliable supplies of the pure product. This has already been used to make a wide range of medicines including insulin to control diabetes, interferon to encourage the body's defense against disease, growth hormone to cure dwarfism, and much more. Vaccines that will immunize us against dangerous diseases can be created by putting genes (encoding the proteins that will give us immunity) into live viruses or bacteria that can be safely given to humans. This technique offers the main avenue of hope toward the development of a vaccine against the AIDS virus. Genetic engineering has been used to give plants foreign genes that make them resistant to pests and disease, or make them taste better, or make them less likely to rot. For example, genetically engineered tomatoes have been grown that will stay fresh and useable for much longer than other varieties. Genetic engineering is being used to help clean up pollution by giving harmless bacteria the genes that allow them to break up toxic chemicals, degrade the oil in oil spills, and so on. And research scientists routinely use genetic engineering to study how human cells react to the loss or addition of genes in a way that is gradually unlocking the secrets of such terrible diseases as cancer. By learning how to remove genes from cells, purify them, study them, and put them into other cells, we are learning how to make life safer, better, more pleasant, and free of dangerous diseases and pests.

most common, a sample of single-stranded DNA is cut with enzymes into portions approximately 500 bases in length. These DNA molecules are mixed with a DNA polymerase enzyme, like the one discussed earlier, plus a supply of nucleotides. The polymerase enzyme links the free nucleotides into new DNA strands complementary to the original strands. The crucial trick, however, is to stop this process at random by mixing in some unnatural nucleotides with the natural ones. This generates many fragments that, if separated properly, form a "nested set," meaning each member of the set of molecules has one more nucleotide than the previous member. All the fragments are separated by using an electric current to draw them through a gel, a process called "gel electrophoresis." As we saw earlier, this relies on the

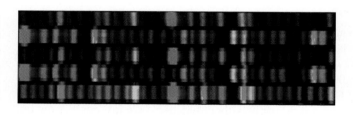

AC C CGGGG N T CCT CT AG N T CG AC CT GC AGG C AT GC AAG C TTG AG TATT CTATA GT GT CAC C TAA
    20   30   40   50   60   70

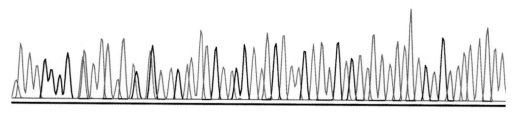

Figure 16.30
Automated DNA sequencer output. The four colors each represent one of the four nucleotides (top). Computerized analysis creates the sequence printout shown (bottom).

con Genetic engineering is a dangerous experiment with nature and one that we may live to regret. When you mess around with nature in ways that you don't fully understand, nature has a nasty habit of biting back. The most common way of using genetic engineering, to make proteins that we want or new vaccines, is to take genes out of animals or plants and switch them into the DNA of bacteria or viruses. We are told that only harmless bacteria or viruses are used, but how can we be sure? There is a danger that we will accidentally create dangerous new bacteria and viruses, and these things are so hard to contain they could escape and multiply out of control. Even seemingly harmless tinkering with plants holds potential hazards. If we start growing vast quantities of crops that are resistant to certain insect "pests," we could upset the "balance of nature" by starving of the beneficial birds and insects that live on these pests. We just don't know enough about the complex ecology of the natural world to go messing around with it by genetic engineering. Then there is the whole issue of the misuse of genetic engineering: the creation of deadly new biological and chemical weapons, for example. We might approve the techniques of genetic engineering for use in beneficial medical or agricultural applications only to find other people and nations adapting these techniques to create selective weapons of mass destruction. Humanity has already made too many mistakes when trying to become masters of nature. Just think about nuclear waste, global warming, and the destruction of the ozone layer. Genetic engineering could eventually prove to be our biggest mistake yet, because the fact that altered animals and plants reproduce and evolve means that we might change things very badly for the worse, and forever . . .

fact that DNA molecules carry negative charges on their phosphate groups, so the molecules are pulled toward positive charge. The shortest molecules travel the farthest, so the nested set is separated by size. Four different dyes are then added to the gel, each one reacting with one of the four bases exposed at the end of the molecules. So one color of dye reacts with exposed adenine (A), another reacts with thymine (T), another with guanine (G), and another with cytosine (C). The base sequence (nucleotide sequence) of the DNA can then simply be read from the gel by looking at the color of the first fragment, then the next, and so on. The whole process has been automated, producing color coded printouts such as the one in Figure 16.30.

The results from the DNA sequencing machines correspond to the real "Book of Life," a bewildering looking code of endless A's, T's, G's, and C's, whose sequence contains the secret of the human genome. Without doubt, many of the biggest scientific, medical, social, and ethical issues of the 21st century will be focused on what we choose to do, and not to do, with our knowledge about the human genome. Consult this book's website to gain access to up-to-date information on the application of this knowledge in medicine, agriculture, and industry.

main points

- Amino acids all share the same basic chemical structure, including an amino group and a carboxylic acid group.

- Proteins are polymers of amino acids.

- Proteins have primary, secondary, and tertiary levels of structure. Multisubunit proteins have quaternary structure.

- Enzymes are the protein molecules that catalyze most of the chemical reactions of life.

- Many of the key structural and functional parts of living things are made largely of protein.

- Proteins have a variety of functions including enzymatic, transport, and structural.

- A DNA molecule consists of a repeating "backbone" of alternating sugar and phosphate groups, with the bases attached to this backbone.

- DNA normally exists in the form of two strands intertwined in a spiral formation, known as a double helix. The strands are held together by hydrogen bonds between the bases.

- The DNA of human cells is found within 46 distinct structures in the cell nucleus known as chromosomes. Within the chromosomes, we find a total of around 50,000 specific regions of DNA known as genes.

- A gene is a long section of a DNA strand that contains a particular sequence of bases able to direct the manufacture of a particular protein.

- Ribonucleic acid, RNA, transfers genetic material to ribosomes as part of the process of protein synthesis.

- The chemical process of DNA replication underpins the ability of humans to have children, of plants to release seeds that can grow into other plants, and of all cells to multiply by splitting in two.

- Carbohydrates are compounds of carbon, hydrogen, and oxygen in which the ratio of H to O is 2 to 1. Carbohydrate is needed principally as a source of body energy.

- Many fats and oils are triglycerides composed of carbon, hydrogen, and oxygen but each fat molecule has only six oxygen atoms. They are examples of lipids.

- ATP is the "energy currency" of life.

- All organisms are made up of cells. Complex multicellular organisms such as humans contain many billions of individual cells, whereas the simplest organisms are single free-living cells.

- Genetic engineering is the artificial manipulation of DNA and genes.

important terms

Adenosine diphosphate (ADP) is the chemical that combines with phosphate to form ATP. (p. 528)

Adenosine triphosphate (ATP) is the main "energy currency" of life, the hydrolysis of which to ADP and phosphate releases the energy to power anabolism. (p. 528)

Amino acids are the monomer building blocks of proteins. (p. 506)

Anabolism is the buildup of chemicals from simpler ones in living things. (p. 528)

Anticodon a set of three bases in tRNA that binds to a specific codon in mRNA. (p. 518)

Carbohydrate class of compound composed of carbon, hydrogen and oxygen (with the H and O in a 2:1 ratio) that includes all sugars, starch and cellulose and glycogen. (p. 521)

Catabolism is the breakdown of chemicals in living things, into simpler ones. (p. 528)

Chromatin is a complex of DNA and protein that forms chromosomes. (p. 516)

Codon is a set of three bases on mRNA that binds to the matching anticodon on tRNA. (p. 517)

Complementary bases are bases in DNA or RNA that can bind together by base-pairing, involving hydrogen bonds. (p. 515)

Deoxyribose, $C_5H_{10}O_4$, is the sugar in DNA. (p. 512)

Dipeptide is a molecule composed of two amino acids linked by a peptide bond. (p. 507)

Disaccharide is a molecule composed of two monosaccharide sugar groups bonded together. (p. 522)

Disulfide bridge is a link between two parts of a protein molecule formed when amino acids with —SH groups react together. (p. 508)

DNA (deoxyribonucleic acid) is a huge molecule that stores genetic information. (p. 512)

Double helix is the interwound double spiral structure of the DNA of genes. (p. 515)

Gene is a section of DNA that encodes a single polypeptide or a functional RNA. (p. 517)

Genetic engineering is the artificial manipulation of DNA and genes. (p. 529)

Genome is the entire complement of genetic information of an organism. (p. 521)

Lipid is a class of fatty water-insoluble substances found in living things. (p. 524)

Messenger RNA (mRNA) is the RNA that carries a gene's "message" to the ribosomes, for it to be "translated" into the amino acid sequence of a protein. (p. 517)

Metabolism is all the chemical reactions of life. (p. 528)

Monosaccharide is a simple "single sugar," form of carbohydrate, usually $C_6H_{12}O_6$. (p. 522)

Mutation is a change in the base sequence of DNA. (p. 529)

Nucleotide is the monomer of DNA or RNA, composed of a sugar, a nitrogenous base and a phosphate group. (p. 512)

Nucleus is the central membrane-bound organelle of a cell, containing the cell's genome. (p. 521)

Polypeptide is a molecule composed of many amino acids linked by peptide bonds. (p. 508)

Primary structure is the amino acid sequence of a protein. (p. 508)

Protein is a polypeptide. (p. 506)

Quaternary structure is the way in which different proteins combine to form multisubunit proteins. (p. 510)

Ribonucleic acid (RNA) is a polymer of ribonucleotides, including mRNA, tRNA, and rRNA. (p. 517)

Ribose, $C_5H_{10}O_5$, is the sugar found in RNA. (p. 517)

Ribosomal RNA (rRNA) is the functional RNA of ribosomes, the sites of protein synthesis. (p. 518)

Ribosomes are the complexes of RNA and protein on which protein synthesis occurs. (p. 518)

Secondary structure is the pattern in which hydrogen bonding holds parts of a protein in distinctive shapes, such as helices and sheets. (p. 508)

Tertiary structure is the way in which protein chains fold into a complex shape. (p. 508)

Transcription is the manufacture of an RNA copy of the DNA of a gene. (p. 517)

Transfer RNA (tRNA) is the class of RNA that carries amino acids to the ribosome, where protein synthesis occurs. (p. 518)

Translation is the process in which the base sequence of mRNA gives rise to the amino acid sequence of a polypeptide. (p. 517)

exercises

1. What three elements are important in both proteins and carbohydrates? What two other elements are important constituents of proteins but not of carbohydrates?
2. What is the name of the bond that forms between two amino acids in building a protein?
3. Draw the structure of an amino acid having —CH_3 as R group.
4. Draw the structure of a dipeptide in which the C-terminal amino acid has —CH_3 as an R group and the N-terminal amino acid has —CH_2SH as an R group.
5. Draw the structure of a tripeptide in which alanine reacts with the acid group of tyrosine, and reacts with the amine group of phenylalanine.

6. What category of bond is responsible for the secondary structure of a protein? What categories of bonds are responsible for the tertiary structure of a protein?
7. Suppose that a protein were made using one molecule of each of the 20 amino acids. What would be its molecular weight?
8. What would be the percent nitrogen in the protein described in Exercise 7? Compare this with the typical nitrogen content of protein which is about 16.0% N.
9. What would be the dimensions of a cubic box that would hold one molecule of each possible protein of 20 amino acids?
10. What makes up the backbone of a DNA chain?
11. What special bond keeps the two DNA strands together in the DNA double helix?
12. What is a chromosome made of? How are these parts arranged to give a chromosome its shape?
13. Describe the process of DNA self-replication.
14. Describe the primary function of each of these:
    a. mRNA
    b. tRNA
    c. rRNA
15. List three differences and similarities between the structures of RNA and DNA.
16. What five elements are primarily responsible for the makeup of DNA and RNA molecules?
17. G-actin is one of the proteins that interact to cause muscle contractions. G-actin has a molecular weight of 42,000 g/mole. Assume that the average molecular weight of an amino acid residue is 120 g/mole.
    a. How many nucleotide bases are in the DNA which codes for this protein?
    b. If this protein were to be transcribed, how many codons would be present in the DNA sequence?
18. This base sequence was found in a newly discovered cancer gene. This gene causes a triplet of base-pairs to repeat in the DNA strand of the tumor cell. The repeated segments are then copied hundreds of times, eventually destroying the cell's ability to control its own growth. Give the complementary repeated triplet that would be found in the replicated DNA strand.

    G A A T C A C C T C A G C A G C A G C A G C A G

19. Indicate the complementary triplets on tRNA for these mRNA triplet sequences.
    a. UAC
    b. CGA
    c. GGU
    d. UAU
20. On a certain location of the DNA strand there is a consensus sequence (particular order of bases) that tells the RNA where to start transcription. One consensus sequence found in the *Escherichia coli* bacterium is TTGACA. Give the complementary sequence of mRNA bases.

21. It has been discovered that DNA directs protein synthesis. However, DNA replication and protein synthesis occur in two different places within the cell. Indicate the location of DNA storage and protein synthesis if it is not located in the same cellular compartment.

food for thought

22. There has been serious discussion of cloning a mammoth from the frozen remains of these prehistoric creatures. In your opinion, would this be a good idea?

23. Genetic testing is currently used in obstetrical care to detect hereditary disorders in the fetus. Gene tests have been and will be used for predicting adult onset of various disorders. However, a positive test does not always correlate with the possibility or severity of occurrence of the disorder. Does the risk of possible "genetic discrimination" by businesses and insurance companies outweigh the benefits of diagnosing an illness before symptoms occur?

24. Throughout this text, we have made the connection between society and technology. The discussion on The Human Genome Project may soon trigger other practical scientific applications. What do you expect will be ramifications of learning our genetic heritage? Remember to consider both risks and benefits.

25. The evaluation of scientific evidence in the courtroom is not always as reliable. In 1923, the court case "Frye v. United States" was the initiator in providing judicial guidance for the handling of scientific evidence. The court wrote in its ruling on Frye that scientific evidence must be "sufficiently established to have gained general acceptance in the particular field in which it belongs." Do you think this rule still adequately applies to the use of scientific evidence in courts today?

26. In the year 2000, small amounts of "Starlink" corn was found in commercial taco shells available for purchase in grocery stores. Starlink corn is a genetically modified corn that was supposed to be used only for animal feed. How do you suppose this corn got into food destined for human consumption? No human health problems were ever traced to this contamination. Should rules be relaxed about this corn in human food?

27. The technique of DNA profiling makes it possible, in principle, for the government to take a sample of DNA from every citizen and store each DNA profile on computer. It might then become very easy for police to identify culprits or victims of many crimes. It would also become much easier to ensure that innocent people are not wrongly accused of such crimes. What do you think about this idea? Would you willingly participate in such a program or vote for a party committed to it? Think of reasons both in support and in opposition to such an idea.

readings

1. Angier, Natalie. Blueprint for a human. *New York Times,* October 6, 1992.
2. Angier, Natalie. A first step in putting genes into action: Bend the DNA. *New York Times,* August 4, 1992.
3. Ayala, Francisco J., and Bert Black. Science and the courts. *American Scientist,* 81, May–June 1993, pp. 230–239.
4. Baum, Rudy M. Progress fitful on understanding AIDS, developing therapies. *Chemical and Engineering News,* August 24, 1992, pp. 26–31.
5. Baum, Stuart J. *An Introduction to Organic and Biological Chemistry,* 4th ed., New York: Macmillan Publ., 1987, pp. 555, 607–642.
6. Brenner, Sydney. That lonesome grail. *Nature,* 358, July 2, 1992, pp. 27–28.
7. Browne, Malcolm W. DNA from the age of dinosaurs is found. *New York Times,* September 25, 1992.
8. Browne, Malcolm W. 40-Million-year-old bee yields oldest gene matter. *New York Times,* September 25, 1992.

websites

www.mhhe.com/kelter   The "World of Choices" website contains web-based activities and exercises linking to: Sandia National Laboratories; National Institute of Justice and their site on DNA evidence; National Center for Biotechnology Information; and much more!

The Chemistry of Food

A Book of Verse beneath the Bough
A Jug of Wine, a Loaf of Bread—and Thou
Beside Me Singing in the Wilderness

From *The Rubáiyát of Omar Khayyám* by 11th century Persian poet Ghiyáthuddin Abulfath Omar bin Ibrá him al-Khayyámi, translated into English by Edward FitzGerald, 1858

A loaf of bread and a jug of wine make a pretty puny picnic compared to today's standard fare, featuring fried chicken, potato salad, hot dogs, baked beans, and cake or cobbler. The shelves of any large food store display a bewildering variety of foods to nourish our bodies and tempt our appetites. Fruits and vegetables of many flavors, shapes, and sizes are piled high. Dozens of different forms of meats and fish are for sale. Pasta, rice, bread, and cakes sit invitingly nearby. There are many rows of cans and bottles containing processed foods and combinations of foods; packaged complete meals (refrigerated, frozen, or dried); hundreds of types of candy; and an amazing variety of beverages, with favorites that fizz! In the developed western world, food companies continually compete to offer us new varieties and combinations of food and drink to the extent that the choices available can sometimes become overwhelming. A single branch of a large food store can have as many as 20,000 different food and drink items on its shelves. What you are really being sold is an astonishing variety of different chemicals. When you enter a food store and contemplate the displays, you are about to decide what chemicals you wish to put into your body. So food shopping is chemical shopping, and, despite the apparently limitless variety, all the foods available contain just six basic categories of chemicals, which perform only three main functions in the body.

The chemicals that the body needs to take in as food and/or drink are classified together as **nutrients**. The six categories of nutrients are

- Water
- Carbohydrates
- Fats (lipids)
- Proteins
- Minerals
- Vitamins

The three essential roles these chemicals perform in the body are

- Serving as raw materials for the growth and maintenance of body tissues.
- Serving as a source of energy.
- Helping to regulate and facilitate the body's chemical and biological processes.

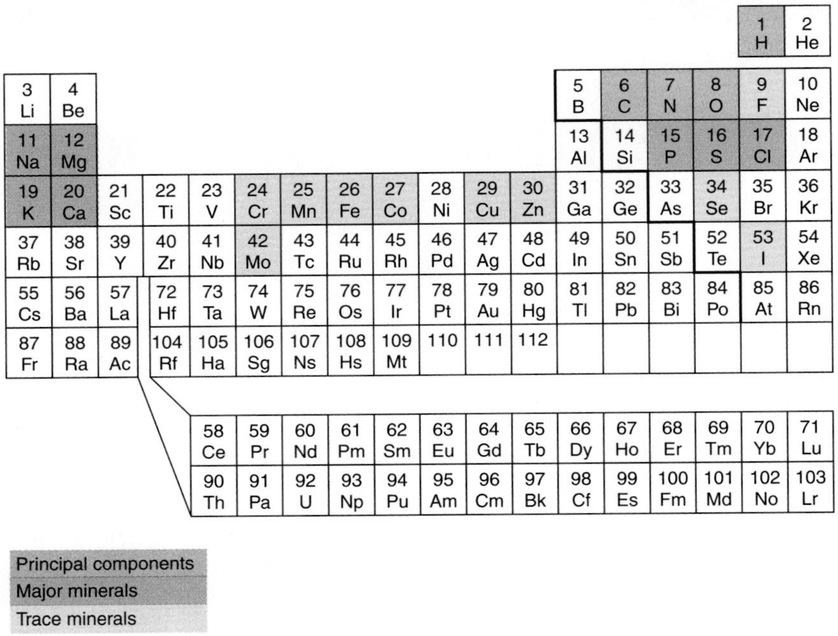

Figure 17.1
Elements essential for human health. Elements highlighted in orange are present in major amounts in the human body, those in blue are present in minor amounts, and those in yellow are needed in trace amounts.

Principal components
Major minerals
Trace minerals

As far as we know, the human body requires 21 different chemical elements, which must be present in the nutrients of our food (see Figure 17.1). In this chapter we look at each category of nutrient and consider the ways in which each contributes to the chemistry that keeps our bodies working and healthy. We also consider the balance of nutrients required for a healthy diet, and some debates and controversies surrounding the food industry.

All of the foods we eat are made from living things, which have all ultimately been sustained by the energy of the Sun. If we eat animals, they themselves will have been sustained by eating plants or by eating other animals that eat plants. So all our food supplies depend on the growth of plants, powered by the energy of the Sun. In summary, the Sun's energy allows water, carbon dioxide, and other chemicals of the Earth (minerals) to become incorporated into plants and the animals that eat the plants. When we eat plants or animals as food we are able to use the chemicals in the food as raw materials and a source of energy.

The chemicals we eat do not remain in us forever, but are continually recycled due to our excretion of solid, liquid, and gaseous wastes. So the process of eating and excreting is another example of the many chemical cycles powered by the energy of the Sun that lie at the heart of everything that happens on planet Earth.

17.1 | Water as a Nutrient

Water (the familiar H_2O) is the most fundamental and simplest nutrient. It is also the most essential in the sense that it is the one we are least able to manage without. Many of us could survive for weeks or even months without eating any carbohydrates, proteins, fat, vitamins, or minerals; but we would die within a few days without water. Around 45–75% of the human body is water. A typical adult contains about 40 L of water: 3 L in blood plasma, 12 L in tissue fluid, and 25 L within blood and tissue cells. This same typical adult takes in 1.30 L of water per day in drink, 0.90 L in food, and gains 0.30 L through oxidation of the nutrients in food. The output values are 1.50 L of water in urine, 0.55 L through skin as perspiration, 0.35 L in exhaled air, and 0.10 L in feces (see Figure 17.2). Notice that the input and output just balance. If this balance is upset significantly for more than a few days, serious health problems will emerge.

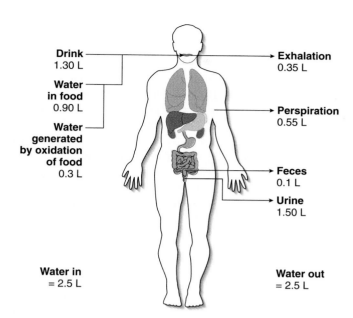

Drink
1.30 L

**Water
in food**
0.90 L

**Water
generated
by oxidation
of food**
0.3 L

Exhalation
0.35 L

Perspiration
0.55 L

Feces
0.1 L

Urine
1.50 L

Water in
= 2.5 L

Water out
= 2.5 L

Figure 17.2
Water taken into the body must be balanced by water excreted by the body, if a healthy water balance is to be maintained. Typical values are shown. Actual values vary.

Fortunately, there are many pleasant ways to get our water other than just drinking the pure compound. Soft drinks, fruit juices, vegetable juices, and alcoholic beverages are common dietary sources of substantial amounts of water. A lot of the water we consume is also contained in the form of seemingly solid foods, which can in fact contain large quantities of water. Potatoes, for example, are nearly 80% water by weight, meat is around 60% water, and bread about 40%. Table 17.1 shows the percentage of water in a variety of common foods.

The key chemical role of water in the body is to serve as the solvent of life. In other words, water is the chemical in which most of the other chemicals of life become dissolved and transported from place to place and in which most of the chemistry of life occurs. The body as a whole is largely built out of many billions of individual cells and each cell is filled with a watery solution. Water-based solutions such as blood, also provide the key transport systems of the body, allowing chemicals to be whisked from place to place as required.

Water is a good solvent for molecules which are small, polar, or form hydrogen bonds. Recall from Chapter 6 that hydrogen-bonding molecules include those that have O—H or N—H groups in their structure. Look for these functional groups in the structures shown in this chapter and in Chapter 16. You will see that carbohydrates have many O—H groups and proteins have many N—H groups (and a few O—H groups). These classes of compounds dissolve well in water. Fats, on the other hand, have no O—H or N—H groups and dissolve very poorly in water. We will describe a very different means by which fats move through the body.

Likewise, some vitamins dissolve well in water—they all have O—H or N—H groups which enhance their water solubility—but others dissolve poorly in water but well in fat. These latter are the so-called "fat-soluble" vitamins. Their structures have few or no O—H or N—H groups so they do not participate in hydrogen-bonding (see Section 6.3) either with each other or with water.

Table 17.1

Water Content of Some Foods

Food	% H$_2$O
Melon	94
Beer	90
Cabbage	88
Milk	88
Apples	84
Jelly	84
Potatoes	76
Eggs	75
Beef steak	67
White bread	39
Cheddar cheese	37
Butter	15
White flour	13
Corn flakes	7
Sugar	0

Table from P. M. Gamon and K. B. Sherrington. 1990. *The Science of Food.* 3rd Edition, Oxford: Pergamon Press, p. 112.

Osmosis

Recall from Chapter 7 (Properties of Water) that we talked about osmotic pressure and the tendency of water molecules to move through membranes. Water can move through tiny pores in the cell membranes in our bodies, but osmotic pressure will only drive water to move through a membrane from an area of low concentration of dissolved solutes to an area of high concentration. In other words, concentrations of solutions on opposite sides of the membrane will tend to become equal as water leaves a dilute solution (thereby raising its concentration) and enters a concentrated solution (thereby lowering its concentration). This flow of water by osmosis governs the transfer of water from cell to cell within the body.

17.2 | Carbohydrate

The cereal box side panel reproduced in Figure 17.3 shows that each serving (30 g) of the cereal contains 24 g of "carbohydrate" (food labels tend to use the singulars—carbohydrate, protein, fat, etc., rather than the chemically more correct plurals). The label indicates that, of each 24 g of carbohydrate, 4 g are simple sugars, 17 g are starch ("other carbohydrate"), and 3 g are fiber. We discussed the chemical structure of all three of these types of carbohydrates in Chapter 16. All are composed of carbon, hydrogen and oxygen, with the general formula $C_x(H_2O)_y$, but they have different effects on the body.

The sugars and starch are digested to glucose ($C_6H_{12}O_6$), which serves as an energy supply for the body. The energy provided by digestible carbohydrates is released by the stepwise oxidation of the glucose they are converted into. This oxidation reaction (see Figure 17.4) is summarized by this overall equation:

$$C_6H_{12}O_6 + 6O_2 \rightarrow 6CO_2 + 6H_2O$$

This is a highly exothermic reaction, releasing 2816 kilojoules per mole of glucose oxidized. Much of that energy can be trapped in the form of ATP, discussed in Chapter 16, then used to power the energy-requiring processes of metabolism. Since the molar mass of glucose is 180 g, we can calculate that 15.6 kJ of energy are released per gram of glucose oxidized.

Typically, the energy content of foods is measured in units of **kilocalories**, usually called **dietary Calories**, or just "**Calories**" with a capital C. So the "110 Calories" listed on the food label of Figure 17.3, really corresponds to 110 kilocalories, that is 110,000 calories. The use of Calories (with a capital C) to mean 1000 calories (with a small c) is unfortunate, but is embedded in usage, so we have to get along with it.

One dietary Calorie (1 Calorie) = 1000 calories (1 kilocalorie) = 4.184 kJ (4184 joules)

One (small c) **calorie** is the amount of heat needed to raise the temperature of 1 g H_2O by 1°C. We will use the abbreviation "kcal" to stand for dietary Calories. The 15.6 kJ available from the oxidation of 1 gram of glucose works out to be 3.74 kcal per gram of glucose, very close to the normally quoted value of four dietary Calories per gram of carbohydrate. The Calorie is well known to most dieters. Each gram of digestible carbohydrate yields about 4 kcal (4 Calories with a capital C) for use by your body, although a significant amount of that will be released as heat. This heat is not all wasted energy—one of the most important functions of food is to release the heat that maintains our bodies at 37°C (98.6°F).

In addition to acting as a source of energy, the carbon, hydrogen, and oxygen atoms within the carbohydrates we eat can also serve as a source of these atoms for building the chemicals of our body.

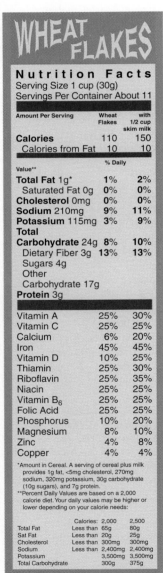

Figure 17.3
Nutrition facts and ingredients are listed on nearly all foods sold in supermarkets.

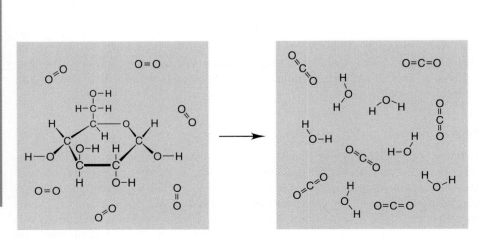

Figure 17.4
Glucose reacts with oxygen to yield carbon dioxide and water.

exercise 17.1

Bond Energy Calculation

Problem

Use bond energy calculations to predict the amount of energy released on oxidation of a mole of glucose.

$$C_6H_{12}O_6 + 6O_2 \rightarrow 6CO_2 + 6H_2O$$

Solution

Bonds Broken				**Bonds Formed**		
C—H	7 mol	7(415) = 2905		C=O	12 mol	12(805) = 9660
C—C	5 mol	5(345) = 1725		O—H	12 mol	12(460) = 5520
C—O	7 mol	7(360) = 2520				
O—H	5 mol	5(460) = 2300				
O=O	6 mol	6(493.6) = 2962				
Total		12412 kJ		Total		15180 kJ

12412 kJ endothermic energy to break bonds is overmatched by the 15180 kJ exothermic energy released in making bonds, so the difference 15180 − 12412 = 2768 kJ is predicted from bond energies to be released for every mole of glucose that is oxidized to carbon dioxide and water.

Recall our discussion of oxidation states in Chapter 15.

 # case in point

Sweetness, Artificial and Natural

Our real need for carbohydrates is as a source of energy, but we also crave sugars for their sweet taste. Carbohydrates in the form of sugars, such as sucrose, fructose, and lactose, are responsible for the natural sweetness of foods. All sugars taste sweet but some taste sweeter than others even though they all deliver the same 4 kcal per gram. Sucrose is used as the standard against which the sweetness of other sweeteners is compared. Table 17.2 lists the relative sweetness of several natural sugars and three artificial sweeteners. The numbers indicate the number of grams of sucrose needed to give the same sweetness as 1 g of the sweetener listed. Thus, a compound with a number greater than one is sweeter than sucrose, and a compound with a number less than one is less sweet than sucrose. Very little is known about why some molecules taste sweet and others do not. Each of the synthetic sweeteners was discovered by accident.

People for whom dietary sugar should be kept to a minimum, such as diabetics, were the original market for artificially sweetened foods. The first successful artificial sweetener was saccharin, which was first produced in

Table 17.2

The Relative Sweetness of Some Sugars and Artificial Sweeteners	
Sweetener	**Sweetness**
(as grams of sucrose equivalent to grams of sweetener)	
Sucrose	1.0
Fructose	1.7
Glucose	0.7
Maltose	0.4
Lactose	0.2
Aspartame	200
Saccharin	500

1879. Saccharin is about 500 times as sweet as sucrose; only a small amount is needed to give the same sweetness as natural sugars. Even this small amount is not metabolized by our bodies; rather it is excreted in urine, having gone through the entire digestive system unchanged. Therefore, a food sweetened with saccharin is lower in calories than one sweetened with sugar. Although artificial sweeteners are still used by diabetics, the largest market for artificially sweetened foods and drinks is for weight conscious dieters. If diet soft drinks were sold only to diabetics, very little would be sold compared to the actual market. Since artificial sweeteners such as saccharin and aspartame are used in such small amounts to provide sweetness and saccharin isn't metabolized by our bodies, the amount of energy released by their digestion is very small. Aspartame is metabolized, and releases about the same amount of energy per gram as carbohydrate, but it is so sweet that very little of it (about 0.2 g in a full can of soda) is needed, so it supplies very little energy (i.e., supplies very few calories). Since sugar is the main source of

calories in soft drinks, the replacement of sugar by noncaloric sweeteners results in these beverages being very low in calories. The sweeteners somehow interact with the portion of our tongue that registers sweet taste. It is not obvious how the structures of aspartame and saccharin are similar to sucrose. It is clear why saccharin provides us no calories, however. Our bodies simply do not have suitable enzymes for oxidizing saccharin the way we do sugar. We do have enzymes suitable for digesting aspartame because aspartame is really just a very small protein molecule, so it is digested just like any other protein and that reaction does release energy. Aspartame is regarded as "calorie-free," however, because so little of it (0.005 g) is required to give as strong a sweet taste as 1.0 g of sucrose.

Saccharine   Cyclamate   Aspartame

Saccharin has a structure very different from that of a sugar but it is very sweet. Other synthetic sweeteners are also very different in structure. Little is known about the relationship of structure to sweetness.

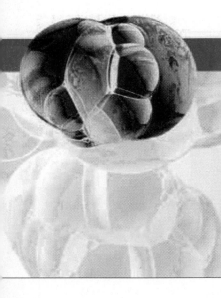

exercise 17.2

Energy from Aspartame

Problem

Calculate the amount of energy available to the body from digesting the amount of aspartame that supplies the same sweetness as 1 g of sucrose.

Solution

Aspartame, like carbohydrate, supplies 4 Calories per gram so 0.005 g of aspartame would supply:

$$0.005 \text{ g} \times 4 \text{ Cal per g} = 0.02 \text{ Calories}$$

This is not zero, but it's pretty close.

■ case in point

Lactose Intolerance

Sucrose, common table sugar, is a disaccharide formed from glucose and fructose, while lactose, the sugar in milk, is formed from galactose and glucose, as shown below.

The difference in structure between sucrose and lactose is very important to the large number of adults around the world who are "lactose-intolerant." In the normal digestion of lactose, the enzyme lactase causes the lactose to split into glucose and galactose; then another enzyme converts the galactose into glucose, which is the main form in which carbohydrate is used in the human body. In those who are lactose-intolerant, the enzyme lactase is not present or is present in too small an amount to convert the milk sugar into glucose and galactose. The undigested lactose therefore passes through the small intestine into the large intestine, where microorganisms devour it and create large volumes of gas. This causes abdominal bloating and other digestive problems. Most lactose-intolerant persons can eat small amounts of dairy products without a problem and can minimize their difficulty by drinking milk to which lactase has been added to convert the lactose into glucose and galactose. This milk is sweeter than regular milk because two molecules of glucose and galactose are sweeter than one molecule of lactose. The lactose in milk is largely lost when it is converted into cheese, so cheese is a good source of protein and other dietary requirements that avoids the digestion difficulties associated with lactose. Likewise, yogurt is made using bacteria that metabolize lactose into lactic acid ($C_3H_6O_3$) and leave it suitable for lactose-intolerant persons.

$$C_6H_{12}O_6 \rightarrow 2C_3H_6O_3$$
$$\text{galactose} \qquad \text{lactic acid}$$

The structure of lactic acid is $H_3C—CHOH—COOH$. This reaction releases 79 kJ per mole of galactose.

Fiber

The "Dietary **Fiber**" listed in the food label in Figure 17.3 is indigestible carbohydrate, largely composed of cellulose. As we discussed in Chapter 16, although made up of glucose molecules, the cellulose cannot be broken down by our digestive enzymes and so it passes through our digestive tract and forms much of the bulk of feces. Although it does not supply any energy or raw materials to us, fiber is believed to play an important role in keeping the digestive tract healthy and our bowels "regular." Part of this role may involve the absorption of toxic compounds, allowing them to be carried along with the fiber, for excretion in feces. There is some evidence linking a deficiency of fiber in the diet with an increased risk of bowel cancer, although recent data have challenged this idea.

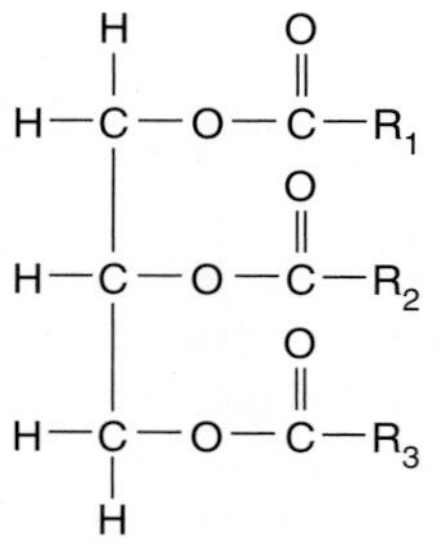

Figure 17.5
Triglycerides share this general structure. R_1, R_2, and R_3 are the remaining portions of each fatty acid chain.

17.3 | Fat

The breakfast cereal label shown in Figure 17.3 shows that a serving contains only a single gram of **fat** in contrast to 24 total grams of carbohydrate. Many of the other items that we put in our carts in the supermarket, however, especially those that we really like to eat, are rich in fat. Examples include hamburgers, ice cream, avocados, bacon, and potato chips. A careful word about nomenclature is needed here, however. When used on food labels, and in many discussions of nutrition, the word "fat" really means all *lipids*. Recall from Chapter 16 that in strict chemical usage a fat is a solid triglyceride, an **oil** is a liquid triglyceride and lipids include fats, oils and all other "fatty" water-insoluble substances found in living things. Figure 17.5 is a reminder of the structure of a triglyceride.

Since this chapter is about the chemistry of food, we often have to use the words fat and fats in their food label context, meaning all lipids.

Deposits of "fat" in animals (including humans) serve as a means of storing energy. Like carbohydrate, fat releases energy when it is oxidized to carbon dioxide and water. The oxidation of 1 gram of fat releases approximately 9 kcal of energy, so, weight-for-weight, fat supplies us with more than twice as much energy as carbohydrate. The fat is stored within "adipose tissue," composed of cells containing many fatty globular deposits. Lipids do not dissolve in water very well so they can be stored a long time in the body without washing out. Fat also plays important structural roles in the body. As we have discussed, each of the cells of the body is surrounded by a membrane based on fat (lipid) molecules, and the "organelles" within cells, such as the cell nucleus, also have their own lipid bilayer membranes. Fat is also a good insulator against the cold. Animals like seals and polar bears, which are active in cold climates, have very thick layers of fat to protect their inner organs from freezing. Fat also surrounds and protects more sensitive parts of the body.

Most fat in plants is associated with the seed of the plant. The fat is used by the germinating seed as a source of energy before photosynthesis begins. Photosynthesis can't begin until the growing plant has sprouted green leaves above the ground. Before that, the plant's growth is powered by energy stored in the seed either in the form of starch or oil.

Fat is a necessary part of the human diet, but the typical American diet now includes too much fat. In response to the health problems of diets too rich in fat, food manufacturers are changing their products to produce healthier foods (and making sure we know it). The difficulty in reducing the amount of fat we eat is that fat adds pleasant sensations to many foods. Fat is soft, especially when warm, and leads to a good "mouth feel," which means just what it says. Also, many compounds that are volatile and add to the aroma and taste of foods dissolve much better in fat than they do in water, so fat contributes strongly to the taste and smell of food.

Fatty Acids—A Closer Look

The difference between different fats and oils found in food is that they contain different fatty acid chains bonded to the common glycerol-derived region. Recall that a fatty acid is a compound that has a long hydrocarbon chain terminating in a carboxylic acid group (see Figure 17.6). In naturally occurring fatty acids within food, the total number of carbon atoms is always an even number, usually in the range of 8–22. Within the long chain of carbon atoms there are sometimes a few double bonds. Names and structures of a few fatty acids found in foods are shown in Table 17.3. A typical fatty acid is stearic acid, which has the formula $CH_3(CH_2)_{16}COOH$. Stearic acid is a **saturated fatty acid** because it contains no carbon-carbon double bonds and therefore carries the maximum number of hydrogen atoms (it is "saturated" in hydrogen). Other fatty acids can contain one or more double bonds, which could, potentially, combine with further hydrogen atoms, and so are known as **unsaturated fatty acids**. One such unsaturated fatty acid (oleic acid) is also shown in Figure 17.7. Fatty acids with two or more double bonds, such as linoleic acid, are known as **polyunsaturated fatty acids**, whereas those with a single double bond are **monounsaturated fatty acids**.

Another important aspect of fatty acid terminology may be familiar to you through terms such as "omega-3 fatty acids" or "omega-6 fatty acids." This is based on the position of the double bond closest to the "omega" carbon atom at the end of the fatty acid chain (see Figure 17.6). If we count the omega carbon as no. 1, and find the first double bond on carbon no. 3, we have an omega-3 fatty acid. Similar reasoning defines an omega-6 fatty acid.

The differences between the triglyceride fats and oils in foods depend on differences in the length and degree of unsaturation of the fatty acids they contain. If the fatty acids are long (typically 18 carbons) the fats tend to be solids; if they are short (typically 12 carbons), the fats tend to be liquids (so technically oils). Even more important is the presence of double bonds. Triglycerides containing unsaturated fatty acids tend to be liquids, while those with saturated fatty acids tend to be solids. This is because a double bond introduces a bend or kink in a fatty acid (see Figure 17.7), making it more difficult for the fatty acids to fit together into the ordered structure of a solid. Table 17.4 compares the melting temperatures of four fatty acids having 18 carbon atoms. Note that the saturated fatty acid melts well above room temperature, so is a solid at room temperature, but the unsaturated fatty acids melt well below room temperature, so are liquids.

Plant oils tend to contain shorter and predominantly unsaturated fatty acids. The main exceptions to this rule are coconut oil and palm oil, which are solids at room temperature because they have such a small number of double bonds in their fatty acids. Coconut oil, palm oil, and palm kernel oil are collectively called "tropical oils" because of the tropical climate in which palm and coconut trees grow. These oils have received some negative publicity due to their high percentage of saturated fatty acids and the related health problems, discussed in the next section.

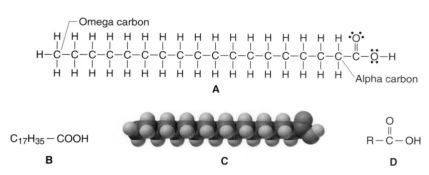

$C_{17}H_{35}$ — COOH

B

C

D

Figure 17.6
A fatty acid can be represented in several different ways.

Oleic acid

Figure 17.7
Oleic acid contains 18 carbon atoms and has a double bond between carbon atoms 9 and 10 (counting the carboxylate carbon as 1). This double bond forces a permanent kink in the molecule.

Table 17.3

Names and Partial Structures of a Few Fatty Acids

Name	Formula
Lauric	C—C—C—C—C—C—C—C—C—C—C—COOH
Myristic	C—C—C—C—C—C—C—C—C—C—C—C—C—COOH
Palmitic	C—C—C—C—C—C—C—C—C—C—C—C—C—C—C—COOH
Stearic	C—C—C—C—C—C—C—C—C—C—C—C—C—C—C—C—C—COOH
Arachidic	C—C—C—C—C—C—C—C—C—C—C—C—C—C—C—C—C—C—C—COOH
Palmitoleic	C—C—C—C—C—C—C=C—C—C—C—C—C—C—C—COOH
Oleic	C—C—C—C—C—C—C—C—C=C—C—C—C—C—C—C—C—COOH
Linoleic	C—C—C—C—C—C=C—C—C=C—C—C—C—C—C—C—C—COOH
Linolenic	C—C—C=C—C—C=C—C—C=C—C—C—C—C—C—C—C—COOH
Arachidonic	C—C—C—C—C—C=C—C—C=C—C—C=C—C—C=C—C—C—C—C—COOH

Note: All carbons have four bonds but these structures show only the carbon atoms and the carboxylate groups to minimize confusion. Hydrogen atoms occupy all other appropriate positions.

Table 17.4

Melting Points of Some Fatty Acids

Fat	Structure*	Melting Temperature
Stearic	C—C—C—C—C—C—C—C—C—C—C—C—C—C—C—C—C—COOH	69.6°C
Oleic	C—C—C—C—C—C—C—C—C=C—C—C—C—C—C—C—C—COOH	13.4°C
Linoleic	C—C—C—C—C—C=C—C—C=C—C—C—C—C—C—C—C—COOH	−5°C
Linolenic	C—C—C=C—C—C=C—C—C=C—C—C—C—C—C—C—C—COOH	−11°C

*These structures show only the carbon atoms and the carboxylate group to minimize confusion. Hydrogen atoms occupy all other appropriate positions.

Table 17.5

A Typical Distribution of Fatty Acids Found in Cow's Milk

Fatty Acid	Number of Carbons	Number Double Bonds	Percent by Weight
Butyric	4	0	3.0
Caproic	6	0	1.4
Caprylic	8	0	1.5
Capric	10	0	2.7
Lauric	12	0	3.7
Myristic	14	0	12.1
Palmitic	16	0	25.3
Stearic	18	0	9.2
Arachidic	20	0	1.3
Lauroleic	12	1	0.4
Myristoleic	14	1	1.6
Palmitoleic	16	1	4.0
Oleic	18	1	29.6
Linoleic	18	2	3.6

Saturated Fats versus Unsaturated Fats

Fats themselves can be described as "saturated" or "unsaturated" or "monounsaturated" or "polyunsaturated," depending on the degree of saturation within the fatty acids they contain. Dietary guidelines suggest that we limit our intake of fats, but especially saturated fats. The saturated fats are related to higher incidence of heart attacks and other health problems. Tropical oils (palm oil and coconut oil) and animal fats are higher in saturation than are other vegetable oils. Table 17.5 shows a typical distribution of fatty acids in cow's milk.

Hydrogenation

Unsaturated fats tend to get rancid. This is the result of oxygen from the air reacting with the double bonds of the fat to make smelly, unpleasant-tasting molecules. Rancid fats make food taste very bad. To minimize the problem of rancidity, food processors often eliminate the double bonds in fats by adding hydrogen to those double bonds. This is done by bubbling hydrogen gas through hot, liquid oil in the presence of fine particles of a nickel catalyst. Margarine is made by hydrogenating some of the double bonds in vegetable oil. This converts the vegetable oil, which is liquid at room temperature, into a solid at room temperature—a solid that does not go rancid even if stored for weeks. Generally, the hydrogenation process removes some, but not all of the carbon-carbon double bonds. The remaining double bonds are often transformed in a subtle way, discussed in the Consider This section on *trans*-Fatty Acids).

consider this:

trans-Fatty Acids

When we look closely at the arrangement of atoms around a carbon-carbon double bond we see two possible arrangements. Remember, when a carbon has a double bond, it is only bonded to three other atoms and these atoms point to the corners of a triangle centered on the carbon atom. When two groups are on the same side of a C=C double bond, they are in the *cis*- configuration, while if they are on opposite sides of the double bond they are in the *trans*- configuration (see the figure). The groups attached to the carbon atoms of C=C bonds are stuck in their relative positions, because the arrangement of electrons in these bonds prevents any rotation around the bond. This is in contrast with C—C single bonds, around which free rotation can occur. So molecules can't convert from cis- to trans- or vice versa under normal circumstances.

Trans-fatty acids are not normally present in natural fats and oils, but they can be formed during the partial hydrogenation process used to harden and stabilize many oils. A growing body of research suggests that *trans*-fatty acids in the diet may be associated with an increased risk of heart attack. Excessive amounts of fats including *trans*-fatty acids seem to lead to a greater risk for heart attacks. It seems that eating *trans*-fatty acids leads to higher amounts of cholesterol being produced in the body than would be produced if we ate only the natural *cis*-fatty acids. The shape of the *cis*- and *trans*- isomers is the cause for the different properties. Our body chemistry is adapted for *cis*-fatty acids.

cis- and *trans*-fatty acids have the same formula but different structures because of the double bond. Fatty acids in our food are virtually all *cis*. Unnatural *trans*-fatty acids have been implicated in heart disease.

Monoglycerides and Diglycerides

If only one of the —OH groups on glycerol is combined with a fatty acid, the resulting molecule is called a **monoglyceride**. We are sure that you can now figure out what a **diglyceride** must be. Both mono- and diglycerides have free —OH groups that can hydrogen-bond to water and they have long carbon chains that are fat soluble. This behavior allows these molecules to dissolve partly in fats and partly in water. Commercially, these compounds are used as **emulsifiers**, compounds that cause fat to become miscible with water. They do this when their long tails dissolve in the fat and their —OH groups dissolve in water. Mono- and diglycerides therefore help fat molecules and water molecules to mingle without the fat technically dissolving. Diet margarines therefore use emulsifiers to keep the oil from separating from the water that is present in 30 to 60% by weight. The water is important because it does not contribute any dietary calories so the overall calorie count is low. These products are not good for frying foods because, on heating, the emulsion breaks and the water separates out from the fat and spatters badly. Look on the label of low-calorie margarines and many other foods and you will find "mono- and diglycerides"; these are there to help blend fat and water, two substances that will not intermingle without an emulsifier.

Cholesterol

Meat, milk, eggs, and other animal-based foods all contain **cholesterol**, surely the most notorious lipid in food. Cholesterol is in these foods because animals (including humans) make cholesterol for their body's use even if there is none in their diet. As we mentioned in Chapter 16, cholesterol is a vital component of animal cell membranes. Plant-based products, no matter how fatty, contain no cholesterol. Cholesterol in the human body arises from the cholesterol we eat directly and the cholesterol our body makes from the food we eat.

Recall our definition that a lipid is a compound which is not soluble in water but is readily soluble in many organic solvents. Cholesterol is, in fact, a steroid, as we discussed in Section 6.3 (page 227). Cholesterol's formula is $C_{27}H_{46}O$ and its structure contains four rings; it is

exercise 17.3

Diglyceride Formation

Problem

Write a balanced equation for the formation of a diglyceride from glycerol and fatty acids.

Solution

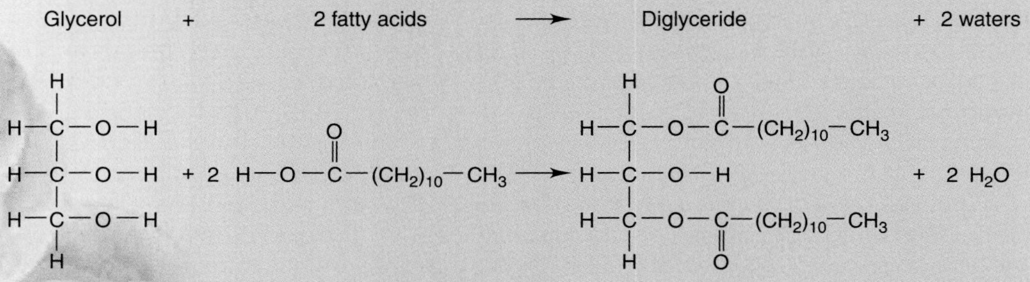

Glycerol   +   2 fatty acids $\longrightarrow$ Diglyceride   + 2 waters

In this case, both fatty acids are lauric acid, $C_{12}H_{24}O_2$.

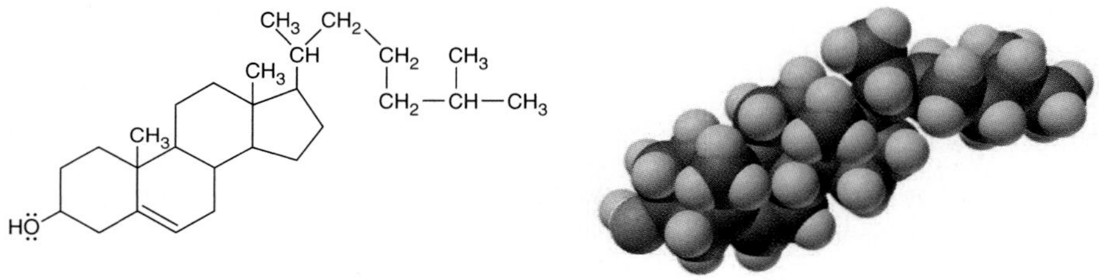

Figure 17.8
Structure of cholesterol.

shown in Figure 17.8. Cholesterol is not a triglyceride but its single —OH group is not enough to help it dissolve well in water but it does dissolve well in nonpolar compounds such as fats.

Dietary Fat, Cholesterol, and Heart Disease

Diets high in fats, especially saturated fats, are correlated with higher incidence of heart disease and strokes. Both these diseases are associated with poor blood circulation. Blood circulation is impaired when plaque containing cholesterol builds up on the walls of arteries. How does cholesterol get there?

Cholesterol enters our circulation both as the cholesterol we eat and as cholesterol synthesized by the liver from other chemicals in food. Cholesterol doesn't dissolve in water so it will not move around through the bloodstream unless it is combined with something else which is soluble. That "something" is a series of "lipoproteins" (see Figure 17.9). Lipoprotein is a combination of lipid (fat and cholesterol) with protein. The protein is quite water soluble, so when combined with lipid it drags the lipid into solution. The more lipid there is in lipoprotein, the lower is its density. Very low density lipoprotein (VLDL) is mostly triglyceride with a bit of cholesterol and protein. VLDLs are formed in the liver, but when they leave the liver, much of the triglyceride is lost, the density increases, and the resultant cholesterol-rich lipoprotein is called LDL, low-density lipoprotein. This material is transported through the bloodstream where much of its cholesterol can be incorporated within deposits on

case in point

Olestra, the Nonfat Fat

For over 30 years, food companies have been trying to invent, improve, market, and sell a product that would behave like fat in foods but that would not supply any calories to the human diet.

In 1996, "Olestra" was approved by the U.S. Food and Drug Administration (FDA) for its use in food. Olestra is a polyester like ordinary fat, but the fatty acids are bonded to a sucrose molecule rather than to glycerol. Before a trade name was assigned, this set of compounds was known as sucrose polyesters. (The figure shows a typical structure of Olestra.)

Remember, fats are formed from the carboxylic acid of three fatty acids each forming an ester linkage through an alcohol group of the trialcohol glycerol. Sucrose has eight —OH groups which could react similarly. When six or all eight of these —OH groups are reacted with fatty acids, the resulting compound looks like a fat, feels like a fat both on our fingers and in our mouths, melts like a fat, and is stable at high temperatures like a fat. Unlike a fat, however, our human enzymes can't digest it, so it passes through our digestive system without being metabolized. As a result, it contributes zero calories to the human diet. All sorts of foods have been made substituting Olestra for triglycerides and they taste just fine. This could be a fine way for calorie-conscious consumers to eat foods that have all the taste and appearance of fatty foods, but do not provide the calories.

There are two clouds to this silver lining. First, the bacteria in our large intestine do have enzymes that can digest this material. That means that if a person eats a large amount of Olestra, it will go directly to the large intestine where it becomes digested by the suite of bacteria living there. These bacteria metabolize foods and release carbon dioxide and water, just as we do. A polite term for the resulting effect is abdominal distress; there are several impolite terms.

The second disturbing feature about Olestra is that, like fats, certain of the fat-soluble vitamins and related compounds dissolve well in it. To the extent that happens, vitamins will be removed from our digestive system and excreted from our bodies. A possible solution to this problem is to enrich Olestra-containing foods with high levels of these vitamins so that even if much of it is lost, plenty will be left behind for use in our bodies.

Most food additives that come before the FDA for approval are used in tiny amounts in foods. Olestra is unusual in that it will be a major ingredient in many foods if it becomes widely used. Thirty years have gone by since the first application for use of this class of compounds. Caution and prudence are surely called for in evaluating a synthetic compound to be used in large amounts in foods with the principle benefit being that it is indigestible.

Structure of Olestra
("R" groups are from fatty acids)

artery walls. These deposits on the inside of blood vessels make it more difficult for blood to flow and so raise blood pressure. If the deposits occur in the heart arteries, preventing adequate amounts of blood from reaching the heart, a heart attack can result. LDL is the so-called "bad cholesterol." "Good cholesterol," HDL, is high-density lipoprotein. HDL is rich in protein and scours cholesterol from where it has been deposited throughout the bloodstream for delivery back to the liver. As an example of how analyzing these components of blood gives an indication of risk, a ratio of cholesterol-to-HDL of *more than* 4.0 significantly increases the risk of heart attack in women.

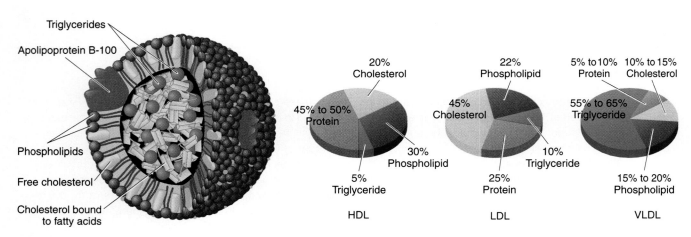

Figure 17.9
Structure of lipoproteins.

exercise 17.4

Assessing Cholesterol Levels

Problem

A woman aged 40 has a cholesterol level of 256 mg/dL and an HDL level of 61 mg/dL. Do these figures suggest the woman is at a significantly increased risk of heart disease?

Solution

The ratio of cholesterol to HDL is 4.2, which is greater than the 4.0 mark at which increased risk occurs. This woman is at a definite increased risk of heart disease, although to an extent that might be resolved by appropriate dietary modification.

Benefits of Cholesterol

That cholesterol brings us any benefit is usually a surprise to most people. The fact that our bodies synthesize cholesterol for themselves, however, is a clue suggesting that, despite the dangers of excessive amounts, cholesterol is important to us. We have already mentioned cholesterol's role in cell membranes. Compare the structure of cholesterol with those of vitamin D, several hormones, and bile acids, all necessary compounds in our bodies (see Figure 17.10). Cholesterol is the starting material for synthesis of these indispensable compounds.

Some of the cholesterol in the body results from our eating cholesterol-containing foods. As we mentioned earlier, only animal products such as meat, milk, and eggs actually contain cholesterol. Vegetable-based foods, no matter how much fat they contain, do not contain cholesterol because only animals make cholesterol. Vegetable fats, however, are good sources of raw material for the internal manufacture of cholesterol. Interestingly, it has been shown that feeding hens a diet containing a lot of flax seed lowers the amount of cholesterol in eggs.

17.4 | Protein

Most of us think of meat when we think of high-protein foods. This is certainly reasonable, but eggs (the white part), milk, beans, nuts, and rice and other grains are also excellent sources of

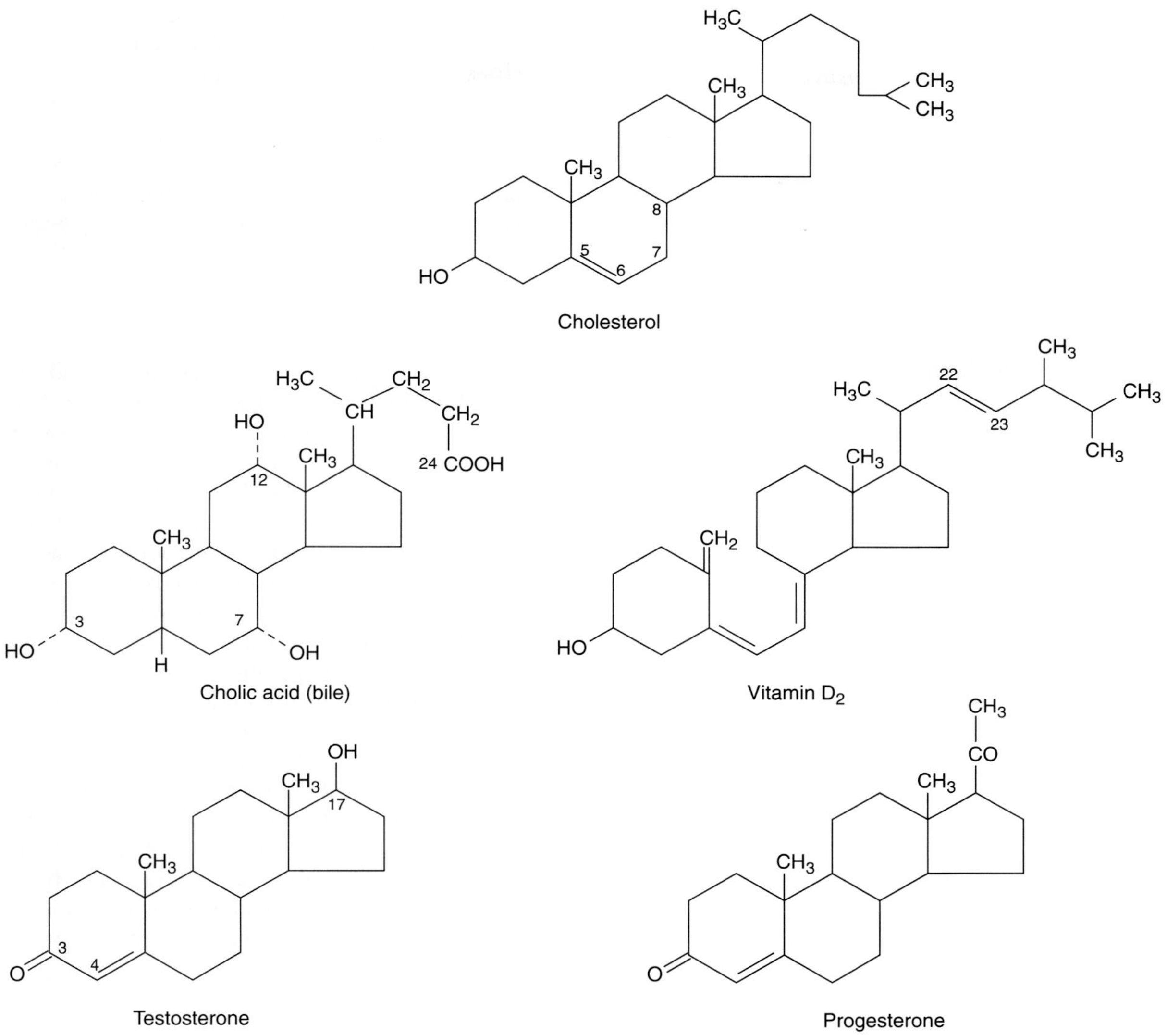

Figure 17.10
Beneficial compounds derived from cholesterol. Cholic acid is a crucial part of bile required in the large intestine; vitamin D_2 is necessary for proper growth; testosterone is the classic male hormone and progesterone the female hormone. Many other steroid compounds are necessary for good health. Look up their structures in any biochemistry book.

protein. Go back to the cereal box label shown in Figure 17.3 and you will find that a serving of this cereal contains 3 g of protein out of the 30-g serving. If we add the normal amount of milk to a serving of cereal, we are up to about 7 g of protein per bowlful. Like carbohydrate and fat, protein can be used as an energy source when it is oxidized to carbon dioxide and water. This releases approximately 4 kcal per gram of protein, much the same as carbohydrate; but it is not the most significant role of protein in our diet. Carbohydrate and fats are our most important energy sources. The most significant function of protein in food is to provide the raw materials (amino acids) needed to build our own proteins, including the protein of muscle and the enzymes that catalyze the chemistry of metabolism.

One obvious way in which protein molecules differ chemically from carbohydrate and lipids is that proteins contain a large amount of nitrogen and a small amount of sulfur in addition to large amounts of carbon, hydrogen, and oxygen.

Denaturing Protein—A Key Step in Cooking

The white part of a raw egg looks very different from that of a cooked egg. In the raw egg, protein called ovalbumin is dissolved in water with each protein molecule (molecular weight about 45,000) held in a soft, globular shape by the secondary and tertiary attractions. As egg white is cooked, the heat from the stove supplies enough energy to break the relatively weak bonds that are responsible for the tertiary and secondary structure of the proteins. These structures break down and the long protein molecule unfolds. Then, when the long, unstructured molecules collide, they can form weak bonds to each other and become entangled. At the same time, water is driven out from between neighboring molecules and the cooked egg becomes drier than the fresh egg. Such a transformation is called "denaturing" because the protein is no longer in its natural form. Most proteins are denatured on cooking and further denatured before digestion.

Digestion of Protein

Our body's digestive system must break proteins down into their constituent amino acids for them to be absorbed from the intestine into the blood. Once absorbed, amino acids can be built back into human proteins or used for the manufacture of other nitrogen-containing compounds in the body. They can also be oxidized (after removal of the nitrogen) as a source of energy. Excess nitrogen from the break-down of proteins is excreted as the chemical that gives urine its name—urea, $CO(NH_2)_2$. The digestion of protein begins in the stomach where the acid conditions in the presence of the digestive enzyme pepsin cause some of the peptide bonds to rupture, breaking a protein into several shorter fragments. These fragments pass into the small intestine where other enzymes break the remaining peptide bonds and release the individual amino acids for absorption through the intestinal wall into the blood. Our bodies can manufacture 11 of the 20 amino acids needed to make proteins, provided sufficient nitrogen is available from any amino acids in the diet. The remaining nine amino acids are called the dietary essential amino acids because our inability to make them makes it essential that they are present in adequate amount in the diet. The essential amino acids are valine, leucine, isoleucine, threonine, methionine, phenylalanine, tryptophan, lysine, and histidine (see Table 16.1, page 506, for the structures). For our diet to be balanced properly, it should contain each of the nine essential amino acids in sufficient amounts. Egg white and meat have nearly perfect proportions for human consumption and are said to be complete proteins. Individual vegetable-based proteins are generally less satisfactory, but if eaten in appropriate mixtures can readily combine to achieve the same effect as a complete protein. For instance, corn and beans together give complete protein whereas either eaten alone would be incomplete.

17.5 | Vitamins

Carbohydrates, fats, and proteins are known as "**macronutrients**" because we need to eat them in substantial amounts as part of a balanced diet. The **vitamins**, by contrast, are a range of compounds which we need only in tiny amounts; hence they are called **micronutrients**. You can appreciate just how little of these nutrients we need by looking at the list of daily values shown in Table 17.6. The technical definition of vitamins is that they are micronutrients that are necessary for health and that either cannot be manufactured in our own bodies, or may not be manufactured in sufficient amounts. This means that vitamins must be included in the diet. These molecules are used as received rather than being broken down and reconstructed. Vitamins were originally discovered when it was realized that people with diets lacking certain categories of foods often developed diseases.

Vitamin C provides us with the best known story about the discovery of a vitamin. Sailors on long ocean trips often became sick with a disease called "scurvy," which was later shown to be due to a deficiency of vitamin C. Scurvy began to be a problem for sailors in the years after 1492 when long voyages away from land became common. Fresh fruit (especially citrus fruit) and vegetables are the best sources of vitamin C. Around 1750 it was proved that eating oranges, lemons, or limes would prevent and cure scurvy. By 1800 the British Navy required limes as part of the diet of sailors (hence the nickname "limey") to keep them from falling sick with this terribly debilitating and often fatal vitamin deficiency.

Table 17.6

"Daily Values" for Vitamins

Vitamin A	5000 IU
Vitamin C	60 mg
Vitamin D	400 IU
Vitamin E	30 IU
Vitamin K	80 μg
Thiamin	1.5 mg
Riboflavin	1.7 mg
Niacin	20 mg
Vitamin B_6	2.0 mg
Folate	400 μg
Vitamin B_{12}	6.0 μg
Biotin	300 μg
Pantothenic acid	10 mg

The chemical structure and therefore chemical name (ascorbic acid) of vitamin C wasn't proved until 1932, almost 200 years after its importance had been demonstrated. The most current studies seem to show that about 200 milligrams of vitamin C per day is a reasonable amount. This would be an increase over the May 1996 recommended dietary allowance (RDA) of 60 milligrams per day.

Vitamins are classified as water-soluble or fat soluble. One of the major reasons we must have some fat in our diet, despite cautions not to consume too much, is because the fat brings essential supplies of fat-soluble vitamins with it.

Water-Soluble Vitamins

Vitamin C

Vitamin C (ascorbic acid) is found mainly in citrus fruits and green vegetables, potatoes and various other types of fruit, though it is found in a wide variety of other foods. It is required to assist in the addition of hydroxyl (—OH) groups to the amino acids proline and lysine. This is an essential step in the production of collagen, a key structural protein in connective tissue, muscles, cartilage, and bone. It is also involved in the production of "neurotransmitters," which carry messages between nerve cells. It also acts as an antioxidant, preventing damaging oxidation reactions, in addition to various other roles. As we said before, deficiency in vitamin C causes scurvy, associated with widespread bleeding. This is due to the rupture of blood capillaries, whose walls normally include much collagen.

Thiamine

Thiamine (vitamin B_1) is essential for carbohydrate metabolism because it acts as a **coenzyme**, needed to bind to various enzymes and assist in their catalytic effects. It occurs in many foods, especially pork, liver, and whole grains. Thiamine deficiency causes the disease beriberi, and initial signs of thiamine deficiency are anxiety, hysteria, nausea, depression, and loss of appetite. The thiamine in whole grain rice is lost during the polishing process that creates white rice.

Riboflavin

Riboflavin (vitamin B_2) is most readily obtained from milk, eggs, liver, and leafy green vegetables. It is needed for proper growth and functioning of the skin, eyes, liver, and nerves. These roles are due to its role within various coenzymes, needed to allow a wide variety of enzymes to perform their tasks. Riboflavin deficiency in humans usually causes cracking of the skin at the corners of the mouth, inflammation of the tongue and dermatitis about the corners of the nose, eyes, and ears.

Vitamin B_6

Vitamin B_6 occurs in three forms: pyridoxine (pyridoxol), pyridoxal, and pyridoxamine. Good sources of vitamin B_6 are fish, chicken, meats, eggs, whole grains, beans, and nuts. The various forms of vitamin B_6 are converted in the body to coenzymes needed for the activity of enzymes involved in the metabolism of amino acids, nucleic acids, and lipids. Symptoms of vitamin B_6 deficiency include immune suppression, depression, convulsions, dermatitis, and anemia.

Niacin

Niacin (nicotinic acid) is another vitamin required to form a variety of important coenzymes. The processes that depend on niacin include the energy-releasing reactions of life, so niacin is needed by all cells. It is found in many foods, but yeasts, wheat germ and meats are particularly rich sources. Niacin-deficiency disease is known as pellagra, the symptoms of which are dermatitis, diarrhea, and depression. Like all vitamin deficiencies, pellagra can eventually cause death, and so is known as "the four D's," dermatitis, diarrhea, depression, and death.

Folic Acid

Folic Acid

Folic acid, and some closely related compounds, serve as coenzymes that transfer fragments of molecules containing single carbon atoms from one chemical to another during the synthesis of amino acids and nucleic acids (DNA and RNA). Folic acid deficiency causes impaired cell division and problems in protein synthesis that affect cell growth. Anemia is an early sign of folic acid deficiency. The vitamin is readily available in liver, yeast, leafy vegetables, peas, beans, and some fruits. Limited folic acid supply in the diet of pregnant women has been linked to serious developmental defects in the fetus. For this reason, women who believe they may become pregnant are recommended to use a multivitamin supplement to ensure adequate supplies of folic acid and other vitamins.

Vitamin B₁₂

Vitamin B₁₂

The term vitamin B_{12} (cobalamin) refers to a variety of complex cobalt-containing compounds that are converted into crucial coenzymes involved in various aspects of metabolism. Animals cannot make vitamin B_{12}, although some, such as cows, can take up supplies of the vitamin made by bacteria in their gut. The major sources of vitamin B_{12} are meats, eggs, and dairy products, although it is also found in yeast extract. Vegetarians and vegans must take special care to consume sufficient vitamin B_{12}. Vitamin B_{12} deficiency causes severe anemia.

Biotin

Biotin

Biotin is a sulfur-containing vitamin that serves as a coenzyme in many reactions involving the transfer of CO_2 groups between molecules. The enzymes that are assisted by biotin are involved in the metabolism of carbohydrates, fats, and proteins. Good sources of biotin are liver, kidney, peanut butter, egg yolk, and yeast. Biotin deficiency causes anorexia, nausea, vomiting, depression, hair loss accompanied by a dry scaly dermatitis, and an increase in blood cholesterol and bile pigment levels. Biotin deficiency is very rare, except in infants.

Pantothenic Acid

Pantothenic acid, usually found in the form of its calcium salt, is part of coenzymes involved in the release of energy from carbohydrate, in fatty acid metabolism, and in the synthesis of steroid hormones. Good sources are liver, kidneys, green vegetables, and egg yolks. Pantothenic acid deficiency has not been clearly recognized in humans, but has been experimentally shown to cause headache, fatigue, impaired coordination, muscle cramps, and gastrointestinal disturbances.

Pantothenic Acid

Fat-Soluble Vitamins

Vitamin A

Vitamin A

Also known as retinol, vitamin A is essential for vision, growth, and the "differentiation" of cells into the specialized types of cells characteristic of various tissues. One of the first signs of vitamin A deficiency is impaired vision in dim light, due to retinol's crucial role in vision. Vitamin A is found in liver, egg yolk, fish, whole milk, butter, and cheese. We can also meet our vitamin A requirements by consuming plants that have "carotenoids," which we can convert into vitamin A. The carotenoids are examples of "provitamins," and are abundant in carrots, yellow vegetables and dark-green leafy vegetables. Retinol and carotenoids are destroyed readily by oxidation, which causes losses during storage and cooking.

Vitamin D

Vitamin D, also known as calciferol, is needed for the proper formation and maintenance of bones. It also promotes the absorption of calcium from the intestines. It can be manufactured in the body from 7-dehydrocholesterol on exposure of the skin to UV light. It is an essential dietary component, however, for people who have insufficient exposure to light to make adequate supplies. Vitamin D deficiency leads to weakening of the bones, and is the cause of the skeletal deformities called "rickets" in children. Rich food sources of vitamin D are eggs, milk, butter, and fish. Years ago children were forced to take a spoonful of cod liver oil each day, especially in the winter, to get an adequate supply of vitamin D. In the United States and many other developed countries, vitamin D is now added to milk to assure that children will get a sufficient supply.

Vitamin E

At least eight naturally occurring forms of vitamin E are known, all coming under the name of "tocopherols." α-tocopherol, shown here, is the most common and active form. One of the first signs of vitamin E deficiency is loss of coordination and balance. Vitamin E's major role is to act as an **antioxidant**, preventing damage to the body from reactive short-lived chemical species generated from oxygen. Vitamin E reacts with and so inactivates these hazardous natural oxidizing agents. Good sources are nuts, seeds, whole grains, leafy vegetables, eggs, and milk.

Vitamin K

There are two naturally occurring forms of this vitamin, menaquinone (vitamin K_1), made by bacteria, and phylloquinone (vitamin K_2), made by plants. It is obtained in the diet from leafy green vegetables and is essential for the formation of proteins involved in the regulation of blood clotting. Unsurprisingly, the major sign of vitamin K deficiency is ineffective clotting of blood and the associated bleeding from wounds.

17.6 Minerals

Minerals are elements usually associated with inorganic (mineral) substances. Many necessary minerals are metals that we require for good health. The most well known necessary metallic elements in our diet are iron (used in hemoglobin), calcium (bones and teeth), and sodium (blood serum). Look at the periodic table in Figure 17.1 to see the mineral elements required for health. After having been consumed, these mineral elements are shuffled around the body, acted on by enzymes, and put to use in forms very different from the form in which they were eaten. Daily Values for key minerals are shown in Table 17.7. As examples of this important nutrient group, we will focus on three mineral elements, iron, calcium, and phosphorus.

Iron

Iron is used in our bodies in a myriad of ways but the two most well known are in the oxygen transport molecules, hemoglobin and myoglobin. Both are protein molecules attached to a ring of atoms containing four nitrogen atoms, which bond very strongly to an iron(II) ion. A fifth bonding position on the iron is taken up by a nitrogen atom, which is part of an amino acid

Table 17.7
"Daily Values" for Minerals

Potassium	3500 mg
Chloride	3400 mg
Sodium	2400 mg
Calcium	1000 mg
Phosphorus	1000 mg
Magnesium	400 mg
Iron	18 mg
Zinc	15 mg
Copper	2.0 mg
Manganese	2.0 mg
Iodine	150 μg
Chromium	120 μg
Molybdenum	75 μg
Selenium	70 μg

from the protein. This leaves the sixth position available to bond to an oxygen molecule. There are four such iron ions in hemoglobin and one in myoglobin (see Figure 17.11). Each hemoglobin molecule picks up four molecules of oxygen, one on each iron, as the red blood cells containing the hemoglobin pass through the capillaries lining the lungs. The oxygen is then transported through our circulation to be released into all the cells that require it for the oxidation of glucose, fat and protein that provides us with energy. When oxygen is plentiful, some of it is transferred to myoglobin molecules, each one similar to a single subunit of hemoglobin, which stores the oxygen until the demand for it is very high. When we start running fast, for example, oxygen will be released from our myoglobin to supplement the usual supplies arriving on hemoglobin.

Iron is not utilized efficiently by our body. Only about 10% of dietary iron is actually absorbed; the rest goes through in solid waste. The chemical form of iron in food can change the utilization efficiency. Iron in meat is utilized most efficiently and powdered iron added to foods is least efficiently utilized.

hands-on

Iron in Breakfast Cereal

Examine the ingredient list on several breakfast cereals. If "iron" is listed as an ingredient, not just as being present in the food, it is present as powdered iron and can be removed with a magnet. Mash up a flake cereal or use a fine granular cereal designed to be added to water and heated. In either case, stir it for several minutes with a magnet that is inside a plastic bag. You will find that a small amount of powdered iron is attracted to the magnet. The plastic bag is to make it easier to get the iron back off the magnet. If the ingredient list says "iron sulfate" or ferrous sulfate, or some other iron compound, it will not be separable by a magnet but will be more efficiently utilized by your body.

Calcium

Calcium is necessary for a host of vital processes such as blood clotting, carrying nerve impulses, and the digestion of fats and proteins. Ninety-nine percent of the body's nearly 3 pounds of calcium is tied up in bone and tooth minerals. The dominant mineral in both bone and tooth structure is called apatite by geologists and has the formula $Ca_5(PO_4)_3OH$. If our diet does not include enough calcium, our body uses some of the large amount of calcium in bone to carry out the other absolutely necessary functions. As you might guess, that makes for weak bones. Older people, especially women, are prone to the disease "osteoporosis." The translation of this word is "porous bones" and the condition results from years of a diet containing too little calcium. It is important to enter older age with strong, calcium-rich bones because our bone mass typically diminishes as we age due to other, more pressing uses of calcium (see Figure 17.12). Dairy products are the best source of dietary calcium so "drink your milk!" (though our discussion on limiting fat intake tells us skimmed or semiskimmed will be best for your heart).

Phosphorus

Phosphorus, along with calcium, is present largely in our bone and tooth mineral but has other, more celebrated uses. DNA and RNA rely on one phosphate group per nucleotide base to provide part of these molecules "backbone." Also, ATP (adenosine triphosphate) contains three phosphate groups (see Figure 16.27). Recall that it is this compound that actually transfers the energy gained from metabolism of our food to power useful chemical reactions. ATP is formed from ADP (adenosine diphosphate) and phosphate.

$$ADP + H_3PO_4 \rightarrow ATP + H_2O$$

ATP is often spoken of as the " energy currency" of life because, like dollar bills, it is earned in one transaction, carried around until needed, and then traded for something else that we find necessary at a later time.

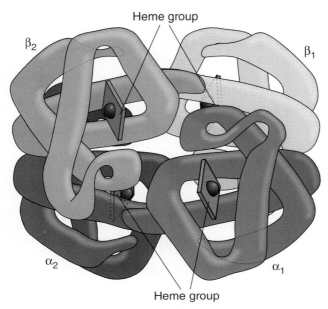

Figure 17.11
Hemoglobin is a large molecule made up of four protein chains, each one of which contains one iron ion which bonds O_2 molecules for transport through the body.

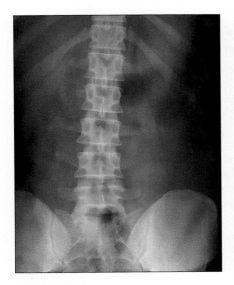 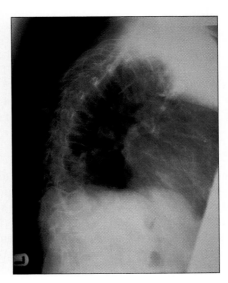

Figure 17.12
A. Normal spine of a healthy person with hard, dense bones resulting from a diet containing enough calcium. **B.** Shows a spinal X-ray from an individual suffering from osteoporosis (a condition resulting from insufficient calcium consumption). The bones lose protein matrix tissue and become brittle and prone to fracture. The colorized area highlights particularly affected areas of bone.

17.7 | Food Additives

Often when we read the ingredients of a food label we see names of substances that appear more like names of chemicals than names of foods. We then remind ourselves that all foods are made of various chemicals but still we suspect that butylated hydroxytoluene (BHT) is not something that grew in a plant and may wonder why it is in our food. Rather than being a natural food ingredient, it is a substance that has been prepared or purified in a laboratory and added to our food. A good working definition of a **food additive** is "a substance added to a food which is not part of the primary nutrients or ingredients of the food." Typically, food additives comprise only a tiny fraction of the weight of a food but may comprise a large fraction of the space in the ingredient section of a food label. A bewildering variety of additives are used in foods for many reasons. A few of the reasons are:

Some reasons for using food additives
- Mineral enhancement
- Vitamin enhancement
- Prevention of spoilage
- Modification of texture and mouthfeel
- Modification of taste and odor
- Modification of appearance and color

The U.S. Food and Drug Administration (FDA) must approve additives in foods sold in the United States. To simplify this process, in 1958 the FDA compiled a list of additives which are "**G**enerally **R**ecognized **A**s **S**afe." These so-called **GRAS substances** can be used with less paperwork than new substances. The entire GRAS list includes things like salt and gelatin as well as many unfamiliar substances. The list is available on the Internet by following the link in this book's website.

Some substances influence more than one of the listed reasons for additive use. For instance, vinegar is added both to change the taste of a food and to change its pH to inhibit the grown of microorganisms.

Food Colors

Most food colors go by names such as "FD&C Red #2." This is because their chemical names are very long and meaningless to most readers. The FD&C means that they are allowed by the Food and Drug Administration (FDA) for use in foods, drugs, and cosmetics. Other nonfood products may contain colors such as "D&C Blue #1," which means the material is suitable for drugs or cosmetics but not foods. The structures of these compounds are complicated (see Figure 17.13) but all of them have long sequences of alternating single and double bonds.

Spices

Spices are used to give improved flavor and odor to foods. They supply virtually no nutritional benefit. Spices are derived from plants, mostly grown in tropical areas. The "Spice Islands," now part of Indonesia, were an important trading station in the early years of transcontinental navigation and trade.

Flavor enhancers such as monosodium glutamate (MSG) are also added to improve flavor of foods.

Table 17.8 shows a few spices and the compounds that are most important in their composition. Only the major compound is shown for each spice. In reality, each consists of several compounds, often of related structure.

Mineral and Vitamin Enhancement

Iron, zinc, and a few other mineral elements are frequently added to foods to improve their nutritional value. Vitamins are often added for similar reasons. As we have already discussed,

FD&C Orange No. 1	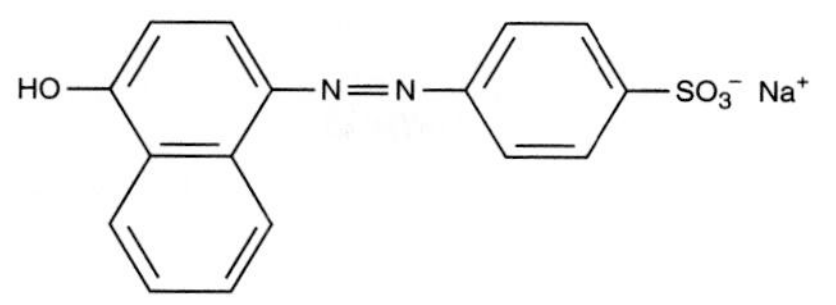
FD&C Yellow No. 5	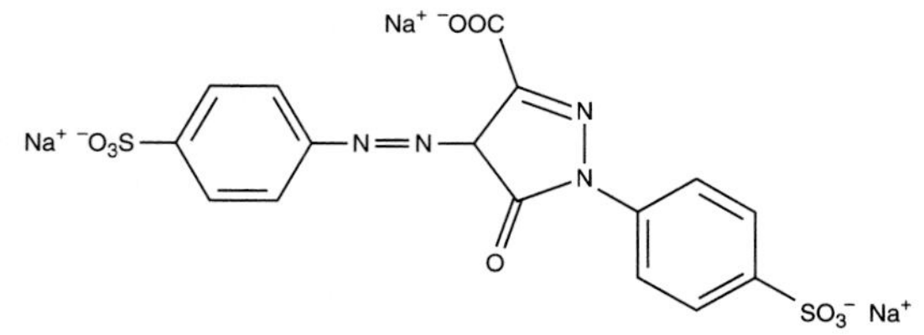
FD&C Red No. 40	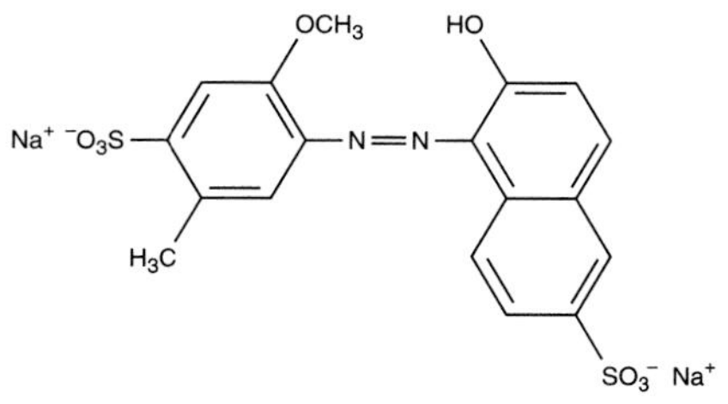
FD&C Blue No. 1	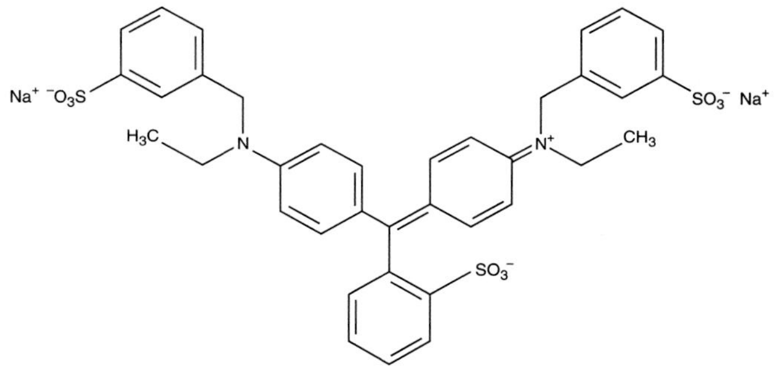
FD&C Green No. 2	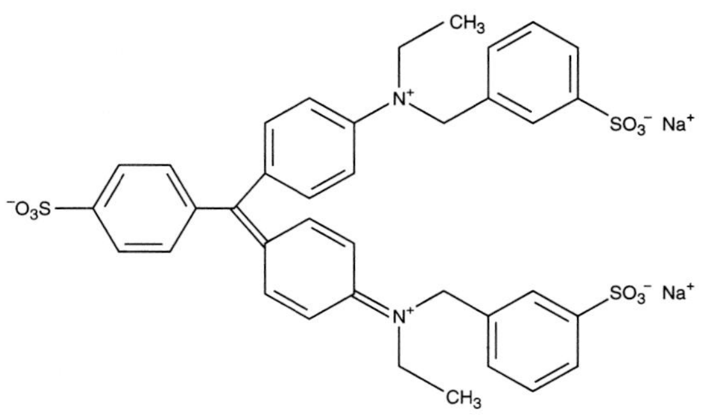

Figure 17.13
Selected food colorings and their important compounds

many foods contain vitamins naturally, but others are added to make the foods supply a more balanced suite of vitamins. Also, some vitamins and minerals are lost in processing foods, so an equal amount can be added back to bring the food back to the nutritional level its un-processed ingredients would have contained. This procedure is called "enrichment" and the best example is wheat flour, into which iron, niacin, riboflavin, and thiamin are added to bring the levels back to those in whole wheat before being ground into flour. Most milk has added

Table 17.8

Selected Spices and Their Important Compounds

Spice	Compound Name	Compound Structure
Cinnamon	Cinnamaldehyde	
Clove	Eugenol	
Nutmeg	Camphene	
Caraway	*d*-Carvone	
Spearmint	*l*-Carvone	
Pepper	Phellandrene	
Coriander	Coriandrol	
Vanilla	Vanillin	
Ginger	Zingiberone	

vitamins A and D to minimize the possibility of deficiencies in these vitamins among children. Phosphate and phosphoric acid are used both as a phosphorous source and as a pH adjuster to prevent microbial growth.

Prevention of Spoilage

The two major types of food deterioration are oxidation and the growth of microorganisms. Oxidation is most damaging to foods containing unsaturated fats. It leads to unpleasant flavors and the term "rancidity" is used to describe unwanted partial oxidation of fatty foods. Ultraviolet light accelerates the development of rancidity and metals such as iron and copper catalyze the rancidity reactions. Susceptible foods are often packed in opaque containers to keep them away from light. Additives such as EDTA (ethylenediaminetetraacetic acid) can also be used to tie up metal ions to prevent their taking part in rancidity reactions. Several antioxidants such as BHT and vitamins C and E are used to prevent these undesirable oxidations. Since vitamin E is fat-soluble and vitamin C is water-soluble, vitamin E is more effective as an antioxidant in many important foods. An antioxidant is really just a reducing agent (see Figure 17.14)

Several strategies are used to prevent growth of microorganisms. Bacteria won't grow in foods having a pH below about 4.3, so acids such as vinegar (acetic acid solution) are added to decrease the pH to such values. Bacteria won't grow when there is insufficient water available, so sugar or salt can be added to tie up the water via hydration to prevent bacterial growth. Foods such as jelly and ham are examples of well-known foods that contain water but have most of it tied up so as to be unavailable to bacteria. Sugar in jelly is usually regarded as a nutrient and flavor component rather than an additive, but one of its purposes is to prevent spoilage.

Vitamin C (asorbic acid)	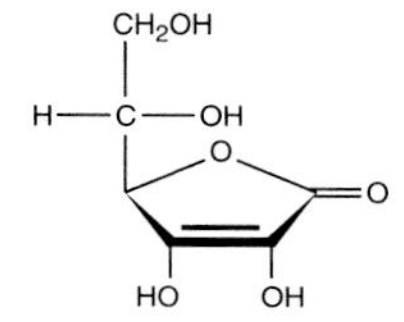
Vitamin E (α-tocopherol)	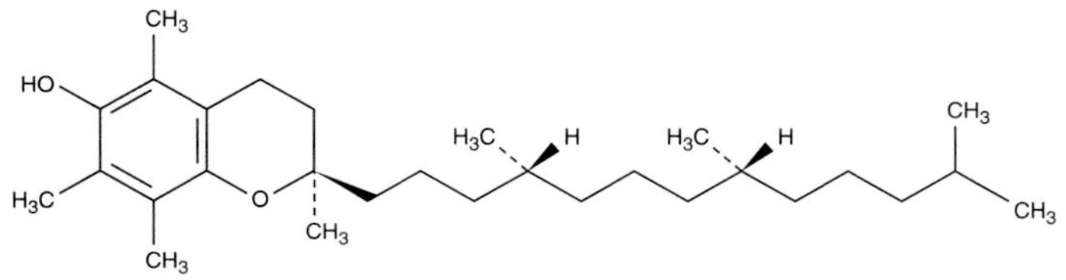
BHT (butylated hydroxytoluene)	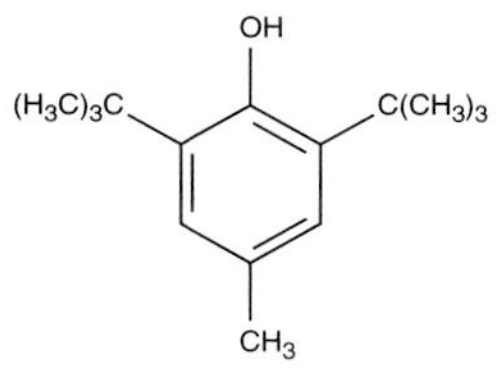
BHA (butylated hydroxyanisole) BHA is a mixture of two isomeric compounds.	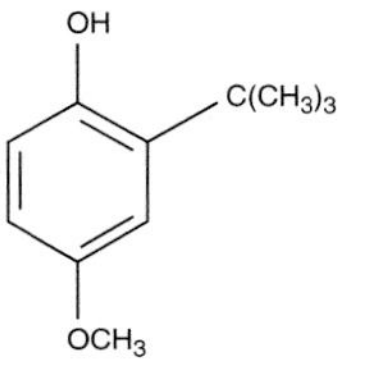 and 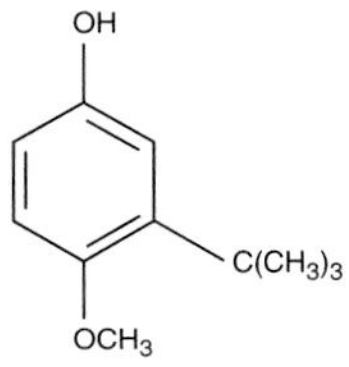

Figure 17.14
Compounds that prevent food spoilage.

Mouthfeel and Texture

The way a food feels in our mouth is important to our enjoyment of it. Compounds like starch and carrageenan, a polysaccharide derived from seaweed, are used as thickeners to make foods smooth. Commercial ice cream is a good example of a compound that often includes thickeners.

Anticaking Agents

Compounds such as calcium silicate, ammonium citrate, magnesium stearate, and silicon dioxide are used in many powdered food products, such as salt and powdered sugar, to keep them from forming lumps. They do this by absorbing water that would cause, for instance, powdered sugar to form lumps after dissolving in the sugar.

17.8 | Choosing a Balanced Diet

A balanced healthy diet must provide us with water, enough energy to maintain a healthy body weight, the raw materials (amino acids, minerals, etc.) to maintain the chemistry of our bodies, and the vitamins to keep the vital processes that rely on them working as they should. A balanced diet must also not contain too much of various components that might be good in moderation. Too much food in general will make us overweight. Too much saturated fat and cholesterol will increase the risk of heart disease and other illnesses. Too much salt will promote high blood pressure and possible damage our kidneys. Too much sugar and acidic soft drinks can promote tooth decay. On the other hand, certain components of food may be especially good for us, such as fish oils rich in specific omega fatty acids, which may promote cardiovascular health; or vegetables rich in antioxidants that may offer protection against cancer.

So many things to consider. So many choices! Yet the basis of a healthy balanced diet is quite easily grasped. In the United States, the simplest and most familiar guidelines on healthy eating are embodied within the Food Guide Pyramid, shown in Figure 17.15. This separates

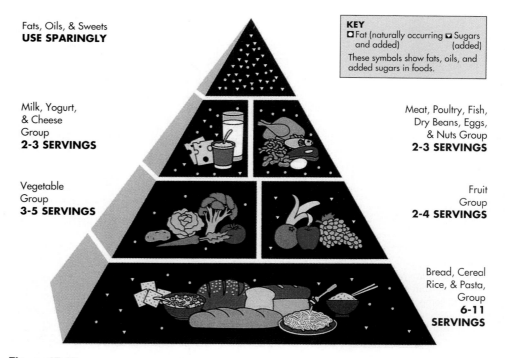

Figure 17.15
USDA's Food Guide Pyramid. The Food Guide Pyramid lists the food groups and the amount to consume from each group.

foods into six key groups, with the relative use we should make of each roughly corresponding to the volume they occupy in the Pyramid. The "Bread, Cereal, Rice, and Pasta" group should form the foundation of our diet. These carbohydrate-rich foods meet our essential energy needs in a health-promoting, low fat way, while also delivering a useful complement of other nutrients. The "Fruit" group and the "Vegetable" group comes next, providing us with food that is rich in vitamins and minerals while low in calories and fat. The "Meat, Poultry, Fish, Dry Beans, Eggs & Nuts" group includes foods that are high in protein, which we must consume. This group should be used with some caution, however, because some of its members are quite rich in fat and cholesterol. Similar reasoning applies to the "Milk, Yogurt & Cheese" group, which supplies needed protein, calcium, and other nutrients but can also be high in fat. Using skimmed milk and low-fat dairy products can give the advantages of this group without the disadvantage of high fat content. At the top of the Pyramid are the "Fats, Oils & Sweets" we should consume only sparingly. As we have seen, the fats and oils supply 9 kcal per gram rather than the 4 kcal per gram available from carbohydrate and protein, so they carry a greater risk of causing overweight. Unfortunately, many of us find food rich in fats, oils, and sugary products to be the most appealing. Why should that be? Part of the reason is that we live in unusual times. We have plentiful supplies of food, while most humans have lived in times when food was scarce. When food is scarce it makes good sense to devour as much energy-rich fats and oils as you can. It makes much less sense when the fridge is full of fatty temptations to enjoy at leisure and when our lifestyles may involve little exercise. What we eat is our choice, but the Food Guide Pyramid helps us identify the choices most likely to maintain good health.

Vegetarians, who, despite the name, generally eat dairy products and eggs in addition to fruits and vegetables, can follow the guidance of the Food Guide Pyramid without difficulty. Vegans, who reject all dairy and other animal products, in addition to meat must take care to consume sufficient protein and also sufficient vitamin B_{12} (available to them in breakfast cereals and yeast extracts, for example). Most vegans and vegetarians are people who take greater care with their food than the average omnivore (a person who "eats everything"— sometimes almost literally, unfortunately!). Hence vegetarians and vegans generally eat healthy diets, although special care is recommended in bringing up children on these restricted diets. Like every chemical issue, there are controversies surrounding the concept of the "ideal diet." We do not propose to enter these controversies, but the general basis of a healthy diet is clear and accepted by most nutritionists.

An alternative way of viewing our dietary needs is to consider each thing we need and outline some good food sources of it. Trying to base a balanced diet on this approach is more complicated, and will in any case just end up duplicating the advice embodied in the Food Guide Pyramid.

17.9 | Energy Needs

How much energy must be released by the oxidation of food in order to keep us healthy, without causing us to become fat? The answer depends on our age, gender and on what activities we choose to perform during the day. Table 17.9 shows average values for young children, males older than 10 years and females older than 10 years. The "resting energy expenditure" is the energy needed to keep us going if we simply rested 24 hours of the day. The "mean activity factor" is the factor by which the resting energy expenditure must be multiplied to allow for the additional energy needed to power our daily activity (walking, exercising, working, etc.). The total estimated energy allowance is shown in kcal per day, kcal per kg of body weight, and megajoules (MJ) per day. The ranges shown in the kcal per day values are very wide, because we vary widely in how active we are.

You can see how activity can greatly influence our energy needs by looking at Table 17.11. Resting in bed we are using approximately 1 kcal per minute. Go for a walk and it rises to around 3.5 kcal per minute, while going for a very fast run can consume in excess of 7 kcal per minute. These figures make clear how increasing the amount of exercise can be a very valuable

Table 17.9

Mean Heights and Weights and Estimated Daily Energy Allowances

	Weight		Height		Resting Energy Expenditure (kcal/day)	Mean Activity Factor	Estimated Energy Allowance (with Range)		
	kg	lb	cm	in			kcal/day	kcal/kg	MJ/day
Children (yr)									
1–1.9	11	24	82	32	600	2.0	1200 (900–1600)	105	4.8
2–3.9	14	31	96	38	700	2.0	1400 (1100–1900)	100	5.9
4–5.9	18	40	109	43	830	2.0	1700 (1300–2300)	92	7.1
6–7.9	22	49	121	48	930	2.0	1800 (1400–2400)	83	7.5
8–9.9	28	62	132	52	1050	1.8	1900 (1400–2500)	69	7.9
Males (yr)									
10–11.9	36	79	143	56	1200	1.8	2200 (1700–2900)	61	9.2
12–16.9	57	126	169	67	1580	1.7	2700 (2000–3600)	47	11.3
17–24.9	70	155	177	70	1750	1.6	2800 (2400–3200)	40	11.7
25–49.9	69	152	176	69	1620	1.6	2600 (2200–3100)	38	10.9
50–69.9	68	149	173	68	1440	1.6	2300 (1900–2700)	34	9.6
>70	66	146	171	67	1310	1.5	2000 (1600–2400)	30	8.4
Females (yr)									
10–13.9	42	96	155	62	1300	1.7	2200 (1700–2900)	50	9.2
14–16.9	56	123	162	64	1410	1.6	2200 (1700–3000)	41	9.6
17–24.9	58	128	163	64	1420	1.6	2300 (1900–2700)	39	9.6
25–49.9	59	130	163	64	1350	1.6	2200 (1800–2600)	37	9.2
50–69.9	59	130	160	63	1220	1.6	2000 (1600–2400)	34	8.4
<70	59	130	158	62	1140	1.5	1700 (1300–2100)	30	7.1

From FAO/WHO/UNU: *Energy and Protein Requirements,* Technical Rep Series 724, Geneva, Switzerland, 1985, World Health Organization.

means of weight control. The exercise can "burn" off the energy stored in our body fat very effectively, provided we don't compensate by eating too much to power the extra activity.

Where does all this energy come from? The proportion of energy supplied by each nutrient varies depending on the proportions in our diet and the amount of stored body fat we have. Average proportions for the typical North American diet (if there is such a thing) are shown in Figure 17.16. This reveals another significant energy source in many people's diet, namely alcohol. The ethanol (C_2H_5OH) in alcoholic beverages is not a recommended nutrient, and we can certainly survive without it! Those who do indulge, however, can gain a significant proportion of their energy from ethanol. Like carbohydrate, fat and protein, ethanol is oxidized to carbon dioxide and water. The oxidation of ethanol releases 7 kcal/gram, second only to fat in the energy league table (see Table 17.10).

The values in Figure 17.16 are consistent with our earlier comments that carbohydrate and fat are the main energy sources for the body, with protein playing a less significant role. Dietary guidelines recommend that our fat intake should supply around 30% of our total calories, so only those "typical" Americans near the lower limit of the

Table 17.10

The Energy Value of the Four Main Energy-Supplying Nutrients

Nutrient	Energy Supplied
Fat	9 kcal per gram
Alcohol	7 kcal per gram
Carbohydrate	4 kcal per gram
Protein	4 kcal per gram

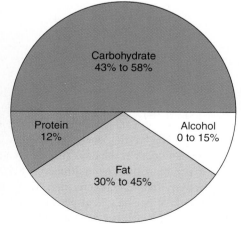

Figure 17.16
Percentages of energy (in kilocalories) contributed by carbohydrate, fat, protein and alcohol in the typical North American diet.

Table 17.11

Energy Expenditure in Specified Activities

	kcal/min	kJ/min
Man (65 kg, or 143 lb)		
In bed asleep or resting	1.1	4.52
Sitting quietly	1.4	5.82
Standing quietly	1.7	7.32
Walking 3 miles/hr (4.9 km/hr)	3.7	15.5
Walking 3 miles/hr (4.9 km/hr) with a 10-kg load	4.0	16.7
Office work (sedentary)	1.8	7.5
Domestic work		
Cooking	2.1	8.8
Light cleaning	3.1	13.0
Moderate cleaning (such as polishing and window cleaning)	4.3	18.0
Industry		
Garage work (repairs)	4.1	17.2
Carpentry	4.0	16.7
Electrical and machine tool industry	3.6	15.1
Laboratory work	2.3	9.6
Construction work	6.0	25.1
Bricklaying	3.8	15.9
Driving tractor	2.4	10.0
Feeding animals	4.1	17.2
Planting	4.7	19.7
Sawing—hand saw	8.6	36.0
power saw	4.8	20.1
Shoveling	6.5	27.2
Recreation		
Sedentary	2.5	10.5
Light (playing pool, bowling, golf, sailing)	2.5–5.0	10.5–21.0
Moderate (such as dancing, horseback riding, swimming, and tennis)	5.0–7.5	21.0–31.5
Heavy (such as athletics, football, and rowing)	7.5+	31.5
Woman (55 kg, or 121 lb)		
In bed asleep or resting	0.9	3.7
Sitting quietly	1.2	4.8
Standing quietly	1.4	5.7
Walking 3 miles/hr (4.9 km/hr)	3.0	12.6
Walking 3 miles/hr (4.9 km/hr) with a 10-kg load	3.4	14.2
Office work (sedentary)	1.6	6.7
Domestic work		
Cooking	1.7	7.1
Light cleaning	2.5	10.5
Moderate cleaning (such as polishing and window cleaning)	3.5	14.6
Light industry		
Bakery work	2.3	9.6
Laundry work	3.2	13.4
Machine tool industry	2.5	10.5
Recreation		
Sedentary	2.0	8.3
Light (playing pool, bowling, golf, sailing)	2.0–4.0	8.3–16.7
Moderate (such as canoeing, dancing, horseback riding, swimming, and tennis)	4.0–6.0	16.7–25.1
Heavy (such as athletics, football, rowing)	6.0+	25+

From Durnin JVGA, Passmore R: *Energy, Work, and Leisure,* London, 1967, as reported in FAO/WHO: *Energy and Protein Requirements,* Technical Rep No 522, Geneva, Switzerland, 1973, World Health Organization.

30–45% range indicated in Figure 17.16 are innocent of the tendency of too many Americans to consume too much fat.

The issue of excess fat consumption leads us into an issue of great concern to many of us—weight control. If we consume too much food, we will grow fat. Each pound of body fat stores about 3500 kcal of energy. So we need to undertake activity that will expend 3500 excess kcal in order to lose one pound of fat. As you can calculate from Table 17.11, each pound of fat could meet all of your energy needs during about 1000 minutes of brisk walking! Walk for nearly 17 hours to lose a pound of fat? Best to choose a diet that prevents you from gaining the excess pound in the first place!

consider this:

What Is "Organic Food"?

Food producers, grocery stores, and the public at large are increasingly using the term "organic food." They generally mean food produced without the use of synthetic pesticides or other chemicals that do not occur naturally, and without the application of inorganic fertilizers (such as ammonia and ammonium salts) to fields. Food cannot be called organic if ionizing radiation, sewage sludge fertilizer, or genetic engineering is used in its production. Also, antibiotics and synthetic growth hormones cannot be used in organic meat production. The basic ethos of organic methods is to produce food "naturally." Some people prefer this food for a variety of reasons, including worries about the health effects of the chemicals used in nonorganic food production, and their effect on the environment.

"Organic food" is a misleading term, chemically speaking, because to a chemist organic chemistry covers all carbon-based chemicals, both natural ones and synthetic ones, and regardless of whether or not we can safely eat them. All foods, however they are grown, are predominantly organic chemicals. Also, synthetic plastics and carbon-based pesticides are just as much organic chemicals as natural proteins, sugars, vitamins, and so on. Nevertheless, use of the term organic food in the sense described above seems to be here to stay.

The U.S. Department of Agriculture (USDA) currently defines organic food as:

> To be sold or labeled as an organically produced agricultural product . . . an agricultural product shall (1) have been produced and handled without the use of synthetic chemicals . . . and . . . (2) not be produced on land to which any prohibited substances, including synthetic chemicals, have been applied during the 3 years immediately preceding the harvest of the agricultural products; and (3) . . . be produced and handled in compliance with an organic plan agreed to by the producer and handler of such product and the certifying agent.

The definition of synthetic chemicals used in that definition is:

> The term "synthetic" means a substance that is formulated or manufactured by a chemical process or by a process that chemically changes a substance extracted from naturally occurring plant, animal, or mineral sources, except that such term shall not apply to substances created by naturally occurring biological processes.

The USDA has also ruled that food labels can use the term "organic" in one of these approved ways:

"100% organic": means that the food contains only organic ingredients. The USDA organic seal, shown here, can be used on such products.

"Organic": means that the food contains at least 95% organic ingredients by weight. Again, the USDA organic seal can be used on such products.

"Made with organic ingredients": means that the food contains at least 70% organic ingredients and up to three of them may be listed on front of the package.

Processed products that contain less than 70% organic ingredients can have organic ingredients listed on the information panel, but the word "organic" cannot be used on the front of the package.

exercise 17.5

Using the Body Mass Index (BMI)

Problem

As of 1999, 27% of Americans are judged by the National Center for Health Statistics to be overweight and an additional 34% to be obese (severely overweight), calculated using the body mass index (BMI), a relative measure of a person's weight to height.

$$BMI = 703 \times \text{weight in pounds}/(\text{height in inches})^2$$

A BMI value of less than 25 is associated with the lowest incidence of obesity-related health problems, while a value over 30 is of greater concern. The BMI does not take into account healthy lifestyle decisions such as long-distance running, which make it less meaningful than for the general, more sedentary population.

One of your textbook authors (PBK) is a long-distance runner who weighs 192 pounds and is 5 feet 11.5 inches tall. What is his BMI, and is this cause for concern?

Solution

$$BMI = 703 \times 192/(71.5)^2 = \mathbf{26.4}$$

Although this value is above the ideal limit, PBK's long-distance running makes the value less meaningful than for those who do not heavily exercise. There are additional measures, such as cholesterol and HDL levels (see Section 17.3) that confirm this assertion.

17.10 | Genetically Modified Food

One of the biggest commercial applications of genetic engineering, which we explored in Chapter 16, is likely to be the production of genetically modified food crops (**GM food**). Such crops are already on our tables, in the form of genetically modified soya, corn and tomatoes, for example. Like all aspects of genetic engineering, this is an issue surrounded by controversy.

First of all, why should we want to alter the genetic make up of food crops? Some of the main incentives are as follows:

Key Reasons for Developing GM Food

- To make plants resistant to specific pests and diseases, which brings the additional benefit of reducing the amount of pesticides and herbicides they must be exposed to.
- To make plants resistant to selected herbicides, so that spraying of the crops with the herbicide will kill all weeds, while leaving the crops unharmed.
- To improve the keeping and processing qualities of foods.
- To make genetically modified plants and animals that grow larger than normal, offering greater yields.
- To make plants that contain less, or none, of certain natural toxins (such as "glycoalkaloids" in potatoes) or allergens (such as allergenic proteins in nuts).
- To make plants and animals with improved nutritional qualities, such as increased content of certain vitamins.
- To develop plants that can grow in areas that are currently too hot, or cold, or wet or dry.

Worries have been raised, however, about the safety and general desirability of using genetic engineering to tamper with our food crops. Next are some of the main concerns that have been raised.

Key Concerns about GM Food

- The general fear that tinkering with nature in ways we do not fully understand will have harmful, but currently unforeseeable, effects on health and/or the environment. This is basically the fear of "nasty surprises," which science and technology have certainly thrown up in the past.
- The possibility that antibiotic resistance genes and viral genes, both of which are used widely in gene transfer technology, will cross species barriers and lead to new strains of antibiotic-resistant bacteria or virulent new viruses.
- The danger of creating new toxins or allergens in the plants or animals whose genetic makeup is modified.
- The risk of genetically modified plants becoming superweeds, able to spread uncontrollably and wipe out more natural competitors.
- The risk of cross-breeding between genetically modified plants and their natural relatives.
- The accidental production of new plant and animal diseases, and new sources of human disease due to unforeseen genetic transfer across species barriers, which some claim may happen much more readily to transferred genes.
- The danger that genetically modified crops will prove harmful to insect species that play crucial roles in "the balance of nature," leading to cumulative unpredictable ecological consequences.

The debate continues as we write, and will be continuing as you read this. You may like to try to find out a bit about the current state of play in this field, by consulting newspapers, magazines, and some of the more popular scientific journals.

17.11 | Eating—A Crucial Choice

There is an old saying "You are what you eat," and to a great extent that is true. It is not precisely true, because in reality you are what your body makes out of what you eat, and the body has a remarkable ability to select what it needs from food, reject what it does not need, and chemically transform the chemicals we eat into the chemicals of the body. The selection and transformation process is far from perfect, however, so toxic chemicals in food can cause great harm. Even the chemicals we know we need can cause us harm if we eat too much of them, or eat the wrong proportions. So choosing what to eat is one of our most crucial chemical choices, and it is a choice we all must make. Eating is not an optional activity! Book stores have high and long shelves packed with all kinds of texts claiming to offer guidance on what we should eat. For a simple, sensible and easy to use guide, however, we need only try to live in accordance with the Food Guide Pyramid (Figure 17.15), make sure we eat plenty of fresh fruits and vegetables, and avoid eating so much that we become overweight. Easy to say. Often not so easy to do. But at least it is our choice.

main points

- Our food is a mixture of chemicals, mostly high molecular weight covalently bonded compounds.

- Water is the single most abundant compound in our food and we need a supply of water every day.

- Carbohydrate, protein, and fat make up the majority of our diet, along with water.

- Carbohydrates are compounds of carbon, hydrogen, and oxygen in which the ratio of H to O is 2 to 1. Carbohydrate is needed in food principally as a source of energy.

- Fats are compounds of carbon, hydrogen, and oxygen. Fats are also valuable sources of energy.

- Proteins are compounds of carbon, hydrogen, oxygen, nitrogen, and sulfur. Protein is used mainly to build body tissue and to manufacture enzymes that catalyze the chemistry of life.

- In digestion, carbohydrates, fats, and proteins are first broken into much smaller molecules, which are reassembled in our bodies for our own purposes.

- Cholesterol is a necessary part of the body but too much is troublesome.

- Various mineral elements, such as calcium, iron, and phosphorus, must also be obtained from our food.

- Food must contain many vitamins, but in tiny amounts.

important terms

Antioxidant is a chemical that protects against damaging oxidizing agents by reacting with these agents. (p. 555)

calorie is the amount of heat needed to raise temperature of 1 g H_2O by 1°C. (p. 540)

Calorie is equal to 1000 calories, 1 dietary calorie, 1 kilocalorie. (p. 540)

Cholesterol is a lipid found only in animal products, necessary for life but a health hazard if present in too high an amount. (p. 547)

Coenzyme is a small molecule needed to bind to an enzyme in order to allow it to work. Many vitamins act as coenzymes or as precursors to coenzymes. (p. 553)

Dietary Calorie (Calorie, kilocalorie) = 1000 calories. (p. 540)

Diglyceride is a compound formed by the reaction of two alcohol groups of glycerol with fatty acids. (p. 547)

Emulsifiers are compounds that interact with both water and fat to allow them to mix. (p. 547)

Fat is a general term for the lipids in food. In this usage, it includes fats and oils. (p. 544)

Fiber is carbohydrate in food that is indigestible by humans. (p. 543)

Food additive is a substance added to a food that is not part of the primary nutrients or ingredients of the food. (p. 558)

GM food is genetically modified food, or food derived from a genetically modified organism. (p. 568)

GRAS substances are food additives classified as generally recognized as safe. (p. 558)

Kilocalorie = 1000 calories = 1 dietary calorie. (p. 540)

Macronutrients are nutrients needed in substantial amounts in the diet, namely, carbohydrates, fats, and proteins. (p. 552)

Micronutrients are nutrients needed in only tiny amounts in the diet, especially all the vitamins and many minerals. (p. 552)

Mineral is a metal ion or nonmetal that appears in the body largely as ionic compounds. (p. 555)

Monoglyceride is a compound formed by the reaction of one alcohol group of glycerol with a fatty acid. (p. 547)

Monounsaturated fatty acid is a fatty acid with only one carbon-carbon double bond. (p. 544)

Nutrient is any substance needed in our diet. (p. 537)

Oil is a liquid fat. (p. 544)

Polyunsaturated fatty acid is a fatty acid with more than one carbon-carbon double bond. (p. 544)

Saturated fatty acid is a fatty acid with no carbon-carbon double bonds. (p. 544)

Unsaturated fatty acid is a fatty acid with one or more carbon-carbon double bonds. (p. 544)

Vitamins are small molecules needed for proper health, which must be in our diet as they are not made, or not made in sufficient amounts, by the body. (p. 552)

exercises

1. What are the six nutrients?
2. Why do we need oxygen to survive?
3. If you drink a can of cola containing 40 g of carbohydrate, how many Calories do you consume?
4. Why do you think artificial sweeteners are used in diet foods rather than sucrose?
5. Explain why both glucose and starch provide 4 Calories per gram.

6. How many grams of fructose are needed to supply the same sweetness as 10.0 g of sucrose? How many grams of saccharine are needed for this same degree of sweetness?
7. How many Calories are supplied by enough aspartame to provide the sweetness of 40.0 g of sucrose?
8. How many Calories are supplied by enough fructose to equal the sweetness of enough sucrose to supply 200 Calories?
9. Draw the structure of a fat molecule having one saturated 12-carbon fatty acid and two monounsaturated 16-carbon fatty acids.
10. If a carbohydrate molecule has 21 oxygen atoms, how many hydrogen atoms does it have?
11. How many water molecules are needed to allow a trisaccharide to form glucose molecules?
12. What monosaccharides are first formed when sucrose is digested?
13. Describe the bonding and structure of a molecule of sucrose. What happens to a sucrose molecule when it is cooked, eaten, and digested?
14. Here is a refresher from Chapter 16 that relates to food. Describe the bonding and structure of a molecule of amylopectin. What happens to an amylose molecule when it is cooked, eaten, and digested?
15. Here is another Chapter 16 food-related refresher. Discuss the similarities and differences in the bonding in amylopectin, and glycogen.
16. Discuss the similarities and differences in the bonding in sucrose, maltose, starch, and cellulose.
17. One french fry weighing 5.0 g is burned in an excess of oxygen and the heat is used to raise the temperature of 300 mL of water. The original water temperature was 20.04°C and the final temperature was 65.04°C. How many dietary calories are available in one serving of 150 g of french fries?
18. Combustion of 1.0 g of sucrose yields 4 kcal. How much sugar is present in a 5.0 ounce piece of candy if it supplies 110 kcal per ounce?
19. Combustion of 1.0 mole of glucose yields 673 kcal. How much heat is released from the burning of 9.0 g of glucose?
20. Balance this reaction.

$$\underline{\quad} \ C_2H_4O_2 + \underline{\quad} \ O_2 \rightarrow \underline{\quad} \ CO_2 + \underline{\quad} \ H_2O$$

21. What is the structural difference between a saturated fat and an unsaturated fat? Why do you think unsaturated fats are healthier for you?
22. Identify the polar and nonpolar portions of the following fatty acid. Which part will dissolve in water?

$$CH_3(CH_2)_{14}COOH$$

23. Can "triglyceride" be accurately used in place of the term "fat"?
24. What is the difference between a fat and an oil?

25. Fats contribute 9 Calories per gram to your diet. How much fat must you eat to get the same number of Calories as eating 45 g of carbohydrate?

26. Why are cholesterol and triglycerides, while structurally very different, both classified as lipids?

27. Draw the structure of a fat molecule in which one fatty acid is stearic acid, one is palmitic, and one is linoleic acid.

28. What is the molar mass of the fat described in Exercise 27?

29. How many grams of hydrogen are necessary to saturate a mole of the fat mentioned in Exercise 27?

30. Suppose you consume 2000 Calories per day and wish 30% of your calories to come from fat. How many grams of fat should you consume each day?

31. The solubility of cholesterol in pure water is about 2 mg/L. How is it possible for an individual to have blood cholesterol level of 200 mg/dL (2000 mg/L)?

32. An average adult male weighing 140 pounds contains about 210 g of cholesterol. Based upon the maximum solubility of 2 mg/L, how much water would be needed to dissolve this much cholesterol?

33. Use bond energy calculations to predict the amount of energy released on hydrogenating a mole of oleic acid.

34. What category of fat is most likely to become rancid?

35. What are the building blocks of proteins? Why are these pieces referred to as "building blocks"?

36. Most people in the United States eat too much protein in their diet. Can you think of a reason why we need only about 12% of our Calories from protein and 55–60% from carbohydrates?

37. People who train for a marathon and do not eat enough carbohydrate sometimes smell like ammonia when exercising. Why do you think this happens?

38. What is an essential amino acid? What will happen if you do not have the essential amino acids in your diet?

39. Examine the structures of each of the amino acids and compare the essential amino acids with the nonessential ones. Can you speculate how the nonessential ones could be made in our bodies from the essential ones?

40. How many dietary calories are available in a food item composed of 5.0 g of sucrose, 8.0 g of starch, 2.0 g of lactose, 6.0 g of fat, 10.0 g of cellulose, 7.0 g of protein, and 9.0 g of water? What fraction of the calories come from fat?

41. Would you expect the compound $C_{18}H_{32}O_2$ to be a fat, a fatty acid, an amino acid, a protein, or a carbohydrate? Explain your answer.

42. Would you expect the compound $C_{18}H_{32}O_{16}$ to be a fat, a fatty acid, an amino acid, a protein, or a carbohydrate? Explain your answer.

43. Protein typically contains about 16.0% nitrogen. Suppose that 8.0 g of a food sample is treated chemically to convert all the nitrogen into ammonia which is then reacted with HCl. It is found that 35 mL of 0.12 *M* HCl is needed to react with all the ammonia. What is the percent protein in the food?

44. What are the differences and similarities of vitamins and minerals?

45. Both vitamin C and vitamin E are necessary for good health but consuming a large amount of vitamin C is not considered harmful but excess amounts of vitamin E are regarded as toxic. Explain the difference.

46. Vitamin D is necessary to store calcium in the bones. If a woman decides to take calcium supplements instead of drinking milk, is she at risk for osteoporosis? If so, how could she lower her risk without ingesting dairy products?

47. One glucose molecule can produce 38 ATP molecules via metabolic processes. How many phosphorus atoms are necessary to allow this to occur?

48. Iodized table salt typically contains about 0.10% KI. If a person consumes 10 g of salt per day, how many days will be needed to consume a mole of iodide ion?

readings

1. Lemonick, Michael D. Are we ready for fat-free fat? *TIME Magazine,* January 8, 1996, p. 53.
2. Ellis, Jerry W. Overview of sweeteners. *J. Chem. Ed.,* 1995, 72, pp. 671–675.
3. Walters, D. Eric. Using models to understand and design sweeteners. *J. Chem. Ed.,* 1995, 72, pp. 680–683.
4. Kinghorn, A. Douglas, and Kennelly, Edward J. Discovery of highly sweet compounds from natural sources. *J. Chem. Ed.,* 1995, 72, pp. 676–679.
5. Karstadt, Myra, and Schmidt, Stephen. Olestra: Procter's big gamble. *Nutrition Action Health Letter,* 1996, 23, pp. 4–5.
6. Brody, Jane. Vitamin C: Is anyone right on dose? *New York Times,* April 16, 1996, p. B5.
7. Weindruch, Richard. Caloric restriction and aging. *Scientific American,* January 1996, pp. 46–52.
8. Barboza, David. As biotech crops multiply, consumers get little choice. *New York Times,* June 10, 2001, pp. 1, 28.
9. Taubes, Gary. The soft science of dietary fat. *Science,* March 30, 2001, pp. 2536–2545.
10. Glausiusz, Josie. The chemistry of fat substitutes—can you stomach it? *Discover,* March 2001, pp. 28–29.

websites

www.mhhe.com/kelter   The "World of Choices" website contains web-based activities and exercises linking to: The Pacific Institute for Studies in Development, Environment and Security's chronology of worldwide water conflicts; the Food and Drug Administration's overview of food additives; NASA's list of Space Shuttle foods; and much more!

Chemistry at Home

Better Things for Better Living—Through Chemistry

—Long-time Dupont Company slogan

Have you ever seen an advertisement for a product that is said to be "Free of Chemicals" or words such as that? By this time in your study of chemistry, you must know that this is incorrect and misleading. All matter on Earth is made of chemical substances so every material thing is a chemical or a mixture of several chemicals. So far in this book we have described many everyday chemicals, including plastics, metals, foods, vitamins, fuels, fabrics, and many others. In this chapter we round things off by looking at some aspects of the chemical products all around us that we rely on in everyday life.

18.1 | Building the House

Homes are made of a lot of things but we will focus on just wood frame and brick construction. Both of these technologies have been around for centuries, way before chemistry became a science. That doesn't keep the chemistry from being interesting just the same.

Wood

Wood is mostly cellulose with smaller amounts of **hemicellulose** and **lignin**. All of these compounds are synthesized to be the structural parts of trees and are made via photosynthesis from carbon dioxide and water. Cellulose (see Chapter 6) is simply glucose molecules linked together in the beta form to make long-chain molecules. Hemicellulose differs from cellulose in that the sugar molecules of which it is composed contain only five carbon atoms rather than six, the missing carbon atom being the one that is not part of the six-membered ring. Lignin is very different, it is a polymer of several aromatic alcohols such as coniferyl alcohol and vanillin (see Figure 18.1).

Glass

Glass is an amorphous mixture of the oxides of silicon (SiO_2), calcium (CaO), and sodium (Na_2O). Remember, mixtures are not compounds because the composition of a mixture is not fixed, but rather can vary over wide ranges. White sand is nearly pure SiO_2 and is used directly in glass manufacturing,

Coniferyl alcohol

Vanillin

Figure 18.1
Lignin comprises a large fraction of wood and is a polymer of several alcohols, including coniferyl alcohol and vanillin.

whereas sodium and calcium oxides are formed by heating sodium carbonate and calcium carbonate, respectively.

$$Na_2CO_3 \rightarrow Na_2O + CO_2$$
$$CaCO_3 \rightarrow CaO + CO_2$$

These ingredients are heated together until they all melt and dissolve together. The mixture is then cooled slowly until the glass stiffens; it does not crystallize, so it is often spoken of as a supercooled liquid rather than as a solid. In ordinary window glass, the mole ratio of the oxides is about 1.6 moles of Na_2O to 1.0 moles of CaO to 6 moles of SiO_2. Glass for other purposes utilize different ratios and often include other oxides. Well-known examples are **Pyrex**™ glass, which includes 12.5% boron oxide; lead crystal, which contains typically 16% PbO; and stained glass, which contains various elements selected from the transition elements from Ti to Cu, depending on the desired color. A wonderful example of stained glass is the York Minster Cathedral in York, England (see Figure 18.2)

Glass for fiberglass insulation is much lower in SiO_2 and Na_2O than window glass but has considerable B_2O_3 and Al_2O_3 as well as CaO. It is sprayed out into very fine fibers as it cools.

Cement and Concrete

People often speak of cement and concrete as if they were the same thing, but they are actually different substances. **Concrete** is the substance we drive on and it is made from cement, sand or gravel, and water. The mixing of these three ingredients is done shortly before the wet concrete is put in place to harden. **Cement**, on the other hand, is made by heating clay (typically $Al_2O_3 \cdot 2SiO_2 \cdot 2H_2O$) and limestone ($CaCO_3$) until they begin to melt at about 1260°C. This melt is called **clinker** and is then cooled and powdered to form what is sold commercially as Portland cement. Although cement is not a pure compound, a typical clinker compound is Ca_3SiO_5, which is more clearly written as $3CaO \cdot SiO_2$.

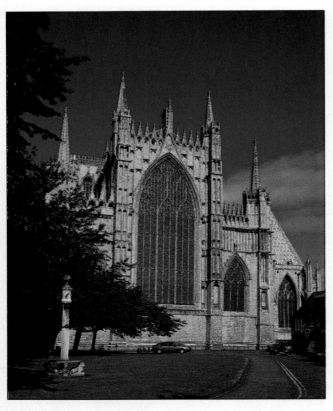

Figure 18.2
The great east window York Minster Cathedral.

Believe it or not, even though concrete and cement have been used for millennia, there is still disagreement about what exactly happens when a mixture of sand, water, and Portland cement is allowed to sit and harden. It is clear that the water reacts with CaO and Al_2O_3 to make $Ca(OH)_2$ and $Al(OH)_3$, which react with calcium silicates to form calcium aluminosilicate crystals, which interlock with each other. Concrete sets in a matter of hours and then continues to harden for years.

Nails

Nails aren't very high-tech, but we all know how necessary they are for successful construction. Most nails are made of steel, which is an alloy containing mostly iron. Look back at Chapter 13 to review iron and steel production. Nails that are never expected to be in contact with water are plain steel, but nails used in installing shingles or external panels must be made resistant to rust. To prevent their rusting, steel nails are often **galvanized**. This means that they are coated with zinc, a good reducing agent. Rather than the iron being oxidized, the zinc instead will oxidize and the steel will not rust. Since zinc melts at a temperature (419°C) much below the melting point of iron (1535°C), steel nails can simply be dipped into molten zinc, removed, and cooled (see Figure 18.3).

 # hands-on

Put several ordinary steel nails in a jar with a damp paper towel and do the same with several galvanized nails. Cover lightly and allow them to sit undisturbed for several days. The difference in corrosion should be very obvious.

Another option for rustless nails is to use aluminum instead of steel. There is a crucial difference between the products formed when iron and aluminum react with oxygen. Iron forms the characteristic red rust, Fe_2O_3, which flakes off and allows more iron to contact the air. Aluminum also forms its oxide, Al_2O_3, but this oxide adheres tightly to the metal and prevents oxygen from reaching the rest of the metal. Aluminum is not as strong as steel so is seldom used for structural purposes, but it is great for adding decorative, lightweight trim work.

Asbestos Siding and Its Replacements

When your authors were young, asbestos siding was regarded as an excellent product. One of us (JDC), remembers his parents' house being sided with asbestos in the late 1940s. These same siding panels are still on the house and have needed no repair or upkeep. They have made the house resistant to fire for all these years and they look as good as new. Now, one never hears of a house being sided with asbestos; rather, vinyl or aluminum siding are preferred. Why this change in attitude? **Asbestos** is an unusual mineral in that it is fibrous. The most abundant form of asbestos is called **chrysotile** and it has the formula $Mg_3(Si_2O_5)(OH)_4$. Its crystals grow very long and slender and pack together like toothpicks in a box as shown in Figure 18.4. In making siding, asbestos fibers are separated and bound into a suitable matrix.

Asbestos can be heated to red hot and it will not burn, melt, change color, shrink, or grow weaker. In short, it makes a wonderful siding material, floor and ceiling tile, and insulation. In recent years, however, it has become common knowledge that breathing asbestos dust can cause severe lung damage and even lung cancer, so very little asbestos is being used any more and old asbestos is being removed, especially from public buildings. A good argument can be made that well-maintained asbestos objects can safely be left in place, but the invisible hazard of asbestos is so worrisome that even tightfisted school boards pay large sums of money to remove old asbestos wherever they find it.

Anyway, back to siding. We've said a lot about aluminum, which is now widely used in siding, but not very much about "vinyl." **Vinyl** is usually **polyvinyl chloride** (PVC) polymer of vinyl chloride, $CH_2{=}CHCl$. Hundreds of vinyl chloride monomers form each molecule of

Figure 18.3
These nails are galvanized, too much so that they have stuck together by the zinc.

Figure 18.4
Asbestos fibers can be several inches long and very thin.

polymer, which then has the structure shown in Table 6.9. The annual U.S. production of PVC is greater than 200,000 tons. In addition to its use in siding, it is widely used in pipes used in plumbing. Even this great stuff has some problems, however. In case of a fire, PVC releases HCl gas, which is dangerous, and the monomer vinyl chloride must be handled carefully because it is a severe carcinogen and has been blamed for several deaths.

Electric Wiring

Very pure copper is used for electrical wiring. All metals conduct electricity well but copper conducts better than any other metal except silver, which is too expensive for common use this way. Possibly as important as the copper, which conducts electricity very well, is the insulation that surrounds the copper and prevents shocks and fires. The insulation on modern wire is a **polysiloxane**, a polymer in which the skeleton of the polymer is alternating silicon and oxygen atoms. The other two bonds to silicon are to either methyl or phenyl groups (see Figure 18.5). This material is much more resistant to temperature extremes, oxidation deterioration, and insect damage than was rubber or the even older fabric covered wires.

Tungsten is also used to conduct electricity but only inside incandescent light bulbs. Tungsten is chosen because it conducts electricity poorly and the electric energy is converted to heat and light when it passes through the very fine tungsten filament. Tungsten is useful because it has a very high melting point ($3410°C$) so it remains solid at the very high temperatures of the filament. If lightbulbs were filled with air, the tungsten at high temperature would burn to form tungsten oxide (WO_3) so lightbulbs with tungsten filaments are filled with inert gas, often argon, krypton, or nitrogen, to prevent oxidation. So-called "halogen" lamps are also tungsten filament bulbs but the bulb also contains a small amount of elemental iodine. As tungsten atoms evaporate off the filament at high temperature, they react with iodine to form WI_2. This compound is a gas at this temperature but when a molecule of it hits the very hot tungsten filament, the molecule decomposes; the tungsten atom redeposits on the filament and the iodine atoms recombine to make an iodine molecule (I_2). This allows the tungsten bulb to operate at higher temperature without burning out and therefore to give a brighter light.

Plumbing

The word "plumbing" comes from the Latin word "plumbum," which means "lead." Lead was known to the ancients and the Romans used it to make pipes for distribution of water. Lead was chosen because it melts at a very low temperature ($327°C$) and can thus be cast and rolled into pipes, which are welded shut with a hot iron. Sadly, lead is also quite poisonous and enough lead dissolved into the water supply from these pipes to cause health problems.

Water pipes are now made of galvanized steel, copper, or PVC, often in the same home (see Figure 18.6). There is still concern about the use of lead-based solders to fasten copper tubes together. Lead-based solders are used because lead melts at a low temperature and adheres well to copper to form water-tight joints.

New construction in the United States uses exclusively copper or PVC pipes. Older homes still have steel pipes and, at least until recently, in the older cities of the eastern United States one could still find lead pipes in use. Where new additions have been added to older homes and copper pipes are connected to older steel ones, there are often corrosion problems due to the very different tendencies of the two metals to become oxidized. Essentially, the iron acts as a sacrificial anode, preventing oxidation of the copper by oxygen in the water but a great deal of iron can dissolve and cause problems. For this reason, a rubber union between the two pipes should be installed to prevent corrosion.

The Air Space

The most pervasive substances in a home are nitrogen and oxygen, the two most abundant components of the air that takes up all the space in the home that isn't taken up by something else. However, the air in one's home is often more polluted than outside air because of all the activities of living. Carpet glue and plywood exude formaldehyde (CH_2O). A hot shower is an excellent way of transferring volatile water pollutants such as chloroform ($CHCl_3$) into the air.

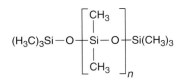

Figure 18.5
Structure of polydimethylsiloxane, used as wire insulation.

Figure 18.6
Iron, copper, and PVC pipes shown together in the plumbing of an older house.

Incomplete combustion in fireplaces and the like can release carbon monoxide (CO), and burned food in the kitchen can release a whole variety of unpleasant substances. Modern buildings are often constructed to be very energy efficient, meaning that they are very airtight and do not allow fresh air to enter from outdoors. Some buildings have been said to be "sick," and the term "sick building syndrome" has entered our vocabulary to describe buildings in which many people feel sick due to high concentrations of indoor air pollutants.

Air also contains water vapor. In many locales there is too much water vapor in the air in summer and too little in winter. Therefore we use humidifiers in winter and dehumidifiers in summer to try to keep a medium humidity. When air gets too dry, wood shrinks and joints loosen, but when the air is too wet, wooden things swell up and make doors difficult to open. People, too, are most comfortable with a relative humidity of about 50%.

18.2 | Keeping the House Clean

In order for a substance to play a role in cleaning, it must remove dirt and grease from clothes, walls, carpet, hands, or whatever. In all means of cleaning, the soil is transferred from the dirty object into a separate phase, either liquid or solid, and then removed from the object. Various products do this in various ways, but generally involve a transfer of the soil into the cleaning product where it can be removed from whatever started out dirty. Often this means that the soil is removed into water, which can be sent down the drain.

Surfactants

Soaps and detergents are both surface active agents (surfactants). A **surfactant** molecule concentrates at a surface between water and nonpolar substances. They lower the surface tension of water and allow it to penetrate into fibers more freely. Their molecules have long chains of CH_2 groups (typically 12–18 carbon atoms) with a very polar or ionic group at one end. Soaps contain the sodium (or potassium) salt of a carboxylic acid (Figure 18.7). Many of the surfactant molecules form **micelles** (Figure 18.8) with the nonpolar part in the center and the ionic part on the outside surface in contact with water. Greasy dirt is quite nonpolar, so it dissolves in the interior of the micelle where it can be washed away into water and down the drain. Detergents and soap as well as most dirt is biodegradable, so when this mess reaches the sewage treatment

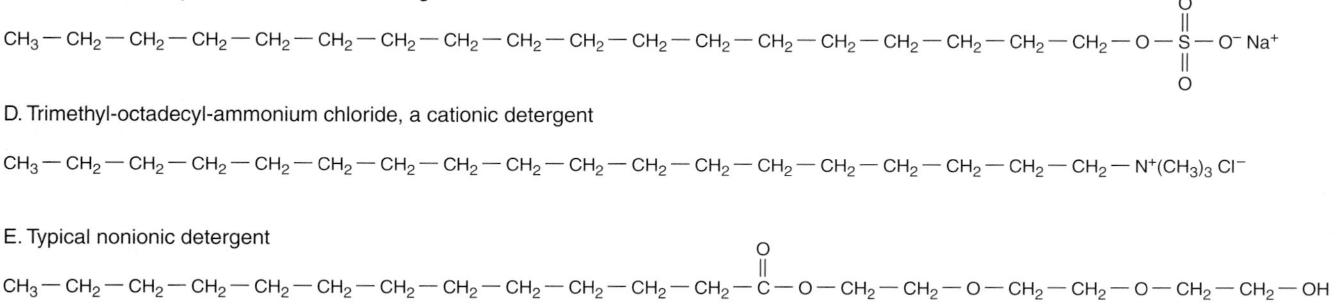

Figure 18.7
Typical structures of cleaning agents: **(A)** soap, **(B)** sulfonate, and **(C)** sulfate detergents and **(D)** cationic and **(E)** nonionic detergents.

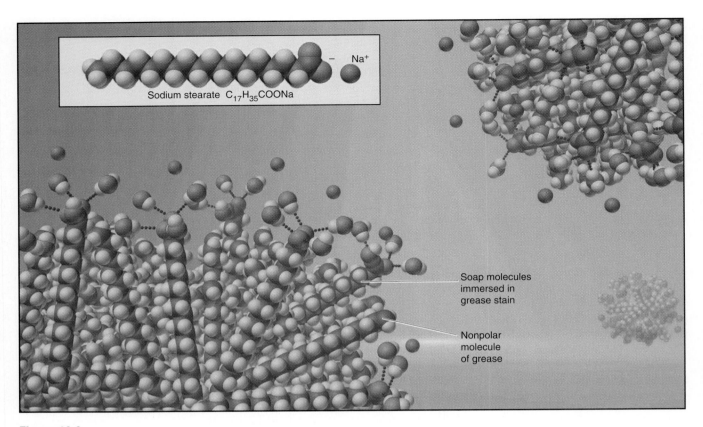

Sodium stearate $C_{17}H_{35}COONa$

Soap molecules immersed in grease stain

Nonpolar molecule of grease

Figure 18.8
A micelle with and without a grease molecule incorporated.

plant, it is converted into either biomass or carbon dioxide and the remaining water is discharged into the environment.

Modern detergents are either anionic, with a negatively charged headgroup; cationic, with a positively charged head group; or nonionic. Anionic detergents are as described in Chapter 7, having a sulfonate or sulfate ionic group. Cationic detergents have a $—N(CH_3)_3^+$ group on the end of a long hydrocarbon chain. Nonionic detergents have a polar end group, which is not ionic. Typical nonionic detergents have a phenolic group, $—C_6H_4(OH)$, a very polar but nonionic group.

Detergent Builders

Commercial detergents now contain much more than just the surfactant material. The most important other compounds are called "builders." A detergent **builder** is something that causes the surfactant material to work more efficiently. It usually involves adjusting the pH and adding something to tie up hardness ions, Ca^{2+} especially. Hardness ions form insoluble precipitates with soap; the familiar but undesirable soap scum is the calcium salt of soap anions. For many years the most widely used builder was sodium tripolyphosphate, $Na_5P_3O_{10}$. This material worked extremely well, but the large amount of phosphate it added to lakes and streams when it slowly hydrolyzed in water, after going down the drain and into the environment, caused objectionable algae blooms.

$$P_3O_{10}^{5-} + 2H_2O \rightarrow 3PO_4^{3-} + 4H^+$$

Algae need several things to grow, but in many natural water settings the limiting reagent is phosphate ion. When large amounts of phosphate become available in rivers and lakes, it acts as a fertilizer and the algae grow tremendously. When they die, they settle to the bottom of the body of water and decompose. This decomposition consumes oxygen and fish die from lack of oxygen. As a result, much less phosphate is now allowed to be used as builder in laundry detergents.

Enzymes that Clean

Another class of substance added to laundry detergents are enzymes (see Figure 18.9). Enzymes are added that cause proteins to be hydrolyzed back into their constituent amino acids. These are called **protease** enzymes. Likewise, enzymes called **lipases** cause the ester linkages in fats to hydrolyze. The enzymes digest the proteins and fats in difficult biological stains such as food, blood, or grass stains. Most enzymes are denatured at high temperature, so detergents using enzymes usually work best in warm water but not in hot water. Enzymes have been isolated from organisms that live in hot springs such as at Yellowstone Park. As you might expect, these enzymes operate fine at higher temperatures and are being developed for use in detergents that will clean at high water temperatures.

Figure 18.9
Box of detergent touting enzyme cleaning.

Brighteners

Laundry detergents frequently advertise that they make the clothes "whiter than white." How can that be, or is it just advertising jargon? Something is perfectly white if it reflects all of the visible light that falls onto it. Therefore, something is whiter than white if it gives off more visible light than what falls onto it (see Figure 18.10).

This is made possible by chemicals called **brighteners.** These compounds absorb ultraviolet radiation, mostly UV-A, and fluoresce visible light, mostly in the blue and violet part of the spectrum. This blue color masks the natural yellowing of many fabrics to make the garment look white. A typical compound used as an optical brightener is blancophor R (see Figure 18.11). The two $SO_3^- Na^+$ groups on the central rings increase the water solubility of the compound.

Figure 18.10
This white shirt and laundry detergent contain brighteners illuminated by only UV light.

Bleach

Sometimes **bleach** is added separately to stained clothes and sometimes mild bleaches are part of the commercial detergent itself. A common bleach material is an aqueous solution of sodium hypochlorite, NaOCl. This is made by dissolving chlorine gas in a dilute solution of sodium hydroxide.

$$Cl_2 + 2NaOH \rightarrow NaOCl + NaCl + H_2O$$

This acts as a bleach because the hypochlorite ion is a modest oxidizing agent that can oxidize the chemicals responsible for many stains. Frequently, the oxidized form is less highly colored than the reduced form so the stain disappears, although frequently the oxidized form remains on the garment.

Newer, less harsh bleaches, are based on oxygen compounds such as peroxides. These actually contain sodium perborate, $NaBO_3 \cdot 4H_2O$. When this is added to hot water, it decomposes into hydrogen peroxide and sodium borate.

$$4NaBO_3 + 5H_2O \rightarrow 4H_2O_2 + Na_2B_4O_7 + 2NaOH$$

The tetraborate ion formed is also useful as a builder as it forms a complex with iron. Sodium tetraborate is known as borax and is sold as a laundry aid.

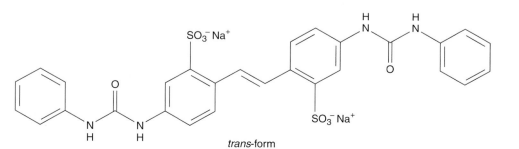

trans-form

Figure 18.11
Structure of blancophor R, a typical brightener in laundry detergents.

Figure 18.12
Solid CO_2 (dry ice) in a closed bottle sublimes to form a gas. When the pressure is great enough, the solid becomes liquid. When the pressure is released, the liquid immediately reforms solid CO_2.

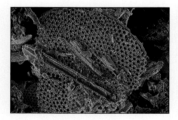

Figure 18.13
Dry cleaning solvents such as tetrachloroethene, C_2Cl_4, are giving way to more environmentally friendly solvents, especially CO_2.

Dry Cleaning

Dry cleaning differs from standard laundry methods which use water as solvent. In dry cleaning, the dirty clothes are submerged in a nonaqueous, volatile solvent that dissolves the grease or dirt. The clothes are then removed from the solvent, but the dirt stays behind in the solvent. The remaining solvent then is easily evaporated away from the clean clothes. Since dirt and grease do not dissolve in water, a nonpolar solvent must be used in dry cleaning. The most used solvent has been tetrachloroethene, C_2Cl_4, (see Figure 18.13) but the use of chlorinated solvents is being discouraged for environmental reasons so replacements are being found. Recently, a few dry cleaning establishments have begun using liquid carbon dioxide as a solvent for dry cleaning. CO_2 will liquify at a pressure of about 5 atmospheres at room temperature (see Figure 18.12). Suitable detergents have been developed for use in this unusual solvent. The detergents used are very different from those used in water-based cleaning. However, they still consist of a region which interacts strongly with the liquid CO_2 and a section which interacts with most greasy stains that are to be removed from the garment. When the garment is cleaned, the CO_2 is allowed to vaporize and is then reliquified away from the clothes. The detergent is repurified and reused.

Adsorbent Cleaners

Wallpaper and carpets are often cleaned by an entirely different process. A dry powder of **diatomaceous earth** is worked into the soiled object; the soil adsorbs into the dry powder, which is then swept or vacuumed away. Diatomaceous earth is nearly pure SiO_2 but in a very porous form (Figure 18.14). It consists of the skeletons of small aquatic organisms called diatoms. This material collected on the bottom of ancient lakes and is now mined and used for many purposes.

Figure 18.14
This microscopic view shows the great porosity of diatomaceous earth.

Ammonia

Aqueous ammonia is often used as a cleaning agent for woodwork and floors. It works because it is basic and dissolves greasy materials but leaves no residue behind because both the ammonia and the water evaporate. Ammonia must never be mixed with hypochlorite bleach. To do so would cause the formation of nitrogen trichloride (also called trichloramine).

$$3Cl_2 + NH_3 \rightarrow NCl_3 + 3HCl$$

Nitrogen trichloride is bad smelling, and poisonous in large amounts.

Dishwashing Compounds

Dishwashing compounds are of two types, those used in automatic dishwashers and those used in hand washing. The compounds used in automatic dishwashers can be more basic than those used in hand washing because, although the dishes can withstand very harsh treatment, one's hands can not. Automatic dishwashing compounds often contain sodium carbonate, sodium silicate, and sometimes sodium phosphate, all of which dissolve in water to give very high pH solution. This high pH at high temperature dissolves and disperses greasy food residue. Materials meant for hand washing dishes are usually medium pH nonionic detergents. They usually form considerable bubbles and foam that most of us regard as a sign of a good cleaner. Foam inside an automatic dishwasher is a bad feature because it is difficult to rinse out of the machine; therefore these products are designed not to foam.

Rust Removers

Rusty iron objects can be cleaned but require fairly vigorous methods. Rust is mostly iron(III) oxide, Fe_2O_3, and related compounds. This must be removed either by physical action of steel wool or chemical methods such as Naval Jelly™ which is a jelly containing phosphoric acid (H_3PO_4). Note its effect on the old hoe shown in Figure 18.15 (see Figure 18.15). All acids react with metallic oxides to form soluble products but phosphoric acid is specially useful because the phosphate ion binds the Fe(III) ion in a complex ion and keeps it in solution as $FeHPO_4^+$ ion.

$$Fe_2O_3 + 2H_3PO_4 \rightarrow 2FeHPO_4^+ + H_2O + 2OH^-$$

Although hydroxide ion is released in this reaction, the solution remains acidic because of the huge excess of phosphoric acid, which reacts with the hydroxide ion as:

$$H_3PO_4 + OH^- \rightarrow H_2PO_4^- + H_2O$$

Products for removal of rust from more delicate things such as fabric often consist of sodium bisulfite ($NaHSO_3$) and sodium hydrosulfite ($Na_2S_2O_4$).

Brass Polish

Brass and copper objects react slowly with air to form CuO and with sulfur compounds to form CuS. These compounds are both black and obscure the beauty of the metals. There are many ways to clean such objects, but commercial products are sometimes just solutions of ammonia in hydrocarbon solvent, often with a bit of very fine diatomaceous earth. The ammonia dissolves any grease but, more importantly, reacts with the CuO or CuS to form the soluble ammonia complex of copper, $Cu(NH_3)_4^{2+}$.

Scouring Powder

Scouring powders contain sodium carbonate and an anionic surfactant along with an insoluble powder, usually either SiO_2 or $CaCO_3$, and disinfectants. Scrubbing with a small amount of water causes the abrasive nature of the SiO_2 to physically remove stains and deposits and organic substances to be adsorbed into the porous powder.

Figure 18.15
Before and after photos of an old hoe cleaned with Naval Jelly™.

Toilet Bowl Cleaner

Toilet bowls are made of ceramic material (as described under the section about plates) with a very tight glaze to prevent water from soaking into the china-like material. This is very resistant to acid and so is usually cleaned with a fairly acidic material. The classic product for this purpose is granular sodium bisulfate, $NaHSO_4$. This is nothing more than sulfuric acid for which one of the ionizable hydrogens has been neutralized by sodium hydroxide. Pure sulfuric acid is too dangerous for use by unsupervised amateurs as it reacts extremely exothermically with water. If care is not taken, the heat of the reaction will cause water to boil and splash up above the liquid mixture. It is even more dangerous to add water to acid because the large amount of acid will always cause the boiling of any small amount of water added to it. Anyway, the addition of sodium bisulfate to water will give a pH of about 1, which cleans away most hard water deposits and iron stains as well as adhering fecal matter. Many common toilet bowl cleaners now contain 7–9% hydrochloric acid, HCl. This is a specially good product for removing hard water deposits ($CaCO_3$) and iron deposits (Fe_2O_3) because of the very high solubility of the chloride salts of calcium and iron.

Sink and Lavatory Trap Cleaners

Most sinks and lavatories are made of steel coated with enamel and the trap beneath the drain is usually made of brass or PVC. Brass is an alloy consisting of mostly copper but with an important fraction of zinc. Copper is quite unreactive toward both acids and bases but zinc reacts readily with acids and slowly with bases. Therefore, the classic cleaning material to get soap scum, hair, and related gunk out of sink traps is sodium hydroxide. The commercial name for NaOH is "lye." Often, commercial trap cleaners are solid NaOH with chips of aluminum. The aluminum metal reacts with concentrated NaOH solution to give hydrogen gas and the bubbling action provides a certain amount of stirring and agitation to help react the material in the trap.

$$2Al + 2OH^- + 6H_2O \rightarrow 2Al(OH)_4^- + 3H_2$$

Alternatively, sulfuric acid solutions are used by professionals as sink trap cleaners. As we just said, it is dangerous to mix concentrated sulfuric acid into water, so this must be done with great care and personal protection. Additionally, the acid also reacts slowly with the zinc in the brass pipe in a process called **dezincification.** That is why manufacturer's instructions always say not to leave the product in contact with the metal trap for a long time. If your pipes are made of PVC instead of brass, they are much more resistant to both acid and base cleaners.

One can also purchase pressurized gases to simply push a plug of stuff out of a trap. Activating the unit allows a burst of gas to be released. Early versions of these products contained Freons (a compound containing only carbon, chlorine, and fluorine, for instance $C_2F_3Cl_3$) but these products have been replaced by tetrafluoroethane (CF_3CH_2F) and dimethoxymethane (CH_3—O—CH_2—O—CH_3). This is just a sophisticated version of the old "plumber's friend" plunger, which rapidly forces surges of air and water down the drain to push the plug out of the trap and into the sanitary sewer system.

Drains that are not really plugged but run slowly, can be cleaned with products that contain enzymes designed to degrade large insoluble molecules into many smaller, water soluble ones. Such products contain enzymes called amylase (breaks starches into glucose units), lipase (breaks fats into glycerol and fatty acids), protease (breaks proteins into amino acids), and cellulase (breaks cellulose into glucose units). These products act slowly so should be added to pipes when water won't be used for several hours.

Sewer Root Remover

The roots of trees and bushes seek water. If a slight crack or hole develops in a sewer pipe, roots grow toward and through it in search of water. This can totally plug a sewer and cause unpleasant water to back up into one's house. In addition to physical removal, copper sulfate pentahydrate, $CuSO_4 \cdot 5H_2O$, is used to kill such roots and slow their return.

18.3 | Chemicals in the Kitchen

The kitchen is a chemical laboratory and cooking is a chemical process. There are so many chemicals in the kitchen that it is difficult to know where to start. Most aspects of food chemistry were discussed in Chapter 17 so this chapter will focus on different aspects of chemistry in the kitchen.

Leavening Agents

Leavening agents are substances that make baked goods such as bread, pancakes, and cookies rise by forming gas bubbles that are retained in the final product. These bubbles cause the final food item to be tender and easy to chew. Unleavened bread is tough and hard to chew and digest.

In thinking about leavening agents, we must first distinguish between yeast and simple chemical leavening agents. Yeast is used in bread and related products. Yeast is a living one-celled organism that is usually stored in a dry, dormant state but when water and sugar are present in a warm bread dough, it begins active respiration and growth and metabolizes the sugar to give carbon dioxide.

$$C_6H_{12}O_6 + 6O_2 \rightarrow 6CO_2 + 6H_2O$$

The CO_2 gas is slowly evolved over perhaps an hour or more and the water is incorporated into the dough. The dough must be stiff enough for the CO_2 gas not to bubble all the way through into the atmosphere. The bubbles are retained in the dough as the bread rises. Then, when the dough is baked, the yeasts are killed by the temperature of the oven but the gas bubbles remain in the bread.

Chemical leavening is also the result of carbon dioxide generation, but by a chemical reaction rather than a biological one. Baking soda (Figure 18.16) is sodium bicarbonate ($NaHCO_3$), which reacts with acidic components of food to produce gas bubbles. Typical acidic ingredients include lemon juice (citric acid), vinegar (acetic acid), and sour milk (lactic acid). Baking powder is a related product that is a mixture of two dry ingredients, sodium bicarbonate and some solid acid. Tartaric acid, often in the form of cream of tartar (potassium hydrogen tartrate, $KHC_4H_4O_6$), is often the solid acid. As a dry mixture, the reactants are unreactive, but when water is added, the acid reacts with the base in a neutralization reaction which releases carbon dioxide, as shown in Figure 18.17.

$$NaHCO_3 + KHC_4H_4O_6 \rightarrow Na^+ + H_2O + CO_2 + K^+ + C_4H_4O_6^{2-}$$

Figure 18.16
Baking soda requires external acid to form CO_2, but baking powder includes its own acid.

Foods that use either baking powder or baking soda are usually more delicate products, such as cakes baked from a batter rather than from a dough. These must be baked right away after adding the leavening agent so that the bubbles form during the baking process and are retained when the product comes out of the oven. If the batter is kept too long, especially at room temperature, the CO_2 will mostly form before baking and bubble right through the batter and leave nothing behind to cause the cake to rise. Recall the Hands On exercise from page 7.

Flour

All baked goods have flour as their major ingredient, but there are a variety of kinds of flour. Flour is the result of grinding grain, usually wheat, although rye, rice, and corn are also used for specialized purposes. In any grain, the important constituents are starch, protein, fiber, oil, and vitamins. From the point of view of the wheat plant, the grain is

Figure 18.17
Baking powder added to soapy water shows vigorous bubbling of CO_2.

the seed necessary for the next generation of wheat. All these constituents are necessary for the seed to germinate and grow successfully. From a human point of view, these are important constituents of a balanced diet.

In the United States, most flour is made from what is called hard red winter wheat, although cake flour is made from soft wheat and pasta from hard, spring wheat called durum wheat. In any case, the wheat grains are ground to a powder by being mashed between wheels of stone (stone ground) or steel. That's all there is to it for whole wheat flour. White flour is further processed to remove much of the fiber from the bran, the outer husk of the seed, and some of the oil. Unsaturated oils in the grain make whole wheat flour go rancid, so white flour stays good much longer. Inadvertently, much of the vitamin content and much of the iron is removed during the processing of the flour. Most flour sold now is enriched. This means that iron and three of the B vitamins, niacin, thiamine, and riboflavin, are added back to the flour so that the amounts of these compounds are the same as it would have been if the processing hadn't occurred. Just now there is serious discussion about enriching flour with a fourth vitamin, folic acid, which is specially important for pregnant women for the health of their babies. Very recent studies seem to show that folic acid also protects against heart disease.

Plates and Dishes

The traditional material for plates is a form of ceramic called "china." All ceramic materials are prepared in similar fashion from clay and water. The composition of the clay and the means of heating determine much of the properties of the finished product. The most common clay to start with is kaolinite, $Al_2(OH)_4Si_2O_5$, which is often written as $Al_2O_3 \cdot 2SiO_2 \cdot 2H_2O$. This is mixed with water to become pliable and easy to mold into shape. Excess water of plasticity is removed by heating to 100°C, at which point the object becomes rigid but would become mud again if it were wetted. Further heat (500°C) removes water from within the crystal and at still higher temperatures (>1000°C) , the kaolinite is converted into mullite ($Al_6Si_2O_{13}$ or $3\ Al_2O_3 \cdot 2SiO_2$) and cristobalite (SiO_2). This mixture of mullite and cristobalite forms the dishes we use on our tables. So-called earthenware dinnerware is usually a darker color of clay that contains traces of iron, but fine china is nearly pure white and has almost no iron content. This description is nearly as valid for making bricks as it is for dishes.

$$3Al_2(OH)_4Si_2O_5 \rightarrow Al_6Si_2O_{13} + 4SiO_2 + 6H_2O$$
$$3\{Al_2O_3 \cdot 2SiO_2 \cdot 2H_2O\} \rightarrow 3Al_2O_3 \cdot 2SiO_2 + 4SiO_2 + 6H_2O$$

Recently, many people have switched to unbreakable plates made of thermosetting plastic. **Thermosetting** means that once the object has been formed, it will not melt upon heating. This is in contrast to **thermoplastic** materials, which will melt upon heating. An example of thermosetting plastics used in dinnerware is melamine-formaldehyde resin. The chemistry of this process is pretty complex but involves the formation of a polymer made from melamine and formaldehyde, which is not just long chains of linkages similar to polyethylene, but rather, long chains that are cross-linked to each other to form a very strong, tough, three-dimensional structure. In making dinnerware, this plastic material is impregnated into a filler usually of pure cellulose.

Silverware

Elegant silverware is still made of silver. Actually, it is made of sterling silver, an alloy of 92.5% silver and 7.5% copper. However, the price and availability of silver have made solid sterling silver flatware nearly unaffordable and very uncommon. A well-respected substitute is silver plate flatware, made by electrically plating a very thin layer of silver on an item made of less expensive metal. Figure 18.18 shows that the silver plating can be worn away in time.

In common use today, is **stainless steel** flatware. Items that frequently come in contact with acidic materials such as tomato paste, fruit juice, and the like must not tarnish or rust, even when allowed to dry overnight in the humid air of a dishwasher.

Figure 18.18
The bottom of an old silverplate spoon shows wear where the silver has been abraded away.

Most steel would not serve well as flatware because it rapidly rusts with ordinary usage. Several formulations of stainless steel are available, but typically they contain up to about 20% chromium, 20% nickel, a fraction of a percent of carbon, possibly a few percent of molybdenum, and the rest iron.

Knives

Knives used for food preparation must be much sharper than knives used at the dinner table and so are made of a different steel. Such knives should be both tough and hard, and to achieve these properties, rust resistance must sometimes be sacrificed. Thus, these knives should not be allowed to sit overnight wet, especially wet with acidic foods or they will rust noticeably. To retain their sharp edge, they should be honed frequently on a sharpening stone made usually of **carborundum,** SiC (silicon carbide). Carborundum takes the same crystal form as diamond and is almost as hard as diamond and therefore can be used to sharpen steel knives made of very hard steel.

Wrapping Materials

Aluminum foil is a chemical substance used to wrap food. An alternate material is **saran** wrap, a copolymer of vinyl chloride (C_2H_3Cl) and vinylidine chloride ($C_2H_2Cl_2$). It is a transparent film used to cover foods in microwave ovens and refrigerators. Flexible sandwich bags and freezer storage bags are usually polyethylene.

18.4 | Personal Care

Now we take a look at some of the chemicals we use to take care of our own personal hygiene. Some are cleaning substances, so are related to issues just discussed, but deserve a place in a separate section focused on personal care.

Toothpaste

We've all been told to brush our teeth after eating and before going to bed. This helps to prevent tooth decay and bad breath, both of which result from food residue being left in the mouth for an extended time. Bacteria in our mouths feed on the food residue and produce acids such as acetic acid, which damage our teeth and produce low-oxidation-state compounds of sulfur and nitrogen, which have unpleasant odors.

Toothpaste consists of a mixture of several compounds to help in oral hygiene. The two major ingredients are abrasives and detergents. The typical detergent used is very similar to the anionic sulfonate or sulfate detergents mentioned in Section 18.3. An abrasive consists of very fine, hard, particles of insoluble material which just scrub resistant stuff off our teeth. The typical abrasive used in toothpaste is hydrated silica. Hydrated silica is SiO_2, which at high temperatures will take on water to form some Si—O—H groups instead of only Si—O—Si bonds. The combination of abrasive action and detergents do a good job of cleaning our teeth.

Many toothpastes are fluoridated to give greater protection against tooth decay. Such toothpastes contain some source of fluoride ion, SnF_2, NaF, Na_2PO_3F, and the like and are effective because the fluoride reacts chemically with tooth enamel to make a more resistant material. Normal tooth enamel contains $Ca_5(PO_4)_3OH$, a mineral the geologists call apatite. The availability of the OH group makes this compound susceptible to action of acids. An exchange reaction occurs in the presence of fluoride ion to form fluoroapatite, $Ca_5(PO_4)_3F$, which is much more resistant to attack by acids generated by bacteria metabolizing food particles.

$$Ca_5(PO_4)_3OH + F^- \rightarrow Ca_5(PO_4)_3F + OH^-$$

A recent improvement in toothpastes involves tartar control. **Tartar** is the hard material deposited on the surface of our teeth as plaque hardens by accreting calcium containing minerals.

Plaque is a thin layer of bacteria that grow in everyone's mouth and accumulates on our teeth. Plaque can be removed by brushing and flossing but tartar can only be removed by a dentist. Therefore tartar control toothpastes do not remove tartar, but do slow its formation. The active ingredient is either sodium or potassium pyrophosphate ($Na_4P_2O_7$), which works by inhibiting certain bacterial enzymes to slow the formation of plaque.

Sodium bicarbonate, $NaHCO_3$, is often used in toothpastes and toothpowders because it will neutralize acids found in the mouth from partial decomposition of food materials and so will bring mouth pH back up to the neutral range.

Some toothpastes also work to minimize the pain of sensitive teeth when their owner eats either hot or cold foods. Active ingredients are either strontium chloride ($SrCl_2$) or potassium nitrate (KNO_3), which block the sensation of pain via nerve transmission from such teeth to the brain.

A new brand of toothpaste is proposed to contain chlorine dioxide (ClO_2), which is an extremely good disinfectant that will kill disease-causing mouth organisms, both bacteria and yeasts.

Shampoo

Shampoo is designed both to clean one's hair and to leave it lustrous, thick, and easy to care for. Hair is protein and is damaged by either low or high pH. Therefore, shampoo is nearly neutral pH (typically pH 4–9) and usually includes a nonionic detergent. It is specially important that shampoo not give insoluble calcium precipitates because that would leave hair sticky and difficult to manage. Shampoos often include conditioners, which are usually pH neutral cationic detergents. These are long-chain ammonium salts such as trimethyl-alkyl-ammonium chloride to make the hair softer and easier to comb. Shampoos often also contain **humectants** such as glycerine to retain moisture to minimize dry hair. Advanced shampoos also contain certain provitamins such as pantothenyl alcohol, which reacts to make panthothenic acid, which is involved in the biosynthesis of fatty acids. This was identified many years ago as an agent useful in strengthening hair.

pantothenyl alcohol is

$$HOCH_2C(CH_3)_2\underset{\underset{\displaystyle OH}{|}}{CH}\overset{\overset{\displaystyle O}{\|}}{C}NHCH_2CH_2CH_2OH$$

Permanent Wave

A permanent wave hair treatment really consists of two treatments, both having to do with the sulfur-sulfur bonds in hair protein. As we mention in Chapter 16, the folded structure of some proteins, especially the keratin of hair, is strengthened by bonds formed between the sulfur atoms of two different cysteine amino acids a considerable distance apart from each other on the primary structure of the polypeptide chain. The first treatment is by a reducing agent such as thioglycolic acid ($HS—CH_2COOH$). This breaks the disulfide bonds so that the folded structure is ruptured. Then the hair is put on curlers and is treated with a weak oxidizing agent such as dilute hydrogen peroxide. This causes the disulfide bonds to reform with partners on polypeptide chains that are in curled positions (see Figures 18.19 and 18.20).

Hair Colorants

The simplest kind of hair coloring is bleaching. Hydrogen peroxide (H_2O_2) is invariably used for this purpose and leads to the term "peroxide blonde." This works because the natural color of dark hair, melanin (brown-black) and phaeomelanin (red-brown), are turned colorless by oxidation. These natural pigments have extremely complex polymeric structures. Bleaching is also usually a step in dyeing one's hair a different color. Hair dyes show up better if they are applied to light-colored hair.

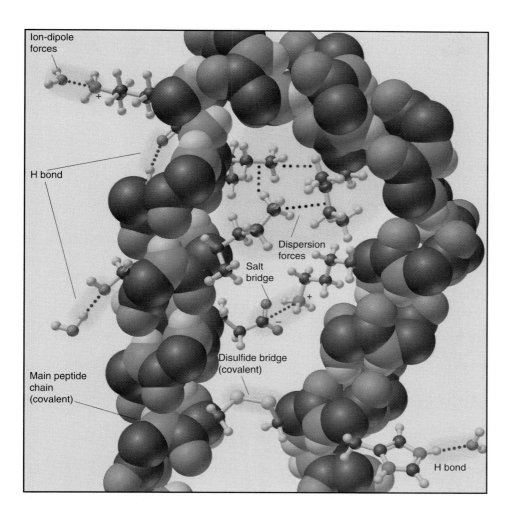

Ion-dipole forces

H bond

Main peptide chain (covalent)

Dispersion forces

Salt bridge

Disulfide bridge (covalent)

H bond

Figure 18.19
Chemistry of breaking and forming disulfide bonds in a permanent wave.

Figure 18.20
Before and after photos of a permanent.

Permanent hair dyes must penetrate the cuticle of the hair so the dye can soak into the hair rather than just adhering to the outside surface. Key ingredients are ammonia, hydrogen peroxide, and the actual components of the dye. Ammonia causes the hair to swell and the cuticle scales to separate to allow the actual dye into the hair shaft. The dyes are applied in a colorless form that soaks into the hair shaft before they are reacted in a redox reaction with hydrogen peroxide to form the final color. Typical dye precursors are diaminobenzene, aminohydroxybenzene, or dihydroxybenzene but mixtures of these and similar chemicals are used to reach a variety of final colors.

$$H_2N-C_6H_4-NH_2 \qquad H_2N-C_6H_4-OH \qquad HO-C_6H_4-OH$$

Some products that color gray hair to return it to a dark color contain lead acetate, $Pb(CH_3CO_2)_2$, which reacts with the sulfur atoms of the amino acid cysteine, which is an important natural part of hair protein. This compound of lead ion bound to sulfur is dark brown-black in color so it very rapidly and easily covers the gray color in hair.

Perfumes

Perfumes have been used for millennia to give people and their possessions desirable fragrances. Many of the early perfumes were derived from flowers. Flowers or their petals were soaked in olive oil, water, or alcohol to dissolve the fragrance and this was daubed onto the person. We still can buy rose-water. The fragrant molecules in nature are organic compounds and most are either esters or **terpenes**. Examples of esters from plants are shown in Figure 18.21

Ester	Structure	Odor, flavor
Ethyl formate	$H-\overset{\overset{\displaystyle O}{\|\|}}{C}-O-CH_2CH_3$	Rum
Isobutyl propanoate	$CH_2CH_3-\overset{\overset{\displaystyle O}{\|\|}}{C}-O-CH_2CHCH_3$ (CH_3)	Rum
Isobutyl formate	$H-\overset{\overset{\displaystyle O}{\|\|}}{C}-O-CH_2-CH-CH_3$ (CH_3)	Raspberry
Methyl butyrate	$CH_3CH_2CH_2-\overset{\overset{\displaystyle O}{\|\|}}{C}-O-CH_3$	Apple
Isopentyl pentanoate	$CH_3CH_2CH_2CH_2-\overset{\overset{\displaystyle O}{\|\|}}{C}-O-CH_2CH_2CHCH_3$ (CH_3)	Apple
Ethyl butyrate	$CH_3CH_2CH_2-\overset{\overset{\displaystyle O}{\|\|}}{C}-O-CH_2CH_3$	Pineapple
Butyl butanoate	$CH_3(CH_2)_2-\overset{\overset{\displaystyle O}{\|\|}}{C}-O-CH_2CH_2CH_2CH_3$	Pineapple
Isopentyl acetate	$H_3C-\overset{\overset{\displaystyle O}{\|\|}}{C}-O-CH_2CH_2CHCH_3$ (CH_3)	Banana
Octyl acetate	$H_3C-\overset{\overset{\displaystyle O}{\|\|}}{C}-O-(CH_2)_7-CH_3$	Orange
Pentyl propionate	$CH_3CH_2-\overset{\overset{\displaystyle O}{\|\|}}{C}-O-(CH_2)_4-CH_3$	Apricot

Figure 18.21
Structures and identities of several esters.

and terpenes are in Figure 18.22. Recall from Chapter 6 that an ester is the condensation product of an alcohol and a carboxylic acid. A terpene is a compound having a formula $C_{10}H_{16}$, which is formally a dimer of isoprene, C_5H_8. Trimers and tetramers of isoprene, called sesquiterpenes and diterpenes and alcohols or ketones derived from these compounds are also frequently found in volatile oils from plants and most have a pleasing odor.

As an example, rose-water is made by soaking rose petals from newly opened blooms in water. Nearly 5000 kg of rose petals are required to give 1 kg of rose oil. Rose oil consists of many compounds, but the two major ones are geraniol and citronellol, both terpene alcohols. The well known pine odor of many cleaners is mostly α-pinene and lemon odor is mostly d-limonene, both cyclic terpenes.

The odor of bananas, raspberries, and cherries are dominantly esters; whose specific structures are shown in Figure 18.21. Perfumes are used in all sorts of products in addition to personal care. Manufactures of foods and cleaning materials all use both natural and synthetic odorants to make their products more attractive.

Perfumes are complicated mixtures of many odorous compounds. Figure 18.23 shows a gas chromatogram of a commercial perfume. Each peak results from a single compound in the mixture. The height of the peak is an indication of the amount of that substance in the mixture.

Suntan Aids

Products that give a person the appearance of a suntan without the person having to spend hours in the sun have become very popular in recent years. These products contain dihydroxyacetone HO—CH_2—CO—CH_2OH, $C_3H_6O_3$) as active ingredient. Notice that this fits our definition of a carbohydrate (Chapter 17), and is half the molecular formula of glucose. Sugars react with amines to give brown products called **melanoidins**. In this case, the part of the sugar that reacts is the oxygen that is double-bonded to the central carbon and the amine is the nitrogen atom on the side chain of the three amino acids arginine, lysine, and histidine. These are fairly abundant in the protein of the outer layer of human skin. Melanoidins are large, cross-linked molecules which are quite stable so the sunless tan lasts quite a while.

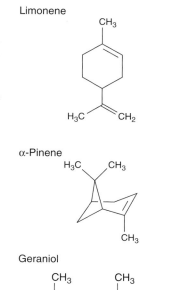

Limonene

α-Pinene

Geraniol

Camphor

Figure 18.22
Structures and identities of several terpenes.

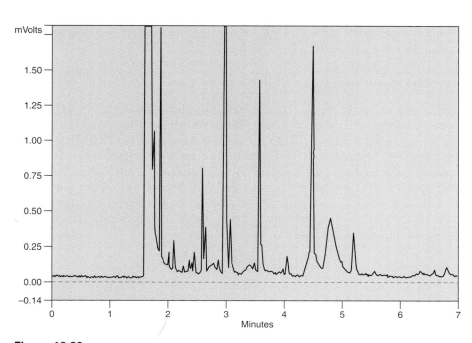

Figure 18.23
This gas chromatogram shows the many components of an expensive perfume.

Sunblock

Products that protect a person from the sun are really protecting us from ultraviolet radiation, specifically UV-B (see Chapter 15). Most of them are designed to allow the less energetic UV-A to pass through and promote a suntan without the harmful effects of the more energetic UV-B radiation. Structures of several sunblock materials are shown on page 483.

Lipstick

Lipstick provides both protection for delicate tissues and color for appearance. It consists of a mixture of waxes, oils, pigments, and emollients. A common wax used for much of the mass of the lipstick is beeswax, an ester in which the alcohol is a long chain of 24 to 36 carbons and the acid also is long chain with up to 36 carbons.

$$CH_3—(CH_2)_{34}—COO—(CH_2)_{35}—CH_3$$

Waxes are too hard to be used without softening, so the wax is mixed with an oil, either some vegetable oil, mineral oil, or lanolin to soften it. Pigment is mixed in to obtain the proper color and often sunscreen or other components to protect the lips are also included.

Pumice

Pumice stones are used to abrade callous tissue, especially from one's feet. **Pumice** is a volcanic stone which is light, hard, and porous. It was formed when complex silicates of aluminum, potassium, and sodium formed with considerable gas dissolved in the molten stone. As the stone cooled rapidly the gas bubbles became trapped in the stone and led to the low-density stone used today.

Mascara

Mascara is a mixture of several powdered polymers and pigments such as TiO_2 and iron oxides held together by oils and waxes.

Artificial Fingernails

These are simply one of several kinds of plastic, often polymethylmethacrylate with a suitable glue such as ethylcyanoacrylate to hold them in place.

Fingernail Polish and Remover

Fingernail polish is a mixture of pigments in a polymer, sometimes nitrocellulose or acrylic in a solvent such as ethyl acetate and butyl acetate ($CH_3COOC_2H_5$ and $CH_3COOC_4H_9$). Fingernail polish remover is simply acetone (CH_3COCH_3) or ethyl acetate ($CH_3COOC_2H_5$). Both of these dissolve the polish and are very volatile. Sometimes one or the other is mixed with mineral oil (saturated hydrocarbons of approximately 10 carbon atoms) to slow down the evaporation of the substance and allow more working time. Care must be taken when using these products because they remove nearly any kind of paint or varnish and dissolve the surface layer of linoleum or most floor tile. You must be very careful not to spill such a product on a vulnerable piece of furniture or floor tile.

Jewelry

The jewelry box is a place that pure elements might be found in the home. Gold, silver, and platinum are frequently used in bracelets and rings. Gold and silver used in jewelry are nearly always actually a mixture. Pure gold (called "fine gold" or 24 karat gold) is too soft to be very useful in jewelry so it is frequently mixed with silver, and silver is usually mixed with some copper, to make the metals hard enough to stand up to the wear and tear of usage (see Figure 18.24).

Figure 18.24
A. The actor Mr. T. made his reputation with tough talk and lots of jewelry. **B.** Jewelry featuring metal and stones.

Table 18.1

Gemstone	Chemical Composition
Diamond	Carbon
Emerald	$Be_3Al_2Si_6O_{18}$ (beryl) + about 2% of Cr
Sapphire	Al_2O_3 (corundum) + traces of Fe and Ti
Ruby	Al_2O_3 (corundum) + traces of Cr
Amethyst	SiO_2 (quartz) + traces of Fe
Aquamarine	$Be_3Al_2Si_6O_{18}$ (beryl) + traces of Cr
Pearl	Smooth layers of nacre produced by oysters around an irritant
Amber	Fossilized tree sap

Gemstones are fairly simple substances. Table 18.1 gives the composition of several well known gemstones.

Several well-known gemstone materials (diamond, ruby, and sapphire among others) have now been synthesized. The impact that these synthetic gems will have on the economics of the jewelry industry is still unclear. Some customers demand that their gems be natural in origin and others are perfectly happy with synthetic stones, which are equally as beautiful as the natural ones.

18.5 | Pharmaceuticals

Thousands of compounds are used as pharmaceutical substances to cure or ameliorate symptoms in a wide variety of human or animal ailments. Many of these compounds are derived from natural sources or are compounds that are simple modifications of natural compounds. Although derived from nature, useful pharmaceuticals are often made in the laboratory because natural sources are insufficient. An excellent example is the recent discovery that a compound known as **paclitaxel,** isolated from the bark of a species of yew tree in the Pacific Northwest, is very powerful in fighting breast, ovarian, and cervical cancers.

When paclitaxel was first announced it was stated that there would be plenty available from natural sources because the tree was plentiful and was thought of as a "weed tree" by people in the lumber industry. It was soon realized, however, that the demand was so great that natural sources couldn't supply enough material. Laboratory syntheses of paclitaxel then began to be studied. Two groups announced successful syntheses in 1994. It has recently been shown that the compound is also found in the needles of yew trees and many parts of hazelnut trees. It is clear, though, that a diet rich in hazelnuts does not supply enough paclitaxel in a form useful as an anticancer product.

Paclitaxel was known by the common name **Taxol** until that name was patented by Bristol-Myers-Squibb Pharmaceutical Co. There is no way that even an experienced cancer researcher could look at the structure of paclitaxel and say, "That looks like something that would fight cancer." In order that useful drugs can be identified, compounds isolated from nature or synthesized in the laboratory are often sent to the National Institutes of Health (NIH), which runs an extensive testing program to determine if compounds have interesting pharmaceutical properties. Only a few compounds show interesting activity, but those that do are subjected to further study and advanced testing.

The structure of paclitaxel is shown in Figure 18.25. This was not an easy structure to decipher. Chemists, biologists, and physicians are working to answer several questions. How can we make this material synthetically so we don't have to chop down every suitable tree in the forest? Are there modifications we can make to the compound to make it even more effective? Of what benefit is the compound to the yew tree? All these are subject of on-going scientific research.

Pharmaceutical products likely to be found in your medicine chest include pain relievers, antihistamines, antiseptics, antibiotics, laxatives, indigestion remedies, anti-diarrhea products, poison ivy remedies and other ointments. Let's see what some of these products are really made of.

Pain Relievers

Aspirin is the oldest pain reliever available today. For centuries, people have known that the willow tree contained a material that was helpful in relieving pain. Similarly, wintergreen leaves were used as a pain reliever for centuries. The relationship of these two substances was made clear in the latter years of the 19th century when the active component of willow bark was shown to be salicin, a compound related structurally to salicylic acid, and wintergreen was shown to contain the methyl ester of salicylic acid. A more palatable compound of salicylic acid, acetylsalicylic acid, was first prepared in 1853. This was later named aspirin and has been sold for well over a century. When either aspirin or oil of wintergreen reaches the intestine, the ester is converted back into salicylic acid, which enters the blood stream and relieves pain and reduces fever. It also acts to diminish blood clotting and so works to prevent heart attacks

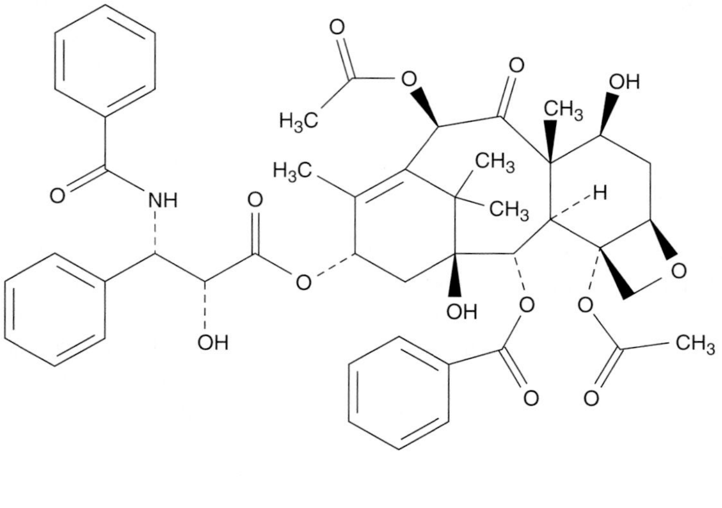

Figure 18.25
Structure of paclitaxel.

caused by poor blood flow through the heart. Aspirin, along with the other pain relievers described here, are shown in Figure 18.26.

Acetaminophen was developed many years after aspirin and is now a major competitor to aspirin for pain and fever relief. Acetaminophen does not have the "blood thinning" properties of aspirin so it is not useful in preventing heart attacks but is superior in cases where bleeding may occur. The anti-inflammatory compound, ibuprofen, is also very popular for relief of similar symptoms.

Codeine is a more powerful pain reliever than either aspirin or acetaminophen and is usually sold in combination with one of these products. It also suppresses coughing (antitussive). Codeine is structurally related to morphine, a pain reliever that is only used in hospital settings.

Antihistamines

Histamine is a compound formed when the amino acid histidine loses a CO_2 group. Histamine is a potent vasodilator. This means that it opens blood vessels wider and causes runny noses in allergic persons. Compounds that are **decongestants** shrink nasal tissues and relieve runny noses. A typical decongestant is pseudoephedrine, a compound which is related to the more well known epinephrine (adrenalin). **Antihistamines** block histamine receptors and so prevent the hay fever symptoms which bother so many people. Many compounds are presently marketed as antihistamines in tablets, syrups, eye drops, and other products. Representative examples are shown in Figure 18.27.

Antiseptics

Antiseptics are compounds that kill microorganisms on contact. The old, tried and true antiseptic, tincture of iodine is still sold today. The word "tincture" means a solution in which alcohol is the solvent. Modern tincture of iodine is about 2% iodine and 2.4% sodium iodide

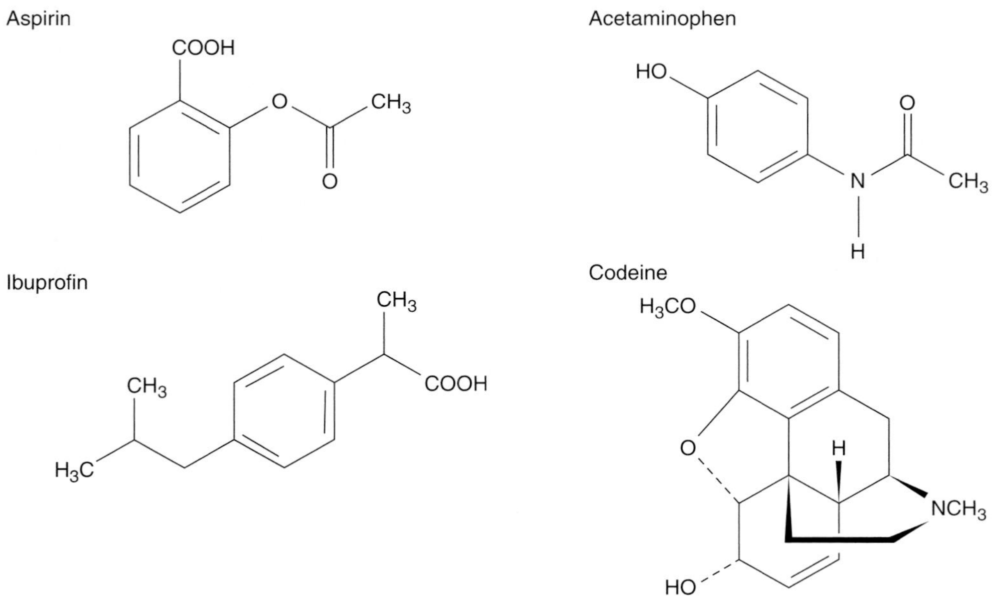

Aspirin

Acetaminophen

Ibuprofin

Codeine

Figure 18.26
Several pain killers are available: aspirin, acetaminophen, ibuprofen, and codeine.

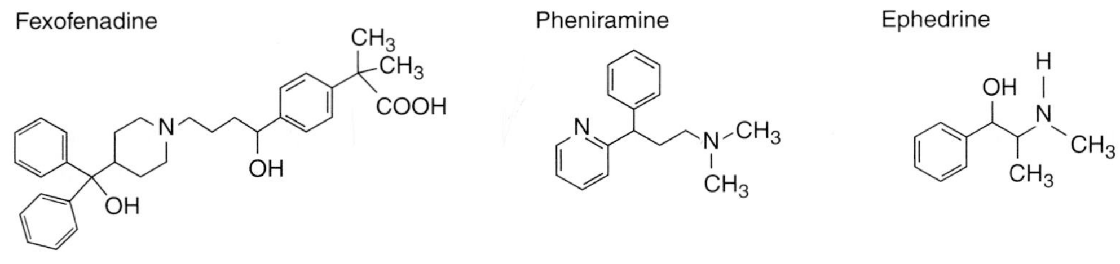

Fexofenadine

Pheniramine

Ephedrine

Figure 18.27
Structures of antihistamines.

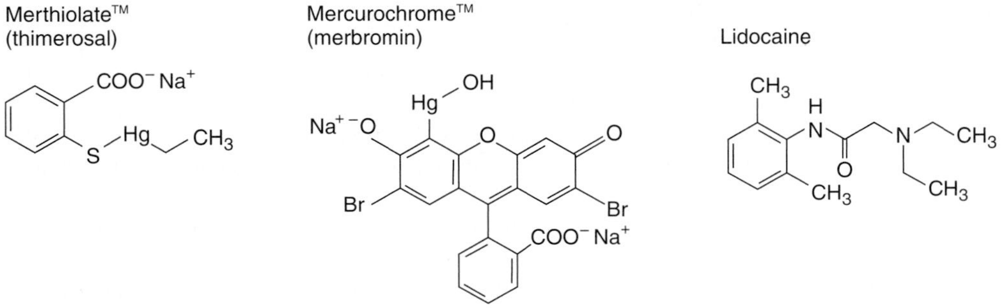

Figure 18.28
Structures of the disinfectants Merthiolate™, Mercurochrome™, and lidocaine.

Merthiolate™ (thimerosal)

Mercurochrome™ (merbromin)

Lidocaine

in alcohol/water solution. Merthiolate and mercurochrome both were formerly sold as dilute solutions in small bottles with glass or plastic applicators as topical antibacterial agents. Either material, painted over minor cuts, scratches, or abrasions, would prevent infection from setting in. All three substances provide lasting antiseptic action even after the solvent evaporates. Note in Figure 18.28 that both merthiolate and mercurochrome contain mercury and to prevent mercury pollution in our environment, these have been taken off the market.

Milder disinfectants without residual effect are simple rubbing alcohol (isopropyl alcohol) and dilute (3% in water) hydrogen peroxide. These compounds kill germs on contact but quickly evaporate (alcohol) or decompose ($2H_2O_2 \rightarrow O_2 + 2H_2O$) so are good at disinfecting minor wounds or tools, but their effect lasts only a few seconds.

To reduce the pain of scratches and abrasions, solutions of lidocaine are used. This material is a local, topical anesthetic.

Indigestion Remedies

Acid indigestion is caused by overproduction of stomach acid, often caused by eating large meals. Neutralization of the acid is one way of treating acid indigestion. Popular remedies therefore contain solid bases such as $NaHCO_3$, $Mg(OH)_2$, $CaCO_3$, $MgCO_3$, and $Al(OH)_3$. Some of these are formulated by mixing with aspirin or another pain reliever. When the stomach is not filled with recently ingested food, its pH is approximately 3 but can drop to as low as 1.5 when the stomach is filled with food. If more acid is generated than is necessary for digestion, discomfort can result, especially if some of this acid finds it way back up into the esophagus. The muscle of the stomach is protected from the acid by being coated with a thick layer of mucus. This protection is not available in either the esophagus or the small intestine, the organs lying directly before and after the stomach. The neutralization reactions (Figure 18.29) of these antacid materials are:

$$NaHCO_3 + HCl \rightarrow NaCl(aq) + CO_2 + H_2O$$
$$Mg(OH)_2 + 2HCl \rightarrow MgCl_2(aq) + 2H_2O$$
$$CaCO_3 + 2HCl \rightarrow CaCl_2(aq) + CO_2 + H_2O$$
$$MgCO_3 + 2HCl \rightarrow MgCl_2(aq) + CO_2 + H_2O$$
$$Al(OH)_3 + 3HCl \rightarrow AlCl_3(aq) + 3H_2O$$

Rather than neutralizing acid, a new prescription medicine that diminishes serious acid indigestion problems contains a compound called rabeprazole sodium (see Figure 18.30), which actually inhibits the secretion of gastric acid.

Laxatives

Laxatives function in more than one way. A bowel movement can be caused by an increase in the amount of water in the large intestine. This can be done by swallowing a large amount of a highly ionic solution. Magnesium citrate and magnesium sulfate (Epsom's salts) are used in this way. The high concentration of highly charged ions in either solution causes a considerable amount of water to pass through the wall of the large intestine, soften the fecal matter in the colon, and create pressure that in a few hours forces a clearance of the bowel.

Figure 18.29
The red cabbage juice indicates neutralization (see pps. 294 and 295) of an acid solution by adding an antacid.

A totally different approach to laxative features is **mineral oil,** which acts as an "internal lubricant." Mineral oil is simply highly purified saturated hydrocarbons having a carbon number in the range of 15–18. The words "mineral oil" is almost a perfect translation of the Greek meaning of "petroleum." To the ancients, "oil" meant olive oil and was called "oleum." The word "petros" means "rock" so the word petroleum was coined to mean oil from rocks or minerals. Mineral oil acts as a laxative because it is not metabolized either by humans or by most of the bacteria which inhabit our large intestines so it moves through the body unchanged and lubricates the passage of the rest of the content of our bowel.

Figure 18.30
Structure of rabeprazole

Ointments and Related Products

Calamine lotion, used as a remedy for poison ivy rash, is simply a suspension of zinc oxide (ZnO) with a small percentage of iron(III) oxide, Fe_2O_3, in water. Petroleum jelly is a mixture of high-molecular-weight saturated hydrocarbons. The number of carbons ranges from approximately 18 to 22. Lanolin is the fat extracted from wool. While on the sheep, lanolin helps to protect the animal from getting soaking wet even in a hard rain. On one's hands, it moistens and softens dry skin.

18.6 | Summary

The home is a virtual chemistry laboratory of fascinating products derived from raw materials found in the water, soil, rock, and biosphere of our planet. This chapter has just scratched the surface of all the hundreds of chemical substances used in every home, but we hope it has opened your eyes to the importance of chemical synthesis, engineering, manufacturing, and distribution of these substances used in everyday life. Our homes and apartments; our schools, offices, parks and trails; our cars, buses and subways; all these places we call 'home' at one time or another—these are all just small parts of the home that we all share—our one, real home—"Spaceship Earth" (see Figure 18.31). And through our study of chemistry, together we appreciate how that one precious home is truly a "World of Chemistry, a World of Choices."

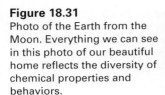

Figure 18.31
Photo of the Earth from the Moon. Everything we can see in this photo of our beautiful home reflects the diversity of chemical properties and behaviors.

main points

- Everything in our homes and environments is composed of chemical substances.

- It is frequently difficult to anticipate specific properties of useful chemicals.

- Some chemicals have hazardous properties even though they are very useful if used as directed.

- Chemicals used in cleaning are often acids or bases.

- In addition to foods, a vast variety of chemicals are present in our kitchens.

- Pharmaceuticals are compounds derived from nature, modified from natural compounds, or completely synthetic compounds. They are subject to extensive testing before commercial distribution is allowed.

important terms

Antihistamine is one of several compounds used to block the action of histamine and prevent hay fever symptoms. (p. 593)

Antiseptic is one of several compounds used to kill microorganisms. (p. 593)

Asbestos is a group of silicate minerals that crystallize in fibers; this is very resistant to fire but harmful if fibers are inhaled. (p. 575)

Bleach is a substance that decolorizes stains by oxidation. (p. 579)

Brighteners are substances that absorb ultraviolet light and emit blue light to make clothes appear whiter than they really are. (p. 579)

Builder is a substance used in major amounts in commercial laundry detergents to allow the surfactant to perform its job, usually is used to raise the pH of laundry water. (p. 578)

Carborundum is silicon carbide, a very hard synthetic material used in grinding wheels and sharpening stones. (p. 585)

Chrysotile is one specific type of asbestos, $Mg_3(Si_2O_5)(OH)_4$. (p. 575)

Clinker is the product formed from heating calcium carbonate and clay to a high temperature; it is powdered to become Portland cement. (p. 574)

Concrete is the rock-like material made by mixing cement, sand, and water and allowing the mixture to harden in place. (p. 574)

Decongestant is a material that is used to clear breathing passages. (p. 593)

Dezincification is selective oxidation of zinc in brass plumbing fittings; it occurs because zinc is much more readily oxidized than is copper. (p. 582)

Diatomaceous earth is a mineral that is the skeletons of large numbers of "diatoms," small water-borne organisms; it is composed mostly of silicon dioxide. (p. 580)

Dry cleaning is cleaning of clothing in which the solvent is some material other than water. (p. 580)

Galvanized refers to a steel object coated with zinc to prevent its rusting. (p. 575)

Hemicellulose is a major component of wood, a polymer of the five-carbon sugar, xylose, $C_5H_{10}O_5$. (p. 573)

Histamine is a powerful vasodilator, formed by removal of CO_2 from the amino acid histidine. (p. 593)

Humectant is a substance added to commercial products such as foods or shampoo to cause retention of moisture. (p. 586)

Leavening agent is a substance that produces carbon dioxide bubbles to form in a baked food. (p. 583)

Lignin is a major component of wood, a polymer of several aromatic alcohols. (p. 573)

Lipases are enzymes that cause ester linkages in fats to hydrolyze. (p. 579)

Melanoidin is a substance that gives a dark color to hair. (p. 589)

Micelle is a cluster of surface active molecules in which the nonpolar ends gather together leaving the polar end in contact with water. (p. 577)

Mineral oil is a mixture of saturated hydrocarbons of approximately 20 carbons, obtained from petroleum. (p. 595)

Paclitaxel is a powerful anticancer drug obtained from yew trees. (p. 591)

Plaque is a thin layer of bacteria that grow in one's mouth and accumulate on teeth. (p. 586)

Polysiloxane is a polymer in which the backbone consists of alternating silicon and oxygen atoms. (p. 576)

Polyvinyl chloride is a polymer of $CH_2{=}CHCl$ that is widely used for many purposes, including water pipes. (p. 575)

Protease is an enzyme that degrades protein into its constituent amino acids. (p. 579)

Pumice is a very porous, low-density stone of volcanic origin. (p. 590)

Saran is a copolymer of $CH_2{=}CHCl$ and $CHCl{=}CHCl$; a copolymer means that the polymer is formed from a mixture of the two monomers. (p. 585)

Stainless steel is a steel alloy that does not rust under normal use, usually includes major amounts of chromium and nickel in addition to iron. (p. 584)

Surfactant is abbreviation for "surface active agent"; it is a substance that has a long nonpolar chain of CH_2 groups and a polar or ionic group at one end. (p. 577)

Tartar is the hard material deposited on the surface of teeth as plaque hardens by accreting calcium containing minerals. (p. 585)

Taxol is a copyrighted name for paclitaxel. (p. 592)

Terpene is any of several compounds having the formula $C_{10}H_{16}$, many of which are natural constituents of plants. (p. 588)

Thermoplastic is a classification of plastic that will soften on heating. (p. 584)

Thermosetting is a classification of plastic that will not soften on heating. (p. 584)

Vinyl literally means a $CH_2{=}CH{-}$ group of atoms; it usually refers to vinyl chloride or polyvinyl chloride. (p. 575)

exercises

1. Why is it environmentally OK to sell iodine as a topical disinfectant but not merthiolate or mercurochrome?
2. Why are gold, silver, and platinum favored in making jewelry instead of nickel, chromium, and tungsten?
3. How can it be that both emerald and aquamarine are chemically the same and even contain the same trace element (chromium) to supply the color?
4. A popular antacid contains 508 mg of $Al(OH)_3$ and 475 mg of $MgCO_3$ per two teaspoons (10 mL). What volume

of pH 2.00 HCl can be neutralized by two teaspoons of this product?

5. Estimate the amount of mercury added to the environment in the United States each year by the use of mercurochrome before it was taken out of the market. It was typically marketed as a solution of 1 gram of mercurochrome per liter of solution.

6. Calculate the percent weight loss when clay objects are fired.

7. What is the mass loss when a 240 gram coffee cup is fired during its manufacture?

8. Of the five antacids, $NaHCO_3$, $Mg(OH)_2$, $CaCO_3$, $Al(OH)_3$, and $MgCO_3$, which can neutralize the most HCl per gram of material?

9. Most bar soaps are made from 80% beef tallow and 20% coconut oil. The table shows the composition of both these sources of fatty acids. Refer to Table 16.3 for structures. Caprylic and capric acids have, respectively, 8 and 10 carbons and are saturated fatty acids. Calculate the average molar mass of soap.

	Caprylic	**Capric**	**Lauric**	**Myristic**
Coconut	8.0	7.0	48.0	17.5
Tallow	0.0	0.0	0.0	2.0

	Palmitic	**Stearic**	**Oleic**	**Linoleic**
Coconut	8.2	2.0	6.0	2.5
Tallow	32.5	14.5	48.3	2.7

10. How many cattle are needed to supply the annual needs of beef tallow in the United States for manufacture of bar soaps? One animal can supply approximately 170 pounds of tallow.

11. Compare the advantages of using bar soap made from tallow and coconut oil versus detergent bars that can be made from petroleum.

12. The unit of mass "carat" is used to describe diamonds and other gemstones. One carat = 200 mg. Calculate the number of carbon atoms in a 5.0 carat diamond. Then, look up the density of diamond and calculate the volume in cubic centimeters of such a diamond.

13. The unit of purity, "karat," is used to describe the purity of gold. What percent gold is 17 karat gold?

14. Aspirin is formed by the reaction of salicylic acid with acetic anhydride. Look up the structures of these two compounds and predict what other compound is formed in this reaction.

15. Aspirin tablets are measured in weight units called "grains," equal to 0.0648 grams. A typical aspirin is 5 grains. What fraction of a mole of aspirin is one tablet?

16. Look up and write down the original definition of the mass measurements "grain" and "carat."

17. Some people have difficulty using aspirin because the acidic tablet causes stomach irritation. A product satis-

factory for them is buffered aspirin. What do you suppose that means?

18. An aspirin tablet requires 36.3 mL of 0.0500 M NaOH to neutralize the acid functional group of the aspirin. How many grams of aspirin are in this tablet? How many grains is this?

19. Aspirin slowly reacts with water to give acetic acid and salicylic acid. Write a balanced equation for this reaction (called hydrolysis) and explain why very old aspirin sometimes smells like vinegar.

20. Suggest some acidic ingredients that might be used in conjunction with baking soda in a recipe for pancakes.

21. In making baking powder, what mass of cream of tartar (potassium hydrogen tartrate) is necessary to exactly react with 250 grams of sodium bicarbonate?

22. Sheetrock ($CaSO_4 \cdot 2H_2O$) is made from plaster sandwiched between two sheets of paperboard. Suggest why this is a very good material at preventing the spread of fire.

23. The monomer in hemicellulose is xylose and each molecule contains 5 carbons and 5 oxygen atoms. How many hydrogens must it contain to be considered a carbohydrate?

24. Archimedes is reputed to have figured out a way to determine whether his king's crown was made of pure gold. His method involved weighing the crown and weighing the amount of water displaced when the crown was submerged. Describe this process in modern terms.

25. Of the gems listed in Table 18.1, which are inorganic and which organic materials?

26. Look up and write down the composition of the semi-precious gems, garnet, cubic zirconia, topaz, and opal.

27. A measuring cup of enriched flour weighs 115 grams and contains 12 grams of protein. What percent protein is flour?

28. A single piece of whole wheat bread weighs 25.0 grams and contains 0.855 mg of iron. Recommended dietary intakes (called DRI for dietary reference intake) of iron for college age females (age 19–30) is 18 mg per day and for males is 8 mg per day. How many slices of bread would supply your daily needs for iron?

29. Suggest why terpenes are invariably volatile compounds.

30. A sample of saran is made from equal masses of vinyl chloride and vinylidene chloride. How much chlorine is in a 500 gram sample of this material?

31. Examine the ingredient list on a commercial cosmetic or cleaning product. Write a report describing each of the ingredients and tell why each is used in the product.

32. Explain why water pipes in new construction have gone from being lead, to steel, to copper, to PVC over the last couple of centuries.

33. Explain why many people think it will be an important change if dry cleaning using carbon dioxide rather than tetrachloroethene becomes widespread.

34. Figure 18.23 shows the structures of several esters that are part of the odor of various fruits. Determine what acid and what alcohol went to make up the compound found in bananas.

35. Find out the environmental consequences of putting incandescent lightbulbs and fluorescent light tubes in the trash.

36. A standard 48″ fluorescent lightbulb contains approximately 25 mg/bulb and approximately 600 million are disposed of each year. How many liters of mercury would that be?

37. A standard 48″ fluorescent lamp contains 25 mg of mercury but a low-mercury fluorescent lamp contains only 16 mg. The standard bulb requires 27 watts of electric power to operate for an hour and the low mercury lamp requires 34 watts per hour. Discuss the costs and benefits of using standard versus low mercury lamps.

38. Explain the difference between cement and concrete.

39. Why is it a simplification to say that wood is made of cellulose?

40. What are the bubbles filled with in a loaf of bread or a pancake?

41. Silver plate on flatware is specified by the mass of silver on a gross of teaspoons according to these criteria.

Specification	Mass of Silver per gross of teaspoons
Federal specific plate	9 troy oz
Quadruple plate	8 troy oz
Triple plate	6 troy oz
Double plate	4 troy oz
Par plate	2 troy oz

Calculate the mass in grams of silver per teaspoon for a triple plated spoon.

readings:

1. "Race to synthesize Taxol ends in a tie" *Science,* 263, 1994, p. 911.
2. ACS web site "What's That Stuff?" (http://pubs.acs.org/cen/whatstuff/stuff/html).
3. Cockburn "Diamonds—The Real Story" *National Geographic,* March 2002, p. 2.

websites

www.mhhe.com/kelter   The "World of Choices" website contains web-based activities and exercises linking to: information on "Green Building"; the diamond exhibit at the American Museum of Natural History in New York City; and much more!

Selected Answers

(Answers to selected odd-numbered questions)

Chapter 1

1. Enlarged belly, enlarged breasts, tiredness, strange eating patterns, morning sickness

3. 2.5×10^{12} J

5. 1.28×10^{16} J

7. 1.7×10^8 atoms

11.

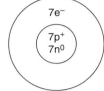

$7e^-$

$7p^+$
$7n^0$

13. Fission of either uranium-235 or plutonium into smaller atoms.

15. ^{99m}Tc in bone scans, ^{125}I in thyroid examinations, many others

17. There is enough hydrogen for several billion more years.

19. This can be made by fusion of $^{204}_{80}$Hg + $^{87}_{37}$Rb. This would require energy.

21. 30 neutrons

23. It requires seven He atoms. This would give off energy.

25. Mass numbers 10 and 11 with ^{11}B being the majority isotope

27. a. Zr
    b. Re
    c. S
    d. C
    e. V
    f. Xe

29. a. 89 protons & 138 neutrons
    b. 31 protons & 39 neutrons
    c. 19 protons & 21 neutrons
    d. 102 protons & 157 neutrons
    e. 93 protons & 146 neutrons
    f. 26 protons & 30 neutrons

31. solids: plants, fish, soil sediment
    liquids: water
    gases: dissolved oxygen, nitrogen, carbon dioxide, sometimes bubbles of methane

33. 9.4 g/mL

35. potassium is positive and bromine is negative

37. From top to bottom: ether, water, carbon tetrachloride, mercury

39. a. 11 protons, 12 neutrons, 10 electrons
    b. 35 protons, 46 neutrons, 36 electrons
    c. 22 protons, 26 neutrons, 20 electrons
    d. 38 protons, 52 neutrons, 36 electrons

41. a. 78 protons, 117 neutrons, 77 electrons
    b. 41 protons, 52 neutrons, 41 electrons
    c. 17 protons, 20 neutrons, 18 electrons
    d. 17 protons, 18 neutrons 10 electrons
    e. 51 protons, 71 neutrons, 49 electrons
    f. 26 protons, 30 neutrons, 24 electrons
    g. 74 protons, 108 neutrons, 74 electrons
    h. 37 protons, 48 neutrons, 36 electrons
    i. 14 protons, 15 neutrons, 18 electrons

43. a. $^{27}_{13}$Al^{3+} 14 neutrons +3 charge
    b. $^{88}_{38}$Sr^{2+} 38 protons, 50 neutrons, 36 electrons
    c. $^{65}_{30}$Zn$^+$ 29 electrons, +1 charge
    d. $^{37}_{19}$K   0 charge
    e. $^{127}_{51}$Sb^{3+} 51 protons, 48 electrons, +3 charge
    f. $^{133}_{55}$Cs$^+$ 55 protons, 78 neutrons, 54 electrons

45. 10 protons, 10 neutrons, 10 electrons

51. The iceberg floated because ice is less dense than water. The ship floated because it was full of air to the extent that its overall mass was less than the mass of water it displaced. When the ship filled with water it sank because the air was displaced by water and the greater density of steel caused it to sink.

Chapter 2

1. 6 carbon atoms, 12 hydrogen atoms, 18 oxygen atoms

3. Different. They are different compounds so are expected to have different properties.

5. Carbon has 6 protons and 6 electrons but nitrogen has 7 of each.

7. a. C
    b. Br
    c. Sn
    d. K
    e. H

9. The **Law of Conservation of Mass** states that matter is neither created nor destroyed during chemical processes. The **Law of Definite Proportions** recognizes that a pure compound always has the same fractions (proportions), by mass, of its constituent elements. The **Law of Multiple Proportions** recognizes that the same elements can combine to form different compounds when the constituents have different ratios of mass.

11. 266 g Cl and 2.52 g of hydrogen

13. H_2O mass of H = 5.58 g; mass of O = 44.4 g
    H_2O_2 mass of H = 2.95 g; mass of O = 47.0 g
    CO mass of C = 21.4 g; mass of O = 28.6 g
    CO_2 mass of C = 13.6 g; mass of O = 36.4 g

15. a. C_2H_4
    b. PF_5
    c. H_2SO_4
    d. Fe_2O_3

17. a. $C_5H_{10}O$
    b. B_2H_6
    c. $Si_3O_2H_8$

19. NO_2 and N_2O_4

21. Elements can be arranged according to columns and rows and so can decks of cards be arranged according to suit and value.

23. Atomic number is more fundamental than atomic mass. The latter depends on the distribution of isotopes of an element but the other is the integer number of protons in the nucleus and hence, for the neutral atom, the number of electrons surrounding the nucleus.

25. Vegetables: carrots, cucumber, potato, green bean
    Fruits: banana, orange, apple, peach
    Meat: chicken, fish, pork chop, beef steak
    Starchy: rice, bagel, pasta, bread
    To further distinguish among the categories one could ask about caloric content, color, size of the item, etc.

27. chlorine and hydrogen

29. a. 19
    b. 16
    c. 18
    d. F, Cl, Br, I
    e. isotopes

31. a. protons . . . electrons
    b. neutrons . . . protons
    c. 8
    d. hydrogen

33. a. boron
    b. oxygen
    c. 65.35 amu on average
    d. 2+
    e. Na
    f. nuclear fusion
    g. 12

35. element 113 will be in group IIIA, under Tl

37. Element number 43 is missing; it was later identified as technetium.

39. C

41. One possibility is: I KNoW He LiKEs US

43. Several possibilities among which are:

B	H		C	E
R	E		F	R

45. In and Ni; Br and Rb

47.

	n = 1	n = 2	n = 3	n = 4
a. K	2	8	8	1
b. Ar	2	8	8	0
c. As	2	8	18	5
d. Mn	2	8	13	2

49. a. IIIB
    b. VIIIB
    c. VIB

51. gold, silver, iron, nickel, copper, manganese, zinc, cobalt, tungsten, mercury, tin, lead, bismuth; These were all discovered and named before modern nomenclature had developed. Tantalum, aluminum, & platinum end in "um" because their pronunciation would be tricky with the extra syllable.

53. a. 0
    b. 2
    c. 6
    d. 7

55. RbBr and don't forget RbOBr (rubidium hypobromite)

57. 2 in the inner circle, 8 in the second, 18 in the third, and 5 in the fourth.

59. Chlorine is more reactive because it does not have a filled p subshell or octet.

61. Sodium: atomic number 11, atomic mass 22.99 amu, 11 protons, 12 neutrons, 11 electrons, 1 valence electron, two fully filled energy levels and one electron in the third level; electron configuration $1s^2\,2s^2\,2p^6\,3s^1$. Magnesium has one more electron than sodium. Potassium has its single electron in the 4s orbital instead of in the 3s.

63. a. Ni in group 8B
    b. Se in group 6A

65. a. Ne
    b. Ar
    c. He

67. a. transition metal
    b. halogen
    c. noble gas
    d. alkaline earth
    e. alkali metal
    f. rare earth (lanthanide)

69. a. transuranic element (actinide)
    b. halogen
    c. transition metal
    d. lanthanide
    e. noble gas
    f. alkali metal
    g. alkaline earth

71. It would probably be a solid much like iodine. No photo exists because it is so rare and radioactive that there has never been enough in one place to photograph.

73. iron—structural use in building things;
    copper—good electrical conductor
    chromium—plating steel to make it shiny and rust resistant
    zinc—coating steel to make it rust resistant

75. Sodium atom will be larger than sodium ion because to form an ion, the outer electron must be lost.

77. Sc^{3+} Ca^{2+} K^+ Cl^- S^{2-} P^{3-}

81. At that time, Bi and Te were known but At was not and element 117 was not so he could only use two neighboring elements rather than four to base his predictions.

83. Each halogen (group 7A) element has an odd number of valence electrons. Only even numbers of atoms will therefore supply an even number of valence electrons necessary to form all the valence electrons in pairs.

Chapter 3

1. Li must lose 1, P must gain 3, I must gain 1, O must gain 2, Mg must lose 2, B must lose 3, H must gain 1

3. Cs^+ is much larger than Na^+ so a larger number of Cl^- can touch it simultaneously.

5. All atoms are non-metals

7. a. ionic
   b. covalent
   c. covalent
   d. ionic

9. Cs (or Fr)

11. Anions are larger than the atom from which they are formed but cations are smaller.

13. a. Na^+
    b. $[:\overset{..}{\underset{..}{Br}}:]^-$
    c. $[:\overset{..}{\underset{..}{Cl}}:]^-$
    d. $[:\overset{..}{\underset{..}{O}} - H]^-$
    e. $[:C \equiv N:]^-$

15. a. $\left[:\overset{..}{O}\ \overset{..}{\underset{C}{O}}: \atop :\overset{..}{\underset{..}{O}}: \right]^{2-}$
    b. $\left[:\overset{..}{O}\ \overset{..}{\underset{N}{O}}: \atop :\overset{..}{\underset{..}{O}}: \right]^-$
    c. $\left[H \atop H-N-H \atop H \right]^+$
    d. $\left[:\overset{..}{O}: \atop :\overset{..}{O}-S-\overset{..}{O}: \atop :\overset{..}{O}: \right]^{2-}$

17. a. CH_3—O—CH_3 CH_3—CH_2—O-H
    b. CH_3—CH_2—CH_2F CH_3—CHF—CH_3
    c. CH_2Cl—CH_2Cl $CHCl_2$—CH_3
    d. CH_2=C=CH_2 $CH_3C \equiv CH$ HC=CH with $\underset{H_2}{C}$

19. a.
    $:\overset{..}{\underset{..}{Cl}}\ \underset{:\overset{..}{\underset{..}{Cl}}.\ .\overset{..}{\underset{..}{Cl}}:}{\overset{|}{\underset{|}{P}}}-\overset{..}{\underset{..}{Cl}}:$
    b.
    $\overset{:\overset{..}{F}:}{\underset{:\overset{..}{F}:}{\underset{:F.}{\overset{.F:}{S}}}}$
    c.
    $\overset{:\overset{..}{F}:}{\underset{:\overset{..}{F}.\ .\overset{..}{F}:}{\overset{:F.\ \ .F:}{Cl}}}$

21. one set of possibilities is: NCl_3 HI $\underset{Cl}{\overset{Br}{F-C-O-H}}$

23. CCl_4 x = 4
    $CHCl_3$ x = 3
    CH_2Cl_2 x = 2

25. 2

27. No, carbon can only form 4 bonds and chlorine can't form a C—Cl—Cl arrangement.

29. $:\overset{..}{F}\ \overset{..}{.F}: \atop \overset{B}{\underset{:\overset{..}{F}:}{|}}$ $\underset{H}{\overset{H}{\underset{..}{O}}}-\overset{..}{O}$ $[K]^+ [:\overset{..}{\underset{..}{Br}}:]^-$ $\underset{S}{\overset{H\ \ H}{..}}$

31. NaCl, because it is an ionic compound.

33. Na_2O, ionic

35. H—C≡N:

37. 0.267 g of As.

39. Closer to Cl because Cl is more electronegative than is H.

41. Many possibilities including $\overset{\delta+}{H}$—$\overset{\delta-}{Cl}$ and H—H or Cl—Cl

43. Certainly choose a metal; silver is the best electrical conductor but it is so expensive that copper is usually used instead.

45. a. $Ca_3(PO_4)_2$
    b. $KAl(SO_4)_2$
    c. $K_2Cr_2O_7$

47. a. CuCl
    b. SnO_2
    c. N_2O_4
    d. $(NH_4)_2SO_4$

49. K_2CO_3 $FeCl_3$ $NaNO_3$ Co_2O_3 CO_2

51. Eight neighboring cubes share each corner atom.

53. 0.741

55. In sharing a piece of cake with an older sibling, the older kid often gets the larger piece of cake in the unequal sharing. Another example is a tug of war between a very strong team and a much weaker team. The stronger team pulls the rope toward its side although both teams are pulling on the rope.

Chapter 4

5. Typically, water falls as rain, runs off into a stream, enters groundwater, is pumped from the ground (or from a lake or river), is treated at a central treatment plant, pumped into a distribution system to the tap where you draw your glass of water.

7. There are six carbons, twelve hydrogens, and six oxygens on each side of the equation.

9. a. not possible
   b. possible
   c. not possible

11. There are seven oxygens in both the reactants and the products.

13. 100 molecules of water are needed. 16.7 (or just 16) molecules of glucose are produced along with 100 molecules of oxygen. A better answer recognizes that only sixteen glucose molecules can be made and that this will require 96 molecules of CO_2 (4 will be left over) and 96 molecules of H_2O will be consumed and 96 molecules of O_2 will be formed.

15. The atomic mass of: Sc is 44.96 amu; Pd is 106.4 amu; O is 16.00 amu; Na is 22.99 amu; and Cl is 35.45 amu

17. 8.6×10^4 mole of Pd

19. 1.0 g of H_2O

21. 4.2×10^{-3} mole of Rb_2O_2

23. 0.77 mol of O_2

25. Assuming that pennies are 100% copper as was nearly the case before 1983: 9.5×10^{23} atoms of Cu
Assuming that pennies are 4% copper as is true now: 3.8×10^{22} atoms of Cu

27. 7.4×10^{22} atoms of Cl

29. 2.4×10^4 g or 24 kg of coal

31. 138 g of H_2O

33. 9.8×10^7 g = 9.8×10^4 kg = 98 metric tons of Al_2O_3

35. $2 H_2 + O_2 \rightarrow 2 H_2O$ 2 mol H_2/mol O_2 2 mol H_2/2 mol H_2O 1 mol O_2/2 mol H_2O

37. 0.044 mol of sucrose

39. 284 aluminum cans (assuming that each can contains 16 g of Al as per problem 29.)

41. 7.3×10^9 mol of CO_2

43. 1.2×10^5 g of sulfuric acid

45. 1.2×10^6 g $TiCl_4$

47. 3.3×10^{23} nm = 3.3×10^{11} km

49. 1.5×10^{23} Pb atoms

51. 9.2 g O_2

53. 1.0×10^{14} molecules

55. You were given 2.36 mol of Cu, which is enough to recover 4.72 mol (or 5.1×10^2 g) of Ag. To recycle 6 mol of Ag, you would need 3 mol of Cu so you don't have enough.

57. 20 g Ca

59. a. $Ca + 2 H_2O \rightarrow Ca(OH)_2$ (aq) + H_2 (g)
b. $Zn + 2 HCl \rightarrow ZnCl_2$(aq) + H_2 (g)
c. $Cl_2 + 2 NaI \rightarrow 2 NaCl$(aq) + I_2(aq)

61. 170 g of Ag

63. Sn^{2+}

67. 695 tons/train

Chapter 5

3. Raised position

5. $32^{\circ}C = 305K$; $100^{\circ}C = 373K$; $0^{\circ}C = 273K$; $150^{\circ}C = 423K$

7. All objects emit radiant energy depending on their temperature. When they are hot, objects emit energy of much shorter wavelength and greater intensity than they emit at room temperature.

9. $2300^{\circ}F = 1260^{\circ}C = 1533K$

11. 1358 kJ is required to break the bonds; 1840 kJ is released during formation of the bonds in 2 mol of H_2O. Therefore, 482 kJ would be given off in formation of 2 mol of H_2O, so 964 kJ of energy would be released in the formation of 4 mol of H_2O.

13. a. endothermic
b. exothermic
c. endothermic
d. exothermic

15. 4.68×10^{-18} kJ. The sun.

17. Heat is clearly released which can be used for many purposes.

19. 1241 kJ is released per mole of acetylene burned (assuming that the combustion products are CO_2 and H_2O, and not CO and H_2O.)

21. $\Delta H = -55$ kJ for two moles or -27 kJ per mole of NBr_3 reacted.

23. $\Delta H = 82$ kJ/mole so the reaction is endothermic.

25. This reaction is endothermic. It only occurs upon heating to a high temperature. Additionally, the reverse reaction occurs readily even at room temperature.

27. a. $2 KClO_3(s) \rightarrow 2 KCl(s) + 3 O_2$ (g)
ΔS is positive because of the release of gaseous oxygen
b. N_2 (g) + 3 H_2 (g) $\rightarrow 2 NH_3$ (g)
ΔS is negative because 4 moles of gas form 2 moles of gas

29. The Cl in ammonium perchlorate is in a very high oxidation state and aluminum is in its lowest oxidation state. The reaction of these reactants is a redox reaction in which aluminum is oxidized and chlorine is reduced in a very exothermic reaction.

31. The second law of thermodynamics states that a process occurs spontaneously in the direction that increases the entropy of the universe.

33. For anthracite: $\Delta H = 485$ kJ/mole CO_2
For lignite: $\Delta H = 453$ kJ/mole CO_2
These energies are both about the same as CH_3COOH so the oxidation number of carbon in coal is about zero. The carbon in anthracite is slightly lower in oxidation number than that in lignite.

35. The lower graph. It has a lower activation energy.

37. Generally, a reaction with a low activation energy will occur faster than one with a high activation energy.

39. a. boiling 10 grams of water
b. warming 200 g from 5° to 35°
c. boiling 25 grams of water
d. equal

41. 1.2×10^{22} molecules I_2

Chapter 6

1. Organic chemistry is the chemistry of compoundsconsisting mainly of carbon.

3. Starch in pasta has (α) alpha linkages. This is important because if they had beta links the starch would be indigestible.

5. The first is a stable molecule but the others are not because they show five bonds to two carbon atoms.

7. The following diagrams omit all the hydrogen atoms to make the images clearer.

9. CH_4 and C_4H_{10} are saturated; C_4H_8 and C_4H_6 are unsaturated

11.

13. saturated C_nH_{2n+2} one double bond C_nH_{2n}

one triple bond C_nH_{2n-2} one ring C_nH_{2n}

two double bonds C_nH_{2n-2}

15. a. double bond
    b. alcohol
    c. carboxylic acid

17.

$$\begin{array}{ccc} H & O & \\ | & \| & \\ H-C-C & -O-H \\ | & & \\ H & & \end{array}$$

19. $H-O-CH_2CH_2CH_2-O-H$ $H-O-CH_2CH_2-O-CH_3$

$$H-O-CH_2-\underset{\underset{H}{\overset{|}{O}}}{\overset{|}{CH}}-CH_3$$

21. $CH_3-C\equiv CH$ or $CH_2=C=CH_2$

23. C, N, and O all have even atomic weights (12, 14, and 16 respectively) so the only way a compound can have an odd molecular weight is to have an odd number of hydrogens. Compounds having only C, O, and H always have an even number of hydrogens because they have an even number of atoms bonded to themselves. Nitrogen, on the other hand, typically has only three bonds so a compound with one nitrogen (as well as C and O) will always have an odd number of hydrogens.

25. From left to right in each compound:

     CH_3COOH C is tetrahedral; C is trigonal planar; O is bent at tetrahedral angle

     C_2H_3Cl both carbons are trigonal planar

     C_4H_{10} all four carbons are tetrahedral

     C_3H_4 C is tetrahedral; C is linear; C is linear.

27. CH_3-CH_2-OH CH_3-O-CH_3

29. Water is very polar and forms hydrogen bonds extensively. Gasoline is a mixture of hydrocarbons that are very nearly nonpolar and do not participate in hydrogen bonding. Therefore, the water molecules attract each other and have very little attraction for the hydrocarbon molecules.

31. Steroids are a class of compounds having three six-membered rings and one five-membered ring in a characteristic pattern. Typically they contain one or two oxygen atoms and a small number of double bonds. Cholesterol fits that description perfectly.

33. a.

$$H_3C-\overset{\overset{O}{\|}}{C}-O-CH_2-CH_3$$

b.

$$C_6H_5-\overset{\overset{O}{\|}}{C}-O-C_4H_9$$

35. Aspirin can react to form salicylic acid and acetic acid. The acetic acid is the active ingredient of vinegar.

37. C_{60} is very nonpolar. If a drug could be coupled to it or located inside it, the entire molecule could bind to the nonpolar lipids of cell membranes.

39.

$$H_3C-CH_2-\overset{\overset{O}{\|}}{C}-O-CH_3 \ + H_2O; \text{ ester}$$

41. 1265 g or 1.27 kg

43. These words describe the position of atoms bonded to double bonded carbons. Trans constituents are on opposite sides of the double bond and cis constituents are on the same sides of the double bond.

45. a. an ester and water
    b. 4; 1; 2; 3.
    c. carboxylic acids and amines

47. 713 molecules

49. The empirical formula is CH_3. This is an impossible molecular formula so the real molecular formula must be C_2H_6. Note that C_3H_9, C_4H_{12} and all higher multiples of CH_3 are also impossible.

51. The formula is C_3H_8O and the structure must be $CH_3-O-CH_2-CH_3$

53. Increasing number of oxygen atoms in an organic molecule causes the average oxidation number of the carbon atoms to increase. Increasing the number of hydrogen atoms in an organic molecule causes the average oxidation number of the carbon atoms to decrease.

Chapter 7

1. Water molecules are bent and carbon dioxide molecules are linear.

3. The fully hydrogen bonded structure of ice has more vacant space than does liquid water.

5. 1489 torr, nearly two atmospheres

7. 20920 J or 20.9 kJ

9. -214 kJ/g

11. 52.0 kJ

13. Since the liquid water has a higher vapor pressure, it would tend to evaporate (endothermically) and condense out on the solid ice (exothermically). The temperature would remain at $-10°C$ because the heat gain would equal the heat loss.

15. Methanol is more water soluble because decanol has a long chain of CH_2 groups which do not interact with water.

17. 8×10^{-3} M

19. ethylene glycol and ammonia

21. 0.341 L or 341 mL water; 150 g caffeine

23. The oil is suspended in micelles of the soap so will disperse in water and come clean.

25. The sodium is at a much higher concentration.

27. Chlorine reacts to form chlorinated compounds which are undesirable in water supply. Ozone forms no such compounds.

29. $Ca(HCO_3)_2 \text{ (aq)} \rightarrow CaCO_3(s) + CO_2 \text{ (g)} + H_2O(l)$

31. 0.020 moles

33. 101.4°C

35. i. b
     ii. b
     iii. b

37. 5.4 molal

39. fraction of molecules = 0.164

41. %glycerine = 84%, freezing point ≈ −10°C.

43. 2.92 molal

45. 6674 g or 6.67 kg

47   100.075 °C; 100.75 °C

49   1.2×10^8 g O_2, area = 6 km^2

51   2.54×10^6 grams or 2.54 metric tons

53. 3.1×10^7 g or 31 metric tons

55. 935 µg/L

57. 5.5×10^{19} mol, this is 1×10^{21} g or 1×10^{18} L; 7.2×10^{-4}

59. 2323 g CO_2

Chapter 8

1. An acid donates hydrogen ions and a base accepts hydrogen ions.
    An acid tastes sour and a base tastes bitter
    An acid turns litmus red and a base turns litmus blue
    An acid vigorously reacts with many metals to produce hydrogen gas. A base forms insoluble hydroxides with many metal ions

3. H^+; H_3O^+

5. A weak acid is one that dissociates into hydrogen ions and the corresponding negative ions only to a small extent.
    A strong acid is one that dissociates into hydrogen ions and the corresponding negative ions completely.
    A concentrated acid is one that has a large number of moles of the acid per liter of solution.
    A dilute acid is one that has only a small number of moles of acid per liter of solution.
    Examples: Weak acid—acetic acid CH_3COOH
                Strong acid—hydrochloric acid HCl
                Concentrated acid—sulfuric acid in drain cleaner
                Dilute acid—acetic acid in vinegar

7. Hydronium ions and bromide ions.

9. a. $H_2SO_4 \rightarrow 2\,H^+ + SO_4^{2-}$
    b. $C_4H_5O_6H \rightarrow H^+ + C_4H_5O_6^-$
    c. $H_2CO_3 \rightarrow H^+ + HCO_3^-$
        $HCO_3^- \rightarrow H^+ + CO_3^{2-}$
    d. $C_7H_5O_3H \rightarrow H^+ + C_7H_5O_3^-$

11. 3×10^{24} ions

13. pH = −0.70

15. Often one finds phosphoric acid, citric acid, and ascorbic acid.

17. 0.977 M

19. 5.69 M

21. donor . . . acceptor

23. HF and F^- are acid and conjugate base respectively

     HCO_3^- and H_2CO_3 are base and conjugate acid respectively

25. acids are CH_3COOH, HI, CO_2
     bases are NH_3, $Ca(OH)_2$, and N_2H_4

27. SO_3, HNO_3, CO_2

29. CO_2 reacts with water to give H_2CO_3 that dissociates into H^+ and HCO_3^-.

31. Closer to 1 because it is acidic.

33. a. 3
     b. 12
     c. 5
     d. 10
     e. 7

35. pH = 1.00

37. a. pOH = 12
     b. pH = 3
     c. pH = 8
     d. pOH = 9.75
     e. pH = 5.02
     f. pOH = 10.33

39. A difference of 2.0 pH units means that the water is 100 times more acidic. Yes, this added acidity could contribute to distruction of a large area of forest downwind from peatlands.

41. This solution is highly basic.

43. a. pOH = 7.0
     b. pOH = 10.0
     c. pOH = 3.0
     d. pOH = 12.0

45. 16.7 g $Ca(OH)_2$

47. HCl + $H_2O \rightarrow H_3O^+ + Cl^-$
     water accepting H^+ so acting as a base

     $NH_3 + H_2O \rightarrow OH^- + NH_4^+$
     water donating H^+ so acting as an acid

49. pH = 4.38

51. Yes, these answers are consistent because
     $(4.2 \times 10^{-5})(2.4 \times 10^{-10}) = 1.0 \times 10^{-14}$

53. They change colors depending on the solution pH.

55. a. $HNO_2 + OH^-$
     b. HPO_4^{2-} and H_3O^+ (but also H_3PO_4 and OH^-)
     c. $Mg(OH)_2$
     d. $CN^- + H_3O^+$

57. 8.1×10^{-4} M CH_3COOH

59. An indicator is a compound the color of which depends on the pH of the solution in which it is dissolved. Each indicator has a characteristic pH range at which it changes color.

61. Use a piece of litmus paper or pH indicating paper. An acid will turn blue litmus to red and a base will change red litmus to blue. A neutral solution will have no effect on either red or blue litmus.

63. Add enough acid to neutralize it before disposing of it. Monitor the approximate pH with either litmus paper or pH indicating paper.

65. $H_3PO_4 + 3\,NaOH \rightarrow 3\,H_2O + 3\,Na^+ + PO_4^{3-}$

67. Concentrated acids are damaging to skin and flesh because they cause protein to depolymerize. Sulfuric acid also reacts strongly with water to remove water from flesh with great release of heat so actual burns can result.

Chapter 9

1. Burning coal releases SO_2, roasting sulfide ores releases SO_2, automobiles form NO. NO_2 and SO_2 react with water to form acid rain.

3. Acid rain is precipitation having a lower pH than would be the case if the only dissolved component in the rain was atmospheric carbon dioxide. Acid rain damages vegetation, damages limestone and marble structures, and dissolves heavy metals.

5. SO_2 can react with water to make sulfurous acid or with oxygen and then water to make sulfuric acid. Sulfurous and sulfuric acids are both stronger acids than H_2CO_3 which is formed when CO_2 reacts with water.

7. Acid can enhance the dissolution of aluminum from the soil and allow it to be transported into the roots where it precipitates as $Al(OH)_3$. Acidic water can be taken directly into the plant through the roots or stomata where damage can result.

9. True and false—it is false because acid-rain has been known for a long time but it is true because much of the added acidity is caused by industrial practices.

11. The initial damage is probably to the smallest organisms, algae being most important. The algae form the basis of the food for higher organisms.

13. No, insects can also cause damage and temperature or rainfall changes can damage trees. Chemical damage can weaken a tree so that it becomes more vulnerable to other stresses. Roads through forests can cause erosion which leads to damage to trees.

15. Sulfuric acid in rainfall damages anything made of calcium carbonate (limestone and marble). The reaction is $CaCO_3 + H_2SO_4 \rightarrow CO_2 + H_2O + CaSO_4$. Calcium sulfate, which is more soluble than calcium carbonate, forms as a principite on the statue or building surface and can be dissolved when exposed to rains or other precipitation. Calcium sulfate is much more water-soluble than is calcium carbonate.

17. White house paint used in the first half of the 20th century often included lead carbonate pigment. Also, lead compounds were used in most gasoline for many years and were deposited along roads. Nitric acid can dissolve lead compounds and the dissolved lead can end up harming animals and even humans that drink the water.

19. 5.0 mol or 200 g of NaOH

21. 0.037 moles = 2.3 g HNO_3

23. pH = 3.28

25. The NO would supply more hydrogen ions than the SO_2.

27. Calcium oxide can react with SO_3 to form $CaSO_4$ to remove the sulfur oxides from the flue before they enter the environment.

29. Calcium sulfate is insoluble and very stable even at high temperatures. Calcium nitrate is very water soluble and melts or decomposes at lower temperatures.

31. The tops of the mountains are more often in clouds than are the bottoms of the mountains. If the clouds are acidic prior to acid rain, the acidic moisture in the clouds would harm the trees.

33. 1.2×10^8 L

35. 6×10^7 grams NO, which is 60 metric tons

37. Assuming that the rain is only slightly acidic, we would want an indicator with a color change in the range of 4–6 so bromocresol green or methyl red would be useful.

Chapter 10

1. Point Sources—sewers which empty into waterways without treatment
   Diffuse Sources—fertilizer & herbicide runoff from agricultural fields or lawns during rainstorms
   Indirect Pathways—rusting of barrels of buried waste materials
   Atmospheric Sources—soot and acidic rain

3. Lack of environmental laws and practices in the former eastern block countries.
   Differing laws among the many countries, although the European Community is regularizing laws throughout the region.

5. Point sources are easier to control.

7. A great many industrial production firms established plants near the great lakes (especially Lakes Michigan, Erie, and Ontario) to allow for easy transport of raw materials and finished products. Much of the waste from these operations was dumped right into the lake and the lake water became highly polluted by the 1960s. Laws and public opinion changed about at that time and corporations responded by greatly improving their waste handling facilities so that the lakes are now much, much cleaner than they were 50 years ago.

11. 40,000 mg. Yes (0.09 lbs).

13. DDT has been shown to damage the eggs of falcons and related meat-eating birds.

15. You want it to last for the duration of the threat from its intended target, but not to remain in the environment longer than necessary.

17. A selective herbicide destroys only a selected group of plants; a nonselective herbicide kills virtually all plants.

19. All nitrates and many phosphates are quite water-soluble. These compounds can dissolve in rainwater, which ultimately becomes groundwater.

21. "On-demand generation" implies that a substance is prepared only when it is needed so it need not be stored or transported. Some industries simply do not have the facilities to prepare all the substances that they need. They count on specialized producers to make these chemicals and ship them when needed.

23. Answers will vary. Some good choices would be: asphalt, gasoline, polyethylene, kerosene, and nylon.

25. Crude oil can be separated into fractions by distillation, which is successful because different compounds in crude oil have different boiling points.

27. Oil is insoluble in water & oil is less dense than water.

Chapter 11

1. Exothermic; energy is released in the form of heat, sound, and light.

3. d = 1.04 g/L

5. A standard atmosphere is a pressure of 760 torr. An atmosphere is whatever the atmospheric pressure is at a given place and time. Atmospheric pressure changes with altitude and weather but a standard atmosphere stays the same.

7. Actual pressure = 25.2 psi

9. The balloon floats because helium is less dense than air. It later sinks because helium can diffuse out through the rubber membrane and decrease the volume such that the mass of the balloon can no longer be lifted by the remaining helium.

11. The most dense gas is the one with the highest molecular weight, SO_3.

13. Approximately 15°C

15. 3.9 moles N_2

17. P_2 = 5.3 atm

19. V_2 = 3.4 L

21. P_2 = 50 psi or 35 psig. If you adjust the pressure in your tires while they are hot, when they cool down the pressure will be too low.

23. V_2 = 7.5 L

25. The pressure is greatest because the volume is smallest when you are hitting the ground.

27. You must put in more air when the temperature is low.

29. Aspen is at a higher altitude than most of the country so the atmospheric pressure is lower. Therefore, to maintain the same gage pressure, one must remove some air from tires when going from low altitude to high altitude. In winter, when the temperatures are lower, one must put in more air to maintain the same pressure.

31. density = 0.286 g/L. This is less than 1.18 g/L for air at this temperature so the balloon will easily rise.

33. Propane has a higher molecular weight than air so if a cylinder of propane leaks, the gas will tend to accumulate in the bottom of the boat where it could become dangerous. Methane has a lower molecular weight than air so if it leaks, it will tend to rise out of the boat and create no danger.

35. The detonator cap is to initiate the decomposition of azide into molecular nitrogen.

37. Water will evaporate from hot pudding and will condense into water droplets on the plastic wrap in the refrigerator. If the pudding is left uncovered, the skin that forms is the pudding that has dried out from the loss of water. Evaporation requires energy so is endothermic.

39. 2.4 mm

41. PV = nRT. P = gas pressure; V = volume of gas; n = number of moles of gas; R = ideal gas constant; T = Kelvin temperature of gas.

43. M = 44.0 g/mole

45. The pressure would be 230/130 = 1.77 times as great and, since air bags don't expand beyond the expected size, the bag would either burst from the pressure or would be so hard that it would harm anyone who was unlucky enough to be in an accident with it.

47. M = 50.2 g/mole

49. 44%

51. A lake at 1000 ft would have a higher concentration of oxygen than at 7000 ft because the partial pressure of oxygen at the lower altitude would be greater.

Chapter 12

1. A thermal inversion occurs when cool air is trapped beneath a layer of warm air.

3. Rocks older than that do not show oxidized forms of metals, especially iron in the form of Fe_2O_3, which are present in rocks younger than 3.5 billion years.

5. If the ozone layer is significantly depleted, more ultraviolet radiation will reach the surface of the earth and damage living organisms of all kinds. Such a depletion will have no significant effect on the amount of O_2 in our atmosphere. If the ozone layer were completely removed, life as we know it could not survive on the surface of the earth.

7. Many responses are possible. Some possible answers are: operating gas powered motorized vehicles, use of CFCs, burning coal for heat, and incineration of trash.

9. nitrogen (N_2) and oxygen (O_2)

11. $3 O_2 \rightarrow 2 O_3$. We don't run out of ozone because it is continually being formed high in the atmosphere.

13. No, volcanoes supply sulfur oxides and particles, oak and pine trees supply hydrocarbon compounds, termites, swamps, and cows generate methane, etc.

15. Oxygen

17. CO_2 is a greenhouse gas because it does not absorb visible radiation but does absorb infrared.

19. forest fires, the ocean, windstorms, volcanoes

21. Droplets of water are sprayed into the air in rough water. If the water evaporates before it reaches the surface of the water, the salt that was in the droplet becomes a small particle that can blow with the wind.

23. Most ozone is found in the stratosphere and small but damaging amounts are found near the surface of the earth in the lower troposphere.

25. There is no chemical difference between the ozone found in the stratosphere and that found in the troposphere. Stratospheric ozone is desirable in that it absorbs much of the ultraviolet radiation from the sun and is formed in the stratosphere under the influence of ultraviolet radiation. Surface ozone is undesirable and is formed as a result of electric sparks and as a result of reaction of NO with O_2.

27. When struck by a photon of ultraviolet of sufficient energy, a chlorine atom from a CFC can break away from the rest of the molecule and serve as a catalyst in the decomposition of stratospheric ozone.

29. We need to be concerned about surface ozone because of the damage to living organisms and items of commerce caused by the extreme oxidizing power of ozone.

31. A health alert is just what it sounds like; a notice that people may be harmed by being outside breathing air containing relatively high concentrations of ozone.

33. A greenhouse gas is one that absorbs infrared radiation but does not absorb visible or ultraviolet radiation. H_2O, CO_2, and CH_4 are good examples.

35. One must consider both the costs and benefits of CFC uses. CFC compounds are extremely useful, and affordable replacements are not fully developed for all uses.

37. Reducing smog contains lots of soot and oxides of sulfur; photochemical smog contains lots of nitrogen oxides and ozone.

39. Photochemical smog requires sunlight which is more intense in summer.

41. Nitrogen oxides are converted to nitrogen, carbon monoxide is converted to carbon dioxide, and unburned hydrocarbons converted to carbon dioxide and water. This is accomplished at fairly high temperature at a catalytic surface.

43. Natural processes, such as volcanoes; high temperature combustion, such as burning of coal; reaction of gaseous compounds to form a solid, such as the formation of droplets of sulfuric acid from sulfur trioxide and water, which deposits on soot.

45. They are reactive because they are so small and have large surface areas.

47. 0.5 ppm = (0.5 mol O_3 /10^6 mol air)(6.02 $\times$ 10^{23} molecules/mole)(1 mol air/24.4 L)(1 L/1000 cm^3) = 1.2 $\times$ 10^{13} molecules per cm^3

49. 6 Fe^{2+} + O_3 + 6 H^+ → 6 Fe^{3+} + 3 H_2O

51. The carbon-chlorine bond is not as strong as the carbon-fluorine bond so is more likely to break to form a free atom.

53. 39.894 g/mole

55. Something must combine permanently with the chlorine atom. A good possibility is water that will react to form HCl which is very soluble in water so will come down in rainfall.

57. N_2 and H_2 are both nonpolar molecules, but NH_3 is polar and a good hydrogen bond former, so NH_3 will dissolve well in water.

Chapter 13

1. The earth neither receives nor loses significant amounts of matter from outside the earth. In an open system we would receive or lose large amounts of matter from (or to) outside the earth.

3. a. yes
   b. no
   c. yes

5. A resource is material of a certain type that exists on earth; a reserve is that part of the resource that can be recovered economically.

7. density = 19.3 g/cm³, melting point = 1063°C, boiling point = 2600°C, atomic weight = 197.0 amu or g/mole, and many others.

9. density = 2.702 g/cm³, melting point = 659.7°C, boiling point = 2057°C, atomic weight = 26.98 amu or g/mole

11. Suppose the can has a mass of 14.5 grams. #electrons = 9.70 $\times$ 10^{23} electrons

13. Cheap gold contains a large fraction of copper which reacts with air and salty moisture to form green oxides of copper.

15. Gold is measured in troy ounces and pounds, but feathers are measured in avoirdupois units. A troy ounce = 31.103 grams, so an ounce of gold is heavier than an ounce of feathers which = 28.349 grams. A troy pound = 373.236 grams, so a pound of gold is less massive than a pound of feathers, which has a mass of = 453.584 grams.

17. 2.0 $\times$ 10^{16} moles of Au would occupy 2.1 $\times$ 10^{17} cm^3. This answer is suspicious because the fraction of gold quoted is for the Earth's crust, not the interior about which nobody knows details.

19. $372

21. As of September 22, 2001, gold price = $291.10/oz and silver price = $4.592/oz. This is a ratio of 63.4

23. Error in cost = (1000.0 oz)(0.0005)($291.10/oz) = $145

25. 3.10 mm

27. Malachite is bright blue in color so would attract anyone who might see it. People might have picked it up and it accidentally fell into a fire or it was used to decorate a piece of pottery before the pottery was fired. Upon firing the previously blue stone would have turned to a brown metal.

29. See section 7.3, water hardness.

31. Impure copper is oxidized to copper ions and the copper ions are plated out onto pure copper metal. Metals less easily oxidized fall to the bottom of the tank as a valuable source of silver and gold. Metals less easily reduced than copper dissolve and remain dissolved in solution.

33. 2098 g

35. The best way to reduce aluminum oxide to aluminum metal is through electricity. A very high voltage is needed and 3 electrons per atom of aluminum, so the process is very costly. Recycling aluminum saves having to prepare more metal from the ore.

37. 1.6 $\times$ 10^5 grams C

39. Gold has oxides of Au_2O and Au_2O_3 (oxidation states +1 and +3). Iron forms FeO, Fe_2O_3, and Fe_3O_4 (oxidation states +2, +3, and +8/3). Copper forms Cu_2O and CuO (oxidation states +1 and +2). Aluminum has only the oxide Al_2O_3 (oxidation state +3). Aluminum's behavior makes sense based on its position in Group 3A. Gold, iron, and copper show multiple oxidation states, which is characteristic of transition elements.

41. A conductor conducts electricity with low resistance, but a superconductor conducts electricity with almost zero resistance.

43. Valence electrons in organic compounds are usually confined to covalent bonds connecting just two atoms. In a metal, the valence electrons are shared among all the atoms in the metal sample.

45. Bronze alloys have more tin than zinc, while brass alloys have more zinc than tin.

47. Most of the organic oxygen became carbon dioxide or water in the process of decay.

49. a. gas
    b. liquid
    c. solid

51. Petroleum is a very complex mixture of hydrocarbons. To be useful, the mixture must be divided into purified fractions. Distillation is a convenient way to separate compounds of different boiling points.

53. Sand is nearly pure SiO_2 and exists in fairly large pieces. Clay is also very small particles of aluminosilicates. Silt is typically deposited by settling out of water and is extremely small particles of broken up rock. Humus is organic material formed by the partial decay of plant material. The three inorganic components of soil are characterized by their particle size; sand (0.06–2 mm), silt (0.006–0.06 mm), and clay (less than 0.002 mm).

55. 3.8%

Chapter 14

1. Radioactive events occur when an atomic nucleus undergoes a change. Naturally unstable nuclei can decay into more stable nuclei by ejecting particles of matter and excess energy.

3. alpha particles, beta particles, and positrons. Positrons are most harmful because they give rise to highly penetrating gamma radiation.

5. Stable ^{31}P ^{35}Cl ^{37}Cl ^{14}N
   Unstable ^{28}P ^{37}P ^{16}N

7. An alpha particle has a mass of 4 amu and a charge of $+2$ whereas a beta particle has a mass of nearly zero (~0.00055 amu) and a charge of -1. An alpha particle is a helium nucleus and a beta particle is a very energetic electron.

9. ^{204}Pb will be formed. This is a stable isotope so it will not decay by emitting another alpha particle.

11. $^{137}Cs \rightarrow {}^{137}Ba + \beta^-$

13. a. β
    b. ^{11}B
    c. ^{218}Po
    d. $3\,^1n$
    e. $2\,\gamma$

15. ^{240}U; γ; ^{142}Xe

17. An alpha particle has a mass of 4 amu, a beta particle has mass of 0.00055 amu and gamma radiation has zero mass.

19. Thorium-234

21. False

23. 2.5 grams

25. 30 minutes

27. 1/16 or 6.25 %

29. 9 hours = 540 minutes, which is 20 half-lives. The original amount of lead would be 1.05×10^6 greater than the analyzed amount, or 5.4 grams of lead. This is an impossibly large amount to have collected from radon emission. Something has gone wrong.

31. 11460 years

35. subcritical, critical, and supercritical For a weapon to explode, it must have a supercritical mass of fissionable material.

37. If ^{238}U absorbs a neutron, it becomes ^{239}U. The ^{239}U emits a beta particle to become ^{239}Np. Finally, ^{239}Np emits a beta particle to become ^{239}Pu.

39. Waste heat is usually dispersed in a river. This means that the temperature of the river goes up. This causes the amount of dissolved oxygen to decrease such that fish may not be able to live in the water.

41. In a breeder reactor, surplus neutrons are used to convert ^{238}U into ^{239}Pu which is a useful fuel for future use in nuclear power generation. ^{238}U (99% abundance) is not fissionable so is, itself, not a useful fuel. Liquid sodium, rather than control rods, is used to keep the reaction under control.

43. Control rods absorb surplus neutrons to keep the reaction under control.

45. The sun's energy comes from the fusion of hydrogen into helium with the release of vast amounts of heat and light.

47. A hydrogen bomb is based on fusion of hydrogen atoms into helium. An atom bomb is based on fission of uranium or plutonium into middle-sized atoms. In order to cause a hydrogen bomb to explode, the hydrogen must be heated to tremendous temperatures. This is carried out by triggering the hydrogen bomb with an atom bomb.

49. The key to controlled fusion is to raise the temperature of a sample of hydrogen to millions of degrees while confining it to a small location.

51. The ^{14}C measurement is meaningless because the bottle can only be about 60 years old and this is too short a time to detect a significant change in ^{14}C levels. The tritium measurement, however, shows that the hydrogen in the alcohol in the gin is only a couple of years old. The bottle is not authentic.

Chapter 15

1. Windmills—kinetic energy of wind turns a turbine to make electricity
   Solar panels—generate electricity directly
   Solar reflectors to generate steam to turn a turbine
   Conversion of biomass into alcohol—burn this alcohol as automobile fuel
   Tidal power—as water moves up and down due to tides, can turn a turbine

3. wavelength, frequency, amplitude

5. $6.0 \times 10^{14}\ sec^{-1}$. Yes, this is in the green region of the spectrum

7. Using the wavelength of 500 nm, E = 3.98×10^{-19} Joules per photon or 239 kJ/mole of photons.

9. 29.6 nm

11. Show your friend that the energy of a UV photon is enough to break chemical bonds. Sunscreen absorbs the UV before it reaches the skin.

13. Chlorophyll absorbs visible light. The energy of this light is used to cause endothermic chemical reactions that result in the synthesis of biomass.

15. Normal constituents of air do not absorb in the visible region or the ultraviolet region but CO_2 and H_2O absorb significantly in the infrared region. High concentrations of CO_2 in the atmosphere absorb energy from both the sun and reflected energy coming from the earth. Larger amounts of CO_2 lead to greater amounts of radiant energy being absorbed and therefore the temperature of the atmosphere is expected to increase.

17. Beta-carotene contains a long series of alternating double and single carbon—carbon bonds causing visible light to be absorbed.

19. starch

21. a. oxidation
    b. oxidation
    c. reduction
    d. reduction

23. Electromotive force is what causes electrons to move from one place to another and measures the tendency for a redox reaction to occur. It is measured in volts.

25. The hydrogen gas and oxygen gas would collect in a single container and the H^+ and OH^- would neutralize each other to make water. $2\,H_2O \rightarrow 2\,H_2 + O_2$

27. Aluminum foil is oxidized and forms the water soluble Al^{3+} while the orange-brown copper metal comes out of solution. Nothing would happen if a copper strip were placed in a solution containing aluminum ions.

29. First is reduction, second is oxidation, third is neither (it is metathesis), fourth is oxidation; Reactions that are neither oxidation nor reduction include acid-base reaction and precipitation reaction; $4\,Fe + 3O_2 + 2H_2O \rightarrow 4\,FeO(OH)$.

31. The salt bridge allows ions to move from one electrode area to the other without allowing the two solutions to mix.

33. When you meet your friends, instead of saying "hi," say "OA," for <u>o</u>xidation <u>a</u>node.

35. The paste-like material is called an electrolyte. It allows charge to move from one electrode to the other.

37. All of its valence electrons are tied tightly to specific pairs of atoms so they can only move with very great difficulty.

39. An intrinsic semiconductor is a material like silicon or germanium which has a small enough band gap that electrons can flow although with high resistance.

41. GaAs can absorb light with wavelengths below 850 nm; this includes the entire visible region.

43. The two single biggest problems are that solar energy is only available when the sun is shining so is unavailable at night and that solar energy is difficult to concentrate at one place. Rather, it is spread evenly over the sunny side of the earth. Fossil fuel is more convenient because coal, oil, and gas can be collected for use any time, and a small volume of any of these can supply a great deal of energy.

Chapter 16

1. carbon, hydrogen, & oxygen; nitrogen & sulfur

3.

5.

7. 2371 g/mole

9. 74.5 cm (29.3 in) on a side

11. hydrogen bonds between the bases; specifically A to T and C to G.

13. Double-stranded DNA begins to separate into two, complementary single-stranded molecules. Once a short section has separated, each single-stranded molecule acts as a template for the formation of a new DNA molecule. This action continues as the original double-stranded molecule "unzips" as new material is formed a short distance away from the "unzipping" portion.

15. Similarities:
    Both contain phosphate, a five-carbon sugar, and a nucleotide base
    Both utilize hydrogen bonding extensively
    Both are polymeric molecules
    Differences:
    RNA contains uracil whereas DNA has thymine
    RNA contains ribose whereas DNA contains deoxyribose
    RNA moves from nucleus to cytoplasm, unlike DNA

17. a. The number of amino acids in G-actin must be approximately $42,000/120 = 350$. Each amino acid requires a sequence of three nucleotide bases so there must be $350 \times 3 = 1050$ bases associated with the DNA code for G-actin.
    b. There is one codon for every amino acid, so 350 codons.

19. a. AUG
    b. GCU
    c. CCA
    d. AUA

21. DNA replication occurs in the nucleus of the cell. Protein synthesis occurs in the cytoplasm, specifically on the surface of a ribosome, under the influence of RNA which "carries the message" from the DNA to the location where protein is synthesized.

Chapter 17

1. water, protein, carbohydrate, fat, vitamins, minerals

3. 160 Cal

5. Both glucose and starch have carbon in the same oxidation state, namely zero, and are both oxidized to CO_2 and H_2O. Therefore, the same amount of energy is released per mole of carbon. Starch is a polymer of glucose monomers.

7. 0.80 Cal. Compare this value to the answer for question 3.

9.

11. A trisaccharide will require two water molecules to form three glucose molecules.

13. Sucrose is a disaccharide formed from a molecule of glucose and a molecule of fructose. Nothing happens to sucrose when it is cooked but during the digestion process it is first converted into glucose and fructose; the fructose is then converted into a glucose, and both glucose molecules are converted into CO_2 and H_2O through reaction with oxygen obtained by breathing. The exothermic energy change of this reaction is coupled via ATP to useful energy for our bodies. The CO_2 is expelled by breathing.

15. Amylose, amylopectin, and glycogen are all polysaccharides made up from a large number of alpha-glucose molecules. The differences are in the amount of branching; amylose has essentially zero branches, amylopectin has moderate amounts of branching, and glycogen is extensively branched.

17. 405 Cal per serving

19. 34 kcal or 34 Cal

21. A saturated fat has no carbon-carbon double bonds whereas an unsaturated fat has one or more carbon-carbon double bonds. Excessive amounts of saturated fats in one's diet can lead to increasing amounts of cholesterol in the blood.

23. No, fat sometimes means all lipids, some of which are not triglycerides.

25. 20 g of fat

27. Structure is drawn without hydrogens for clarity. Remember, all carbons have four bonds

29. Stearic and palmitic acids have no C=C but linoleic has two C=C double bonds. Therefore, two moles of H_2 or 4.0 grams of H_2 are needed to fully saturate a mole of this fat.

31. Cholesterol in the blood is not dissolved as a simple molecule. Rather, it is imbedded in a lipoprotein which is very nonpolar on the inside but presents a polar, hydrogen-bonding exterior to the blood.

33. −30.6 Cal or −128 kJ

35. Amino acids are the building blocks of proteins because they polymerize to make protein.

37. The body of a person who does not consume enough carbohydrate to supply energy needs uses protein as an energy source. The nitrogen atoms are not used to build body protein or enzymes but rather are converted into useless ammonia which is lost in perspiration and breathing.

39. The essential amino acids tend to be more complicated in structure. The nonessential ones can be formed in our bodies through bond-breaking to make the simpler substances.

41. $C_{18}H_{32}O_2$ must be a fatty acid. It can't be a protein or amino acid because it has no nitrogen; it can't be a fat because a fat always has six oxygen atoms; it can't be a carbohydrate because a carbohydrate always has twice as many hydrogens as oxygens. This compound must be a fatty acid having two carbon-carbon double bonds because if there were no double bonds (or rings) it would have $(18 \times 2) + 2 = 38$ hydrogen atoms.

43. 4.6% protein

45. Vitamin C is highly water soluble so any excess will be excreted in urine. Vitamin E is fat soluble and so is not excreted but rather stored in fatty tissue.

47. 114 P atoms

Chapter 18

1. Mercurochrome and merthiolate both contain mercury which is a severe poison; iodine, however, rapidly becomes iodide in the environment and causes no problems.

3. The amount of chromium governs the intensity of the color. More Cr gives the darker green color of emerald whereas lower concentrations lead to the more delicately colored aquamarine.

5. Assuming that each person in the U.S. used 1 mL of mercurochrome solution each year, 6.7×10^4 g Hg/year or 67 kg/year

7. 33.5 g of water loss

9. This can be done accurately or approximately. The approximate method assumes that coconut oil fatty acids can be represented as being 100% lauric acid, $C_{12}H_{24}O_2$, and that the soap from it is $C_{12}H_{23}O_2Na$ with a molar mass of 222 g/mol and that tallow fatty acid be represented as being 100% oleic acid, $C_{18}H_{34}O_2$ and its soap is $C_{18}H_{33}O_2Na$ with a molar mass of 304 g/mol. Therefore soap has an average molar mass of $(0.20 \times 222) + (0.80 \times 304) = 44.4 + 243.2 = 287.6$ or 288 g/mol.

The accurate method accounts for each fatty acid. Calculate the total amount of each fatty acid in soap by multiplying the amount in coconut by 0.20 and adding that to the tallow amount multiplied by 0.80. Then multiply each of those combined fractions by the molar mass of each compound and take the total of these values. This method leads to a value of 271.3.

11. Tallow and coconut oil are renewable resources whereas petroleum is a finite resource. Detergent bars, however, have better performance in that they do not form soap scum because the calcium salts of the detergent anions are much more water soluble than those of the fatty acids.

13. 70.8 % gold

15. 0.00180 mol

17. For something to be buffered, ingredients are added that prevent the solution pH from changing dramatically when the substance dissolves or is added as a liquid. Buffered aspirin is prepared by mixing the aspirin with the solid basic components

such as $Al(OH)_3$ or $MgCO_3$ to speed the dissolving of the acidic aspirin to minimize stomach discomfort.

19. $C_9H_8O_4 + H_2O \rightarrow C_7H_6O_3 + C_2H_4O_2$
    (aspirin)　　　　　(salicylic acid)　(acetic acid)

    Vinegar is 5% acetic acid in water so the smell of vinegar is really the smell of acetic acid.

21. 560 g $C_4H_5O_6K$

23. Carbohydrates have twice as many hydrogens as oxygens so there are 10 hydrogen atoms.

25. Emerald, ruby, sapphire, and aquamarine are definitely inorganic. Pearl and amber are definitely organic. Most chemists would classify diamond as inorganic even though it is 100% carbon.

27. 10.4 % or, rounded to correct significant figures, 10%

29. Terpenes are volatile because they have no strongly polar groups that would act to hold the molecules together in condensed phase. Compounds with polar or ionic functional groups are much less volatile than similar size compounds that lack such groups. For instance, α-pinene has a boiling point of 156°C but decanoic acid ($CH_3(CH_2)_8COOH$), which has the same number of carbons as pinene, has a boiling point of 270°C.

33. Volatile chlorinated liquids such as tetrachloroethene are hazardous in the environment but CO_2 is a normal constituent of the air. The CO_2 used in dry cleaning is purified from any of several processes that form CO_2, which would otherwise be allowed to escape into the atmosphere. Therefore dry cleaning is not responsible for introducing more CO_2 than would be the case in the absence of this use.

35. Incandescent light bulbs consist of a tungsten filament inside a glass bulb filled with inert gas and capped with a brass or aluminum base but with a lead based "button" at the very bottom of the base. This lead is environmentally hazardous. Fluorescent tubes are a source of mercury. A small amount of mercury vapor fills the interior of the bulb; this emits ultraviolet light when excited by a high voltage. The UV light strikes the phosphor, which emits white light. Burned out fluorescent bulbs should be returned to a location where the mercury can be recovered.

37. The biggest "cost" of operating a standard mercury lamp is that the standard bulb contains a greater amount of mercury than the low-mercury lamp. Thus, appropriate disposal of standard lamp bulbs involves the transfer of more toxic waste to landfills than would be the case with the low-mercury bulbs. The biggest benefit of using the standard bulbs is that they require less energy to operate than is required by the low-mercury bulbs.

39. Wood is composed of cellulose, hemicellulose, lignin, and water.

41. 1.3 grams Ag

glossary

Absolute temperature is the temperature measured with the coldest theoretically possible temperature set to be zero on Kelvin scale.

Absolute zero is the lowest temperature possible, at which atoms, molecules, and ions stop moving.

Absorbance range is the range of wavelengths that are absorbed by a sample.

Absorption spectroscopy examines the extent to which light from a known source is absorbed by a sample as a measure of its concentration.

Acetylcholine is a compound that transmits nerve signals.

Acetylcholinesterase inhibitor is a compound that interferes with the functioning of acetylcholine in transmitting nerve signals.

Acid is a substance that increases hydronium concentration in aqueous solution.

Acid anhydride is a nonmetal oxide that forms an acid when combined with water.

Acid deposition is the total effect of acid falling in rain, snow, etc. or dry solids.

Acid-base indicator is a substance that has different appearance in acid and base.

Acid-neutralizing capacity (ANC) describes the ability of water or soil to neutralize falling acids.

Actinide series is a series of 14 inner-transition elements (atomic numbers 90-103) that are all similar in properties and result from filling of the 5f orbitals.

Activation energy is the energy needed to start a reaction going.

Addition polymer is a polymer formed by adding together many monomers containing double bonds.

Addition reaction is a reaction in which a reagent X-Y adds to a multiple bond so X is attached to one of the carbons of the bond and Y to the other.

Adenosine diphosphate (ADP) is the chemical that combines with phosphate to form ATP.

Adenosine triphosphate (ATP) is the main "energy currency" of life, the hydrolysis of which to ADP and phosphate releases the energy to power anabolism.

Aerobic, requiring oxygen.

Aerosol, tiny particles suspended in air.

Air pollution is the result of something entering the atmosphere usually as a result of human activity, that has a harmful effect on some aspect of the environment.

Air quality is the description of deviation of air composition from that of pure air.

Alchemy was the study of chemical and physical properties of substances with goals that included conversion of base metals into gold.

Alkali is a substance that furnishes hydroxide ion in water by dissociation of a metal hydroxide.

Alkanes are hydrocarbon that only contains single bonds. Alkanes not in a ring share the general formula C_nH_{2n+2}.

Alkenes are hydrocarbons that contain a carbon-carbon double bond.

Alkynes are hydrocarbons that contain a carbon-carbon triple bond.

Allotopes are alternate forms of the same element that have significantly different chemical and physical properties.

Allotropes are different molecular or crystalline forms of a pure element.

Alloy is a homogeneous mixture of metals, sometimes containing small amounts of nonmetals.

Alpha particle is the helium nucleus emitted from certain heavy radioactive nuclei.

Amino acids are the monomer building blocks of proteins.

Amphiprotic is a substance that can behave either as an acid or a base.

amu is the abbreviation for **a**tomic **m**ass **u**nit, also called "dalton."

Anabolic steroid is a steroid that aids in building body mass.

Anabolism refers to processes in the body associated with building new tissue

Anaerobic, not requiring oxygen.

An **anion** is an atom or group of atoms having a negative charge.

Anode is an electrode at which oxidation occurs.

Anthocyanin is a class of substances responsible for the color of many flowers, fruits, and vegetables.

Anti-oxidant is a chemical that protects against damaging oxidizing agents by reacting with these agents.

Anticodon a set of three bases in tRNA that binds to a specific codon in mRNA.

Antihistamine is one of several compounds used to block the action of histamine and prevent hay fever symptoms.

Antiseptic is one of several compounds used to kill microorganisms.

Area is the two-dimensional size of an object.

Asbestos is a group of silicate minerals that crystallize in fibers; this is very resistant to fire but harmful if fibers are inhaled.

Atmosphere is the unit of pressure of air at the Earth's surface.

Atom is the smallest part of an element to retain the properties of the element.

Atom economy involves having as much of the reactant as possible appear in the product.

Atomic mass is the mass of one atom in amu or 1 mol of an element in grams.

Atomic mass unit (amu) is approximately the mass of one hydrogen atom; exactly 1/12 the mass of one atom of carbon having mass number 12.

Atomic number is an integer equal to the number of protons in an atom's nucleus.

Atomic theory stresses that matter is composed of atoms that combine in simple proportions.

Base is a substance that furnishes hydroxide ion in water.

Basic anhydride is a group 1A or 2A metal oxide that forms a base when combined with water.

Basin environment is a geographical region surrounded at least partially by hills.

Bauxite is an ore of aluminum, consisting of aluminum oxide and water along with impurities.

Benzene (C_6H_6) is a planar six-membered ring (hexagon) of carbon atoms. The carbon atoms are also bonded to hydrogen atoms.

Beta particle is the electron emitted from nuclei having too many neutrons for stability.

Big Bang is the event, occurring 12-15 billion years ago, when the universe was created.

Binary compounds are compounds composed of only two elements.

Binary salts are compounds containing one metal and one nonmetal.

Biomagnification is the process by which the concentration of pollutant increases as predators eat plants and animals that originally have taken on pollutant substances.

Biomass is plant material such as wood, grasses, or leaves.

Bleach is a substance that decolorizes stains by oxidation.

Boiling point is the temperature at which a liquid boils and at which vapor pressure equals external applied pressure.

Boiling point elevation is the increase of the boiling point by dissolved solutes in a liquid.

Bond angle is the geometric angle at which two atoms are attached to a third atom.

Bond energy is the energy required to break a given bond in a mole of gaseous molecules.

Bonding is a collective term used to describe ways in which atoms are held together in compounds or crystals.

Brighteners are substances that absorb ultraviolet light and emit blue light to make clothes appear whiter than they really are.

Buffer is a chemical system, typically composed of conjugate acid-base pairs that resist change in pH.

Builder is a substance used in major amounts in commercial laundry detergents to allow the surfactant to perform its job, usually is used to raise the pH of laundry water.

Calorie is equal to 1000 calories, 1 dietary calorie, 1 kilocalorie.

calorie is the amount of heat needed to raise temperature of 1 g H_2O by 1°C.

Calorimeter is a device used to measure the amount of heat released or consumed during a chemical reaction.

Carbohydrate class of compound composed of carbon, hydrogen and oxygen (with the H and O in a 2:1 ratio) that includes all sugars, starch and cellulose and glycogen.

Carbon cycle is the sequence of reactions characteristic of carbon atoms in the natural environment—the material cycle wherein carbon atoms are cycled through many different compounds.

Carbonate, CO_3^{2-} ion, is an anion of limestone and hard water.

Carborundum is silicon carbide, a very hard synthetic material used in grinding wheels and sharpening stones.

Carboxyl acid group are the four atoms, COOH, carbon double bonded to one oxygen and single bonded to an —OH group resulting in an acidic H atom.

Carcinogen is a substance which causes cancer.

Catabolism is the breakdown of chemicals in living things, into simpler ones.

Catalyst is a substance added to lower the activation energy of a chemical reaction, thus causing a faster reaction.

Catalytic converter is a device used on automobiles to react CO and NO into CO_2 and N_2.

Cathode is an electrode at which reduction occurs.

A **cation** is an atom or group of atoms having a positive charge.

Celsius is the name of a temperature scale, formerly called "centigrade," in which melting ice is defined as 0° and boiling water defined as 100° at sea level.

CFC is the abbreviation for chlorofluorocarbon.

Chalcocite is a copper ore composed of copper(I) sulfide.

Chalcopyrite is a copper ore composed of copper/iron mixed sulfide.

Charles' Law describes how the volume occupied by a gas is related to temperature.

Chemical bond involves electrons holding two or more atoms together in a molecule or crystal.

A **chemical formula** is a shorthand method of describing the composition of pure substances, giving the identity and number or ratio of each element in the substance.

Chemical equilibrium is a state in which chemical reactions appear to have come to completion.

A **chemical equation** is a formal method for describing reactants and products of a chemical change.

Chemical kinetics is the study of the speed at which chemical reactions occur.

A **chemical reaction** is a process in which one set of chemicals is converted into different substances.

Chemistry, at the atomic level, is study of the interaction of atoms as they react to form different combinations and make new substances.

Chlorinated hydrocarbon is a compound containing only carbon, hydrogen, and chlorine.

Chlorofluorocarbon is a small molecule containing only carbon, chlorine, and fluorine.

Chloroplast is a substructure of a plant cell in which photosynthesis occurs.

Cholesterol is a lipid found only in animal products, necessary for life but a health hazard if present in too high an amount.

Chromatin is a complex of DNA and protein that forms chromosomes.

Chrysotile is one specific type of asbestos, $Mg_3(Si_2O_5)(OH)_4$.

Clay consists of fine particles of aluminum silicate in soil.

Clinker is the product formed from heating calcium carbonate and clay to a high temperature; it is powdered to become Portland cement.

A **closed system** is a collection of matter shut off such that no other matter can enter or leave.

Coagulation is the process of making insoluble solids from dissolved materials.

Codon is a set of three bases on mRNA that binds to the matching anticodon on tRNA.

Coenzyme is a small molecule needed to bind to an enzyme in order to allow it to work. Many vitamins act as coenzymes or as precursors to coenzymes.

Colligative property is a property of a solution that depends only on the concentration of a solute.

Colloid is a suspension of a material in a matrix in which it is not soluble but in which the suspended material is so finely dispersed that it does not settle out.

Combined gas law $\dfrac{P_{start} V_{start}}{T_{start}} = \dfrac{P_{end} V_{end}}{T_{end}}$

Complementary bases are bases in DNA or RNA that can bind together by base-pairing, involving hydrogen bonds.

Composting is a process of allowing leaves, grass, etc. to react with water, air, and microorganisms to form a soil-like solid while giving off carbon dioxide.

Compound is a substance made of specific ratios of different chemical elements in a particular arrangement.

Concentration is the measure of the amount of solute dissolved in a solution.

Concrete is the rock-like material made by mixing cement, sand, and water and allowing the mixture to harden in place.

Condensation is the process of a gas becoming a liquid, a physical change.

Condensation polymer is a polymer formed by reacting molecules that split out other small molecules, usually water.

Conductor is a substance capable of conducting electric current.

Conjugate base is that which remains when hydrogen ion is removed from an acid.

Conjugate pair is an acid and its conjugate base.

Constitutional isomers are the same as structural isomers.

Conversion factor is a ratio of quantity expressed in two different units, used to convert a quantity in one unit to the corresponding quantity in the other unit.

Cooling tower is a large building in which warm water is allowed to cool.

Copper is an elemental metal, chemical symbol Cu.

Covalent bond is a chemical bond formed by the sharing of electrons between two atoms.

Critical is the mass of fissionable nuclide that will just sustain a chain reaction.

Cross-over method is a strategy for writing the correct formula for an ionic compound.

Curie is a measure of the amount of radioactivity equal to 37 billion disintegrations (emissions) per second (DPS). This is the number of radium-226 nuclei in a 1-gram sample of the isotope undergoing radioactive decay in a second.

Current is the flow of electrons through a medium, usually a wire.

A **d orbital** is one of a set of five orbitals (four are shaped like three-dimensional four-leaf clovers, one like an hourglass surrounded by a doughnut); it appears in level $n = 3$ and higher.

Dalton's law of partial pressures says that the total pressure of gases is the sum of the partial pressures of each component gas of the mixture

Decay series is the series of nuclides formed on successive radioactive decay events.

Decongestant is a material that is used to clear breathing passages.

Density is the ratio of the mass of a substance to its volume.

Deoxyribose, $C_5H_{10}O_4$, is the sugar in DNA.

Detergent is a synthetically prepared cleaning agent.

Dezincification is selective oxidation of zinc in brass plumbing fittings; it occurs because zinc is much more readily oxidized than is copper.

Diatomaceous earth is a mineral that is the skeletons of large numbers of "diatoms," small water-borne organisms; it is composed mostly of silicon dioxide.

Dietary Calorie (Calorie, kilocalorie) = 1000 calories.

Digestion is a process that breaks down food into nutrients that can be absorbed into the bloodstream.

Diglyceride is a compound formed by the reaction of two alcohol groups of glycerol with fatty acids.

Dimensional analysis is a technique for solving numerical problems in which the units of the quantities guide the solution and serve as a check for the solution.

Dipeptide is a molecule composed of two amino acids linked by a peptide bond.

Direct evidence is the result of a measurement or observation.

Disaccharide is a molecule composed of two monosaccharide sugar groups bonded together.

Dissolve is to disperse one substance at the molecular level in another.

Distillation is the process for purifying a liquid by evaporating it, condensing the vapors, and collecting the liquid in a clean container.

Disulfide bridge is a link between two parts of a protein molecule formed when amino acids with —SH groups react together.

DNA (deoxyribonucleic acid) is a huge molecule that stores genetic information.

Double bond is a covalent bond formed by two pairs of shared electrons.

Double helix is the interwound double spiral structure of the DNA of genes.

Dry cleaning is cleaning of clothing in which the solvent is some material other than water.

Electrolysis is the passage of electricity through a substance to cause chemical change.

Electrolyte is a substance that on dissolving, allows current to flow through its solution.

Electromagnetic force is the force associated with light, electricity, and magnets.

Electromagnetic radiation is the collective term including visible light, ultraviolet, infrared, microwave, radio, X-rays, and gamma rays.

Electromotive force measured in volts, measures the tendency for a redox reaction to occur.

Electron is the subatomic negatively charged particle surrounding the atomic nucleus.

Electron cloud is the region of space surrounding a nucleus where the electrons are located.

Electron configuration is the electron arrangement and pattern of orbitals in an atom.

Electron shells are sets of energy levels corresponding to a single value of n.

Electronegativity is a measure of the ability of an atom to attract available electrons to itself.

Electrostatic force is the attraction of opposite electric charges and mutual repulsion of like charges.

Element is one of over 92 naturally occurring substances of which the matter of the solar system is made.

Emission spectroscopy studies the intensity of light given off by a sample as a measure of its amount.

Emulsifiers are compounds that interact with both water and fat to allow them to mix.

Emulsion is a colloidal dispersion of one liquid in another.

Endergonic is a process that takes in energy.

Endothermic is a process that takes in heat.

Energy is the capacity for doing work or causing a change, either chemical or physical.

Energy levels are the discrete energies of electrons in an atom.

Enriched means having a greater fraction than normal of an unusual isotope.

Entropy is a measure of the disorder of a system.

Environmental estrogens are pollutants that make male animals develop some female characteristics.

Environmentally benign means having no impact on the environment.

Enzyme is a large molecule which acts as a catalyst in biological reactions.

Equilibrium is the state of a reaction at which there is no change in the concentration of the reactants and products (that state of a reaction in which the rates of the forward and reverse processes are equal).

Estrogens are female hormones.

Ethanol (C_2H_5OH) is ethyl alcohol, the alcohol of beverages and gasohol.

Ethyne (C_2H_2) includes a triple bond.

Evaporation occurs when molecules of a liquid become gaseous.

Excess deaths, statistically, more deaths than are expected in a given time period.

Excited state, an atom or a molecule possessing more than the minimum amount of energy is said to be in an excited state.

Exergonic is a process that gives off energy.

Exothermic is a process that gives off heat.

An *f orbital* is one of seven orbitals of complex shape; it appears in level $n = 4$ and higher.

Fat is a general term for the lipids in food. In this usage, it includes fats and oils.

Fatty acid is a constituent of a fat, an acid with a long hydrocarbon chain.

Fertilizer is a substance used to increase the growth rate of plants.

Fiber is carbohydrate in food that is indigestible by humans.

First Law of Thermodynamics states that heat and work are interconvertible and together constitute energy of a system— in essence this is a restatement of the Law of Conservation of Energy.

Fission is a nuclear reaction in which one nucleus fragments into two smaller ones.

Flue-gas desulfurization (FGD) is a process for removing sulfur from combustion products of sulfur-containing fuel, usually coal.

Food additive is a substance added to a food that is not part of the primary nutrients or ingredients of the food.

Force is the mass of an object multiplied by its acceleration.

Formula mass is the mass of one formula unit or molecule in amu or the mass of 1 mol of a compound in grams.

Formula unit is the number of atoms of each kind in a molecule or the simplest formula of an ionic compound.

Fossil fuels are residues such as coal or petroleum left from the decomposition of ancient plant life.

Fractional distillation is a process involving numerous vaporization-condensation steps, used to separate two or more volatile components.

Freezing point is the temperature at which a liquid changes to a crystalline solid (freezes).

Freezing point depression is the decrease of the freezing point by dissolved solutes in a liquid.

Frequency is the number of electromagnetic waves that pass a point in a second.

Functional group is collections of atoms that give special properties and reactivities to molecules.

Fundamental forces are the four forces responsible for all interactions between matter in the universe: strong and weak nuclear forces, electromagnetism, and gravity.

Fusion is a nuclear reaction in which two nuclei combine (fuse) into one.

Galvanized refers to a steel object coated with zinc to prevent its rusting.

Galvanizing is coating iron with zinc to prevent it from oxidizing.

Gamma rays, high-energy electromagnetic radiation emitted from many radioactive isotopes.

Gas (g) is one of three states of matter; it fills the volume of the container that holds it regardless of the shape and size of the container.

Gene is a section of DNA that encodes a single polypeptide or a functional RNA.

General contact herbicide is a substance that kills plants after contacting the plant's surface.

Genetic engineering is the artificial manipulation of DNA and genes.

Genome is the entire complement of genetic information of an organism.

Geometric isomers are compounds differing only in that one is cis and one trans about a double bond.

Global warming is the process by which the atmosphere and the surface of the earth show an increased temperature.

GM food is genetically modified food, or food derived from a genetically modified organism.

Gold is an elemental metal, chemical symbol Au.

Gram formula mass is the mass of 1 mol of a compound or element in grams.

GRAS substances are food additives classified as generally recognized as safe.

Gravity is the force of gravitational attraction between two bodies.

Green Chemistry is concerned with making the practice of chemistry as environmentally benign as possible by the reduction or elimination of hazardous wastes and efficient use of energy.

Ground state, an atom or molecule possessing the minimum amount of energy is said to be in a ground state.

Group, as used in the periodic table, refers to the chemical elements found in a vertical column; the elements all share the same outer-shell electron configuration.

Haber process is the Nobel prize-winning process for making ammonia from hydrogen and nitrogen.

Half-life is the length of time necessary for half of an amount of a radioactive isotope to undergo decay.

Half-reaction is any reaction that is written with electrons as either reactants or products.

Hall-Heroult process is the process for the economical, commercial production of aluminum.

Halogen is any one of the elements in Group 7A (F, Cl, Br, I, At).

Halon is a small molecule containing carbon and bromine in addition to chlorine and/or fluorine but no hydrogen.

Hard water is water that contains significant concentrations of calcium ion.

Heat is the energy of a system caused by the continuous motion of its particles.

Heat capacity is the amount of heat needed to raise the temperature of an object by one degree Celsius (or by one kelvin).

Heat of fusion is the amount of heat needed to melt 1 mol or 1 g of a solid.

Heat of vaporization is the amount of heat needed to vaporize 1 mol or 1 g of a liquid.

Hemicellulose is a major component of wood, a polymer of the five-carbon sugar, xylose, $C_5H_{10}O_5$.

Hemoglobin is a protein (mol wt = 56,000) that carries oxygen through the blood system from the lungs to the rest of the body.

Herbicide is a compound used to kill selected plants.

Hexane (C_6H_{14}) is a saturated hydrocarbon, a constituent in gasoline.

Histamine is a powerful vasodilator, formed by removal of CO_2 from the amino acid histidine.

Homologous series is a group of organic chemicals that share the same general formula and similar chemical characteristics.

Hormone is a compound made in our bodies to carry out a function at another location in the body.

Humectant is an edible substance added to foods to maintain their moisture level; glycerol, $C_3H_8O_3$, is a common example.

Humus is the decayed remains of plants in topsoil.

Hydrocarbon is a compound made up of only carbon and hydrogen atoms.

Hydrochloric acid is an aqueous solution of hydrogen chloride, $HCl(aq)$.

Hydrogen bond is a moderately strong intermolecular attraction caused by the partial sharing of electrons between an atom of F, O, or N and the polar hydrogen atom in an F—H, O—H, or N—H bond.

Hydronium ion, H_3O^+, is a representation describing species found when water accepts a hydrogen ion from an acid.

Hypertonic is having a salt concentration greater than blood serum.

Hyperventilation is breathing too rapidly, prefix "hyper" means "too much."

Hypotonic is having a salt concentration less than blood serum.

Hypoventilation is breathing too slowly, prefix "hypo" means "too little."

Ideal gas constant, R = 0.08206 L atm/K mol.

Ideal gas equation, $PV = nRT$, describes relationship of pressure, volume, and temperature of a given quantity of gas.

Imaging is the use of radiation to gain a picture of hidden objects, often internal organs of living persons.

Immutable is unchangeable.

Incinerate is to burn to ashes in an excess of oxygen.

Indicator is a substance that has a different appearance in one set of circumstances than in another.

Indirect evidence is inferred from observation of related phenomena.

Induced fission is the nuclear fission triggered by a neutron striking an unstable nucleus.

Inhibitor is a substance that slows or prevents a chemical reaction.

Initiator is a substance that starts a reaction between two other molecules.

Inorganic refers to a compound made up primarily of atoms other than carbon.

Insecticide is a compound used to kill selected insects.

Intermediate is a substance formed in a chemical reaction which then reacts further so that it is not a final product of the reaction.

Intermolecular forces cause the attraction of one molecule for another, but not such that the two react.

Intrinsic semiconductor is a material that, when very pure, conducts electricity less well than a conductor but better than an insulator.

Ion is an atom or molecule possessing an electric charge.

Ion exchange is a process in which one kind of ion in solution is exchanged for another.

Ionic bond is a bond formed by attraction of positively and negatively charged ions.

Iron is an elemental metal, chemical symbol Fe.

Isoelectronic means having the same electron arrangement.

Isotonic is having a salt concentration equal to that of blood serum

Isotope effect, different isotopes in reactants cause very minor but detectable

changes in the reactivity of these substances.

Isotopes are atoms with the same number of protons but a different number of neutrons.

Kelvin is a temperature scale the zero of which is set to be absolute zero.

Kilocalorie = 1000 calories = 1 dietary calorie.

Kinetic energy is the energy of motion.

Lachrymator is a compound whose vapors cause copious tears to form.

Landfill is a place where garbage and trash are buried in such a way that they will interact with the rest of the environment as little as possible.

Lanthanide series is a series of 14 inner-transition elements (atomic numbers 58-71) that are similar in properties and result from filling of the $4f$ orbitals.

A **lattice** is a tightly packed regular array of particles.

Law of Conservation of Energy is the scientific law that states that energy is neither created nor destroyed.

Law of Conservation of Mass states that matter (mass) is neither created nor destroyed although matter is frequently transformed into different substances.

Law of Definite Proportions recognizes that a pure compound always has the same fractions (proportions), by mass, of its constituent elements.

Law of Multiple Proportions recognizes that the same elements can combine to form different compounds when the constituents have different ratios of mass.

Leaching is the slow dissolving of metals by passing solvent through granular solid.

Leavening agent is a substance that produces carbon dioxide bubbles to form in a baked food.

Le Châtelier's principle states that when a system at equilibrium is subjected to a stress, the system responds to partially alleviate that stress.

Lewis structure is a diagram showing atoms and valence electrons in an ion or molecule.

Lignin is a major component of wood, a polymer of several aromatic alcohols.

Lipases are enzymes that cause ester linkages in fats to hydrolyze.

Lipid is a class of fatty water-insoluble substances found in living things.

Liquid (*l*) is one of three states of matter; it fills the bottom of the container that holds it to the extent of its own volume.

Lone pair describes a pair of valence electrons which are not shared but rather retained by a single atom.

Lye is solid sodium hydroxide, NaOH.

Macronutrients are nutrients needed in substantial amounts in the diet, namely, carbohydrates, fats, and proteins.

Main group, as used in the periodic table, consists of groups of elements whose column is headed by the letter A.

Main-group elements are those that belong to the "A" groups in the periodic table.

Marble is calcium carbonate transformed geologically from limestone by high pressure and temperature.

Mass number is the integer which is the sum of the number of protons and neutrons in a nucleus.

Mass is a measure of an amount of matter.

Material cycle is a description of the transformations of substances in the environment, usually implying that elements undergo continuous chemical change but often reappear in forms they have been in before.

Matter is anything that occupies space and has mass.

Mechanism is the sequence of molecular events that occur during a chemical reaction.

Melanoidin is a substance that gives a dark color to hair.

Melting point is the temperature at which a substance melts; above this temperature the substance is a liquid.

Messenger RNA (mRNA) is the RNA that carries a gene's "message" to the ribosomes, for it to be "translated" into the amino acid sequence of a protein.

Metabolism is all the chemical reactions of life.

Metallic bonding is chemical bonding in metals, neither ionic nor covalent, in which atoms share their valence electrons communally.

Metalloids are semimetals, elements that have properties intermediate between metals and nonmetals.

Metals are elements that tend to form positive ions, conduct electricity, be malleable, and form ionic compounds with nonmetals.

Methane (CH_4) is a natural gas.

Micelle is a cluster of molecules of soap or similar substance in which a long nonpolar group of atoms intermingle to leave a small polar group extending into surrounding water.

Micronutrients are nutrients needed in only tiny amounts in the diet, especially all the vitamins and many minerals.

Mineral is a metal ion or nonmetal that appears in the body largely as ionic compounds.

Mineral oil is a mixture of saturated hydrocarbons of approximately 20 carbons, obtained from petroleum.

Model is a representation of reality (an image) used to relate unfamiliar phenomena to something familiar for ease of understanding.

Molality is the measure of concentration, number of moles of solute per kilogram of solvent.

Molar mass is the mass of one mole of a compound or element in grams.

Molarity is the number of moles solute per liter of solution.

Mole-mole conversion factor is the ratio of moles of two substances in a chemical reaction.

Mole is Avogadro's number of atoms, molecules, or formula units; an amount of any substance equal to its molar mass expressed in grams.

Molecular mass is the mass of 1 mol (Avogadro's number) of molecules of a compound.

Molecule is a group of atoms bonded together by shared electrons in a specific arrangement, which behaves as an independent unit.

Monoglyceride is a compound formed by the reaction of one alcohol group of glycerol with a fatty acid.

Monomer is a compound of which many molecules will react together to form one much larger molecule.

Monosaccharide is a simple "single sugar," form of carbohydrate, usually $C_6H_{12}O_6$.

Monounsaturated fatty acid is a fatty acid with only one carbon-carbon double bond.

Mutable is capable of being changed.

Mutagenic is causing change in the DNA genetic material of an organism.

Mutation is a change in the base sequence of DNA.

Mutual annihilation, electron and positron collide and are both converted into two gamma rays.

Natural gas, mostly methane (CH_4), is used as a fuel and raw material.

network covalent compound is a three-dimensional lattice of covalently-bonded atoms.

Neurotransmitter is a chemical that transmits a chemical message used in the nervous system.

Neutral pH means neither acid nor base.

Neutralization is the process of adding acid to a base or vice versa till neither is in excess.

Neutron is a subatomic nuclear particle that has no electrical charge and so contributes only to the mass of the nucleus.

NO$_x$ is the combination of several oxides of nitrogen, especially NO and NO$_2$.

Nonmetals are elements that tend to form negative ions, don't conduct electricity, and form covalent bonds with other nonmetals and ionic bonds with metals.

Nuclear fusion is the formation of a larger nucleus by the merging of two smaller nuclei.

Nuclear reactions are those reactions in which the makeup of an atomic nucleus changes and new elements or isotopes are produced. *Chemical* reactions, in comparison, are changes in which one set of substances is converted into another set but the nuclei of atoms involved remain unchanged.

Nucleotide is the monomer of DNA or RNA, composed of a sugar, a nitrogenous base and a phosphate group.

Nucleus (of a cell) is the central membrane-bound organelle of a cell, containing the cell's genome.

Nucleus (of an atom) is the positively charged center of an atom.

Nuclide is a nucleus of a specific proton and neutron number.

Nutrient is any substance needed in our diet.

Octet is a set of eight valence electrons associated with an atom.

Oil is a liquid fat.

On-demand generation is a strategy of making dangerous compounds just before they are needed.

Optical Isomers are nonsuperimposable mirror images of each other.

Orbitals are regions of space in which electrons of certain energy levels are usually found.

Organic chemistry is the study of most carbon-based compounds.

Organophosphate is an organic compound which includes a phosphate (PO$_4$) group.

Osmosis is the process in which solvent molecules move through a semipermeable membrane.

Osmotic pressure is the pressure developed as solvent undergoing osmosis expands.

Oxidation is the loss of electrons by an atom or group of atoms.

Oxidation state is the actual charge of an ion or the apparent charge on a covalently bonded atom.

Oxide is a compound of oxygen with any other element.

Oxygenate is a gasoline additive that contains an oxygen atom.

Ozone hole is the dramatically lowered concentration of ozone in regions of the upper atmosphere.

Ozone is an allotrope of oxygen having three atoms per molecule, O$_3$.

Ozone layer is the region in the stratosphere in which O$_3$ is formed by the reaction of O$_2$ with sunlight and which diminishes the amount of UV radiation reaching the Earth's surface from the Sun.

A *p orbital* is one of three orbitals shaped like an hourglass; it appears in level $n = 2$ and higher.

Paclitaxel is a powerful anticancer drug obtained from yew trees.

Partial pressure, the portion of the total pressure which is exerted by a gas.

Particulate matter are small pieces of solid or liquid suspended in the air.

Period, as used in the periodic table, refers to the elements arranged in horizontal rows.

Periodic table is a systematic arrangement of elements based on electron configuration and periodic behavior.

Pesticide is a substance that kills a selected set of plants or animals.

pH is the negative logarithm (base 10) of hydrogen (hydronium) ion molarity.

Phloem is tissue that transports food materials in plants from where they are produced to where they are needed.

Photochemical smog is a mixture of chemicals formed from interaction of sunlight with unburned gasoline and nitrogen monoxide formed mostly in automobile exhaust.

Photoelectrolysis is the decomposition of water into hydrogen and oxygen with light supplying the energy for the decomposition.

Photon is the smallest amount of light, a "particle" of light.

Photosynthesis is a sequence of reactions in green plants which results in water and carbon dioxide being converted into oxygen gas and carbohydrate under the influence of sunlight.

Photovoltaic device (PV) uses the conversion of light energy into electrical energy.

Phytoplankton are very tiny plants that live in water and convert sunlight into biomass.

Planar means all the atoms of the molecule or group are in a single plane.

Plaque is a thin layer of bacteria that grow in one's mouth and accumulate on teeth.

Plastic describes a high molecular-weight synthetic polymer.

pOH is the negative logarithm (base 10) of hydroxide ion molarity.

Polar means the negative charge of a molecule is concentrated at one position.

Polar covalent bond is a bond in which the electrons are shared unevenly between two atoms.

Polar molecule is a molecule in which the center of negative charge is not at the same place as the center of positive charge.

Pollutant is a substance present in the environment, usually resulting from human activity, that decreases the quality or value of its surroundings.

Polyatomic ions consist of ions composed of more than one atom.

Polyethylene is the addition polymer made from ethylene (C_2H_4).

Polymer is a molecule formed by a large number of identical smaller molecules (monomers) reacting together to form a much larger molecule.

Polypeptide is a molecule composed of many amino acids linked by peptide bonds.

Polysiloxane is a polymer in which the backbone consists of alternating silicon and oxygen atoms.

Polyunsaturated fatty acid is a fatty acid with more than one carbon-carbon double bond.

Polyvinyl chloride is a polymer of $CH_2\!\!/\!\!CHCl$ that is widely used for many purposes, including water pipes.

Positron is a short-lived particle with the mass of an electron but a positive charge.

Potential energy is energy possessed by virtue of location or situation.

Power is the rate at which work is done or energy is transferred; electrical power is the voltage multiplied by the current. The units of power are watts.

Precipitate (noun) is a solid substance that comes out of solution to settle to the bottom of a solvent.

Precipitate (verb) is a process of coming out of solution.

Pressure is the force of gas pushing per area of surface.

Primary structure is the amino acid sequence of a protein.

Product is a substance formed in a chemical reaction.

Protease is an enzyme that degrades protein into its constituent amino acids.

Protein is a very large biological molecule containing C, H, N, O, and S and is a polymer of amino acids.

Proton is a positively charged subatomic particle in the nucleus of atoms; its number determines atomic number.

psi, "pounds per square inch" is a unit of pressure.

psig, "pounds per square inch gauge" means the gas pressure in excess of the surrounding atmospheric pressure.

Pumice is a very porous, low-density stone of volcanic origin.

Pyrite is a shiny golden-colored iron ore, FeS_2.

Quaternary structure is the way in which different proteins combine to form multisubunit proteins.

R is the ideal gas constant = 0.08206 L atm/K mol.

Radiation, energetic particles emitted from atoms undergoing radioactive decay.

Radical is an atom or molecule having an unpaired electron.

Radioactivity is the spontaneous change in the nucleus of an atom.

Rate is the speed of a chemical reaction, sometimes expressed in moles per second.

Reactant is a substance consumed in a chemical reaction.

Reactive describes elements or compounds that readily undergo chemical reactions.

Reactivity series describes a ranking of chemical elements in order of their ability to oxidize or reduce each other.

Receptor is the location on a biological molecule that receives another molecule and transmits a chemical signal.

Recycle is to utilize waste materials rather than throw them away.

Redox reaction is an abbreviation for "reduction and oxidation," a class of chemical reactions in which both reduction and oxidation occur.

Reducing smog is smog containing high concentrations of SO_2 and CO.

Reduction is the gain of electrons by an atom or group of atoms.

Relative atomic mass is the average mass of one atom of an element.

Relative humidity, the ratio of the actual partial pressure of water vapor in the air to the maximum possible pressure, expressed in percent.

rem is the unit of measure for the effect on humans of the energetic emissions from radioactive elements. An exposure of 1 rem results in 0.01 J absorbed per kilogram of body tissue.

Reserve is a deposit of minerals that can be economically utilized.

Residual herbicide is an herbicide that kills selected plants a considerable time after its application.

Resource is a known deposit of a useful mineral.

Respiration is the reaction carried out in living organisms in which oxygen and carbohydrate are converted into water and carbon dioxide so that energy useful to the organism is obtained.

Reverse Osmosis is a process in which greater than osmotic pressure is applied to the solution side, so solvent (often water) is forced to the pure solvent side. The solution becomes more concentrated as a result.

Ribonucleic acid (RNA) is a polymer of ribonucleotides, including mRNA, tRNA, and rRNA.

Ribose, $C_5H_{10}O_5$, is the sugar found in RNA.

Ribosomal RNA (rRNA) is the functional RNA of ribosomes, the sites of protein synthesis.

Ribosomes are the complexes of RNA and protein on which protein synthesis occurs.

An ***s orbital*** is a spherical orbital centered on the nucleus; it appears in all levels, starting with $n = 1$.

Saponification is the process of making soap from fat.

Saran is a copolymer of $CH_2\!\!/\!\!/CHCl$ and $CHCl\!\!/\!\!/CHCl$; a copolymer means that the polymer is formed from a mixture of the two monomers.

Saturated (solution) is when the maximum possible concentration of solute is present in solution.

Saturated (hydrocarbon) means containing as many hydrogen atoms as possible with no multiple bonds.

Saturated fatty acid is a fatty acid with no carbon-carbon double bonds.

Science literacy is the ability to read and comprehend scientific articles and arguments.

Second Law of Thermodynamics states that a process occurs spontaneously in the direction that increases the entropy of the universe.

Secondary structure is the pattern in which hydrogen bonding holds parts of a protein in distinctive shapes, such as helices and sheets.

Selective, only harms certain organisms and not others.

Semimetals are metalloids, elements that have properties intermediate between metals and nonmetals.

Slaked lime is calcium hydroxide.

SO_X is the combination of oxides of sulfur, especially SO_2 and SO_3.

Soft water is water having only very low concentrations of calcium ion.

Solid (*s*) is one of three states of matter; it has a fixed shape.

Solubility is the amount of a solute that will dissolve in a standard amount of a given solvent.

Solute is a substance (solid, liquid, or gas) dissolved in a larger amount of another substance (the solvent).

Solvent is the substance, usually a liquid and often water, in which substances dissolve.

Soman ($C_7FH_{16}O_2P$) is a nerve gas originally created for use in World War II.

Space-filling diagram is a molecular diagram that shows how the electrons of all the atoms within a molecule occupy space.

Specific heat is the amount of heat needed to raise the temperature of 1 g of substance by one degree Celsius or kelvin (see Table 7.3).

Spontaneous is occurring without continuous input of energy.

Spontaneous fission is nuclear fission that occurs without the nucleus being struck by a neutron.

Stainless steel is a steel alloy that does not rust under normal use, usually includes major amounts of chromium and nickel in addition to iron.

Standard atmosphere is the atmospheric pressure at sea level on a nice day, equaling 760 torr.

Standard pressure, 1 bar = 100,000 pascals = 0.9869 atmospheres = 750.6 torr

Standard temperature, 0°C or 273.15 K.

States of matter are the three forms in which matter appears: gas, liquid, and solid.

Stereoisomers are isomers that contain atoms bonded in the same order, but arranged differently in space.

Steroid is a class of biologically active molecules that include a set of four rings fused together.

Stoichiometry is a means of comparative measuring that uses the fact that chemical reactions occur in ratios of moles and that calculations of amounts of reactants and products are possible.

STP is the "standard temperature and pressure."

Strong acid is an acid that completely dissociates in water.

Strong nuclear force is the force that holds protons and neutrons together in an atomic nucleus.

Structural formula is a molecular formula that shows how atoms are bonded together.

Structural isomers are compounds that have the same molecular formula but different arrangements of their atoms.

Subcritical is when not enough neutrons are available to maintain fission.

Sublimation is the conversion of a solid directly into a gas.

Sulfide ore is a mineral in which the anion is a sulfide or disulfide ion.

Sulfurous acid (H_2SO_3) forms when sulfur dioxide (SO_2) comes in contact with water (H_2O).

Superconductor is a material(*s*) that conducts electricity with zero resistance at temperatures typically below 90K.

Supercritical is a mass of fissionable nuclide that is greater than a critical amount so a chain reaction occurs.

Surfactant is abbreviation for "surface active agent"; it is a substance that has a long nonpolar chain of CH_2 groups and a polar or ionic group at one end.

Synapse is a gap between adjacent nerve cells.

Synergistic describes two (or more) processes that increase each other's effects such that the total effect is greater than the two effects taken separately.

Tartar is the hard material deposited on the surface of teeth as plaque hardens by accreting calcium containing minerals.

Taxol is a copyrighted name for paclitaxel.

Temperature is the measure of the average speed of molecules or atoms of a substance.

Terpene is any of several compounds having the formula $C_{10}H_{16}$, many of which are natural constituents of plants.

Tertiary structure is the way in which protein chains fold into a complex shape.

Testosterone is a male hormone.

Tetrahedron is a three-dimensional figure having four corners and four faces.

Thermal inversion is when cool air cannot rise through the warm air above

Thermal pollution are the damaging effects of too much heat being added to a body of water.

Thermodynamics is the study of the interactions of energy with chemical or physical systems.

Thermoplastic is a classification of plastic that will soften on heating.

Thermosetting is a classification of plastic that will not soften on heating.

torr is the unit of pressure equal to pressure needed to support a column of mercury 1 mm high; short for Torricelli.

Trace metal ions are metal ions present in very low concentrations in water or soil.

Transcription is the manufacture of an RNA copy of the DNA of a gene.

Transfer RNA (tRNA) is the class of RNA that carries amino acids to the ribosome, where protein synthesis occurs.

Transition elements are elements that belong to the "B" groups in the periodic table.

Translation is the process in which the base sequence of mRNA gives rise to the amino acid sequence of a polypeptide.

Translocated herbicide kills plants only after being taken into the plant through the roots or leaves.

Transmutation is the conversion of one element into another.

A **triad** is a set of three elements having similar properties.

Triple bond is a covalent bond involving the sharing of three pairs of electrons.

Unsaturated means it can take on more hydrogen atoms and has one or more multiple bonds.

Unsaturated fatty acid is a fatty acid with one or more carbon-carbon double bonds.

Vacuum distillation is the distillation done under vacuum to lower the boiling point of a liquid.

Valence refers to the number of bonds which an atom can form.

Valence electrons are the electrons in the outermost energy level of the atom (highest n value), those most likely to be involved in bonding.

Vinyl literally means a $CH_2\!\!/\!\!/CH-$ group of atoms; it usually refers to vinyl chloride or polyvinyl chloride.

Vitamins are small molecules needed for proper health, which must be in our diet as they are not made, or not made in sufficient amounts, by the body.

Vitrified is to change or make into glass, especially through heat.

Volatile is easily evaporated.

Voltage, a measure of the electromotive force, measures the extent to which a given redox reaction will occur.

VSEPR is the Valence Shell Electron Pair Repulsion Theory that guides prediction of molecular structure.

Water cycle describes the sequence of forms characteristic of water in the environment.

Wavelength is a measure of the distance between crests in a wave of electromagnetic radiation.

Weak acid is an acid that only partially dissociates in water (typically $< 5\%$).

Weak nuclear force is responsible for the transformations of sub-subatomic particles in the nucleus of some atoms.

Weight is a measure of force equal to the product of an object's mass times the local gravitational acceleration.

Work is the energy transferred when an object is moved by a force.

Xylem is tissue which transports water and mineral nutrients in plants.

Credits

Index

The Elements

Element	Symbol	Atomic Number	Atomic Mass†	Element	Symbol	Atomic Number	Atomic Mass†
Actinium	Ac	89	(227)	Molybdenum	Mo	42	95.94
Aluminum	Al	13	26.98	Neodymium	Nd	60	144.2
Americium	Am	95	(243)	Neon	Ne	10	20.18
Antimony	Sb	51	121.8	Neptunium	Np	93	(237)
Argon	Ar	18	39.95	Nickel	Ni	28	58.69
Arsenic	As	33	74.92	Niobium	Nb	41	92.91
Astatine	At	85	(210)	Nitrogen	N	7	14.01
Barium	Ba	56	137.3	Nobelium	No	102	(259)
Berkelium	Bk	97	(247)	Osmium	Os	76	190.2
Beryllium	Be	4	9.012	Oxygen	O	8	16.00
Bismuth	Bi	83	209.0	Palladium	Pd	46	106.4
Bohrium	Bh	107	(267)	Phosphorus	P	15	30.97
Boron	B	5	10.81	Platinum	Pt	78	195.1
Bromine	Br	35	79.90	Plutonium	Pu	94	(242)
Cadmium	Cd	48	112.4	Polonium	Po	84	(209)
Calcium	Ca	20	40.08	Potassium	K	19	39.10
Californium	Cf	98	(251)	Praseodymium	Pr	59	140.9
Carbon	C	6	12.01	Promethium	Pm	61	(145)
Cerium	Ce	58	140.1	Protactinium	Pa	91	(231)
Cesium	Cs	55	132.9	Radium	Ra	88	(226)
Chlorine	Cl	17	35.45	Radon	Rn	86	(222)
Chromium	Cr	24	52.00	Rhenium	Re	75	186.2
Cobalt	Co	27	58.93	Rhodium	Rh	45	102.9
Copper	Cu	29	63.55	Rubidium	Rb	37	85.47
Curium	Cm	96	(247)	Ruthenium	Ru	44	101.1
Dubnium	Db	105	(262)	Rutherfordium	Rf	104	(261)
Dysprosium	Dy	66	162.5	Samarium	Sm	62	150.4
Einsteinium	Es	99	(252)	Scandium	Sc	21	44.96
Erbium	Er	68	167.3	Seaborgium	Sg	106	(263)
Europium	Eu	63	152.0	Selenium	Se	34	78.96
Fermium	Fm	100	(257)	Silicon	Si	14	28.09
Fluorine	F	9	19.00	Silver	Ag	47	107.9
Francium	Fr	87	(223)	Sodium	Na	11	22.99
Gadolinium	Gd	64	157.3	Strontium	Sr	38	87.62
Gallium	Ga	31	69.72	Sulfur	S	16	32.07
Germanium	Ge	32	72.61	Tantalum	Ta	73	180.9
Gold	Au	79	197.0	Technetium	Tc	43	(98)
Hafnium	Hf	72	178.5	Tellurium	Te	52	127.6
Hassium	Hs	108	(269)	Terbium	Tb	65	158.9
Helium	He	2	4.003	Thallium	Tl	81	204.4
Holmium	Ho	67	164.9	Thorium	Th	90	232.0
Hydrogen	H	1	1.008	Thulium	Tm	69	168.9
Indium	In	49	114.8	Tin	Sn	50	118.7
Iodine	I	53	126.9	Titanium	Ti	22	47.88
Iridium	Ir	77	192.2	Tungsten	W	74	183.9
Iron	Fe	26	55.85	Uranium	U	92	238.0
Krypton	Kr	36	83.80	Vanadium	V	23	50.94
Lanthanum	La	57	138.9	Xenon	Xe	54	131.3
Lawrencium	Lr	103	(262)	Ytterbium	Yb	70	173.0
Lead	Pb	82	207.2	Yttrium	Y	39	88.91
Lithium	Li	3	6.941	Zinc	Zn	30	65.39
Lutetium	Lu	71	175.0	Zirconium	Zr	40	91.22
Magnesium	Mg	12	24.31			110	(269)
Manganese	Mn	25	54.94			111	(272)
Meitnerium	Mt	109	(268)			112	(285)
Mendelevium	Md	101	(258)			114	(289)
Mercury	Hg	80	200.6				

†All atomic masses are shown to four significant figures. Values in parentheses represent the mass number of the most stable isotope of radioactive elements.